Waste Heat Recovery

Principles and Industrial Applications

Waste Heat Recovery

Principles and Industrial Applications

Chirla Chandra Sekhara Reddy

Gade Pandu Rangaiah

National University of Singapore, Singapore

World Scientific

NEW JERSEY • LONDON • SINGAPORE • BEIJING • SHANGHAI • HONG KONG • TAIPEI • CHENNAI • TOKYO

Published by

World Scientific Publishing Co. Pte. Ltd.
5 Toh Tuck Link, Singapore 596224
USA office: 27 Warren Street, Suite 401-402, Hackensack, NJ 07601
UK office: 57 Shelton Street, Covent Garden, London WC2H 9HE

British Library Cataloguing-in-Publication Data
A catalogue record for this book is available from the British Library.

WASTE HEAT RECOVERY
Principles and Industrial Applications

ISBN 978-981-124-839-9 (hardcover)
ISBN 978-981-124-840-5 (ebook for institutions)
ISBN 978-981-124-841-2 (ebook for individuals)

For any available supplementary material, please visit
https://www.worldscientific.com/worldscibooks/10.1142/12588#t=suppl

Typeset by Stallion Press
Email: enquiries@stallionpress.com

Preface

Thermal energy (or heat) is required for the manufacture of products in process plants (such as desalination, chemicals, food and beverage, mineral processing, oil and gas, paper and pulp, petroleum refining, petrochemicals and pharmaceuticals) and power plants. Some of the energy used in these plants is wasted for one reason or another. Hence, waste heat recovery (WHR) and its effective reuse are essential for improving energy efficiency and sustainability of process and power plants. This has encouraged the existing/operating plants to implement WHR projects. Despite these projects and efforts, further WHR is needed in existing and new plants due to the increasing concerns related to carbon (dioxide) emissions and climate change as well as because of developments in techniques, technologies and equipment for WHR.

This book covers fundamentals, state-of-art technologies and industrial applications of WHR in process and power plants. Simple and effective WHR techniques are illustrated with practical examples to help readers to identify, estimate potential of and develop projects for WHR in their plants. Key benefits of WHR projects, which are useful for developing successful WHR business cases, are demonstrated. Special emphasis is placed on major technical risks and mitigation plans for implementing sound WHR projects. In general, WHR projects involve proven technologies and, hence, have minimal technical risks. Strategies for reaping benefits of WHR projects for longer periods are also outlined. It is hoped that this book will help to increase WHR and energy efficiency in process and power plants.

This comprehensive and valuable book has 20 chapters, organized in three parts. After the Introduction chapter, Part I consists of Chapters 2–7 on heat transfer, heat integration, pinch analysis, heat exchangers, heat pumps, cost estimation, economic evaluation, and estimation of energy requirements and CO2 emissions. These chapters present the fundamentals, description and merits of available equipment, equations/procedures for analysis, design and costing for WHR. They provide the essential background for understanding WHR applications in Part II.

Chapters 8–17 in Part II are on diverse industrial applications of WHR, which is the significant and unique feature of this book. They cover WHR in vacuum systems, distillation, power generation, flue gas systems, compression systems, use of waste heat for desalination and water recovery, WHR using a heat transfer fluid, cooling systems for waste heat reduction, supporting systems for WHR, steam and condensate return systems. Each of these chapters presents descriptions and alternatives of the specific system, opportunities and processes for WHR, and relative merits of available choices. All these are illustrated with industrial examples and scenarios.

Part III, the last part in this book, is on the implementation aspects of WHR projects. Chapters 18, 19 and 20 in this part are, respectively, on sustainability of, project management for and process safety of WHR projects. These chapters are useful for any WHR application and are important for developing and implementing sustainable WHR projects.

This book is based on the authors' extensive experience on WHR in industry, research and teaching. Specifically, their experience from teaching a short course on WHR, a number of times in the last 4 years, to practitioners from different industries, was valuable in planning, organizing and preparing this book. Each chapter is organized similarly starting with an overview section on the relevance, importance, contents and learning outcomes of that chapter. It has a number of sections for carefully chosen contents on the topic, organized for easy reading and learning.

Finally, each chapter ends with a summary of key points (or takeaways) and important books/book chapters for more details. Many exercises are provided at the end of each chapter; they include qualitative questions to test understanding, quantitative problems to perform calculations and analysis, and yes/no questions to confirm concepts and principles learned. Depending on the reader's interest, each chapter in this book can be read on its own.

This book can be used as the main text for a course on WHR for students in chemical, process and related engineering programmes at undergraduate or postgraduate level. For this, the instructor can choose applicable chapters in this book according to the students' background and course syllabus. Alternatively, several chapters are relevant to courses such as those on plant utilities, distillation and heat pumps, and so the book can be employed for supplementary reading in such courses. Almost all chapters in this book are useful to engineers and managers in process and power plants, and in design and engineering companies, for improving energy efficiency of their plants via WHR.

Each chapter was prepared by one author, reviewed by the other author and then revised; this cycle was repeated several times. Finally, all chapters were read by Sharad Sharma, a chemical engineer (B. Tech IITM and M. Tech IITK) with over 40 years of experience in the fertilizer, power, alumina and software industries in India and Australia. The authors are very grateful to Sharad for his constructive suggestions and careful reading of all chapters, which helped us to improve the presentation and correct typographical errors in the book. However, the authors take the full responsibility for any remaining errors. They request the readers to send their comments and corrections to enhance the book, through email to chegpr@nus.edu.sg and/or cchandrasreddy@gmail.com. The authors thank Associate Professor A.F.A. Hoadley of Monash University for his thoughtful comments on the outline of this book as well as their industry contacts, who provided inputs for Table 1.1. They also acknowledge Sandhya Devi M.G. of *World Scientific* for her suggestions and assistance during the preparation of this book.

I thank my guru, Prof. G.P. Rangaiah for his inspiration and valuable research guidance, especially on WHR. I acknowledge the support of many colleagues and managements in petro-chemicals and refinery plants, in developing and implementing numerous successful WHR projects. I am very grateful and indebted to my parents (Rama Krishna Reddy and Kanaka Ratnam) and my family members for their support and sacrifices during the preparation of this book.

Chirla Chandra Sekhara Reddy
Singapore

I thank C.C.S. Reddy and other graduate students, who contributed to my research on WHR, as well as the National University of Singapore for supporting my research in process systems engineering. On the personal front, I am very grateful to all my family members for their deep affection and unwavering support. I am indebted to my dearest wife (Krishna Kumari) for taking care of me and our family, thus enabling me to pursue my research and teaching interests.

Gade Pandu Rangaiah
Singapore

Contents

Chapter 1

Introduction

1.1 Overview

Energy is essential not only for all living beings but also for all industries. It should be utilized efficiently for sustainability. If not, today's wastage of energy will be tomorrow's problem. This can be seen in the serious concerns on global warming and climate change. Industrial usage of energy is spread across several types such as chemical, electrical, nuclear and thermal energy. Significant quantities of these energy types are used in process and power plants. Here, process plants include industries such as desalination, chemicals, food and beverage, mineral processing, oil and gas, paper and pulp, petroleum refining, petrochemicals and pharmaceuticals. Many power plants generate electricity using fuels and nuclear materials whereas other plants use water in dams, solar energy, wind, etc.

Thermal energy (i.e., heat) is used extensively in process and power plants. Some of this energy is wasted in the plant for one reason or other. Recovering and reusing this waste heat (WH) in the plant will increase energy efficiency of the system, improve economics and reduce harmful emissions (e.g., carbon dioxide, CO_2). Hence, WH recovery (WHR) is critical for the sustainability of process and power plants. It is relevant for both new plants (greenfield projects) and existing/operating plants (brownfield projects). Technological advances in equipment, materials and techniques occur continually, and they create new and attractive opportunities for WHR. In other words, there will always be new possibilities for increasing WHR in and energy efficiency of existing plants.

This book describes in detail the fundamental principles and applications of WHR in process and power plants. This Introduction chapter begins with WHR examples in various industries in Section 1.2. Drivers for WHR and suggestions for identifying WHR opportunities are outlined in Sections 1.3 and 1.4, respectively. Then, role of simulation and optimization in WHR is discussed in Section 1.5. Common utilities in process and power plants, and effect of their costs on WHR are presented in Section 1.6. Scope and use of chapters in this book are described in Section 1.7. Finally, main points of this chapter are summarized in Section 1.8.

Learning outcomes of this Introduction chapter are as follows.

1. Describe the need for, examples of and drivers for WHR in industries.
2. Identify the WHR opportunities in any process/power plant.
3. State the common utilities and their effect on WHR in plants.
4. Outline the scope and use of chapters in this book.

1.2 Waste Heat Recovery in Industries

WHR is often possible in all process and power plants. This can be seen from many WHR projects in diverse industries, summarized in Table 1.1. These are compiled with inputs from our industry contacts, based on some websites and authors' experience. Many of these projects are already implemented whereas some are under consideration, which indicate continued importance of and need for WHR in plants. The projects in Table 1.1 are for process and power plants in various regions of the World.

The projects in Table 1.1 are just some examples to motivate the readers to learn and identify WHR opportunities in their plants. There are numerous such projects implemented in or being evaluated for existing and new plants throughout the World. WHR is often involved in improving energy efficiency, and industry initiatives and projects are recognized with energy efficiency awards in many countries. More details on such awards and projects in Singapore are available at https://www.nea.gov.sg/programmes-grants/grants-and-awards/eenp-awards (accessed on 26th September 2021). A couple of these are included in Table 1.1. Various

Table 1.1: Brief details and benefits of some WHR projects in process and power industries

Industry (Chapters*)	Brief Details of WHR	Benefits	Remarks and/or References
Alumina Refining (Chapter 4)	Heat in blow-off vapour (from a digestion unit), that would have been vented to atmosphere, is recovered and reused by installing a heater to heat wash water returned to the process. Aspen Plus was used for process simulations.	Recovery of 50–80% of WH, permits higher temperature of feed liquor and, therefore, reduces steam required for heating the feed liquor.	Chatfield and Sharma[1]
Alumina Refining (Chapter 5)	Low-pressure waste vapour stream in an alumina refinery will be turned into high-pressure steam for process heating, by mechanical vapour recompression (MVR). Electricity for powering the compressor will be sourced from renewable energy.	Reduction of fossil fuel used to produce steam, and de-carbonise operations. Alumina refining is energy and emissions intensive (14.3 million tonnes of CO_2 emissions in Australia in 2019). The technology has potential to reduce these by up to 70% as well as significantly reduce water use by capturing water vapour that would otherwise be lost to the atmosphere.	Currently, conducting technical and commercial feasibility studies. If successful, a 3-MW MVR will be installed by 2023, at a refinery in Australia, to test the technology. https://www.alcoa.com/global/en/what-we-do/alumina/mvr-project, accessed on 6 October 2021.

(*Continued*)

Table 1.1: (*Continued*)

Industry (Chapters*)	Brief Details of WHR	Benefits	Remarks and/or References
Alumina Refining (Chapter 18)	WHR of heat exchangers (HEs) for liquor to digestion was reduced by scale deposited on tube (heat transfer) surface. A study guided an appropriate heater maintenance schedule, which was implemented successfully.	More frequent cleaning to reduce the approach temperature by 1°C in the first HE was estimated to result in energy saving equivalent to $100,000/year, through better recovery of heat from flash vapour and thus reduce steam required for heating in digestion.	Chatfield and Sharma[1]
Biochemical (Chapter 4)	The column overhead vapour stream is used to generate low-pressure steam (LPS), which is then used in evaporators upstream of the plant.	Recovery of ≈2 MW energy for use in evaporators. Increase in investment of ≈$2.5 Millions for steam generation (instead of using cooling water) in the condenser. Payback period is 1.5 years.	This plant is in North America.
Cement (Chapter 9)	Potential use of WH from high-temperature effluent gas from the rotary kiln, for electric power generation using organic Rankine cycle (ORC) and/or for drying raw material (limestone)	Different effluent gas temperatures and working fluids in ORC are analysed. Best alternative is for power generation using ORC. Payback period is 6 years.	Study is for a cement plant in Colombia. See Fierro *et al.*[2] for more details.

Chemical (Chapter 16)	The process reactions generate significant amount of steam. The plant still operates two small boilers for reliable steam supply. Proposed to recover excess steam for power generation in a two-stage steam turbine.	Reduction in electric power import to the site. Payback period of investment for steam turbine is 6 years	Proposal is based on energy efficiency study executed in the Asia-Pacific region. Project is yet to be implemented.
Chemical (Chapters 13 and 9)	Saline wastewater treatment by using heat pump (HP) to recover latent heat of condensation in the condenser, and use recovered heat in the reboiler.	Reduction in the steam consumption at the reboiler of a distillation column. Payback period is about 2 years.	This project is already implemented in a plant in the Asia Pacific region.
Chemical — specialized (Chapter 9)	High pressure overhead vapour stream of a column is used in its reboiler via a HP.	Minimizes steam usage in the reboiler. Capital investment is required for installing HP.	This plant is in North America.
Chemical (Chapter 17)	Additional steam condensate recovery from the plant condensate system. This reduced the demand for LPS at the deaerator.	Reduction in fuel burned at the utility boiler. Payback period is 2 years	Already implemented in Asia.

(*Continued*)

Table 1.1: (*Continued*)

Industry (Chapters*)	Brief Details of WHR	Benefits	Remarks and/or References
Chemical (Chapter 4)	Troubleshooting and upgrading of feed effluent HE on the outlet of a reactor. This HE heats reactant (ethylene dichloride) against reactor output. Aspen Exchanger Design and Rating (EDR) was used for HE analysis.	The new HE has replaced an old, inefficient HE. The new unit is more efficient and is designed for higher flows. Improved efficiency means lower heat addition into the reactor during steady state.	HE installed in a chemical plant in Louisiana, USA.
Food and Beverage (Chapter 13)	Retrofit a multi-effect evaporator with a thermo vapour re-compressor (steam jet ejector) to compress, and reuse the compressed vapour at a higher temperature.	Reduction in fresh steam consumption by 6.6%.	Implemented in Singapore (https://drive.google.com/drive/folders/1TC89RyqoM4B9tUxSpMxnD7dFR4Ccv-zl, accessed on 26 September 2021)
Oil and Gas — Offshore (Chapter 14)	WHR from flue gas (FG) to heat pressurised water (in closed loop)**, which is used for process heating such as crude heating or stabilisation, amine reboiler and fuel gas superheating.	Reductions in power consumption, space, weight and capital cost (by decreasing number and size of gas turbines for power generation). This WHR is widely used on process platforms and FPSOs.	https://doczz.net/doc/1283067/offshore-whru-brochure---clyde-bergemann-power-group; https://www.bihl.com/waste-heat-recovery-units/

Paper and Pulp (Chapters 2 and 4)	Heat recovery from the exhaust of the paper machine dryer, to heat the pocket ventilation air and process water.	Reduction in steam usage. This WHR was already implemented in some plants around the World, and others are considering it.	See Ghosh[3] for details.
Paper and Pulp (Chapter 9)	Recovery of low temperature (75–85°C) heat from a wastewater steam, to generate power by ORC with NH_3 as the working fluid.	Analysis was performed at various heat source conditions and capacity, for a Swedish pulp mill. For reasonable return, value of produced power should be at least ≈0.11 $/kWh	See Öhman[4] for details.
Petroleum Refining (Chapter 4)	Improve the heat recovery in the feed-effluent HE of hydro-desulfurization reactor, by using a more efficient HE.	Reduction in fuel consumption at furnace. Payback period is about 4 years.	Project under evaluation for implementation.
Petroleum Refining (Chapter 15)	Product coolers using air are normally provided with water-cooled trim coolers to achieve required temperature for storage. Demineralised (DM) water (instead of cooling water) may be used in trim coolers.	Reduction in steam consumption in deaerator due to higher DM water temperature. Payback period is about 1 year.	Already implemented in Asia.

(*Continued*)

Table 1.1: (*Continued*)

Industry (Chapters*)	Brief Details of WHR	Benefits	Remarks and/or References
Petroleum Refining (Chapter 14)	Recover low-grade WH from multiple sources in a petroleum refinery, into a hot water circuit, which serves to transfer recovered heat to other process streams (i.e., heat sinks).	Reduction in steam demand from utility boilers, leading to fuel savings. Payback period is about 5 years.	See Moorthy *et al.*[5] for details on a similar case study, which includes both new design and revamp scenarios.
Petroleum Refining (Chapter 11)	WHR from to preheat combustion air, which increases combustion air temperature to a fired heater.	Reduction in fuel supplied to the fired heater. Payback period is about 3 years	Already implemented in Asia.
Petrochemicals (Chapter 9)	In the separation of propylene and propane mixture by distillation, HP is used to save both steam and cooling water. High-purity overhead stream is compressed to increase pressure and temperature. Then, it is condensed in the reboiler to supply heat.	Net reduction of energy for the separation as overhead stream is used in the reboiler, which avoids a separate condenser. However, compressor and power are required for compressing the overhead stream. Payback period is 3–4 years.	Already implemented in Asia

Pharmaceuticals (Chapter 10)	Installation of a tri-generation plant and downsizing boilers: this system includes a gas turbine to generate electricity, WHR boiler to generate steam, absorption chiller to produce chilled water (from WH), economizer and high-efficiency boiler (with turndown ratio of 10).	Reduction in total energy required to generate steam and energy used in the plant. Payback period of the project is less than 7 years.	Implemented in Singapore (https://drive.google.com/drive/folders/1TC89RyqoM4B9tUxSpMxnD7dFR4Ccv-zl, accessed on 26 September 2021)
Power Plant (Chapter 10)	Optimal heat integration of power plant with carbon capture system and CO_2 compression train. This optimization was by mixed integer nonlinear programming.	Reduction in 'penalty' on power plant output (due to carbon capture and compression).	Project is yet to be implemented.
Waste Plastic (Chapter 14)	Hot reactor vapour at ≈540°C is used to heat oil that later runs the reboiler of a stripper in the plant.	Recovery of energy (≈0.44 MW). Investment is required for adding one small column and hot oil recovery HEs.	This plant is in North America.

(*Continued*)

Table 1.1: (*Continued*)

Industry (Chapters*)	Brief Details of WHR	Benefits	Remarks and/or References
Wastewater (Chapter 13)	In a novel chemicals-free treatment of wastewater (like multi-effect distillation), the challenge was to recover as much heat as possible from the overhead stream of the treatment unit whilst ensuring that the stream is split into a clean water stream and a vapour stream with pollutants (which is routed to a boiler for burning).	Simulations show that about 70% reduction in heat input can be achieved in steady state operation.	This will be the first plant of its kind in the world. Plant operation is expected to commence in early 2022.
Cement, Glass, Oil and Gas, and Steel (Chapter 9)	Potential of ORC for WHR in energy intensive industries in 27 countries of European Union is evaluated.	Benefits from 2705 MW of power from ORC are cost savings of 1.95 billion Euros and reduction of 8.1 million tons of CO_2 emissions.	Details are available in Campana *et al.*[6]

* Chapters most relevant to each project are given in brackets.

** The other option for process heating on a FPSO (Floating Production, Storage and Off-loading) is electrical, which requires additional power generation and fuel consumption. Electric heater is used only for small loads and start-up. Steam generation from WHR requires more equipment and large maintenance, and hence it is not common in offshore applications.

chapters of this book cover fundamentals, technical details and implementation aspects of many WHR projects. To help readers in this, most relevant chapters for each WHR project are also given in the first column of Table 1.1.

Research and technological advances in methods, techniques, equipment and applications of WHR continue to occur. These are reported and described in journal and conference papers. For an indication of these, SCOPUS database is searched for journal/conference papers, book chapters and books containing WHR (together as a phrase) in title, keywords and abstract; this search is limited to publications from the year 2000 to (and including) the year 2021, subject area of chemical engineering and in English. As shown in Fig. 1.1, number of these papers published in each year has been increasing from around 40 in early 2000s to nearly 250 in the year 2021, i.e., by about six times. There are probably many more papers on WHR than those shown in this figure, for two reasons: some papers on WHR may not have used WHR in title, keywords or abstract, and the subject area of energy is not included (in order to focus on WHR in chemical and related processes). Total number of papers in Fig. 1.1 is 2388, whereas it is 9425 if the subject area of energy is also included.

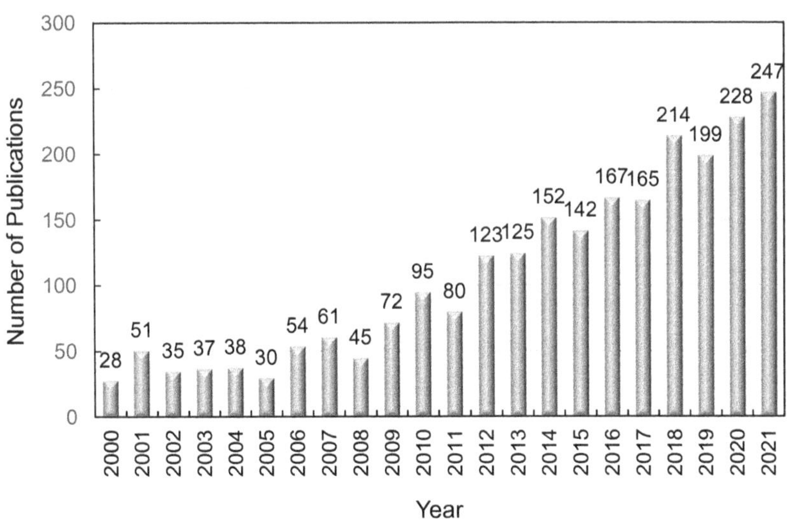

Fig. 1.1: Number of publications with WHR in the title, keywords and abstract, published in the Chemical Engineering subject area from the year 2000 to 2021.

Papers on WHR identified by SCOPUS search (Fig. 1.1) are from more than 80 countries with 45 of them contributing 10 or more papers. In particular, 442, 340, 172, 157 and 118 papers came from investigators in China, United States, United Kingdom, India and Japan, respectively. International journals, which published more than 50 papers are *Chemical Engineering Transactions* (205), *Desalination* (123), *International Journal of Heat and Mass Transfer* (120), *Industrial and Engineering Chemistry Research* (86), *Fuel* (62), *Bioresource Technology* (56) and *Energy and Fuels* (55).

In summary, research and development are occurring in WHR techniques and equipment as well as their applications. These will lead to more and more WHR projects in industry to improve energy efficiency and sustainability of process and power plants.

1.3 Drivers for WHR Projects

Key drivers and considerations for WHR projects are outlined as follows:

- **Sustainable development:** WHR projects decrease use of fossil fuel and consequently decrease the carbon footprint of the plant. Heating and/or cooling (using chilled water generated by lithium bromide chiller) can be provided by WHR. Further, electric power and freshwater can be produced using WH. Overall, WHR projects achieve economic, environmental and social benefits. Hence, they lead to sustainable development in process plants (covered in Section 18.2 of Chapter 18).
- **Little technology risk:** WHR generally requires conventional equipment used in process plants, such as HEs, pumps, compressors, fans, steam turbines, steam jet ejectors and cooling towers. Hence, WHR projects pose very little technology risk provided appropriate equipment with right metallurgy and/or damage mitigation methods (covered in Section 18.3 of Chapter 18) are selected.
- **Enablers for carbon capture technologies:** Many of technologies for industrial CO_2 capture require FG to be cooled (from more than

200°C to less than 100°C) before CO_2 is absorbed by a solvent. Use of metal and polymer air preheaters (covered in Section 18.4 of Chapter 18) in series can reduce FG temperature to <100°C and, hence, reduce the capital cost for FG cooler, before sending FG to CO_2 capture system.

- **Reduces the use of fossil fuels and emissions:** As WHR projects recycle the recovered WH to the process plant, they reduce the fuel consumption and, hence, carbon emissions (covered in Section 7.5 of Chapter 7).
- **Increases energy efficiency and reduces energy cost:** WHR projects increase energy efficiency of the process, and hence reduces the energy and operating costs of the plant (covered in Section 6.5 of Chapter 6).
- **May reduce the capital cost for WH disposal equipment:** WHR projects reduce temperature of WH streams. For example, use of economizer and/or air preheater can reduce FG temperature at boilers, fired heaters and furnaces (covered in Sections 11.3–11.5 of Chapter 11), which helps to reduce the height of FG stack. This WHR feature reduces the capital cost of FG stack for new projects. Having shorter FG stack may improve the company's image in public as shorter stacks may not attract attention of the public.
- **Reduces cooling duty:** WHR projects recover WH and, hence, reduce the cooling duty. For example, hot wastewater streams require cooling before their treatment. WHR from a wastewater stream can eliminate the need for cooling it. Similarly, WHR from boiler blowdown eliminates the need for blowdown cooler (covered in Section 16.3 of Chapter 16).
- **May increase productivity and competitiveness:** In processes that are limited by heating duty, WHR projects can help to increase the plant capacity. For example, crude distillation unit in petroleum refineries is often constrained by the fired heater duty. Increasing WHR in crude preheat train reduces the heating duty demand at the fired heater and, hence, can help to increase the crude processing capacity.
- **Cheap generation of power:** WH-powered ORC (organic Rankine cycle) or KC (Kalina cycle) can produce electric power at a fraction

of the cost of power generation by Rankine cycle (covered in Section 9.6 of Chapter 9).

- **Cheap production of chilled water (at or above 5°C):** WHR can be used to produce cheaper chilled water using lithium bromide absorption chiller (covered in Section 5.5 of Chapter 5), compared to chilled water generated by a mechanical chiller using a compressor.
- **Recycle and reduce the amount of freshwater required:** WHR projects often increase water recycle and/or produce desalinated water (by thermal desalination covered in Section 13.3 of Chapter 13). This reduces the amount of freshwater intake and can result in substantial operating cost savings. Using WH for thermal desalination, freshwater can be produced cheaply, compared to other desalination methods. Steam condensate recovery (covered in Section 17.2 of Chapter 17) not only recycles heat/energy but also water.
- **May improve product quality:** For example, reflux rate and, hence, reboiler duty of a distillation column have to be increased for a better product quality. In such a scenario, LPS generated by WHR or a HP can be very useful to support higher reboiler duty.
- **Contribution to circular economy:** WH in one process plant can be recovered and reused in another process plant, if the recovered WH is more than that can be consumed within the original plant. For example, in petroleum refineries and petrochemical plants, WHR can be used to produce LPS and export part (or all) of it to nearby plants (covered in Section 14.6 of Chapter 14).
- **Better corporate image:** WHR leads to better reputation among shareholders and society at large. This is becoming increasingly important under ESG (Environmental, Social and Governance) framework in the finance sector.
- **Better legal compliance:** WHR projects help to meet government regulations such as on emissions to the environment.

1.4 Identifying WHR Opportunities

Process and power plants offer many WHR opportunities. Given the large range and complexity of operations in these plants, it may not be straightforward and requires extensive knowledge of the plant, to find suitable WH streams for WHR and potential WH sinks. Identifying and evaluating opportunities for WHR in manufacturing facilities can be time consuming, for selecting the most attractive outcome (using suitable criteria such as payback, internal rate of return and CO_2 emissions). Availability of many WHR techniques and technologies makes this task even more complex.

Knowledge and skills required for identifying and evaluating WHR opportunities are (a) familiarity with the entire plant (especially locations at which heating and cooling duties are involved) including utilities production and availability, (b) good understanding of business drivers for WHR projects, (c) comprehensive understanding of WHR techniques and technologies, (d) in-depth knowledge about WHR equipment and their relative merits, and (e) skills in simulation and optimization of WHR processes. Chapters in this book are especially useful for (b)–(d).

A comprehensive approach for identifying WH sources and WHR potential is to conduct a plant-wide energy audit or detailed review of data in the database of the distributed control system (DCS). Such an audit or data review should identify WH sources, their temperatures, pressures, flow rates and composition, as well as potential WH sinks in the plant. Once WH source and sink data are available, WHR opportunities can be evaluated by systematic screening (covered in Section 19.3 of Chapter 19). Often a quick scan of the process plant may reveal attractive WH opportunities. Some ideas for this quick scan are given in the following paragraphs.

Steam and condensate systems may have many WHR opportunities as described in Chapters 16 and 17. A quick way to evaluate WHR potential in these systems is to check boiler feed water (BFW) make-up to the deaerator. If this make-up is more than 30% of steam generation (at equipment such as boilers and heat recovery steam generators), there can be attractive WHR projects for recovering vented/leaking steam and/or condensate. For exploring WHR projects in the combustion system, one should review FG stack temperature; if this temperature is more than 100°C, there is a potential for WHR from FG covered in Sections 11.3–11.9 of Chapter 11).

If the process plant has a condensing steam turbine, a WHR project may be developed for converting such a turbine to back pressure type and use LPS generated by the back pressure steam turbine for process heating. Alternatively, a WHR project can be developed for using flash steam (at 1–2 bar) for thermal desalination (covered in Section 13.6 of Chapter 13).

Any steam venting or steam trap/piping leakages from the steam system can be easily spotted, and necessary WHR opportunity or repair plan can be established. A steam system survey by plant operators or energy consultants can be very useful. If a process plant is generating chilled water using electricity and there is a WH source (with sufficient amount of energy at >100°C), the mechanical chiller can be potentially replaced by lithium bromide absorption chiller using WH (covered in Section 5.5 of Chapter 5).

Check the quantity and temperature of blowdown at boilers and other steam generation equipment. If sufficient blowdown flow rate (above 10 m^3/hr) is available at or above 100°C, WHR project can be developed to use WH in the blowdown to heat water or process streams.

Other than steam and condensate system, a quick way to establish WH discarded from a process plant is by summing the cooling duties of cooling towers and other cooling systems (such as seawater cooling systems and fin fan coolers). Then, the next step is to list the hot process and utility streams (requiring cooling at these coolers and condensers) as well as cold streams (requiring heating). Such streams can be screened for attractive WHR projects. Hot streams at greater than 70°C are more suitable for WHR applications. Often, one or two major hot streams can be quickly identified for heat exchange with a few cold streams by analyzing their proximity within the process unit, heat duties and temperatures. If the plant has many hot and cold streams, a more beneficial HE network can be developed by pinch analysis (covered in Chapter 3) and/or mathematical optimization.

Even if WH sources/streams are identified, they may not be suitable for WHR projects for several reasons such as the following. WHR from a stream containing highly corrosive substances and solid material may not be economical. A WH stream may not be suitable for WHR if it has components that can leak to and contaminate the process stream to be heated (e.g., in case WHR HE tube ruptures). WHR from a WH stream may be limited by acid dew point corrosion or salt formation; then, it needs careful and comprehensive assessment of potential benefits. WH streams such as FG may be at close to or below atmospheric pressure; then, WHR from them may require induced draft fans. Similarly, some process liquid streams may be at low pressures and, hence, may require additional pumps for WHR. Maximum temperature of a WH stream is important for certain WHR equipment. For example, temperature of a WH stream used for WHR in heat pipes (covered in Section 11.7 of Chapter 11) should not exceed the critical temperature of the working fluid. A WH stream available on batch wise (i.e., intermittently) is generally not suitable for heating a continuous process stream.

Shell and tube HEs may be modified (covered in Section 4.5 of Chapter 4) for increasing WHR by using heat transfer enhancements (covered in Section 4.8 of Chapter 4). For revamp type WHR projects, conventional shell and tube HEs with plain tubes can be replaced by plate HEs (covered in Section 4.6 of Chapter 4) or heat pipes.

Many WHR opportunities exist in distillation columns ranging from easy to complex options (covered in Sections 9.4–9.7 of Chapter 9). For distillation systems handling separation of close boiling substances, heat pumps can be very attractive for WHR from the overhead stream and using it in the reboiler (covered in Section 9.8 of Chapter 9).

Steam ejector–assisted vacuum systems offer significant WHR opportunities by converting them to liquid ring and/or dry vacuum pumps (covered in Section 8.6 of Chapter 8). Multistage compression systems offer low-temperature WHR opportunities (covered in Sections 12.4–12.6 of Chapter 12).

A quick look at power generation systems may reveal opportunities for co-, tri- or quad-generation (covered in Sections 10.3–10.5 of Chapter 10). For example, power generation based on a simple steam Rankine cycle is inefficient as latent heat of steam is discarded to cooling water. Replacing such a system with co-, tri- or quad-generation can be very beneficial. Chilled water generated by tri-generation can be used for producing more electric power at a gas turbine by cooling its inlet air (covered in Section 10.6 of Chapter 10).

Distance between WH source and sink streams strongly influence the economic attractiveness of a WHR project. Proximity within the same process unit makes WHR economical. If WH sources and sinks are scattered in the process plant, then a HE network, using water or hot oil as a heat transfer fluid, may be beneficial to develop an attractive WHR project (covered in Section 14.5 of Chapter 14).

1.5 Simulation and Optimization for WHR Projects

WHR projects often require process simulation, optimization, and detailed thermal and hydraulic analysis, for developing the most beneficial WHR scope. For analysing WHR projects involving only water/steam streams, Microsoft Excel file with built-in steam properties can be used. Excel spreadsheets can be used for pinch analysis, steam network (covered in Section 16.6.1 of Chapter 16) and vacuum system analysis/optimization (covered in Section 8.5 of Chapter 8). Process plant data from its DCS can be directly imported into Excel. Even after implementation of the WHR project, Excel-based models are easy to implement and understand by users, and are also very useful for real-time monitoring and adjusting process variables, for sustaining WHR benefits. HEs can be simulated, designed and rated using software tools such as Heat Transfer Research, Inc. (HTRI) and Aspen EDR (Exchanger Design and Rating). For estimating WHR opportunities in reaction and separation systems, process simulators such as Aspen Plus and PRO/II are generally used; they are also useful for detailed hydraulic analysis, thermal analysis and real-time optimization. Aspen Energy Analyser is convenient for performing pinch analysis.

Process simulations are effective only if they represent process operations closely. Hence, before using them for assessing WHR potential, a simulation model should be validated to match the base operating data of the process. Pressure, temperature and flow rate from plant operating data are essential for validating the process simulation model. The validated model can then be used for developing or evaluating WHR project scope. Experience of the process engineer is beneficial for selecting good initial values for process variables, which help to converge the simulation.

Such process experience is also useful for reviewing and deciding the simulation results.

For many linear programming (LP) and nonlinear programming (NLP) problems, Solver tool in Microsoft Excel may be sufficient. A simple NLP optimization model is presented in Section 8.5 of Chapter 8, for minimizing steam consumption; it can be easily solved by Solver tool. Another NLP model is presented in Section 12.2.1 of Chapter 12, for minimizing the compression power. Again, this model can be easily solved using Solver tool. For bigger and highly complex optimization involving NLP or mixed integer linear/nonlinear programming (MILP or MINLP) problems, software such as GAMS[a] or FrontlineSolvers[b] can be used. A MILP model presented for maximizing power generation in steam network (Section 16.6.1 of Chapter 16) can be solved by Frontline Solver or GAMS. Practical constrains such as forbidden matches between heat source and heat sink streams (e.g., due to their location, shutdown and safety issues such as to avoid contact of shell and tube side fluids due to chemical reaction, overpressure damage and atmospheric release) must be clearly specified in optimization models. Frontline Solver works with Microsoft Excel, and so it is very useful for monitoring and sustaining WHR benefits.

As stated in Section 1.3, there are many drivers for WHR projects, which achieve economic, environmental and social benefits. Hence, optimization of WHR projects is likely to involve multiple criteria or objectives, either maximization or minimization type. Such problems require multi-objective optimization (MOO), for which many techniques and programs are now available. See Rangaiah *et al.*[7] for an overview to techniques and software for MOO as well as their applications in chemical engineering.

1.6 Common Utilities and their Effect on WHR

WHR projects generally involve HEs, which may require hot utility (e.g., steam for heating) and/or cold utility (e.g., cooling water for cooling).

[a]GAMS, https://www.gams.com/, accessed on 2 October 2021.

[b]FrontlineSolvers, https://www.solver.com/, accessed on 2 October 2021.

WHR equipment, other than HEs, may require electricity, fuel, steam, cooling water, etc. Utility requirements for many WHR equipment are compiled in Table 1.2.

Utility costs are covered in Section 6.5 of Chapter 6. They can significantly influence the best WHR option for a given WHR project, and hence their reliable estimation is important. Utility costs generated from WHR can be lower than those based on (fossil) fuel cost. True cost of steam is

Table 1.2: WHR equipment, their utility requirements and uses

WHR Equipment or Process	Utilities Required	WHR Uses
Absorption HP	Medium-pressure steam (MPS) and electricity (for pumps)	Upgrades low-temperature WH to a medium temperature (between WH and MPS temperatures) and nearly doubles the amount of usable heat.
Absorption Heat Transformer (AHT)	Cooling water and electricity (for AHT pumps)	Upgrades nearly half of WH at low temperature to a higher usable temperate level.
Back Pressure Steam Turbine (BST)	High-pressure steam (HPS) or MPS	Provides co-generation by producing electricity/shaft power and steam (MPS and/or LPS).
Compressor	Electricity, HPS/MPS or fuel if motor, steam turbine or gas turbine/reciprocating engine is the driver, respectively.	WHR from flash steam or low pressure vapor/gas streams in steam, distillation and evaporation systems, by compression.
Gas Turbine	Fuel	Provides co-generation by producing electricity/shaft power and steam. Can be used for tri-generation to generate electricity, steam and chilled water/fresh water using thermal desalination.
Reciprocating Engine	Fuel, cooling water and electricity (for lube oil pumps if this engine is to provide shaft work)	Provides co-generation by producing electricity/shaft power and steam. Can be used for tri-generation to generate electricity, steam and chilled water, and also for quad-generation to generate electricity, steam, CO_2 and chilled water.

Table 1.2: (*Continued*)

WHR Equipment or Process	Utilities Required	WHR Uses
Micro Turbine	Fuel and cooling water	Provides co-generation by producing electricity/shaft power and steam.
ORC and KC	Cooling water	Generates electricity.
Pumps, Liquid Ring and Dry Vacuum Pumps	Electricity	Used in WHR systems such as heat pumps, ORC, KC, heat transfer fluid and vacuum systems.
Steam Jet Ejector	HPS, MPS or LPS	Upgrades low-pressure flash steam (at ≈1 bar) to higher pressure steam that is suitable for process heating.
Heat Transfer Fluid System	DM water or hot oil	Provides more opportunities for WH recovery and reuse between heat sources and heat sinks.
Low-Temperature Thermal Desalination	Flash steam and cooling water	Uses WH to produce desalinated water.

illustrated in Section 16.6.9 of Chapter 16. For example, co- and tri-generation can reduce the cost of steam and/or electricity. Sometimes, MPS, LPS, flash steam and electricity required by WHR equipment in a project can be generated by the same project or other WHR projects. This increases the attractiveness of WHR projects by reducing the operating costs. Co-generation using BSTs provides electricity/shaft power, required for driving WHR equipment at very low operating cost, as long as the outlet/extraction steam from BST can be used for process heating. Reciprocating engines, gas and micro turbines also provide such co-generation benefits and make WHR more attractive.

1.7 Scope and Use of the Book and Chapters

Comprehensive knowledge and experience are required for selecting and optimizing the most appropriate from available techniques, technologies, equipment and designs in order to implement a successful WHR project and maintain its sustainability. Careful design, right equipment selection and appropriate operating conditions as well as operations and maintenance strategy are required to address issues during production such as corrosion, fouling and clogging for sustaining benefits from WHR. These key requirements should be developed during the project design stage itself.

This book would assist in achieving sustained benefits of WHR, and comprehensively covers fundamentals, latest technologies and industrial applications of WHR in process and power plants. Simple and effective WHR techniques are illustrated with industrial examples, to help readers to identify, estimate potential of and develop projects for WHR in their plants. Key benefits of WHR projects, useful for developing successful WHR business cases, are demonstrated. Special emphasis is given towards major technical risks and mitigation plans, for implementing sound WHR projects. Strategies for reaping benefits of WHR projects for longer periods are also outlined. In short, this book brings together essential material and industrial applications on WHR, and presents them coherently and concisely.

This book has 20 chapters in total. After this Introduction chapter, Chapters 2–20 are organized into three parts: Part I on fundamentals and principles for WHR, Part II on many industrial applications of WHR and Part III on implementation aspects of WHR projects. Part I has 6 chapters (Chapters 2–7) on heat transfer, heat integration, pinch analysis, HEs, heat pumps, cost estimation, economic evaluation, and estimation of energy and CO_2 emissions. These chapters present fundamentals, description and relative merits of available equipment, and equations and procedures for analysis, and design and costing for WHR. The chapters in Part I provide necessary background for understanding WHR applications in Part II. Depending on their background and interest, readers can choose and study one or more chapters in Part I.

Part II is on many industrial applications of WHR, and it is the major and significant aspect of this book. Chapters 8–17 in this part cover WHR in vacuum systems, distillation, power generation, FG systems, compression systems, WH use for desalination and water recovery, WHR using a heat transfer fluid, cooling systems for WH reduction and supporting WHR equipment (such as AHT, ORC and KC), steam systems and condensate return systems. In general, each of these chapters present description and alternatives of the specific system, opportunities and processes for WHR, and relative merits of available choices. All these are illustrated with industrial examples and scenarios. One can choose to study one or more chapters in Part II, which are relevant to her or his interest and application.

Part III, the last part in this book, is on the implementation aspects of WHR projects. Chapters 18, 19 and 20 in this part are, respectively, on sustainability of, project management for and process safety of WHR projects. These chapters are applicable and useful for any WHR application such as those described in the various chapters of Part II. Chapters 18–20 are important for developing sustainable WHR projects. Hence, design/practicing engineers should consider many aspects covered in these chapters, before developing any WHR project.

Many of the chapters in this book can be studied on their own without much difficulty. Each chapter is organized similarly starting with an overview section on the relevance, importance, contents and learning outcomes of that chapter. Learning outcomes are new capabilities and skills that one should achieve after studying and solving exercises at the end of a chapter. Each chapter has a number of sections for carefully chosen contents on its topic, organized for easy following. Finally, it ends with a summary of key points (i.e., takeaways) and most relevant books and/or book chapters for more details. Many exercises are provided at the end of each chapter; they include qualitative type to test understanding, quantitative type to perform calculations and analysis, and yes/no type on fundamentals and principles gained from that chapter.

Many of the chapters in this book can be used as the textbook for a course on WHR for students in chemical, process and related engineering at undergraduate and/or postgraduate level. Alternatively, students can choose and read chapters relevant to courses such as on plant utilities, distillation and heat pumps. Some or all chapters in this book are useful for engineers and managers in process and power plants, and in design and engineering companies, for improving energy efficiency of their plant via WHR and reuse. For identifying the chapters of interest, first browse the title and sections of chapters in the Table of Contents (at the beginning of the book), to shortlist some chapters of interest; then, read the Overview (i.e., the first) section in the shortlisted chapters to select the chapters for reading and learning.

1.8 Summary

Main points in this Introduction chapter of the book on the important subject of Waste Heat Recovery: Principles and Industrial Applications, are summarized as follows.

1. WHR is essential for energy efficiency and sustainability of process and power plants. As presented in Section 1.3, there can be several business drivers for a WHR project.
2. Many WHR projects are already implemented and in operation in diverse process industries. Yet, many new WHR projects are possible because of advances in WHR equipment and techniques as well as changes in energy costs and priorities.
3. Of the knowledge and skills required for identifying potential WHR projects, this book provides thorough understanding of business drivers, methods, techniques, equipment and technologies along with their relative merits, for WHR projects. Many suggestions for identifying potential WHR projects in any process/power plant are given in Section 1.4.
4. Process simulation and optimization are generally required for identifying, analysing and developing WHR projects. They can be performed using a spreadsheet (e.g., Excel), process simulators (e.g., Aspen Plus) and optimization software (e.g., GAMS and Solver tool in Excel), depending on the project.
5. Hot utilities (e.g., steam), electricity and/or cold utilities (e.g., cooling water) are often involved in WHR projects. Hence, their availability and reliable costs are important for a WHR project.
6. A comprehensive and systematic procedure is required for achieving the maximum benefits from the implementation of a WHR project throughout the plant life.
7. Chapters in this book are organized into three parts on principles, various industrial applications and implementation aspects of WHR projects.
8. Depending on the background and requirements, one or more chapters of interest can be identified. Many chapters are independent of others; thus, each of them can be read on its own.

9. Chapters in this book can be used by undergraduates (in upper years) and postgraduates, in courses related to WHR, either as a textbook or supplementary book, and by engineers and managers for improving energy efficiency of process and power plants.

References

1. Chatfield R, Sharma SC. (2006). The use of process models to resolve capital and operating cost trade-offs in Bayer Digestion. Proc. Seventeenth Intl. Symp. I.C.S.O.B.A. **33:** 37; Vol 1. Proceedings of the Bauxite and Alumina related papers, OMBKE (Hungarian Mining and Metallurgical Society), Budapest, K. Solymar, Ed., 114–127.
2. Fierro JJ, Escudero-Atehortua A, Nieto-Londono C, *et al.* (2020) Evaluation of waste heat recovery technologies for the cement industry. *Int J Thermofluids* **7–8:** 100040.
3. Ghosh AK. (2011). Fundamentals of Paper Drying – Theory and Application from Industrial Perspective, Chapter 25 in Evaporation, Condensation and Heat Transfer, Edited by A. Ahsan, INTECH Open Access. https://www.intechopen.com/chapters/19429 (accessed on 26th September 2021).
4. Öhman H. (2012) Implementation of a low temperature waste heat recovery power cycle using NH_3 in an organic Rankine Cycle. *Energy* **48:** 277–232.
5. Moorthy AN, Reddy CCS, Rangaiah GP. (2014) Optimization of heat exchanger networks for the utilization of low-temperature process heat. *Ind Eng Chem Res* **53:** 17989–18004.
6. Campana F, Bianchi M, Branchini L, *et al.* (2013) ORC waste heat recovery in European energy intensive industries: Energy and GHG savings. *Energy Convers Manag* **76:** 244–252.
7. Rangaiah GP, Feng Z, Hoadley AF. (2020). Multi-objective optimization applications in chemical process engineering: Tutorial and Review. *Processes* **8:** 508. doi:10.3390/pr8050508.

Acronyms and Notation

Acronyms

AHT	Absorption Heat Transformer
BFW	Boiler Feed Water
BST	Back Pressure Steam Turbine

CO_2	Carbon Dioxide
DCS	Distributed Control System
DM	Demineralized
EDR	Exchanger Design and Rating
ESG	Environmental, Social and Governance
FG	Flue Gas
FPSO	Floating Production, Storage and Off-loading
HE	Heat Exchanger
HP	Heat Pump
HPS	High-Pressure Steam
HTRI	Heat Transfer Research, Inc.
KC	Kalina Cycle
LP	Linear Programming
LPS	Low-Pressure Steam
MILP	Mixed Integer Linear Programming
MINLP	Mixed Integer Nonlinear Programming
MOO	Multi-Objective Optimization
MPS	Medium-Pressure Steam
NLP	Nonlinear Programming
NH_3	Ammonia
ORC	Organic Rankine Cycle
WH	Waste Heat
WHR	Waste Heat Recovery

Exercises

1.1 Identify one or two potential WHR projects in your plant or a process in a book or paper. Briefly describe each of them and its benefits.

1.2 Choose a recent paper (say, published in the last 2 years) on WHR, from journals or conferences, and carefully study it. Outline scope of WHR in this paper, technology and equipment employed, benefits of WHR, new contribution in the paper and further potential for improving energy efficiency of the process/system in the paper.

1.3 Among the drivers for WHR outlined in this chapter, which one is the most important in your view? Justify your selection.

1.4 What is the easiest way to identify the WHR potential at the steam and condensate system?

1.5 Is a quick look at the total cooling duty of the plant useful to assess WH rejected? Briefly discuss.

1.6 Answer the following with yes (right) or no (wrong), and give brief reasoning, where applicable.
  (A) Many WHR projects are possible in process and power plants, and only a few of them may not be economical or practical.
  (B) There is only one driver (motivation) for a WHR project.
  (C) Utility cost is an important factor for deciding the best WHR project.
  (D) WHR improves sustainability of the plant.
  (E) Good understanding of the entire plant is not required for developing a WHR project.
  (F) Process simulators are very useful for developing a WHR project. So, it is not necessary to validate the base case simulation (i.e., plant simulation model before WHR project) with the actual plant data.
  (G) A systematic plant-wide survey of WH sources and WH sinks is not compulsory for estimating maximum potential of WHR at the site.

Chapter 2

Principles of Heat Transfer and Heat Integration

2.1 Overview

Heat transfer and heat integration are central in waste heat recovery (WHR) and reuse. This chapter covers principles of heat transfer, heat exchangers (HEs), heat integration and exergy analysis, all with illustrative examples. First, energy, enthalpy and heat capacity are introduced in Section 2.2. Next, heat transfer mechanisms, namely, conduction, convection and radiation are presented in Section 2.3. The subsequent section covers basics and governing equations of HEs. Section 2.5 discusses heat transfer coefficients and their effect on HEs. Section 2.6 describes heat integration and its benefits. The subsequent section describes temperature-enthalpy (T-H) plots, which are useful in heat integration. Sections 2.8 to 2.10 present exergy concepts, exergy analysis and their application to HEs. Finally, this chapter ends with summary in Section 2.11.

Learning outcomes of this chapter on heat transfer and heat integration are:

1. Describe heat transfer mechanisms
2. Perform HE calculations
3. Explain heat integration and its benefits
4. Describe exergy and its merits
5. Conduct exergy analysis of WHR systems

This chapter may be a refresher for readers having a chemical engineering background, but it will be useful for other readers in order to be able to understand subsequent chapters in this book. Heat integration and exergy basics covered later in this chapter, will be new to many readers. In particular, heat integration basics will provide a sound background for Chapter 3 on pinch analysis for heat integration.

2.2 Energy, Enthalpy and Heat Capacity

Energy is omnipresent; although it cannot be seen, its effects can be seen and/or felt. It exists in either of the two forms: potential energy (i.e., stored energy due to position or structure) and kinetic energy (due to moving of objects or particles). Examples of potential energy are water behind a dam, any object at an altitude and chemical energy (due to molecular structure or bonds of a component). Kinetic energy examples are flowing fluid, moving vehicles and electrical energy (due to motion of electrons).

Common units or dimensions of energy in engineering practice are Joule (J, named after the famous physicist: James Prescott Joule), calorie (= 4.184 J) and BTU (British Thermal Unit = 1055.06 J = 252 calories). One Joule is equal to the work done by a force of 1 Newton over a distance of 1 m. Since the energy unit Joule and power unit Watt (= J/s) are small quantities, kilo Joule (kJ) and kilo Watt (kW) are often used and also employed in this chapter. For example, capacity of a domestic water heater is 2 to 4 kW whereas that of a typical power plant is in the range of a few hundred to a few thousand mega Watt (MW = 1,000 kW).

Energy can be transformed from potential to kinetic energy and vice versa, and also from one type to another (e.g., electrical energy to shaft power in motors, shaft power to electrical energy in generators, electrical energy to thermal energy in electric heaters and thermal energy to shaft power in gas turbines). Living things are excellent examples, where one type of energy is transformed to other forms; for example, plants convert solar energy to chemical energy, and human bodies transform chemical energy in food into heat and work for survival and daily activities. In general, energy is conserved and cannot be destroyed, but it can be transferred from one substance to another; for example, heat transfer can occur from a hot stream to a cold stream.

In this book, we are mostly concerned with thermal energy, which is sometimes termed as heat energy. Thermal energy of a substance is due to vibration of molecules and atoms in the substance; hence, it is in the form of kinetic energy at microscopic level. Vibration of molecules/atoms increases with temperature. Hence, thermal energy of a substance increases with its temperature. In other words, thermal energy and temperature are positively correlated.

Enthalpy (H, kJ) of a substance is a common quantity used in thermodynamics and heat transfer as well as in this book. It is the sum of two terms, namely, internal energy (u, kJ, which includes thermal energy) and product of pressure (P, bar) and volume (V, m^3). Of these two terms in enthalpy, u is generally more than PV; this can be seen from u and H values (per kg) for water/steam at selected temperatures and pressures given in Table 2.1; these values are from the steam table calculator at https://www.steamtablesonline.com/steam97web.aspx. As expected, both u and H increase with temperature. For water (in liquid state), u and H are nearly the same, and not much affected by pressure; on the other hand, they differ and are affected by pressure for steam (in vapour state). The main reason for this is that the liquid volume is almost unaffected by pressure. Since water/steam occurs in many chapters of this book, it is used as an example in this chapter.

Table 2.1 has values of u and H (as well as other quantities to be discussed later) for water (both subcooled and saturated) and steam (both saturated and superheated) at three pressures, namely, 0.006117, 1.0 and 10.0 bar. Saturated water and steam are respectively water and steam at the boiling point corresponding to the specified pressure. Subcooled water refers to water below its boiling temperature at the given pressure whereas superheated steam is at a temperature above boiling point, at the given pressure.

Table 2.1 shows that u and H increase substantially when water (saturated) changes to steam (saturated) at the same temperature, for a given pressure. Latent heat of vapourization is the difference between enthalpy of saturated steam and saturated water at a given pressure. It is a large quantity, and significant in WHR. These details are particularly useful for readers new to the study of heat transfer. Applications in process industries often involve mixtures of diverse components (and not just water/steam), and their values of u and H will differ from those in Table 2.1.

Table 2.1: Internal energy, enthalpy and heat capacity of water/steam at selected temperatures and pressures

State	P (bar)	T (°C)	u (kJ/kg)	H (kJ/kg)	C_v (kJ/kg.K)	C_p (kJ/kg.K)
Water (saturated)	0.006117	0.010967	0.0040824	0.0046943	4.2174211	4.2198942
Steam (saturated)	0.006117	0.010967	2374.9125	2500.9128	1.4221438	1.8882228
Water (subcooled)	1.000000	30.00000	125.73207	125.83251	4.1174826	4.1800238
Water (saturated)	1.000000	99.60592	417.33217	417.43649	3.7696997	4.2161494
Steam (saturated)	1.000000	99.60592	2505.5474	2674.9496	1.5526970	2.0759380
Steam (superheated)	1.000000	200.0000	2658.2256	2875.4751	1.4992365	1.9756881
Water (subcooled)	10.00000	30.00000	125.64921	126.65317	4.1146249	4.1775896
Water (saturated)	10.00000	179.8856	761.55561	762.68284	3.3967480	4.4051120
Steam (saturated)	10.00000	179.8856	2582.7707	2777.1195	1.9303382	2.7149848
Steam (superheated)	10.00000	200.0000	2622.2639	2828.2675	1.7526191	2.4288462

Internal energy and enthalpy values are with respect to a reference state. For water/steam data such as those in Table 2.1, this reference state is saturated liquid at the triple point of water (0.01°C and 0.006117 bar), where solid, liquid and vapour co-exist, and is the state at which the specific internal energy and specific entropy are taken as equal to zero. The first row of Table 2.1 refers to water near this state, with extremely small values of u and H.

Note that the reference state/value can be different. For example, reference enthalpy value in Aspen Hysys is that of ideal gas enthalpy of formation at 25°C; thus, enthalpy value in Aspen Hysys includes both chemical energy and thermal energy. This is important if enthalpy values from different sources are used. In heat transfer, heat integration and

WHR, enthalpy change (and not enthalpy itself) and thermal energy are the relevant properties that are employed. Further, enthalpy change is often calculated using heat capacity (described below). Hence, reference state/value for enthalpy is not important, unless a chemical reaction also takes place.

Another quantity of importance in heat transfer is the specific heat or heat capacity of a substance, which is the energy required for increasing temperature of unit mass of the substance by one degree. This property can be measured at constant pressure or volume, and correspondingly there exist heat capacity at constant pressure (C_p kJ/kg.K) and heat capacity at constant volume (C_v kJ/kg.K). Values of these two quantities for water/steam at selected Ts and Ps are included in Table 2.1. As shown by these values, C_p and C_v for a liquid have comparable values whereas C_p is more than C_v for a vapour mixture. Often, C_p is used since heating/cooling occurs at constant P. As can be seen from Table 2.1, heat capacity varies with T; this variation is small for water whereas it is large for steam. Further, heat capacity of water is about twice that of steam. For quick (back of the envelope) calculations, C_p of water and steam can be taken as 4.2 (or even 4.0) and 2.0 kJ/kg.K, respectively.

Example 2.1: A typical shower requires hot water at 45°C and 0.0025 m^3/min (2.5 litres/min) for 10 min. Assuming water is available at 30°C, how much energy is required for producing hot water for the shower? How will this energy change if water is available at 20°C in winter? State your assumptions, if any.

Solution: Assume density and C_p of water are respectively 1000 kg/m^3 and 4.2 kJ/kg.K.

Mass of (hot) water = (0.0025 m^3/min) × (1000 kg/m^3) × 10 min = 25 kg

Energy required for producing hot water = (25 kg) × (4.2 kJ/kg/K) × (45 − 30) = 1575 kJ

If electric heating is used for producing hot water for 10 min, power required = $\frac{1575\,\text{kJ}}{600\,\text{s}}$ = 2.625 kW. Recall that power unit is Watt, W (= J/s).

Assuming efficiency of electric heating including heat losses to pipes and surroundings is 0.8, actual power required = 2.625/0.8 = 3.281 kW. Compared to this, power rating of energy-saving light bulbs is less than

20 W (0.02 kW). These values give a feel for energy and its unit of kJ. Finally, energy required will be more, if water is available at 20°C. The reader is encouraged to calculate this quantity and compare it with that required if water is at 30°C.

2.3 Heat Transfer Mechanisms

Heat is thermal energy in transition from one substance to another under the influence of temperature driving force (i.e., energy flow from one material at a higher temperature to another material at a lower temperature). It can occur by three mechanisms, namely, conduction, convection and radiation.

Heat transfer by conduction occurs via molecular interactions and free electrons in materials. It is particularly significant in metallic solids. Ability of a material to transfer heat by conduction is characterized by its thermal conductivity, k (kW/m.K), which varies with material and temperature. Heat transfer rate, q (kW) across a rectangular wall (Fig. 2.1) is given by:

$$q = k A \frac{T_h - T_c}{t} \tag{2.1}$$

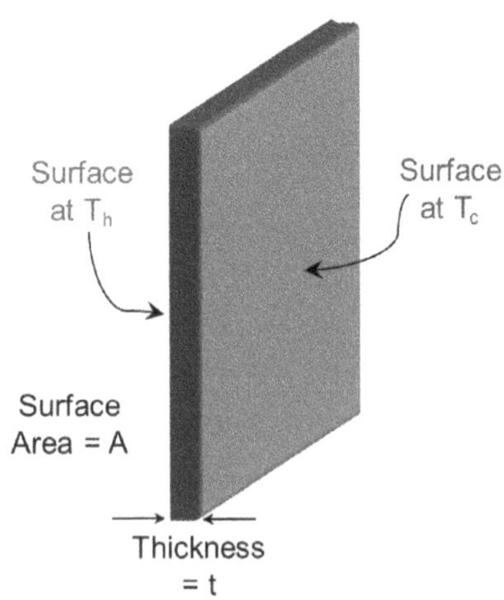

Fig. 2.1: Conductive heat transfer across a rectangular wall.

Here, A is the surface area in m^2, T_h and T_c are temperature (in K or °C) on hot and cold sides of the wall, respectively, and t is the wall thickness in m.

Thus, conductive heat transfer increases with increase in k, increase in temperature driving force or difference (= $T_h - T_c$) and decrease in t. Typical k values of copper, aluminium, carbon steel, water, rock wool (insulating material) and air are 0.385, 0.230, 0.050, 0.0006, 0.00004 and 0.000026 kW/m.K. In other words, copper and aluminium are good conductors of heat (as well as of electricity) whereas rock wool and air are poor conductors of heat.

Heat transfer by convection is the heat transfer between a surface and adjacent fluid (i.e., liquid or gas), and it occurs by motion of fluid. Consider hot water at T_h flowing through a pipe (of inner diameter, d_i, outer diameter, d_o and length L, all in m) in ambient air at T_a (Fig. 2.2). Pipe inner and outer surface temperatures are $T_{c,i}$ and $T_{c,o}$, which are slightly different due to heat flow (heat loss) through the pipe wall.

Heat transfer rate, q by convection from hot water to pipe inner surface is given by:

$$q = h_i\, A_i\, (T_h - T_{c,i}) \tag{2.2}$$

Here, A_i (m^2) = $\pi\, d_i\, L$ is the lateral surface area (and not cross-sectional area) of the pipe based on inner diameter and h_i ($kW/m^2.K$) is the convective heat transfer coefficient inside the pipe. Note that units of h are different from those of k (kW/m.K); this is because we have temperature *gradient* (across the wall thickness) on the right-hand side of Eq. 2.1, whereas we have temperature *difference* on the right-hand side of Eq. 2.2.

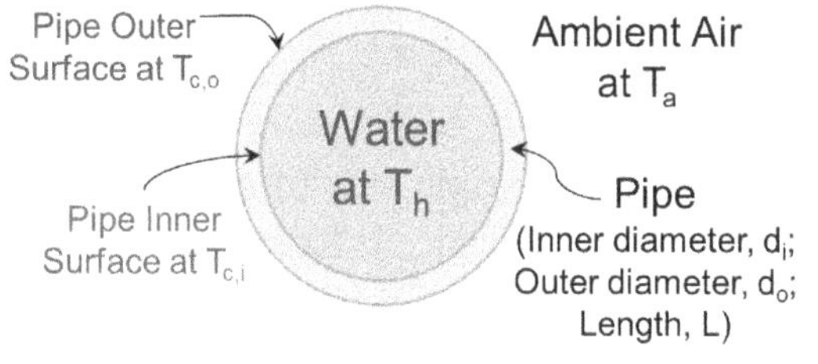

Fig. 2.2: Cross-section of a pipe with hot water flowing inside.

Defining resistance to heat transfer inside the pipe, $R_i = \frac{1}{h_i A_i}$, Eq. 2.2 can be written as:

$$q = \frac{T_h - T_{c,i}}{R_i} \tag{2.3}$$

The value of the heat transfer coefficient, h, varies significantly (by a factor of 10 to 100) depending on the fluid, its velocity, and phase change (boiling or condensation). Typical h values are given later. Equation 2.3 shows that the heat transfer rate is equal to temperature difference (driving force) divided by corresponding (thermal) resistance; this is similar to electric current flow being equal to voltage difference divided by (electrical) resistance.

Similarly, heat transfer rate from pipe outer surface to ambient air is given by:

$$q = h_o\, A_o\, (T_{c,o} - T_a) \tag{2.4}$$

Here, $A_o = \pi\, d_o\, L$ is the lateral surface area of the pipe based on outer diameter, and h_o is the convective heat transfer coefficient outside the pipe. In terms of resistance to heat transfer outside the pipe, $R_o = \frac{1}{h_o A_o}$, Eq. 2.4 becomes:

$$q = \frac{T_{c,o} - T_a}{R_o} \tag{2.5}$$

Heat transfer from pipe's inner surface to outer surface (i.e., through the pipe wall) is by conduction, and it is given by:

$$q = k\left(\frac{2\pi L}{\ln\left(\frac{d_o}{d_i}\right)}\right)(T_{c,i} - T_{c,o}) \tag{2.6}$$

The first two terms on the right side of this equation are somewhat different from those in Eqs. 2.2 and 2.4. Reasons for this are heat transfer is by conduction rather than convection, and lateral surface area increases slightly from inner diameter to outer diameter of pipe. Defining wall resistance to heat transfer, $R_W = \frac{\ln\left(\frac{d_o}{d_i}\right)}{2\pi kL}$, Eq. 2.6 can be written as:

$$q = \frac{T_{c,i} - T_{c,o}}{R_W} \tag{2.7}$$

At steady state, q in Eqs. 2.2–2.7 is identical. Equations 2.3, 2.5 and 2.7 can be combined as:

$$q\,(R_i + R_w + R_o) = (T_h - T_{c,i}) + (T_{c,i} - T_{c,o}) + (T_{c,o} - T_a) \tag{2.8}$$

Defining total resistance, $R = R_i + R_w + R_o$ and simplifying the above equation,

$$q = \frac{T_h - T_a}{R} \tag{2.9}$$

In other words, resistances to heat transfer in series can be added (similar to electric resistances in series), but heat transfer coefficients can't be added. As in Eq. 2.3, we see in Eqs. 2.5, 2.7 and 2.9, that heat transfer rate is equal to temperature difference (driving force) divided by the corresponding thermal resistance (similar to electric current flow being equal to voltage difference divided by electrical resistance).

Example 2.2: Water at 55°C is flowing through a carbon steel pipe (d_i = 0.108 m, d_o = 0.114 m and k = 0.050 kW/m.K). The pipe is not insulated and is in ambient air at 30°C. Convective heat transfer coefficient is 1.0 kW/m^2.K on water side (inside the pipe) and 0.010 kW/m^2.K on air side (outside the pipe). Find $T_{c,i}$, $T_{c,o}$ and q per metre length of pipe.

Solution: First, calculate R_i, R_w and R_o using the given values and taking L = 1 m. Then, use Eq. 2.9 to find q and finally use Eqs. 2.2 (or 2.3) and 2.4 (or 2.5) to find $T_{c,i}$ and $T_{c,o}$.

$$R_i = \frac{1}{h_i A_i} = \frac{1}{1.0 \times \pi \times 0.108 \times 1} = 2.947 \text{ K/kW}$$

$$R_w = \frac{\ln\left(\frac{d_o}{d_i}\right)}{2\pi k L} = \frac{\ln\left(\frac{0.114}{0.108}\right)}{2 \times \pi \times 0.050 \times 1} = 0.1721 \text{ K/kW}$$

$$R_o = \frac{1}{h_o A_o} = \frac{1}{0.010 \times \pi \times 0.114 \times 1} = 279.2 \text{ K/kW}$$

$$R = 282.3 \text{ K/kW}$$

From Eq. 2.9, $q = \frac{55-30}{282.3} = 0.08858 \text{ kW}$

From Eq. 2.3, $T_{c,i} = T_h - \frac{q}{h_i A_i} = T_h - q \times R_i = 55 - 0.08858 \times 2.947$
$= 54.74°C$

From Eq. 2.5, $T_{c,o} = T_a + \frac{q}{h_o Ao} = T_a + q \times R_o = 30 + 0.08858 \times 279.2$
$= 54.73°C$

The above example brings out three important points. First, resistance to heat transfer is small when heat transfer coefficient is large. Second, temperature difference (driving force) is small when the corresponding resistance is small (e.g., difference of only 0.01°C from the pipe's inner surface to outer surface because R_w is very small compared to R_o and R_i). Third, the largest resistance, R_o is the controlling (deciding) factor of q. These observations can be used to simplify calculations (as in Exercise 2.1 at the end of this chapter) and also identify suitable modifications to improve heat transfer (e.g., by decreasing the largest/controlling resistance, R_o by increasing h_o and/or A_o through the use of fins, inserts etc.).

Heat transfer by (thermal) radiation occurs by emission and transmission of electromagnetic waves by any object at above zero absolute temperature. Maximum heat transfer rate (q kW) by radiation emitted from a surface at temperature, T is given by:

$$q = \sigma A T^4 \tag{2.10}$$

Here, σ is Stefan–Boltzmann constant (= 5.67×10^{-5} kW/m^2.K^4), A is area (m^2) of the emitting surface and T must be in K (and not °C). It is clear from Eq. 2.10 that radiative heat transfer increases with temperature and is significant at high temperatures. Thus, radiative heat transfer is important in the combustion (radiation) zone of furnaces (fired heaters), where temperatures are well above 300°C.

In summary, heat transfer can occur by conduction, convection and/or radiation. Of these, resistance to conductive heat transfer is negligible compared to other resistances, particularly in the context of WHR, and radiative heat transfer is significant at high temperatures. Hence, convective heat transfer is involved and important in WHR. Accordingly, Eq. 2.2–2.9 are the most relevant in WHR.

2.4 Heat Exchanger Basics

WHR generally requires one or more HEs for transferring thermal energy from a hot stream to a cold stream, as required in the production process. HEs come in many types and sizes, which are covered in Chapter 4. This section focuses on HE basics including design equations considering the double-pipe HE, which is the simplest type; these are required for understanding heat integration covered later in this chapter and also in Chapter 3.

A double-pipe HE consists of two concentric pipes, as shown in Fig. 2.3. Here, the hot stream is flowing from left to right inside the inner pipe whereas the cold stream is flowing from right to left in the annulus between the two pipes. Thus, the hot and cold streams are flowing in opposite directions. This configuration is known as counter-current flow, and is common in the process industry. Alternatively, in a co-current flow configuration, both hot and cold streams flow in the same direction, which is less efficient and consequently rare in practice.

In Fig. 2.3, the hot stream is shown as flowing in the inner pipe and the cold stream in the annulus between the two pipes. In general, it could be either way — hot inside and cold outside, or hot outside and cold inside — depending on process conditions, nature of fluids, material of construction, scaling and other considerations.

Typical temperature profiles in the double-pipe HE, are also shown in Fig. 2.3, which is also used to illustrate the calculations in Example 2.3, later. The difference between hot stream temperature (HST) and cold stream temperature (CST) (i.e., vertical distance between the two curves in the plot) at any distance along the HE is the driving force at that

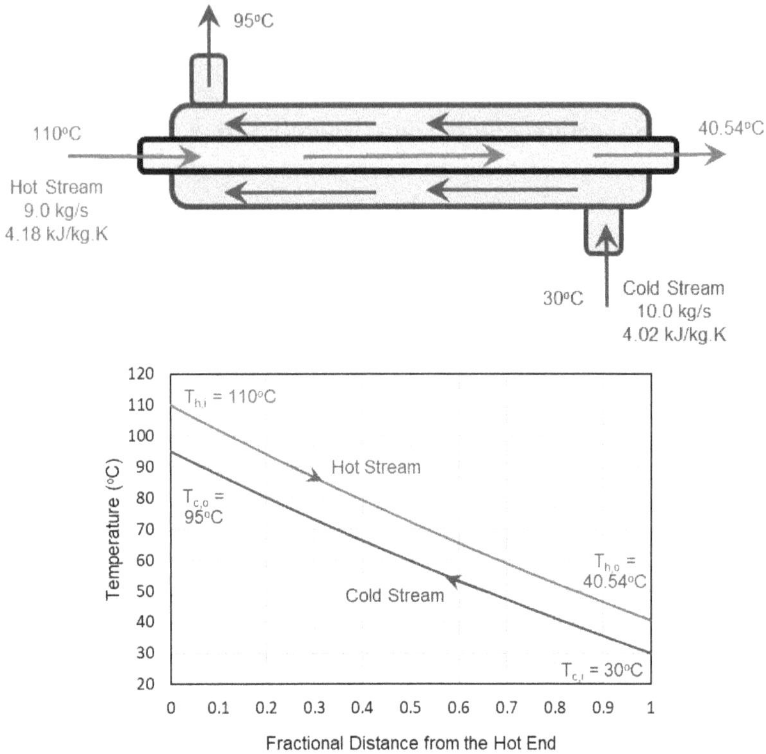

Fig. 2.3: Double-pipe HE with hot stream flowing inside the inner pipe and cold stream flowing in the annulus: schematic and temperature profiles.

location. In Fig. 2.3, driving force decreases gradually from 15°C (at the left, hot end) to 10.54°C (at the right, cold end). It cannot be zero or negative (i.e., the two temperature profiles cannot meet or cross). However, outlet temperature of hot stream (here, 40.54°C) can be lower than outlet temperature of cold stream (here, 95°C) in counter-current flow configuration as in Fig. 2.3 (but not in co-current flow configuration).

Temperature driving force at either end of an HE is termed approach temperature (or temperature approach). There are two approach temperatures: hot-end approach temperature (15°C in Fig. 2.3) and cold-end approach temperature (10.54°C). The lower of these two is the minimum approach temperature. Design should be based on a specified minimum approach temperature, optimized to balance capital and operating costs and for feasible operation. This would typically be 5°C–20°C for above-atmospheric temperatures in the HE, or ≈2°C in case of cryogenic temperatures in the HE.

The slope (ΔT) of the temperature profiles in Fig. 2.3 is equal to the heat transferred (q) divided by the product of mass flow rate (M, kg/s) and C_p of the stream. For steady-state operation, q for the hot stream is equal to that for the cold stream. Here, MC_p of hot and cold streams is respectively 37.62 (= 9 × 4.18) and 40.2 (= 10 × 4.02) kW/K; thus, the slope of the HST curve is more than that of the CST curve. In this case, minimum approach temperature occurs at the cold end. In the case where the slope of the HST curve is less than that of the CST curve, minimum approach temperature occurs at the hot end. Thus, minimum approach temperature in an HE can occur at the cold or hot end.

Governing or design equations of HEs in this chapter are developed based on the following assumptions:

- Continuous process
- Steady state operation
- No heat loss from the HE to the ambient
- No leakage of material from one stream to another or to outside
- No reaction or separation in the HE
- Constant heat capacity (specific heat)
- Constant heat transfer coefficient
- Potential and kinetic energy changes are negligible.

The above assumptions are valid for many HEs. Although heat loss from the HE is not zero, it is small at less than 5% of heat transferred from hot stream to cold stream. For this and/or for any other invalid assumption, design equations can be modified suitably; some modifications (e.g., for heat loss) are simple. However, this is outside the scope of this chapter. Pressure of cold and hot streams decrease due to friction as they flow through the HE; heat transfer analysis in this chapter does not involve this pressure drop of cold/hot streams.

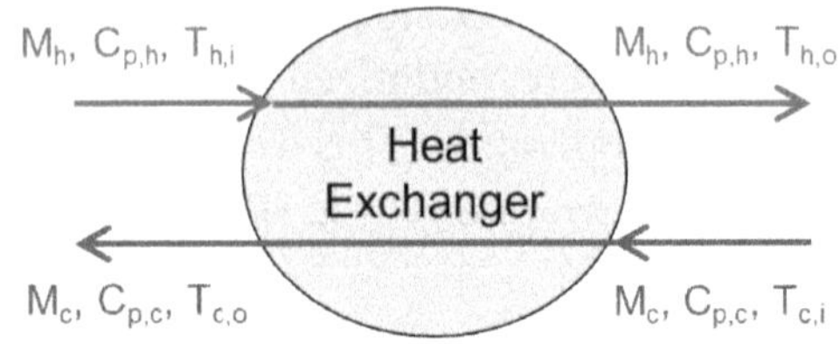

Fig. 2.4: Schematic of a counter-current HE showing hot/cold streams and their variables of interest.

The simple schematic of an HE in Fig. 2.4 can be used for calculations and analysis. It is valid for any type of HE having counter-current flow configuration. In this schematic, hot stream is placed above the cold stream to emphasize that HST should be higher than CST in the HE. Direction of hot stream is shown from left to right whereas that of cold stream is from right to left. Alternatively, the hot stream may be presented as from right to left and the cold stream from left to right. In Fig. 2.4, mass flow rate, heat capacity, inlet and outlet temperatures of the hot stream are respectively M_h, $C_{p,h}$, $T_{h,i}$ and $T_{h,o}$. Corresponding quantities of the cold stream are respectively M_c, $C_{p,c}$, $T_{c,i}$ and $T_{c,o}$.

Based on the definition of heat capacity at constant pressure, thermal energy change of the hot stream per unit time is:

$$Q = M_h\, C_{p,h}\, (T_{h,i} - T_{h,o}) \tag{2.11}$$

Since no heat loss is assumed, the above energy is transferred to the cold stream. Hence,

$$Q = M_c\, C_{p,c}\, (T_{c,o} - T_{c,i}) \tag{2.12}$$

The above two equations involve nine variables (namely, M_h, $C_{p,h}$, $T_{h,i}$, $T_{h,o}$, M_c, $C_{p,c}$, $T_{c,i}$, $T_{c,o}$ and Q). For a unique solution of them, 7 (= 9 − 2) variables must be specified. Usually, M_h, $C_{p,h}$, M_c, $C_{p,c}$, $T_{h,i}$, $T_{c,i}$ and either $T_{h,o}$ or $T_{c,o}$ are given. Then, the unknown temperature and Q can be calculated, as illustrated in the following example:

Example 2.3: A hot stream (C_p = 4.18 kJ/kg.K and M = 9.0 kg/s) at 110°C is entering an HE for heating a cold stream (C_p = 4.02 kJ/kg.K and

M = 10.0 kg/s) from 30°C to 95°C. Find the HST at the outlet of the HE and also heat duty of the HE.

Solution: First, use Eq. 2.12 to find Q:
Q = 10.0 × 4.02 × (95 − 30) = 2,613 kW
Now, calculate $T_{h,o}$ by re-arranging Eq. 2.11.

$$T_{h,o} = T_{h,i} - \frac{Q}{M_h C_{p,h}} = 110 - \frac{2613}{9.0 \times 4.18} = 40.54^\circ C$$

Data and results for this example are shown in Fig. 2.3.

For an HE operating continuously at steady state, the third and last equation accounts for heat transfer rate, and it is as follows:

$$Q = U\,A\,(\Delta T)_{LM} \tag{2.13}$$

Here, U (kW/m^2.K) is the overall heat transfer coefficient accounting for all resistances to heat transfer from hot stream to cold stream in the HE, A (m^2) is the surface area available for heat transfer in the HE and $(\Delta T)_{LM}$ is the log mean temperature difference. More details on U are presented later in this chapter. Many HEs (e.g., shell-and-tube type covered in Chapter 4) are not fully counter-current; even then, for calculations, they are assumed to be counter-current and a factor (F < 1.0) is included on the right side of Eq. 2.13 to account for deviations from counter-current flow.

Log mean temperature difference (driving force) is defined as:

$$(\Delta T)_{LM} = \frac{(T_{h,i} - T_{c,o}) - (T_{h,o} - T_{c,i})}{\ln\left[\frac{(T_{h,i} - T_{c,o})}{(T_{h,o} - T_{c,i})}\right]} \tag{2.14}$$

In the above, $(T_{h,i} - T_{c,o})$ and $(T_{h,o} - T_{c,i})$ are the approach temperatures at hot and cold ends, respectively. Note that $(\Delta T)_{LM}$ is less than arithmetic mean temperature difference $= \frac{(T_{h,i} - T_{c,o}) + (T_{h,o} - T_{c,i})}{2}$. It is exactly equal to arithmetic mean temperature difference in the special case of $(T_{h,i} - T_{c,o}) = (T_{h,o} - T_{c,i})$, when $(\Delta T)_{LM}$ becomes indeterminate $\left(= \frac{0}{0}\right)$.

Eq. 2.11–2.14 involve 12 variables: M_h, $C_{p,h}$, $T_{h,i}$, $T_{h,o}$, M_c, $C_{p,c}$, $T_{c,i}$, $T_{c,o}$, Q, U, A and $(\Delta T)_{LM}$. For their unique solution, 8 (= 12 − 4) variables must be specified. Usually, M_h, $C_{p,h}$, M_c, $C_{p,c}$, U and three of the four temperatures are given. Then, one can calculate the unknown temperature, Q, $(\Delta T)_{LM}$ and A. This is illustrated in the following example, also referring to Fig. 2.3.

Example 2.4: A hot stream (C_p = 4.18 kJ/kg.K and M = 9.0 kg/s) at 110°C is entering an HE for heating a cold stream (C_p = 4.02 kJ/kg.K and M = 10.0 kg/s) from 30°C to 95°C. Given U for this situation is 1.0 kW/m^2.K. Find the HST at the outlet of the HE, and Q and A of the HE.

Solution: HST at the outlet of the HE and Q are already found in Example 2.3. They are 40.54°C and 2,613 kW. Now, use Eq. 2.14 to calculate $(\Delta T)_{LM}$.

$$(\Delta T)_{LM} = \frac{(110-95)-(40.54-30)}{\ln\left[\frac{(110-95)}{(40.54-30)}\right]} = 12.64^{\circ}C$$

From Eq. 2.13, $A = \frac{2613}{1.0 \times 12.64} = 206.7\,m^2$

Calculations in this and previous examples have to be performed many times in heat integration (covered later in this chapter) and HE networks (covered in Chapter 3). Hence, readers new to them should try Exercises 2.2 and 2.3 at the end of this chapter.

In many HEs, hot and cold streams are in separate zones and do not mix. They are mixed in certain situations such as quenching (sudden cooling) to stop reactions and produce hot water by mixing cold water and steam (e.g., see Example 16.8 in Chapter 16). This type of direct contact heat transfer also occurs in cooling towers, wherein returning/warm cooling water (cold utility) is sprayed directly into a flowing stream of air in order to cool the warm cooling water for recycling to the plant. Another example is a deaerator (Fig. 16.9 in Chapter 16), where demineralized water is heated and degassed by direct contact with steam.

In some applications, there may be a phase change in the HE: the hot stream may condense or the cold stream may vapourize. For example, in reboilers, steam condenses while process fluid partially vapourizes. Phase change involves large amount of heat (i.e., latent heat of condensation/vapourization) but relatively small change in temperature. If the stream is a pure component (e.g., steam condensation for heating a process stream), its temperature during phase change is practically constant.

2.5 Heat Transfer Coefficients

Now, let's discuss U in Eq. 2.13. In an HE (Fig. 2.3), thermal energy has to travel outwards from bulk of hot stream to inner surface of inner pipe (by convection), through inner pipe wall (by conduction) and then from outer surface of inner pipe to bulk of cold stream (by convection). As seen in Example 2.2, resistance to conductive hear transfer in the pipe wall is small and can be neglected. Hence, total resistance for heat transfer is:

$$\frac{1}{UA} = \frac{1}{h_i A_i} + \frac{1}{h_o A_o} \tag{2.15}$$

Here, h_i and h_o are the convective heat transfer coefficient inside the pipe (here, hot stream) and in the annulus (here, cold stream), respectively; and A_i and A_o are heat transfer area based on inner and outer surface area of the inner pipe, respectively. Often, A_i and A_o are comparable, and $A = A_i = A_o$ is assumed for process design (and not for detailed design of HEs, which is beyond the scope of this book). With this, Eq. 2.15 simplifies to:

$$\frac{1}{U} = \frac{1}{h_i} + \frac{1}{h_o} \tag{2.16}$$

During operation, fouling of (particle deposits on) heat transfer area is likely. These deposits pose resistance to heat transfer, which is referred to as fouling factor or resistance. Fouling coefficient is the reciprocal of fouling factor/resistance, and is similar to heat transfer coefficient of convective heat transfer. For an HE with fouling, U is given by:

$$\frac{1}{U} = \frac{1}{h_i} + \frac{1}{h_{i,F}} + \frac{1}{h_o} + \frac{1}{h_{o,F}} \tag{2.17}$$

Here, $h_{i,F}$ and $h_{o,F}$ are the fouling (heat transfer) coefficient inside the pipe and in the annulus, respectively.

Heat transfer coefficient, h, depends on the condition (liquid, gas, two-phase flow or phase change) and velocity of the stream, as well as the dimensions/design of HE (e.g., inner pipe diameter and presence of inserts or fins to increase heat transfer). Text-books on heat transfer present dimensionless correlations for estimating h (based on experimental data). Fouling coefficients are given mostly based on industrial experience.

The range and mean values of h and h_F for different fluids and scenarios are compiled in Table 2.2. These are based on values reported in Smith[1] and are useful for preliminary design of double-pipe and shell-and-tube HEs (and not intensified HEs such as plate HE). Range of h for a fluid/scenario may not be the same in different sources/books. For example, see Table 9.4 in Kakaç *et al.*[2] for alternative ranges of h to those in Table 2.2.

Chapter 6 in the book by Kakaç *et al.*[2] is on fouling of HEs. It describes effects of fouling on heat transfer, required heat transfer area, pressure drop and costs, fouling types, design and operation of HEs, and techniques to control fouling. Specifically, Chapter 6 of Kakaç *et al.*[2] includes several tables on fouling factor/resistance values (in SI units of m^2.K/W) of different process streams; these tables are based on the standards of the Tubular Exchanger Manufacturers Association (TEMA).

Using mean values of h and h_F in Table 2.2, typical U for different fluids without and with phase change in double-pipe and shell-and-tube HEs is calculated and compiled in Table 2.3. A small value of U implies

Table 2.2: Heat transfer and fouling coefficients for different fluids, boiling and condensing

	h (kW/m².K)		hF (kW/m².K)	
Fluid and Scenario	**Range**	**Mean**	**Range**	**Mean**
Gases (no phase change)	0.01–0.5	0.255	5–11	8.0
Organic Liquid (high viscosity, no phase change)	0.1–1	0.55	1–3	2.0
Organic Liquid (low viscosity, no phase change)	1–3	2.0	3–11	7.0
Water (no phase change)	2–6	4.0	3–6	4.5
Organic Liquid Boiling (high viscosity)	0.1–0.5	0.3	1–3	2.0
Organic Liquid Boiling (low viscosity)	0.5–2	1.25	3–11	7.0
Boiling Water	2–10	6.0	6–11	8.5
Organic Vapour Condensing (high viscosity)	0.5–1	0.75	1–3	2.0
Organic Vapour Condensing (low viscosity)	1–2.5	1.75	3–11	7.0
Condensing Steam	5–15	10.0	5–11	8.0

Table 2.3: Typical value of U of double-pipe and shell-and-tube HEs for different situations

Heat Transfer Situation	**U ($kW/m^2.K$)**
Gases to/from Gases (no phase change)	0.124
Organic liquid (high viscosity) to/from air (no phase change)	0.157
Organic liquid (high viscosity) to/from water (no phase change)	0.358
Organic liquid (low viscosity) to/from Organic liquid (low viscosity)	0.631
Water to/from water (no phase change)	1.059
Organic vapour condensing (high viscosity) to/from water	0.434
Organic vapour condensing (low viscosity) to/from water	0.843
Condensing steam to/from high viscosity organic liquid boiling	0.246
Condensing steam to/from low viscosity organic liquid boiling	0.856
Condensing steam to/from water	1.434

the need for large A and consequently higher capital cost. It can be seen that U is low ($\approx$ 0.14 $kW/m^2.K$) for heat transfer situations involving gases, moderate ($\approx$ 0.35 $kW/m^2.K$) for situations involving a high-viscosity organic liquid (with or without phase change) and high (> 0.5 $kW/m^2.K$) for scenarios involving low-viscosity organic liquids and/or water/steam. In particular, U for water-to-water heat transfer is quite high at about 1.0 $kW/m^2.K$.

Values of U such as those in Table 2.3 here, and in Table 9.5 in Kakaç *et al.*,[2] are adequate for analysing WHR projects. However, for HE design, it is better to consider the values of convective (or film) heat transfer coefficients (h) and fouling coefficients (h_F) involved in order to recognize the controlling resistance (i.e., lowest of these values) for heat transfer in the HE. The lowest coefficient value should be as accurate as possible for realistic sizing and costing of the HE.

2.6 Heat Integration

Many processes have a number of hot streams that have to be cooled and a number of cold streams that need to be heated. It is often possible to transfer some or all of the thermal energy recovered in cooling hot streams to some of or all the cold streams, thus heating them either partially or fully as required. This is essentially the idea and intention of heat integration, to recover and reuse thermal energy, and is very important in WHR. Similarly, energy of high-pressure gases and liquids can be recovered and reused via expansion (using power recovery turbines), thus producing shaft power. However, this is outside the scope of this book on WHR.

In this section, heat integration is introduced considering the simple situation of one hot stream and one cold stream. For this, consider bioethanol separation from a dilute mixture of water and ethanol, by distillation. Feed to this distillation column typically contains 10 wt% ethanol, is at 30°C, and its flowrate and C_p are 10 kg/s and 4.02 kJ/kg.K, respectively. It is desirable to heat this feed to 98°C in order to reduce the (heat) duty of the column reboiler (not shown separately in Fig. 2.5). This pre-heating of feed can be achieved using low-low pressure (LLP) steam at 2 bar with saturation temperature of 120.2°C and latent heat of condensation of 2201.5 kJ/kg (Fig. 2.5). Amount of LLP steam required can be found as follows.

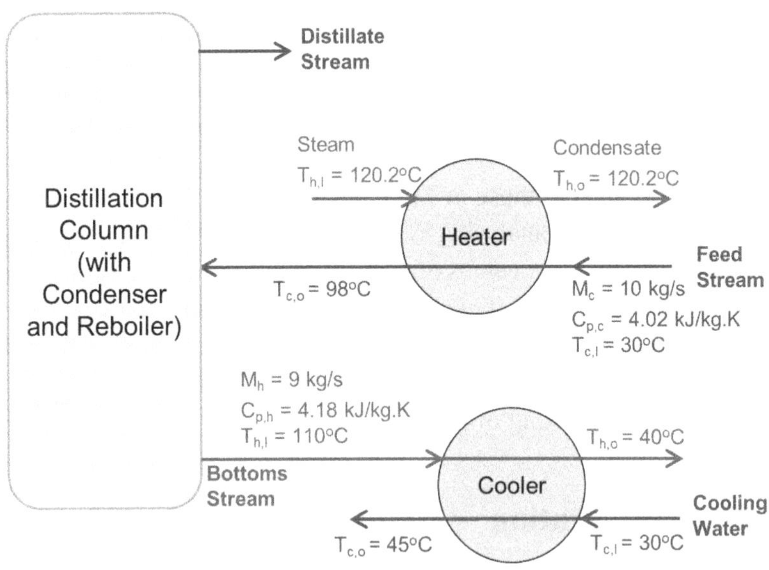

Fig. 2.5: Distillation column with pre-heating feed in a heater and cooling bottoms stream in a cooler.

First, energy required for feed pre-heating is Q = 10 × 4.02 × (98 − 30) = 2733.6 kW. This is assuming that feed before and after pre-heating will be in liquid state (i.e., no phase change). For heating, LLP steam is usually taken to be saturated steam although it is often slightly superheated to avoid condensation in the supply pipeline due to small heat loss. It condenses in the HE and leaves HE as saturated liquid (sub-cooling to below saturation temperature is small with negligible contribution to heating). In essence, only latent heat of condensation (in phase change from saturated steam to saturated water) is available for heating. Accordingly, LLP steam required for pre-heating the feed to distillation column is $\frac{2733.6}{2201.5} = 1.242$ kg/s or 4.47 ton/hr. Assuming LLP steam cost is US$ 20/ton and 8,500 hr of operation per year, annual cost of feed pre-heating is 4.47 × 20 × 8,500 = US$ 759,920. Obviously, this is substantial.

An HE is required for heating the feed using LLP steam, and it is often termed a heater because a process stream is being heated using a hot utility. Its area can be calculated using Eqs. 2.13 and 2.14; this requires a U value. In the heater, steam is condensing and feed is a dilute mixture of ethanol and water. For this situation, U for condensing steam to water (= 1.434 kW/m^2·K) in Table 2.3 can be used. For this feed heater, calculated $(\Delta T)_{LM}$ = 48.5°C, and area is 39.3 m^2.

Bottoms stream of the distillation column is liquid at a higher temperature and often needs to be cooled for storage or further processing. This stream from the ethanol–water column is essentially water (>99.5 wt%) at 110°C (and pressure of 2–3 bar) with flow rate of 9 kg/s and C_p of 4.18 kJ/kg.K. It needs to be cooled to between 35°C and 40°C for returning to upstream fermentation or disposing off as wastewater. For this cooling to 40°C, energy to be removed from the bottoms stream is 9 × 4.18 × (110 − 40) = 2633.4 kW. This cooling can be accomplished using cooling water at 30°C to be returned at around 45°C to the cooling tower for recycling (Fig. 2.5). Assuming cooling water is to be returned at 45°C and its C_p is 4.18 kJ/kg.K, cooling water required is $\frac{2633.4}{4.18 \times (45-30)} = 42.0$ kg/s or 151.2 ton/hr or 1,285,200 ton/a for 8,500 hr of operation per year. Thus, the amount of water required is huge. However, the cost of cooling water is often low at US$ 0.11/ton. Accordingly, annual cost of cooling water is 1,285,200 × 0.11 = US$ 141,372/a.

An HE is required for cooling the bottoms stream of the bioethanol column using cooling water, and it is often termed a cooler (wherein a process stream is being cooled using a cold utility). Its area can be calculated using Eqs. 2.13 and 2.14 along with a suitable U value. Heat exchange in this cooler is between the bottoms stream (mostly water) and cooling water. For this situation, U for water to water (= 1.059 $kW/m^2 \cdot K$) in Table 2.3 can be used. For this cooler, calculated $(\Delta T)_{LM}$ = 29.4°C, and area is 84.6 m^2.

The above results show that steam quantity is only 4.47 ton/hr, much lower than that of cooling water (151.2 ton/hr). On the other hand, annual cost of steam is US$ 759,920, which is about five times cooling water cost of $ 141,372/a. This is the typical situation of cost of steam (hot utility) versus cost of cooling water (cold utility). Overall, is it possible to reduce cost of utilities for feed pre-heating and bottoms product cooling of a distillation column? Indeed, this is possible. One option is by heat exchange between the feed and bottoms product streams. This is practical since both hot and cold streams are in proximity within the same section/unit of the plant. In addition to the bottoms product stream, some other hot streams may be available for pre-heating the feed to the distillation column. If energy required from steam is reduced by heat integration, energy removal by cooling water will be lowered by the same amount because of the overall energy balance.

Let's explore using the bottoms product stream to pre-heat the feed to its target temperature of 98°C. This heat integration is intuitive and an effective example of WHR. It is illustrated in Fig. 2.6 along with values of variables involved. Using Eqs. 2.11 and 2.12, HE duty, Q, is 2,733.6 kW, $T_{h,o}$ is 37.34°C (less than 40°C achieved in Fig. 2.5), approach temperatures are 7.34°C (at cold end) and 12°C (at hot end) and $(\Delta T)_{LM}$ = 9.48°C. Taking U = 1.059 $kW/m^2.K$ from Table 2.3, area of this HE is 272.4 m^2. For practice, reader should calculate and verify all these values.

Minimum approach temperature of the HE in Fig. 2.6 is 7.34°C. This may or may not be acceptable, particularly for a shell-and-tube HE. Considering minimum approach temperature and requirement on the final temperature of bottoms product stream, three possible solutions are discussed below. Of them, Solution A or B is more likely to be acceptable for implementation although it may deviate slightly from the original design

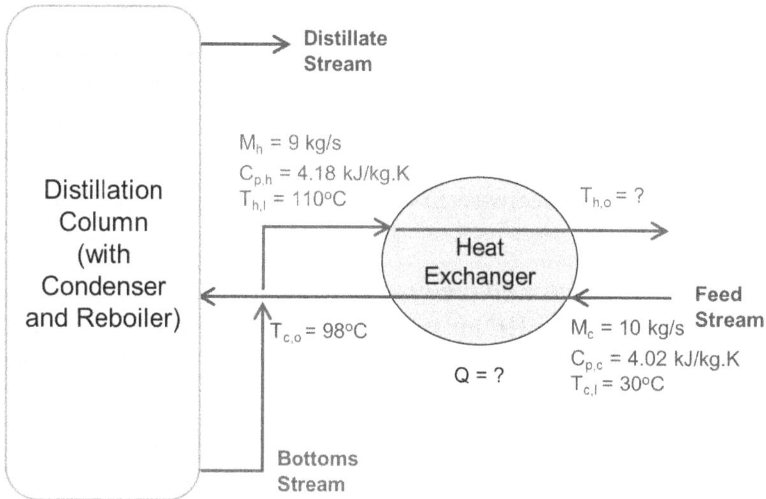

Fig. 2.6 Heat integration (exchange) between the feed and bottoms streams of a distillation column.

requirements. In any case, the following discussion indicates the need to explore possible alternatives for heat integration.

Solution A: Minimum approach temperature of 7.34°C and $T_{h,o}$ of 37.34°C in the heat integration in Fig. 2.6, are acceptable. A plate HE can achieve this minimum approach temperature. In this case, use of heat integration between column bottoms stream and feed stream will eliminate cost of steam and cooling water (i.e., operating cost) and also capital cost of heater and cooler. But it will require an HE with area of 272.4 m^2. Since HE cost is correlated with area, capital cost for this HE is likely to be more than the capital cost for heater (of 39.3 m^2) and cooler (of 84.6 m^2). Even then, operating cost saving of US$ 901,292 (= 759,920 + 141,372) from not using steam and cooling water is substantial, and can recover HE capital cost over a short period (i.e., payback period of one to two years). Moreover, not using steam will substantially decrease CO_2 emissions.

Solution B: It is required to have a minimum approach temperature of 10°C for the HE. This can be met by setting bottoms stream temperature at HE outlet ($T_{h,o}$) to 40°C. Then, for the HE, Q will be $9 \times 4.18 \times (110 - 40) = 2633.4$ kW, $T_{c,o} = 30 + 2633.4/(10 \times 4.02) = 95.51$°C and

A is 205.4 m^2. Feed pre-heating to 95.51°C (instead of 98°C specified earlier) may be acceptable as it will lead to only a slightly smaller decrease in reboiler duty, compared to solution A. In effect, the process requirement of feed pre-heating to 98°C is relaxed marginally. Such a possibility should always be explored in WHR projects. Operating cost savings, capital cost required and decrease in CO_2 emissions will be comparable to those in Solution A.

Solution C: Here, both feed pre-heating to 98°C and minimum approach temperature of 10°C for HE must be met. These can be achieved by adding a heater using LLP steam to heat feed from 95.51°C (at the outlet of HE in Solution B) to 98°C; this solution will be like Fig. 2.7 without the cooler. In this Solution C, duty of this heater and LLP steam required are $10 \times 4.02 \times (98 - 95.51) = 100.1$ kW and 0.1643 ton/hr or 1,396.9 ton/a. Cost of this LLP steam is US$ 27,938/a, which is less than 4% of US$ 759,920 without heat integration. Heat transfer area of this small heater is just 2.99 m^2, which is less than 8% of the heater area of 39.3 m^2 without heat integration. Due to small area requirement, this heater can be double-pipe type.

Thus, steam cost and capital cost of the heater in Solution C will be quite low, and benefits of Solution C will be marginally lower than those of Solutions A and B. Instead of a heater and LLP steam, the possibility of using another hot stream available in or near the distillation unit should be explored for feed heating from 95.51°C to 98°C in another HE. Such a

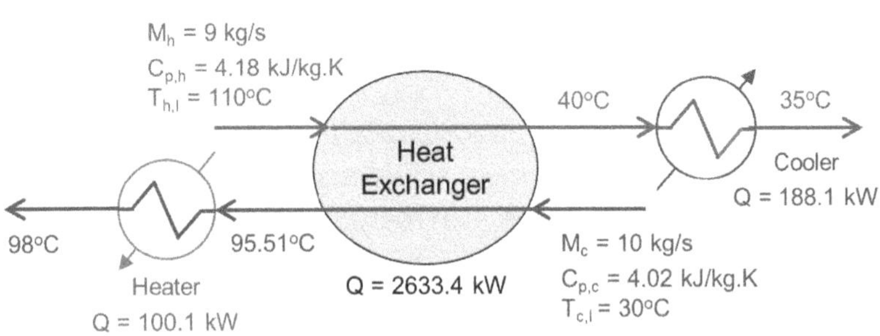

Fig. 2.7: Heat integration between feed and bottoms streams of a distillation column, meeting minimum temperature approach of 10°C for the HE and target temperatures of cold and hot streams.

case involves heat integration among more than one hot and cold streams, which is the topic for Chapter 3.

In Solutions B and C, it is assumed that bottoms stream temperature of 40°C at the outlet of HE is acceptable. In case, the bottoms stream must be cooled to 35°C, a small cooler is required. Solution C with this small cooler is shown in Fig. 2.7. Here, minimum approach temperature of 5°C is assumed to be acceptable for this cooler (e.g., using a plate HE); otherwise, chilled water, which is substantially more expensive than cooling water, will be required. This cooler duty is 188.1 kW, its area is 35.5 m^2 and it requires 9 kg/s of cooling water at a cost of US$ 30,294/a (for 8,500 hr of operation per year and cooling water cost of US$ 0.11/ton). In this case, shown in Fig. 2.7, heat integration involves a small heat exchanger network (HEN) consisting of one HE, one cooler and one heater. Heat integration and design of HEN can be facilitated by pinch analysis, which is covered in Chapter 3.

In summary, heat integration to recover and reuse waste heat is feasible and beneficial to reduce operating cost, energy required and CO_2 emissions although additional capital may be required. It often involves trade-off between operating and capital costs. However, operating cost reduction is likely to be substantial, resulting in a short payback period of a few years.

2.7 Temperature-Enthalpy Plot

A T-H plot presents temperature against enthalpy of a cold/hot stream. It is simple in concept and yet provides a visual picture of temperature driving forces and potential for heat integration. It is somewhat similar to Fig. 2.3 where stream temperature is plotted against fractional distance from hot end of a double-pipe HE. In a T-H plot, x-axis is enthalpy of the stream (instead of fractional distance). T-H plot assumes counter-current flow configuration and is valid for any type of HE. T-H is described below for Solution A in Fig. 2.6 and for HEN in Fig. 2.7.

Figure 2.6 has only one HE and two streams flowing counter-currently. Temperature and enthalpy of the hot stream decrease as heat is transferred out of it whereas temperature and enthalpy of the cold stream increase as it receives the heat. T-H plot in Fig. 2.8 displays this variation of temperature with enthalpy of cold/hot streams; here, the two curves are straight lines (linear) because C_p is assumed to be constant and enthalpy change (ΔH, kW) is given by:

$$\Delta H \text{ of hot stream} = M_h\, C_{p,h}\, (T_{h,o} - T_{h,i}) \quad (2.18)$$

$$\Delta H \text{ of cold stream} = M_c\, C_{p,c}\, (T_{c,o} - T_{c,i}) \quad (2.19)$$

In the above, ΔH of hot stream is negative since HST is decreasing as hot stream gives up heat in the HE whereas ΔH of cold stream is positive since CST is increasing.

The sign of ΔH in Eqs. 2.18 and 2.19 follows the sign convention for enthalpy changes, namely, negative sign for heat removal from the system (hot stream) and positive sign for heat supply to the system (cold stream). Eqs. 2.18 and 2.19 are similar to HE Eqs. 2.11 and 2.12 except for a few differences. First, the former set of equations is for cold/hot streams, which may flow through more than one HE (as in Fig. 2.7), whereas the latter set of equations is in the context of a singe HE (Fig. 2.4). Second, Eq. 2.11 for hot stream does not follow the sign convention for enthalpy changes (i.e., Q in both Eqs. 2.11 and 2.12 is positive).

According to Eqs. 2.18 and 2.19, and since C_p is assumed to be constant here, cold/hot stream curve in T-H plot in Fig. 2.8 is a straight line

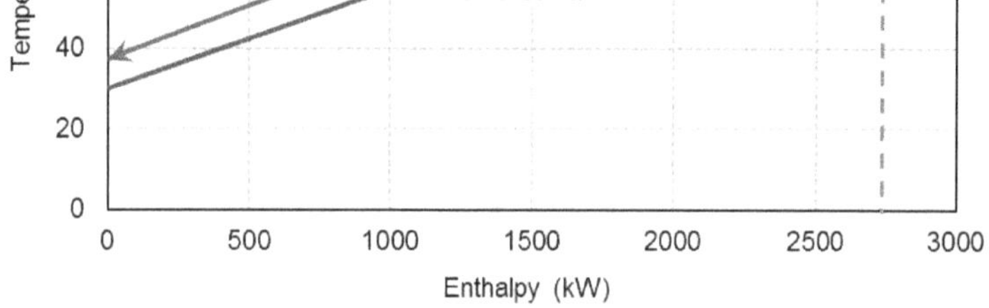

Fig. 2.8: T-H plot for heat transfer in the HE in Fig. 2.6.

with slope equal to reciprocal of MC_p. The downward arrow is used for the hot stream whereas the upward arrow refers to the cold stream. Enthalpy change of the hot stream is equal to that of the cold stream, and hence the two lines span the same range of enthalpy (as emphasized by the vertical dashed line in Fig. 2.8). In other words, some energy of the hot stream is transferred to increase enthalpy/temperature of the cold stream. This transfer is possible because HST is higher than CST at any enthalpy value. The difference, i.e., the vertical distance between the red and blue lines, is the driving force for heat transfer. Obviously, the cold and hot stream lines in the T-H plot must not touch or intersect for feasible heat transfer.

For an HE, the T-H plot is similar to the plot of temperature profiles (Fig. 2.3) but for a few differences. First, the x-axis variable is enthalpy in a T-H plot, whereas it is the fractional distance from one end of HE, in the temperature profile plot. Second, temperature profiles in Fig. 2.3 are slightly curved because temperature difference, and consequently heat transfer at a fractional distance changes from one end to another. Third, the slope of the temperature profile can be decreasing (as in Fig. 2.3) or increasing in case the hot stream enters at the right (instead of left) end of HE and the cold steam enters at the left (instead of right) end of HE (thus, retaining counter-current flow configuration). Slope of CST and HST in T-H plot (Fig. 2.8) is always positive (or zero in case of phase change).

It is possible to present streams, utilities and heat transfer in HEN (Fig. 2.7) in a T-H plot. Recall that this HEN involves only one hot stream with T changing from 110°C to 35°C and $\Delta H = 9 \times 4.18 \times (110 - 35) = 2821.5$ kW, and one cold stream with T range of 30°C to 98°C and $\Delta H = 10 \times 4.02 \times (98 - 30) = 2733.6$ kW. Fig. 2.9(a) shows these streams in the T-H plot. Here, ΔH of hot stream is more than that of cold stream and so the hot stream line is wider than the cold stream line. In Fig. 2.9(a), minimum temperature driving force is at the left end and it is 5°C only whereas minimum temperature approach is 10°C in the HEN in Fig. 2.7 (except in cooler to avoid using chilled water). Can Fig. 2.9(a) be modified for minimum driving force of 10°C? If yes, how?

The enthalpy values (x-axis) on the T-H plot in Fig. 2.9 are taken as relative to the reference enthalpy, usually taken as zero at the left end of the hot stream, for convenience. However, it is the change in enthalpy value (e.g., from the left side of the stream to the right side in Fig. 2.9) and

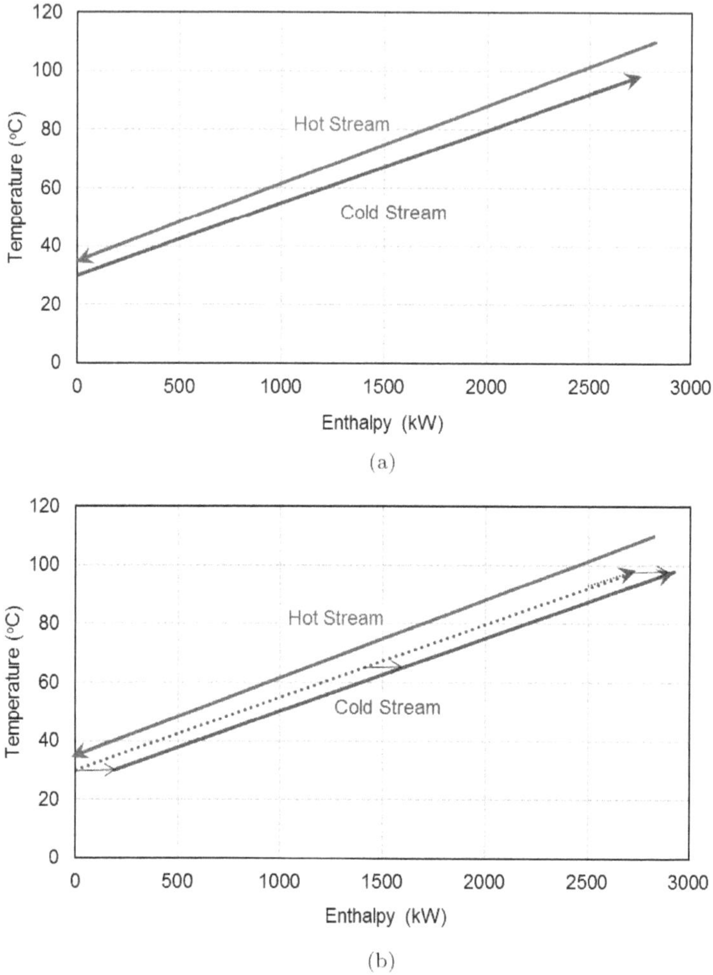

Fig. 2.9: One cold stream and one hot stream on T-H plot: (top plot, a) lines for cold/hot stream start/end at zero enthalpy, and (bottom plot, b) cold stream shifted to right for increasing temperature driving force.

not enthalpy value itself, that is important in WHR and heat transfer (beside temperature change and driving force.[a])

---

[a] Temperature change refers to that of any one stream whereas temperature driving force is the difference between HST and CST.

The cold stream line needs to begin at a positive (non-zero) enthalpy to satisfy the given minimum temperature driving force. Fig. 2.9(a) can thus be modified for a specified minimum driving force, by shifting the cold stream line horizontally, without violating the energy balance. Note that the line cannot be shifted vertically since given Ts cannot be changed. Also, shifting of the hot stream line, although possible, is not preferable, as that would take it into negative territory on the plot.

Shifting of the cold stream line is illustrated in Fig. 2.9(b), wherein the dotted blue line is as in Fig. 2.9(a) and the continuous blue line is after shifting. Note that the entire line for the cold stream is shifted horizontally to the right and hence T range and enthalpy change of the cold stream remains unchanged. Extent of cold stream shifting can be calculated by simple analysis. In Fig. 2.9(b), minimum temperature driving force is at the left end and we want it to be 10°C. For a CST of 30°C, therefore, HST should be 40°C. Enthalpy difference of hot stream between 35°C and 40°C is $9 \times 4.18 \times (40 - 35) = 188.1$ kW. Hence, the cold stream line should be shifted horizontally to the right by 188.1 kW.

As can be seen in Fig. 2.9(b), the shifting has increased the vertical distance between the two lines (i.e., temperature difference between HST and CST). But, the cold stream line goes beyond the hot stream line on the top right, which means hot utility is required for this extended part of the cold stream. Also, there is no cold stream below some part of the hot stream line at bottom left, which requires cold utility to remove some energy from the hot stream.

Figure 2.10 shows the T-H plot with the addition of cold and hot utilities: double blue line for cooling water at the bottom left end and double red line for steam at the top right end. Hot utility line is horizontal since steam temperature does not change during condensation for heating a process stream. Also, temperature driving force for this heating is more than 10°C, as indicated by vertical distance between hot utility and cold stream lines. On the other hand, cold utility line has some slope because cooling water temperature increases as heat is removed from the hot stream, and temperature driving force for this cooling is 5°C, as indicated by smaller vertical distance between hot stream and cold utility lines.

Enthalpy change from y-axis to the first vertical dashed line is the cold utility required (188.1 kW, which is the value used for shifting the cold

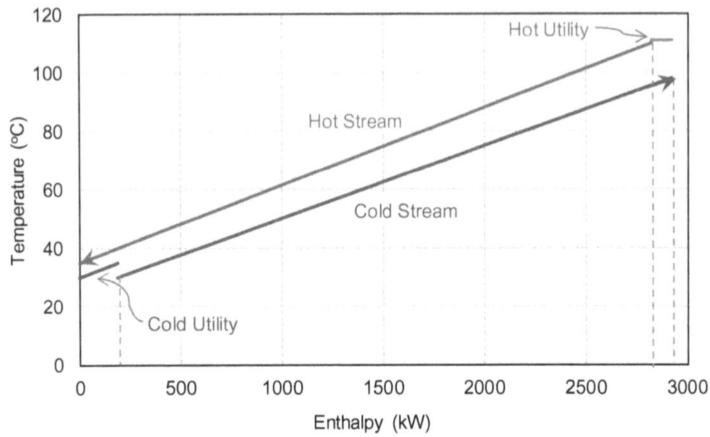

Fig. 2.10: T-H plot for one cold stream and one hot stream, and minimum temperature driving force of 10°C; cold and hot utilities are respectively at bottom left and top right of the plot.

stream line), between the two middle vertical lines is the energy recovery from hot stream to cold stream (= 9 × 4.18 × (110 − 40) = 2633.4 kW), and between the last two vertical dashed lines is the hot utility required (= 10 × 4.02 × (98 − 30) − 2633.4 = 100.2 kW). These can be read approximately from T-H plot or calculated accurately by energy balance of relevant streams/utilities. As shown in Fig. 2.7, one cooler (for cooling the hot stream at the bottom left end of T-H plot), one HE (for heat recovery) and heater (for heating the cold stream at the top right end of T-H plot) are required.

T-H plot such as that in Fig. 2.9(b) or 2.10 indicates heat recovery potential and available temperature driving forces, which determine size and consequently cost of HEs required for heat recovery. In this section, it is introduced for one cold stream and one hot stream only. In fact, T-H plot can be prepared for any number of cold/hot streams well before designing the HEN that can involve any number of HEs, heaters and coolers. These will be covered in Chapter 3.

2.8 Exergy Concepts

Why is exergy (i.e., another quantity) required for energy analysis of engineering systems? As stated in the beginning of this chapter, energy occurs in different forms such as chemical, electrical, mechanical (shaft work) and thermal energy, which are all measured in the same SI unit of J or kJ. However, quality[b] (or value or usefulness) of these different forms of energy is not the same. Thermal energy in steam at a high temperature is more valuable than that in steam at a lower temperature. Difference in the (economic) value of different forms of energy can be seen in their unit prices. According to Table 8.3 in Turton *et al.*,[3] unit cost of electrical energy, high pressure (41 barg) steam, medium pressure (10 barg, with power generation credit) steam and low pressure (5 barg, with power generation credit) steam are US$ 16.8/GJ, US$17.7/GJ, US$ 14.19/GJ and US$13.28/GJ, respectively. Unit costs and economic analysis can account for different value of energy forms employed in industrial systems including WHR system. However, these unit prices of electricity, steam and other utilities vary with time and location.

Exergy (denoted here as Ex) accounts for different values of energy forms in a fundamental way on the basis of thermodynamics instead of monetarily. According to Dincer and Rosen,[4] exergy of a system (which can be a substance or energy) is the maximum possible (shaft) work that can be done by the interaction between (i.e., composite of) the system and the chosen environment. In other words, exergy is the maximum possible work produced by bringing the system into thermodynamic equilibrium with the environment (which can be chosen to be some desired or ambient conditions). Maximum possible work in this definition implies the processes involved for bringing the system into thermodynamic equilibrium should be reversible and ideal. Further, the processes involve interactions between the system and environment only. Finally, a system that is already in thermodynamic equilibrium with the environment has zero exergy, and units of exergy is the same as that of energy (kJ).

Electrical energy can be converted into mechanical energy (shaft work) using a motor and maximum possible work thus obtained is the same as the electrical energy in the best case of a 100% efficient motor. Conversely, shaft work can be converted into electrical energy in a

[b] Energy quality is different from steam quality, which is the mass fraction of vapor in a vapor–liquid mixture of steam/water. However, high values of both are better and valuable.

generator (opposite of motor), and both are equal in the case of a 100% efficient generator. Thermal energy such as that in combustion products and steam at high temperature can be converted into shaft work in turbines. This conversion is limited by the ideal Carnot heat engine or cycle, whose efficiency is $\left(1-\frac{T_c}{T_h}\right)$; here, T_h and T_c are temperature (in K and not in °C) of energy source and sink, respectively.

Thus, maximum possible work from Q kJ of thermal energy (e.g., latent heat of condensation in steam) at constant T_h is only a fraction $\left(=1-\frac{T_c}{T_h}\right)$ of Q and the remaining energy $\left(=\frac{T_c}{T_h}Q\right)$ is rejected at constant T_c. Hence, exergy of Q kJ of thermal energy at constant T_h is $Ex=\left(1-\frac{T_c}{T_h}\right)Q$. For example, assuming T_c is 300 K, exergy of 250 kW of latent heat in high pressure steam at 254.7°C (= 527.85 K) is 107.9 kW whereas exergy of 250 kW of latent heat in low pressure steam at 111.4°C (= 384.55 K) is only 54.97 kW. These exergy values depend on T_c or on the assumed environment, and they will be more in case T_c is lower (e.g., in cold places). In case T_h or T_c is not constant, then it should be replaced by thermodynamic average temperature; see Section 2.2.4 in Ghannadzadeh *et al.*[5].

The above exergy is the physical or thermomechanical exergy since only physical processes are involved in bringing the system of thermal energy into equilibrium with the environment. On the other hand, chemical exergy is the maximum possible work from a substance when it is brought into equilibrium (in terms of temperature, pressure and composition as well) with the chosen environment (called dead state) through chemical reactions. For example, hydrocarbon fuels at the environment temperature and pressure have chemical exergy, which is nearly equal to their higher heating value (HHV).

In summary, exergy of energy or substance is the maximum possible (shaft) work that can be obtained from the energy/substance. It can be equal to energy itself (as in the case of mechanical and electrical energy) or less than energy (as in the case of thermal energy). It is interesting to recall that the energy/exergy unit of 1 J is the work done by a force of 1 Newton over a distance of 1 m. Other names of exergy include work capability (or potential), available (or usable) energy, and essergy.

The first law of thermodynamics states the conservation of energy. Second law of thermodynamics has several formulations; one is the Clausius' statement that heat transfer occurs in the direction of temperature decrease, and another is Kelvin–Planck's statement that it is impossible to convert all of the heat (say, 100 kW) from a high temperature source into equal amount (i.e., 100 kW) of work. Exergy is based on both first and second law of thermodynamics. It brings different forms of energy to the same basis, namely, maximum possible work (i.e., mechanical energy or shaft work). This is particularly relevant and required for thermal energy or steam, which may be available at different temperatures. Although exergy does not depend on economic value, it depends on the assumed environment for final equilibrium.

It is well known that energy is conserved; it can be transformed but it cannot be created or destroyed. On the other hand, exergy is usually not conserved since destruction of exergy occurs in real processes due to the irreversibility involved. Exergy destruction or exergy consumption or internal exergy loss implies reduction in maximum possible work, quality and usefulness of energy. Hence, it should be minimized as far as possible. This requires minimizing irreversibility in the unit/process since exergy destruction and irreversibility are positively correlated. Low exergy destruction in a system (which can be an equipment or process) indicates closeness of its operation to reversibility (i.e., ideality).

Phrases such as energy consumption and energy efficiency are used by the public and also often encountered in technical communications. Scientifically, both these phrases are questionable since energy is always conserved. Probably, energy consumption in day-to-day communications refers to consumption of valuable energy resource(s), and energy efficiency refers to efficient use of valuable energy for the desired purpose. For example, the reboiler of a distillation column uses or consumes (valuable) steam energy; in reality, energy in steam supplied to the reboiler is mostly transferred to vapour produced (desired purpose) and the rest to the ambient as heat loss (not the desired purpose). The modern LED (light emitting diode) lights are energy efficient compared to earlier light types, to provide the same amount of lighting (desired purpose) using less electrical power.

In technical communications, it is better to employ exergy, exergy destruction (or consumption) and exergy efficiency instead of energy consumption and efficiency, for the following reasons. First, exergy has sound basis from thermodynamics and quantifies maximum potential work of all energy forms and substances, on a common basis. Second, although it depends on the chosen environment, exergy is independent of prices of energy and substances, which vary with both location and time. Third, exergy destruction in an equipment or process indicates extent of deviation from reversibility and consequently potential to improve the equipment/process. Fourth, since exergy is not conserved, it is meaningful to define exergy efficiency to quantify exergy destruction (irreversibility) in the equipment/process.

2.9 Exergy Analysis

Consider a continuous equipment or process (such as HE or HEN or separation process) operating at steady state. Energy analysis involves the following equation based on energy conservation:

$$\text{Energy In} = \text{Energy Out} \tag{2.20}$$

Here, Energy In must include all energies entering and Energy Out must include all energies leaving the equipment/process. These energies can be in any form such as potential, kinetic, chemical, electric, work and heat, and some of them can be associated with material streams. It is possible to ignore some of these if they are absent or negligible but there should be clear understanding and justification (e.g., negligible heat loss) for ignoring any energy. See the assumptions stated in Section 2.4 for developing HE Eqs. 2.11 and 2.12.

Exergy balance of the same equipment/process is given by:

$$\text{Exergy In} = \text{Exergy Out} + \text{Exergy Destruction} \tag{2.21}$$

Here, Exergy In and Out are similar to Energy In and Out of energy balance. They must include all material and energy flows, and each exergy flow should include physical (due to temperature and pressure), chemical, potential and kinetic components, whichever are applicable. Chemical, potential and/or kinetic (due to fluid motion) exergy of a material stream can be ignored if they do not change when that stream enters and leaves the equipment/process. In other words, this material stream is undergoing only temperature and pressure changes (and not in composition, level/head and velocity) in the equipment/process.

The last term (Exergy Destruction or consumption) in Eq. 2.21 accounts for irreversibility of the unit/process. For example, heat transfer from HST to CST involves finite temperature gradient, which contributes to irreversibility and to exergy destruction that can be reduced by decreasing temperature gradient. However, smaller temperature gradient means larger heat transfer area and consequently larger HE or intensified HE with a higher heat transfer coefficient.

If there are some waste streams (e.g., heat loss, flue gas leaving a furnace or wastewater leaving a separation unit) besides desired streams leaving the system, then Exergy Out in Eq. 2.21 is the sum of exergy in desired outlet streams and exergy in the waste stream(s). External loss of exergy refers to exergy lost in the waste stream(s); it is in contrast to internal loss of exergy, which is due to irreversibility of the process in the system. Internal and/or external loss of exergy may have to be reduced to improve the performance of the system.

Exergy efficiency can be defined in different ways considering exergy destruction (i.e., internal loss of exergy) and/or external loss of exergy; see Section 2.3 in Ghannadzadeh *et al.*[5] for an overview of several definitions. Simple exergy efficiency is the ratio of Exergy Out to Exergy In. This is easy to understand. But it may be close to unity in case most of Exergy In is unaffected by the system (e.g., chemical exergy is unaffected in an HE). Simple exergy efficiency of close to unity may indicate little scope for improving the system, which may not be the case (e.g., there is potential to reduce destruction of physical or thermomechanical exergy in an HE).

A better alternative to simple exergy efficiency is the rational exergy efficiency, which is defined as the ratio of desired exergetic effect to exergy used. Here, exergy used is the sum of desired exergetic effect, external loss of exergy and internal loss of exergy. For a heater or cooler, desired exergetic effect is the exergy change in the process stream and exergy used is the physical exergy of a utility. Understanding of the equipment/process and its purpose is required to identify desired exergetic effect and exergy used. Table 4 in Ghannadzadeh *et al.*[5] presents rational exergy efficiency of common unit operations.

Analysing rational exergy efficiency of each equipment in the process (instead of the entire process) is a better way to understand how to improve the performance of each equipment and consequently the entire process. If the exergy efficiency of any equipment is found to be low, then identify reason(s) for the irreversibility of the equipment and explore ways to reduce internal loss of exergy. Table 3 in Ghannadzadeh *et al.*[5] summarizes sources of irreversibility and possible improvements for many unit operations. For an HE, irreversibility is due to large and/or non-uniform driving force as well as pressure drop. So, these should be addressed to improve the performance of an HE that has low exergy efficiency. Temperature driving force is uniform or similar throughout an HE having counter-current flow (Fig. 2.3) whereas it changes from a large to a small value in case of co-current flow, which is less efficient. Exergy analysis and its usefulness are illustrated with a few examples in Section 2.10.

Although exergy analysis is useful, its application in industrial practice is very limited. Reasons for this are (a) exergy concepts and analysis are new to practitioners, (b) exergy values of streams/substances are required and (c) exergy values depend on the environment state and so can change from one source to another. However, researchers have been applying exergy analysis to processes. Exergy values are becoming available in process simulators such as Aspen Plus. Ghannadzadeh *et al.*[5] implemented an exergy calculator in ProSimPlus®. Hence, exergy analysis is likely to find use in industrial practice in the coming years. Considering the very limited use of exergy analysis in practice currently and the industrial focus of this book, exergy analysis is covered only in this chapter.

2.10 Application of Exergy Analysis

Consider heating a stream, C1 from 30°C to 98°C with flow rate of 10 kg/s using LLP steam (at 2 bar), ST1, and also cooling a hot stream, H1 from 110°C to 35°C with flow rate of 9 kg/s using cooling water, CW1. These are shown in Fig. 2.11 along with stream data. Suitable pressures of streams and pressure drops in equipment are assumed. For illustration, specific enthalpy and specific entropy of each stream are found using steam/water tables, which are readily available at different temperatures and pressures (e.g., at, https://www.steamtablesonline.com/steam97web.aspx). Alternatively, these values of a process stream can be found from a

Stream	C1	C3	ST1	ST2	H1	H3	CW1	CW2
T (°C)	30.0	98.0	120.2	119.4	110.0	35.0	30.0	45.0
P (bar)	2.00	1.80	2.00	1.95	3.00	2.80	2.00	1.90
Flow Rate (kg/s)	10.0	10.0	1.292	1.292	9.0	9.0	45.18	45.18
Enthalpy (kJ/kg)	125.92	410.73	2706.24	501.29	461.48	146.89	125.92	188.59
Entropy (kJ/kg.K)	0.43673	1.2843	7.12686	1.52147	1.41854	0.50507	0.43673	0.63855
Exergy (kJ/kg)	6.6256	59.9085	759.4692	85.6837	73.9897	8.9259	6.6256	14.1693
Enthalpy Flow (kW)	1259.2	4107.3	3495.5	647.5	4153.3	1322.0	5688.8	8520.1
Exergy Flow (kW)	66.256	599.085	980.973	110.674	665.907	80.333	299.323	640.122

Fig. 2.11: Heater for heating stream C1 (left image) and cooler for cooling hot stream H1 (right image) with stream data.

commercial simulator such as Aspen Plus. It is best to take these values from the same source for consistency and to avoid differences due to reference conditions for enthalpy and entropy used in alternative sources.

In the stream data in Fig. 2.11, enthalpy flow (kW) is the product of mass flow rate (kg/s) and specific enthalpy (kJ/kg), and exergy flow (kW) is the product of mass flow rate and specific exergy (kJ/kg). Steam tables often give specific enthalpy and specific entropy, and not specific exergy. Here, exergy is calculated using the following relation, which is valid for physical (thermomechanical) exergy due to changes in T and P (and not chemical exergy due to composition changes).

$$Ex(T,P) = \{H(T,P) - H(T^o,P^o)\} - T^o \{S(T,P) - S(T^o,P^o)\} \qquad (2.22)$$

Here, H and S are specific enthalpy and entropy of the stream at T and P given in brackets, and T^o and P^o are the reference conditions. In many steam tables, reference state is saturated liquid at the triple point of water (namely, 0.01°C and 0.006117 bar), where specific internal energy and the specific entropy are equal to zero. Hence, $T^o = 273.16$ K, $H(T^o,P^o) = 0$ and $S(T^o,P^o) = 0$. Equation 2.22 for (physical) exergy of streams in Fig. 2.11 simplifies to:

$$Ex(T,P) = H(T,P) - T^o\, S(T,P) \qquad (2.23)$$

This equation, specific enthalpy and specific entropy (from steam tables), and $T^{o} = 273.16$ K are used to calculate specific exergy in Fig. 2.11.

With the exergy flow values in Fig. 2.11, exergy analysis for the heater can be performed as follows:

Ex In = 66.256 + 980.973 = 1047.229 kW
Ex Out = 599.085 + 110.674 = 709.759 kW
Ex Destruction = 1047.229 − 709.759 = 337.47 kW

Simple exergy efficiency of heater is $\frac{709.759}{1047.229} = 0.6777$. This indicates scope to improve heater performance. One possibility for improving the heater is using lower temperature heat source instead of LLP steam, to decrease temperature gradient and hence irreversibility in the heater but this will decrease temperature driving force and increase heat transfer area required. Similar analysis gives exergy efficiency of cooler = $\frac{80.333 + 640.122}{665.907 + 299.323} = 0.7464$, which is higher than that of heater. This efficiency can be improved by increasing outlet temperature of cooling water.

Exergy efficiency of both the heater and cooler in Fig. 2.11 can be improved substantially by heat integration, as shown in Fig. 2.12, to transfer major portion of thermal energy in H1 to C1. Duties of heater, HE and cooler in Fig. 2.12 are respectively 126.3, 2721.8 and 109.5 kW. The heater and cooler duties in Fig. 2.12 are much lower than heater duty of 2848.0 kW and cooler duty of 2831.3 kW in Fig. 2.11. Minimum approach temperature of cooler in Fig. 2.12 is only 2.9°C, which may or may not be acceptable. If it is not, heat recovery in HE can be reduced to increase H3 temperature to 40°C, which will affect performance of HEN. The following discussion, however, is assuming that minimum approach temperature of 2.9°C is acceptable.

Using the stream data in Fig. 2.12, simple exergy efficiency of heater, HE and cooler individually are respectively 0.9906, 0.9012 and 0.9866. These high values are due to smaller and uniform temperature gradient in these units. Consequently, simple exergy efficiency of HEN (i.e., complete process) in Fig. 2.12 is also high at 0.9016. It can be seen that the exergy efficiency of HEN is closer to that of the HE (which has a much

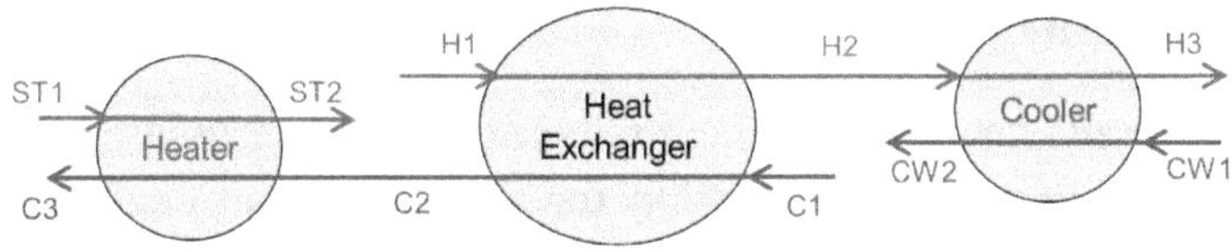

Stream	ST1	ST2	C3	C2	C1
T (°C)	120.2	119.4	98.0	95.0	30.0
P (bar)	2.00	1.95	1.80	1.90	2.00
Flow Rate (kg/s)	0.057275	0.057275	10.0	10.0	10.0
Enthalpy (kJ/kg)	2706.2	501.2895	410.73	398.10	125.92
Entropy (kJ/kg.K)	7.126856	1.521474	1.284299	1.250107	0.436733
Exergy (kJ/kg)	759.4692	85.68372	59.90853	56.61985	6.625625
Enthalpy Flow (kW)	154.9987	28.71112	4107.278	3980.99	1259.237
Exergy Flow (kW)	43.49824	4.907494	599.0853	566.1985	66.25625

Stream	H1	H2	H3	CW1	CW2
T (°C)	110.0	37.9	35.0	30.0	35.0
P (bar)	3.00	2.90	2.80	2.00	1.90
Flow Rate (kg/s)	9.0	9.0	9.0	5.2435	5.2435
Enthalpy (kJ/kg)	461.48	159.0603	146.89	125.92	146.81
Entropy (kJ/kg.K)	1.418537	0.544207	0.50507	0.43673	0.50510
Exergy (kJ/kg)	73.98971	10.40476	8.92589	6.62562	8.83641
Enthalpy Flow (kW)	4153.296	1431.543	1322.02	660.281	769.801
Exergy Flow (kW)	665.9074	93.64285	80.333	34.7415	46.3337

Fig. 2.12: HEN for heat integration of one hot stream and one cold stream, and stream data.

larger heat duty compared to both heater and cooler). Hence, exergy efficiency of HE and consequently HEN can be improved by decreasing temperature driving force and increasing heat transfer in the HE (although both these changes will increase heat transfer area of the HE). Minimum approach temperature of the HE in Fig. 2.12 is 7.9°C. This may be reduced to 5°C or even lower considering operation, equipment and capital cost requirements.

2.11 Summary

Basics of energy, heat transfer, heat integration and exergy are covered in this chapter to provide a foundation for the subsequent chapters in this book. Key points in this chapter are as follows:

- Thermal (or heat) energy of a substance is one of many forms of energy. It increases with temperature and with phase change from liquid to vapour.
- Enthalpy (relative to a reference state/value) and heat capacity of a substance are important quantities in WHR.
- Heat transfer occurs by conduction, convection and radiation, and driving force for it is the temperature difference. Of these, convective heat transfer and heat transfer coefficient (inversely proportional to resistance) are important in WHR.
- HEs are equipment for transferring thermal energy (enthalpy change) from a hot stream to a cold stream. Their analysis involves Eqs. 2.11–2.14.
- Heat transfer in an HE involves several heat transfer coefficients (i.e., resistances). The largest (controlling) resistance has significant effect on heat transfer area (i.e., size of HE) and hence capital cost of HE. This should be addressed for increasing heat transfer in an HE.
- Heat integration between a hot stream and a cold stream decreases utilities required and operating cost. It can involve one or more HEs (including heaters and coolers).
- T-H plot shows curves of temperature versus enthalpy of cold and hot streams involved in heat integration. It gives a visual picture of available temperature driving forces and scope for heat integration.
- Exergy of a system (which can be a substance or energy) is the maximum possible (shaft) work that can be done by the composite of the system as it is brought into equilibrium with the chosen environment.
- Exergy is often not conserved (unlike energy that is always conserved), and there can be exergy destruction due to irreversibility in an equipment or process.
- Exergy analysis can be used to assess and find ways to improve the performance of an HE, HEN or process. This requires exergy values of streams in the process.

Fundamentals and important equations of heat transfer, HEs, heat integration and exergy analysis covered in this chapter are sufficient for understanding the remaining chapters in this book. Detailed coverage of HEs and their design is available in the book by Kakaç *et al.*[2] whereas Dincer and Rosen[4] describe thermodynamic background, exergy concepts and exergy analysis of a variety of processes including heat pump systems, steam power plants, cryogenic and liquefaction systems. The paper by Ghannadzadeh *et al.*[5] is excellent for a concise description of exergy concepts and exergy analysis using a process simulator.

References

1. Smith R. (2005) *Chemical Process Design and Integration.* John Wiley.
2. Kakaç S, Liu H, Pramuanjaroenkij A. (2012) *Heat Exchangers: Selection, Rating, and Thermal Design*, 3rd ed. CRC Press.
3. Turton R, Bailey RC, White WB, *et al.* (2013) *Analysis, Synthesis and Design of Chemical Processes*, 4th ed. Prentice Hall.
4. Dincer I, Rosen MA. (2013) *Exergy: Energy, Environment and Sustainable Development*, 2nd ed. Elsevier.
5. Ghannadzadeh A, Thery-Hetreux R, Baudouin O, *et al.* (2012) General methodology for exergy balance in ProSimPlus® process simulator. *Energy* **44**: 38–59.

Acronyms and Notation

Acronyms

BTU	British Thermal Unit
CST	Cold Stream Temperature
HE	Heat Exchanger
HEN	Heat Exchanger Network
HHV	Higher Heating Value
HST	Hot Stream Temperature
LED	Light Emitting Diode
LLP	Low-Low Pressure
TEMA	Tubular Exchanger Manufacturers Association
T-H	Temperature-Enthalpy
WHR	Waste Heat Recovery

Notation

A	Heat transfer area in the heat exchanger (m^2)
C_p	Heat capacity at constant pressure (kJ/kg.K = kJ/kg.°C)
C_v	Heat capacity at constant volume (kJ/kg.K = kJ/kg.°C)
Ex	Exergy (kJ/kg) or exergy flow (kW)
h	Heat transfer coefficient ($kW/m^2.K = kW/m^2.°C$)
H	Enthalpy (kJ/kg) or enthalpy flow (kW)
k	Thermal conductivity (kW/m.K = kW/m.°C)
M	Mass flow rate (kg/s)
P	Pressure (bar)
Q	Duty of heat exchanger (kW)
q	Heat transfer rate (kW)
R	Resistance to heat transfer (K/kW = °C/kW)
T	Temperature (°C or K)
t	Thickness of wall (m)
U	Overall heat transfer coefficient ($kW/m^2.K = kW/m^2.°C$)
u	Internal energy (kJ/kg)

Greek Symbols

σ	Stefan–Boltzmann constant ($= 5.67 \times 10^{-5}$ $kW/m^2.K^4$)

Subscripts/Superscripts

a	Air
c	Cold
F	Fouling
h	Hot
i	Inner or inlet
o	Outer or outlet
w	Wall (of pipe)

Exercises

2.1 Consider two simplifications in Example 2.2. (a) Assuming R_w is negligible and so $T_{c,i} = T_{c,o}$, find q and $T_{c,o}$. Compare these values with those found in Example 2.2. (b) Assuming both R_i and R_w are negligible and so $T_h = T_{c,o}$, find q and compare it with that found in Example 2.3. Which of these simplifications is acceptable?

2.2 In Examples 2.3 and 2.4, hot stream flow rate increases to 11 kg/s (from 9 kg/s) and all other data remain unchanged. Find the HST at the outlet of the HE, heat duty, approach temperatures and A of the HE. Does the minimum approach temperature occur at the hot or cold end of the HE?

2.3 In Examples 2.3 and 2.4, outlet temperature of hot stream is specified as 35°C (instead of cold stream outlet temperature of 95°C) and all other data remain unchanged. Find the CST at the outlet of the HE, heat duty, approach temperatures and A of the HE. Does the minimum approach temperature occur at the hot or cold end of the HE?

2.4 A process has one hot stream and one cold stream, and their data are as follows. Hot stream flow rate, heat capacity and temperature range are 5 kg/s, 4.1 kJ/kg.K and 140°C to 40°C, respectively. Cold stream flow rate, heat capacity and temperature range are 4.2 kg/s, 3.8 kJ/kg.K and 30°C to 160°C, respectively. Present these two streams on a T-H plot for a minimum driving force of 10°C. From this plot or otherwise, state the possible heat recovery, range of temperature driving force for heat recovery and the (minimum) cold/hot utilities required.

2.5 Develop an HEN for heat integration of the two streams in the previous exercise. Available hot utility is medium-pressure steam at 186°C and cooling water at 30°C (to be returned at or below 40°C).

2.6 Investigate the effect of using low-pressure steam at 4.5 bar (instead of LLP steam at 2 bar), on the exergy efficiency of the HEN in Fig. 2.12. Use enthalpy and entropy values of all streams from the same source for steam/water tables.

2.7 Answer the following with yes (right) or no (wrong), and give brief reasoning, where applicable.
   (A) Does enthalpy decrease with increase in temperature?
   (B) Can there be heat transfer from a cold surface to a hot surface? If so, what are the requirements?

(C) In a counter-current HE, can the HST curve intersect the CST curve?
(D) In an HE, should heat transfer coefficient be small or large? Why?
(E) Is heat integration between two cold streams meaningful? Why or why not?
(F) In a typical T-H plot, enthalpy scale on x-axis starts from zero. Instead, can enthalpy scale start from any negative or positive value? Why or why not?
(G) Should hot stream line be above or below cold stream line in a T-H plot? Why?
(H) What is the relationship between exergy destruction and irreversibility?
(I) For improving exergy efficiency of an HE, heater or cooler, should the temperature driving force be reduced or increased? Why?

Chapter 3

Fundamentals of Pinch Analysis for Heat Integration

3.1 Overview

Heat integration (HI) of cold and hot streams in the process is a key strategy in waste heat recovery (WHR) and reuse. It results in a heat exchanger network (HEN) with a number of heat exchangers (HEs), heaters and coolers for heat transfer, designed taking into consideration, together, all relevant process streams and cold/hot utilities. Although an HEN can be developed based on experience, and improved by iterative design, systematic methods such as optimization and/or pinch analysis (PA) will usually result in more comprehensive, plantwide heat recovery looking at all options. Optimization is mathematical oriented and can find the optimal HEN (without much interaction with the user); however, it does not provide insights into HI and HEN development. On the other hand, PA provides sound basics, compelling concepts and effective procedures for quantifying HI benefits, analysis of existing/proposed HENs and HEN design.

This chapter covers the fundamentals of PA for HI and HEN design, all with illustrative examples. Section 3.2 introduces PA for HI. Stream data required for PA and guidelines for choosing streams for HI are covered in Section 3.3. Next, composite curves for finding minimum utilities (i.e., utility targets), pinch and its significance are covered in Section 3.4. The procedure for finding the corresponding target on heat transfer area is described in Section 3.5. Target on minimum number of units (i.e., HEs, heaters and coolers together) is covered in Section 3.6. Finding optimum minimum driving force, $(\Delta T)_{min}$ based on targets is presented in Section 3.7. Subsequently, Section 3.8 describes HEN and its representations. Section 3.9 illustrates analysis of a given HEN whereas Section 3.10 describes HEN design. Finally, this chapter ends with summary in Section 3.11.

The learning outcomes of this chapter on PA for HI are:

1. Outline PA for HI
2. Find targets for utilities, heat transfer area and number of units
3. Distinguish between pinched and threshold problems
4. Describe the procedure for finding optimum $(\Delta T)_{min}$
5. Identify and present HENs in different representations
6. Analyse HENs for finding the potential to improve them

This chapter is essential for learning PA, analysing HENs and WHR projects. It is useful as a refresher to those who have already learnt PA for HI. Depending on the reader's background, it may be better to read this chapter after studying heat transfer and HI presented in Chapter 2.

3.2 What is Pinch Analysis?

HI is performed between a given set of cold and hot streams in the process. PA consists of a number of techniques/procedures for this HI and analysis/design of HENs. It has two main steps: targets, followed by HEN development. In the first step, targets, namely, minimum utilities, total heat transfer area and minimum number of units are found for a given $(\Delta T)_{min}$. These targets can be used to estimate the operating cost (OC) and capital cost (CC) of the HEN (without actually designing it). By repeating this procedure for a number of $(\Delta T)_{min}$, the optimal value of $(\Delta T)_{min}$ that would minimize total annual cost (TAC, which includes both CC and OC), can be found. The second step of PA consists of the systematic development of an HEN for the given cold/hot streams and a chosen/optimal $(\Delta T)_{min}$. Several techniques/procedures are available for the design of an HEN meeting the targets and then for evolving it to reduce

CC, which may increase OC. Also, the performance of a given HEN (e.g., designed by an engineer or already existing/operating in the plant) can be analysed using PA concepts, to assess the potential for its improvement.

The development of PA began around 1975, driven by the sudden and substantial increase in crude petroleum (oil) prices in 1973. The first application of PA was in revamping an existing HEN for (pre-)heating crude petroleum, which is the feed to a crude distillation column in a petroleum refinery. The feed is heated using a number of intermediate and product streams of the column in an HEN consisting of many HEs. This complex and large HEN is known as the crude pre-heat train. The objective of this revamping application in the early 1980s was to increase the plant capacity by 25%. The HEN designed by PA, with the increased capacity, achieved energy savings of 10% over the existing plant with the original capacity and 25% over contractor design as well as payback of the capital investment within months. For more details of this application, see Chapter 9 in Kemp.[1]

The above significant and successful application of PA in the early 1980s attracted lot of attention from both academia and practitioners, and also led to quick acceptance and adoption of PA in industry. Thus, PA has been employed in design/consulting companies and in process industries since 1980s. Software for PA is now readily available within process simulators such as Aspen Plus. Some free programs for PA are also available (e.g., THEN at https://the-heat-exchanger-network.software.informer.com/1.0/). An MS Excel based pinch program available with Kemp[1] is easy-to-use and effective, particularly for utility targets. Although capabilities of these programs vary, they are adequate for PA in many applications.

The strengths of PA include sound basis from thermodynamics, compelling concepts, simple and effective methods/procedures, the engineer is actively involved (compared to solution using optimization, which is more automated), provides deeper insights into the HI problem under investigation and is useful for identifying potential improvements of HENs. This chapter describes many PA concepts, methods and procedures, which can be applied to WHR projects involving HI.

3.3 Stream Data for Pinch Analysis

HI is performed between the cold and hot streams in the process. For this, PA requires mass flow rate (M kg/s), heat capacity (kJ/kg.°C), supply or initial temperature (T_S in °C), target or final temperature (T_T in °C) and heat transfer coefficient including fouling (h kW/m^2.°C) of each of these cold/hot streams. These data are similar to that required for analysing an HE as in Section 2.4. Table 3.1 shows the stream data required for PA of the process in Fig. 3.1. This process and its stream data will be referred to as the example process and example stream data, respectively, in the rest of this chapter. It is the product of mass flow rate and heat capacity (MC_p, kW/°C) that is required for PA, and not M and C_p separately; so, this is what is presented in Table 3.1. In some literature, symbol for this product is CP, which may be confused with C_p for heat capacity, and hence it is not used in this book. The pressure of streams is not required for PA, unless pressure drop in HEs, and pumping requirements, are to be studied.

Assumptions made in compiling the stream data in Table 3.1 for the example process are: the process is at steady state; C_p is constant; only T is changing for each cold/hot stream and phase and pressure of a stream may change, but composition remains the same. In short, HI is between cold/hot streams whose temperature and/or phase change from supply to

Table 3.1: Stream data required for PA of the process in Fig. 3.1

Stream No.	MC_p (kW/°C)	T_S (°C)	T_T (°C)	h (kW/m^2.°C)
C1	20	30	180	0.43
C2	35	60	180	1.6
H1	45	200	70	1.6
H2	30	120	50	2.1

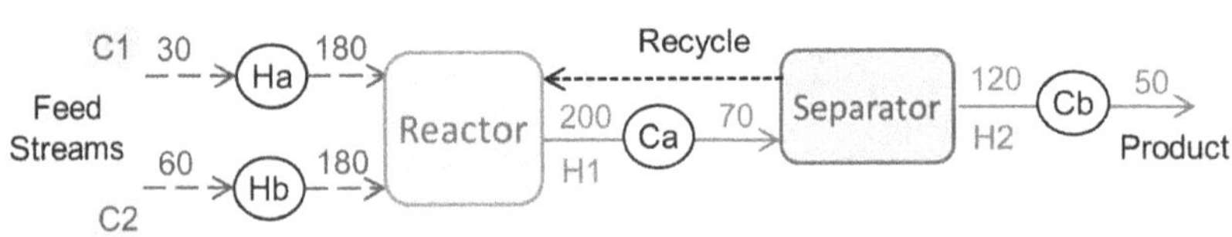

Fig. 3.1: A simplified process showing cold and hot streams (with temperatures in °C), heaters (Ha and Hb) and coolers (Ca and Cb).

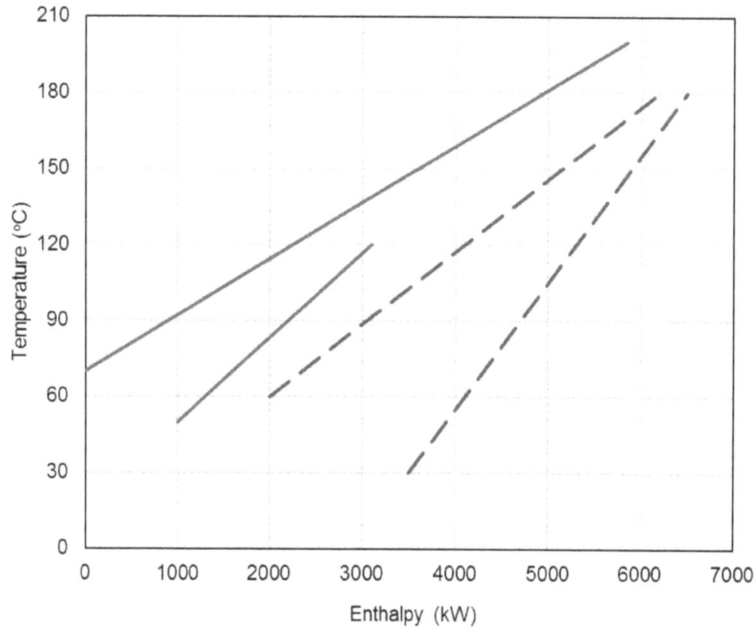

Fig. 3.2: T-H plot of streams in Table 3.1.

target conditions. The temperature and enthalpy change (ΔH) of each of the streams in Table 3.1 can be represented as a straight line (with slope = $\frac{1}{MCp}$) in the temperature-enthalpy (T-H) plot (Fig. 3.2). Recall this plot introduced in Chapter 2 and match the lines in the plot with the corresponding streams in Table 3.1.

The temperature (and consequently enthalpy) of streams in Table 3.1 changes significantly. However, temperature may not change much (despite significant change in enthalpy) for a stream undergoing phase change either partially or fully. For example, this occurs in the case of a condenser and reboiler of a typical distillation column (Fig. 3.3). Here, the condenser has one vapour inlet stream (at 105.5°C) and two liquid streams (both at 105.1°C), and 3,300 kW of energy is removed using a cold utility (CU). For HI, this condenser can be taken as one hot stream from 105.5°C to 105.1°C, and its $MC_p = \frac{3{,}300}{(105.5-105.1)} = 9{,}250\ kW/°C$. Note that this MC_p is fictitious to account for energy removed from (i.e., duty of) the condenser, and is related to mass flow rate and latent heat of condensation (and not heat capacity). Similarly, the reboiler in Fig. 3.3 is

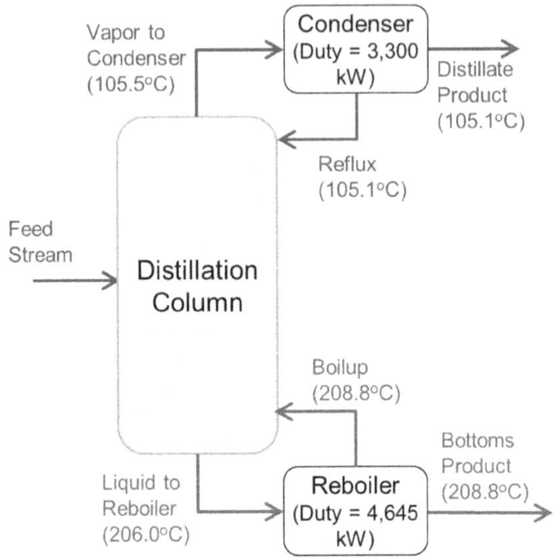

Fig. 3.3: Typical distillation column showing a condenser and a reboiler along with their duties, and temperatures of relevant streams for HI.

equivalent to a cold stream (since 4,645 kW of thermal energy is supplied); it can be taken as a cold stream from 206.0°C to 208.8°C, and its $MC_p = \frac{4,645}{(208.8-206.0)} = 1,659$ kW / °C.

In summary, the distillation process in Fig. 3.3 has one hot stream (with T_S = 105.5°C, T_T = 105.1°C and MC_p = 9,250 kW/°C) and one cold stream (with T_S = 206.0°C, T_T = 208.8°C and MC_p = 1,659 kW/°C). As stated earlier, these are sufficient for HI and PA, and M, C_p and latent heat are not required separately. MC_p of these two streams is substantially larger than those in Table 3.1 due to large latent heat and small temperature change.

As seen by the cold and hot stream temperatures (HSTs) in the distillation process (Fig. 3.3), HST can be lower than cold stream temperature (CST). In HI, a hot stream means it requires heat removal for cooling from the supply to target temperature; on the other hand, a cold stream requires heating (heat supply) to increase its temperature from supply to target value.

The following points should be considered for choosing cold/hot streams for HI.

- Enthalpy change: choose those streams with large ΔHs.
- Location: if possible, choose those streams located close by (in order to avoid piping cost for to-and-fro transfer, safety and/or operational issues if a stream is located far away in the plant).
- Heat transfer coefficient: low values of heat transfer coefficients increase HE sizes and consequently CC. Hence, choose streams with high heat transfer coefficient values (e.g., streams with phase change, and low-viscosity liquid streams).
- Partial HI: this may be present in the existing plant or new process design. For greater heat recovery and reuse, ignore existing HI and start afresh for choosing cold/hot streams.
- Accuracy: stream data may be from the design reports or operating plant. In either case, stream data should be as accurate as possible for reliable results from PA.
- Heat capacity: if this changes significantly with temperature for a stream, then temperature range of that stream should be divided into several temperature intervals such that the heat capacity within each interval can be assumed to be a constant. Then, such a stream should be taken as several (sub-)streams for PA.
- Phase change: for a stream undergoing partial or complete phase change, MC_p value is based on ΔH (including latent heat of vapourization/condensation) and not based on heat capacity of the stream. Such an MC_p is often large because of large latent heat over small change in temperature.
- Phase change along with cooling/heating of a stream: in a process, a stream may be heated before vapourization or may be cooled after condensation. In either case, ΔH of the stream is due to phase change for one temperature interval and due to temperature change (i.e., sensible heat) in another temperature interval. Then, such a stream should be considered as two or more (sub-)streams for PA such that MC_p of each of them can be taken to be constant.

3.4 Target on Utilities and Pinch

The minimum cold/hot utilities required for a set of cold/hot streams and for a given minimum driving force, $(\Delta T)_{min}$ can be found by two methods.

One is graphical, based on a T-H plot (known as composite curves) and another is computational (known by two names: temperature-interval analysis and problem table procedure). The graphical method provides a visual picture of potential heat recovery and driving force for it, as well as brings out concepts such as pinch and its significance. Hence, it is described here considering the example stream data in Table 3.1. See Chapter 3 in Rangaiah and Morrison[2] for details on the computational method: temperature-interval analysis (problem table procedure).

The energy balance on the given cold/hot streams can be performed as presented in Table 3.2. Here, ΔH of each stream follows the convention: negative sign for heat removal from the system (i.e., a hot stream) and positive sign for heat supply to the system (i.e., a cold stream). Net ΔH is the algebraic sum of ΔHs of all cold/hot streams involved; it is $3{,}000 + 4{,}200 - 5{,}850 - 2{,}100 = -750$ for the streams in Table 3.2. This means 750 kW of energy needs to be removed from the set of C1, C2, H1 and H2 streams for achieving the respective target temperature. This assumes heat transfer from hot to cold streams as per energy available in them; for example, C1 receives 3,000 kW from H1, and C2 receives the remaining 2,850 kW of H1 and 1,350 kW of H2 with the remaining 750 kW of H2 to be removed using a CU. In this, the temperature driving force for heat transfer is not considered. Hence, minimum CU of 750 kW based on energy balance may or may not be feasible.

To find the minimum utilities, an energy balance is not sufficient and temperature driving force should also be considered. This can be done using a T-H plot. Figure 3.2 is the T-H plot of the two cold and two hot streams in Table 3.1. It is possible to represent the (two) hot streams together into one curve, known as the hot composite curve (HCC), by

Table 3.2: Energy balance for the streams in Table 3.1

Stream No.	MC_p (kW/°C)	T_S (°C)	T_T (°C)	ΔH (kW)
C1	20	30	180	$20 \times (180 - 30) = +3{,}000$
C2	35	60	180	$35 \times (180 - 60) = +4{,}200$
H1	45	200	70	$45 \times (70 - 200) = -5{,}850$
H2	30	120	50	$20 \times (30 - 180) = -2{,}100$
			Net ΔH (kW)	−750

Table 3.3: Temperature intervals and ΔHs for plotting the HCC

Temperature (°C)	Interval No.	Streams Present	ΣMC_p (kW/°C)	ΔH in the Interval (kW)	Enthalpy at a Temperature (kW)
200					7,950
	1	H1	45	45 × (120 − 200) = −3,600	
120					7,950 − 3,600 = 4,350
	2	H1, H2	45 + 30	75 × (70 − 120) = −3,750	
70					4,350 − 3,750 = 600
	3	H2	30	30 × (50 − 70) = −600	
50					600 − 600 = 0

considering temperature intervals and hot streams present in each temperature interval. Required data and calculations for HCC for the hot stream data in Table 3.2 are shown in Table 3.3; calculations in this table are outlined below, starting from the left column to the right column.

In Table 3.3, there are three temperature intervals: first from 200°C to 120°C, second from 120°C to 70°C and third from 70°C to 50°C. These are chosen based on hot stream data (Table 3.2) and such that same set of hot streams occurs throughout an interval. Here, interval 1 has one hot stream (H1), interval 2 has two hot streams (H1 and H2) and interval 3 has one hot stream (H2). Then, ΔH in each interval (in column 5 in Table 3.3) is calculated by multiplying the sum of MC_p (in column 4) of streams present in that interval by the temperature range of that interval. In other words, ΔHs of all hot streams in an interval are added,[a] which gives the total enthalpy change or enthalpy available for transfer to cold streams in that interval. The lowest temperature (50°C) in the HCC in Table 3.3 is taken as the reference for enthalpy (i.e., enthalpy = 0) for convenience, and the enthalpy at the highest temperature (200°C) is the sum of absolute ΔHs of all hot streams from their supply to target temperatures (= 5,850 + 2,100 = 7,950 kW). Starting from this sum, enthalpy at a temperature is calculated as shown in the last column in Table 3.3.

[a] This addition is for developing HCC only, and it does not mean hot streams are mixed.

The HCC is a plot of temperature and corresponding enthalpy (first and last columns in Table 3.3); this is shown as a continuous/red curve in Fig. 3.4. Similar to Table 3.3 for HCC, required data and calculations for the cold composite curve (CCC) for cold stream data in Table 3.2, are shown in Table 3.4. Observe that the temperature is decreasing for the hot

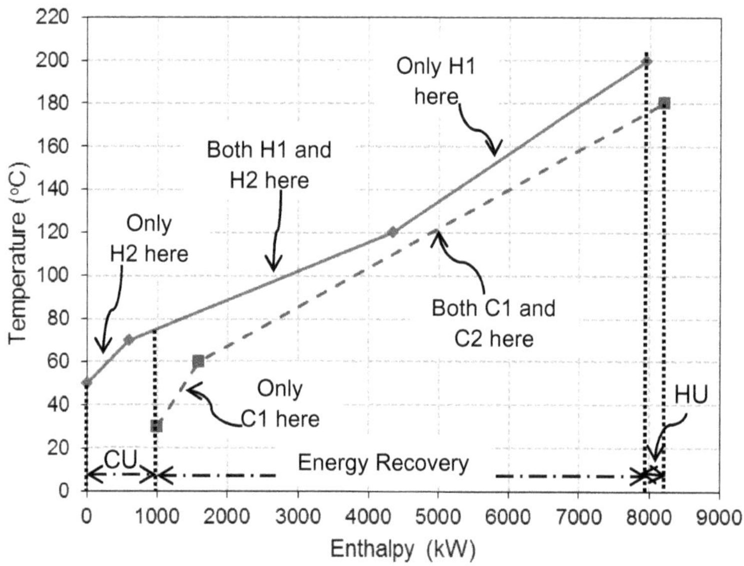

Fig. 3.4: Composite curves for the stream data in Tables 3.1 and 3.2 and $(\Delta T)_{min} \approx 10°C$, arising from arbitrary reference enthalpy (of 1,000 kW) for CCC.

Table 3.4: Temperature intervals and ΔHs for plotting the CCC

Temperature (°C)	Interval No.	Streams Present	ΣMC_p (kW/°C)	ΔH in the Interval (kW)	Enthalpy at a Temperature (kW)
30					1,000
	1	C1	20	$20 \times (60 - 30) =$ + 600	
60					1,600
	2	C1, C2	20 + 35	$55 \times (180 - 60) =$ + 6,600	
180					8,200

streams, and the (reference) enthalpy at the lowest temperature is zero in Table 3.3. On the other hand, in Table 3.4, the temperature is increasing for the cold streams, and the (reference) enthalpy at the lowest temperature is arbitrarily chosen. The CCC is shown as a dashed/blue curve in Fig. 3.4.

Compared to the T-H plot for one hot stream and one cold stream in Fig. 2.8, the HCC/CCC is a series of straight lines whose number is equal to number of temperature intervals in Tables 3.3 and 3.4. Hot/cold streams involved in each straight line can be identified based on Tables 3.3 and 3.4. In Fig. 3.4, energy from the hot streams can be transferred to the cold streams in the middle portion (between the middle two dotted vertical lines), and it is approximately 7,000 kW. The temperature driving force for this is the vertical distance between the HCC and the CCC, and the minimum driving force is about 10°C (at HST of 120°C). Some hot utility (HU) is required for heating the cold streams at the top right (between the last two dotted vertical lines) whereas CU is needed for cooling the hot streams at the bottom left (between the first two dotted vertical lines).

How can minimum utilities be found for $(\Delta T)_{min}$ equal to, say, 20°C from the composite curves in Fig. 3.4? This requires shifting the CCC such that the minimum driving force between HCC and CCC is $(\Delta T)_{min}$ = 20°C. Like shifting a cold stream on the T-H plot in Fig. 2.8, the entire CCC can be shifted horizontally to left or right (and not vertically). (Although the HCC can also be shifted horizontally, this is generally avoided since the HCC ends at zero enthalpy, convenient for reading enthalpy values from the plot of composite curves.) In Fig. 3.4, the CCC should be shifted to the right to increase the driving force to 20°C. This can be done approximately by drawing a curve parallel to the CCC or accurately by changing the arbitrarily chosen enthalpy (1,000 kW in Table 3.4) at the starting point of the CCC, as described in the next paragraph. Like shifting the cold stream horizontally in Fig. 2.8, shifting of the CCC horizontally from the position in Fig. 3.4 to that in Fig. 3.5 does not affect any energy balance because enthalpy change (ΔH) of all cold streams is preserved.

The starting value of enthalpy for the CCC for the desired $(\Delta T)_{min}$ can be calculated by energy balance in a suitable section. For example, it can be observed from Fig. 3.4 that the temperature of the HCC at the

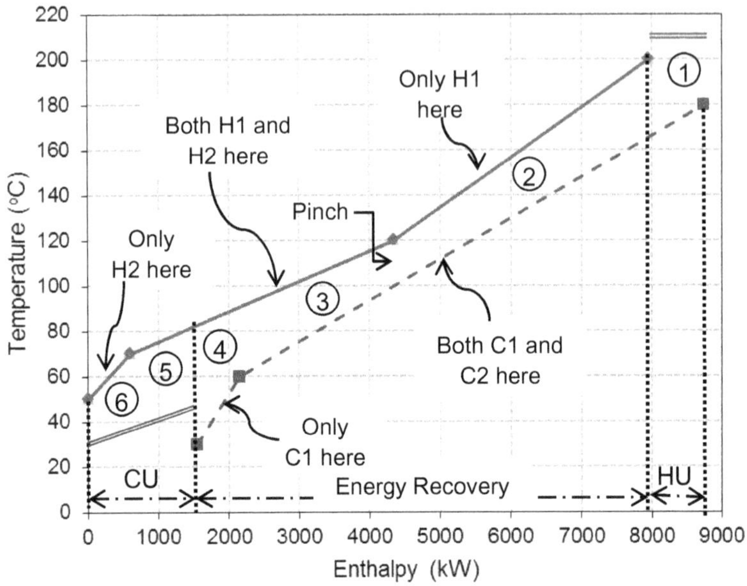

Fig. 3.5: Composite curves for the stream data in Tables 3.1and 3.2 and $(\Delta T)_{min} = 20°C$; numbers in circles refer to segments, used later for area calculations.

minimum driving force is unaffected by shifting and remains at 120°C. So, for $(\Delta T)_{min} = 20°C$, the corresponding temperature of the CCC is 120 − 20 = 100°C. ΔH of cold stream(s) from 30°C to 100°C can be calculated based on Table 3.4 as: 20 × (60 − 30) + 55 × (100 − 60) = 2,800 kW. Noting that the enthalpy of the HCC at 120°C is 4,350 kW, the starting enthalpy of the CCC should be 4,350 − 2,800 = 1,550 kW.

Figure 3.5 presents composite curves for the stream data in Table 3.1 and $(\Delta T)_{min} = 20°C$. The point, where the driving force is $(\Delta T)_{min} = 20°C$, is known as the pinch, and is very important in PA. In Fig. 3.5, HST is 120°C and CST is 100°C at the pinch and so pinch is said to be at 120/100°C. Minimum CU (= 1,550 kW) is at the bottom left. Using the enthalpy of HCC at the highest temperature (i.e., 7,950 kW) from Table 3.3, energy recovery in the wide middle portion is 7,950 − 1,550 = 6,400 kW. Then, using the enthalpy change in the two intervals (i.e., 600 and 6,600 kW) in Table 3.4 for cold streams, minimum HU (= 1,550 + 600 + 6,600 − 7,950 = 800 kW) at the top right. Instead of these calculations, all these values can be read approximately from Fig. 3.5.

The CU is usually cooling water with supply and return temperatures of 30°C and 45°C; this is shown by the double line at the bottom left of Fig. 3.5. The CU line has a gradient because cooling water temperature increases as heat is transferred to it. The HU is often saturated steam at a high temperature, say, at 210°C, which is shown by the double line at the top right of Fig. 3.5. The HU line is a horizontal line (i.e., parallel to x-axis) since steam temperature does not change during condensation.

At the beginning of this section, from the energy balance alone (without considering the temperature driving force constraint), we calculated the minimum CU of 750 kW and no HU. Compared to this, the composite curves (Fig. 3.5) for $(\Delta T)_{min}$ = 20°C, indicate minimum CU of 1,550 kW and HU of 800 kW. In other words, both CU and HU increased by 800 kW. Based on this and Figs. 3.4 and 3.5, both CU and HU, and consequently OC, increase with increasing $(\Delta T)_{min}$. On the other hand, CC (related to heat exchange area required) of an HEN decreases with increasing driving force resulting from increasing $(\Delta T)_{min}$. This is the expected trade-off between CC and OC.

Composite curves such as those in Fig. 3.5 show the pinch (in the middle portion), CU (at the bottom left), energy recovery (in the middle part) and HU (at the top right), as well as available temperature driving force for energy recovery. All these can be grasped easily from a single plot. Figure 3.5 is a typical plot of composite curves but there can be many variations. Pinch can be at the left or right end, there can be more than one pinch, and CU or HU can be zero. In rare situations, both CU and HU are zero.

Significance of pinch, where driving force is the minimum, in PA is as follows.

- No CU should be used above the pinch, for minimum utilities.
- No HU should be used below the pinch, for minimum utilities.
- No heat transfer should take place across the pinch (i.e., from a hot stream above pinch to a cold stream below pinch), for minimum utilities.

The above are known as golden rules. Both CU and HU increase if any of them is violated. For example, use of 10 kW of CU for cooling a hot stream above pinch increases HU by 10 kW. Heating a cold stream below pinch using 25 kW of HU increases CU by 25 kW. If a hot stream above pinch transfers 15 kW to a cold stream below pinch, then both CU and HU increase by 15 kW.

A given HEN can be analysed for improvement potential by verifying whether the three golden rules are satisfied or not. This will be illustrated later in this chapter. Golden rules also provide guidance on type of new streams to be added (besides those already included in PA) in order to reduce CU or HU. A new hot stream at a temperature above pinch will reduce HU whereas a new cold stream at a temperature below pinch will reduce CU.

Moreover, the pinch divides HEN design into two sub-problems: HEN design above pinch and HEN design below pinch. This facilitates HEN design since two sub-problems are generally easier to solve than a single, larger problem. Further, each of the two sub-problems can be tackled by two persons independently. Generally, driving force is low near the pinch compared to that away from the pinch (e.g., 20°C at pinch increasing to around 40°C at either end of the energy recovery region in Fig. 3.5). HEN design is expected to be more intricate when driving force is low. Both these provide an effective and simple guideline on HEN design above/below pinch, namely, start HEN design from pinch and proceed to away from pinch.

Many problems like the example problem require both CU and HU, as in Fig. 3.5. They are known as pinched problems. In contrast, there are a number of HI problems, which require only one utility (either CU or HU). These problems do not have pinch and are known as threshold problems. In such problems, minimum driving force for heat recovery is often more than the specified $(\Delta T)_{min}$, and it can occur either at one end or between the two ends of composite curves. See Chapter 6 in Rangaiah and Morrison[2] for examples of threshold problems.

3.5 Target on Heat Transfer Area

The total heat transfer area required for energy recovery and use of minimum utilities (e.g., given by composite curves) can be found using stream data, composite curves and HE equations described in Chapter 2.

Table 3.5: Process streams, utilities and their data including heat transfer coefficients for the example process

Stream No.	MC_p (kW/°C) or CU/HU (kW)	T_S (°C)	T_T (°C)	h (kW/m^2.°C)
CU (Cooling Water)*	1,550 kW	30	45	2.1
C1	20 kW/°C	30	180	0.43
C2	35 kW/°C	60	180	1.6
H1	45 kW/°C	200	70	1.6
H2	30 kW/°C	120	50	2.1
HU (Steam)	800 kW	210	210	4.4

*If cooling water goes from 30°C to 45°C, required MC_p of CU is $= \frac{1,550}{45-30} = 103.33$ kW / °C.

The systematic procedure for estimating this total heat transfer area consists of four steps. It requires process stream and utility data including heat transfer coefficient, which are summarized in Table 3.5 for the example process.

First step is to divide the composite curves into a number of segments such that cold/hot streams and CU/HU do not change within a segment. The composite curves for the example stream data can be divided into six segments, as indicated by circles with numbers in Fig. 3.5; these segments cover those using utilities as well. Streams and/or utilities present in each of these segments are presented in column 2 of Table 3.6. As can be seen, a segment may have more than one hot stream or HU, and/or more than one cold stream or CU.

Second step is to compile temperatures of cold/hot streams, CU and HU at the two ends of each segment. These temperatures can be found from stream data and/or read-off from composite curves (Fig. 3.5). Reading from the plot is approximate; to avoid this, unknown temperature at the end of a segment can be accurately calculated based on energy balance of streams/utilities present in that segment. These calculations are illustrated in column 4 of Table 3.6.

Third step is to calculate the area required for heat transfer from hot streams/HU to cold streams/CU within each segment. This area

Table 3.6: Details of calculations for heat transfer area in the segments of the composite curves in Fig. 3.5 for the example stream data

Segment No.	Streams Present	Temperature Range (°C)	Temperature Calculations	$(\Delta T)_{LM}$ and Area
1	Steam C1, C2	210 ← 210 165.45 → 180	T of HU (steam) is unchanged. Using minimum HU = 800 kW, $T = 180 - \frac{800}{20+35} = 165.45°C$	$(\Delta T)_{LM}$ = 36.80°C A = 32.0 m^2
2	H1 C1, C2	120 ← 200 100 → 165.45	Pinch is at 120/100°C	$(\Delta T)_{LM}$ = 26.62°C A = 252.7 m^2
3	H1, H2 C1, C2	90.667 ← 120 60 → 100	$T = 120 - \frac{(20+35)\times(100-60)}{45+30} = 90.667°C$	$(\Delta T)_{LM}$ = 24.95°C A = 159.5 m^2
4	H1, H2 C1	82.667 ← 90.667 30 → 60	$T = 90.67 - \frac{20\times(60-30)}{45+30} = 82.667°C$	$(\Delta T)_{LM}$ = 40.68°C A = 42.6 m^2
5	H1, H2 CU	70 ← 82.667 35.806 → 45	$T = 45 - \frac{(45+30)\times(82.667-70)}{103.33} = 35.806°C$	$(\Delta T)_{LM}$ = 35.90°C A = 27.6 m^2
6	H2 CU	50 ← 70 30 → 35.806	Confirmation using MC_p of CU (=103.33 kJ/°C): 30 × (70 − 50) = 103.33 × (35.806 − 30) = 600 kW	$(\Delta T)_{LM}$ = 26.47°C A = 21.6 m^2

calculation for each segment utilizes Eqs. 2.13 and 2.16 (in Chapter 2) for an HE. Details of this major step are described in the following paragraphs. For each segment, there is a single area calculation based on stream data for all the streams in that segment, but this does not mean there is only one HE in that segment. The minimum number of HEs and hot/cold stream matching for HEN design are covered later.

The area calculation requires heat transfer coefficient (h $kW/m^2.°C$) of each cold/hot stream, CU and HU, and also temperatures of CU and HU. All these data for the example stream data are summarized in Table 3.5. Equations 2.13 and 2.16 from Chapter 2 can be combined and re-arranged to:

$$A = \frac{Q}{U(\Delta T)_{LM}} = \frac{1}{(\Delta T)_{LM}}\left[\frac{Q}{h_i} + \frac{Q}{h_o}\right] \quad (3.1)$$

Here, Q is the duty, $(\Delta T)_{LM}$ is the log mean temperature difference (driving force) and h_i and h_o are the heat transfer coefficient of the cold and hot streams involved. It does not matter whether the cold or hot stream is on the inside or outside of tubes of a shell-and-tube HE, but it is important to use a suitable h value. Each of h_i and h_o can include fouling, if any; then, there is no need to consider fouling coefficient/factor separately.

The HE duty, Q, is the same as the heat given by the hot stream, say, inside stream; denote this as q_i. It is also the same as the heat transferred to the cold stream, say, outside stream; denote this as q_o. Both q_i and q_o are positive and equal in magnitude. Thus, Eq. 3.1 can be re-written as:

$$A = \frac{1}{(\Delta T)_{LM}}\left[\frac{q_i}{h_i} + \frac{q_o}{h_o}\right] \tag{3.2}$$

Observe that each term in the square brackets is the heat (either given or taken up) divided by the corresponding heat transfer coefficient.

Interestingly, Eq. 3.2 can be extended to more than one cold stream and/or one hot stream involved in the heat transfer, say, in one HE or in one segment in the composite curves. The extended equation for nhu (≥ 1) hot streams and/or HU, and ncu (≥ 1) cold streams and/or CU in a segment is:

$$A = \frac{1}{(\Delta T)_{LM}}\left[\sum_{j=1}^{nhu}\left(\frac{q_{h,j}}{h_{h,j}}\right) + \sum_{k=1}^{ncu}\left(\frac{q_{c,k}}{h_{c,k}}\right)\right] \tag{3.3}$$

Here, the first summation on the right side is for all hot streams/HU, the second summation is for all cold streams/CU, $q_{h,j}$ is the heat given by (= absolute ΔH of) jth hot stream/HU, $q_{c,k}$ is the heat taken by (= absolute ΔH of) kth cold stream/CU and $h_{h,j}$ and $h_{c,k}$ are heat transfer coefficient of jth hot stream/HU and kth cold stream/CU, respectively. Energy balance of all streams and utilities involved in one segment, gives:

$$\sum_{j=1}^{nhu} q_{h,j} = \sum_{k=1}^{ncu} q_{c,k} \tag{3.4}$$

Use the above equation to confirm calculations/values.

Equation 3.3 is applied to each of the segments in Fig. 3.5 and the final results are in the last column of Table 3.6. Note that calculations in the fourth and fifth/last columns require data from Table 3.5. Area calculations for segments 1 and 3 are described below. Exercise 3.2 at the end of this chapter is on verification of the results for other segments in Table 3.6.

Heat transfer in segment 1 involves HU, C1 and C2. Based on end temperatures of this segment in column 2 of Table 3.6,

$$(\Delta T)_{LM} = \frac{(210-165.45)-(210-180)}{\ln\left[\frac{210-165.45}{210-180}\right]} = 36.80°C \tag{3.5}$$

Using the data in Table 3.5, q_{HU} = 800 kW, q_{C1} = 20 × (180 − 165.45) = 291 kW and q_{C2} = 35 × (180 − 165.45) = 509 kW. These values satisfy the energy balance: $q_{HU} = q_{C1} + q_{C1}$. Finally, using Eq. 3.3 and heat transfer coefficients from Table 3.5, area required for heat transfer in segment 1 is:

$$A = \frac{1}{36.8}\left[\frac{800}{4.4} + \frac{291}{0.43} + \frac{509}{1.6}\right] = 32.0 \text{ m}^2 \tag{3.6}$$

Heat transfer in segment 3 involves H1, H2, C1 and C2. Based on end temperatures of this segment in column 2 of Table 3.6,

$$(\Delta T)_{LM} = \frac{(90.667-60)-(120-100)}{\ln\left[\frac{90.667-60}{120-100}\right]} = 24.95°C \tag{3.7}$$

Using the data in Table 3.5, q_{H1} = = 45 × (120 − 90.667) = 1,320 kW, q_{H2} = = 30 × (120 − 90.667) = 880 kW, q_{C1} = 20 × (100 − 60) = 800 kW and q_{C2} = 35 × (100 − 60) = 1,400 kW. These values satisfy the energy balance: $q_{H1} + q_{H2} = q_{C1} + q_{C1}$. Finally, using Eq. 3.3 and heat transfer coefficients from Table 3.5, area required for heat transfer in segment 3 is:

$$A = \frac{1}{24.95}\left[\frac{1{,}320}{1.6} + \frac{880}{2.1} + \frac{800}{0.43} + \frac{1{,}400}{1.6}\right] = 159.5 \text{ m}^2 \tag{3.8}$$

Fourth and final step is to find the total heat transfer area by summing areas required for all segments in the composite curves. Thus, total area required for all six segments for heat transfer among all process and utility streams in Fig. 3.5 is 32.0 + 252.7 + 159.5 + 42.6 + 27.6 + 21.6 = 536 m^2. The above four steps are for calculating the total heat transfer area required for heat transfer from hot streams/HU to cold streams/CU while using minimum CU and HU. They do not require the number of HEs in the HEN yet to be designed. The number of segments may or may not be the same as the number of HEs (to be determined in Section 3.6).

Assumptions in the above four-step procedure for heat transfer area estimation are no heat loss, counter-current configuration, negligible wall resistance for heat transfer and heat transfer is from hot to cold streams within each segment (and not across adjacent segments). These assumptions are reasonable. The heat transfer area thus estimated (before HEN design) is for the HEN using minimum utilities, and it is expected to be within ±10% of that calculated after HEN design. Note that HEN design requires experience and may have several solutions. A major advantage of the four-step procedure is that the number and details of HEs in the HEN are not required. The calculations, as seen above, are numerous but they are systematic and can be implemented in a computer program.

3.6 Minimum Number of Units

The word 'units' in this section heading refers to all HEs, heaters and coolers in the HEN. For example, the example process in Fig. 3.1 has an HEN consisting of two heaters (Ha and Hb) and two coolers (but no HEs), and so this HEN has four units. From stream data in Tables 3.1 and 3.2, we see that the (heat) duties of Ha and Hb (on C1 and C2 streams) are 3,000 kW and 4,200 kW. Let us suppose both these are supplied by only one HU (say, high pressure steam). In other words, there is only one HU but two heaters with each supplied by the required amount of the same utility. The total amount of heat added by the HU used in the HEN in Fig. 3.1 is 3,000 + 4,200 = 7,200 kW. Similarly, duties of Ca and Cb (on H1 and H2 streams) are 5,850 kW and 2,100 kW. Often, there is only one CU (e.g., cooling water) removing this much energy in the two

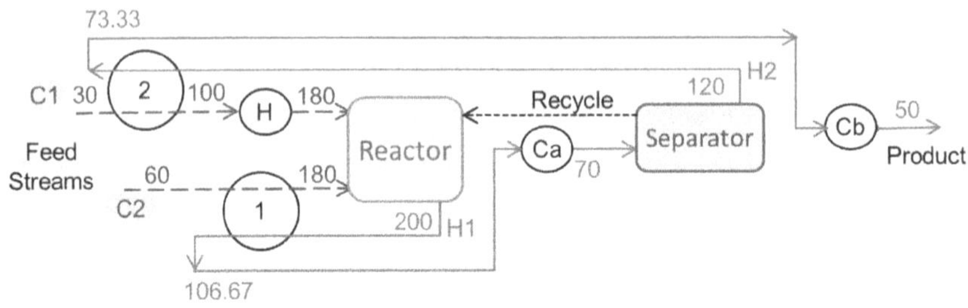

Fig. 3.6: Example process with an HEN for HI.

coolers, each using the required amount of cooling water. The total amount of heat removed by the CU used in the HEN in Fig. 3.1 is 5,850 + 2,100 = 7,950 kW.

Consider the example process with some HI in Fig. 3.6, which is developed using stream data, $(\Delta T)_{LM} = 20°C$ (same as that assumed for utility target, for comparison) and design experience.[b] The main features of the process (e.g., feed streams, reactor, separator, product and recycle) are identical in Figs. 3.1 and 3.6. The difference is in the HEN. In Fig. 3.6, the HEN has two HEs (shown as big circles with numbers inside), one heater (H) and two coolers (Ca and Cb). Hence, in this HEN, the number of units is 5, heat added by HU is 20 × (180 − 100) = 1,600 kW, heat removed by CU is 2,350 kW (with 45 × (106.67 − 70) = 1,650 kW in Ca and 30 × (73.33 − 50) = 700 kW in Cb).

In the above paragraphs, number of units is counted after HEN is designed as in Figs. 3.1 and 3.6. In fact, minimum number of units (U_{min}) can be found *without* HEN design, based on number of cold streams (nc), number hot streams (nh) and number of utility (nu) streams. For the example data (Table 3.1 or Fig. 3.6), nc = 2, nh = 2 and nu = 2 (i.e., one CU and one HU). The formula for U_{min} is:

$$U_{min} = nc + nh + nu - ns \tag{3.9}$$

Here, ns is the number of subsets among cold/hot/utility streams involved. It can be one or more, as discussed in the following paragraphs.

[b] Several solutions for heat integration and HEN are possible for the example process. This is the case for many processes. Systematic procedure for HEN design in PA is outlined later in this chapter. Even then, a few solutions are often possible for HEN design.

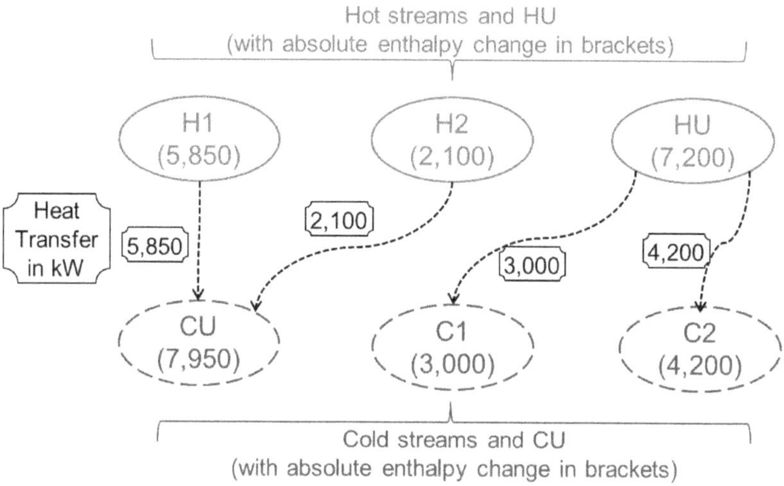

Fig. 3.7: ΔHs and heat transfer in the example process (Fig. 3.1).

To understand the above equation, its validity and applications, it is convenient to show ΔHs of streams and heat transfer in the example process (Fig. 3.1) as in Fig. 3.7. Here, ΔHs (kW) of hot streams and HU are shown in the top ovals whereas those of cold streams and CU are in the bottom ovals. Heat transfer from hot streams/HU to cold streams/CU is shown by arcs/curves, and the quantity of heat transferred (kW) is in the rectangular box. Observe that the energy balance is satisfied (i.e., sum of ΔHs of hot streams/HU = sum of ΔHs of cold streams/CU = sum of heat transferred). Heat transfer for each arc requires one unit, and four units are required in Fig. 3.7. Note that temperatures are neither shown nor considered for heat transfer in this figure.

In Fig. 3.7, H1, H2 and CU form one subset whereas HU, C1 and C2 form another subset. There is energy balance among the streams/utility within each subset, and so each subset is like a separate problem. Thus, there are two subsets (i.e., ns = 2) besides two hot streams, two cold streams and two utility streams in Fig. 3.7. Using these values in Eq. 3.9, $U_{min} = 2 + 2 + 2 - 2 = 4$, which is consistent with the heat transfer shown in Fig. 3.7. Different configurations of transferring heat from hot streams/HU to cold streams/CU within each subset are possible (although this is not the case in Fig. 3.7 because it has heaters and coolers only). Even then, number of units required is at least 4.

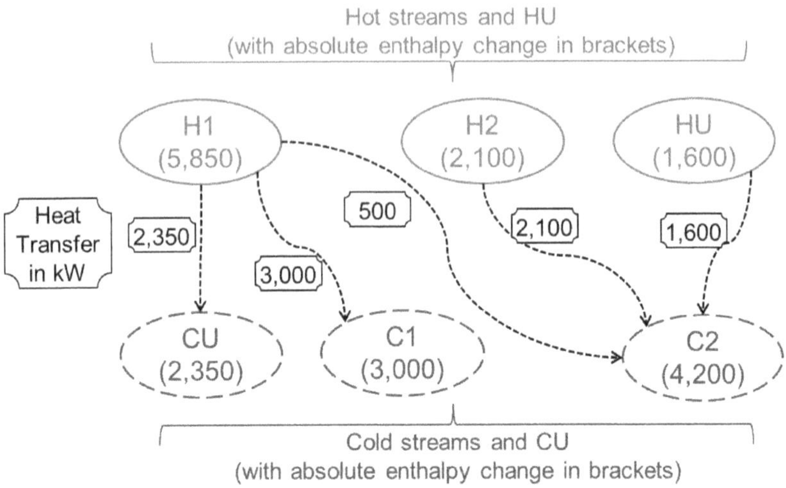

Fig. 3.8: ΔHs of streams and heat transfer in the example process with HI in Fig. 3.6.

The example process with HI (Figs. 3.6 and 3.8) requires five units. Here, it is possible to transfer heat from hot streams/HU to cold streams/CU in different configurations. Even then, number of units required is at least four. The streams/utilities in Fig. 3.8 together satisfy the energy balance, which can be used for confirmation of all values or to calculate if any one value of ΔH or utility is unknown. In Fig. 3.8, there is only one subset (ns = 1); that is, the energy balance among the streams/utility is satisfied only if the whole problem is taken together. There are five units, which is consistent with $U_{min} = 2 + 2 + 2 - 1 = 5$ (since nc = 2, nh = 2, nu = 2 and ns = 1).

The above two examples indicate that the U_{min} formula (Eq. 3.9) is correct; it can be proven using graph theory but that is beyond the scope of this chapter. The formula is simple but requires caution in its application. First, ns should be found carefully. In industrial problems, it is unlikely that ns > 1, as ΔHs will not be round numbers (i.e., ending with zero like those in Fig. 3.7) to satisfy the energy balance for a subset of streams and utilities. However, ΔHs of all streams/utilities together must satisfy the energy balance.

Second, very importantly, the U_{min} formula should be applied separately to above and below the pinch, since pinch divides the HEN design into two problems (i.e., above and below pinch). Hence, for HEN of pinch problems,

$$U_{min,MER} = U_{min,above} + U_{min,below} \quad (3.10)$$

Here, MER subscript denotes maximum energy recovery (i.e., using minimum utilities) for the given $(\Delta T)_{min}$.

Let's apply Eq. 3.10 to the example problem using minimum utilities for $(\Delta T)_{min}$ = 20°C. The HEN is not yet designed. However, from the analysis for utility target, composite curves are available in Fig. 3.5. In this example, streams (and their ΔHs in kW) above the pinch are HU (800), H1 (3,600), C1 (1,600) and C2 (2,800), and below the pinch are H1 (2,250), H2 (2,100), C1 (1,400), C2 (1,400) and CU (1,550). These satisfy the energy balance above and below the pinch separately (i.e., 800 + 3,600 = 1,600 + 2,800, and 2,250 + 2,100 = 1,400 + 1,400 + 1,550). Review of ΔHs shows that ns = 1 above and also below the pinch. Applying Eqs. 3.9 and 3.10 for this example:

$$U_{min,above} = 2 \text{ cold streams} + 1 \text{ hot stream} + 1 \text{ HU} - 1 \text{ subset} = 3 \quad (3.11)$$

$$U_{min,below} = 2 \text{ cold streams} + 2 \text{ hot streams} + 1 \text{ CU} - 1 \text{ subset} = 4 \quad (3.12)$$

$$U_{min,MER} = 3 + 4 = 7 \quad (3.13)$$

Without HI and pinch considerations, utility consumption (and therefore OC) may be more, but U_{min} is only 5, as seen in Fig. 3.8, rather than 7 with HI. Thus, the number of units (and hence CC) without considering pinch (i.e., U_{min}) can be less than the number of units (and hence CC) with pinch considerations (i.e., $U_{min,MER}$). In short, there is a trade-off between CC and OC of HEN design for HI.

The U_{min} formula (Eqs. 3.9 and 3.10) give the minimum number of units in the HEN yet to be designed. The reader may be wondering whether it is possible to design an HEN meeting the minimum utilities, $U_{min,MER}$ and specified $(\Delta T)_{min}$. This is indeed possible with the design procedures available as part of PA. Some of these will be covered later in this chapter. In fact, several HEN designs may be possible. As can be expected, difficulty of HEN design increases with the number of cold/hot streams involved.

3.7 Optimum Minimum Driving Force

A certain value of $(\Delta T)_{min}$ was assumed and used in finding minimum utilities (Section 3.4), required heat transfer area (Section 3.5) and $U_{min,MER}$ (Section 3.6). However, what is the optimum value of $(\Delta T)_{min}$ for minimizing TAC? Can this be found before HEN design? This is indeed possible, and it is described in this section. OC is mostly related to CU and HU used in the HEN; pumping and maintenance costs are minor and nearly unaffected by HEN design. So, it can be taken that OC is equal to cost of CU and HU. Then, OC can be calculated knowing minimum CU and HU (e.g., from composite curves) and unit prices of CU and HU.

CC of a unit (i.e., HE, heater or cooler) depends on its heat transfer area, material of construction and pressure. It should be the installed cost and not just the purchase cost of the unit. These are covered later in Chapter 6. For the present, assume that CC of a unit is given by:

$$CC = C + D\,(\text{heat transfer area})^{E} \tag{3.14}$$

Here, C, D and E are coefficients with appropriate values for the unit; of these, E is generally less than 1.0.

The heat transfer area in each unit of an HEN is unavailable until the HEN is designed. Only the total heat transfer area in the network and $U_{min,MER}$ are available, from the procedures described earlier. For estimating the CC of the HEN, it may be assumed that the heat transfer area in each unit is the same, and is equal to total heat transfer area divided by $U_{min,MER}$. Then, CC of HEN is given by:

$$CC = U_{min,MER} \times \left[C + D\left(\frac{\text{Total Heat Transfer Area}}{U_{min,MER}} \right)^{E} \right] \tag{3.15}$$

Experience shows that this estimate of CC assuming same heat transfer area in each of the units is reasonable and is often within ±10% of that calculated after the HEN is designed. Considering this and the approximate nature of the cost correlation (Eq. 3.14), the CC of an HEN estimated by Eq. 3.15 is sufficient for finding the optimum value of $(\Delta T)_{min}$.

The OC is incurred daily/yearly during the operation whereas CC is a one-time cost. For combining these two, the CC can be annualized by simply dividing it with the payback period, which is typically 3–5 years. The payback period is different and lower than the life of the equipment/plant. Annualized CC, ACC is $\frac{\text{CC}}{\text{Payback Period}}$ and TAC = ACC + AOC, where AOC is the annual OC.

Thus, ACC, AOC and TAC can be calculated based on the utility targets, total heat transfer area and $U_{min,MER}$, found for a particular value of $(\Delta T)_{min}$. The calculations can be repeated for a series of $(\Delta T)_{min}$ values, say, from 1°C to 40°C in steps of 1, 2 or 5. Then, ACC, AOC and TAC can be plotted against $(\Delta T)_{min}$, as illustrated in Fig. 3.9. In this plot, the following trends are typical as $(\Delta T)_{min}$ increases from small to large values.

- ACC decreases significantly and then gradually; this is mainly due to increasing driving force and consequently decreasing heat transfer area.

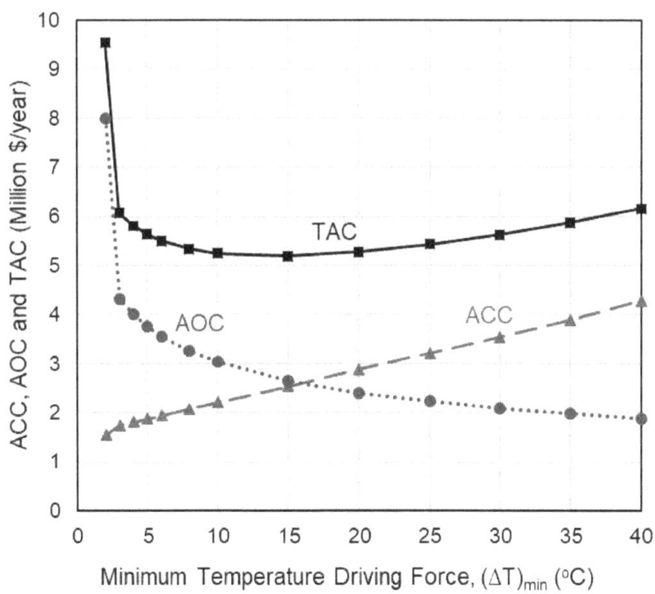

Fig. 3.9: Variation of annualized capital, operating and total costs with minimum temperature driving force.

- AOC increases steadily because of increasing utilities.
- TAC decreases initially (like ACC) and then slowly increases due to increasing AOC.
- TAC curve is nearly flat around its minimum and so any value of $(\Delta T)_{min}$ in this flat region (e.g., between 10°C and 20°C in Fig. 3.9) can be taken as the optimum.

The procedure outlined in this section for finding the optimum value of $(\Delta T)_{min}$ is known as supertargeting, in PA. It is systematic, does not require the HEN design and provides a sufficiently accurate optimum. However, it does involve numerous calculations although these can be automated in a computer program. One such program is the Energy Analyzer in Aspen Hysys/Plus.

3.8 Heat Exchanger Network Representation

Figures 3.1 and 3.6 are simple process flow diagrams (PFDs), which include HEN. Only heaters and coolers are used in Fig. 3.1, and they are shown without the utility streams. Fig. 3.6 includes HI among cold and hot streams in the process. Both these figures show process equipment (reactor and separator) besides the HEN with HEs, heaters and coolers. Thus, they provide more information about the process. However, comparison of Figs. 3.1 and 3.6 indicates that a PFD with the HEN becomes complicated with HI. Further, the example process is simple with only two process equipment, namely, reactor and separator. Many processes have more equipment, and their PFDs with HEN will be large and complex. Hence, PFDs are not convenient for developing HI and HEN. Therefore, an HEN representation, known as grid diagram or representation, was developed for HEN design and analysis.

The grid diagram of the HEN in Fig. 3.6 is presented in Fig. 3.10. Its key features and explanation are as follows.

- Hot streams are shown as horizontal lines with temperature decreasing from left to right and arrow direction to the right whereas cold streams are shown by horizontal lines with temperature increasing

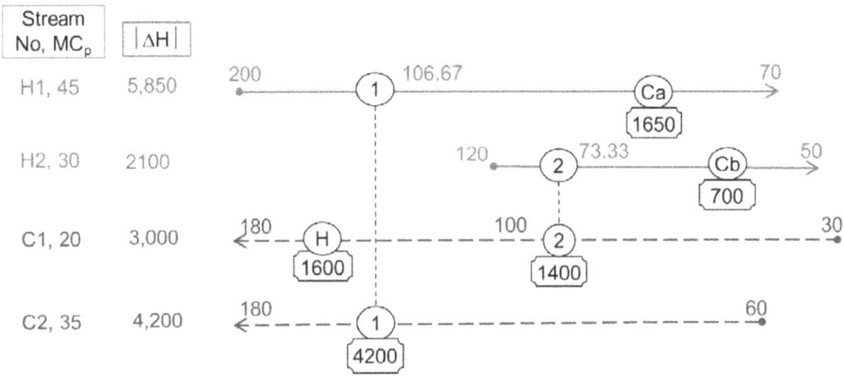

Fig. 3.10: Grid diagram or representation of HEN in PFD in Fig. 3.6.

from right to left and arrow direction to the left. This ensures counter-current configuration for heat transfer in HEs.

- Supply, some intermediate, and target temperatures are given just above each stream.
- All hot streams are in the top portion whereas all cold streams are in the bottom portion of the grid diagram. This serves as a reminder or a check that HST at the inlet or outlet of an HE should be higher than the corresponding CST.
- Name, MC_p and $|\Delta H|$ of each stream are placed on the left side of the diagram so that these data are readily available for designing the HEN, calculating intermediate temperatures of a stream and verifying values shown on the HEN.
- Only HEs, heaters and coolers (i.e., units in HEN) are shown in the grid diagram, and all other equipment (e.g., reactors and separators) in the process are excluded. Thus, some details including location of cold/hot streams are missed out. However, the advantage of the grid diagram is that the engineer can focus on HEN design without being distracted by other details.
- For simplicity, heaters and coolers are shown without the utility streams.
- The HEN in Fig. 3.10 has two HEs (1 and 2). Each HE is represented by two circles: one on a hot stream and another on a cold stream; both

Fig. 3.11: Schematic representation (employed in Chapter 2, e.g., Fig. 2.4) of HEs 1 and 2 in Fig. 3.10.

circles have the same number, and are connected with a dashed line. For illustration, Fig. 3.11 is a schematic representation, introduced in Chapter 2 (Fig. 2.4), of HEs 1 and 2, which are shown in grid representation in Fig. 3.11.

- Duty of (i.e., heat transferred from the hot to cold stream in) each unit is given just below its bottom circle. For example, duties of HE 1 and cooler Ca are 4,200 kW and 1,650 kW, respectively.
- For simplicity, dimensions of temperatures, MC_p, ΔH and duties are not included. They are assumed to be in consistent units.
- Calculated temperatures with two decimal digits are sufficient to avoid minor discrepancies due to round-off errors.

Some references employ slightly different grid diagrams (such as vertical lines instead of horizontal lines for streams; stream name, MC_p and/or ΔH on the right side; and include utility streams as well). A grid diagram that has CU and HU streams along with cold/hot streams, is known as a balanced grid diagram.

The grid diagram in Fig. 3.10 is that of an HEN without a pinch. A grid diagram for an HEN with a pinch has more details, as presented in Fig. 3.12. This is the HEN using minimum utilities for the example process stream data (whose composite curves are in Fig. 3.5). This grid diagram includes data on targets, namely, pinch and a vertical line for it, minimum HU, $U_{min,above}$, minimum CU and $U_{min,below}$. Further, on the left side, ΔH of a stream is given as two parts — ΔH above and below pinch. For example, for C1 stream, ΔH above pinch = 20 × (180 − 100) = 1,600 kW and ΔH below pinch = 20 × (100 − 30) = 1,400 kW.

In Fig. 3.12, stream H1 with MC_p = 45 is split into two sub-streams with MC_p of 35 and 10 (as shown in the left column), for heat transfer to C1 and C2 in HEs 1 and 2 while meeting the specified $(\Delta T)_{min}$ = 20°C. The

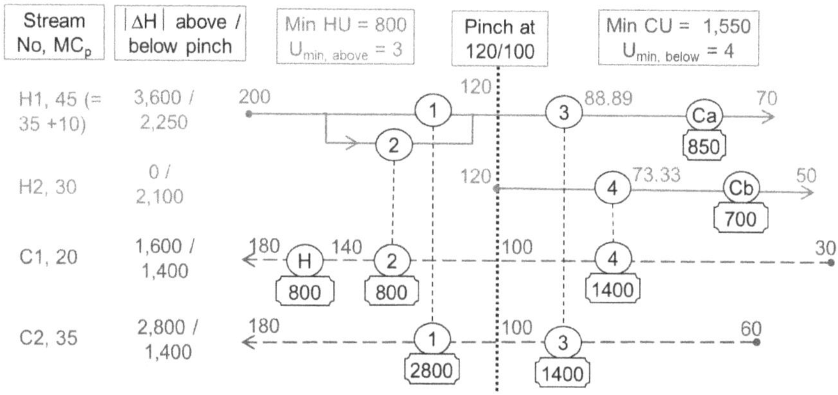

Fig. 3.12: Grid diagram of the HEN with pinch and using minimum utilities, for the example process stream data.

upper branch of H1 with $MC_p = 35$ is flowing through HE 1 whereas the lower branch with $MC_p = 10$ is flowing through HE 2. Both the branches of H1 are mixed after they exit HEs 1 and 2. Stream splitting is an accepted practice in industries, and it does not affect the target temperature of the stream leaving the HEN and consequently the rest of the process.

Depending on the problem, one or more cold and/or hot streams may have to be split, and the number of branches of each split can be two or more. In some cases, there can be two or more splits on one stream. Stream splitting greatly facilitates HEN design meeting the minimum utilities, $(\Delta T)_{min}$ and $U_{min,MER}$. However, it requires additional piping and instrumentation (to maintain the required split in the presence of fluctuations during the operation), and hence increases CC slightly. Stream splitting should therefore be avoided as far as possible.

It is possible to incorporate the HEN in Fig. 3.12 within the corresponding PFD. Figure 3.13 demonstrates this for the simple process in Figs. 3.1 and 3.6. In Fig. 3.13, each HE is shown by two separate circles with the same number (same as those in Fig. 3.12). For example, for HE 4, H2 is the hot stream (from 120°C to 73.33°C) and C1 is the cold stream (from 30°C to 100°C). This is unlike the schematic representation shown in Fig. 3.11; if each HE was to be shown with counter-current flow of cold/hot streams as in that figure, the PFD in Fig. 3.13 would become

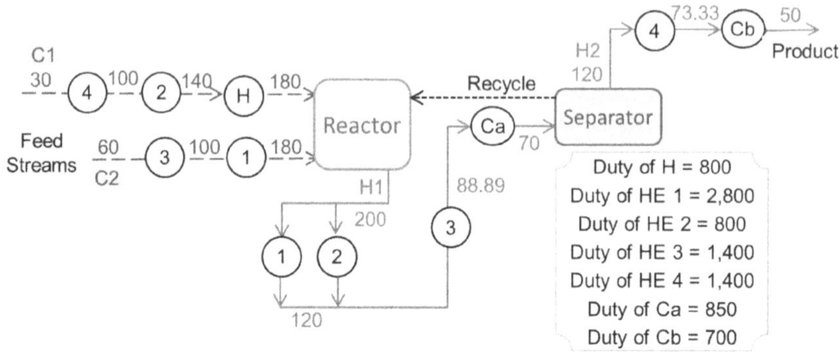

Fig. 3.13: PFD incorporating the HEN using minimum utilities in Fig. 3.12.

extremely complex with streams criss-crossing, and also make it challenging to incorporate correctly the HEN from the grid diagram into the PFD. For convenience and ready reference, duties of all units are listed besides the PFD in Fig. 3.13.

3.9 Analysis of a Given Heat Exchanger Network

A given HEN (currently operating, designed by a consultant or from some publication) can be analysed using the grid diagram (presented in Section 3.8), utility targets, pinch and golden rules (namely, no CU above pinch, no HU below pinch and no heat transfer across pinch) described in Section 3.4. This analysis indicates potential for improving the given HEN.

Consider an existing HEN in a chemical plant, shown on the grid diagram in Fig. 3.14. It involves two cold streams and three hot streams, and has three coolers, two HEs and three heaters. Among the approach temperatures of the two HEs, the minimum is 60°C on the right side of HE 2. Temperatures of CU and HU are not given, and so it is not possible to find the approach temperatures of coolers and heaters. Assuming $(\Delta T)_{min}$ = 60°C, minimum CU, minimum HU and pinch can be found for the two cold streams and three hot streams in Fig. 3.14, and they are 2,420 kW, 5,429 kW and 123/63°C, respectively. On the other hand, the HEN in Fig. 3.14 is using 1,236 + 530 + 2,215 = 3,981 kW of CU and 2,165 + 4,500 + 325 = 6,990 kW of HU, which are more than the minimum. Thus,

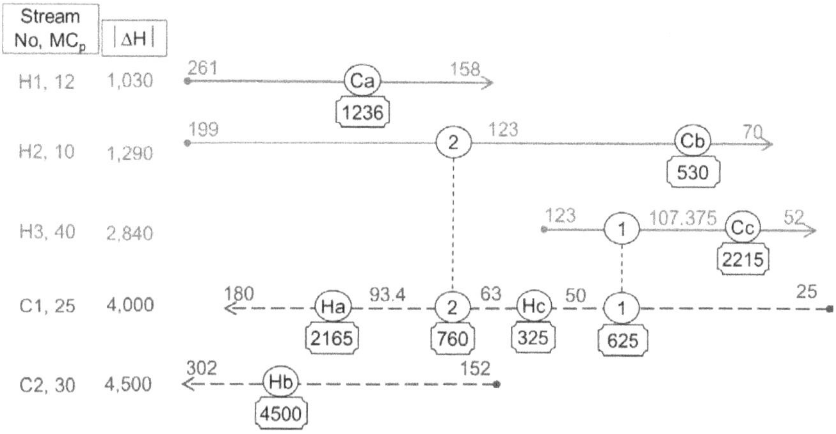

Fig. 3.14: Existing HEN in a chemical plant.

finding the utility targets and comparing with the utilities used in the given HEN indicates the potential for decreasing the utilities used in the HEN. For a fair comparison, utility targets should be found for the minimum approach temperature in the given HEN because they will be lower for a lower $(\Delta T)_{min}$.

What are the reasons for the use of more than the minimum utilities, in the existing HEN? How can the energy efficiency of the existing HEN be improved? Both these can be answered as follows. Scrutiny of Fig. 3.14 shows violation of two golden rules: CU of 1,236 kW is used above the pinch and HU of 325 kW is used below the pinch. Both these violations increase CU and HU by 1,236 + 325 = 1,561 kW in total. Therefore, it is possible to improve the HEN design in Fig. 3.14 in order to decrease both CU and HU by 1,561 kW. Besides this, driving forces in this HEN are large with $(\Delta T)_{min}$ = 60°C. It is possible to reduce $(\Delta T)_{min}$ to 30°C, find corresponding utility targets and pinch, and then analyse the existing HEN. This is given as Exercise 3.6 at the end of this chapter.

The example process with HI, given in Fig. 3.6, can be analysed similarly. HEN in this figure is using 2,350 kW of CU and 1,600 kW of HU compared to the targets of minimum CU = 1,550 kW and minimum HU = 800 kW for $(\Delta T)_{min}$ = 20°C. Thus, there is potential to decrease both CU and HU by 800 kW, by a better design of HEN. The HEN in Fig. 3.6 is using more utilities because of violation of one or more golden rules. For finding the golden rules satisfied or violated, present the HEN on the grid diagram, as already given in Fig. 3.10. Visualize the locus of pinch temperatures (120°C/100°C) on this grid diagram; here, it is through the HE 1. Keeping this locus in view, it can be seen that no CU is used above pinch and no HU is used below pinch (i.e., two of the three golden rules are satisfied). Since more utilities are used in the given HEN, it must be because the golden rule of no heat flow across pinch is violated. How can this be confirmed?

Within HE 1 in Fig. 3.10, stream H1 is transferring 45 × (200 − 120) = 3,600 kW above the pinch, whereas cold stream C2 can receive only 35 × (180 − 100) = 2,800 kW above the pinch. In other words, there is heat flow of 800 (= 3,600 − 2,800) kW across the pinch from above to below the pinch. This can also be confirmed by considering heat transfer below the pinch, as follows. Within HE 1, below the pinch temperatures, stream H1 can transfer only 45 × (120 − 106.67) = 600 kW but stream C2 is receiving 35 × (100 − 60) = 1,400 kW. This means C2 is receiving 1,400 − 600 = 800 kW of energy from above the pinch.

It is possible to use the HEN in the PFD in Fig. 3.6 for identifying violation of golden rules, if any. However, this requires careful examination of the HEN and temperatures in the PFD, and the difficulty increases with number of equipment and streams in the PFD. On the other hand, the grid diagram of the HEN (Fig. 3.10) is better for finding violation of golden rules, if any, and one is less unlikely to overlook any violation.

3.10 Heat Exchanger Network Design

After finding utility targets, pinch and $U_{min,MER}$, the HEN can be designed systematically using the grid diagram. The starting point of HEN design is the preparation of the grid diagram with the stream data and targets found. Figure 3.15 shows this grid diagram for the stream data in the example process (Fig. 3.1) along with targets found for $(\Delta T)_{min}$ = 20°C (Fig. 3.5). Main features/steps of HEN design in PA are as follows:

- Pinch divides the design problem into two sub-problems: design above and below pinch, which can be performed independently and

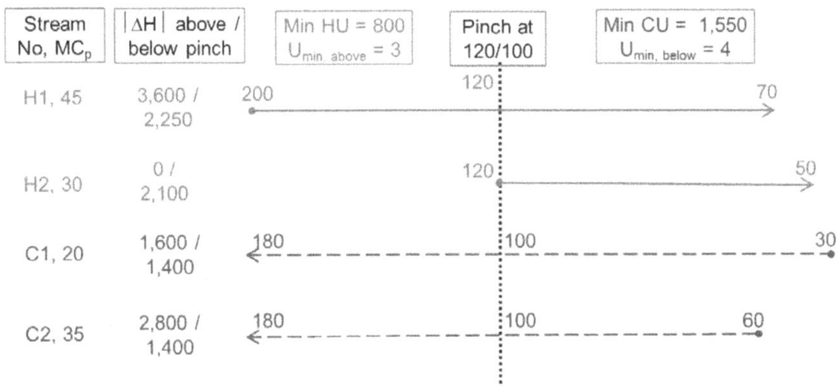

Fig. 3.15: Stream data and targets on the grid diagram for designing HEN.

sequentially (i.e., design above pinch followed by design below pinch or vice versa, whichever is convenient).

- HEN design either above or below pinch must follow the golden rules (no CU above pinch, no HU below pinch and no heat flow across pinch).
- Driving force for heat transfer, as can be seen in Fig. 3.5, is small at and near the pinch compared to that away from the pinch on both right and left sides. Hence, HEN design is more constrained and challenging near the pinch. In PA, this is tackled first.
- Based on the previous remark, HEN design (either above or below pinch) begins at the pinch, and then proceeds away from the pinch.
- For heat exchange, a hot stream is matched with a suitable cold stream ensuring that approach temperature on either side of the HE (also known as match in PA) is $\geq (\Delta T)_{min}$.
- Guidelines for stream matching at the pinch are the so-called MC_p and number criteria:

MC_p of hot stream $\leq$ MC_p of cold stream for matches above pinch (3.16)

MC_p of hot stream $\geq$ MC_p of cold stream for matches below pinch (3.17)

Number of hot streams $\leq$ Number of cold streams above pinch (3.18)

$$\text{Number of hot streams} \geq \text{Number of cold streams below pinch} \quad (3.19)$$

- If MC_p and/or number criteria are not satisfied, either a hot or cold stream is split into two or more branches. Experience and foresight (on design yet to be done) are required for selecting the best way for stream splitting.
- Heat duty of (i.e., heat transferred from hot to cold stream in) each match (HE) is maximized in order to satisfy enthalpy change requirement of cold/hot streams involved. This heuristic is to minimize the number of units and consequently CC of the HEN.
- After matching cold/hot process streams for heat exchange, remaining heating (of cold streams) required above pinch is achieved by adding heater(s) on cold streams, and remaining cooling (of hot streams) required below pinch is achieved by adding cooler(s) on hot streams.
- Stream splitting provides great flexibility for HEN design, facilitating HEN design using minimum CU/HU and $U_{min,MER}$.
- In general, several solutions for HEN design are possible. Each of them may have certain advantage.

Detailed procedure and application of the above steps for HEN design are beyond the scope of this single chapter on PA. They can be found in several chapters in books such as Rangaiah and Morrison,[2] Kemp[1] and Smith.[3] The above steps are for HEN design of pinched problems. Some of them can be adopted for HEN design of threshold problems introduced in Section 3.5. Chapter 5 in Rangaiah and Morrison[2] has a few examples on HEN design of threshold problems.

Specifically, the design for the example process with the stream data and targets in Fig. 3.15, is as follows. HEN design above pinch is described first, and it is followed by HEN design below pinch; alternatively, HEN below pinch can be designed first and then HEN above pinch. At the pinch, driving force is low at or above $(\Delta T)_{min}$. Above pinch, since no CU is allowed, cooling hot streams to pinch temperature is more difficult and it is by heat transfer to cold streams only. In other words, focus on how to cool hot streams to pinch temperature (because heating cold streams has more flexibility by using HU). Accordingly, design the matches at the pinch first.

Above pinch, there are one hot (H1) and two cold (C1, C2) streams. Hence, the number criterion: Number of hot streams ≤ Number of cold streams above pinch (Eq. 3.18), is satisfied. However, MC_p of hot stream (H1) is 45, MC_p of cold stream (C1) is 20 and MC_p of cold stream (C2) is 35; therefore MC_p criterion: MC_p of hot stream ≤ MC_p of cold stream for matches above pinch (Eq. 3.16), is not satisfied for H1 with either cold stream. Hence, H1 needs to be split.

Since C2 has the maximum of the ΔH's (2,800, 1,600) of the cold streams, we can try to match that with available enthalpy in H1, in order to maximize heat duty of that match. This would require a branch of H1 with $MC_p = 35$ (since $\Delta T = 200 - 120 = 80$°C) such that $MC_p \times \Delta T = 2{,}800$, giving us Match 1. That would leave $MC_p = 10$, with $\Delta T = 200 - 120 = 80$°C and $MC_p \times \Delta T = 800$ for the other branch of H1 to transfer to C1 (Match 2). The resulting outlet temperature of C1 would be 140°C $(= 120 + \frac{800}{10})$. We can thus confirm that approach temperature on either side of the HE is $\geq (\Delta T)_{min}$ of 20°C. Figure 3.16 shows HEN design above pinch until this point.

The remainder of the heating of C1 to 180°C, can be done by placing a heater (H) with duty of 800 kW (indicated by ? in Fig. 3.16). Complete

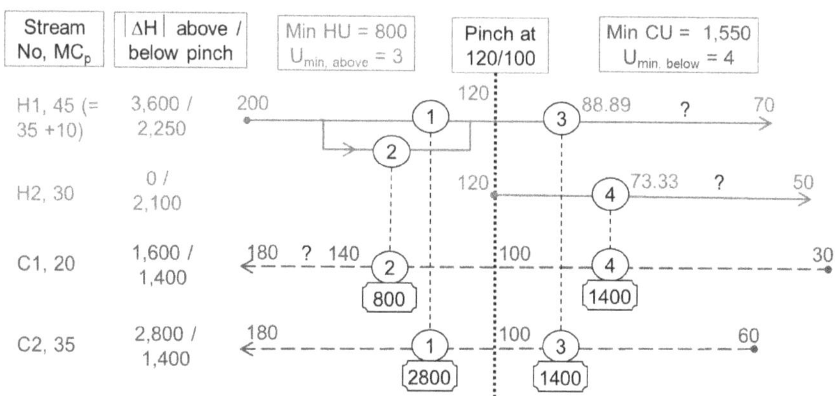

Fig. 3.16: HEN design at the pinch for the example stream data and targets (Fig. 3.15): above and below pinch.

HEN design above pinch meeting targets on HU and number of units is presented in Fig. 3.12.

Note two important points on stream splitting. First, splitting of H1 stream is absent initially in Fig. 3.15 but later it is required for HEN design above pinch because MC_p criterion (Eq. 3.16) is not met by original H1, C1 and C2 streams. In some problems, stream splitting may be required for satisfying number criterion or both MC_p and number criteria; this can be for HEN design above and/or below pinch. Second, stream splitting does not affect $U_{min,above} = 3$ in Fig. 3.15. In other words, the U_{min} equation must use the number of streams originally present (without stream splitting).

HEN design below pinch, where there are two cold and two hot streams (Fig. 3.15), is now performed. Below pinch, design for heating cold streams to pinch temperature is more difficult since no HU is allowed. It must be done by heat transfer from hot streams only. Hence, matches at the pinch are designed focusing on heating cold streams to pinch temperature (and not on cooling hot streams where there is more flexibility because CU can be used).

Both MC_p and number criteria (Eqs. 3.17 and 3.19) are satisfied below pinch. Matches satisfying these criteria in Fig. 3.16 are H1 with C2 and H2 with C1, which are matches 3 and 4, respectively. Duty of match 3 is maximized (i.e., minimum of 2,250 and 1,400 kW) to satisfy enthalpy change of C2 below pinch; then, H1 temperature at the outlet of match 3 is calculated as $120 - \frac{1,400}{45} = 88.89$°C. Similarly, duty of match 4 is maximized (i.e., minimum of 2,100 and 1,400 kW) to satisfy enthalpy change of C1 below pinch; then, H2 temperature at the outlet of match 4 is calculated as $120 - \frac{1,400}{30} = 73.33$°C. HEN below pinch until this point is shown in Fig. 3.16. Remaining HEN design below pinch is cooling of H1 from 88.89°C to 70°C and H2 from 73.33°C to 50°C (indicated by ? in Fig. 3.16). Both these can be achieved by placing coolers. Fig. 3.12 shows the complete HEN design below pinch.

A final HEN design for a certain set of industrial data with more streams is presented in Fig. 3.17. This is for $(\Delta T)_{min}$ = 20°C and using minimum utilities. This HEN, particularly above pinch, is more complex. Above pinch, each of C1 and C2 streams is split into two sub-streams; MC_p of these sub-streams are not given in Fig. 3.17 but it can be

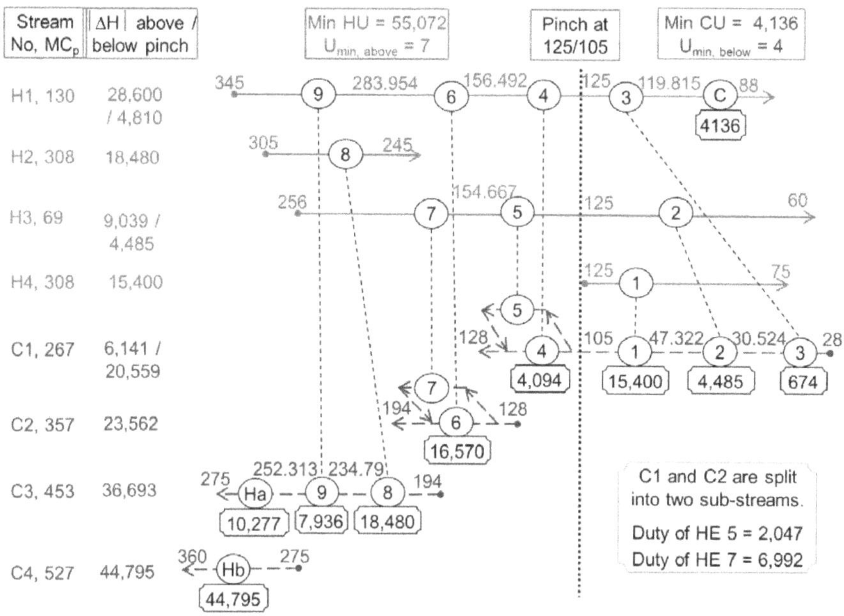

Fig 3.17: HEN design for HI of four cold streams and four hot streams in a plant.

calculated based on given duties of HEs 4, 5, 6 and 7. HEN above pinch has one unit more than the minimum number; this is due to the splitting of C1 and C2, required for matching streams, satisfying $(\Delta T)_{min}$. This does not mean stream splitting leads to more units in all problems; for example, see HEN in Fig. 3.12 for the example stream data. Returning to Fig. 3.17, HEN below pinch does not require stream splitting and has minimum number of units.

In the stream data used in Fig. 3.17, all four cold streams together correspond to only one cold stream with supply and target temperatures of 28°C and 360°C. This wide range is divided into four parts to account for temperature dependence of MC_p, thus creating four cold streams. With this process knowledge, it is possible to merge C1 from 105°C to 128°C (i.e., above pinch) with C2, and find MC_p of the merged stream to achieve the same enthalpy changes as in the original data. Note that this merging above pinch changes the number of streams and consequently $U_{min,above}$. HEN design using only six units above the pinch is possible for the stream data with the merged stream. In addition, the two heaters Ha and Hb can be realized in a single unit, since C3 and C4 have the same material flow. Thus, process knowledge is beneficial to improve HEN design.

HEN design (e.g., Fig. 3.12) obtained by the above steps, uses minimum utilities for the $(\Delta T)_{min}$ employed. It is not possible to reduce utilities without reducing $(\Delta T)_{min}$. Hence, HEN designed by the above steps has the minimum OC. Possible improvement of HEN design is with respect to CC, which is related to number of units; this is because HEN without a pinch (e.g., Fig. 3.10) often requires fewer units than $U_{min,MER}$. Therefore, HEN design using the minimum utilities may be improved to reduce number of units and CC, but CU/HU is likely to increase. Procedure for exploring this is referred to as HEN evolution in PA. Selection between HEN using minimum utilities and evolved HEN is generally based on TAC. Procedure for HEN evolution including this selection is available with an example in Chapter 8 of Rangaiah and Morrison.[2]

HEN design for a threshold problem (i.e., without a pinch) is presented in Fig. 3.18. Here, there are six hot and two cold streams from a petroleum refining process. Of the hot streams, H1 and H2 are the same material stream taken as two sub-streams (one from 332°C to 288°C, and another from 288°C to 200°C) to account for large variation in MC_p. The eight cold/hot streams require only CU and no HU for $(\Delta T)_{min}$ = 30°C, which is thus a threshold problem; it becomes a pinched problem for a larger $(\Delta T)_{min}$. HEN for the threshold problem is designed by following some of the above steps/features of HEN design in PA.

In the resulting HEN (Fig. 3.18) for the threshold problem, HE 2 duty could not be maximized as per heuristic,[c] and so it leads to one unit more than the minimum number. Further, the design requires splitting of C2 stream with MC_p of sub-streams chosen carefully to satisfy $(\Delta T)_{min}$ on the left side of HEs 4 and 5, and to use minimal number of units. Temperature of sub-streams of C2 on the left side of HEs 4 and 5 is not the same but it is acceptable. Temperature after mixing of sub-streams of C2 can be calculated by energy balance as $\frac{40.908 * 193.97 + 12.789 * 187.0}{40.908 + 12.789} = 192.31°C$.

[c] If HE 2 duty is maximized to 7,452 kW to cool H2 from 288°C to 200°C, then there is no hot stream to heat C1 to 212.794°C while satisfying $(\Delta T)_{min}$ = 30°C. Different values of HE 2 duty are possible.

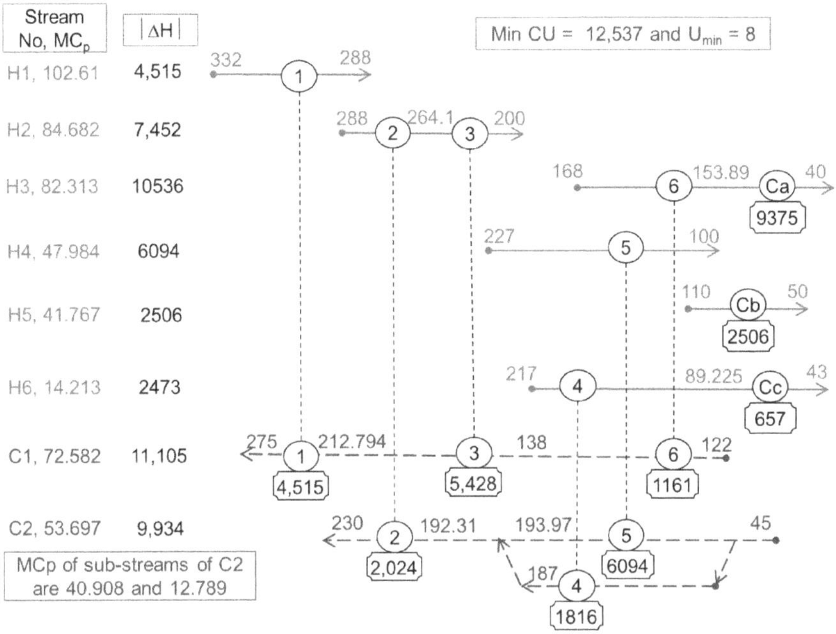

Fig. 3.18: HEN for a threshold problem with $(\Delta T)_{min}$ = 30°C for stream data from a petroleum refining process.

In conclusion, HENs in Figs. 3.17 and 3.18 are realistic and somewhat complex. They demonstrate the applicability of PA for both pinched and threshold problems.

3.11 Summary

This chapter describes fundamentals of PA for HI, which is critical for WHR. Main steps and features of PA are as follows.

- PA consists of many methods/procedures for HI.
- Targets on utilities, number of units and heat transfer area for a given $(\Delta T)_{min}$ are found before HEN design.
- Composite curves give minimum utilities and pinch information as well as range of available driving force for heat recovery.

- Pinch divides HEN design into two sub-problems: above and below pinch, and also provides golden rules for HEN design.
- Total heat transfer area can be found from composite curves (without/before HEN design).
- Formula for minimum number of units is simple.
- CC and OC can be estimated based on targets found for a given $(\Delta T)_{min}$.
- By repeating the calculations for a range of $(\Delta T)_{min}$ values, optimum $(\Delta T)_{min}$ minimizing TAC can be found.
- Grid diagram is very convenient and useful for both analysis of existing/given HEN and design of a new HEN.
- HEN design is systematic using grid diagram, golden rules, two criteria for matching cold/hot streams and a few heuristics.
- HEN design using minimum utilities and minimum number of units is often possible, and it may involve stream splitting. For some applications, design for minimum utilities may require a few units more than the minimum.

PA concepts, methods and procedures described in this chapter are particularly useful for finding targets, interpreting composite curves and HEN analysis. There are other concepts, methods and procedures in PA (e.g., grand composite curve, stream splitting, threshold problems, multiple utilities and HEN evolution) that are not covered in this chapter. These are presented with illustrations in Chapters 5–8 in Rangaiah and Morrison.[2] PA applications in industry and a number of case studies are available in Chapters 8 and 9 in Kemp.[1]

References

1. Kemp IC. (2007) *Pinch Analysis and Process Integration: A User Guide on Process Integration for Efficient Use of Energy*, 2nd ed. Butterworth-Heinemann.
2. Rangaiah GP, Morrison A. (2019) *Reference Manual on Energy Recovery and Reuse Under Singapore Certified Energy Manager Programme*. https://scemrefmanuals.wordpress.com/
3. Smith R. (2005) *Chemical Process Design and Integration*. John Wiley.

Acronyms and Notation

Acronyms

ACC	Annualized Capital Cost
AOC	Annual Operating Cost
CC	Capital Cost
CCC	Cold Composite Curve
CST	Cold Stream Temperature
CU	Cold Utility
HCC	Hot Composite Curve
HE	Heat Exchanger
HEN	Heat Exchanger Network
HI	Heat Integration
HST	Hot Stream Temperature
HU	Hot Utility
MER	Maximum Energy Recovery
OC	Operating Cost
PA	Pinch Analysis
PFD	Process Flow Diagram
TAC	Total Annual Cost
T-H	Temperature-Enthalpy
WHR	Waste Heat Recovery

Notation

A	Heat transfer area in the heat exchanger (m^2)
C_p	Heat capacity at constant pressure (kJ/kg.K = kJ/kg.°C)
h	Heat transfer coefficient ($kW/m^2.K = kW/m^2.°C$)
H	Enthalpy (kJ/kg) or enthalpy flow (kW)
k	Thermal conductivity (kW/m.K = kW/m.°C)
M	Mass flow rate (kg/s)
MC_p	Product of mass flow rate and heat capacity (kW/°C = kW/K)
nc	Number of cold streams (-)
ncu	Number of cold streams and cold utility (-)
nh	Number of hot streams (-)
nhu	Number of hot streams and hot utility (-)
ns	Number of subsets among the streams (-)

nu	Number of utility streams (-)
P	Pressure (bar)
Q	Duty of heat exchanger (kW)
q	Heat transferred to a cold stream or from a hot stream (kW)
T	Temperature (°C or K)
U	Overall heat transfer coefficient ($kW/m^2{\cdot}K = kW/m^2{\cdot}°C$)
U_{min}	Minimum number of units (-)

Greek Symbols

ΔH	Enthalpy change (kW)
ΔT	Temperature difference or driving force (°C or K)

Subscripts/Superscripts

c	Cold
h	Hot
i	Inner or inlet
min	Minimum
o	Outer or outlet

Exercises

3.1 As part of HI for the stream data in Table 3.1, 500 kW of HU is used in the HEN. For this, find CU required by energy balance. Calculations are similar to those in Table 3.2 but they should include HU (similar to a hot stream in giving heat).

3.2 For the stream data in Table 3.1, prepare composite curves for $(\Delta T)_{min}$ = 10°C, and then find the minimum utilities and pinch. Verify your results using any computer program for PA. Compare the minimum utilities and pinch found for $(\Delta T)_{min}$ = 10°C with those given in Section 3.4 for $(\Delta T)_{min}$ = 20°C.

3.3 Perform calculations to verify values of temperatures, $(\Delta T)_{LM}$ and area of each segment in Table 3.6.

3.4 Find the total heat transfer area based on the composite curves prepared in Exercise 3.2 for $(\Delta T)_{min}$ = 10°C. Compare this with that found in Section 3.5 for $(\Delta T)_{min}$ = 20°C, and then comment on the relationship between total heat transfer area and $(\Delta T)_{min}$.

3.5 Develop another configuration using minimum number of units (i.e., arcs) for transferring heat from hot streams/HU to cold streams/CU in Fig. 3.8. Compare the number of units in the developed configuration with U_{min} given by the formula.

3.6 For the stream data in the existing HEN in Fig. 3.14, minimum CU, minimum HU and pinch for $(\Delta T)_{min}$ = 30°C are found to be 1,670 kW, 4,679 kW and 123/93°C. For these targets, which golden rules are violated in the existing HEN in Fig. 3.14? Compared to CU and HU used in the existing HEN, how much HU can be reduced and how much CU can be reduced?

3.7 HEN design using minimum utilities and minimum number of units is given in Fig. 3.12. Using HE Eqs. 2.13, 2.14 and 2.16 in Chapter 2, and the data on CU and HU from Table 3.5, find the heat transfer area of each unit (i.e., heaters, HEs and coolers) in Fig. 3.12. What is the range of heat transfer areas of all units? Find the total heat transfer area of all units, and then compare it with that estimated from the composite curves in Section 3.5.

3.8 Study the HEN in Fig. 3.18. Calculate duties and approach temperatures of each and every unit in this figure. Is $(\Delta T)_{min}$ = 30°C satisfied by all units in this HEN?

3.9 Answer the following with yes (right) or no (wrong), and give brief reasoning, where applicable.

(A) In PA, HEN can be designed before finding targets.

(B) Minimum cold and hot utilities increase with increasing $(\Delta T)_{min}$.

(C) There should be no heat flow across the pinch for minimizing utilities.

(D) Total heat transfer area increases with increasing $(\Delta T)_{min}$.

(E) For a given set of process streams, only one HEN design is possible.

(F) Number of units in an HEN can be more than that given by U_{min} formula.

(G) Supertargeting is the procedure to find the optimum value of $(\Delta T)_{min}$.

(H) Pinch information is not required for assessing potential to reduce utilities in a given HEN.

(I) Utility streams are shown on the balanced grid diagram.

(J) Grid diagram is not essential for assessing potential to reduce utilities in a given HEN.

(K) HEN design in PA is based on pinch and golden rules.

Chapter 4

Heat Exchangers for Waste Heat Recovery

4.1 Overview

Waste heat recovery (WHR) techniques require heat exchangers (HEs) for transferring waste heat (WH) to process and/or cold utility streams. Feasibility and sustainability of WHR projects strongly depend on factors such as HE technology, capital cost, operating and maintenance costs, process safety, reliability and plot space requirements. Use of thermally efficient HEs helps to maximize WHR and minimize the size of HE for WHR. Many types of HEs are applicable and are used in process industries. They include direct contact HE (discussed in Chapter 16), double pipe HE (discussed in Chapter 2), shell and tube HE (STHE), enhanced surface HE, plate HE (PHE), heat pipe and regenerative HE (covered in Chapter 11).

Often, selection of the most appropriate HE is not easy. It requires consideration of many factors. Good understanding of HEs is necessary for correctly choosing the HE for an application. Very often, WHR projects require debottlenecking/retrofitting the existing plant HEs for increasing the heat transfer coefficient (HTC) and/or heat transfer area, mainly to increase the economic attractiveness of the WHR project. Debottleneck/retrofit options are different for different type of HEs. Hence, good understanding of these options is vital for developing an economically attractive and sustainable WHR project.

This chapter outlines HE types and then focusses debottlenecking/retrofitting options for STHEs and PHEs. It begins with Section 4.2 focusing on HE classification, HE design standards, advantages and disadvantages of HE types. Section 4.3 presents TEMA classification for STHEs, advantages and disadvantages of various STHE types. Sections 4.4 and 4.5 discuss key considerations for design and options for increasing heat transfer in STHEs, respectively. The subsequent section covers PHEs. Section 4.7 describes WHR steam generators (WHRSGs). Section 4.8 discusses the heat transfer enhancement (HTE) techniques. Finally, Section 4.9 summarizes the salient points for selection, design and retrofitting of HEs.

Learning outcomes of this chapter on HEs are as follows:

1. Describe and discuss various types of HEs.
2. Discuss advantages and disadvantages of various STHEs.
3. Explain useful design guidelines for STHE and PHE.
4. Discuss the debottleneck/retrofit options for STHE and PHE.
5. Outline various types of WHRSGs.
6. Expound various HTE options.

4.2 Classifications of Heat Exchangers Used for WHR

HEs can be broadly classified into four service categories. Following are the main functions of these HEs.

- **Cooler:** Cools a fluid (liquid or gas such as a reactor outlet stream) with water or any other fluid.
- **Condenser:** Condenses vapour (consisting of one or more components) or steam with or without some non-condensing gases. A common example is the distillation overhead vapour condenser.
- **Heater:** Heats liquid (usually with no phase change) or gas stream using a heating medium. For example, a heater is used to heat a reactor inlet stream to the reaction temperature.
- **Vapourizer/steam generator:** Generates vapour/steam using a heating medium; examples are WH boilers, reboilers or WHRSGs.

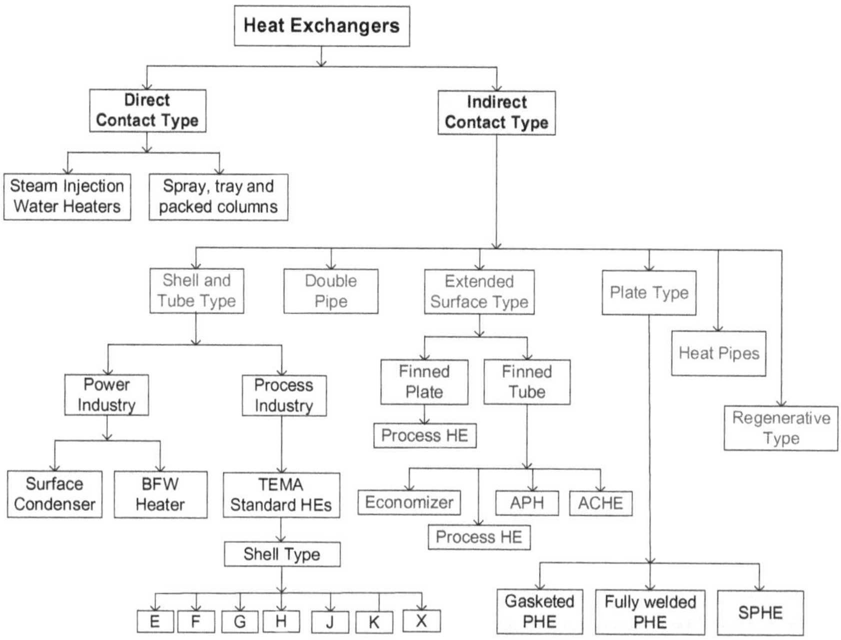

Fig. 4.1: Detailed classification of HEs.

Based on the type of HE used, HEs can be further classified into direct and indirect contact type, as shown in Fig. 4.1. Direct contact HEs (wherein cold and hot streams mix and leave as a single stream) are used to produce hot water and for heating slurries in processes such as Bayer process. Examples are direct injection of steam into water, using a specially designed mixing device. Another example is the direct contact of low-pressure steam (LPS) and demineralized water in deaerator. Deaerator and direct contact heating are covered in Chapter 16.

Indirect contact type HEs (wherein hot and cold streams remain and leave as separate streams) include double pipe HE, STHE, extended surface HE, PHE, heat pipe and regenerative HEs. STHEs for power industry such as boiler feed water (BFW) heater and surface condensers are designed using Heat Exchange Institute standards. On the other hand, STHEs for process industries are typically designed using ASME Section VIII, TEMA (Tubular Exchangers Manufacturers Association), American Petroleum Institute 660 (API 660) and API 662 standards.[9,10] There are three main TEMA classes. Class R is for the generally severe requirements of petroleum and related processing applications. Class C is for moderate requirements for commercial and general process applications. Class B is for chemical process services. Class R has more stringent design requirements compared to Classes B and C. Detailed specifications for TEMA Class R, C and B type HEs are available in TEMA standard.[1]

Double pipe HEs are outlined in Chapter 2, and detailed classification of STHEs is covered in Section 4.3.

Extended surface HEs are of two types: finned tube and finned plate. In finned tube HE, tubes may have fins on their inner and/or outer surfaces. Examples of finned tube HEs are economizer (covered in Chapter 11), air preheater (APH, covered in Chapter 11), air cooled HE (ACHE, covered in Chapter 15) and STHEs used for heat exchange between process streams. Finned plate HEs are used mainly for WHR from process streams. PHEs can be further classified into three categories: gasketed PHE (GPHE), fully welded PHE (FWPHE) and spiral PHE (SPHE). GPHEs are mainly used for heat exchange in low-pressure and low-temperature (<160°C) applications. FWPHEs can be used for high-pressure (up to 40 bar) and high-temperature (up to 400°C) applications. SPHEs are used mainly for WHR from highly viscous fluids. Heat pipes are used in process and flue gas (FG) WHR. Regenerative HEs are mainly used for FG WHR. Both heat pipes and regenerative HEs are discussed in Chapter 11. Advantages and disadvantages of double pipe HE and STHE are summarized in Table 4.1 whereas those of PHE, SPHE and direct contact HE in Table 4.2.

4.3 STHE Details and Classification

STHE is the most common WHR equipment in process industries. Main components of STHE are shown in Fig. 4.3. According to TEMA classification (in Fig. 4.4), STHE is differentiated by three main parts[a]:

- Front end head: Tube side fluid enters and leaves the HE from this head, except for single tube pass arrangement.

[a] Shell and Tube Heat Exchangers (thermopedia.com), accessed 10-05-2021. Main parts of STHEs and detailed explanation of TEMA classification can be found at this link.

Table 4.1: Advantages and disadvantages of double pipe HE and STHE

HE Type	Advantages	Disadvantages
Double Pipe	• Very economical for heating area up to 50 m^2. • Can be designed to achieve pure counter or co-current flow. • Flexible for increasing or decreasing the heating duty by addition or removal of some double pipe portions, connected in series and/or parallel. • Low inventory of spare parts is required due to design standardization. • These are installed in the form of hair pins, and so there is no problem with thermal expansion.	• For large heat transfer areas, many double pipe HEs are required. Piping connections get cumbersome for such double pipe HEs. • Occupy lot of plot space. • Risk of leakage due to large number of flange joints used in double pipe HEs. This is a serious concern for handling hazardous fluids.
STHE	• Practically there is no limit on its use for higher operating temperatures and pressures. • Pressure drop (PD) across an STHE is lower. • Rugged mechanical construction that can withstand more physical and process abuse. • Tube leaks can be easily detected by hydro tests, and leaking tubes can be plugged. • Can be used to hold inventory (e.g., kettle-type HE). • Use of sacrificial anodes in HE protects the whole process system against corrosion. • Long service life.	• Lower heat transfer efficiency compared to PHEs. • HE removal requires plot space since a horizontal STHE requires enough clearance (more than the tube bundle length) at one end to remove the tube bundle for hydro jet cleaning. Plot space can be reduced to some extent by using a vertical STHE. • Contains stagnant (dead) zones on the shell side, and they can lead to plugging and/or corrosion. • Flow maldistribution on shell side can happen especially with two-phase inlet streams.

(*Continued*)

Table 4.1: (*Continued*)

HE Type	Advantages	Disadvantages
		• Difficult to increase the capacity of an STHE. • Potential of flow-induced vibration can lead to equipment failure, unless proper care is taken in design. • Safety risks with large inventory. • Cannot handle temperature cross[b] unless multiple units in series are used.

Table 4.2: Advantages and disadvantages of PHE, SPHE and direct contact HE

HE Type	Advantages	Disadvantages
PHE	• GPHE is competitive in cost whereas FWPHE and SPHE are more expensive, compared to STHE. • Higher HTCs compared to STHE, and hence smaller heat transfer areas are required. • Closer temperature approach. • Temperature cross can be handled easily. • Lighter weight. • Reduced fouling due to high turbulence between the plates (and so self-cleaning effect). • Can be easily cleaned, especially GPHE.	• Plates are thinner than STHE tubes; hence, they require proper material selection. • Plates can be damaged by differential pressure reversal (i.e., ΔP frequently changes from positive to negative value and vice versa). • Narrower range of allowable pressures and temperatures compared to STHE. • Gasket material between plates limits operating temperature of GPHE. • Finding leakage is difficult in FWPHE since pressure test is not easy as compared to STHE.

[b]Temperature cross is a condition where the outlet temperature of the hot stream is lower than the outlet temperature of the cold stream.

Table 4.2: (*Continued*)

HE Type	Advantages	Disadvantages
	• Significantly smaller space required for installation and maintenance. • Maintenance is relatively simple compared to STHE. • GPHE area can be increased by introducing plates in pairs. But this is not possible for FWPHE or SPHE. • Leaking plates can be removed in pairs in GPHE, without replacement of the whole unit.	• PD in PHE is higher than that in STHE due to close clearance between plates. • Subject to plugging/fouling due to very narrow flow path, unless proper filters are used for process streams with solids. • Maintenance requires careful dismantling and assembling to avoid damage to plates and/or gasket. • Over tightening of clamping bolts result in increased PD across GPHE. • Titanium is a noble metal; if it is used, other parts of the cooling system are susceptible to corrosion and hence they require corrosion protection. • GPHE is not generally recommended for handling hazardous fluids. • Short service life. PHEs are quickly clogged. Maximum service life without cleaning is ~3 years for relatively clean services.
SPHE	• Single flow paths reduce the rate of scale formation. • Ability to work with hot and cold streams that are both highly viscous liquids. • No dead zones and no accumulation of solid particles inside SPHE. • Counter-current flow heat transfer. • Low PD.	• Designs are the intellectual property of a few companies, and a limited number of manufacturers. • Generally, more expensive than other types of HEs.

(*Continued*)

Table 4.2: (*Continued*)

HE Type	Advantages	Disadvantages
Direct contact type	• Large flow rates • Low PD • High efficiency • Low fouling	• Mixing of hot and cold streams, which is not allowed in many applications. • Large size • Requires makeup water, chemical treatment and wastewater treatment

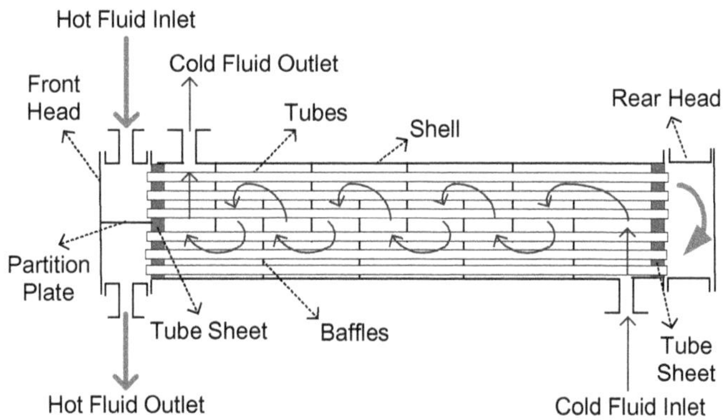

Fig. 4.2: Main components/parts of an STHE with one shell pass and two tube passes.

- Rear end head: Tube side fluid leaves from this end in single tube pass arrangement of HE; it is usually returned to the front end head in HEs with even number of tube side passes (as in Fig. 4.2).
- Shell: This contains the tube bundle.

Shell Selection: The single-pass shell, type E in Fig. 4.3, is widely used in process industry for general services. Two-pass shell with longitudinal baffle (type F) is mainly used if counter-current flow configuration is required. If shell side PD is the main limiting factor, double-split flow shell (type H), divided flow shell (type J) or cross flow shell (type X) is considered. For horizontal shell-side thermosiphon reboilers, split flow shell (type G or H) is selected. The kettle type shell (type K) is selected for boiling or vapourization on shell side and where vapour liquid separation is required.

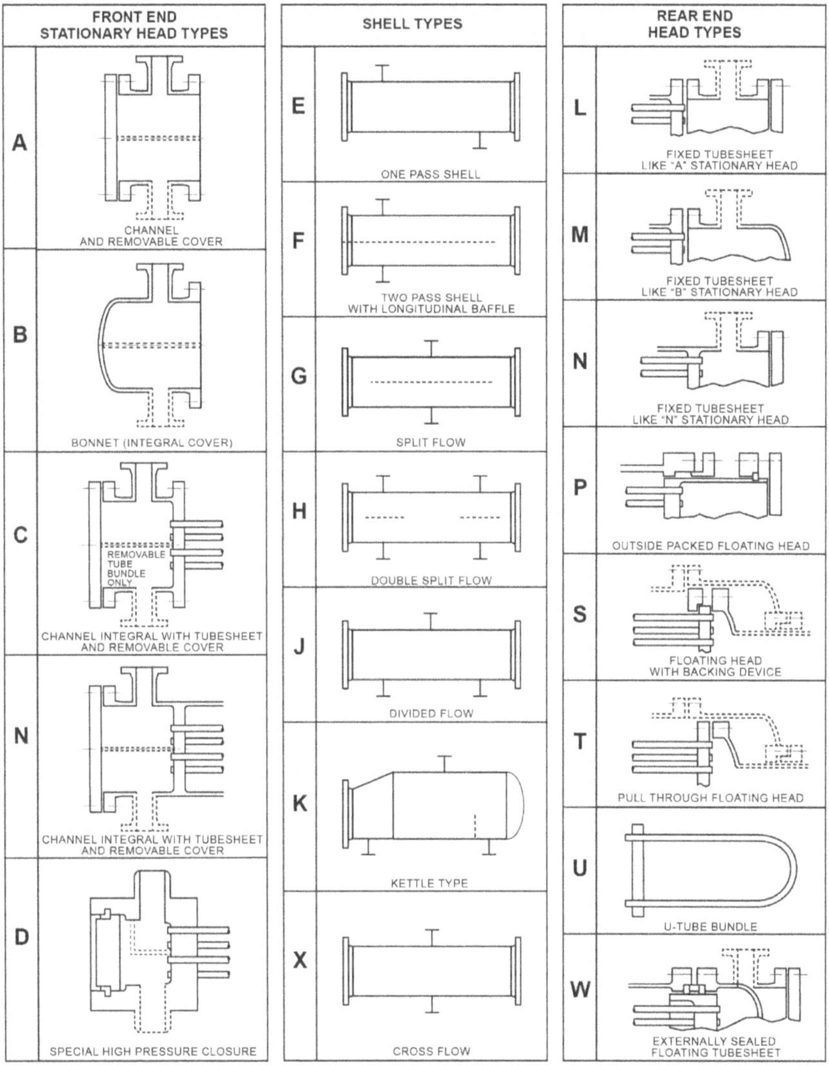

Fig. 4.3: TEMA classifications of HEs (Courtesy of Tubular Exchanger Manufacturers Association Inc., TEMA).

Front End and Rear End Head Selection: Type A head is selected for applications, where frequent tube side cleaning is anticipated, and the tube design pressure is low. Front end bonnet type B head is generally used for HEs, where frequency of cleaning on the tube side is expected to be low. Type A and B heads are used with fixed tube sheet, U-tube, and floating head HEs. For high-pressure and/or very toxic or hazardous service, where it is desirable to limit the number of external joints, stationary head B, C or N is considered for the front end, and L, M or N type head for the rear end. If very high pressure on tube side is required, front head type D is used.

Rear end type S head is used for floating head HEs whereas rear end types L, M and N are used for fixed tube sheet design. The outside packed floating head type P is not used for hydrocarbons and toxic fluid services on shell side. Externally sealed floating tube sheet type W head is not used for hydrocarbons and toxic fluid services on either tube or shell side as both shell and tube side fluids can potentially leak. Advantages and disadvantages of TEMA front end, shell type and rear end heads are summarized in Tables 4.3, 4.4 and 4.5, respectively.

Table 4.3: Advantages and disadvantages of TEMA front head types

Front End Head Type	Advantages	Disadvantages
A-type: It has a flange joint at both ends.	• Tubes can be cleaned without opening the piping. • Easy to repair and replace.	• Higher cost compared to B type head due to two flange joints requiring good seals (one between the tube sheet and head, and another between the head and the end plate). • Safety concern of leaks for handling hazardous substances.
B-type: One side of the head is flanged, and the other side is welded in semi-elliptical head.	• Cheapest front head • Preferred for clean tube side fluids and high-pressure applications as it has only one seal.	Tube cleaning access requires dismantling the piping.
C-type: One side of the head is flanged whereas the other side is welded to tube sheet.	• Allows cleaning access to tubes, without dismantling the piping. • Used for high-pressure applications (>100 bar) and handling hazardous substances on tube side.	Tube bundle is an integral part of the front head; hence, it is difficult to be repaired or replaced.

Table 4.3: (*Continued*)

Front End Head Type	Advantages	Disadvantages
N-type: Similar to C type head except that the integral tube sheet is welded to the shell.	• Allows cleaning access to tubes, without dismantling the piping. • Used for very high-pressure applications (>150 bar) and handling hazardous substances on tube side.	Tube bundle is an integral part of the front head and shell. Hence, it is difficult to be repaired or replaced.
D-type: Used for very high pressures (>150 bar).	Allows cleaning access to tubes, without dismantling the piping.	• Most expensive type of front head. • Tube bundle is an integral part of the front head and shell. Hence, it is difficult to be repaired or replaced.

Table 4.4: Advantages/applications and disadvantages of TEMA shell types

Shell Type	Advantages/Applications
E type (one pass shell)	Most common shell type, and suitable for most duties and applications.
F type (two pass shell)[#]	It is used when pure counter-current flow is required.
G type (split flow)	Mainly used for horizontal thermosyphon reboilers and for applications requiring low PD on shell side.
H type (double split flow)	Similar use as G shell but for larger and longer HE compared to G shell.
J type (divided flow)	This is used if the shell side PD of E shell is high even after implementing double or triple segmental baffles. This type can also reduce tube vibration due to divided flow.
K (kettle) type	This shell provides sufficient disengagement space for minimizing shell side liquid carryover. Hence, it is mainly used for WHRSGs, reboilers and some refrigerated coolers.
X type (cross flow)	This shell provides least PD among all shell and baffle combinations. It is mainly used in vacuum applications, condensers and gas coolers.

[#]Disadvantage of F type shell is thermal and hydraulic leakage around longitudinal baffle, especially at temperatures above 175°C.

Table 4.5: Advantages and disadvantages of TEMA rear head types; note that L, M and N are used for fixed tube HEs whereas P, S, T, U and W are used for floating head HEs

Rear Head Types	Advantages	Disadvantages
L type: used for HE with fixed tube sheets only.	• Tubes inner surface can be cleaned easily without dismantling the tube side piping. • Tube bundle to shell clearance is small. • This type of head is cheaper compared to floating head.	• Tube sheet is welded to the shell; hence, access to and cleaning of shell side are difficult. • Expansion bellows on shell side are required to handle large thermal expansions/ contractions.
M type: used for HE with fixed tube sheets only.	This type of head is cheaper compared to L type.	• Head and piping should be removed for inspection and cleaning of the tube inner surface. • Expansion bellows on shell side are required to handle large thermal expansions/ contractions.
N type: used for HE with fixed tube sheets only.	Tubes can be accessed without dismantling the tube side piping.	• Head and tube sheet are an integral part of the shell. Hence, it is difficult to maintain and replace the tube bundle. • Expansion bellows on shell side are required to handle large thermal expansions/ contractions.
P type: it is an outside packed floating head.	• It is generally a low-cost floating head design. • It allows easy access to the inside of tubes for inspection and cleaning. • It allows the entire tube bundle removal for cleaning.	• Bigger shell due to large bundle to shell clearance. • Shell side fluid may leak if the packing ring leaks; so, its use is limited to low pressure, non-hazardous fluids. • Only small thermal expansions are permitted.

Table 4.5: (*Continued*)

Rear Head Types	Advantages	Disadvantages
S type: it is a floating rear head with backing device.	• It allows removal of the bundle. • Unlimited thermal expansion is possible. • It has smaller shell to bundle clearance compared to other floating head types.	• It is the most expensive of the floating head types. • Bundle pulling for cleaning is difficult. • Shell diameter and bundle to shell clearance are larger compared to fixed head type HEs.
T type: it is a pull through floating head.	• It is cheaper and easier to remove the bundle compared to S type rear head. • It allows for unlimited thermal expansion.	• It has the largest bundle to shell clearance of all floating head types. • It is more expensive.
U type	• Cheapest of all types of HEs that can handle thermal expansion/contraction or have removable tube bundles. • It allows easy removal of the tube bundle for cleaning. • It has the tightest bundle to shell clearance. • Uses a single tube sheet only.	• It is comparable or slightly more expensive than a fixed tube sheet type. • It cannot normally have pure counter-current flow unless an F-type shell is used. • U-tube designs are limited to even number of tube passes.
W type: it has a packed floating tube sheet with lantern ring.	• It is the cheapest of the floating head designs. • It allows for unlimited thermal expansion/ contraction. • It allows easy removal of the tube bundle for cleaning.	• Large tube bundle to shell clearance; hence, it requires bigger shell diameter. • Limited to low pressure non-hazardous fluids due to the possibility of both shell and tube side fluids to leak via the packing rings.

There are three main types of STHEs, namely, fixed tube sheet, U-tube and floating head HEs, which are briefly described in the following paragraphs. Advantages and disadvantages of fixed tube, U-tube and floating head STHEs are summarized in Table 4.6.

Table 4.6: Advantages and disadvantages of fixed tube sheet, U-tube and floating head STHEs

Type	Advantages	Disadvantages
Fixed tube sheet	• Generally, this is the cheapest (as long as no expansion joints on shell are required) and most common type HE in process industries, as they have simple construction features. • Fewer gaskets compared to floating head STHE. • No restriction on number of tube passes, either even or odd. • Individual tubes are replaceable.	• Their application is limited to clean fluids on shell side. • Shell side cannot be inspected or mechanically cleaned as the tube sheets are fully welded to the shell. Chemical cleaning can be done but complete cleaning cannot be guaranteed. Hence, maintenance is difficult. • Not economical for large difference between the shell and tube side metal temperatures, which requires an expansion bellow on shell side. Expansion bellows are expensive and pose safety hazards due to the possibility of their failure.
U-tube	• Lower cost compared to floating head STHE. • Internal gasket is eliminated; hence, there is less concern on internal leakage. • Only one tube sheet is required. • Tube bundle can be easily pulled out for inspection, cleaning or replacement. • Bypassing of shell side fluid is minimal due to small clearance between the shell and outermost tubes.	• Tube inner surface cannot be thoroughly cleaned, especially at the U bends, by mechanical means such as hydro jetting. • Chemical cleaning is required. • Difficult to replace individual tubes. • Owing to U bends, fewer tubes can be fixed on tube sheet, compared to other STHE types. • High tube side velocity may cause erosion-corrosion at bends. • Only even number of passes is possible. Pass partition plates are built in front head only.

Table 4.6: (*Continued*)

Type	Advantages	Disadvantages
Floating head	• Tube bundle can be removed easily for inspection, repair or cleaning. • Easy to replace damaged tubes. • Mechanical cleaning of inner and outer surfaces of tubes is easy. • Either odd or even number of tube passes is possible for P type rear head. S type can have any number of even passes whereas W type can have only one or two passes.	• It is very costly. • It has many gasket joints, which may lead to leakages.

Fixed tube sheet STHE: This has straight tubes that are secured at both ends to tube sheets, which are welded to the shell. It may have removable or bonnet-type channel covers. Steel STHE can take a maximum of 25°C temperature difference between shell and tube metal surfaces, when the tube side is the hottest. When the shell side is the hottest, the maximum temperature difference is 85°C.[2] Fixed tube sheet HEs are used in services where differential expansion between the tubes and the shell does not give rise to unacceptable stresses, shell side fluid is non-fouling or shell side fouling can be removed by chemical cleaning and/or even slightest mixing of shell and tube side fluids is not allowed. Tube side can be easily cleaned by hydro jet cleaning.

U-tube STHE: In this HE, the tube bundle consists of U-shaped tubes with their both ends fixed to the same tube sheet. Tubes can expand or contract freely due to any thermal expansion, resulting from temperature difference between shell and tube side fluids. Hence, thermal expansion or contraction is not a problem in U-tube HE. U-tube STHEs are mainly used if thermal expansion/contraction of tubes is significant and tube side fouling can be removed by chemical cleaning without requiring thorough mechanical cleaning. U-tube bundle performs the same function as floating head tube bundle. Also, U-tube STHE has fewer joints, which can leak.

Floating head STHE: These are used in applications, where fixed tube sheet or U-tube HEs are not recommended. STHEs installed with TEMA rear head types P, S, T and W allow use of floating heads. Compared to P and W types, S type floating head STHE is commonly used in process industries. There are two variations of floating tube sheet HEs: pull-through and non-pull-through. In the pull-through unit, the entire floating tube sheet and cover assembly can be drawn through the shell without disassembly. In the non-pull-through unit, both the shell cover and the floating tube sheet cover must be removed before the bundle can be taken out of the shell.

In floating head STHE, straight tubes are fixed at both ends in stationary (at front end side) and floating (at rear end side) tube sheets. This allows tubes to expand or contract freely due to thermal expansion. The tube sheet at the front end is larger in diameter compared to the shell diameter. It is fixed to shell by a flange connection. The floating tube sheet at rear end is smaller in diameter compared to the shell diameter. Smaller diameter of rear end tube sheet permits the entire tube bundle to be removed from the front end side for mechanical cleaning. Floating head HE is expensive compared to fixed tube sheet or U-tube STHE. Floating head STHEs are suitable for both clean and fouling services with high temperature, high pressure and large temperature difference between the shell and tube side fluids.

STHEs in order of increasing cost are fixed tube sheet, U-tube, fixed tube sheet unit with an expansion or packed joint, and floating tube sheet (pull-through and non-pull-through). Commonly used HE types in process industries are fixed tube sheet: AEL, AEM, AEN, BEL, BEM, BEN; U-tube: AEU, CEU, DEU and floating head: AES and BES. Here, the first, second and third letter is for the front end stationary head type, shell type and rear head type, respectively, as per TEMA classification in Fig. 4.3.

4.4 Design Considerations for Heat Exchangers

HEs are typically designed for 20 years of operation. For maximizing WHR, new HE should be selected based on life cycle cost assessment, which involves consideration of initial cost, operating cost, maintenance cost, operation down time losses due to potential failures of HE and replacement cost (if the unit fails and at the end of its life). HEs can be designed using the methods in books such as Serth[3] or using software such as HTRI and Aspen EDR.

Key considerations for selection of an HE are as follows:

- Type: Selection criteria for STHE, PHE and SPHE are presented in Tables 4.7 and 4.8. These are based on first author's experience.
- Flow rates of both streams: Minimum, normal and maximum operating flows should be used to check fluid velocities and PD.
- Physical properties of both the streams: Viscosity, thermal conductivity, density and specific heat/latent heat over the entire operating range are required. For this, process simulators such as Hysys or inbuilt property databases in HTRI can be used.
- Operating pressure and temperature: While normal operating pressures and temperatures are used for HE design, it is recommended to check the design for minimum and maximum operating conditions also.
- Design pressure and temperature: These are established based on maximum and minimum conditions of pressure and temperature, with

Table 4.7: General recommendations for HE selection

Process Application	Suitable HE and Comments
Low viscosity liquid on both sides or low viscosity liquid to steam	For low pressures and non-hazardous substances, GPHEs are preferred. STHE or FWPHEs are preferred otherwise.
Medium viscosity fluid on both sides	For low pressures and non-hazardous substances, GPHEs are preferred. SPHE or FWPHEs are preferred otherwise.
High viscosity fluids/slurries	SPHE is suitable since spiral path without baffles ensures less fouling.
Fouling fluids	Use STHE with removable tube bundle or PHE.
Heat sensitive fluids	PHE that can maintain low-approach temperatures.
Gases under pressure	High gas pressure is better handled by STHE.
Process stream cooling with air	ACHE as it requires only ambient air for cooling.

Table 4.8: Selection criteria of STHE, GPHE and SPHE

Criterion	STHE	GPHE	SPHE
Pressure, bar	<551.6	<20.7	<17.2
Temperature, °C	<537.8	<260	<399
Maximum area, m^2	2,787/shell	1,486/frame	279/unit
Compactness	Very poor	Good	Good
Mechanical cleaning	Fair	Very good	Very good
Cost per unit area	Poor	Good	Fair
Corrosion risk	Poor	Good	Good
Fouling risk	Poor	Very good	Good
Leakage risk	Fair	Very poor	Very poor
Temperature cross	Poor	Good	Good
Viscous flow	Poor	Good	Very good
Heat sensitive fluids	Poor	Good	Good
Hazardous fluids	Good	Very poor	Fair
Fluids with solid	Very poor	Poor	Very good
Gases	Good	Very poor	Fair
Phase change	Good	Very poor	Good

appropriate safety margin. Design pressure and temperature are very important for material selection of HE components.

- Heat duty
- Fouling factors: Suitable fouling factors should be used. Typical fouling coefficients (= reciprocal of fouling factors) are given in Table 2.2 of Chapter 2. It is good to minimize fouling as much as possible by using HE fouling mitigation methods, described in Section 18.7 of Chapter 18.
- Allowable PDs for STHEs: For liquids and gases, typically PDs are 0.5–0.7 kg/cm^2 and 0.05–0.2 kg/cm^2, respectively.
- Tube size for STHEs: Tube outer diameters (ODs) of ¾″ and 1″ are popular. For fouling applications, up to 2″ OD is used.
- Tube length for STHEs: Commonly used tube lengths are 2,438, 3,048, 3,658, 4,877 and 6,096 mm. Typical tube length to shell diameter ratio is 5–10.

- Shell diameter for STHEs
- Nozzle sizes
- Materials of construction: These used in HEs depend on the fluid nature, operating temperature and pressure. Selection of materials involves careful consideration of initial cost, longevity, maintenance, performance and corrosion resistance. General material selection for all types of HEs is covered in Section 18.6 of Chapter 18.
- Cleanability, maintenance and repair considerations: For STHEs, tube pitch is important for easy cleanability. This is outlined in a following section with reference to Fig. 4.4. GPHEs and SPHEs are easier to clean and maintain.
- Future expansion: GPHEs are easier to expand their capacity/heat duty by addition of plates within the same HE frame. STHEs and other types of HEs generally require addition of new units in parallel.
- HE cost, weight and plot space requirements.
- Overdesign
- Turndown Ratio: This information is required for designing proper control methods to avoid excessive fouling and HE oversize. For example, if the turndown is 50%, two HEs each serving 50% of the maximum duty can be installed in parallel. During turndown operation, one of the HEs can be isolated.
- Special considerations such as thermal stability of fluids, presence of solids, crystallization, corrosion and erosion problems, tubes vibrations, thermal expansion and contractions of shell and/or tubes are discussed in Chapter 18. Another consideration is requirement of steam cleaning. In such cases, HE is also required to be designed for full vacuum. Further, even if the temperature difference between the shell and tube side fluid is very small, an expansion bellow may still be required on the shell. Occasionally, some applications may require removable tube bundles or double tube sheets.

Important considerations for STHE design are described in the following paragraphs:

Tube Pitch: Tube pitch is the shortest center to center distance between the adjacent tubes. Minimum tube pitch should be 1.25 times the OD of

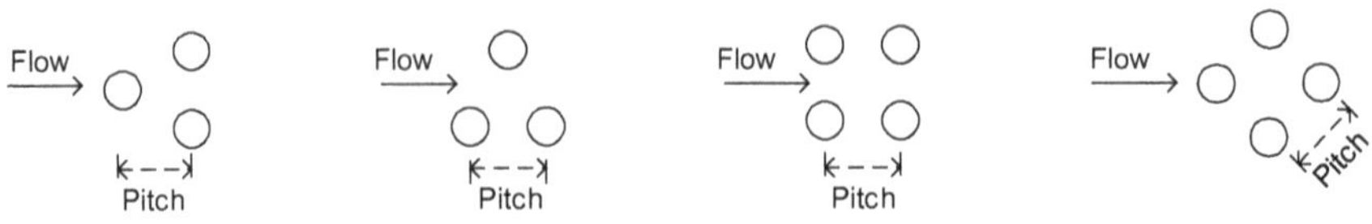

(a) Triangular Pitch (b) Rotated Triangular Pitch (c) Square Pitch (d) Rotated Square Pitch

Fig. 4.4: Tube pitch arrangements (cross-sectional view).

the tube. For mechanical cleaning, minimum 6.4 mm cleaning lanes are required (as per TEMA guidelines). Commonly used tube pitch configurations are illustrated in Fig. 4.4. Triangular and rotated triangular pitch creates high PD, turbulence and provides high HTC on shell side. But, it poses problem for STHE cleaning. Hence, (rotated) triangular pitch is generally used in clean services. Square or rotated square pitch facilitates easy cleaning and results in low PD. But, its disadvantage is that it creates less turbulence and hence results in lower HTC on shell side.

Baffles: There are two types of baffles: longitudinal and transverse. Longitudinal baffles are used in F, G and H shells to establish the desired flow pattern for the respective shell; see dashed lines in these shells in Fig. 4.3. Transverse or segmental baffles support the tube bundle against bending and vibrations (Fig. 4.5), and direct the fluid flow towards tubes repeatedly for efficient heat transfer. Single segmental baffles are most widely used. The spacing between baffles should be between 0.2 and 1.0 of shell diameter (but not less than 51 mm). Baffle cut (explained below and Fig. 4.6) should be between 15% and 45%. For single-phase flow, baffle cut of 20%–35% is recommended.[3]

Single segmental baffles force the fluid across the entire tube count. This can result in excessive pressure loss, especially in high velocity gas streams. If high PD is not acceptable, double segmental baffles (Fig. 4.5) can be used. This provides the necessary structural support for the tube bundle and allows the fluid to flow between alternating sections of tube count, thereby reducing the PD. This approach takes full advantage of the available tube surface but results in lower thermal performance due to reduction in shell side HTC. Double segmental baffles reduce the shell side velocity by half, compared to single segmental baffles. This results in approximately one quarter of the PD in a single segmental baffle arrangement, for the same tube count. If further PD reduction is required,

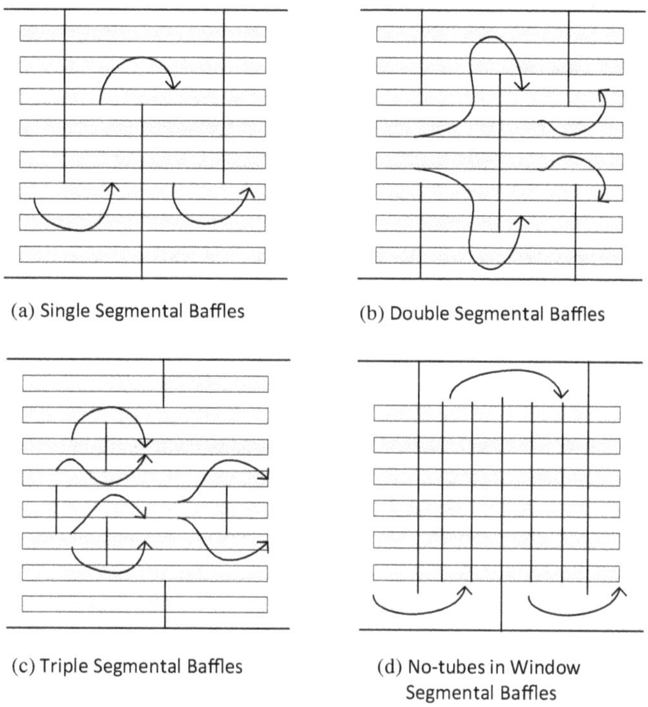

Fig. 4.5: Different baffle types.

triple segmental baffle or no tubes in window arrangement can be used (Fig. 4.5).

Baffle cut is the ratio of the height of the cut-out segment to the inner diameter (ID) of the shell. It can be horizontal or vertical (as shown in Fig. 4.6) or inclined. Horizontal baffles are the most commonly used segmental baffles. For single-phase fluids on the shell side, a horizontal baffle cut is recommended, because it minimizes accumulation of deposits at the bottom of the shell and also allows mixing of warmer fluid (at the top of the HE) with colder fluid (at the bottom of the HE). However, in the case of a two-pass shell (TEMA F type), a vertical cut is preferred for ease of fabrication and bundle assembly. Vertical baffles are also preferred for vapour condensation on shell side as it helps to quickly drain the condensed liquid, or when low PD is required on the shell side.

No tubes in window baffles (Fig. 4.5) are used mainly to reduce flow-induced vibrations. Their main disadvantage is the requirement of a larger

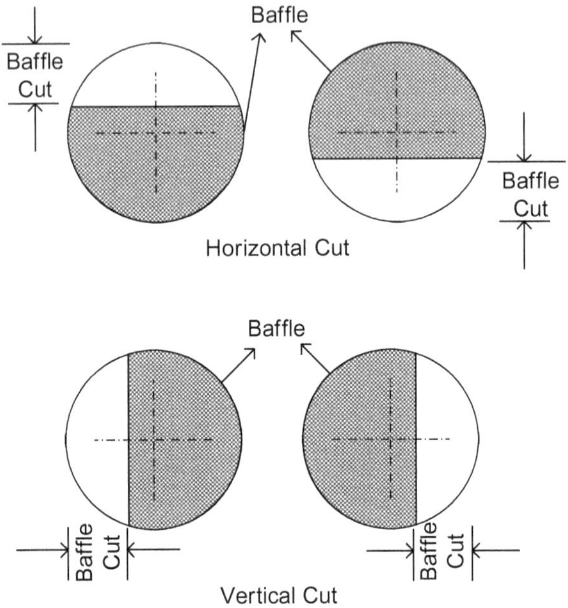

Fig. 4.6: Horizontal and vertical baffle cut arrangements.

diameter shell to provide the same amount of heat transfer area. Other options of reducing shell side PD are the use of rod baffles, EMBaffle,[c] twisted tubes, helical baffles and different shell types such as divided flow, double split flow and cross flow shells (Fig. 4.3). In case of rod baffles, baffles are replaced by solid metal rods that support every tube and hence remove any dead zones and minimize the possibility of tube vibrations. EMBaffles are expanded metal baffle grid made of plate material that has been slit and expanded; as the metal grids support all tubes at many points along the tube length, tube vibrations are minimized; further, there are no plate baffles used, and so shell side PD is low. Twisted tubes[d] do not require any baffles, support each other at numerous points along their length and hence reduce PD and tube vibrations. Further, due to many directional changes of fluid while flowing inside the tubes, they

[c]Technology — EMBaffle — World leader in innovative heat transfer solutions, accessed on 15-05-2021.

[d]Twisted Tube™ Heat Exchanger Technology — Koch Heat Transfer, accessed on 15-05-2021.

Table 4.9: Comparison of different baffle arrangements

Characteristic	Segmental Baffle	Rod Baffle/ EMBaffle	Helical Baffle	Twisted Tube
High shell side HTC	Yes	No	Yes	No
Tube side HTC enhancement	Possible with tube inserts	Possible with tube inserts	Possible with tube inserts	Yes, even without tube inserts
Requirement of segmental baffles	Yes	No	No	No
Minimize dead zones / fouling on shell side	No	Yes	Yes	Yes.
Mechanical cleaning	Yes, with square pitch	Yes	Yes, with square pitch	Yes. Chemical cleaning may be required.
Tube vibrations	Low, only with use of proper preventive design solutions such as impingement devices and closer baffle spacing	Yes	Yes, with double helix design	Yes

increase the tube side HTC by ~40%. Helical baffles (e.g., Helixchanger[e]) uses a series of baffles arranged at an angle (usually ~40°) to the tube axis. This makes shell side fluid to flow in helical pattern. It increases the HTC on shell side, reduces fouling, tube vibrations and PD. Different baffle arrangements are compared in Table 4.9, which is prepared from the first author's experience and Table 1.1 in Shah and Sekulic.[4]

Baffle Space: Segmental baffles should not be spaced closer than 10% of the shell ID or 51 mm, whichever is higher. As per TEMA standard, maximum baffle spacing is given in Table 4.10.

Shell Side Flow Streams: Depending on the shell type and baffle type, cut and spacing, there are five shell side flow streams as follows.

[e] HELIXCHANGER® Heat Exchanger — Koch Heat Transfer, accessed on 15-05-2021.

Table 4.10: Maximum baffle spacing for different tube ODs and materials of construction (Reprinted with permission from the Tubular Exchanger Manufacturers Association)[1]

	Maximum Baffle Spacing (mm) for Materials of Construction	
Tube OD, mm	**Carbon Steel, High Alloy Steel, Low Alloy Steel, Nickel–Copper, Nickel, Nickel–Chromium-Iron**	**Aluminium, Aluminium Alloys, Copper and Copper Alloys, Titanium Alloys**
6.4	660	559
9.5	889	762
12.7	1,118	965
15.9	1,321	1,143
19.1	1,524	1,321
22.2	1,753	1,524
25.4	1,880	1,626
31.8	2,235	1,930
38.1	2,540	2,210
50.8	3,175	2,794
63.5	3,175	2,794
76.2	3,175	2,794

- Stream A: Flow through annuli between tubes and baffles. This is good for heat transfer as the fluid is in contact with tubes. However, it is not as efficient as Stream B.
- Stream B: Crossflow through bundle between baffle tips. This is the main cross flow across the tube bundle. It is the most effective for heat transfer.
- Stream C: Flow through annulus between bundle and shell wall.
- Stream E: Flow through annulus between baffle and shell.
- Stream F: Flow through tube pass partition lanes parallel to crossflow direction.

See Fig. 11 in Mukherjee[5] for a schematic of the above flow streams.

Streams C and F contact the tubes only partially. Stream E does not contact tubes at all. Thus, relative effectiveness towards heat transfer is B > A > (C, F) > E. Hence, for good HE design, special attention is required to increase B stream and reduce C, F and E streams. Based upon the values of each of these streams, the shell side HTC is estimated.

Mukherjee[5] discussed these flow streams in detail. HE design software such as HTRI estimates the flow stream values and gives them in the output file. User can examine these values and adjust the specifications of HE (such as baffle spacing, baffle cut, clearance between the tube and baffle hole, tube layout angle, tube pitch, number of tube lanes in the flow direction and lane width, clearance between the shell ID and baffle plate, location of sealing strips and sealing rods), for achieving a good design of HE.

Impingement Protection Devices: As per TEMA, an impingement device should be provided to protect tube bundle against impinging fluids, when the entrance nozzle value of ρV^2 exceeds 2,232 and 744 $\frac{kg}{ms^2}$, for non-abrasive single-phase fluids and all liquids, including liquids at boiling points, respectively. Here, ρ is the fluid density ($\frac{kg}{m^3}$) and V is the fluid velocity (m/s) in the entrance pipe, respectively. For gases, vapours, vapour–liquid mixtures or if ρV^2 is higher than the allowed value (for liquid services), one of the impingement protection devices in Fig. 4.7 can be used. Figure 4.7(a) shows a thick impingement plate at the shell inlet nozzle, which handles the impingement safely. Figure 4.7(b) uses solid metal rods instead of tubes at the HE inlet nozzle. This method is

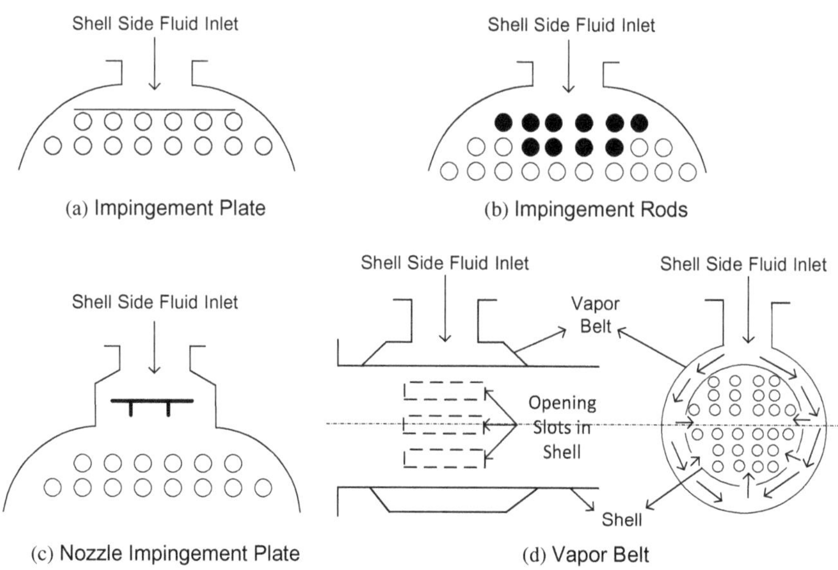

Fig. 4.7: Details of impingement devices.

commonly used as it is very robust. For very large HEs, arrangements in Fig. 4.7(c) (nozzle impingement plate) and Fig. 4.7(d) (vapour belt) are used. For tube side, ρV^2 value should not exceed 8,928 $\frac{kg}{ms^2}$. This can be achieved by using a bigger tube side entrance nozzle.

Shell and Tube Passes: E shell has single shell pass whereas F shell has two shell passes. Several shells can be arranged in series to provide more shell passes. The tubes are arranged within the shell of an STHE in passes. Tube side passes can be one, two, four, six or eight.

Number of Tubes: This depends on the fluid flow rates and available PD in STHE. Number of tubes is selected such that the tube side velocity for liquids ranges from 2 to 3 m/s and shell-side velocity ranges from 0.6 to 1.5 m/s. For vapour service, the velocity limits are 50–70 m/s, 10–30 m/s and 5–10 m/s for vacuum, atmospheric pressure, and high-pressure conditions, respectively. A lower velocity leads to fouling, and very high velocity leads to erosion. When sand and silt are present, velocity is kept sufficiently high to prevent settling.

Minimum Approach Temperature (ΔT_{min}): High ΔT_{min} leads to lower capital cost; however, it results in higher operating cost. On the other hand, low ΔT_{min} leads to higher capital cost; however, it results in lower operating cost. Hence, for achieving economical WHR, proper selection of ΔT_{min} is very important. Typical values of ΔT_{min} for various industrial sectors and process applications are presented in Tables 4.11 and 4.12, respectively.

Velocity Criteria for Shell and Tube Sides: For designing or revamping an HE, one should take note of velocity and PD criteria. HTC is inversely proportional to viscosity to the power of 0.47, and it is directly proportional to thermal conductivity to the power of 0.67. For turbulent flow, tube side HTC varies with tube-side velocity to the power of 0.8, whereas tube side PD varies to the square of velocity. Velocity and PD criteria for shell and tube side fluids are summarized in Table 4.13, which is based on Treese *et al.*[6] (Part V, Table 27) and the first Author's experience.

Placement of fluids on Shell and Tube Sides: Placing a fluid on tube versus shell side is driven by the capital, maintenance and operating cost considerations. Maintenance costs include HE cleaning costs whereas

Table 4.11: Typical ΔT_{min} values in various industrial sectors

Industrial Sector	Typical ΔT_{min} Values	Comments
Petroleum refining	20°C–40°C	High fouling and lower HTC situations
Petrochemicals/ Chemicals	10°C–20°C	Situations of moderate or low fouling and high heat transfer rates
Low temperature (cryogenic) applications	2°C –5°C	Mainly economic reasons as the energy requirements for refrigeration is very high

Table 4.12: Typical ΔT_{min} values for selected process applications

Process Application	Typical ΔT_{min} Values	Comments
Steam generation or condensation by heat exchange with a process stream	10°C–20°C	Good HTCs for boiling and condensation
Process stream cooling by a refrigerant	3°C–5°C	High cost of refrigeration
Process stream heating by FG	50°C	Low HTC of FG
Steam generation by heat exchange with FG	25°C–40°C	Good HTC for boiling and low HTC of FG
Combustion air heating by FG	50°C–100°C	Low HTC of both air and FG. ΔT_{min} depends on acid gas dew point temperature for FG
Process stream cooling by cooling water	5°C–20°C	
Liquid to liquid heat transfer	10°C–20°C	
Heater	10°C–50°C	
Superheater	50°C–100°C	

operating costs include the utility cost. Typical criteria for fluid placement in HE are given in Table 4.14.

Overdesign: Process designers prefer to use overdesign for heat transfer area, for the following considerations:

- Deviation of operating conditions from design conditions
- To meet transient conditions during process upsets

Table 4.13: Typical velocity and PD criteria for shell and tube sides

	Tube Side		Shell Side	
Stream Detail	**Velocity, m/s**	**PD, bar**	**Velocity, m/s**	**PD, bar**
Non-viscous liquid	2–3	0.7	0.6–1.5	0.7
Viscous liquids	1.8	1–2	0.9 max	1–1.4
Clean cooling water	2–3	0.7–1	—	—
Dirty cooling water	0.9	0.7	—	—
Suspended solids in liquid	0.6–0.9	0.7	0.45	0.9
Gases and vapours	$\frac{100}{\sqrt{\rho}}$ max	0.1–0.4	5–30	—
Condensing vapours	—	—	—	0.9–1.5
Steam condensation under vacuum	50–70	—	—	—
Steam condensation at 1–7 bar	50	—	—	—
Steam condensation at above 7 bar	40	—	—	—

Table 4.14: Selection of fluids in STHE

Process Stream Property	**Shell Side**	**Tube Side**	**Comments**
High fluid temperature		√	Lower heat loss
High stream pressure		√	Reduces shell thickness and so capital cost
High flow rate stream		√	High velocity and high heat transfer
High fouling stream		√	Can maintain high velocity to minimize fouling
Cooling tower water		√	Can maintain high velocity to minimize fouling. Better effectiveness of corrosion inhibitors inside tubes. Cleaning is easier.
Require alloy materials for corrosion control		√	Less metal quantity and cost
Requires low ΔP	√		More cross-sectional area for flow
Viscosity above 2 cP	√		Staggered tube layout induces good heat transfer at low Reynolds number.

Table 4.14: (*Continued*)

Process Stream Property	Shell Side	Tube Side	Comments
Lower HTC expected	√		Finned tubes improve the limiting HTC
Boiling	√		Easy vapour–liquid separation on shell
Very low system pressure or ΔP available	√		Can use J or X shell type to shorten flow path and reduce PD.
Large temperature change of streams (> 100°C)	√		Excessive ΔT in stationary tube sheet, if placed on tube side.
Condensing	---	---	Determined by the coolant nature. Mostly shell side is used.
Normal fouling	---	---	Other factors will decide on the shell and tube sides fluid selection

- Future retrofits for higher duty
- Allowance for unexpected fouling
- For possible plugging of some tubes if they leak during the operation

With the use of recommended velocities on shell (cross flow velocity of 0.6–1.5 m/s) and tube sides (of 2–3 m/s), design safety factor should be as small as possible. It is not recommended to use overdesign higher than 30%, under any consideration as it decreases the tube side velocity and causes aggressive fouling.

4.5 Options for Increasing Heat Transfer in STHEs

For new or retrofit WHR projects, increasing the heat transfer duty in an STHE by increasing HTC and/or heat transfer surface area is one of the main goals. Various options for achieving it are compiled in Fig. 4.8. Tube side HTC increases by increase in the number of tube passes, decrease in the tube OD, use of tube inserts (explained in Section 4.8), use of extended surfaces such as fins (on inner surface of tube) and use of twisted tubes. For retrofit designs increasing the number of tube passes is generally the cheapest option if sufficient PD is available and tube side velocity is low. Use of twisted tube is relatively expensive (as it requires tube bundle

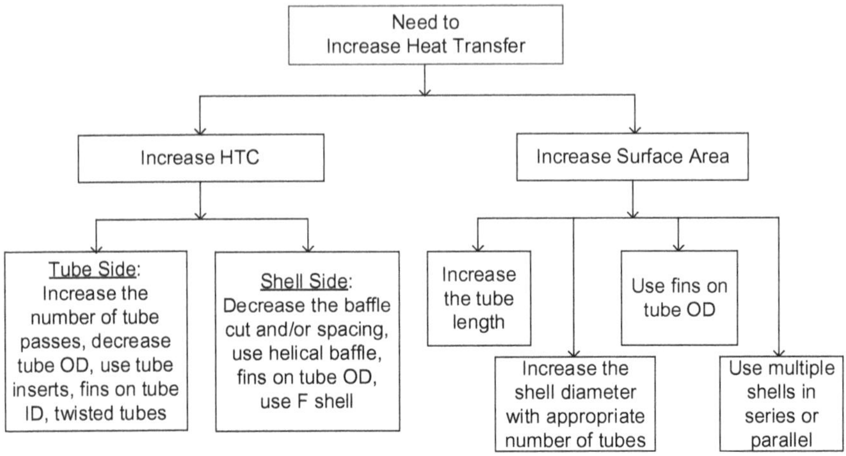

Fig. 4.8: Options to increase heat transfer in STHE.

replacement with twisted tubes, within the same shell) but still it is a good retrofit option if other options do not meet the desired increase in HTC.

Shell side HTC increases by decrease in baffle cut and/or spacing. Other options include conversion of E shell to F shell (i.e., increasing shell side passes from one to two), use of triangular or rotated triangular tube pitch (applicable only for new designs involving clean fluids) and use of helical baffles. If shell side fluid is clean and adequate PD is available, decreasing baffle cut and/or spacing is the cheapest retrofit option for increasing shell side HTC. Shell side fouling can be reduced by use of twisted tube bundle, special baffles such as EmBaffle, rod or helical baffles. These options can be used for sustaining shell side heat transfer.

Surface area for heat transfer in an STHE can be increased by increasing the tube length, increasing number of tubes along with the use of bigger shell, using extended surfaces such as tube fins (on outer surface of tubes) and using multiple shells in series or parallel. Except for use of fins on outer surface of tubes, which can be used for both new and retrofit designs, all other options are applicable for new designs only. For retrofit design, involving use of low fouling fluid on the shell side, use of finned tubes is the cheapest option. For new designs, use of longer tubes is generally the cheaper option. Some of STHE debottleneck/retrofit options are compared in Table 4.15. Sometimes, changing from STHE to compact HE such as PHE can be beneficial, especially if the site has space constraint.

Table 4.15: Comparison of STHE duty debottleneck/retrofit options

Technology	Capital Cost	Additional Operating Cost	Benefits	Disadvantages
Online cleaning	Low	Low to Medium	Less downtime and related expenses than offline cleaning	Foulant still builds up and reduces efficiency.
Anti-foulant additives	Low	Medium	Substantially decreases fouling if mechanisms are well understood	• Fouling mechanisms must be understood. • Continual expense
Tube inserts	Medium	Low	• Increases turbulence and HTC on tube side. • Decreases fouling on tube side. • Lasts 2–3 times longer between cleanings on tube side.	Inserts must be replaced every 2 years, due to wear and tear.
Increase tube passes	Low	Medium	• Increases and sustains HTC on tube side, if tube side velocity can be maintained at 2–3 m/s.	High tube side PD
Increase the number of baffles	Medium	Low	• Increases turbulence and HTC on shell side. • Decreases fouling on shell side. • Lasts 2–3 times longer between cleanings on shell side.	Some baffles make cleaning more difficult.

(*Continued*)

Table 4.15: (*Continued*)

Technology	Capital Cost	Additional Operating Cost	Benefits	Disadvantages
Special (e.g., EMBaffle, rod or helical) baffles	Medium	Low	Reduces shell side fouling and PD	Do not increase shell side HTC significantly
Twisted	Medium	Low	• Increases HTC on tube side • Decreases fouling on both shell and tube sides	High PD
Finned tubes on inner and/or outer surface of tubes	Medium	Low	Increases HTC on tube and/or shell sides	Applicable to low fouling services
Coatings on inner and outer surfaces of tubes	Medium	None	Reduces corrosion and fouling	Must be designed to match specific application.
Anti-vibration technology	Medium	Low	• Reduces vibration from high velocities. • May decrease fouling by allowing higher velocities. • Possibly longer time between cleanings.	Only small impact on fouling (from increased velocity).

Utilizing the PD: Maximizing the use of available PD, within velocity limits (to avoid erosion and tube vibration), maximizes HTC and hence improves the heat transfer. Options to increase the shell and tube side PD are given with comments, in Table 4.16. Although it is beneficial to increase the shell and tube side PD, certain brownfield WHR projects may require designing HE within certain PD constraints. Hence, PD reduction options are also important. Options for reducing the shell and tube PD of STHEs are summarized in Tables 4.17 and 4.18.

Table 4.16: Options for increasing the shell and tube side PDs of STHEs

Detail	Option Followed by Comments
Shell side only	• **Reduction in baffle spacing** increases the crossflow velocity and consequently the shell side PD. However, it leads to an increased fraction of shell-side flow by-passing the tube bundle, thereby, lowering heat transfer. Therefore, it can be done in a limited way only.
Tube side only	• **Increasing the number of tube passes** increases the tube side velocity and PD. However, with the TEMA E shell, use of two or more tube passes lowers the effective mean temperature difference (EMTD), which means required heat transfer area is more. Hence, combined impact of increased HTC and reduced EMTD on the required heat transfer area should be evaluated. • **Reduction in tube diameter** increases PD as it increases tube side velocity due to decreased flow area.
Both shell and tube sides	• **Reducing shell diameter** increases shell side velocity and hence PD. It also leads to increased tube side velocity and PD due to reduction in number of tubes. • **Changing E shell to F shell** increases both shell and tube side PDs. • **Shells in series** can be used to maximize the available PDs on both shell and tube sides. However, it results in increased cost due to increased number of shells, channels/bonnets, tube sheets, nozzles, flanges, etc.

Table 4.17: Options for reducing the shell side PD of STHEs

Option	Remarks
Baffle type change from single to double segmental or other types (e.g., rod or EMBaffle)	This reduces the shell side flow velocity and, thereby, lowers the shell side PD.
Increase baffle spacing, within maximum unsupported tube span, recommended by TEMA	It increases the crossflow area and, hence, decreases crossflow velocity.

(*Continued*)

Table 4.17: *(Continued)*

Option	Remarks
Increase the baffle cut (should not exceed 35% of shell ID)	It reduces window velocity and hence reduces PD. However, it requires bigger shell diameter and more capital cost. It also reduces shell side HTC.
Change the shell type	TEMA J shell requires less PD than E shell as the shell-side flow is divided and flow velocity becomes half. Further PD reduction can be achieved in the G and H shells. PD is least in TEMA X shell, which has crossflow, largest flow area and least velocity.
Increase the tube pitch ratio	This reduces the crossflow velocity and, thereby, reduces PD. However, tube pitch ratio is generally 1.25, 1.33 or 1.50; therefore, it can be varied in a limited way only.
Increase the nozzle size	If the PD across the nozzles is excessive relative to the total PD, the nozzle size can be increased reasonably to lower the PD.
Use shells in parallel	Multiple shells can be used in parallel so that total shell-side flow is split, and flow velocity is reduced. However, it increases the cost due to more number of shells, tube sheets, channels/ bonnets, nozzles, flanges, etc.

Table 4.18: Options for reducing the tube side PD of STHEs

Options	Remarks
Increase shell diameter	This permits the use of a greater number of tubes, reduces tube flow velocity and hence reduces tube side PD.
Reduce tube length	This reduces PD.
Increase tube diameter	This reduces tube side velocity and hence reduces PD. Tube OD more than 1 inch is generally not desirable as it means larger shell diameter to accommodate the required number of tubes due to increased tube pitch, which, in turn, means higher cost.
Increase the nozzle size	If nozzles are too small in diameter, their diameter can be increased reasonably to lower the PD.
Multiple shells in parallel	Multiple shells can be used in parallel so that total tube side flow is split, and flow velocity is reduced. Consequently, the PD is reduced. However, it increases the cost.
Decrease number of tube passes	This reduces flow path length and hence reduces the PD.

4.6 Plate Heat Exchangers

A PHE contains a series of corrugated metal plates in mutual contact. Each plate has two inlet and two outlet ports, and gasket seals are designed to direct the fluids in alternate flow passages, formed by adjacent plates, so that hot and cold streams exchange heat through the separating plate. In a PHE, high levels of turbulence and swirl or vortex flows exist and hence leads to higher shear stress at the plates and hence lower fouling. It also leads to high film and overall HTCs. The corrugated plates in a PHE create many contact points and form a strong rigid matrix that can withstand high pressure difference between channels at relatively small plate thickness. High HTCs of PHEs enable smaller weight and volume compared to STHEs. Hence, PHEs usually require less capital cost (for purchasing, transportation, foundation and installation), are compact and need a small area for maintenance. They can operate at a small temperature approach (down to 1°C) owing to high HTCs and true counter-current flow arrangement. Technical details of PHEs are given in Table 4.19.

PHEs can be gasketed,[f] fully welded, spiral or plate and fin type. GPHEs are used mainly for handling non-hazardous substances at low pressure and temperature. Welded or gasket-less PHEs (e.g., Compabloc[g]) are used for handling hazardous services with high temperature (up to 400°C) and high pressure (up to 42 bar). Spiral[h] PHEs are used mainly for handling high viscosity fluids whereas finned PHEs in cryogenic

[f]Typical parts and working animation for GPHE can be seen at Sondex Plate Heat Exchanger — Working Principles — YouTube, accessed on 16-05-2021.

[g]Animation for Compabloc with four passes each for hot and cold fluid, can be seen at Alfa Laval Compabloc — Liquid-to-liquid — YouTube, accessed on 16-05-2021.

[h]Parts and working animation of Spiral PHE can be viewed at (29) TRANTER Spiral Heat Exchanger — YouTube, accessed on 16-07-2021.

Table 4.19: Typical details of GPHEs. 'Courtesy of WIT Press from the book [Wang l., Sundén, B., and Manglik R.M., Plate Heat Exchangers: Design, Applications and Performance, 2007][7]

Quantity	Value
Maximum operating pressure	25 bar
Maximum operating temperature	160°C
Maximum unit flowrate	3,600 m^3/h
Overall HTC	Up to 7,500 W/m^2.K
Heat transfer area per unit	Up to 2,200 m^2
Maximum nozzle size	450 mm
Minimum approach temperature	1°C
Maximum heat recovery[i]	93%
Maximum number of plates	700
Plate size	0.3–3.5 m
Plate thickness	0.4–1.2 mm
Plate spacing	1.5–5.4 mm
Corrugation depth	1.5–5.4 mm

applications. GPHE and STHE are compared in Table 4.20. Typical fouling factors and overall HTCs in PHEs for some services are given in Tables 4.21 and 4.22, respectively. Values in these two tables are extracted from Table 9.4 in Cao.[8] PHEs are available in many metallurgies. Typical material of construction for PHE plates and PHE gasket materials are presented in Tables 4.23 and 4.24, respectively. In the past, asbestos gaskets were used for temperatures above 180°C, but they are now banned in many countries.

[i] Maximum heat recovery is expressed as the percentage of heat in the hot stream that can be transferred to the cold stream (with minimum approach temperature and assuming zero heat loss).

Table 4.20: Comparison of GPHE and STHE

Quantity	GPHE	STHE
Minimum approach temperature, °C	~1	~5
Multiple duty[j]	Possible	Not possible
Maximum operating temperature	Limited to 180°C, by the gasket material	No limit up to 537.8°C, with suitable metallurgy
Safety threats for handling hazardous substances	High	low
Handling liquids with solids	Not suitable unless proper filtration is used to remove solids	Suitable
PD	Higher	Lower
Piping connections	From one direction	From many directions
Operating weight ratio	1	3–10
Hold-up volume	Low	High
Internal welds	None	Many
Vibration damage	Possible with pressure fluctuations on shell and tube side fluids	Possible if baffle spacing is very high and/or nozzle velocities are high
Gaskets	On every plate	On each bonnet
Access for inspection	On both sides of plates	Limited
Disassemble time	Low	High
Repair	Easy to replace gaskets and plates	Relatively difficult to replace tubes
Heat duty increase	Easy by increasing number of plates	Difficult
Relative fouling tendency	0.1–0.25	1

[j] Multiple duty refers to heat exchange between more than two (i.e., one hot and one cold) streams. It is mainly used in cryogenic applications.

Table 4.21: Fouling resistances of PHE

Process Stream	Fouling Resistance, $m^2.K/W$
Fresh water, condensate	$(0.172–0.344) \times 10^{-4}$
Sea water, Acetic acid, acetone, benzene, carbon tetrachloride, Dowtherm. ethyl acetate, ethanol, ethylene glycol, n-heptane, methanol, methyl ethyl ketone, wine and n-octane	$(0.43–0.86) \times 10^{-4}$
Beer, milk	$(0.086–0.258) \times 10^{-4}$

Table 4.22: Typical overall HTCs of PHE

	Overall Heat Transfer Coefficient, W/m.K	
Service	**Transversal Corrugation Plate**	**Herringbone Corrugation Plate**
Water to water or steam	3,100–3,900	3,000–3,700
Viscous aqueous solution to water or steam	1,000–1,200	700–800
Mineral oil to water or steam	450–580	300–350
Mineral oil to mineral oil	210–270	120–190
Organic solvent to water or steam	1,850–2,100	1,500–1,950
Vegetable oil to water or steam	870–1,000	810–930

Table 4.23: Typical materials for plates of PHE and their applications

Material	Typical Applications
Stainless steel	Fresh water, non-saline cooling water, food products, pharmaceutical products, copper sulfate solutions, dilute chloride solutions (<200 ppm chloride)
Nickel	Caustic (50–70 wt. %) solutions
Incoloy	Hydrogen gas/water vapour with mercury carryover and acids
Hastelloy	Sulfuric and nitric acids
Titanium	Sea or brackish water, dilute acids (<70°C), chloride solutions (>200 ppm chloride)
Titanium–Palladium alloy	Dilute nitric acids and dilute sulfuric acids (<70°C and 10% concentration)

Table 4.24: Typical PHE gasket materials and their applications

Material	Maximum Operating Temperature, °C	Typical Applications
Rubber	70	Oxygenated solvents, acids and alcohols
Neoprene	70	Alcohols, acids, alkalies, hydrocarbon solvents
Synthetic butadiene rubber	85	Aqueous solutions, alkalies, oxygenated solvents and acids
Nitrile rubber	135	Dairy products, beverages, pharmaceutical and food products, gasoline, oils, alkalies and organic solvents
Fluoroelastomers	150	Oils
Butyl rubber	155	Acids, alkalies, oils, aldehydes, ketones, phenols and esters
Ethylene propylene rubber	155	Alcohols, sodium hypochlorite, wide range of chemical media
Fluorinated rubber	180	Oils, gasoline, aqueous solutions and organic solvents
Silicone	180	Corrosive liquids

4.7 Unfired Waste Heat Recovery Steam Generators

Steam generation from WH streams is very common in process industries. Typical vertical WHRSG, shown in Fig. 4.9, is generally used for WHR from plug flow reactors, where exothermic chemical reaction occurs on catalyst packing inside the tubes. In this arrangement, BFW is introduced to the steam drum, which provides for water storage, steam–water separation and the static head for natural circulation. BFW is heated in the steam drum by direct mixing with saturated steam and water mixture generated at the WHRSG. Hot saturated water flows down to shell side of the WHRSG, through downcomer pipes under gravitational driving force. Hot process fluid flows inside the WHRSG tubes and its heat is transferred to water on the shell side. This causes 5%–10% of the shell side water to vapourize into steam. Resulting steam and hot water mixture

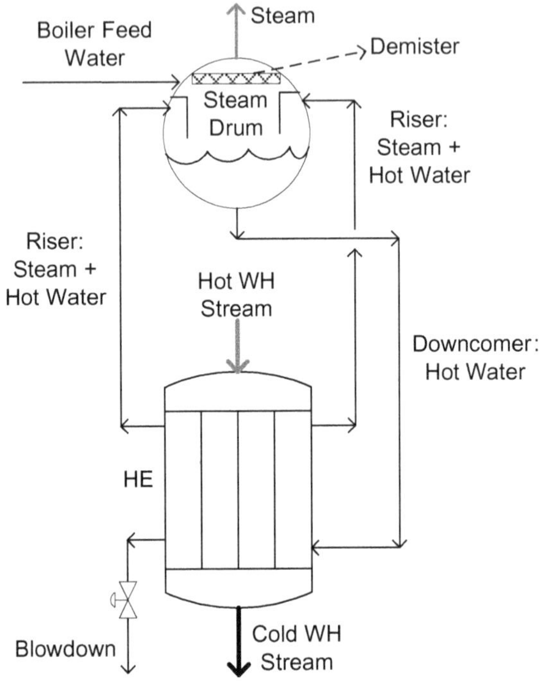

Fig. 4.9: Steam generation in a vertical WHRSG using natural circulation.

flows upwards in the WHRSG and then through riser pipes into the steam drum, due to density difference.

Natural circulation flow rate should be designed such that the actual heat flux in HE is lower than the critical heat flux. It is calculated by equating static head (i.e., elevation difference between the steam drum water level and boiling water surface in HE) to sum of the PDs due to friction losses in downcomer pipes, friction losses in riser pipes and HE system, gravity losses in HE and riser pipes, acceleration losses due to phase change, and losses in steam drum internals. Natural circulation flow rate increases by increase in static height difference between steam drum and HE, and density difference between water and steam (i.e., by lowering the steam generation pressure). Water and steam are separated in the steam drum due to their density difference. Steam is sent to steam header after passing through a demister or other type of water droplet removal system (e.g., vane, cyclone or combination of them), installed inside the steam drum. Blowdown, either on continuous or intermittent basis, is conducted based on the water quality control requirements.

Large process WHRSGs generally use horizontal steam generators shown in Figs. 4.10 and 4.11. The arrangements in Figs. 4.9 and 4.10 use an external steam drum while that in Fig. 4.11 uses an integrated steam drum. Due to large shell diameter, kettle type HE (Fig. 4.11) provides integrated steam drum. In kettle type WHRSG, a minimum vapour space height of 500 mm should be maintained above the liquid surface.

Design of WHRSG depends on many factors presented in Fig. 4.12. Of these, nature of the WH fluid is very important. Corrosive fluids require corrosion resistant shell or tube material. Fouling fluids are recommended to be used inside the tubes, at a velocity of 2–3 m/s. Selection of natural or forced circulation mainly depends on site specific factors such as plot space and feasibility of installing steam drum at an elevated platform. A forced circulation system uses a water pump to pump hot water from steam drum to HE. Hence, in this system, steam drum elevation does not dictate water and water-steam circulation rates. More discussion on natural and forced circulation systems is given in Section 11.9 of Chapter 11. Unfired WHRSGs generally produce saturated steam at a

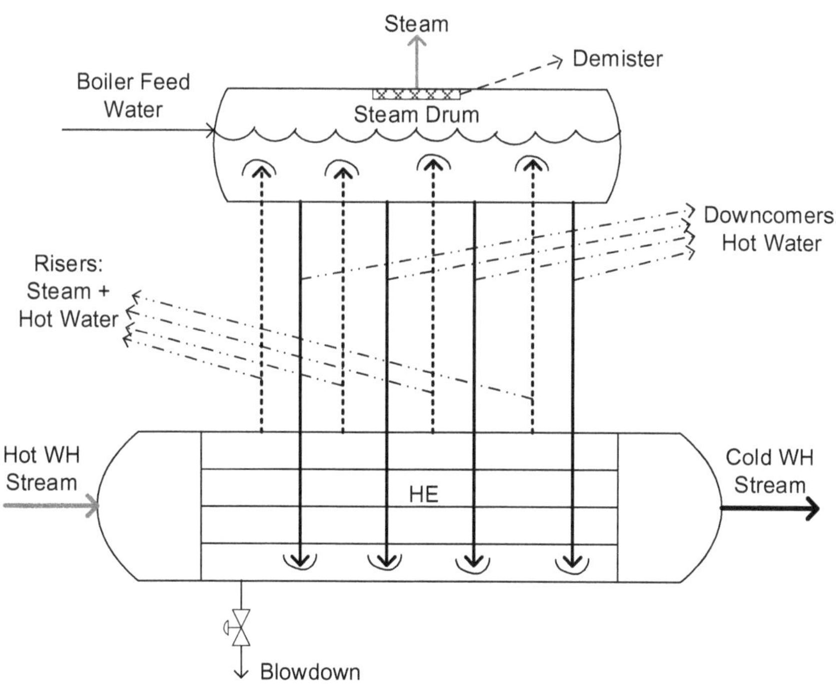

Fig. 4.10: Steam generation from a horizontal WHRSG using natural circulation.

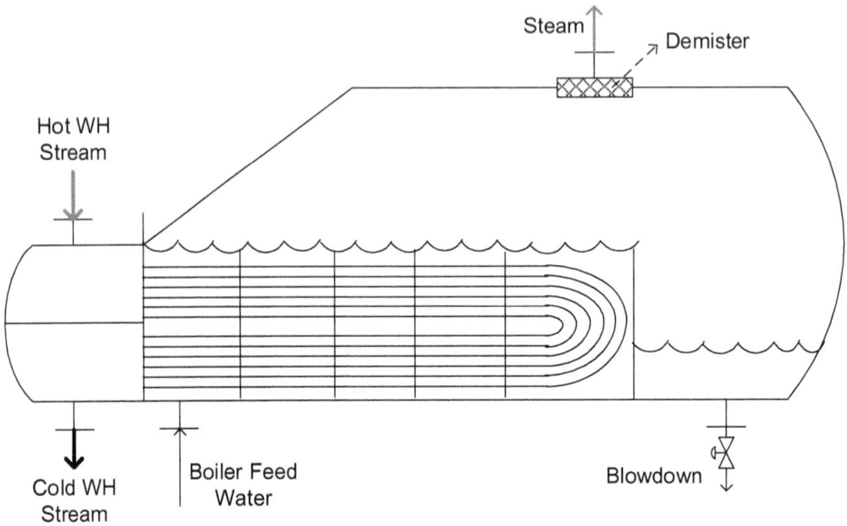

Fig. 4.11: Steam generation from a kettle type WHRSG.

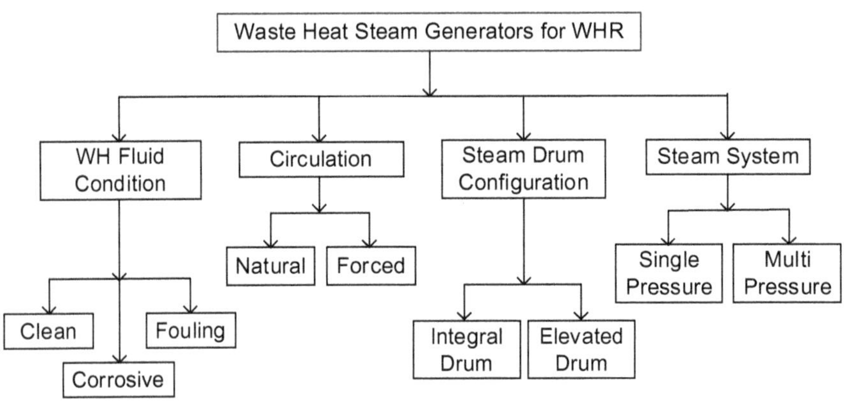

Fig. 4.12: Classification of WH steam generators.

single pressure. However, multilevel steam generation is sometimes possible by installing WHRSGs in series, with each WHRSG generating steam at a different steam pressure, by using WH from one or more process streams. Steam drum can be elevated as in Figs. 4.9 and 4.10, or integral drum as in Fig. 4.11.

4.8 Heat Transfer Enhancement Techniques

The goal of HTE or augmentation is to improve the HTC. HTE techniques generally reduce the thermal resistance either by generating turbulence in the fluid flowing inside the device or by increasing the effective heat transfer surface area. Hence, they are widely used in HEs to reduce their size, capital and operating costs. Rough or extended surfaces are utilized for increasing the effective surface area whereas inserts, winglets and turbulators are employed for increasing/generating the turbulence.

HTE techniques are divided into three categories: passive, active and compound techniques, as presented in Table 4.25. Active techniques employ supplementary power supply. Their examples are use of

Table 4.25: Classification and details of passive and active techniques for HTE

Passive Techniques	Details
Treated or rough metal surfaces	Heat transfer surfaces can be specially treated to create a fine-scale alteration to them or rough surfaces. These techniques increase turbulence at heat transfer surface; however, there is no appreciable increase in the heat transfer surface area. They are mainly used in STHEs such as reboilers and condensers for increasing heat transfer duties. Heat transfer surfaces can also be coated with other metals or polymers such as Teflon.
Extended surfaces	Extended surfaces such as fins increase the surface area and hence increase the heat transfer. Plain fins are commonly used in double pipe HE, STHE, finned PHE and ACHEs. Fins can be in different shapes to improve HTCs (e.g., low fins on tube inner and/or outer surface).[k,l]
Swirl flow devices	These use tube inserts to produce swirl flow inside the tubes and hence increase turbulence for single and two-phase flows. Some examples are static mixer, twisted tape, helical vane and cored screw-type tube inserts in STHEs.

(Continued)

[k] Low-finned tubes — Wieland Thermal Solutions (wieland-thermalsolutions.com), accessed on 15-05-2021.

[l] NEOTISS — High performance tube — NEOTISS™ HPT Finned Tubes, a finned tubular solution which yields numerous benefits for shell and tubes heat exchangers, accessed on 15-05-2021.

Table 4.25: (*Continued*)

Passive Techniques	Details
Surface tension devices (applicable for all HE types)	These utilize surface tension forces to drain or transport liquid film. They consist of wicking or grooved surfaces, which direct and improve the liquid flow to boiling surfaces and condensate drainage from condensing surfaces (e.g., fluted tubes).
Additives for fluids (applicable for all HE types)	Solid particles or gas additives can be added to liquid streams. Solid particles or liquid droplets are used for gas streams.
Coiled tubes	The tube coil curvature produces changes in direction of flow and creates secondary flow, which increases HTC in both single and two-phase flows.
Displaced enhancement devices	Displaced enhancement devices are inserts, used inside the tubes to improve mixing in the main flow and at the wall regions in STHEs (e.g., spaced disk device, rod inserts and wire coil insert such as hiTRAN® wire matrix[m]).
Active Techniques	**Details**
Surface/fluid vibrations	Surface vibration is applied in single-phase flows to obtain higher convective HTCs. It can use pulsations of frequencies ranging from 1 Hz to ultrasound (~20,000 Hz). Flow-induced vibrations can be used, with proper design, to enhance the HTC in STHEs.
Electrostatic fields	Electric, magnetic fields or combination of the two can be used to cause greater bulk mixing of dielectric fluids, near the heat transfer surface.
Jet impingement	Jet impingement forces a single-phase fluid normally or obliquely towards the heating surface. Single or multiple jets may be used.
Mechanical aids	These stir the fluid by mechanical means or by rotating the surface. Some examples are rotating tube HE and scraped surface HE.
Injection	Injection, used only in single-phase flow, involves injection of the same or a different fluid into the main bulk fluid either through a porous heat transfer interface or upstream of the HE. This technique is generally used in STHEs and ACHEs
Suction	Suction involves vapour removal, in nucleate or film boiling, or fluid withdrawal in single phase flow through a porous heated surface.

[m] hiTRAN® Thermal Systems (calgavin.com), accessed on 15-05-2021.

mechanical aids, surface fluid vibration and electrostatic fields. Passive techniques break the thermal boundary layer to increase HTC, and they include use of twisted tapes, helical screw tape inserts, ribbons, wire coils, rough surfaces, extended surfaces and additives for liquid and gases. Their implementation generally requires lower cost compared to the active techniques, and hence they are commonly used in industry. Two or more of the active and/or passive techniques can be utilized simultaneously to produce an enhancement that is larger than each of them operating separately. This is called compound enhancement of HTC. One example is rough tube outer surface with tube insert inside the tube.

Other than commonly used HTE techniques, for example, extended surfaces, typical applications of passive techniques used in process industries are summarized in Table 4.26. When specifying tube inserts, special care should be taken to check their suitability for a given service. In some cases, inserts may break and plug the tubes and hence cause very high PD and/or tube cleaning problems.

Table 4.26: Typical application of passive techniques in process industries

Type	Description	Enhancement Principle	Typical Applications
Rifled tubing	A helical groove or rib of low projection or depth. Geometry limited to those of manufacturers.	Interrupt the boundary layer.	• WH boilers with non-turbulent flow due to high process gas temperatures. • Viscous streams with solids and fibers.
Twisted tape	A thin ribbon of metal twisted into a helical shape. The geometry is fully customizable.	Bulk mixing between the centre and wall regions.	• Diverse applications due to lower cost and well documented calculation methods. • Often used for retrofitting existing HE.

(*Continued*)

Table 4.26: (*Continued*)

Type	Description	Enhancement Principle	Typical Applications
Helical wire wrapped core or twisted tube	Wire wrapped around a rod or twisted tube. Geometry can be customized.	Bulk mixing between the centre and wall regions, blocking the tube centre to reduce hydraulic diameter.	• Viscous flows with low velocities. • Can tolerate solids and fibers.
Wire matrix	A matrix of looping wires around a central core wire. Wire size, spacing and matrix density can be tailored to suit.	Interrupt the boundary layer.	Viscous fluids that do not contain fibers. Can tolerate some small solids. Commonly used for retrofitting existing HE.
Helical wire (Stationary)	Simple helical wire with interference fit into tube. Wire diameter and pitch can be tailored to suit.	Interrupt the boundary layer.	Viscous streams with solids and fibers. Often used in retrofitting to increase heat duty or control fouling.
Rotating or vibrating helical wire inserts	Simple helical wire with non-interference fit into tube. Wire diameter and pitch can be tailored to suit.	Mechanical cleaning of soft fouling layers, interrupt boundary layer.	Heavy fouling due to high HE tube wall temperature (e.g., in crude oil preheat train).
Static Mixer	Ribbon and bladed inline static mixers.	Bulk mixing between the centre and wall regions	High viscosity materials such as plastic injection molding heating and adhesive heating.

4.9 Summary

This chapter is on types of, selection guidelines for and retrofit/revamp options for HEs. Main points of this chapter are as follows:

- Many types of HEs are available for WHR. Understanding of their relative advantages and disadvantages is important for selecting the right type of HE.
- Fixed tube sheet STHEs are the cheapest among STHEs for those applications handling clean fluids and minimal thermal expansion (i.e., not requiring floating head). U-tube HEs are the cheapest option for applications requiring tube bundle expansion and handling clean services. S-type floating head HE is commonly used for applications involving fouling fluids and requiring floating heads.
- There are many options for increasing HTC (on tube and/or shell side), heat transfer area, increasing or decreasing PD, in STHEs. They are useful for academicians as well as practicing engineers.
- Steam generation is commonly used for WHR from process streams and other WH sources such as FG. Several useful WHRSG options are presented in Section 4.7.
- Active, passive and their combination techniques are available for HTE, which ultimately improve STHE performance. Among them, passive options are commonly used in process industries.

Chapters 3–7 in Serth[3] are recommended for process design of HEs. Chapter 10 in Shah[4] is recommended for detailed coverage of HE selection.

References

1. TEMA (Tubular Exchanger Manufacturers Association). (2019) *Standards of the Tubular Exchanger Manufacturers Association (TEMA)*, 10th ed.
2. Raju KS. (2011) *Fluid Mechanics, Heat Transfer, and Mass Transfer: Chemical Engineering Practice*. John Wiley.
3. Serth RW. (2007) *Process Heat Transfer: Principles and Applications*, 1st ed. Elsevier.
4. Shah RK, Sekulic DP. (2003) *Fundamentals of Heat Exchanger Design.* Hoboken, NJ: John Wiley.
5. Mukherjee R. (February 1998) *Effectively Design Shell-and-Tube Heat Exchangers Chemical Engineering Progress*. http://www.torr-engenharia.com.br/wp-content/uploads/2011/05/exchanger.pdf
6. Treese SA, Pujado PR, Jones DSJ. (2015) *Handbook of Petroleum Processing*, 2nd ed. Netherlands: Springer.

7. Wang l, Sundén B, Manglik RM. (2007) *Plate Heat Exchangers: Design, Applications and Performance*, Wit Press.
8. Cao E. (2010) *Heat Transfer in Process Engineering*. McGraw-Hill Education.
9. API Std. 660. (August 2007) *Shell and Tube Heat Exchangers*, 8th ed.
10. API Std. 662. (February 2006) *Plate Heat Exchanger for General Refinery Services, Part I — Plate and Frame Heat Exchangers,* 1st ed.

Acronyms and Notation

Acronyms

ACHE	Air Cooled Heat Exchanger
APH	Air Preheater
API	American Petroleum Institute
ACHE	Air Cooled Heat Exchanger
BFW	Boiler Feed Water
DM	Demineralized Water
EMTD	Effective Mean Temperature Difference
FG	Flue Gas
FWPHE	Fully Welded Plate Heat Exchanger
GPHE	Gasketed Plate Heat Exchanger
HE	Heat Exchanger
HTC	Heat Transfer Coefficient
HTE	Heat Transfer Enhancement
ID	Inner Diameter
LPS	Low-Pressure Steam
OD	Outer Diameter
PD	Pressure Drop
PHE	Plate Heat Exchanger
SPHE	Spiral Plate Heat Exchanger
STHE	Shell and Tube Heat Exchanger
TEMA	Tubular Exchangers Manufacturers Association
WH	Waste Heat
WHR	Waste Heat Recovery
WHRSG	Waste Heat Recovery Steam Generator

Notation

V	Velocity (m/s)

Greek Symbols

ΔT	Temperature difference or driving force (°C or K)
ρ	Density (kg/m^3)

Subscript

min	Minimum

Exercises

4.1 List important considerations for selection of shell, front and rear heads of an STHE.

4.2 Which TEMA type of STHE is economical for handling clean and non-hazardous fluids (on both shell and tube sides) at low operating pressure and temperature?

4.3 Which TEMA type of STHE is economical for handling hazardous fluid on tube side, with low operating pressure and requiring thermal expansion of the tube bundle?

4.4 What are the important uses of longitudinal and transverse baffles?

4.5 What tube pitch is recommended for handling heavy fouling substances on shell side?

4.6 What is the importance of impingement devices?

4.7 What is the significance of shell side streams? What are the strategies for maximizing B stream in an STHE?

4.8 What are the advantages and disadvantages of PHEs over STHEs?

4.9 What are the considerations for using HTE techniques? Are they equally recommended for both new STHEs and retrofitting existing

STHEs. List the reasons for the recommended use of HTE techniques.

4.10 Answer the following with yes (right) or no (wrong), and give brief reasoning, where applicable.

(A) STHEs are always the best choice.
(B) Designing an STHE for low PD is recommended to save on pump/compressor energy.
(C) Placing fouling fluid on the shell side is the preferred option.
(D) It is economical to place high-pressure fluid on the tube side.
(E) E shell is the most used in the industry.
(F) F shell is always preferred over E shell.
(G) Since U-tube STHE is the cheapest among removable tube bundle STHEs, it is always the preferred removable tube bundle type design.
(H) It is not possible to increase the heating duty of STHE, without replacing the STHE.
(I) Increasing tube side velocity increases HTC faster than tube side PD.
(J) Square pitch is always the best option.
(K) Triangular pitch provides more HTC compared to square pitch.
(L) GPHEs can be used for handling high-pressure hazardous fluids.
(M) For a given service, overall HTC is the same for both STHE and PHE.
(N) PHE has generally lower PD compared to STHE.
(O) For a given duty, PHE is always cheaper than STHE.
(P) Use of extended surfaces like fins is always beneficial.
(Q) Active HTE options are generally cheaper compared to passive HTE options.

Chapter 5

Heat Pumps for Waste Heat Recovery

5.1 Overview

Process industries use heat pumps (HPs) for providing process heating and cooling (via chilled water). Traditionally, these duties are met using plant utilities, which consume primary and/or secondary energy (covered in Chapter 7), thus contributing to energy and environmental costs. On the other hand, millions of HPs are used for space heating or cooling in residential and commercial buildings. Application of HPs using waste heat (WH) is now growing in process industries. HPs reduce fuel and/or electricity consumption, and consequently reduce greenhouse gas/acid rain generation and air pollution. Hence, they contribute significantly to the creation of a sustainable, low emission industry.

An HP can upgrade heat from a low temperature level to a high temperature level with the aid of an external energy source. It can maximize WH utilization; hence, it is economical for providing heating duties in some low temperature applications such as light hydrocarbon separations, process heating and HVAC (heating, ventilation and air conditioning) applications. HPs are also very attractive for producing chilled water using WH. Those HPs used for refrigeration are commonly known as chillers; they can be driven by electricity (e.g., mechanical chiller) or WH (e.g., absorption chiller). A mechanical chiller is used in chilled water production described in Section 7.3.4. The present chapter covers chillers using WH only.

WH driven HPs maximize the plant's WHR in a sustainable way by increasing energy efficiency, reducing environmental emissions and operating costs. Although electric power is consumed, mechanical compression (also known as mechanical vapour recompression, MVR) can be beneficial for upgrading flash steam (FS) to higher temperature/pressure (covered in Chapter 16) and reducing energy consumption in distillation system (covered in Chapter 9). Absorption HPs (AHP) can utilize WH to provide process heating and chilled water; they can be used for producing both chilled and hot water. Chemical HPs (using a suitable chemical reaction) can be used for achieving higher temperature lift of WH compared to AHPs.

HPs are widely used in almost all process industries such as food, pharmaceutical, pulp, paper, metal, water, power, chemical, petrochemical and petroleum refineries. Many types of HPs are available, and each of them has its own merits and demerits. Hence, comprehensive knowledge of HPs is required for correct selection of an economical HP for a particular WHR application. This chapter outlines classification, working principles and main uses of industrial HPs. It begins with Section 5.2 on the classification of HPs. Section 5.3 presents the operating principle and applications of MVR and thermal vapour recompression (TVR). Section 5.4 outlines the working principles of lithium bromide (LiBr) HPs and chemical HPs, for amplifying temperature of heat. Then, Section 5.5 describes operating principles of LiBr and adsorption HPs, for producing chilled water. Section 5.6 presents selection strategies for HPs. Finally, Section 5.7 summarizes the salient points for maximizing WHR using HPs.

The learning outcomes of this chapter on industrial HPs are as follows:

1. Outline the classification of HPs using WH.
2. State operating principles of MVR, absorption (LiBr) and chemical HPs for process heating.
3. Discuss the operating principle and applications of absorption (LiBr) and adsorption chillers for producing chilled water.
4. Discuss selection strategies for application of HPs using WH.

5.2 Classification of Heat Pumps

HPs are used for process heating and/or cooling work by performing the following basic functions:

- Receiving low temperature WH
- Raising the temperature/heat content of WH
- Delivering useful heat at a higher temperature

Under cooling mode of HPs, low temperature WH from chilled water return is raised to a high temperature/heat content and rejected to cooling water (i.e., a heat sink). Classification of industrial HPs for process heating and/or cooling is shown in Fig. 5.1.

HPs consume external energy (WH or mechanical energy) to increase the temperature of WH and ultimately reduce the fuel consumption. The common types of industrial HPs are as follows:

- **Open-Cycle MVR** uses mechanical (re)compression to increase the pressure and temperature of process gas or water vapour. It is employed in evaporation and distillation processes (e.g., by upgrading

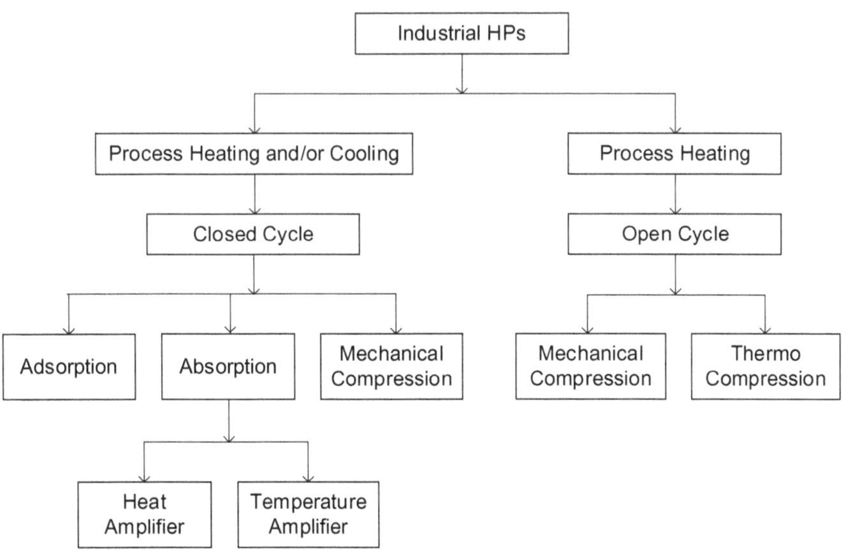

Fig. 5.1: Classification of HPs for heating and/or cooling.

FS to low-pressure steam [LPS]), commonly found in process industries.

- **Open-Cycle Thermocompression (TVR)** uses medium-pressure steam (MPS) or high-pressure steam (HPS) to increase the pressure/temperature of FS. It is utilized in evaporators, FS recovery systems and paper dryers. Open-cycle TVR has both higher coefficient of performance (COP) and lower capital cost. Hence, its selection for FS recovery, where product steam can be sent to plant's steam header, is generally more cost-effective.
- **Closed Cycle MVR** uses mechanical compression of FS or process vapour, produced by heat exchange with WH stream. This produces steam/process vapour at a higher temperature and pressure, which can be used for heat exchange with a process cold (sink) stream. It is mainly used in distillation, heating of water/process liquids and drying processes.
- **Closed Cycle AHP** uses working fluid mixtures such as LiBr and water or ammonia and water. Boiling point elevation and heat of absorption are used to provide a high temperature heat source. There are two types of AHPs: Type 1 or AHP and Type 2 or absorption heat transformer (AHT), which are explained in Section 5.4. They have the ability to serve both cooling (using chilled water) and heating duties, and are typically used in distillation, process heating and chilling applications.
- **Closed Cycle Adsorption HP** uses adsorbent (such as silica gel or zeolite) and working fluid such as water. Adsorption, desorption, condensation and evaporation steps are involved to produce both chilled and hot water.

Note that, when absorption or adsorption HPs are configured for simultaneous chilled and hot water production, chilled water production is the main objective while hot water is produced at condenser and absorber sections of the HP. Usually, hot water temperature of 70°C is achievable.

Open-cycle HPs can be subdivided into MVR and TVR type. TVR (e.g., steam jet ejector, SJE) increases WH stream temperature to a higher and useful temperature level. It is covered in Section 5.3. Open-cycle HPs are not applicable/economical for producing process cooling or chilled water. SJEs can be used for producing chilled water by vacuum evaporation of water. However, all the steam at the outlet of SJE needs to be condensed, which results in loss of latent heat of steam to cooling water or air. This leads to energy wastage. Hence, SJE for chilled water production is not a WHR application and not further discussed in this chapter.

Closed cycle HPs can be subdivided into absorption, adsorption and MVR types. They can be configured to produce chilled water also. Chemical HPs are absorption Type-2 HPs or AHTs. They utilize reaction of acids, their salts and water to upgrade WH to a temperature higher than that achieved by LiBr HPs. They are covered in Section 5.4. Adsorption HPs uses WH and solid adsorbents to provide both chilled water (typically 5°C–15°C) and hot water (generally at <90°C). They commonly employ water, ammonia or methanol as the working fluid. They are described in Section 5.5.

HPs are widely used in process industries. Their typical applications in process industries are summarized in Table 5.1.

Table 5.1: Typical application of HPs in process industries

Industry	Process Details	HP Details
Petroleum refining and petrochemicals	Separation of close boiling mixtures such as propane and propylene, and i-butane and n-butane	MVR (open cycle)
	FS recovery	TVR (open cycle)
	Production of chilled water	Closed cycle MVR, absorption or adsorption chiller
Chemicals	Concentration of liquids including wastewater streams	MVR (open cycle)
	FS recovery	TVR (open cycle)
	Production of chilled water	Closed cycle MVR, absorption or adsorption chiller
Power and Utilities	Cooling of gas turbine inlet air using chilled water produced by flue gas WHR	Absorption chiller

(*Continued*)

Table 5.1: *(Continued)*

Industry	Process Details	HP Details
	Concentration of blowdown water from cooling towers	MVR (open cycle)
Water	Producing fresh water from thermal desalination of sea water or treated wastewater	MVR (open cycle)
Pharmaceuticals	Process/water heating	Closed cycle MVR or AHP
Pulp and paper	Concentration of black liquor	MVR (open cycle)
	Process water heating	Closed cycle MVR or AHP
Food and Beverage	Concentration of (a) waste liquids in alcohol manufacture, (b) waste beer in breweries, (c) solutions in starch and sugar industry, (d) milk, cheese and whey in dairy industry, (e) juices in beverage industry and (f) wastewater in soft drink/food industry	MVR (open cycle)
	Water heating	Closed cycle MVR or AHP
Textiles	Process water heating and concentration of effluent	Closed cycle MVR or AHP
Miscellaneous	Solvent recovery, process/water heating and cooling	All types

5.3 Mechanical and Thermal Vapour (Re-)Compression

Compression of the working fluid/refrigerant can be achieved with a mechanical compressor or SJE. HPs, using mechanical compressor or SJEs, are commonly known as MVR and TVR, respectively. They can be open or closed cycle type. In open-cycle MVR/TVR, mechanical compressor or SJE compresses the FS or process vapour. This increases the steam or vapour temperature. The high-pressure/temperature steam or vapour can then be used in other equipment in the plant (Fig. 5.2).

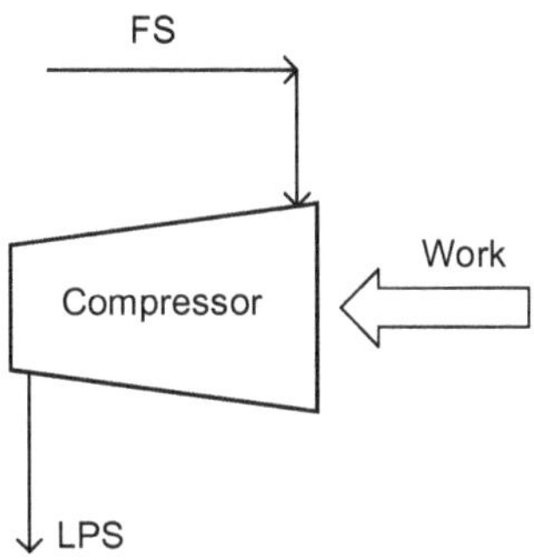

Fig. 5.2: Typical open-cycle MVR arrangement.

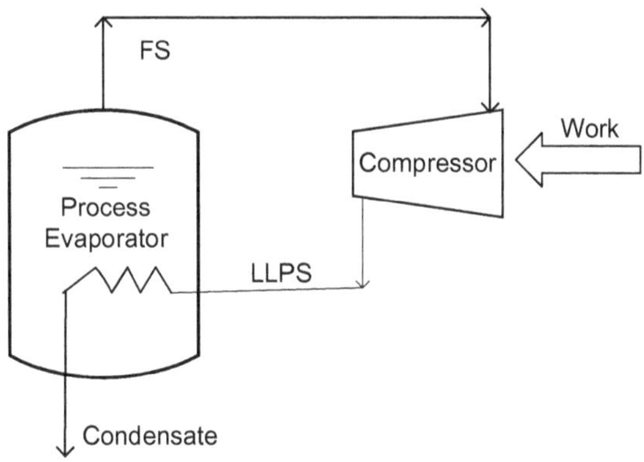

Fig. 5.3: Typical semi-open-cycle MVR arrangement.

In semi-open cycle, heat from the recompressed FS/vapour is transferred back to the same process equipment, producing the FS/vapour, via a heat exchanger (Fig. 5.3). COP of MVR is the ratio of heat rejected at higher temperature (in the condenser) to the energy input to the compressor. MVRs have high COP (3–9), do not use any chemicals and simple to operate and maintain.

The open-cycle TVR is similar to open-cycle MVR except that no mechanical energy input is required. TVR uses the kinetic energy of a vapour jet, normally steam, to compress a lower pressure vapour stream. Its equipment (i.e., SJE) is simple, has no rotating parts and hence very reliable. It is also most economical if the steam produced by TVR can be used for process heating. However, in most cases, the motive steam required in the TVR is much greater than FS being compressed. This restricts TVRs to systems with a smaller quantity of FS source and a large amount of low-low-pressure steam (LLPS) or LPS demand. TVR operates efficiently at the design conditions. Its performance can fall dramatically at off-design conditions making it more suitable for processes with steady operating conditions. Using a control scheme such as one illustrated in Section 8.2.1 of Chapter 8, TVR can be operated at off-design conditions.

Open-cycle TVR (i.e., SJE) uses HPS or MPS as motive fluid, to compress FS (suction fluid) at typically 1–3 bar, to usually LLPS or LPS. Typical arrangement of a TVR is shown in Fig. 5.4.

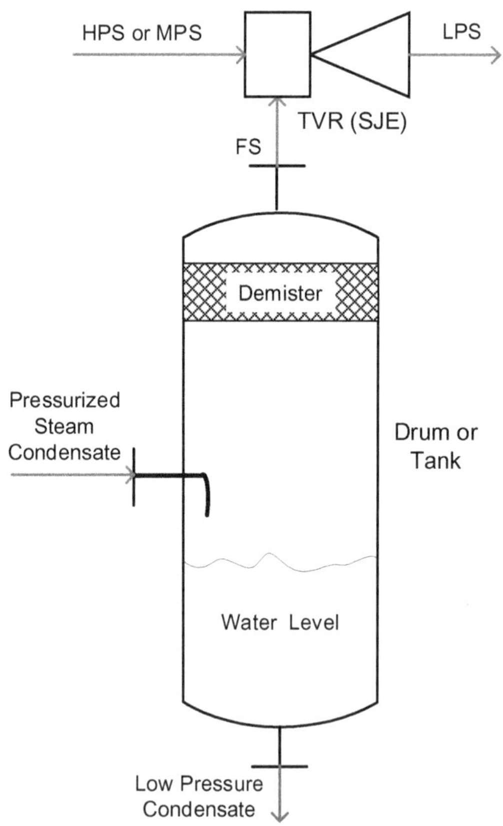

Fig. 5.4: FS recovery drum, installed with TVR (SJE).

MPS consumption for the TVR can be estimated using RM ratio, defined by:

$$\text{RM} = 0.4 \times e^{\frac{4.6 \times \ln\left(\frac{P_D}{P_L}\right)}{\ln\left(\frac{P_M}{P_L}\right)}} \tag{5.1}$$

Here, RM is the ratio of mass flow rate of motive steam to mass flow rate of load steam (FS or LLPS at lower pressure), P_M is the absolute pressure of motive steam, P_L is the absolute pressure of load steam and P_D is the target pressure of discharge steam. Application of this equation is illustrated in the following exercise.

Exercise 5.1: A TVR is required to compress 2 ton/hr of saturated LLPS at 2 bar to LPS at 4.5 bar, using motive MPS (saturated) at 12.7 bar. Assume discharge stream is saturated LPS. Estimate the MPS consumption and COP of TVR.

Solution: In this exercise, LLPS is the load steam, MPS is the motive steam and LPS is the discharge steam.

Substituting P_D = 4.5 bar, P_L = 2 bar and P_M = 12.7 bar in Eq. 5.1, RM = 3.009

Amount of MPS required = 3.009 × 2,000 = 6,018.927 kg/hr

Amount of LPS generated = Amount of LLPS + Amount of MPS = 2,000 + 6,018.927 = 8,018.927 kg/hr

Enthalpy of saturated LPS at 4.5 bar (from steam tables) = 2,743.386 kJ/kg

Enthalpy of saturated MPS at 12.7 bar (from steam tables) = 2,785.712 kJ/kg

Hence, $\text{COP} = \frac{\text{Amount of useful LPS energy}}{\text{MPS energy used}} = \frac{8{,}018.927 \times 2{,}743.386}{6{,}000 \times 2{,}785.712} = 1.316.$

Although COP of 1.316 is not very high, there is no wastage of WH or motive steam if the LPS can be used for process heating. Hence, TVR is very attractive as it is very reliable and a low-cost solution.

Closed cycle MVRs are generally used for upgrading WH at temperatures less than 100°C. A working fluid (water, ammonia, hydrocarbon-based or refrigerant) is heated by WH in the evaporator (Fig. 5.5). Vapour thus generated is compressed and then condensed to release heat at a

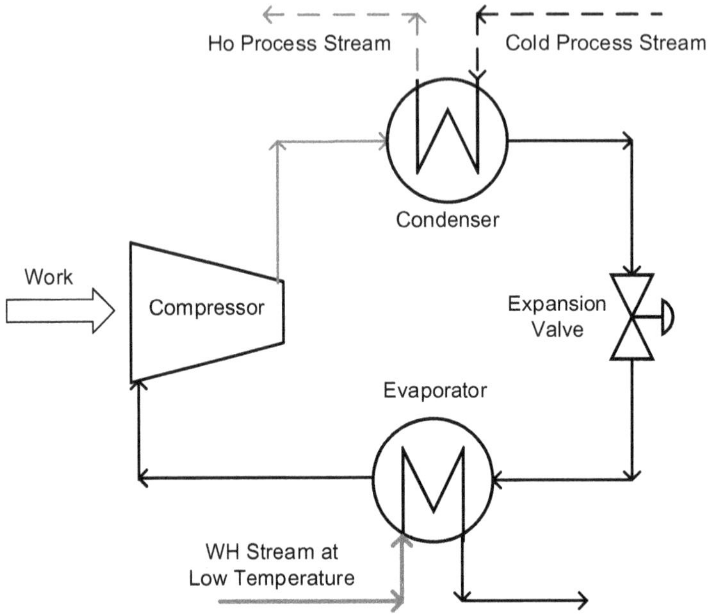

Fig. 5.5: Schematic of closed cycle MVR.

higher temperature, compared to the temperature of WH stream. Condensed working fluid is finally returned to the evaporator via an expansion valve. Closed cycle MVR in Fig. 5.5 uses a mechanical compressor (e.g., centrifugal or reciprocating compressor, or rotary blower), for increasing the pressure and temperature of the vapour. HPs can use shell and tube (with plain or finned tubes), plate or finned coil heat exchangers (Chapter 4). Available MVRs from some industrial suppliers are listed in Table 5.2.

Compressor in MVR can be driven by electric motor or steam turbine or diesel/gas engines. Electric motors are simple, compact, highly efficient, incur lower capital and maintenance costs; hence, they are commonly used. They have additional advantage if cheap electricity is available. Back pressure type steam turbines (described in Chapter 16) increase economic attractiveness of MVR if plant has demand for MPS/LPS and sufficient HPS/MPS is available for use in the steam turbine. Gas engine generally uses natural gas as the fuel.

Table 5.2: Some of the industrial MVRs with high heat sink temperature of 90°C and above (courtesy of Arpagaus *et al.*[1])

Manufacturer	Product	Refrigerant	Max. Heat Sink Temperature, °C	Heating Duty, kW	Compressor Type
Kobe Steel (Kobelco steam grow HP)	SGH 165	R134a/R245fa	165	70–660	Twin screw
	SGH 120	R245fa	120	70–370	
	HEM-HR90, 90A	R134a/R245fa	90	70–230	
Vicking Heating Engines AS	Heat Booster S4	R1336mzz(Z) R245fa	150	28–188	Piston
Ochsner Energie Technik GmbH	IWWDSS R2R3b	R134a/ÖKO1	130	170–750	Screw
	IWWDS ER3b	ÖKO (R245fa)	130	170–750	
	IWWHS ER3b	ÖKO (R245fa)	95	60–850	
Hybrid Energy	Hybrid HP	R717/R718 (NH3/H2O)	120	0.25–2,500	Piston
Mayekawa	Eco Sirocco	R744 (CO2)	120	65–90	Screw
	Eco Cute Unimo	R744 (CO2)	90	45–110	
Combitherm	HWW 245fa	R245fa	120	62–252	Piston
	HWW R1234ze	R1234ze(E)	95	85–1,301	
Dürr thermea GmbH	thermeco2	R744 (CO2)	110	51–2,200	Piston (up to 6 in parallel)
Friotherm	Unitop 22	R1234ze(E)	95	0.6–3,600	Turbo
	Unitop 50	R134a	90	9–20,000	(two stage)
Star Refrigeration	Neatpump	R717 (NH3)	90	0.35–15,000	Screw (Vilter VSSH 76 bar)
GEA Refrigeration	GEA Grasso FX P 63 bar	R717 (NH3)	90	2,000–4,500	Twin screw (63 bar)
Johnson Controls	HeatPAC HPX	R717 (NH3)	90	326–1,324	Piston (60 bar)
	HeatPAC Screw	R717 (NH3)	90	230–1,315	Screw
	Titan OM	R134a	90	5–20,000	Turbo
Mitsubishi	ETW-L	R134a	90	340–600	Turbo (two stage)
Viessmann	Vitocal 350-HT Pro	R1234ze(E)	90	148–390	Piston (2–3 in parallel)

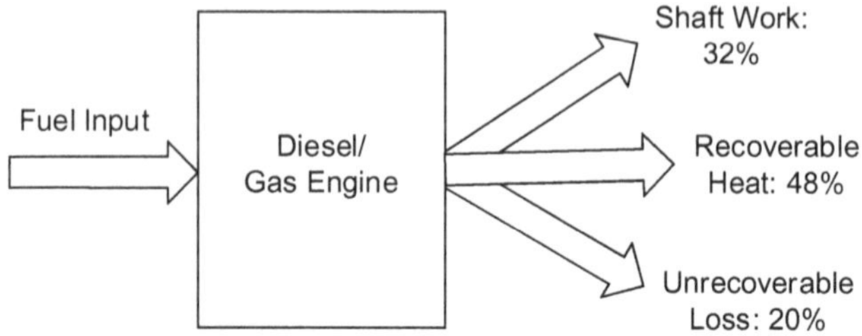

Fig. 5.6: Typical energy flows in a diesel/gas engine using diesel or natural gas as fuel.

In addition to producing shaft power, gas engine drivers reject most of the input energy in the form of WH in exhaust gases and jacket cooling water. The exhaust gases are at a relatively high temperature (200°C–500°C), while the engine cooling circuit produces useful heat in the range of 70°C–110°C. As shown in Fig. 5.6, approximately 32% of fuel energy goes into the compressor shaft energy. Balance 68% of fuel energy generates WH. Majority (~48%) of the fuel energy can be recovered economically and used for process or utility heating. Gas/diesel engines are generally expensive to purchase, install and maintain compared to motor of same power output. However, they can be attractive if fuel is very cheap compared to the electricity and WH from the flue gas can be recovered and reused.

COP of an HP is given by:

$$COP_H = \frac{Q_{out}}{W} \quad \text{or} \quad \frac{Q_{out}}{Q_{in}} \tag{5.2}$$

Here, Q_{out} is the useful heat released, Q_{in} is the WH input and W is the work done by the compressor. However, when considering the performance of a refrigeration system, cooling duty (Q_C) is important and not the heat output. Hence, refrigeration COP is calculated as:

$$COP_C = \frac{Q_c}{W} \quad \text{or} \quad \frac{Q_c}{Q_{in}} \tag{5.3}$$

Care must be taken in using COP as a performance measurement as it does not include auxiliary power for supporting equipment like cooling water circulating pumps and fans. These supporting equipment (e.g., cooling water circulating pumps in Type-2 AHP and absorption chillers) often consume significant energy and add to the operating cost. Also, COP does not consider any WHR from steam turbine driver outlet steam or diesel/gas engine drive's exhaust flue gas. A slightly better performance indicator for MVR type HP performance is primary energy ratio (PER) of useful heat delivered to total energy consumed, which includes all the energy used for driving the auxiliary equipment such as cooling water circulating pimps and fans. However, both COP and PER make no allowance for different value/cost of the heat supplied and the power used, and do not give any indication of capital cost.

If MVR type HP is driven by a diesel/gas engine with a thermal (= ratio of work done by the diesel/gas engine to the total heat energy input) efficiency of η, PER can be calculated as:

$$\text{PER} = \text{COP} \times \eta \tag{5.4}$$

If X% of fuel energy to the diesel/gas engine can be recovered by WHR, PER improves to:

$$\text{PER} = (\text{COP} \times \eta) + \frac{X}{100} \tag{5.5}$$

Exercise 5.2: An HP with COP of 5 is driven by a diesel engine with an efficiency of 0.3. Estimate PER and then PER if 40% of the energy input to the diesel engine can be recovered.

Solution: PER without WHR from diesel engine = 5 × 0.3 = 1.5.

This means, without WHR from diesel engine, HP only generates useful heat, which is 1.5 times the energy used at the diesel engine.

PER with WHR from diesel engine = $5 \times 0.3 + \frac{40}{100} = 1.9$.

As expected, PER increases to 1.9 from 1.5, for this case.

Specific work done by blower type MVR can be estimated by:

$$W = \frac{(P_{\text{d}} - P_{\text{s}}) \times V_{\text{s}}}{\eta_{\text{MVR}}} \tag{5.6}$$

Here, W is specific work done (kJ/kg), P_{d} and P_{s} are the discharge and suction pressures (kPa), respectively, V_{s} is the specific Inlet volume (m^3/kg) and η_{MVR} is the efficiency of the MVR.

Exercise 5.3: An MVR is installed in a multi-effect evaporator to compress 3,000 kg/hr of FS, generated from an aqueous solution in the first stage. The inlet pressure and temperature of MVR are 1 bar and 100°C, and the compression ratio is 1.3. Since the compression ratio is small, a blower is sufficient rather than a compressor. Assuming a blower efficiency of 0.6, estimate the electric power required for the MVR. Then, assuming electric power to heat ratio of 3, estimate the COP and PER of MVR.

Solution: Specific volume of steam at 1 bar (from steam tables) = 1.694 m^3/kg

Using Eq. 5.6, specific energy input $= \frac{(1.3-1)\times 100\times 1.694}{0.6} = 84.7\,\text{kJ/kg}$

From steam tables, latent heat of vapourization of water at 100°C and 1 bar is 2,257.513 kJ/kg.

Hence, in this case, 84.7 kJ/kg of electrical energy is sufficient to recover 2,257.513 kJ/kg of thermal energy from FS steam.

$$\text{COP} = \frac{2{,}257.513}{84.7} = 26.65$$

$$\text{PER} = \frac{2{,}257.513}{3\times 84.7} = 8.88$$

Exercise 5.4: A centrifugal compressor is used to compress 5,000 kg/hr of saturated FS at 100°C to 2.975 bar, which is then used to heat a process stream in a heat exchanger. This condenses and cools the compressed steam to 2.484 bar and 110°C. Condensate is depressurized to 1.504 bar. Finally, condensate is heated with WH stream in a heater to fully vapourize it. Outlet condition of heater is saturated steam at 100°C. Estimate the electric power and COP required for the MVR. Assume polytropic efficiency of 0.75 for the centrifugal compressor, efficiency of 0.95 for the centrifugal compressor's drive, and thermal to electrical energy ratio of 0.33. Estimate PER.

Solution: Calculation of MVR's power consumption requires the estimation of the value of polytropic exponent. To avoid lengthy procedure for this, Aspen Hysys simulation (Fig. 5.7) of centrifugal compressor type MVR can be used. From the simulation, 403.378 kW of power is required for the compressor and 3,478.856 kW of useful heat is generated at the condenser

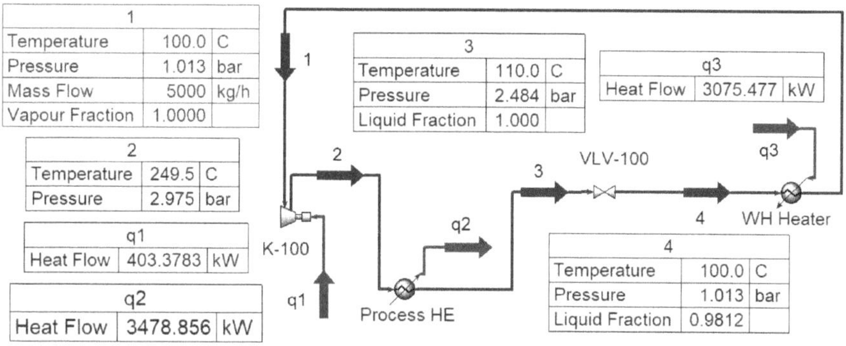

Fig. 5.7: Aspen Hysys simulation for Exercise 5.4. Pressure drop in WH heater is included in the pressure reduction of VLV-100 itself; this avoids possible warning in simulating WH Heater with temperature decrease of a process stream.

(Process HE) from 3,075.477 kW of WH utilized at the evaporator (WH heater in Fig. 5.7). Note that 3,075.477 + 403.378 = 3,478.855 kW. Hence,

COP of the MVR $= \frac{3{,}478.855}{403.378} = 8.624.$

PER = COP × η_m × ratio of thermal to electrical energy = 8.624 × 0.95 × 0.33 = 2.703

In case the process simulation in Fig. 5.7 is not available, approximate power consumption of MVR and discharge temperature of the compressor can be estimated by the following equations, respectively:

$$W = \left(\frac{mZRT_S}{M\eta_p}\right) \times \left(\frac{n}{n-1}\right) \times \left(\left(\frac{P_D}{P_S}\right)^{\frac{(n-1)}{n}} - 1\right) \tag{5.7}$$

$$T_D = T_S \times \left(\frac{P_D}{P_S}\right)^{\frac{(n-1)}{n}} \tag{5.8}$$

Here, W is the power required for compression (kW), Z is the compressibility constant, R is the universal gas constant (8.314 J/K.mol), T_S is the compressor inlet temperature (K), P_S and P_D are the compressor inlet and outlet pressures (bar), m is the mass flow rate of gas/vapour (kg/sec), η_p is the polytropic efficiency, M and n are molecular weight and polytropic exponent (i.e., polytropic compression obeys

PV^n = constant), respectively. Isentropic and polytropic exponents are related by:

$$\frac{n}{(n-1)}=\frac{k}{(k-1)}\times\eta_{\mathrm{p}} \tag{5.9}$$

Here, k is the isentropic exponent (equal to the ratio of specific heats, C_p/C_v). Using the above equations and steam tables, Exercise 5.4 can be solved as follows.

From steam tables, Z = 0.985. C_p = 2.098 kJ/kgK and C_v = 1.556 kJ/kgK.

M=18 (molecular weight of steam).

MVR suction pressure = saturation pressure of steam at 100°C = 1.01418 bar

Note that steam tables provide slightly higher steam pressure (1.01418 bar) at 100°C saturation temperature, compared to Aspen Hysys (1.013 bar). This difference is very small and will not significantly affect the solution.

$$k=\frac{2.08}{1.556}=1.337$$

$$m=\frac{5,000}{3,6000}=1.389\frac{\text{kg}}{\text{s}}$$

$$T_S = 100 + 273.15 = 373.15 \text{ K}$$

$$\frac{P_D}{P_S}=\frac{2.975}{1.01418}=2.9334$$

Using values of k and η_p (= 0.75) in Eq. 5.9, n = 1.505.

Using Eq. 5.7, power consumption of the MVR =

$$=\left(\frac{1.389\times0.985\times8.314\times373.15}{18\times0.75}\right)\times\left(\frac{1.505}{1.505-1}\right)\times\left((2.9334)^{\frac{(1.505-1)}{1.505}}-1\right)$$

$$=407.5\,\text{kW}$$

Using Eq. 5.8, discharge temperature of MVR =

$T_D=373.15\times(2.9334)^{\frac{(1.505-)}{1.505}}$ = 535.499 K = 535.499 − 273.15 = 262.29°C

From steam tables, enthalpy of steam at the outlet of MVR (at discharge conditions of 2.975 bar and 262.29°C) = 2,992.974 kJ/kg, and enthalpy of steam condensate at the outlet of process HE (at 2.484 bar and 110°C) = 461.4398 kJ/kg.

Recalling m = 1.389 kg/sec, heat transferred in the process HE = (2,992.974 − 461.4398) × 1.389 = 3,516.3 kW

COP of MVR $= \frac{3{,}516.3}{407.5} = 8.629$

PER = COP × η_m × ratio of thermal to electrical energy = 8.629 × 0.95 × 0.33 = 2.705

From steam tables, enthalpy of saturated steam at the suction of MVR (at suction temperature of 100°C) = 2,675.572 kJ/kg

Amount of WH required at the WH heater = (2,675.572 − 461.4398) × 1.389 = 3,075.18 kW

These values of COP, PER and amount of WH required are close to those found using Aspen Hysys simulation.

5.4 Absorption Heat Pumps for Heat/Temperature Upgrade

AHPs use WH, high temperature hot utility (such as MPS or HPS) or direct gas firing, to deliver higher temperature or quantity of heat, for process usage. There are two types of AHPs: Type 1 or AHP or heat amplifier, and Type 2 or AHT or temperature amplifier, as shown in Fig. 5.8.

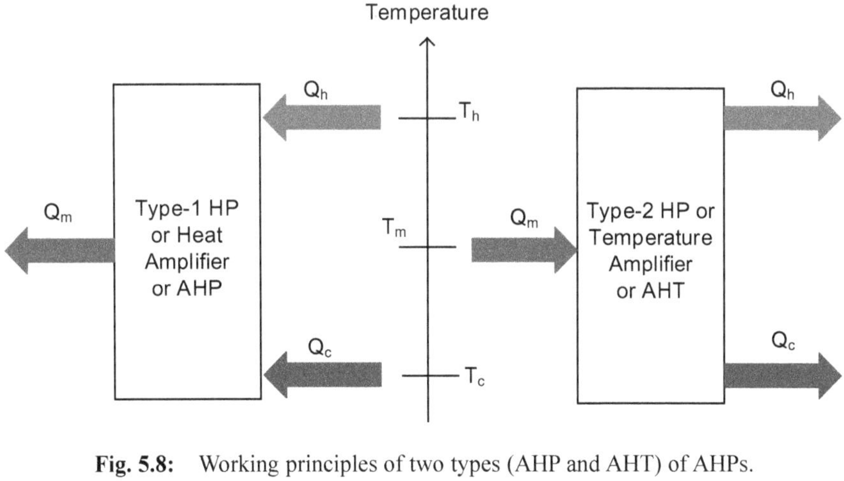

Fig. 5.8: Working principles of two types (AHP and AHT) of AHPs.

APH/AHT differs from MVR in that the former requires low mechanical energy input for achieving higher quantity of heat/temperature, and the principal external energy supply to APH/AHT is in the form of WH, LLPS or LPS/MPS.

Type-1 AHP has two heat inputs: a low temperature WH source and a high temperature energy source (e.g., MPS, HPS or direct gas firing). Heat output is at a temperature between the WH source and high temperature energy; thus, WH is upgraded to a higher temperature whereas heat from the energy source is downgraded to a lower temperature (Fig. 5.8). Electrical energy required for pumps in AHP/AHT is very small. Typical COP of single stage AHP is 1.2–1.7 whereas a two-stage AHP can achieve a COP of 2.3.[a]

Type-2 AHP operates in a cycle opposite to that of AHP. WH is supplied at a medium temperature level. AHT splits the total heat into two parts at different temperatures: one at a temperature higher than the WH source and the other rejected at a lower temperature, commonly using cooling water. AHTs are heat splitters and do not require any other energy input. Approximately 50% of the WH can be upgraded to a higher and useful temperature level. Typical COP of AHT is 0.3–0.5.

AHP/AHT/absorption chillers use two working fluids: a refrigerant and an absorbent. Various refrigerant and absorbent pairs are used in industrial practice. Of these, water–LiBr combination is most found in industrial applications, where water is the refrigerant. LiBr operates under very low pressure, has high efficiency, and can achieve temperature lift of up to 50°C from WH sources at 80°C–100°C. AHT can have output temperature up to about 150°C. Apart from LiBr–water mixture, ammonia–water mixture is sometimes used. Comparison of these working fluid mixtures is presented in Table 5.3. The temperature lift (typically 30°C–50°C) in AHT is limited by the potential crystallization of the absorbent. Ammonia/water AHPs can be operated below 0°C. The main drawback for ammonia–water mixture is the high operating pressure level and hence high energy consumption for the circulation pump that requires special design. Also, a costly distillation column is required to serve as the generator for this system. Moreover, ammonia is toxic and can create safety hazard if it leaks.

---

[a]Absorption heat pump Type1 (hitachiaircon.com), accessed on 10-06-2021.

Table 5.3: Absorption system working fluids

Property	Ammonia–Water	Water–LiBr
Refrigerant	Ammonia	Water
High latent heat	Good	Excellent
Low vapour pressure	Too high	Too low
Freezing temperature	−33°C	above 0°C
Low viscosity	Good	Good
Absorbent	Water	LiBr
Low vapour pressure	Poor	Excellent
Low Viscosity	Good	Good
Refrigerant and Absorbent Mixture		
Avoid Crystallization	Excellent	Only under a controlled operating range
Low toxicity	Poor	Good
High affinity between refrigerant and absorbent	Good	Good
Vacuum pump	Not required	Required
Corrosion	Good	Poor

5.4.1 *Absorption Heat Pump or Type-1 AHP*

The pressure–temperature profile of AHP is shown in Fig. 5.9, where low temperature WH (Q_e) is accepted in the evaporator to vapourize water. Water vapour from the evaporator is then absorbed in concentrated LiBr solution in the absorber, thus diluting LiBr solution and also liberating heat (Q_a), which can be transferred to a process heat sink (cold) stream. Diluted LiBr solution is pumped through an economizer, to heat it by heat exchange with hot concentrated LiBr solution. It then flows to the generator, where high temperature utility (Q_g) such as LPS is used to generate saturated steam at the generator pressure. As water evaporates, LiBr solution concentrates and is returned to the absorber, after exchanging heat at the economizer. Steam generated at generator is condensed in AHP's condenser, to release heat (Q_c) to the process heat sink stream. Steam condensate (water) is sent through an expansion valve to the evaporator, to complete the cycle. Evaporator temperature (T_e) is fixed according to WH stream temperature whereas condenser temperature (T_c) and absorber temperature (T_a) are fixed according to the process heat sink (cold stream) temperatures.

COP of the AHP system is given by:

$$\mathrm{COP_{AHP}} = \frac{Q_c + Q_a}{Q_g} \tag{5.10}$$

This equation ignores power required for the pumps in AHP, which is negligible (~2% of the total heat input) as liquid mixture leaving the absorber is nearly incompressible. Typical range of COP of AHP is 1.2–1.7. A large COP for an AHP indicates better heat recovery by using minimum external utility heat. In AHP, high temperature heat input (Q_g) to the generator is the main energy requirement, unlike the MVR that mainly requires compressor power.

Exercise 5.5: Estimate COP of an AHP operating under the following conditions. With reference to Fig. 5.9, outlet temperature of absorber ($T1$),

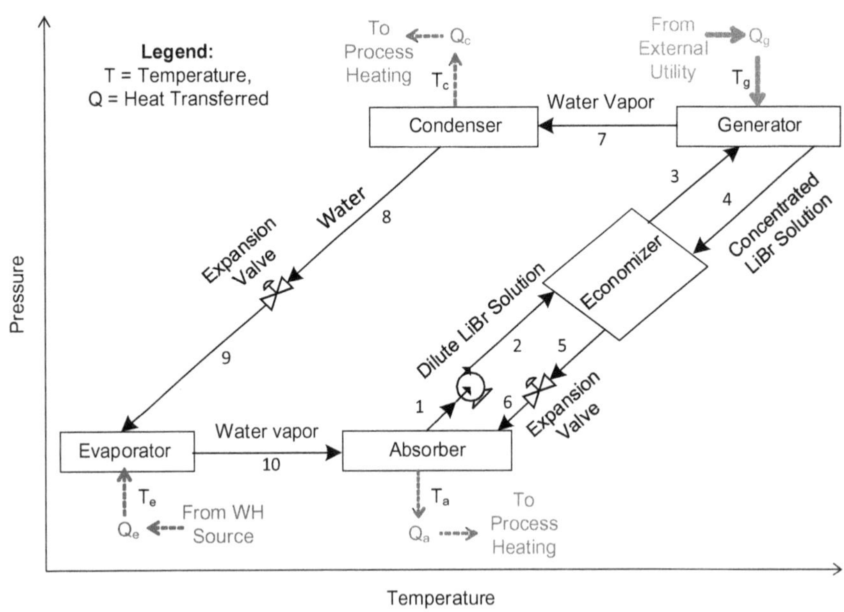

Fig. 5.9: Operational details of AHP on temperature versus pressure plot.

condenser ($T8$), generator ($T4$) and evaporator ($T10$) are 120°C, 120°C, 208.4°C and 83.56°C, respectively. Heat released at condenser (Q_c) and absorber (Q_a) are 519.23 kW and 819.3 kW, respectively. WH supplied to evaporator (Q_e) is 485.79 kW, and high-temperature utility supplied to generator (Q_g) is 852.74 kW.

Solution: $\text{COP} = \dfrac{Q_c + Q_a}{Q_g} = \dfrac{519.23 + 819.3}{852.74} = 1.57$

5.4.2 *Absorption Heat Transformer or Type-2 AHP*

Similar to AHP, AHT operates mainly with four heat exchangers but at different temperatures and pressures as shown in Fig. 5.10. Here, WH is used in the generator to generate steam from dilute LiBr solution, and so it is denoted as Q_g. This concentrates LiBr solution, which is pumped to the absorber, after heat exchange with hot dilute LiBr solution at the economizer. Steam from the generator is sent to the AHT's condenser for condensation, using cooling water (Q_c). Condensate from the condenser is pumped to the evaporator, where water is evaporated by heat exchange

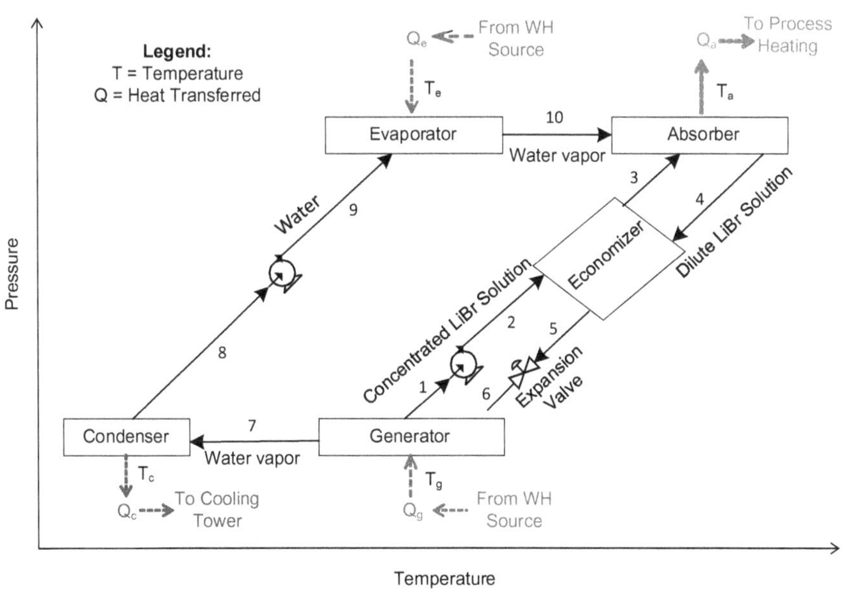

Fig. 5.10: Operational details of AHT on temperature versus pressure plot.

with WH stream (Q_e). Thus, WH is used in both the generator and evaporator. Steam generated in the evaporator is absorbed in concentrated LiBr solution, in the absorber. It generates high temperature heat (Q_a), which is utilized for process heating. Hot diluted LiBr solution from the absorber is sent to the generator after cooling at the economizer, to complete the cycle. AHT upgrades nearly half of the low temperature WH to a high temperature heat, and the other half of WH is rejected to cooling water.

COP of the AHT system is given by:

$$\mathrm{COP}_{\mathrm{AHT}} = \frac{Q_{\mathrm{a}}}{Q_{\mathrm{g}} + Q_{\mathrm{e}}} \tag{5.11}$$

Like COP for AHP, the above equation ignores power required for the pumps in AHT, which is negligible (~2% of the total heat input). Typical COP of AHTs is 0.5. A higher COP for AHT indicates upgrading of a larger portion of WH to a higher temperature level, and hence better performance. A low temperature lift gives a high COP, and a large amount of heat is upgraded per unit heat input. Note that AHT requires only cold utility whereas AHP requires only hot utility.

Exercise 5.6: Estimate COP of an AHT operating under the following conditions. With reference to Fig. 5.10, outlet temperature of generator ($T1$), Absorber ($T4$), condenser ($T8$) and evaporator ($T10$) are 80°C, 130°C, 25°C and 80°C, respectively. Heat released at absorber (Q_a) is 272 kW. WH supplied at evaporator (Q_e) and generator (Q_g) are 289.1 kW and 273 kW, respectively. Heat rejected to cooling water at condenser (Q_c) is 290 kW.

Solution: $\mathrm{COP} = \dfrac{Q_{\mathrm{a}}}{Q_{\mathrm{g}} + Q_{\mathrm{e}}} = \dfrac{272}{(273 + 289.1)} = 0.484$

5.4.3 *Chemical Heat Pumps*

Chemical HPs are essentially AHT; they release useful heat at a higher temperature by heat of reaction (and not by absorption). They can generate a higher temperature lift of WH compared to the AHPs. Chemical HPs use reversible chemical reactions between a vapour/gas and a salt that is part of a porous solid material inside a reactor (which replaces absorber in Fig. 5.10). WH at 75°C–150°C can be upgraded to 150°C–241°C by using liquid phosphates and water.[b] Typical COP of a chemical HP is ≈0.5. Chemical HPs are now available for industrial application.

In the first step of a chemical HP, water is removed at low pressure from phosphoric acid and its salts, by evaporation using WH; this occurs in the generator (in Fig. 5.10). By removing water and heating up phosphoric acid and its salts with WH, an endothermic polymerization takes place in the generator. Evaporated water is condensed with cooling water in the condenser (Figure 5.10). The condensed water is pumped to a higher pressure and vapourized using WH, in the evaporator. The polymerized-oligomerized product is pressurized using another pump and contacted with water vapour in the absorber. (Economizer can be added between generator and absorber, for increasing the COP of the chemical HP). The condensation of water in the liquid acid and its salts results in a highly exothermic hydrolysis reaction and a big temperature increase. This will depolymerize the working medium. The temperature increase is relatively large due to the latent heat of water vapour in combination with the hydrolysis energy. High temperature lift of up to 100°C is possible.[c] After the hydrolysis, the acids and its salts in water are recycled back from the absorber to the generator (endothermic reactor), to complete the cycle (Figure 5.10).

5.5 Heat Pumps as Chillers

Water boils and evaporates at 100°C, at standard atmospheric pressure of 101.325 kPa. When the pressure is reduced, water boils at a lower temperature. Water evaporation at sufficiently low pressure provides cooling duty; for example, to produce chilled water by heat exchange. Figure 5.11 shows variation of water boiling temperature with pressure (e.g., water boils at 5°C and 10°C at a pressure of 0.9 kPa and 1.2 kPa, respectively).

AHP can also be used as a chiller, which uses thermal energy rather than mechanical/shaft energy for operation. Absorption chiller operates at lower pressures compared to AHP. Otherwise, operating principle is same for both. Operating principle of LiBr absorption chiller is shown in

---

[b] https://www.borealisgroup.com/news/borealis-announces-start-up-of-heat-recovery-unit-based-on-revolutionary-qpinch-technology, accessed on 8-06-2021.

[c] https://www.qpinch.com, accessed on 8-06-2021.

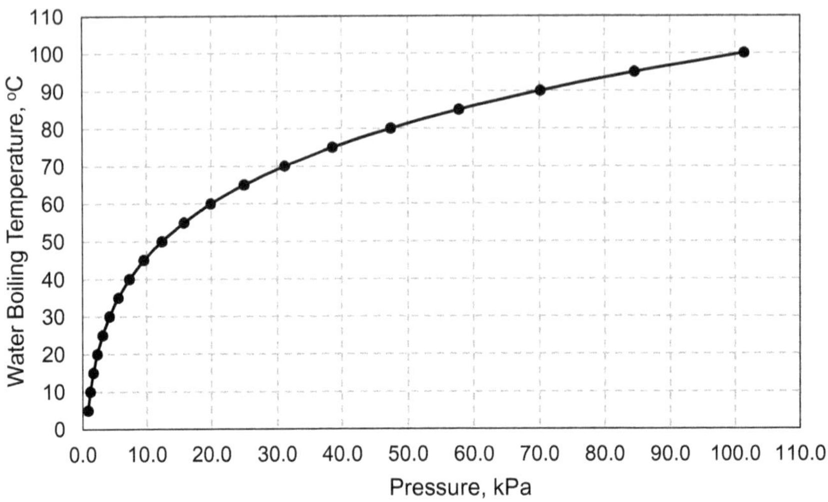

Fig. 5.11: Variation of water saturation/boiling temperature with pressure.

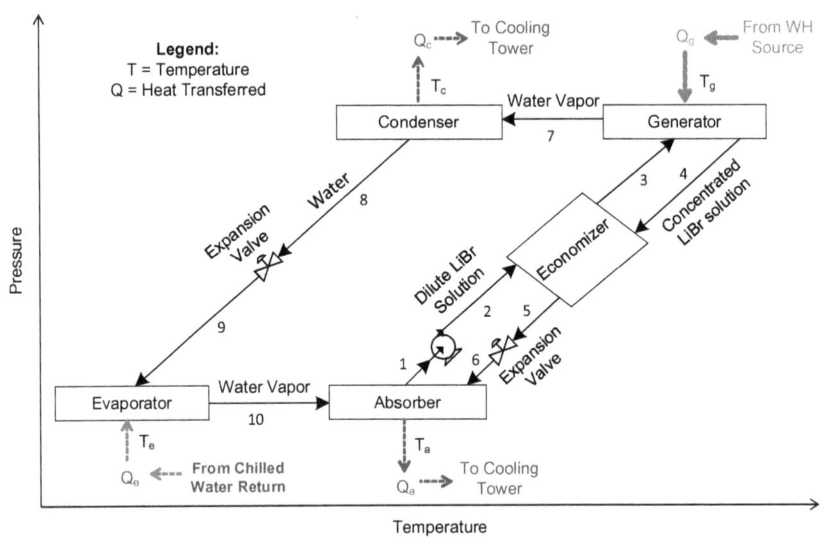

Fig. 5.12: Operational details of LiBr absorption chiller on temperature versus pressure plot.

Fig. 5.12, which is similar to that of AHP in Fig. 5.9. Absorption chillers use either LiBr or ammonia absorption in water. LiBr–water systems are limited to evaporation temperatures above water freezing temperature as water is used as the refrigerant. The main advantage of these systems is that its operation is at lower pressures compared to ammonia–water system. But this is also a drawback because pressures are below atmospheric, causing air infiltration into the system, which must be purged periodically. Due to corrosion problems, special inhibitors must be used in LiBr–water systems. Ammonia absorption systems require high-pressure distillation for regeneration of ammonia as water is also volatile. Refrigerant ammonia requires much higher pressures of 1,100–2,100 kPa in the condenser. The ammonia–water system can achieve evaporating temperatures below 0°C. Many industrial applications of absorption chillers are presented by Herald et al.[2]

COP for absorption refrigeration (COP_{AC}) is the ratio of refrigeration rate to heat input at the generator. With reference to Fig. 5.12, it is given by:

$$COP_{AC} = \frac{Q_e}{Q_g} \tag{5.12}$$

COP of an absorption chiller decreases with an increase in the cooling water temperature (for given chilled water and heat source temperatures). For given cooling and chilled water temperatures, COP drops with decrease in the heat source temperature. COP of an absorption chiller increases with increase in chilled water supply temperature.

Exercise 5.7: Estimate COP of an absorption chiller operating under the following conditions. With reference to Fig. 5.12, outlet temperature of absorber ($T1$), condenser ($T8$), generator ($T4$) and evaporator ($T10$) are 30°C, 30°C, 90°C and 4°C, respectively. Cooling duty of evaporator (Q_e) is 498.34 kW, and high temperature utility supplied to generator (Q_g) is 606.99 kW.

Solution: $COP = \frac{Q_e}{Q_g} = \frac{498.34}{606.99} = 0.821$

Typical utility requirements and performance of absorption and mechanical chillers are compared in Table 5.4. Ammonia–water and LiBr–water chiller systems have similar heating/cooling duty requirements and comparable COP. Capital cost of an absorption chiller rises sharply as the temperature of the heat source falls, making WHR uneconomical. Compared to mechanical chillers, absorption chillers have similar heating/cooling duty requirements and low COP.

Table 5.4: Typical operating requirements of absorption and mechanical chillers

Type of Chiller	Main Driver	Requirement for One kW of Refrigeration: Steam and Electric Power	Requirement for One kW of Refrigeration: Cooling Duty	COP
Single-stage LiBr–water absorption	LLPS (≈2 bar), hot water or WH stream at > 90°C	1.48–1.51 kW of steam, and 0.003 – 0.011 kW of electric power for pumping	≈ 3 kW	0.6–0.75
Two-stage LiBr–water absorption	MPS (9 bar or more) or WH stream at >170°C	0.78–0.81 kW of steam, and 0.003– 0.011 kW of electric power for pumping	≈ 2.4 kW	1.1–1.35
Mechanical compression (R134a)	Motor or steam turbine or gas/ diesel engine driven compressor	Electric power of 0.18 kW	≈ 1.8 kW	5.68

*Ammonia–water and mechanical chillers can be used to achieve sub-zero temperatures, whereas LiBr–water chillers are generally used for chilled water temperatures above 5°C.

Nonetheless, they can substantially reduce operating cost because they are energized by low-grade WH (often available at no cost), while mechanical chillers require a motor or engine for vapour compression. The WH source for an absorption chiller can be LLPS, a hot gaseous or liquid stream.

The COP of the system can be increased significantly by more stages. However, addition of each stage increases the required heat source temperature. For example, two-stage LiBr–water chiller has higher COP compared to single stage chiller (Table 5.4). However, the former requires MPS as heat source, which incurs cost; hence, two-stage chiller is generally not a WHR option.

Main problems with LiBr–water AHP systems for heating or cooling are as follows:

1. Crystallization: Crystallization problem can be avoided by operating away from the crystallization zone, controlling the condenser pressure

sufficiently high (by regulating cooling water flow rate) and/or use of chemical additives to inhibit crystallization.
2. Air leakage: As the LiBr–water chiller operates under vacuum, outside air can leak into the system. Leaked air is generally removed by dry vacuum pump or steam ejector system.
3. Pressure drops: The operating pressures are very low, and so specific volume of vapour is very high. These result in relatively high-pressure drops in vapour piping, which is minimized using twin drum arrangement (i.e., condenser and generator in one drum, and evaporator and absorber in another drum).

Compared to mechanical/compression chillers, absorption chillers contain very few moving parts, have less noise and vibration problems, are compact and require little maintenance, but they have a low COP. Further, as absorption chillers use WH, they incur substantially reduced energy costs. Moreover, compared to mechanical chillers, COP of absorption chiller is not sensitive to load variations (i.e., it does not reduce significantly at part loads).

Use of chilled water (generated by utilizing WH in a single-stage absorption chiller) at pre-condensers, can substantially reduce the fixed and operating costs of multi-stage SJE systems, by condensing most of the suction vapour at the entrance to the vacuum system. This is described in detail in Chapter 8. Use of absorption chillers driven by WH from flue gases, can increase the gas turbine power output in cogeneration plants, which is covered in Chapter 10. WH driven absorption chillers can be used to debottleneck compression system capacity. This is especially true if the compressor suction gas has significant condensable vapour. Further, reduction in suction temperature reduces the energy required for compression systems. These applications of chilled water for compression system are covered in Chapter 12.

Exercise 5.8: In a plant, saturated LLPS at 1.015 bar is vented to the atmosphere, and a mechanical chiller with a COP of 5.0 is used to produce 1,000 tons of refrigeration. Calculate annual operating savings by replacing the mechanical chiller by an absorption chiller (using LLPS) with a COP of 1.5. The cost of electricity at the plant is \$0.1/kWh. Also, estimate the amount of LLPS required to support the new absorption chiller.

Solution: Noting that one ton of refrigeration is 3.517 kW, required cooling duty = 1,000 × 3.517 = 3,517 kW

From Eq. 5.3, $COP_C = \frac{Q_C}{W}$. Hence, power consumed by mechanical compressor = $\frac{3{,}517}{5}$ = 703.4 kW

For 365 days of operation per year, annual electricity savings = 703.4 × 0.1 × 24 × 365 = $616,178.4

From Eq. 5.12, $COP_{AC} = \frac{Q_e}{Q_g}$ and so $Q_g = \frac{3{,}517}{1.5}$ = 2,344.67 kW

From steam tables, latent heat of LLPS at 1.015 bar = 2,256.41 kJ/kg

Amount LLPS required = $\frac{2{,}344.67}{2{,}256.41}$ = 1.039 kg/sec = 3,740.81 kg/hr

5.5.1 *Adsorption Heat Pump as Chillers*

The adsorption HP chiller cycle, like the absorption chiller cycle, uses thermal compression rather than mechanical compression. As adsorbent (solid material such as silica gel) cannot be pumped, it does not use any pump like LiBr absorption chiller. Instead, adsorption HPs use valves to connect absorber and desorber to condenser and evaporator.

Unit operations in an adsorption HP are evaporation, adsorption, desorption and condensation. Accordingly, an adsorption HP consists of two or more adsorbent/desorbent beds, a condenser, an evaporator and an expansion valve, as shown in Fig. 5.13. Here, first bed and second bed are shown to be operating in adsorption mode and desorption mode, respectively (as per the first option for valve positions in Fig. 5.13). These beds switch between adsorption and desorption modes alternatively by the programmed operation of on–off valves, to provide continuous flow of refrigerant to the condenser. The adsorption/desorption vessel is filled with adsorbent (e.g., zeolite, silica gel, alumina, active carbon or metal salts), which in its neutral state adsorbs the refrigerant (in most cases water). When heated, the solid desorbs refrigerant vapour, which subsequently is cooled and liquefied at high pressure/temperature in the condenser. This liquid refrigerant is depressurized to a lower pressure; then, it vapourizes in the evaporator to provide the cooling duty (i.e., absorb heat from the returning chilled water). In the final stage, the refrigerant vapour is (re)adsorbed onto the solid. After the adsorbent is saturated, supplying WH to it leads to desorption of refrigerant.

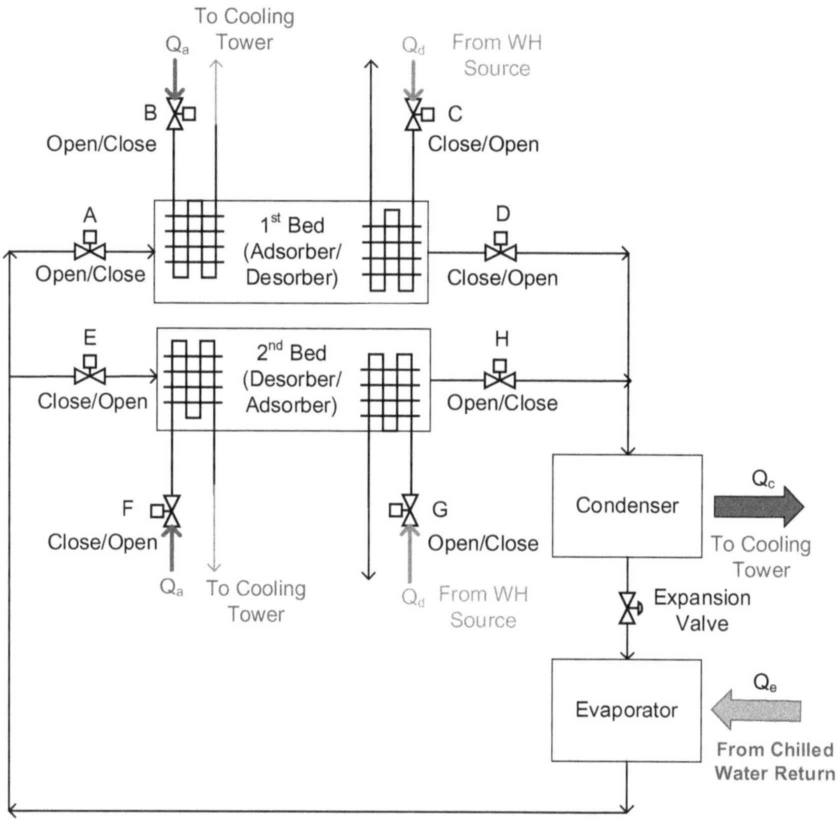

Fig. 5.13: Details of an adsorption chiller with two beds (in alternating operation of one as an adsorber and another as desorber).

Basically, adsorption HP operates by cycling refrigerant between adsorber, desorber, condenser and evaporator. In this cycle, adsorption phenomenon plays the same role of mechanical power, so that the working fluid can be circulated in the cycle without any mechanical power. In fact, adsorption HP is a discontinuous process with adsorption and desorption occurring successively in the same vessel by special arrangement of valves on the inlets and outlets of the vessel. To ensure a reasonably continuous heating or cooling effect, two vessels (each of them switches service from adsorption to desorption and back to adsorption, repeatedly) are usually used.

Usual adsorbent and refrigerant pairs for using low temperature WH sources, are silica gel–water, zeolite (alkaline rock crystal)–water, activated carbon–methanol and activated carbon–ammonia. Combinations suited to high temperature WH sources include zeolite–water, activated carbon–ammonia, metal chlorides–ammonia, and composite adsorbents–ammonia. More details on some of these pairs are given in the following paragraphs.

Like in an absorption chiller, water is often used as the refrigerant in adsorption HPs. It is adsorbed in the pores of a solid adsorbent. Silica gel–water and activated carbon–methanol are used for low temperature WH (<120°C). COP of adsorption chiller can be calculated using equation (5.3). Silica gel–water adsorption chiller can be used to obtain chilled water at 14°C using WH at 75°C, and its COP is 0.6. Silica has low adsorption capacity (about 0.2 kg/kg), and hence it requires large adsorber/desorber vessels. As water is the refrigerant used with silica gel, it cannot produce evaporation temperature below 0°C. Silica gel cannot be used with WH source temperatures above 120°C as it is destroyed beyond this temperature.

Activated carbon–methanol pair serves applications, which require cooling below 0°C, and it can be driven by WH sources at <120°C. Adsorption chillers with silica gel–water, activated carbon–methanol and zeolite–water operate under vacuum conditions, and leakage of atmospheric air can deteriorate the system performance. Hence, similar to LiBr absorption chillers, leak-proof design, installed with dry vacuum pumps or SJEs, is essential for maintaining this type of adsorption HP performance. Systems that utilize ammonia–metal pair operate under positive pressure, and hence they do not require vacuum pumps.

Zeolite–water is very stable and does not break down at high (WH) temperatures. It can serve both heating and cooling duties. The adsorption heat for this pair is higher (3,300–4,200 kJ/kg) than that of the silica gel–water pair (~2,500 kJ/kg).[3]

Adsorption HPs can use WH or other heat sources to produce chilled and/or hot water. They are environmentally friendly as they do not use any fuel, compressor or pumps. Moreover, they do not have any crystallization and corrosion issues like LiBr AHPs. However, large size (compared to mechanical or AHPs), higher capital cost and lower efficiency (typically 0.3–0.6) are restricting the wide application of adsorption HPs.

5.6 Selection of Heat Pumps

The COP of an HP must be as high as possible for it to be economically attractive. Most of the HPs operate with temperature lifts of less than 55°C. Selection of an HP for an application is based on many factors such as nature of the service, nature of process sink, temperature requirement at the process sink, plot location and size, availability and cost of utilities (such as cooling water, LLPS, LPS, MPS or HPS), cost of energy (e.g., fuel and electricity) and payback period/time. Some of the general selection criteria for HPs based on nature of service, process sink and its temperature requirements are summarized in Table 5.5, whereas

Table 5.5: HP selection guidelines for various process applications

No.	Process Detail	Comment
1	Process involves multi-stage evaporation.	MVR is favoured for complete utilization of WH.
2	Process involves WH stream at ≥90°C.	MVR or absorption chiller can be used to generate LLPS or chilled water.
3	Process streams are heated with LLPS to 100°C–120°C.	MVR or Type-1 AHP can easily provide the heat in this temperature range, using WH at 80°C–90°C.
4	WH streams are available at >200°C	AHP can be used to produce more LPS.
4	Vented FS or LPS	These heat sources are very suitable for driving AHT/AHP.
5	Distillation columns with low temperature difference (10°C–30°C) between reboiler and condenser	MVR can be used for simultaneous reduction of cooling (at condenser) and heating (at reboiler) duties.
6	WH source is clean fluid, available continuously and recoverable WH is more than 150 kW.	Economy of scale and simple construction materials can help to justify appropriate and cost-effective HP, based on heating or cooling needs of the plant.
7	Electricity is cheaper than fuel.	Motor driven MVR can be economical.
8	Both fuel and electricity costs are high.	AHPs are generally favourable.
9	Sufficient availability of WH and demand for LPS or LLPS	Installation of AHP or AHT is favourable.

Table 5.6: Guidelines for selection of MVR, AHP and AHT based on heat source and sink characteristics

Temperature Lift	Heat-Source Type	Heat-Sink Type	Suitable HP
Low at 10°C–30°C	Sensible cooling of liquid or partial condensation of liquid from vapour stream	Sensible heating of fluids, and boiling liquids	Closed cycle MVR or AHP
	Condensing steam	Evaporation of water	Open-cycle MVR or TVR
	Condensing vapour	Sensible heating of fluids or boiling liquids	Semi open-cycle MVR
>30°C	Hot WH streams	Process heat sinks	AHT with high temperature lift or Multi-stage MVR
	WH or LPS	MPS	Open-cycle MVR, Multi-stage MVR or AHT

guidelines for selection of HPs based on heat source and sink characteristics are summarized in Table 5.6.

Generally, MVR is considered first as they do not waste WH and can be used for almost every heating duty, provided WH energy upgraded to higher temperature is substantial compared to the electrical energy consumed. Reciprocating and rotary compressors are highly efficient, less costly and hence widely used for lower capacities (up to 1,700 m^3/hr at inlet flow conditions). However, reciprocating compressors require more space and maintenance costs compared to centrifugal compressors. Centrifugal compressors are well suited for higher capacities (>1,700 m^3/hr at inlet flow conditions).

Choice of refrigerant in absorption/adsorption HP involves many considerations. Water is the most favoured refrigerant for production of chilled water. Very often, process industries use a fluid that needs to be cooled as a refrigerant. Ammonia is very good and cheap, but it is toxic in nature, requires high-pressure operation and hence costly equipment. Its use is generally restricted to remote locations away from populated areas. Halocarbons have many advantages (non-toxic, not explosive and odourless), but environmental issues are the main concern. R134a (1,1,1,2-tetrafluoroethane) is an attractive alternative for CFCs (chlorofluorocarbons) and HCFCs (hydrochlorofluorocarbons).

AHPs can be considered when low-cost FS, LPS or WH is available, and temperature lift and refrigeration load are relatively high. Typical application range is for water chilling at 7°C–15°C, and capacities from 300 kW to 5 MW in a single unit.[4] LiBr absorption chillers operate under vacuum, and hence require leak-free system, installed with a dry vacuum pump or SJE; also, corrosion, crystallization and high-pressure drop of vapour pipes are the other main problems with these chillers.

5.7 Summary

This chapter is on the types, working principles, main uses of and selection guidelines for industrial HPs. Main points of this chapter are as follows

- MVRs have a wide range of applications in industries. MVR is widely used for low temperature lift applications such as multi-stage evaporation, distillation, and process heating. It has a high COP of 3–9.
- TVR is useful for low temperature lift applications as it is cheap and there is no wastage of heat if steam generated at the outlet of TVR can be used for process heating. For example, open-cycle TVR is very useful for upgrading FS.
- Compared to COP, PER is a better measure for comparing HP performance as it considers power consumed by auxiliary equipment such as cooling water pumps and fans. However, both do not include capital cost.
- Type-1 AHP or AHP amplifies the amount of WH but achieves lower temperature lift. It can achieve a COP of up to 1.7.
- Type-2 AHP or AHT can upgrade nearly half of the WH quantity to a higher temperature. However, it achieves a lower COP (~0.5) only.
- Chemical HPs can achieve higher temperature lift compared to AHT.
- LiBr chillers employ the same working principle as AHP. They are very useful for production of chilled water in process industries as they can work with WH at around 90°C.

- Adsorption chillers can use WH and do not involve any rotating equipment. They are generally bigger in size and achieve lower COP (0.3–0.5) compared to absorption chillers (with COP 0.6–0.75).
- Many types of HPs are available for WHR. Good understanding of their advantages and disadvantages is important for selecting the economical HP. Considerations discussed in this chapter are useful for correct selection of an HP for a particular WHR application.

Chapters 6 and 7 in Herald *et al.*[2] are recommended for comprehensive coverage of absorption chillers.

References

1. Arpagaus C, Bless S, Uhlmann M, Schiffmann J, Bertsch S. (2018) *High Temperature Heat Pumps: Market Overview, State of the Art, Research Status, Refrigerants, and Application Potentials (purdue.edu)*. https://docs.lib.purdue.edu/cgi/viewcontent.cgi?article=2875&context=iracc.
2. Herald KE, Radermacher R, Klein SA. (2016) *Absorption Chillers and Heat Pumps*, 2nd ed. CRC Press.
3. Wang R, Wang L, Wu J. (2014) *Adsorption Refrigeration Technology: Theory and Application.* John Wiley & Sons.
4. Green DW, Perry RH. (2008) *Perry's Chemical Engineers' Handbook*, 8th ed. McGraw-Hill.

Acronyms and Notation

Acronyms

AHP	Absorption Heat Pump
AHT	Absorption Heat Transformer
CFC	Chlorofluorocarbons
COP	Coefficient of Performance
FS	Flash Steam
HCFC	Hydrochlorofluorocarbon
HP	Heat Pump

HPS	High-Pressure Steam
HVAC	Heating, Ventilation and Air Conditioning
LiBr	Lithium Bromide
LLPS	Low-Low-Pressure Steam
LPS	Low-Pressure Steam
MPS	Medium-Pressure Steam
MVR	Mechanical Vapour Recompression
PER	Primary Energy Ratio
SJE	Steam Jet Ejector
TVR	Thermal Vapour Recompression
WH	Waste Heat
WHR	Waste Heat Recovery

Notation

k	Isentropic exponent
m	Mass flow rate of gas/vapour (kg/sec)
M	Molecular weight
n	Polytropic exponent
P	Pressure (bar or kPa)
Q	Heat exchanged or duty (kW)
R	Universal gas constant (8.314 J/K.mol)
RM	Ratio of mass flow rate of motive steam to mass flow rate of load (flash) steam
V_s	Specific volume (m^3/kg)
W	Work done by the compressor (kW) or specific work done by the blower (kJ/kg)
X	% of the fuel input to the diesel generator that can be recovered by WHR
Z	Compressibility constant

Greek Symbols

η	Efficiency

Subscripts

a	Absorber
AC	Absorption chiller
AHP	Absorption heat pump
AHT	Absorption heat transformer
c	Condenser
C	Cooling
d	Desorber
D	Discharge
e	Evaporator
g	Generator
H	Heating
L	Load or suction
m	Motor
M	Motive
MVR	Mechanical vapour recompression
p	Polytropic
S	Suction

Exercises

5.1 An HP is required to compress 3 ton/hr of saturated LLPS at 1.1 bar to LPS at 4 bar, which can be sent to the steam header for consumption throughout the plant. The plant has sufficient capacity of electricity and saturated MPS at 13 bar. Suggest the most beneficial HP for this application. Assume discharge condition of LPS from the HP to be saturated. Estimate the COP of the selected HP and utility consumption.

5.2 What are the situations where open-cycle mechanical compression is preferred over closed cycle mechanical compression?

5.3 What are the advantages and disadvantages of LiBr absorption chillers compared to mechanical chillers?

5.4 In a plant, 5 ton/hr of saturated LLPS at 2 bar is currently vented to the atmosphere. This can be utilized in a new LiBr absorption chiller with a COP of 0.7, to produce chilled water. Estimate the amount of chilled water that can be produced with chilled water supply and return temperatures of 15°C and 21°C, respectively. Assume heat capacity of chilled water is 4.184 kJ/kg.K.

5.5 An MVR with a COP of 6 is driven by a steam turbine with an efficiency of 0.3. This steam turbine uses HPS at 42 bar and 360°C and generates 4 ton/hr of LPS at 4 bar and 200°C, which is used in a process heat exchanger for latent heat transfer. Assume 0.5 bar pressure drop in the heat exchanger. Saturated condensate from the heat exchanger is drained as there is no condensate recovery equipment. Estimate PER of the MVR.

5.6 An AHP with a COP of 1.4 is used to produce 400 kW of useful heat. Estimate the amount of hot utility required?

Chapter 6

Cost Estimation and Economic Evaluation

6.1 Overview

Approval for waste heat recovery (WHR) projects is mostly based on the required investment and attractive return on the investment. Other considerations in this are reduction in environmental impact and safety risks. Hence, WHR project approval requires good estimates of fixed capital investment (FCI) required, operating cost (OC) of WHR and/or revenue from WHR, and then economic evaluation to determine the profitability of the project. Other terms for FCI are capital cost (CC) and, simply, investment. Note that WHR may reduce OC and/or generate revenue (e.g., from steam generation). This chapter covers both cost estimation and economic evaluation with focus on WHR equipment costs and economics.

First, different procedures, methods and correlations for estimating FCI for and OC of WHR equipment are covered in Sections 6.2–6.5. Then, time value of money and several profitability criteria for economic evaluation are presented in Sections 6.6 and 6.7. Illustrative examples throughout this chapter are on WHR; some of them are taken from later chapters.

Data required for estimating FCI, classification of and accuracy of FCI estimates are described in Section 6.2. Next, Section 6.3 presents (plant) cost index for escalating historical cost of process equipment to present/future time. Section 6.4 describes and discusses selected methods for estimating equipment cost and FCI. This major section includes six-tenths rule and module costing technique (MCT) for FCI estimation as well as discussion on FCI estimation. Section 6.5 presents OC of and revenue from WHR. Subsequently, Section 6.6 describes time value of money, which is important for sound economic evaluation of any investment, be it in the plant or personal investment. Section 6.7 presents common profitability criteria, which are used in later chapters. Finally, this chapter ends with summary of key points in Section 6.8.

Learning outcomes of this chapter on cost estimation and economic evaluation are as follows:

1. Describe required data, procedures and methods for estimating process equipment cost.
2. Estimate FCI for WHR projects.
3. Explain and estimate OC of and/or revenue from WHR.
4. Describe time value of money, payback period (PBP), net present value (NPV) and internal (or investor's) rate of return (IRR).
5. Perform economic evaluation of WHR projects.

6.2 Data for FCI Estimation, and Accuracy and Classification of FCI Estimates

WHR requires new equipment such as heat exchangers (HEs), pumps, compressors and pressure vessels.[a] These equipments have to be purchased, brought to the site, their supports need to be built, they have to be connected via pipes with other equipment, required instrumentation have to be added, preliminarily tested for their performance and so on before WHR becomes operational. All these incur costs. The sum of these for any equipment is its installed cost (IC) or total module cost (TMC), which is often 2–5 times its purchase cost (PC). It includes material, labour, engineering and transportation costs, contingencies and contractor fee incurred at various stages of equipment fabrication, transportation and installation. The sum of TMC of all equipment in a (WHR) project is the required FCI

[a] In addition to these for WHR, a process plant requires other equipment such as separation columns, reactors, evaporators, mixers and filters. Obviously, it requires many more equipment than WHR projects.

or CC. In this chapter, FCI is employed instead of CC, to emphasize that FCI is a one-time cost, it is fixed (i.e., not easy to recover) and acceptable return on it is expected.

For estimating PC, IC and then FCI, many details about the equipment in the proposed process or WHR are required. Minimum data required for estimating PC and IC of any equipment are type, size or capacity, material of construction (MOC), and operating pressure (OP) of the equipment. For example, HE can be double-pipe, fixed tube-sheet, floating head, U-tube, kettle, plate or spiral type; size of an HE is the required heat transfer area; and MOC can be carbon steel (CS), stainless steel (SS) or titanium. These details should be available in the screening and feasibility study stage of a WHR project.

More details of a shell-and-tube HE (STHE) include tube inner/outer diameter and length, number of tubes, configuration of tubes, shell diameter and length, number of baffles and shell cover. Determination of these requires detailed design of STHE, and so these details will be available as the project progresses from screening, feasibility to implementation stage. Not surprisingly, accuracy of the cost estimate improves (i.e., error or uncertainty in the estimate decreases) with the availability of increasing details of the equipment. On the other hand, cost of finding a cost estimate increases with improvement in the accuracy of the estimate. As shown in Fig. 9.41 in Holland and Wilkinson,[1] cost of an estimate with ±40% accuracy is about 0.1% of total project cost whereas that of an estimate with ±5% accuracy increases by 40 times to about 4% of total project cost.

Five types of FCI estimates with their accuracy, purpose and details required are summarized in Table 6.1. These are based on Couper *et al.*[2] and Turton *et al.*[3] Probable accuracy of −25 to +30% means the true cost of installing/building the equipment, process unit or plant is likely to be between 25% lower and 30% higher than the estimate. Accuracy range given in Table 6.1 is different from one reference to another, which indicates variations in data sources for, varied experiences and therefore difficulties in FCI estimation. Owing to uncertainty in cost estimation, PC, IC and FCI are generally rounded off and given to the nearest thousands or millions. See Dysert[4] or AACE[5] for detailed description of cost estimation classifications.

Table 6.1: Types of FCI estimates and their accuracy, purpose and design details required

Type of Estimate	Probable Accuracy (%)	Purpose	Design Details Required and Comments
Order of Magnitude, Ratio or Feasibility	−30 to +50	For screening alternatives	Least accurate estimate but requires minimal effort using only plant capacity. For a plant, it is found by updating historical cost of a similar plant for inflation and capacity using a plant cost index (Section 6.3) and six-tenths rule (Section 6.4.1), respectively.
Study, Factored or Major Equipment	−25 to +30	For feasibility studies	This estimate requires type, size (capacity), MOC and OP of all equipment starting from pumps and bigger equipment, typically shown on process flow diagram (PFD). It can be found by MCT described in Section 6.4.2.
Preliminary Design, Budget Authorization or Scope	−20 to +25	For approval and budget	This estimate requires more details from detailed design of all equipment on PFD and piping and instrumentation diagrams (P&IDs). Vendor budgetary quotes for expensive equipment are required.
Definite or Project Control	−10 to +15	For bid/tender	In addition to design details used for earlier estimates, this estimate requires equipment specifications/drawings, isometric diagrams, P&ID, pipe lengths and sizes, electrical and instrumentation and detailed bill of materials.
Detailed or Firm	−5 to +10	Check cost estimate or bids/ tender for sub-contracts	Besides design details, specifications and drawings used for earlier estimates, this estimate requires vendor quotes, final PFD, P&ID and site surveys for equipment installation.

6.3 Plant Cost Index

Many of FCI estimates in Table 6.1 are based on historical cost data of the same type of equipment or plant. PC and IC of equipment required for FCI estimate vary with time due to inflation (e.g., changes in prices of material and labour) and technological developments. Hence, a suitable plant cost index is required for escalating PC, IC and FCI from the past to present time. Examples of such a cost index are Marshall & Swift Equipment Cost Index (MSECI) started in 1926, Nelson-Farrar Refinery Construction Index from 1946, Chemical Engineering Plant Cost Index (CEPCI) started in 1957 and Engineering News-Record Construction Index from 1967. These indices are updated regularly and published; for example, CEPCI is published (e.g., www.chemengonline.com/pci).

Of the plant cost indices, CEPCI is commonly used in process industries. Hence, it is presented in some detail in this chapter and is employed in other chapters of this book. Similar to other indices, CEPCI is a composite[b] of components/items involved in purchasing and installing diverse equipment in a process plant. Specifically, components/items and their respective weights in CEPCI are HEs and tanks (0.171); process machinery (0.0649); pipes, valves and fittings (0.0963); process instruments (0.0532); pumps and compressors (0.0324); electrical equipment (0.0355); structural supports and miscellaneous (0.0532); buildings (0.0458); engineering and supervision (0.158) and construction labour (0.29). These weights indicate the relative contribution of various components/items to FCI. More details on CEPCI and these components/items can be found in Vatavuk.[6]

Yearly CEPCI from 1981 to 2020 is given in Table 6.2; it is useful in updating historical data on FCI of a plant or IC of any equipment to the latest year under consideration using simple proportionality. For example, FCI of a plant built in the year 2000 (when CEPCI = 394.1) was US$150 million; FCI of the same plant and capacity next year (when CEPCI is estimated to be 600) is given by:

---

[b] This is similar to CPI, which is a composite of components/items such as food, housing, clothes, transportation, entertainment and so on for our living.

Table 6.2: Yearly variation of CEPCI from 1981 to 2020

Year	CEPCI	Year	CEPCI	Year	CEPCI	Year	CEPCI
1981	297.0	1991	361.3	2001	395.0	2011	586.0
1982	314.0	1992	358.2	2002	396.0	2012	584.6
1983	317.0	1993	359.2	2003	402.0	2013	567.3
1984	322.7	1994	368.1	2004	444.0	2014	576.1
1985	325.3	1995	381.1	2005	468.0	2015	556.8
1986	318.4	1996	381.7	2006	500.0	2016	541.7
1987	323.8	1997	386.5	2007	525.4	2017	567.5
1988	342.5	1998	389.5	2008	575.4	2018	603.1
1989	355.4	1999	390.6	2009	521.9	2019	607.5
1990	357.6	2000	394.1	2010	551.0	2020	595.0

$$\begin{aligned}\text{FCI in next year} &= \text{FCI in 2000} \times \left(\frac{\text{CEPCI of next year}}{\text{CEPCI of 2000}}\right) \\ &= 150 \times \frac{600}{394.1} = \text{US\$228.4 million}\end{aligned} \tag{6.1}$$

Use of another plant cost index will give a slightly different estimate (and not US$228.4 million based on CEPCI) because of differences in the components, their weights and inflation of their costs in different indices.

For example, consider use of MSECI, whose value in the year 2000 was 1,089 and its value for next year is estimated to be 1,700; then, FCI of the same plant and capacity next year is:

$$\begin{aligned}\text{FCI in next year} &= \text{FCI in 2000} \times \left(\frac{\text{MSECI of next year}}{\text{MSECI of 2000}}\right) \\ &= 150 \times \frac{1{,}700}{1{,}089} = \text{US\$234.2 million}\end{aligned} \tag{6.2}$$

Thus, FCI estimates by two different indices in this example differ by less than 3%. Similarly, Example 7.6 in Turton *et al.*[3] on escalating PC of an HE from 1996 to 2011 using CEPCI and MSECI has a difference of 7%.

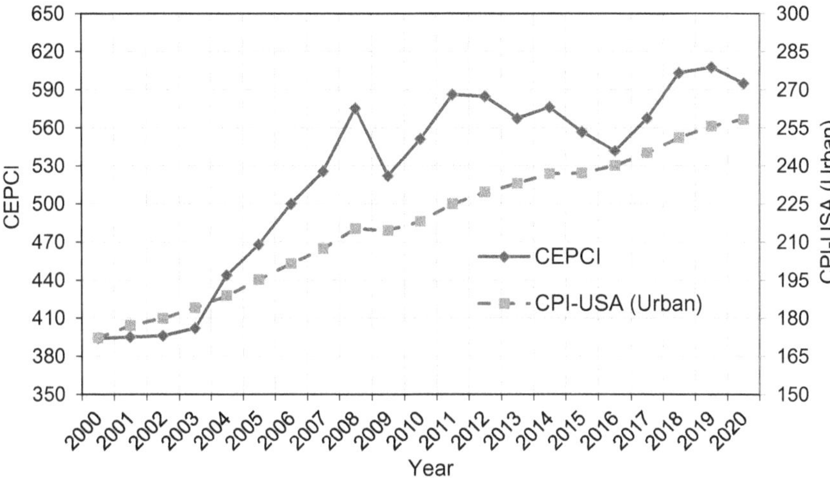

Fig. 6.1: Variation of CEPCI and CPI of the United States (urban) from the year 2000–2020.

Yearly variation of CEPCI from the year 2000 and its comparison with consumer price index (CPI) of USA for urban consumers are presented in Fig. 6.1. Curiously, overall increase over two decades from 2000 to 2020 is around 50% in both CEPCI and CPI of the United States. However, CEPCI has greater fluctuation compared to steady rise in CPI of the United States. As shown in Fig. 6.1, CEPCI rose quickly from 2003 to 2008 (due to substantial increase in prices of MOC due to higher oil prices and interest rates) and nearly constant around 550 from 2008 to 2016 and about 600 from 2018 to 2020. The difference in the trends of CEPCI and CPI of the United States is expected because components in these two indices are very different.

6.4 Estimation of Equipment Cost and Capital Cost

A new process may require many types of equipment with different specifications and operating conditions. Further, it may be for building immediately or after a few years in some country/region of the world. In general, PC, IC and consequently FCI depend on characteristics and considerations such as (a) time/year of construction, (b) plant location (i.e., country/region), (c) type, size/capacity and MOC of each and every equipment, (d) pressure and temperature of operation of each equipment, and so on. Finding the cost of each equipment from vendor quotations may not be possible due to time and effort involved. Hence, PC, IC and FCI are often estimated based on historical cost data available in design books and/or company's records. How can the above characteristics/considerations be properly accounted in cost estimation from historical data? This requires sound procedures for estimating PC, IC and FCI on a consistent basis.

As described in the previous section, a cost index such as CEPCI can be used to account for time/year of equipment purchase/installation. PC and IC vary with plant location due to differences in material, transportation and labour costs and labour productivity. However, historical cost data available in the open literature are limited to a few countries such as the United States and United Kingdom. These data can be used for other plant locations by multiplying with a location (site) factor. Value of this factor reported by Guthrie,[7] who termed it as area correction factor, and by Seider *et al.*[8] are summarised in Table 6.3. In addition, Table 9.55 in

Table 6.3: Location or site factors for CC estimation

Country or Region	Reported by Guthrie[7]	Reported by Holland and Wilkinson[9]	Reported by Seider *et al.*[8]
United States	0.89–1.17	1.0	0.95–1.25
Mexico	NA	NA	0.95
South America	0.86–1.0	1.36	0.95
Western Europe	0.76–0.98	NA	1.2
Germany and United Kingdom	NA	0.76	NA
Australia	1.07–1.11	1.04	NA
Pacific Rim	NA	NA	1.0
Japan	NA	1.46	1.15
India	NA	0.8 for imported element; 0.25 for indigenous element	0.85
Malaysia	NA	0.42	NA

Holland and Wilkinson[1] has location factor values (for 1993) for about 30 countries around the World; some of these are also included in Table 6.3. Location factor values in different references are limited, and their accuracy is unknown. Moreover, they will change with time and industrial development of a country/region. Hence, as far as possible, find and use cost data for the region/country under consideration.

Procedures to consider equipment type, size (capacity), MOC and operating conditions are often based on MCT of Guthrie,[7] who presented many procedures and extensive data in plots. Some of these procedures and data have been adopted in later books. In cost estimation for process plants by different techniques, PC of equipment is the basis for IC and FCI. Historical PCs of various equipment are presented in plots and/or as correlations. Often, these are for equipment made of CS and operating at ambient pressure, which are referred to as base conditions. MCT is outlined and illustrated in Section 6.4.2.

In the current section, a simple and handy technique is presented. It is based on the observation that costs of two similar equipment/plants but of different sizes are related by:

$$\frac{C}{C_R} = \left(\frac{S}{S_R}\right)^n \tag{6.3}$$

Here, C and C_R are cost (US\$) of an equipment/plant of size S and S_R, respectively. The exponent, n is often (but not always) in the range of 0.4–0.8, as can be seen from its values for many equipment in Table 9.50 in Holland and Wilkinson.[1] Subscript R denotes that C_R and S_R are the reference values, which are known. Equation 6.3 can be re-written as:

$$C = K\,(S)^n \tag{6.4}$$

By comparing Eqs. 6.3 and 6.4, $K = \frac{C_R}{(S_R)^n}$. Since n is often less than unity, Eq. 6.4 implies economy of scale (i.e., equipment cost per unit size/capacity decreases with increasing size/capacity). For example, when the size doubles, cost of the plant/equipment does not double but increases by a factor of $(2)^n$; for n = 0.6, this factor is 1.58 or by approximately 60% (instead of 100%).

For estimating PC (free-on-board, i.e., excluding transportation costs and CEPCI = 500) of equipment involved in WHR, data on *K*, *S* and n in Eq. 6.4 are given in Table 6.4. These are extracted from Table 22.32 in Seider *et al.*[8] which contains many other equipment employed in process

Table 6.4: Exponential equation for estimating PC (free-on-board and CEPCI = 500) of equipment involved in WHR

Equipment Details (MOC* and Remarks)	Size/Capacity Parameter with Its Units	Range of Size Parameter	PC (US$)
Plate and frame HE (SS)	Heat transfer area, A ft^2	150–15,000	$8{,}800\ A^{0.42}$
Spiral plate HE (SS)	Heat transfer area, A ft^2	20–2,000	$6{,}200\ A^{0.42}$
Air-cooled fin-fin cooler (CS)	Bare tube heat transfer area, A ft^2	40–150,000	$2{,}500\ A^{0.4}$
Gas expanders (pressure discharge, CS)	Power extracted, P HP	20–5,000	$530\ P^{0.81}$
Gas expanders (vacuum discharge, CS)	Power extracted, P HP	200–8,000	$1{,}190\ P^{0.81}$
Steam turbines (non-condensing, CS)	Shaft power, P HP	250–10,000	$9{,}400\ P^{0.41}$
Steam turbines (condensing, CS)	Shaft power, P HP	250–10,000	$25{,}000\ P^{0.41}$
Gas turbines (CS)	Shaft power, P HP	100–10,000	$2{,}500\ P^{0.76}$
Internal combustion engines (CS)	Shaft power, P HP	100–4,000	$1{,}400\ P^{0.75}$
One-stage steam jet ejector, SJE (CS)#	Suction flow rate/Suction pressure, S (lb/hr)/torr	0.1–100	$1{,}690\ S^{0.41}$
Liquid-ring vacuum pump (SS with sealant recirculation)	Flow at suction, S ft^3/min	50–350	$8{,}250\ S^{0.35}$
Three-stage lobe vacuum pump (includes intercoolers)	Flow at suction, S ft^3/min	60–240	$7{,}120\ S^{0.41}$
Three-stage claw vacuum pump (includes intercoolers)	Flow at suction, S ft^3/min	60–270	$8{,}630\ S^{0.36}$
Screw compressors (with protective controls)	Flow at suction, S ft^3/min	50–350	$9{,}590\ S^{0.38}$

* For data given for MOC of CS, use a multiplying factor of 2 for PC of equipment made of SS.

\# See Table 22.31 in Seider *et al.*[8] for other types and condensers.

industries. For STHE (floating-head type), not available in Table 6.4, PC = US$ 1,496 $A^{0.59}$ where A is the heat transfer area from 1.9 to 1,860 m^2 (Holland and Wilkinson)[1]; further, PC of fixed tube-sheet HE, U-tube HE and kettle reboiler is respectively 0.85, 0.87 and 1.35 that of floating-head HE. This cost of STHE is delivered (i.e., including transportation) when MSECI = 1,000 or CEPCI = 398 in the year 1994. Note that PC, IC and FCI given in this chapter are all in US$ and for a location in the United States.

Data in Table 6.4 and for STHE given above can be used for estimating PC of an equipment. However, note the following points.

1. Data given are specific for PC of a particular type of equipment and near ambient pressure of operation. They cannot be used for another equipment type or conditions.
2. The stated units or dimensions (and not others) must be used for size/capacity parameter. If necessary, convert size in other units to the stated units and then use the data and equation.
3. Estimated PC corresponds to the stated CEPCI for a location in the United States. It should be adjusted to the current year using the corresponding CEPCI and, if required, to another location using the location/site factor.
4. Equation 6.4 with the data should be used within the range given in Table 6.4. If any equipment is smaller than the lower limit of the range, then estimate PC as that of size equal to the lower limit. If any equipment is larger than the upper limit of the range, then assume multiple units of same size within the range and estimate PC of multiple units. For example, PC of a plate and frame HE (SS) of 20,000 ft^2 (which is outside the range in Table 6.4) should be estimated as PC of two plate and frame HEs (SS), each of 10,000 ft^2. These assumptions provide conservative (i.e., higher) estimates.
5. Data in Table 6.4 and for STHE given above provide PC only. It has to be multiplied by suitable factors (e.g., bare-module factors in Table 22.11 in Seider *et al.*[8] and then 1.18 for contingency and contractor fee) to find IC or TMC, which is often 2–5 times PC.

Example 6.1: Find the PC of a spiral plate HE of 300 m^2 and SS in the near future when CEPCI is estimated to be 600. Assume location factor of 0.9. Then, find IC of this spiral plate HE.

Solution: The required area of 300 m^2 is equal to 3,229 ft^2, which is outside the range of 20–2,000 ft^2 for spiral plate HE in Table 6.4. Hence, assume two units of spiral plate HEs, each of 3,229/2 = 1,614.5 ft^2. Then, PC of these two units is:

PC (for CEPCI = 500) = 2 × 6,200 $(1{,}614.5)^{0.42}$ = 275,929 US$

The above value is now adjusted for CEPCI and location as follows:

PC (for CEPCI = 600, location factor = 0.9) = 275,929 × $\frac{600}{500} \times 0.9$ = 298,000 US$

To find IC of the spiral HE, assume bare-module factor of 1.8 for double-pipe HE (in Table 22.11 in Seider *et al.*[8]) and another factor of 1.18 for contingency and contractor fee, are applicable. Here, the assumed bare-module factor is for double-pipe HE because this factor for spiral plate HE is not available. Then, IC = 298,000 × 1.8 × 1.18 = 633,000 US$.

6.4.1 *Six-Tenths and Seven-Tenths Rules*

Six-tenths rule is essentially Eq. 6.3 or 6.4 with n is 0.6 (i.e., six-tenths). It is based on observation that average value of n in Eq. 6.4 for many equipment is 0.6. Six-tenths rule is particularly useful for FCI of plants with diverse equipment. Instead of 0.6, some prefer using 0.7, which gives seven-tenths rule. Six- or seven-tenths rule can be used when data are limited and n in Eq. 6.4 is unknown. This is illustrated in the following example.

Example 6.2: FCI of a plant built in 1994 (CEPCI = 368.1), for producing 100 thousand tons/year of styrene from benzene and ethylene, was $40 million. Estimate the CC for setting up a similar plant to produce 200 thousand tons/year of styrene, next year when CEPCI is projected to be 600. Investigate the effect of using six- and seven-tenths rule. Is there a need to adjust for location?

Solution: First, use cost index to find FCI for setting up the same plant/capacity next year as: 40 × 600/368.1 = $65.2 million. Next, estimate FCI for a larger plant as follows:

For six-tenths rule: $65.2 \times \left(\frac{200}{100}\right)^{0.6} = 98.8\ \$\text{million}$

For seven-tenths rule: $65.2 \times \left(\frac{200}{100}\right)^{0.7} = 105.9$ \$million

The difference in the estimates by six- and seven-tenths rules is 7%. Note that economy of scale is more when n is smaller. The above two steps can be inter-changed (e.g., estimate FCI for a larger plant and then adjust for cost index). Adjustment for location is required only if the location of the new plant is different from that of the earlier plant. The above estimates are valid if the new plant is using the same raw materials and process technology. In case the new plant incorporates latest technologies such as better catalyst and process intensification, then the above estimates may be poor.

Example 6.3: A company purchased an STHE of CS and heat transfer area of 220 m^2, and installed it in the plant. IC of this HE purchased/installed in the year 2014 was US\$ 750,000. The company is now planning to purchase and install the same type of HE but with area of 110 m^2 for WHR. Estimate IC of this new HE. State your assumptions, if any.

Solution: Assume MOC and OP of the new HE are same/similar to those of HE purchased/installed in 2014. Further, CEPCI and six-tenths rule are applicable, and CEPCI now is 600. Instead of six-tenths rule, one can assume seven-tenths rule. However, one of these is required since value of exponent n is not given. In case, IC of another HE of same type is available, then the exponent can be found. See and solve Exercise 6.1 at the end of this chapter.

Using six-tenths rule, IC of the new HE, if it were purchased/installed in 2014, is $750{,}000 \times \left(\frac{110}{220}\right)^{0.6} = 494{,}800$ US\$. Notice that IC of the HE of half the size/capacity is about 66% (and not 50%) of the bigger HE. This is due to economy of size from the exponent less than 1.0. Then, using CEPCI of 576.1 in 2014 (Table 6.2) and assumed CEPCI of 600 for the new HE, IC is 494,800 × (600/576.1) = 515,300 US\$.

6.4.2 *Module Costing Technique*

In MCT of Guthrie,[7] PC of equipment with a different MOC (such as SS) and OP is estimated by multiplying PC at base conditions with suitable factors for MOC and OP. As stated earlier, IC of an equipment is much more than its PC for the following additional costs: material and labour for installation; freight, insurance and taxes for equipment transport from its fabrication location to plant site; construction overhead and contractor engineering expenses. These are estimated as certain factors of PC, which depend on equipment. Turton *et al.*[3] presented all these factors combined together as bare-module factor for each equipment. For example, bare-module factor for a floating-head STHE increases from 3.29 for base conditions (i.e., CS shell and tubes, and ambient pressure operation) to 7.90 for SS shell and tubes, and 100 barg pressure. This increase is due to costlier MOC, higher metal thickness and higher pressure.

Product of PC at base conditions with the bare-module factor gives bare-module cost (BMC). Contingencies and contractor fee at, respectively, 15% and 3% of BMC are added to find TMC = 1.18 BMC. For additions such as WHR and expansions to existing plants, the sum of TMC for all new equipment can be taken as the required FCI. For grass-roots or green field projects (where there is no existing plant or facilities on the site), additional investment is required for site development, auxiliary buildings, off-sites and utilities. This is estimated as 0.5 times sum of TMC of all equipment at base conditions (and not at specified MOC and OP). Grassroots cost or FCI for grassroots plant is the sum of TMC of all equipment for specified MOC and OP, and this additional investment. For more details on MCT, see Chapter 7 in Turton *et al.*[3]

Working capital to pay for raw materials, fuels, salaries and so on, before receiving revenue from product sales, is required for grassroots projects. It may be required for large additions to existing plants but is not needed for WHR projects, and hence it is not discussed further in this chapter. Working capital, if required, is taken as 15%–20% of FCI. Different from FCI, it can be fully recovered at the end of plant life.

The following examples illustrate application of MCT for estimating PC and TMC (i.e., FCI) for some equipment, which are often encountered in WHR. Procedure for MCT and cost data/correlations used in these examples, are mainly from Chapter 7 and Appendix A in Turton *et al.*[3] Equipment in these examples are for Cases B and C described in Chapter 16 of this book.

Example 6.4: A proposed WHR project requires a new reboiler with the following specifications: STHE of U-tube type; area = 354 m^2, CS for both shell and tubes, design pressure is 21 barg for tubes and 5 barg for shell. Find PC, BMC and TMC for this reboiler. Assume CEPCI = 600. State other assumptions, if any.

Solution: PC, BMC and TMC are estimated in four steps as follows:

Step 1 — PC of equipment at base conditions of CS and ambient pressure operation: Cost correlation for PC of U-tube STHE at base conditions and CEPCI = 397 from Appendix A of Turton *et al.*[3] is:

$$\mathrm{Log}_{10}(\mathrm{PC}) = 4.1884 - 0.2503\ \log_{10}(A) + 0.1974\ [\log_{10}(A)]^2$$

Here, A is heat transfer area in m^2 and the correlation is valid for area from 10 to 1,000 m^2. Carefully note the terms such as logarithm to base 10 and use them correctly. Substitution of A = 354 m^2 in the above correlation gives PC of U-tube STHE at CEPCI = 397 is 68,073 US$.

Step 2 — Update calculated PC for cost index: CEPCI for the cost correlation in Step 1 is 397 whereas cost for CEPCI = 600 is required. The escalated PC of U-tube STHE at base conditions is 68,073 × 600/397 = 102,881 US$.

To confirm this, the CAPCOST program, which comes with the book by Turton *et al.*[3] was used. While one might expect that the program would have separate options for fixed tube-sheet and U-tube STHE's, the Equipment Cost Data worksheet of this program combines the two in a 'fixed tube-sheet or U-tube' option, implying that there is a common cost calculation for both.

For CEPCI = 600, the CAPCOST program calculated the PC of a 'fixed tube-sheet or U-tube' STHE and a floating-head STHE as US$62,100 and US$81,600, respectively. These are lower than the PC of US$102,881 calculated above for U-tube STHE at base conditions, escalated to CEPCI = 600. However, it is expected that the cost of a fixed tube-sheet STHE or U-tube STHE is less than that of a floating-head STHE. Hence, the cost calculated by CAPCOST is preferred; it is suspected that there are minor typographical errors in the cost correlation used in Step 1, from Appendix A in the book by Turton *et al.*[3] The PC of a U-tube STHE is thus taken as US$ 62,100.

Step 3 — Find the relevant equation for bare-module factor, and then factors for pressure and MOC: For HEs, this equation is BMC = (PC at base conditions) × ($B_1 + B_2 F_M F_P$) where $B_1 = 1.63$, $B_2 = 1.66$, F_M (material factor) = 1 for shell and tubes made of CS, and pressure factor is given by (from Appendix A of Turton *et al.*[3]):

$$\text{Log}_{10}(F_P) = -0.00164 - 0.00627 \log_{10}(P) + 0.0123 [\log_{10}(P)]^2$$

The above equation is valid for 5 < P, pressure < 140 barg. Using this equation for tube pressure of 21 barg gives $F_P = 1.027$. In this example, shell pressure of 5 barg is lower than tube pressure.

Step 4 — Calculate BMC and then TMC:

$$\text{BMC} = 62100 \times (1.63 + 1.66 \times 1.0 \times 1.027) = 207{,}092 \text{ US\\$}$$
$$\text{TMC} = 1.18 \times \text{BMC} = 244{,}369 \text{ US\\$}.$$

Useful pointers from the above example are summarised below:

- Perform sanity (common sense) check of calculated costs for their reasonableness, since cost calculations involve numerous numbers and some typographical errors may occur.
- Perform cost calculations in a spreadsheet (e.g., MS Excel), which is convenient for checking and doing cost calculations for different cases.
- Sequence of calculations can be slightly different (e.g., adjusting for CEPCI can be at the end for TMC instead of in Step 2).
- BMC and TMC are several times PC of equipment at base conditions.
- Use a program such as CAPCOST for quick and consistent cost calculations.

For practice, implement the above calculations in a spreadsheet and then estimate cost of another STHE in Exercise 6.3 at the end of this chapter.

Example 6.5: A proposed WHR project requires a new steam turbine with 196 kW output. Find PC, BMC and TMC for this steam turbine. Assume CEPCI = 600. State other assumptions, if any.

Solution: Cost correlation for PC of a steam turbine from Table A.1 in Turton *et al.*[3] is:

$$\text{Log}_{10}(\text{PC}) = 2.6259 + 1.4398\ \log_{10}(\text{P}) - 0.1776\ [\log_{10}(\text{P})]^2$$

This correlation is valid for P in the range 70–7,500 kW, and it gives PC at CEPCI = 397. Substituting P = 196 kW, PC of steam turbine is US$ 94,424 (at CEPCI = 397) and US$ 148,752 (at CEPCI = 600). Bare-module factor for a steam turbine is 3.5 (Fig. A.19 of Turton *et al.*[3]). Hence, BMC = 3.5 × 148,752 = 520,632 US$ and then TMC = 1.18 × BMC = 613,346 US$. PC and BMC are close to those using CAPCOST (namely, PC of US$ 149,000 and BMC of US$ 521,000).

Instead of cost correlations from Turton *et al.*,[3] PC of steam turbine can be estimated using the following exponential equation for steam turbines (non-condensing, CS) in Table 6.4.

$$\text{PC} = 9{,}400\ \text{P}^{0.41} \tag{6.5}$$

This correlation is valid for P for 250–10,000 HP and gives PC at CEPCI = 500. Converting P = 196 kW to 262.84 HP and substituting in the above equation, PC = 9,400 × $(262.84)^{0.41}$ = 92,299 US$. Updating this estimate for change in CEPCI, PC of steam turbine (at CEPCI = 600) = 92,299 × 600/500 = 110,760 US$. Bare-module factor for pumps and drivers is 3.3 in Table 22.11 in Seider *et al.*[8] With this factor, BMC = 3.3 × 110,760 = 365,506 US$.

Notice that PC and BMC of steam turbine (at CEPCI = 600) using the cost correlations in Seider *et al.*[8] — listed in Table 6.4 — are lower by 29% and 35% respectively, compared to those estimated using cost correlations in Turton *et al.*[3] This is expected and further discussed in Section 6.4.3.

Example 6.6: A proposed WHR project requires a new SJE with the following specifications: suction flow rate of 23,501 lb/hr at suction pressure of 8,740 torr, and MOC is SS. Find PC, BMC and TMC of this SJE. Assume CEPCI = 600. State other assumptions, if any.

Solution: Correlation of PC of one-stage SJE in Table 6.4 is used. One reason for this is that cost of SJE is not available in Turton *et al.*[3]

First, calculate S = 23,501/8,740 = 2.689 (lh/hr)/torr, and then PC of ejector made of SS = 1,690 × $(2.689)^{0.41}$ = US$ 2,535. This is at CEPCI of 500. Adjusting for CEPCI = 600, PC of SJE = 2,535*600/500 = US$3,042.

Bare-module factor of SJEs is not available. Our experience indicates that PC of ejectors is low but cost of installation is high. Hence, assume a bare-module factor of 5 on PC of ejector made of SS (which translates to a factor of 10 on PC of SJE made of CS for F_M of 2 for SS). Thus, BMC of SJE made of SS = 5 × 3,042 = 15,211 US$, and TMC = 1.18 × 15,211 = 17,949 US$. This estimate is lower than that ($26,071) in Chapter 16 because of different reference/cost data used for estimation.

6.4.3 *Discussion on FCI Estimation*

Procedures, correlations and data for cost estimation of process equipment and plants are available in several books such as Ulrich and Vasudevan[9], Seider *et al.*[8] and Turton *et al.*[3] One must use these carefully and correctly as given in the chosen reference; using a computer program, already tested by its developers, is better, in order to avoid calculation and/or typographical errors. Even then, cost estimates using two or more different references may not be comparable.

Feng and Rangaiah[10] employed five different sources (such as CAPCOST, DFP and CCEP) for cost estimation of equipment in seven different processes, and compared the estimates found thus. CCEP program implements cost correlations and procedures in Seider *et al.*[8] whereas DFP program is based on cost correlations and procedures in Sinnott and Towler.[11] Both these and CAPCOST are in MS Excel and easy to use. Here, total PC of all equipment (at base conditions) and TMC of each process by CAPCOST, DFP and CCEP programs are compared in Table 6.5. In all, seven processes from different industry sectors are considered; equipment involved in these processes include HEs, fired heaters (furnaces), pumps, compressors, distillation columns and vessels.

Table 6.5 presents total PC of all equipment in each process, given by CAPCOST in Million US$. Each of these estimates by DFP and CCEP is given as percent difference from that by CAPCOST:

Table 6.5: Estimates of total PC and TMC of several processes, by CAPCOST, DFP and CCEP programs

	Total PC by			TMC by		
Process (Reference)	**CAPCOST (Million $)**	**DFP (Δ%)**	**CCEP (Δ%)**	**CAPCOST (Million $)**	**DFP (Δ%)**	**CCEP (Δ%)**
Alkylate splitter in a petroleum refinery[12]	0.8	2	15	4.3	−6	−4
Mono-chlorobenzene separation process[8]	0.3	−4	33	1.0	16	71
Crystallization process[8]	1.3	0	−61	2.5	6	−11
Formalin production from methanol[13]	0.9	86	8	4.1	79	19
Styrene production process[13]	9.8	−3	−13	33.3	27	−23
Maleic anhydride synthesis from benzene[13]	3.8	82	59	17.0	94	14
Beta-galactosidase batch process[9]	2.2	22	16	9.4	34	7
Average Absolute Difference		**28**	**29**		**37**	**21**

$$\Delta = \frac{(\text{Cost by DFP or CCEP} - \text{Cost by CAPCOST})}{\text{Cost by CAPCOST}} \times 100 \qquad (6.6)$$

Hence, positive (negative) value of Δ means the cost estimate by that method is more (less) than that by CAPCOST. This way of presenting many estimates is to facilitate comparison and bring out the differences. However, there is no implication whatsoever on the (relative) accuracy of estimates by any of the three programs compared here.

Results in Table 6.5 indicate that differences in estimates by different programs can vary significantly from small to large values, which is due to differences in cost estimation of different equipment by the program employed. These differences are illustrated for the common STHE considering U-tube type made of CS with shell/tube pressure of 5 barg (i.e., similar to base conditions). Figure 6.2 shows the variation of PC of this U-tube STHE by CAPCOST, DFP and CCPE with heat transfer area, all at CEPCI = 600. In this figure, cost estimate by CAPCOST is that of fixed tube-sheet STHE since it does not have U-tube STHE separately;

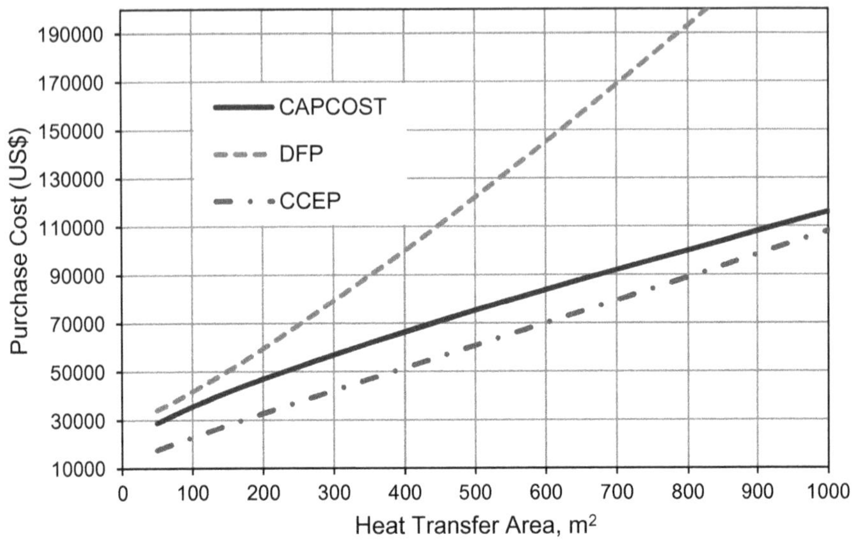

Fig. 6.2: Effect of heat transfer area on PC (at CEPCI = 600) of U-tube STHE by CAPCOST, DFP and CCEP programs.

however, this is reasonable since costs of U-tube and fixed tube-sheet STHE are nearly same.

Over the heat transfer area from 50 to 1,000 m^2, PC estimate of U-tube STHE by CCEP is lower than that of CAPCOST whereas that of DFP is higher than the other two estimates. In particular, DFP estimate of PC of U-tube STHE is significantly higher and also increases faster with area. This may be due to the data used and/or some discrepancy in the reported correlation. In general, U-tube STHE is quite common and so relatively more cost data might have been employed for developing its cost correlation. Despite this, Fig. 6.2 shows variations in PC estimate of U-tube STHE based on different cost data/programs. Hence, it emphasizes the potential differences in cost estimates based on different programs (references or data sources) employed.

As shown by more positive Δ values in Table 6.5, estimate of TMC by DFP is often more than that by CAPCOST whereas TMC estimate by CCEP can be lower or higher than that by CAPCOST. Overall, average absolute difference is around 30%, which is comparable to the probable accuracy (−25–+30%) of study/factored/major equipment type of estimate given in Table 6.1. Readers should particularly note that the difference/accuracy in estimates in Fig. 6.2 and Table 6.5 is despite using the procedures, cost correlations and data carefully and correctly using a computer program.

Besides CAPCOST (which comes with the book by Turton *et al.*[3]), CCEP and DFP (both available at the website of World Scientific for this book), other programs are available. They include EconExpert (web-based program available with the book by Ulrich and Vasudevan[9]), Aspen Economic Analyzer (available in the commercial software: Aspen Hysys and Aspen Plus) and Matche (both free and subscription versions available at https://www.matche.com/equipcost/Default.html). Some of these are updated regularly. Based on open literature, it is not possible to recommend one of these programs over another, for finding more realistic cost estimates. The user should choose one of them based on his/her experience and some comparison with available cost data within his/her company.

WHR projects may require additional capital besides FCI estimated for purchasing and installing the equipment involved. They may require additional piping/ducting and instrumentation (on top of those required for normal installation of equipment). Additional piping may be required for connecting to utilities and other process units whereas additional instrumentation may be required for maintaining the design conditions and improving the safety. HAZOP and LOPA studies usually recommend protection layers involving both instrumented (e.g., additional level/pressure/temperature transmitters, safety instrumented system), and non-instrumented (e.g., pressure relief valves). There will be additional piping and effluent disposal costs for pressure relief fluid and any wastewater generated by WHR equipment. If the electrical substation does not have enough capacity, expanding it requires additional cost. Electrical cable laying costs can be significant (i.e., more than equipment installation cost), if buried cables are required and the substation is far away from the equipment.

6.5 Estimation of Operating Cost and Revenue

Cost of manufacture (COM) or manufacturing cost is comprehensive and includes various OCs for manufacturing one or more (chemical) products from a process plant. According to Turton *et al.*,[3] it is the sum of direct manufacturing cost, fixed manufacturing cost and general manufacturing expenses. Direct manufacturing cost, as the phrase suggests, is related to plant operation for producing products from raw materials. It includes costs of raw materials, waste treatment, utilities, operating labour and supplies, supervision, maintenance and repairs, laboratory tests and patents and royalties. If the plant is not in operation, these costs are absent. On the other hand, fixed manufacturing cost is incurred even if there is no operation and no product is produced. It includes depreciation of FCI, local taxes, insurance and plant overheads (such as costs for accounting, fire, safety, medical, canteen and general engineering). Finally, general manufacturing expenses are costs of administration, heating/cooling/maintenance of buildings, distribution and selling of products and research and development.

Obviously, all components/items in COM vary from one plant to another. Some items in direct/fixed/general manufacturing costs are estimated as a fraction of other costs (e.g., maintenance and repairs as 0.02–0.1 of FCI, and distribution and selling of products as 0.02–0.2 of COM itself). Taking mid-value of these ranges, assuming depreciation of 10% of FCI per year, and re-arranging (since COM occurs on both left and right sides of the equation), gives:

$$\text{COM} = 0.28 \times \text{FCI} + 2.73 \times \text{COL} + 1.23 \times (\text{CRM} + \text{CWT} + \text{CUT}) \quad (6.7)$$

Without depreciation, COM is 0.18 × FCI + 2.73 × COL + 1.23 × (CRM + CWT + CUT). Here, COL, CRM, CWT and CUT are respectively cost of operating labour, raw materials, waste treatment and utilities. See Chapter 8 in Turton *et al.*[3] for more details on COM, many components contributing to it and the above equation.

There are two important points in COM outlined above. First, CRM, CWT and CUT in estimating COM by Eq. 6.7 are multiplied by a factor 1.23. This is mainly due to estimating patents and royalties, distribution and selling of products and research and development as certain fractions of COM itself and later re-arrangement to find COM. In other words, contribution of CRM, CWT and CUT to COM is more than their values.

Second, terminology on COM in different books may not be the same. For example, in Seider *et al.*,[8] COM is the sum of direct and indirect manufacturing costs (i.e., does not include general manufacturing expenses) whereas total product cost is the same as COM in Turton *et al.*[3]

WHR implementation in a plant is likely to change CUT only, and all other components/items in COM remain unaffected. Hence, economic evaluation of WHR requires changes in the value of CUT; it can be expected that WHR reduces CUT or generates utilities for use elsewhere (thus getting credit or revenue for the utilities produced). Hence, this section focuses on unit prices of utilities, which are common in WHR. Using these and knowing changes in utility quantities used in the plant due to WHR, CUT can be estimated. Calculation of CUT per annum requires number of operating hours (typically 8,000–8,500) per year. Further, instead of COM, the phrase, OC is used in this and other chapters of this book because it is more common and WHR affects CUT only.

In plants, utilities are purchased from an outside source (e.g., a power plant generating/selling electricity and a company specialising in producing/selling steam and other utilities) or produced internally within the plant. Utility prices vary for this reason, due to location/region, availability of cheap source of energy and with time. To show possible variation, utilities commonly encountered in WHR and their unit prices from Seider *et al.*[8] and Turton *et al.*[3] are summarized in Table 6.6. Scrutiny of utility prices in this table shows some similarities and significant differences. For example, natural gas price differs by a factor 2, electricity price is same, prices of steam at 10 barg and cooling water deviate by nearly 30%, and chilled water price differs by around 10%. These differences are expected because of assumptions and data used for them in each reference. Unit prices of utilities in other references can be similar or different from those in Table 6.6.

Unit prices of utilities such as those in Table 6.6 are often based on OC for utility generation; that is, they do not include contribution of FCI for utility plants, which is assumed to be accounted elsewhere (e.g., in FCI for a grassroots plant). Hence, one may have to increase utility prices in Table 6.6 by 10%–20% to account for FCI. On the other hand, steam generated from a process given in the second column of Table 6.6 is about

Table 6.6: Utilities common in WHR and their unit prices from two references

Utility and Its Unit Price in Brackets from Seider *et al.*[8]	Utility and Its Unit Price in Brackets from Turton *et al.*[3]
Natural Gas ($0.156/SCM or $4.19/GJ)	Natural Gas ($11.1/GJ)
Electricity ($0.06/kWh)	Electricity ($16.8/GJ or $0.06/kWh)
Steam at 31.0 barg ($14.5/1,000 kg)	Steam at 41 barg ($17.7/GJ or $29.97/1,000 kg)
Steam at 10.3 barg ($10.5/1,000 kg)	Steam at 10 barg after including credit for power ($14.19/GJ or $28.31/1,000 kg)
Steam at 3.45 barg ($6.60/1,000 kg)	Steam at 5 barg after including credit for power ($13.28/GJ or $27.70/1,000 kg)
	Steam Generated from a Process ($12.33/GJ)
Cooling Water ($0.02/m^3)	Cooling Water supplied at 30°C and returned at 40 or 45°C ($0.354/GJ or $0.0148/m^3)
Chilled Water at 4.44°C ($4.0/GJ)	Chilled Water supplied at 5°C and returned at 15°C ($4.43/GJ)
Refrigerant at −34.44°C ($7.90/GJ)	Refrigerant at −20°C ($7.89/GJ)

10% cheaper than steam at 5 barg obtained from the utility plant (i.e., steam generator using fuel such as natural gas). The reader may be wondering whether one should use this for steam generated by WHR or not. Where possible, check and follow the practice in the company.

In short, estimation of CUT and its change for WHR projects are straightforward but require reliable and reasonable unit prices of utilities involved. These can be based on the company's internal sources and/or a text-book such as Seider *et al.*[8] and Turton *et al.*[3] In any case, readers should be aware of significant variation in utility prices and should analyse sensitivity of WHR benefits to expected range of utility prices. This is illustrated in the following and later examples in this chapter.

Example 6.7: In a plant, WH of 2,000 kW is currently rejected to cooling water. It is proposed to recover and use this thermal energy to generate low-pressure steam (LPS), thus reducing cooling water required to remove 2,000 kW. Estimate the credit/revenue from generating LPS and reduction in the cooling water cost, both on annual basis. State your assumptions.

Solution: The following assumptions are required.

1. Number of operating hours per year is 8,250 (= mean of 8,000 and 8.500). However, any justifiable number of hours can be assumed.
2. LPS pressure is not given, and so it is assumed to be 5 barg. The corresponding saturation temperature of steam is 158.92°C (from steam tables or Google search).
3. LPS can be produced from boiler feed water (BFW) at room temperature (which requires both sensible heat to increase water temperature to saturation temperature and latent heat to vapourise) or from BFW at saturation temperature (which requires only latent heat). For the purpose of this example, assume LPS is produced from BFW at room temperature, and is saturated (and not superheated) at 5 barg. Then, from steam tables, energy required for LPS production is equal to 2,756.3 (enthalpy of saturated steam at 5 barg) −125.73 (enthalpy of water at 30°C) = 2,630.57 kJ/kg.
4. Assume cooling water is supplied at 30°C and returned at 45°C, and its heat capacity and density are respectively 4.2 kJ/kg.K and 1,000 kg/m^3. Any reasonable values for these quantities are satisfactory.
5. Assume LPS cost is \$27.70/1,000 kg of steam and cooling water cost is \$0.0148/m^3 of water (i.e., as in the second column of Table 6.6).

It is given that 2,000 kW of WH is used to produce LPS. Using this and energy required for LPS production stated above, LPS produced = $\frac{2{,}000\ \text{kJ/s}}{2.630.57\,\text{kJ/kg}} = 0.76029\,\text{kg/sec}$ or 2,737 kg/hr.

For the assumed 8,250 hours of operation, LPS produced per annum = 2,737 × 8,250 = 22,580,250 kg/year.

Using LPS cost of \$27.7/1,000 kg, credit/revenue from LPS generated = $\frac{22{,}580{,}250 \times 27.7}{1{,}000} = \$625{,}473/\text{year}$. This will be about 10% lower in case unit price of \$12.33/GJ for steam generated from a process (in the second column of Table 6.6) is used.

With the use of WH for LPS production, cooling water is not required to remove 2,000 kW. Using the assumed cooling water data, cooling water not required = $\frac{2{,}000}{4.2 \times (45-30)} = 31.75\,\text{kg/sec}$ or 0.03175 m^3/sec or 114.3 m^3/hr.

Cooling water saved per annum = 114.3 × 8,250 = 942,975 m^3/year.

Using cooling water cost of $0.0148/m^3, annual savings in cooling water cost = 942,975 × 0.0148 = $13,956/year.

Total gain from using WH for LPS generation is the sum of credit/revenue for LPS and reduction in cooling water cost, which is equal to 625,473 + 13,956 = $639,429/year. This is substantial; however, some FCI is required for implementing the proposed WHR for LPS generation. Most of the gain is due to LPS generation and contribution from cooling water reduction is small.

To understand the effect of variation in unit prices of LPS and cooling water, the above calculations can be performed for LPS and cooling water price of $6.60/1,000 kg (for Steam at 3.45 barg) and $0.02/m^3, respectively, from the first column in Table 6.6. Results obtained from calculations in MS Excel are as follows. Credit/revenue for LPS generated is $149,032/year, which is about one quarter of $625,473/year using LPS price from Turton *et al.*[3] On the other hand, reduction in cooling water cost is $18,857/year, which is about 50% more than $13,956/year using cooling water price from Turton *et al.*[3] As expected, these differences are proportional to difference in unit price of the utility in the two references. Total gain from using WH for LPS generation is $167,889/year compared to $639,429/year in the previous paragraph. These emphasize the importance of using reliable unit of prices of utilities, particularly for expensive utilities such as steam, for estimating benefits of WHR projects.

6.6 Time Value of Money

Economic evaluation (i.e., assessment of economics of investments/costs and gains/benefits) of any project requires one important concept, namely, time value of money. Many people understand and are aware of investment, interest rate (or rate of return) and inflation. For example, they expect interest on a deposit in a bank and want to grow their savings by investing. They are also aware of inflation or purchasing power of currency. In other words, prices of daily needs have generally been increasing with passing time. CPI captures this variation with time, and its trend for the United States (urban) can be seen in Fig. 6.1.

Time value of money is related to interest (or return) on investment (and not to inflation or purchasing power). In the present context, interest and return are synonymous, and return on investment is mainly used in this chapter. Like individuals, companies invest and expect attractive return on their investments. There may be risk in the investment (e.g., failure of the plant to produce product(s), and product demand/price fluctuation in the future). To offset this, companies expect higher return on their investment (compared to what banks offer to customers on their deposits and what banks charge on loans given to their borrowers). Note that customers' deposits in banks are not completely free from risk although it is very low.

Return on investment can be according to simple, annual compound and/or continuous compound interest (e.g., daily interest on bank deposits). If P \$ is invested at present time and annual return of i (e.g., 0.1/year) is expected, then accumulated amount (F) after n years is:

$$F = P\,(1 + n \times \mathrm{i}) \quad \text{for simple interest} \tag{6.8}$$

$$F = P\,(1 + \mathrm{i})^{n} \quad \text{for annual compound interest} \tag{6.9}$$

$$F = P\,e^{(n \times \mathrm{i})} \quad \text{for continuous compound interest} \tag{6.10}$$

The first two equations are known to many readers. All the three equations can be developed through simple analysis of each case. Interested readers can try on their own or read it in books. The following example shows the effect of three types of interest. Note that value of i should be positive.

Example 6.8: A sum of \$100,000 is invested for 5 years and rate of return of 6% (i.e., i = 0.06) per year. Find the amount after 5 years for three cases: simple, annual compound and continuous compound interest. Repeat this for i = 0.08/year.

Solution: The future amount (F) after 5 years is as follows:

Interest Type	Amount for i = 0.06 (6%)	Amount for i = 0.08 (8%)
Simple	$F = 100{,}000 \times (1 + 5 \times 0.06)$ = 130,000	$F = 100{,}000 \times (1 + 5 \times 0.08)$ = 140,000
Annual Compound	$F = 100{,}000 \times (1 + 0.06)^5$ = 133,823	$F = 100{,}000 \times (1 + 0.08)^5$ = 146,933
Continuous Compound	$F = 100{,}000 \times e^{(5 \times 0.06)}$ = 134,986	$F = 100{,}000 \times e^{(5 \times 0.08)}$ = 149,182

The above results show that the amount in the future (F) increases from simple interest to annual compound interest and to continuous compound interest. Further, F increases with i (as shown in the above table); it also increases with n; confirm this by solving Exercise 6.4 at the end of this chapter. Although continuous compound interest gives higher amount to the investor, annual compound interest is commonly used in economic evaluation of industrial projects. Hence, it is assumed and used in the rest of this chapter.

Present value (P) of the amount at a future time (F) after n years can be found by re-arranging Eq. 6.9 to:

$$P = \frac{F}{(1+\mathrm{i})^n} \tag{6.11}$$

This equation is assuming investment return of i/year and annual compound interest. For example, consider future amount of \$100,000 after 5 years; its present value for return of 0.08/year is $P = \frac{100{,}000}{(1+0.08)^5} = 68{,}058.$

Bringing the future amount to present value/time (Eq. 6.11) is referred as discounting. Hence, rate of return (i) is also known as discount rate. Typically, a project involves investment initially and then profit each year over the project life. In other words, it involves money/amounts (i.e., cash flows, CFs) at different times. It is necessary to bring them to the same time (e.g., initial or final) for economic evaluation. This requires use of Eq. 6.11 for discounting future amount to find the present value and/or Eq. 6.9 for finding future value of present amount. This will be further described in Section 6.7.

Example 6.9: For a WHR project, a company borrowed \$800,000 from a bank and agreed to re-pay in instalments of \$190,000 at the end of each year for the next 5 years. Assuming annual compounding, find and comment on the interest rate that the bank is charging for this loan for WHR project.

Solution: From the perspective of the company, there is inflow of \$800,000 from the bank at time = 0 and outflows of \$190,000 at the end of each year for 5 years. These CFs are shown in the first two columns of the following table; by convention, inflow is positive whereas outflow is negative. Present value of each CF is calculated for i = 0.05/year using

Eq. 6.11; it is given in the third column. This is required because CFs occur at different years and for including time value of money.

Year	CF ($)	Present Value, $ (for i = 0.05)	Present Value, $ (for i = 0.06)	Present Value, $ (for i = 0.06016)
0	+800,000	+800,000	+800,000	+800,000
1	−190,000	−180,952	−179,245	−179,218
2	−190,000	−172,336	−169,099	−169,048
3	−190,000	−164,129	−159,528	−159,455
4	−190,000	−156,313	−150,498	−150,407
5	−190,000	−148,870	−141,979	−141,872
NPV		−22,601	−349	0

Sum of present values in all 5 years is NPV in the last row of the above table. The annual compound interest that the bank is charging on the loan is that value of i which makes NPV equal to zero. It can be found by trial and error; for example, calculated present values and NPV for i = 0.06 are shown in the fourth column of the table. NPV is close to zero for i = 0.06, which can be taken as the interest that the bank is charging. To find precise value, NPV can be made equal to zero using Goal Seek tool in MS Excel. The corresponding i is 0.06016/year. Instead of present values and NPV, future values and net future value can be used to find bank's interest on the loan. This is given as Exercise 6.5.

From the perspective of the company, calculated interest rate on the loan from the bank seems to be reasonable. WHR should generate revenue and/or save CUT to cover this interest as well as the principal of loan (i.e., $800,000). Although WHR project life can be more than 5 years, assume that the principal should be recovered in 5 years. In this case, WHR should generate annual revenue and/or CUT savings of at least $190,000 (i.e., around 24% of the investment).

From the perspective of the bank, the entire loan amount and annual compound interest at 6.016% are recovered in 5 years. For economic sustainability, the bank can only give much lower interest (say, 3%–4%) on customers' deposit, that is, with sufficient margin to cover their operating expenses including salaries for employees, some loan defaults and reasonable profit for shareholders of the bank.

In summary, time value of money refers to different values of investment (i.e., money) at different times due to the earning capability of money. As seen from the above simple examples, money (say, \$10,000) at present time is more valuable than the same amount (i.e., \$10,000) at a future time. Time value of money does not consider inflation (purchasing power) and risk in investment.

6.7 Profitability Criteria

Economic evaluation of a proposed plant/project is essentially application of chosen profitability criteria to relevant cost data of the plant/project such as FCI, working capital, COM, revenue from products produced, depreciation of FCI, plant lifetime, salvage value of equipment, tax on profit and gross/net profit. Many profitability criteria are available; some of them do not consider time value of money whereas others include time value of money. Here, commonly used profitability criteria, namely, PBP (also termed as payback time), NPV (or net present worth in some references) and IRR (also known as discounted CF rate of return) are described with examples related to WHR.

WHR projects do not affect many components of COM, raw materials used, products produced, revenue, working capital, plant lifetime and so on. They require some FCI for new equipment and/or changes in the existing equipment/plant. This is often required initially (i.e., time = 0) since a WHR project, once approved, can be implemented in 1–2 years. Thus, there is a cash outflow or negative CF at time = 0. In terms of benefit or gain, WHR reduces CUT and/or generates valuable utilities (e.g., steam, electric power and chilled water), thus getting credit or revenue for the generated utilities. Both these are equivalent to revenue from WHR. Although this revenue may vary from year to year, it is reasonable to assume that same benefit/revenue is obtained in each year of WHR plant operation. Thus, there will be cash inflow or positive CF in each year of WHR operation.

By definition, PBP is the number of years (time), starting from project operation, required to recover FCI for the (WHR) project from the benefit/

revenue. Companies/investors prefer to recover FCI as early as possible, and so they favour a low value of PBP (say, a few years). PBP is simple to calculate and easy to understand but it does not consider time value of money. A related criterion, namely, discounted PBP is based on discounted CFs, which includes time value of money. However, its usage is limited and so it is not described further in this chapter.

NPV of a project is equal to the sum of yearly discounted CFs in all years; thus, it accounts for time value of money. Hence, its estimation requires a value for rate of return (or discount rate), i for calculating discounted CFs. The value of i used is essentially acceptable rate of return. It can be based on company's experience with their investments, should be positive and will be more than bank's interest on loans and deposits, as seen in Example 6.8. NPV is a commonly used profitability criterion although it requires the assumption of a value for rate of return.

NPV is the net amount on top of recovering FCI along with return at the rate i/year. Hence, large positive NPV is better and preferred by companies/investors. However, NPV can be negative in some projects and depending on i value. Negative NPV means the project is not generating sufficient revenue for recovering investment and return on it at the rate of i/year. Hence, projects having negative NPV are not preferred. PBP and NPV are illustrated in the following example:

Example 6.10: For a WHR project to improve the steam network in a petroleum refinery (Case C in Chapter 16), FCI required is \$766,256. Benefit (from increased power generation) is \$676,723/year. (Note that this is referred to as OC savings in Chapter 16.) Find PBP and NPV (for i = 0.15, 0.2, 0.25 and 0.3) for this project. Assume a plant lifetime of 10 years for the equipment and changes made for WHR (using FCI). In other words, benefit will be for each of 10 years.

Solution: Assuming FCI is incurred at time = 0 (i.e., initially), CFs and discounted CFs in this project are given in the following table. Sum of CFs in all years (\$6,000,974) is known as cumulative cash position; it is the extra CF generated by the project after recovering the entire investment.

		Discounted CFs ($) for			
Year	**CF ($)**	***i* = 0.15**	***i* = 0.20**	***i* = 0.25**	***i* = 0.30**
0	−766,256	−766,256	−766,256	−766,256	−766,256
1	676,723	588,455	563,936	541,378	520,556
2	676,723	511,700	469,947	433,103	400,428
3	676,723	444,956	391,622	346,482	308,021
4	676,723	386,919	326,352	277,186	236,940
5	676,723	336,451	271,960	221,749	182,261
6	676,723	292,566	226,633	177,399	140,201
7	676,723	254,405	188,861	141,919	107,847
8	676,723	221,222	157,384	113,535	82,959
9	676,723	192,367	131,153	90,828	63,815
10	676,723	167,276	109,295	72,663	49,088
Sum	6,000,974	2,630,060	2,070,886	1,649,986	1,325,860

Sum of CFs in years 1 and 2 is 1,353,446, which is well over FCI, whereas CF in year 1 is less than FCI. Hence, PBP is between 1 and 2 years and it can be calculated as $\frac{766{,}256}{676{,}723} = 1.13$ years. This equation for PBP is valid if CF is same in all years (except initially), as in this example. If CF varies from year to year, then PBP can be calculated by linear interpolation in the year, when sum of CFs until that year exceeds FCI (here, year 2); interpolation in this example is $1 + \frac{766{,}256 - 676{,}723}{676{,}723}$ (i.e., year 1 plus fraction of year 2) = 1.13 years. Note that 1.13 years is from the time of plant operation. In other words, FCI will be recovered from year 1 and 0.13 of year 2 (after 2.13 years of starting investment due to 1 year of construction).

Discounted CF (i.e., present value of CF) in each year is calculated using Eq. 6.11 and assumed i value; this is repeated for each of the required i values. For example, discounted CF in year 4 with i = 0.2 is $\frac{676{,}723}{(1+0.2)^4} = 326{,}352\$$. Notice the steady decrease of discounted CF with years although CF is the same in each of years 1–10. In other words, present value of future amount in later years becomes less due to the term $(1 + i)^n$ in the denominator of Eq. 6.11.

NPV for one i value is the sum of discounted CFs in all years of the project, and it is given in the last row of the above table. NPV for i = 0 is the same as sum of CFs ($6,00,974) or cumulative cash position, at the bottom of column 2. Calculations in the above table can be performed in a spreadsheet. It is possible to use the built-in function for NPV in the spreadsheet; remember to understand its inputs and give them correctly. Results in the above table show that NPV decreases as i increases. NPV can become negative for large values of i.

Finally, low PBP of 1.13 years and large NPV indicate that WHR project is economically very attractive. In general, low PBP corresponds to large NPV.

The above example did not consider depreciation of FCI for WHR project and tax on benefit from power generation. It is possible to account for these with suitable assumptions. Since equipment value decreases with time, tax regulations in many countries allow companies to depreciate equipment cost (i.e., FCI), thereby recover investment made and reduce tax payable on profit. They also specify allowed depreciation for different equipment. For preliminary study, straight-line depreciation of FCI over plant life and zero salvage value of equipment at the end of plant life are reasonable. Then, depreciation per year is simply FCI ($) divided by plant life (years). Other ways of depreciation are sum-of-the years and double declining balance; details on these can be found in Chapter 9 in Turton *et al.*[3]

Although generated power is used internally, it can reduce CUT and correspondingly increase (gross) profit and tax payable. Hence, benefit/gain from generation may be interpreted as revenue on which tax is payable. In general, tax payable is on gross profit (equal to the difference between revenue and COM of a plant) at the local tax rate (TR); recall COM includes depreciation. So, after-tax CF is:

$$(\text{Revenue} - \text{COM}) \times (1 - \text{TR}) + \text{Depreciation} \qquad (6.12)$$

For economic evaluation of a WHR project, revenue is the benefit/gain and FCI for WHR depreciates (which is an expense). This is assuming WHR does not affect revenue from products and COM of the plant (except for depreciation of FCI for WHR). Hence, depreciation of FCI for WHR should be subtracted from the benefit/gain, and after-tax CF in case of a WHR project is simply:

$$(\text{Benefit} - \text{Depreciation of FCI for WHR}) \times (1 - \text{TR}) + \text{Depreciation} \qquad (6.13)$$

The following example considers tax and depreciation for the WHR project analysed in Example 6.10.

Example 6.11: For a WHR project to improve the steam network in a petroleum refinery (Case C in Chapter 16), FCI required is \$766,256. Benefit from increased power generation is \$676,723/year. Find PBP and NPV (for i = 0.15, 0.20, 0.25 and 0.3) for this project. Assume a plant lifetime of 10 years, straight-line depreciation and zero salvage value for the equipment and changes made for WHR (using FCI), and TR of 30% (0.3) on benefit.

Solution: After-tax CF calculated using Eq. 6.13 is given in the second column of the following table. Discounted CFs and NPV are computed as for the previous example. With the consideration of tax and depreciation, CF in each of years 1–10 has decreased from 676,723 to 496,694 (i.e., by around 27%) as compared to example 6.10. Further, the PBP is $\frac{766{,}256}{496{,}694} = 1.54$ years, and NPV decreases from 1.7 to 0.77 million \$ as i increases from 0.15 to 0.30. As expected, these are less than those in the previous example. Overall, the proposed WHR project is economically very attractive even with consideration of tax and depreciation.

		Discounted CFs (\$) for			
Year	**CF (\$)**	***i* = 0.15**	***i* = 0.20**	***i* = 0.25**	***i* = 0.30**
0	−766,256	−766,256	−766,256	−766,256	−766,256
1	496,694	431,908	413,911	397,355	382,072
2	496,694	375,572	344,926	317,884	293,902
3	496,694	326,584	287,439	254,307	226,078
4	496,694	283,986	239,532	203,446	173,906
5	496,694	246,945	199,610	162,757	133,774
6	496,694	214,734	166,342	130,205	102,903
7	496,694	186,726	138,618	104,164	79,156
8	496,694	162,370	115,515	83,331	60,889
9	496,694	141,191	96,263	66,665	46,838
10	496,694	122,775	80,219	53,332	36,029
Sum	4,200,682	1,726,535	1,316,119	1,007,191	769,292

As discussed earlier, there is uncertainty in FCI estimate and benefit because of variations in cost and price data in different sources, location and time. Calculations like in the above example can be performed for other scenarios such as increase in FCI (say, by 20% or 40%) and/or decrease in benefit/revenue (say, by 25% or 50%). This is given as Exercises 6.6 and 6.7 at the end of this chapter. In general, consider both more and less favourable scenarios for comprehensive assessment of a proposed project.

As stated above, one drawback of NPV is that some i value should be given for its calculation. This is avoided in the profitability criterion, namely, IRR. Recall that NPV decreases as i increases. Hence, NPV can be expected to become zero for certain value of i, which is termed IRR. Significance of IRR is similar to NPV; that is, the plant/project and its CFs under study recover the investment along with return at the rate of IRR. IRR is generally positive, and a high IRR is desirable. One limitation of IRR is that its calculation requires iterations. However, these are easy with the current computational tools such as spreadsheets.

Example 6.12: Consider the following slightly different scenario (compared to that in the previous exercise). For a WHR project to improve the steam network in a petroleum refinery (Case C in Chapter 16), FCI required is \$766,256. Benefit (from increased power generation) is \$676,723/year. The project construction requires 2 years with half of FCI spent initially (time = 0) and the remaining half of FCI at the beginning of year 1. WHR will take place from year 2 for 7 (shorter life of 6 years). Salvage value is zero, depreciation is straight-line type over 6 years (from years 2 to 7), and TR is 30% on benefit. Find PBP, NPV (for i = 0.4 and 0.5) and then IRR.

Solution: CFs and discounted CFs are presented in the following table. Observe CF initially and in year 1 are outflows (negative) for WHR project construction. For 6 years plant life, depreciation is $\$\,\frac{766{,}256}{6} = 127{,}709$. Using Eq. 6.13, after-tax CF from years 2 to 7 is (676,723 − 127,709) × (1 − 0.3) + 127,709 = \$512,019. It is slightly (3%) more than that in the previous example due to higher depreciation and consequently lower tax payable. PBP can be calculated from CFs; it is 1.5 years $\left(=\frac{766{,}256}{512{,}019}\right)$, comparable to that in the previous example. Note that 1.5 years is from the time of plant operation. In other words, FCI will be recovered from year 2 and half of year 3 (after 3.5 years of starting investment due to 2 years of construction).

		Discounted CFs ($) for			
Year	**CF ($)**	***i* = 0.4**	***i* = 0.5**	***i* = 0.49**	***i* = 0.487659**
0	−383,128	−383,128	−383,128	−383,128	−383,128
1	−383,128	−273,663	−255,419	−257,133	−257,538
2	512,019	261,234	227,564	230,629	231,355
3	512,019	186,596	151,709	154,784	155,516
4	512,019	133,283	101,140	103,882	104,538
5	512,019	95,202	67,426	69,720	70,270
6	512,019	68,001	44,951	46,792	47,235
7	512,019	48,572	29,967	31,404	31,751
Sum	2,305,857	136,098	−15,789	−3,051	0

NPV values are in the last row of the above table. For i = 0.4, NPV is positive at $136,098 whereas, for i = 0.5, it is negative −15,789 $. This indicates IRR is slightly less than 0.5. By linear interpolation of NPV for i = 0.4 and 0.5, IRR can be estimated as 0.49. Corresponding discounted CFs are given in column 5 of the above table. As can be seen, NPV for i = 0.49 is not zero due to IRR estimate by interpolation. Exact value of IRR can be found by a few trials or using Goal Seek. Results of using Goal Seek are shown in the last column of the table. Exact value of IRR is 0.4877 (or 48.77%). In other words, proposed WHR project can recover the investment of $766,256 and can give 48.77% return.

6.8 Summary

Approval for implementing any project including WHR projects requires cost estimation to find the investment required and then economic evaluation to assess return on investment. Common methods and procedures for cost estimation, time value of money and popular profitability criteria are described and illustrated in this chapter. Key points in this chapter are summarized below.

- FCI (or TMC) is required for purchasing and installing the equipment in the proposed (WHR) project. It is estimated based on historical cost data of process equipment, available in books, software and/or company's database.
- Cost of process equipment varies with time (inflation) and location (due to different costs for materials and labour). Plant cost index (such as CEPCI) is employed to adjust past cost to current/future time.
- Cost estimation is subject to substantial uncertainty (e.g., −30% to +50%) for several reasons including diverse equipment and their designs, unavailability of all details about process equipment and plant layout, limitations and sources of historical cost data, and use of a gross cost index.
- Accuracy of cost estimate increases to the range −5% to +10% with the project progress and availability of more details. Accordingly, cost of finding the cost estimate increases too.
- Depending on the data available and desired accuracy, six-tenths rule and/or MCT can be used for TMC. MCT accounts for equipment type, MOC and OP explicitly.
- COM is comprehensive with FCI contribution and various OCs for manufacturing product(s) in a process plant. It is the sum of direct manufacturing cost, fixed manufacturing cost and general manufacturing expenses.
- WHR projects mainly require estimation of CUT and its changes due to reduction in utilities such as steam, power, chilled water and cooling water, and/or production of steam, power and/or chilled water from WHR.
- Estimation of CUT and its changes requires unit prices of utilities involved, which depend on location and time. Approximate unit prices of utilities can be taken from books and/or other sources. Note that they can be significantly different from one source to another.

- Important concept of time value of money is related to rate of return on investment (popularly known as interest rate), and not to inflation or purchasing power.
- Time value of money (i.e., earning capability of money invested) implies present value of money is more than its value at a future time (i.e., future value). It is required since a project involves CFs in different years of project construction and operation (i.e., plant life).
- WHR projects require FCI for new equipment and/or changes in the existing plant, which incurs outflows (i.e., negative CFs) in the first 1–2 years of a project. Their benefit/gain (or profit) is from savings in utilities used and/or credit for utilities generated for use elsewhere. These are inflows (i.e., positive CFs) starting from the plant operation to end of plant life.
- PBP, NPV and IRR are the profitability criteria commonly used in industry. PBP is simple to understand but does not include time value of money. NPV and IRR include time value of money; the former requires the input of expected rate of return whereas the later computes rate of return from CFs in the project.
- PBP and IRR are generally positive whereas NPV can be negative, zero or positive depending on CFs in the project and rate of return. Small values of PBP correspond to high values of NPV and IRR.
- Projects with low PBP, high NPV and/or high IRR are economically attractive.
- Owing to substantial uncertainty in cost estimation, sensitivity of PBP, NPV and IRR to variation in FCI and unit prices of utilities should be analysed comprehensively.

Concepts, techniques and procedures for cost estimation and economic evaluation covered in this chapter are sufficient for assessing economics of WHR projects. They are relatively simple, and calculations can be performed using a calculator or a spreadsheet. However, lot of data is required for cost estimation, and some cost data may have substantial uncertainty. There are other cost estimation methods, types of depreciation, profitability criteria and so on, which are beyond the scope of this chapter. Chapters 7–10 in Turton *et al.*[3] are recommended for all these topics. Relatively more details of cost estimation and economic evaluation are described in Chapters 22 and 23 in Seider *et al.*[8] Terminology and presentation are slightly different in this book compared to those in Turton *et al.*[3]

References

1. Holland FA, Wilkinson JK. (1997) Process economics, section 9. In: RH. Perry, DW. Green, JO. Maloney (eds), *Perry's Chemical Engineers' Handbook*, 7th ed. McGraw-Hill, New York.
2. Couper JR, Hertz DW, Smith FL. (2008) Process economics, section 9. In: DW. Green, RH. Perry (eds), *Perry's Chemical Engineers' Handbook*, 8th ed. McGraw-Hill, New York.
3. Turton R, Bailie RC, Whiting WB, Shaeiwitz JA, Bhattacharyya D. (2013) *Analysis, Synthesis, and Design of Chemical Processes*, 4th ed. Pearson Education International, Singapore.
4. Dysert LR. (2004) Estimating, chapter 9. In: S. Amos (ed), *Skills & Knowledge of Cost Engineering*, 5th ed. AACE International.
5. AACE. (2005) *Cost Estimate Classification System — As applied In Engineering, Procurement, and Construction for the Process Industries,* AACE International Recommended Practice No. 18R-97. AACE. Available at https://web.aacei.org/docs/default-source/toc/toc_18r-97.pdf (accessed on 13 December 2021).
6. Vatavuk WM. (January 2002) Updating the CE Plant Cost Index. *Chemical Engineering*, pp. 62–70.
7. Guthrie KM. (1974) *Process Plant Estimating Evaluation and Control.* Craftsman Book Company of America, California.
8. Seider WD, Seader JD, Lewin DR, Widagdo S. (2010) *Product and Process Design Principles: Synthesis, Analysis, and Evaluation*, 3rd ed. International Student version, John Wiley.
9. Ulrich GD, Vasudevan PT. (2004) *Chemical Engineering Process Design and Economics — A Practical Guide*, 2nd ed. Process Publishing, New Hampshire.
10. Feng Y, Rangaiah GP. (August 2011) Evaluating Capital Cost Estimation Programs. *Chemical Engineering*, pp. 22–29.
11. Sinnott RK, Towler G. (2009) *Chemical Engineering Design*, 5th ed. Butterworth, Heinemann.

12. Ulrich GD. (1984) *A Guide to Chemical Engineering Process Design and Economics*. John Wiley, New York.
13. Turton R, Bailie RC, Whiting WB, Shaeiwitz JA. (2009) *Analysis, Synthesis, and Design of Chemical Processes*, 3rd ed. Prentice Hall, New Jersey.

Acronyms and Notation

Acronyms

BFW	Boiler Feed Water
BMC	Bare-Module Cost
CC	Capital Cost
CEPCI	Chemical Engineering Plant Cost Index
CF	Cash Flow
COL	Cost of Operating Labour
COM	Cost of Manufacture
CPI	Consumer Price Index
CRM	Cost of Raw Materials
CS	Carbon Steel
CUT	Cost of Utilities
CWT	Cost of Waste Treatment
FCI	Fixed Capital Investment
HE	Heat Exchanger
IC	Installed Cost
IRR	Internal or Investor's Rate of Return
LPS	Low-Pressure Steam
MCT	Module Costing Technique
MOC	Material of Construction
NPV	Net Present Value
OC	Operating Cost
OP	Operating Pressure
PBP	Payback Period
PC	Purchase Cost
PFD	Process Flow Diagram
P&ID	Piping and Instrumentation Diagram
MSECI	Marshall and Swift Equipment Cost Index
NPV	Net Present Value

SJE	Steam Jet Ejector
SS	Stainless Steel
STHE	Shell-and-Tube Heat Exchanger
TMC	Total Module Cost
TR	Tax Rate
WH	Waste Heat
WHR	Waste Heat Recovery

Notation

A	Heat transfer area in the HE (m^2)
F_M	Material factor (–)
F_P	Pressure factor (–)
P	Power extracted or shaft power (kW or HP) or pressure (barg)

Greek Symbols

Δ	Percent difference defined in Eq. 6.5 (–)

Subscripts/Superscripts

R	Reference

Exercises

6.1 A company's records show that two STHEs of CS were purchased/installed in the past. IC for one STHE with heat transfer area of 220 m^2 installed in 2014 was US$ 750,000 whereas IC for another STHE with heat transfer area of 150 m^2 installed in 2018 was US$ 610,000. The company plans to purchase and install the same type of HE with heat transfer area of 230 m^2 for a proposed WHR project. Estimate IC of this new HE assuming current CEPCI = 600. State other assumptions, if any.

6.2 Based on historical cost data given in the previous exercise, is it possible to estimate IC of a new STHE with heat transfer area of 180 m^2 and each of the three variations: (a) MOC is SS instead of CS; (b) OP is less than that of STHEs purchased in the past and (c)

OP is more than that of STHEs purchased in the past? Give reasoning for your answer to each variation.

6.3 A proposed WHR project requires two reboilers; PC, BMC and TMC of one of them are computed in Example 6.4. Specifications of another reboiler are STHE of U-tube type, area = 796 m^2, CS for both shell and tubes, design pressure is 21 barg for tubes and 12 barg for shell. Find PC, BMC and TMC for this reboiler. Assume CEPCI = 600. State other assumptions, if any.

6.4 A sum of $100,000 is invested for 7 years at an annual return of 6% (i = 0.06). Find the amount after 7 years for three cases: simple, annual compound and continuous compound interest. Repeat this for i = 0.8. Compare the obtained results with those in Example 6.8.

6.5 Earlier in this chapter, Example 6.9 is solved using present values. Instead, use future values to find the interest that the bank is charging on the loan, and compare it with that found in Example 6.9.

6.6 For the WHR project to improve steam network in a petroleum refinery (Case C in Chapter 16), FCI has increased by ≈20% to $920,000 (from $766,256) and benefit (from increased power generation) has decreased by ≈25% to 507,540/year (from $676,723/year). Find PBP, NPV (for i = 0.2) and IRR for this project. Assume a plant life of 10 years, straight-line depreciation and zero salvage value for the equipment and changes made for WHR (using FCI), and TR of 30% (0.3) on benefit.

6.7 For the WHR project to improve steam network in a petroleum refinery (Case C in Chapter 16), FCI has increased by ≈40% to $1,073,000 (from $766,256) and benefit (from increased power generation) has decreased by ≈50% to 338,362/year (from $676,723/year). These changes are more than those in the previous exercise. Find PBP, NPV (for i = 0.2) and IRR for this project. Assume a plant life of 10 years, straight-line depreciation and zero salvage value for the equipment and changes made for WHR (using FCI), and TR of 30% (0.3) on benefit. Compare PBP, NPV and IRR found in this exercise with those in the previous exercise.

6.8 Addition of a mechanical vapor re-compressor (MVR) is proposed for improving energy efficiency of a batch distillation unit. Two options are available for this. Option 1 is for a compressor with FCI of \$454,000 to reduce CUT by \$66,000/year. Option 2 is for a compressor with FCI of \$ 297,000 to reduce CUT by \$\$54,000/year. Assume FCI is incurred initially at time = 0 years, compressor life and salvage value are respectively 7 years and zero, straight-line depreciation and TR = 0.3. Find PBP, NPV (for $i = 0.1$) and IRR for each option. Which option is better for each of these profitability criteria? Does the chosen option change with the profitability criterion?

6.9 Answer the following with yes (right) or no (wrong), and give brief reasoning, where applicable.

(A) PC of an equipment is comparable to its IC.
(B) Uncertainty in cost estimation of an equipment is substantial.
(C) COM does not depend on the amount of product(s) produced.
(D) COM is comparable to cost of utilities consumed.
(E) COM includes depreciation of equipment.
(F) Depreciation of an equipment is desirable for companies/investors.
(G) Time value of money is due to inflation (or purchasing power).
(H) All CFs are positive in all projects.
(I) PBP is independent of expected rate of return on investment.
(J) NPV can be negative, zero or positive.
(K) Large PBP corresponds to large positive NPV.
(L) Large positive NPV indicates large IRR.

Chapter 7

Estimation of Energy and CO_2 Emissions

7.1 Overview

Process industries use primary energy (i.e., original/unconverted fuels such as natural gas [NG], coal, biomass, solar, wind and nuclear energy) and secondary energy (i.e., converted or stored energy such as electricity, fuel oil and heat) for producing valuable products for the society. Waste heat recovery (WHR) reduces primary and/or secondary energy used in and CO_2 emissions from process industries. Estimation of this reduction in energy used and CO_2 emissions requires holistic and correct assessment of energy used, which is the focus of this chapter.

Various units/equipment in process industries require utilities such as electricity, steam, cooling water (CW) and/or chilled water (ChW), which are produced either internally and/or purchased from suppliers. Often, steam is produced internally, whereas electricity is purchased from power generation companies via electricity grid. In either case, energy is required for producing the utilities. Often, waste heat (WH) can be utilized to produce electricity (covered in Chapter 10), steam and/or ChW (covered in Chapter 5). This contributes to reducing primary and/or secondary energy used in the plant and consequently CO_2 emissions from the plant.

The common utilities in process industries are electricity (mainly for motors of pumps and compressors) and steam (mostly for heating), and CW and ChW (for cooling purposes). Other utilities include instrument air, plant air (for purging, ventilation and cleaning) and nitrogen (for purging). Production of these utilities requires primary and/or secondary energy. This chapter outlines the production of common utilities, and then describes primary and secondary energy required for and CO_2 emissions in their production. It begins with Section 7.2 on the primary and secondary energy sources utilized in the World. Section 7.3 outlines the production process of electricity, steam, CW and ChW. Sections 7.4 and 7.5 describe, respectively, the energy required for and CO_2 emissions in producing unit quantity of utilities. Finally, Section 7.6 summarizes the main points for correctly estimating primary and secondary energy required and CO_2 emissions due to consumption of utilities in the plant.

The learning outcomes of this chapter on estimation of primary and secondary energy and CO_2 emissions are as follows:

1. Outline the production of electricity, steam, CW and ChW.
2. Describe primary/secondary energy required for producing common utilities.
3. State CO_2 emissions in producing common utilities.
4. Calculate total energy and CO_2 emissions for given quantities of utilities.

7.2 Primary and Secondary Energy Sources

Process industries, power plants, commercial/residential buildings, transportation and other sectors utilize one or more primary and/or secondary energy sources for different purposes including production of utilities. Table 7.1 presents sources of energy consumed in the world, in 2018 and 2019.[a] Note that EJ stands for Exajoules (= 1×10^{18} Joules = 1×10^{12} MJ). Although changes in energy consumption and sources are small from 2018 to 2019, contribution of renewables (currently at ~5%) such as solar and wind power will increase in the coming years, whereas contribution of coal is expected to decrease. Currently, oil (crude oil/petroleum), NG and coal account for nearly 85% of the energy consumption in the world.

[a] https://www.bp.com/en/global/corporate/energy-economics/statistical-review-of-world-energy.html (accessed on 6th February 2021)

Table 7.1: World energy consumption in 2018 and 2019 by energy source

	Energy Consumption in 2018		**Energy Consumption in 2019**	
Energy Source	**Quantity (EJ)**	**Contribution (%)**	**Quantity (EJ)**	**Contribution (%)**
Oil (crude oil/petroleum)	191.45	33.2	193.03	33.1
Natural gas (NG)	138.66	24.1	141.45	24.2
Coal	158.79	27.6	157.86	27.0
Hydro	37.3	6.5	37.66	6.4
Renewables (excluding hydro)	25.83	4.5	28.98	5.0
Nuclear	24.16	4.2	24.92	4.3
Total	**576.19**	**100.0**	**583.90**	**100.0**

Oil, NG and coal have several uses. Most of the oil is currently used for transportation fuels; and the rest of it (e.g., in the form of fuel oil and liquefied petroleum gas) for other purposes such as heating, cooking and utilities production. In 2019, NG uses in United States are for electric power generation (~36%), in industries for heating and as raw material to produce chemicals, fertilizer and hydrogen (~33%), and in residential and commercial sectors for heating and cooking (~27%).[b] Of the coal used in the world, about 60% is for electricity generation,[c] and 20% is in steel, cement and other industries.

In summary, based on Table 7.1 and different uses of oil, NG and coal, the two main sources of primary energy for the production of electricity and steam are NG and coal. The production of CW and ChW utilizes electricity produced from NG, coal or other energy sources (e.g., burning waste streams in plants, renewable energy sources).

Electricity is the most common secondary energy in the world. Primary energy sources for its generation in 2018 and 2019 are compiled

[b] https://www.eia.gov/energyexplained/natural-gas/use-of-natural-gas.php (accessed on 6th February 2021).

[c] https://www.iea.org/reports/global-energy-review-2020/coal#abstract (accessed on 6th February 2021)

Table 7.2: Primary energy source for electricity generation in the world in 2018 and 2019; one terra Watt hour, TWh = 1 × 10^9 kWh = 0.0036 EJ

	Electricity Generation in 2018		Electricity Generation in 2019	
Source of Primary Energy for Electricity	**Quantity (TWh)**	**Contribution (%)**	**Quantity (TWh)**	**Contribution (%)**
Oil	890.4	3.3	825.3	3.1
Natural gas (NG)	6082.5	22.8	6297.9	23.3
Coal	10,091.3	37.9	9824.1	36.4
Hydro	4171.4	15.7	4222.2	15.6
Renewables (excluding hydro)	2468.0	9.3	2805.5	10.4
Nuclear	2700.4	10.1	2796	10.4
Others	248.9	0.9	233.6	0.9
Total	26,652.9	100.0	27,004.6	100.0

in Table 7.2.[d] Similar to variation in energy consumption and sources in Table 7.1, changes in the contribution of different energy sources for electricity generation from 2018 to 2019 are small. However, in the coming years, electricity generation from renewable sources (i.e., solar and wind power) will increase, whereas that from coal will decrease, due to environmental concerns and technological developments. Currently, NG and coal account for around 60% of the total electricity generation in the world (Table 7.2).

Other secondary energies are thermal energy for heating, oil products (e.g., liquefied petroleum gas, gasoline, diesel, jet fuel and fuel oil), biofuels (e.g., bioethanol and biodiesel) and hydrogen. Hydrogen can be produced from oil and NG (e.g., by steam reforming) as well as by electrolysis using electrical energy.

7.3 Production of Common Utilities

The following sub-sections describe the production of electricity, steam, CW and ChW. They also cover (primary) energy required for producing

[d] https://www.bp.com/en/global/corporate/energy-economics/statistical-review-of-world-energy.html (accessed on 6th February 2021)

these utilities; this energy includes that in the fuel used (e.g., coal, NG or biomass) but excludes that involved for building utility production facilities and for manufacturing other materials (e.g., chemicals for water treatment in steam and CW production) used in producing the utilities. Thus, the stated energy for producing the utilities is a slight under-estimation (by <5%) of all (both direct and indirect) energy required.

7.3.1 *Generation of Electricity*

As stated earlier, around 60% of total electricity generation in the world is using coal and NG as fuel in thermal power plants. In these plants, chemical energy in fuels is converted into thermal energy in a combustion chamber (known as boiler or furnace), which is then converted into mechanical energy (or shaft work) in a turbine and finally into electrical energy in a generator. In coal-based power plants, steam is produced and then used in a steam turbine to rotate the electromagnetic shaft (known as rotor) of the generator, which produces electricity. This is also the case in some (older) NG-based power plants.

NG-based power plants often use a combined cycle (e.g., Brayton cycle in a gas turbine and Rankine cycle in a steam turbine) for conversion of thermal energy to shaft work. Schematic of a combined cycle power plant is shown in Fig. 7.1[1]; this is a co-generation plant using NG to

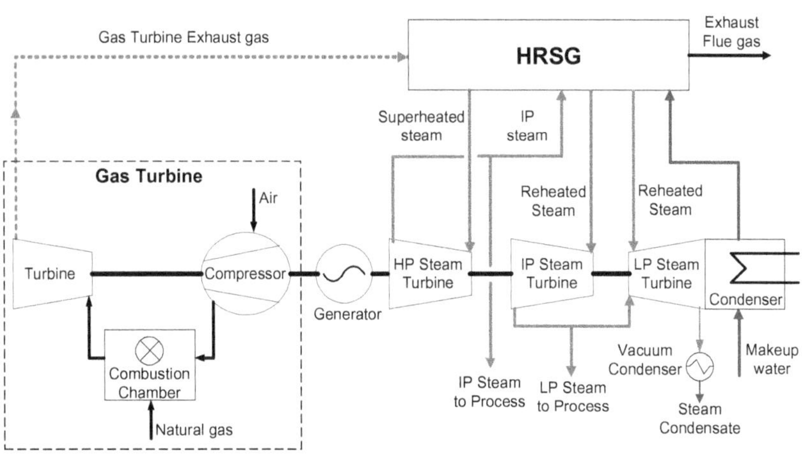

Fig. 7.1: Schematic of a combined cycle co-generation plant.

produce electric power and steam. This plant consists of a gas turbine, compressor (for combustion air), generator (producing electricity) and several steam turbines, all on a common shaft. Gas/steam turbines produce shaft work to drive both the compressor and generator on the common shaft. In addition, the power plant has several heat exchangers to recover thermal energy from hot flue gases leaving the gas turbine and use it for producing high-/medium-/low-pressure steam; these exchangers are part of the heat recovery steam generator (HRSG) block in Fig. 7.1. See Huichao *et al.*[1] for more details of this combined cycle power plant including its operating conditions, modelling and validation.

Efficiency of thermal power plants has been improving marginally over the years. Based on the heat rate (i.e., amount of thermal energy used by a power plant to generate *net* electricity of 1 kWh *after* deducting electricity consumed to operate fuel and boiler water pumps, cooling equipment, pollution control devices, etc.),[e] efficiency of coal- and NG-based power plants in the year 2020 is, respectively, 32.3% and 44.1%. These efficiency values are based on net electricity produced at the generation plant, and do not include losses in the subsequent transmission and distribution of electricity.

Assuming 5% of losses in transmission and distribution of electricity (from the power plant to consumers), we calculate the following efficiencies of electrical energy reaching the plant (i.e., consumer):

$$\text{Efficiency from a coal-based power plant} = 0.95 \times 32.3 = 30.7\% \quad (7.1)$$

$$\text{Efficiency from an NG-based power plant} = 0.95 \times 44.1 = 41.9\% \quad (7.2)$$

This means only 30.7% or 41.9% of the chemical energy in coal or NG, respectively, ends up as electrical energy used by the plant. The rest of the chemical energy goes out in the form of various losses (in flue gas, steam generation, gas/steam turbines converting thermal energy to shaft work, generator producing electrical energy from shaft work, electricity transmission and distribution) and for operating the power plant.

---

[e] https://www.eia.gov/tools/faqs/faq.php?id=107&t=3 (accessed on 18th April 2021), and Table A6 from https://www.eia.gov/totalenergy/data/browser/index.php?tbl=TA6#/?f=A&start=1949&end=2019&charted=5-6-7-8

In summary,

Primary energy in coal required for one GJ of electrical energy = $\frac{100}{30.7}$ = 3.257 GJ (7.3)

Primary energy in NG required for one GJ of electrical energy = $\frac{100}{41.9}$ = 2.387 GJ (7.4)

Subsequent discussion in this chapter assumes that these values of primary energy and efficiency of electrical energy supplied are based on lower heating value (LHV, also known as net calorific value) of the fuel.

It is necessary to be clear on the assumption of heating value used. As the terms indicate, higher heating value (HHV, also known as gross calorific value) is greater than LHV by ≈5% and ≈10% in the case of coal and NG, respectively.[f] Industry practice for reporting efficiency of boilers, furnaces and power plants is based on LHV of fuel used, perhaps because it leads to larger values for efficiency – though this is not universal, as can be seen from Gulen.[2] However, as noted by Gulen, the supply of NG is based on HHV, which is better for sellers. Readers should consistently use one or the other carefully for interpreting and comparing efficiency values reported in various sources, as well as for analysis of their own application. Finally, it is not clear whether the heat rate data given at https://www.eia.gov/tools/faqs/faq.php?id=107&t=3 (accessed on 18 April 2021) are based on HHV or LHV of fuel.

7.3.2 *Steam Production*

In industries, steam is produced in the utilities section (also known as offsites) of the plant and then distributed to process consumers (e.g.,

---

[f] HHV is the amount of heat released in the process of complete combustion of fuel initially at 25°C and cooling combustion products to 25°C. Hence, it includes latent heat of condensation of water vapor formed by combustion of hydrogen present in the fuel. On the other hand, LHV is the amount of heat released by the complete combustion of fuel initially at 25°C and after cooling combustion products to 150°C only. So, water vapor formed does not change phase and LHV excludes latent heat of condensation of water vapor formed. Both HHV and LHV vary with fuel and its composition (including carbon and moisture contents); their typical values are in given later in Table 7.3.

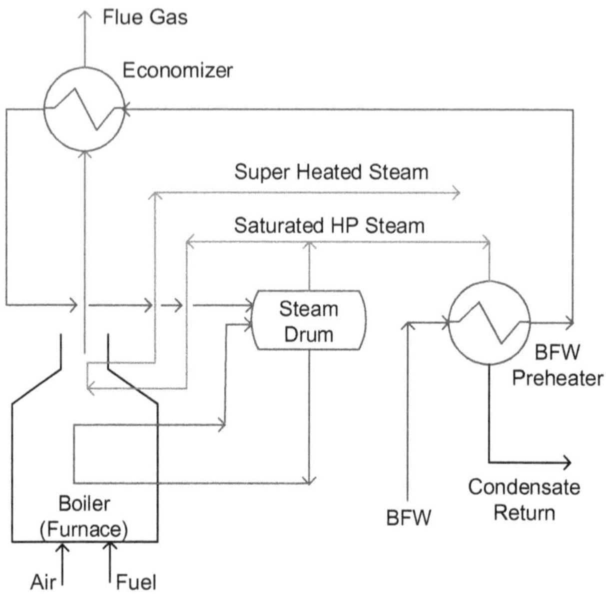

Fig. 7.2: Typical utility boiler (furnace) with economizer, BFW preheater and using NG.

reboilers, strippers, for atomizing oil, heaters requiring steam for heating, velocity steam to reduce residence time in fired heaters, steam jet ejectors and cleaning purposes) as required. Generally, superheated steam at high pressure (above 40 bar) is produced in boilers/furnaces using fuel, which can be primary energy (e.g., NG or coal) or secondary energy (e.g., fuel oil or waste streams). WH boilers also generate steam in some process plants. In the boiler (Fig. 7.2), boiler feed water (BFW), which is specially treated water to remove dissolved solids and gases that can damage boiler tubes, is heated, vaporized and superheated to produce superheated steam. A plant may have more than one boiler depending on the required capacity, and some plants may produce very high pressure (above 100 bar) steam. The boiler shown in Fig. 7.2 is using NG fuel; usually, it has an economizer (and not an air preheater). In one installation of a NG-fired boiler, the economizer improves the boiler efficiency from 84% to 93%. On the other hand, an air preheater (and not an economizer) is common in coal-fired boilers, as described in Section 11.5.

The efficiency of a boiler (furnace or fired heater) refers to the fraction of energy of fuel that goes into steam (i.e., for water heating, vaporizing and steam superheating). It depends on the fuel itself and the use of WHR equipment such as air preheater and economizer in the boiler. Efficiency of a boiler using NG fuel is 92%–94%, whereas that of a boiler using coal as the fuel is 88%–90%. These values are based on LHV, as per industry practice, and they will be lower by 5%–10% in case HHV is used instead of LHV. Reasons for ≈4% difference in the efficiency of NG and coal-fired boilers are as follows. First, HHV/LHV of coal is about half of that of NG (Table 7.3) and so more coal (mass) is required per GJ of energy; although air (mass) required for complete combustion of unit mass of coal is lower than that for NG, overall more combustion air (mass) is required per GJ of energy from most coals. Further, higher percentage excess air is used in a coal-fired boiler for ensuring complete combustion. Consequently, there will be more flue gas and heat loss with it. Second, some heat is lost with the fly/bottom ash leaving the boiler. Third, energy is needed for the operation of pollution control devices in case of coal-fired boilers.

A typical plant uses steam at high pressure (HP at ~40 bar), medium pressure (MP at ~12 bar) and low pressure (LP at ~5 bar). Medium- and low-pressure steam (MPS and LPS) are produced by lowering the pressure of steam generated at high pressure in the boiler(s), through steam turbines (to produce electric power) or through let-down valves (which

Table 7.3: HHV and LHV of selected fuels (https://h2tools.org/hyarc/calculator-tools/lower-and-higher-heating-values-fuels, accessed on 18th April 2021)

Fuel	LHV (MJ/kg)	HHV (MJ/kg)	Difference between HHV and LHV (%)
Hydrogen	119.96	141.88	16.74
Natural gas	47.13	52.21	10.23
Propane	46.28	50.22	8.17
Butane	45.27	49.20	8.32
Crude oil	42.68	45.53	6.46
Bituminous coal (wet basis)	26.12	27.26	4.27
Coal (wet basis)	22.73	23.96	5.27

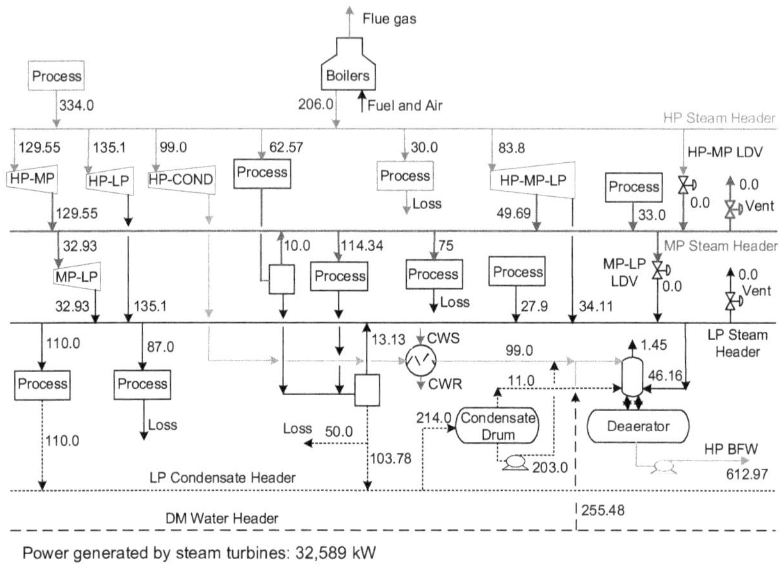

Fig. 7.3: Schematic of the steam network; HP-MP-LP is the steam turbo-generator and values (without units) are steam flow rates in ton/h. A blowdown rate of 2% (of steam generated at the boilers and process steam generators) is considered for calculating HP BFW supply from the Deaerator.

are not economical). A typical steam network in a plant is a complex system as shown in Fig. 7.3. Besides boilers (whose details are shown in Fig. 7.2), it has (i) HP steam (HPS), MPS and LPS headers, (ii) HPS produced by boilers and process units, (iii) MPS produced by some other process units, (iv) many turbines producing MPS and/or LPS from HPS or MPS, respectively, (v) HPS/MPS/LPS supply to diverse process consumers, (vi) deaerator and (vii) condensate drum. In particular, Fig. 7.3 has a HP-MP-LP turbo-generator producing electric power, MPS and LPS from HPS. Additionally, if the site has a cogeneration plant (such as that in Fig. 7.1), HRSG can produce HPS, MPS and/or LPS. Some steam turbines produce electric power for use in the plant, whereas others drive compressors/pumps within the plant.

A deaerator uses LPS to remove dissolved gases in the returned condensate and makeup demineralized (DM) water to produce BFW. Steam condensate (SC) recovery from process consumers is important for energy efficiency since significant thermal energy is lost with SC not returned for steam production. In general, SC returned is at ~115°C and not all SC is returned to the steam network for several reasons (e.g., steam losses in steam traps and relatively high cost of SC return from process consumers located far away from the boiler in the utilities section). SC return of more than 70% is considered good. In addition, some water and so thermal energy is lost with the blowdown from boiler mud drums, to remove solids/sludge in the recirculating SC. Required quantity of DM water makeup (to deaerator) is supplied at room temperature. More details on the steam network system, condensate recovery, WHR from them and their optimization are covered later in Chapters 16 and 17.

Typical energy inflows and outflows (all in GJ) of a steam production and distribution system are given in Table 7.4. Basis for these values is 1 GJ of LHV (chemical energy) of fuel to the boiler/furnace. Two cases, namely, NG and coal as the fuel, are covered in Table 7.4. Main difference between these two cases is in more heat loss in the boiler itself (i.e., efficiency of 92.5% for NG-fired boiler compared to 88.5% for coal-fired boiler due to items 2–5 in Table 7.4). Reasons for this are already stated earlier.

In addition to heat losses in the boiler (items 2–5), there are many other energy/heat losses (items 7–12 in Table 7.4) in the steam distribution and SC return system. These losses are applicable for process industries having a centralised steam production (in utilities section of the plant) and steam distribution (requiring headers, piping, flanges and steam traps) to many process operations, and are unaffected by the fuel used in the boiler. Energy losses in items 7–12 of Table 7.4 are estimated using percentages from Paffel[3] and/or are based on the authors' experience. These percentages assume good design (e.g., sufficient insulation and correct sizing) and regular maintenance of steam production and distribution system. Obviously, steam losses vary from one plant to another depending on the scope of the steam network, process operations involved, design and maintenance practices. For example, percentage values for items 7–12 of Table 7.4 may be lower or negligible for power plants and for smaller plants with a few process operations consuming steam.

It is clear from Table 7.4 that energy losses in the steam distribution and condensate recovery system are not small, and that boiler efficiency does not account for them. For example, 92.5% of primary energy in NG

Table 7.4: Energy supply/loss in large steam generation and distribution systems

Item No.	Energy/Heat Supply/Loss	Energy, GJ NG Fuel	Energy, GJ Coal Fuel	Comments
1	**Primary energy (in fuel)**	1.000	1.000	LHV of NG or coal
2	Loss in flue gas	0.070	0.090	Boiler is assumed to have an economizer or air preheater
3	Loss in fly/bottom ash		0.010	In case of coal as fuel
4	Loss due to pollution control devices		0.010	In case of coal as fuel
5	Loss to ambient from boiler surface	0.005	0.005	Estimated as 0.5% of energy in fuel
6	**Energy in steam produced (ESP)**	**0.925**	**0.885**	This is the boiler efficiency. It is also the energy in steam going to the steam distribution system.
7	Loss in boiler blowdown	0.0185	0.0177	Estimated as 2% of ESP
8	Loss to ambient from insulated piping carrying steam and SC	0.0139	0.0133	Estimated as 1.5% of ESP
9	Loss due to steam leaks in steam distribution piping and flanges	0.0093	0.0089	Estimated as 1% of ESP
10	Loss due to stream trap issues	0.0093	0.0089	Estimated as 1% of ESP
11	Loss due to unrecovered condensate	0.0352	0.0336	Estimated as 3.8% of ESP
12	Loss in steam vented* to atmosphere	0.0093	0.0089	Estimated as 1% of ESP
13	**Energy in steam used by (available to) process units**	**0.8297**	**0.7938**	Estimated by deducting sum of losses (items 7–12) from ESP
14	Electrical energy for combustion air blower	0.0171	0.0180	Based on Example 8.6 in Turton *et al.*[4] for a boiler using NG. Slightly more in case of coal as fuel
15	Electrical energy for BFW pump	0.0021	0.0021	For pressure increase of 45 bar

*In condensate/flash tanks, deaerators, automatic steam vents, etc.

goes out as energy in steam produced (ESP); in other words, boiler efficiency is 92.5%. Owing to various energy losses in the steam distribution and condensate recovery system, only 82.97% of primary energy in NG is used in process units, predominantly as latent heat of condensation of steam. This is the overall efficiency of steam generation, distribution and condensate recovery system.

Considering NG fuel in Table 7.4:

Primary energy in NG required to make one GJ of steam energy available to process units

$$= \frac{1.0}{0.8297} = 1.205\,\text{GJ} \tag{7.5a}$$

Electrical energy required (items 14 and 15 in Table 7.4) to make one GJ of steam energy available to process units

$$= 1.205 \times (0.0171 + 0.0021) = 0.02314\ \text{GJ} \tag{7.5b}$$

From Eq. 7.4, primary energy in NG required to make 0.02314 GJ of electrical energy

$$= 0.02314 \times 2.387 = 0.0552\ \text{GJ} \tag{7.5c}$$

Total primary energy in NG required to make one GJ of steam energy available to process units

$$= 1.205 + 0.0552 = 1.2602\ \text{GJ} \tag{7.5d}$$

Considering coal fuel in Table 7.4:

Primary energy in coal required to make 1 GJ of steam energy available to process units

$$= \frac{1.0}{0.7938} = 1.260\,\text{GJ} \tag{7.6a}$$

Electrical energy required (items 14 and 15 in Table 7.4) to make one GJ of steam energy available to process units

$$= 1.260 \times (0.018 + 0.0021) = 0.0253\ \text{GJ} \tag{7.6b}$$

From Eq. 7.3, primary energy in coal required for to make 0.0253 GJ of electrical energy

$$= 0.0253 \times 3.257 = 0.0825 \text{ GJ} \tag{7.6c}$$

Total primary energy in coal required to make one GJ of steam energy available to process units

$$= 1.260 + 0.0825 = 1.3425 \text{ GJ} \tag{7.6d}$$

The above values are applicable for HPS, which is generally produced in steam boilers. MPS and LPS are produced from HPS by two different ways: using HPS through steam turbines, which produce shaft/electric power, and by pressure reduction of HPS through let-down valves. The values in Eqs. 7.5 and 7.6 can be used for MPS and LPS as well with the assumption that MPS and LPS are produced from HPS through let-down valves since no energy is lost in the pressure reduction step.

Table 7.4 includes all energy supplies/losses in a steam production/distribution/condensate network. However, percentage values used for losses in Table 7.4 may vary from one plant to another. Hence, readers can update percentage values in Table 7.4, as per their plant conditions and experience, to estimate primary energy required for unit quantity of steam consumed in process operations of the plant.

Many studies assume efficiency of 90% for steam and estimate primary energy. However, this percentage ignores energy/steam losses in the steam distribution and condensate recovery system, which amount to about 10% of ESP. Hence, one should use overall efficiency of ~80% for a typical steam network such as that shown in Fig. 7.4. Moreover, steam production requires electrical energy for combustion air blower and BFW pump (items 14 and 15 in Table 7.4). This electrical energy is about 0.02 GJ per GJ of ESP. Recall from Eqs. 7.3 and 7.4, 1 GJ of electrical energy requires 2.4 to 3.3 GJ of primary energy. This contribution of electrical/primary energy to steam used in the plant is often neglected. All these lead to under-estimation of primary energy used and consequent CO_2 emissions due to steam used in the plant. CO_2 emissions is described with examples later in this chapter.

Based on steam balance, Gadalla *et al.*[5] presented the following equation for steam production and its use in the process.

$$Q_F = Q_P \times \frac{(h_p - 419)}{\lambda_p} \times \frac{(T_{FT} - T_0)}{T_{FT} - T_{stack}} \quad (7.7)$$

Here, Q_F is the amount of fuel energy (say, kW) required to provide heat duty of Q_P (kW) to the process, h_P and λ_P are, respectively, the enthalpy and latent heat of steam delivered to the process, and T_{FT}, T_0 and T_{Stack} are the temperature of flame in the boiler/furnace, ambient and flue gas going to the stack, respectively. The ratio of Q_P to Q_F is the fraction of fuel energy finally used in the process via steam, and so it is the efficiency of the steam generation and distribution network (and not that of the boiler producing steam alone).

Enthalpy of BFW at 100°C is taken to be 419 kJ/kg. Suggested values for T_{FT} and T_{Stack} are 1800°C and 160°C, respectively.[5] For MPS at 10.5 barg and 220°C (superheated), h_P and λ_P are, respectively, 2868 kJ/kg and 1992 kJ/kg (from steam tables). Assume ambient temperature, T_0 of 30°C. For these values, the ratio of Q_P to Q_F is

$$\frac{Q_P}{Q_F} = \frac{1992}{(2868 - 419)} \times \frac{(1800 - 160)}{(1800 - 30)} = 0.7537 \quad (7.8)$$

In other words, steam generation/distribution network efficiency is 75.37%. This value is lower than 82.97% (of primary energy in NG used in process units) based on energy supply/losses in Table 7.4. The latter becomes ≈80% after including electric energy required for blower and pumps. On the other hand, the estimated efficiency of 75.37% depends on the values assumed for several quantities in Eq. 7.7. In summary, steam network efficiency found via detailed energy supply/losses in Table 7.4 is comparable to that by Eq. 7.7.

The focus of this sub-section is on energy losses in a steam generation/distribution network (Fig. 7.4) and not their cost implications. Operating costs for energy (boiler fuel and electricity), makeup BFW and chemicals for BFW treatment, in steam production are illustrated in Example 8.6 in Turton *et al.*[4] This example indicates that energy cost for fuel and electricity is about 98% of total operating cost. It does not consider other operating costs such as for labor and maintenance. Further, capital cost for setting up steam production unit is not included, and it is assumed to be included with that for the rest of the plant.

7.3.3 *Cooling Water Production*

In industries, the CW circuit is often closed loop for both economics and water conservation. It consists of one or more cooling towers, CW pumps and coolers (i.e., heat exchangers for cooling process streams), as shown in Fig. 7.4. CW supply at ~30°C is pumped by motor-driven pump(s) through many coolers (in parallel and/or series) in the plant, where it picks up thermal energy from process streams. Warm CW at 40 to 45°C is returned to the cooling tower, where ambient air is used to cool the water, to close to the surrounding temperature, for recirculation. A small amount of water (evaporated for cooling the warm CW and carried away by the wind) is lost in the cooling tower. Further, a small amount of CW is purged from the cooling tower basin, to avoid accumulation of dissolved solid components in water. To compensate for both these losses, makeup

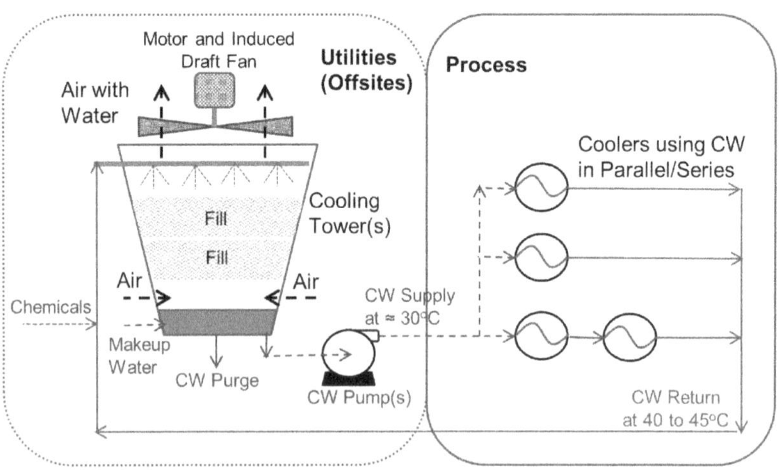

Fig. 7.4: Recirculating CW system with induced-draft cooling tower; there will be many coolers but only four are shown for simplicity.

water is required (Fig. 7.4). Chemicals to reduce fouling in the cooling tower and coolers are also added to the recirculating CW.

In the CW circuit, thermal energy from process streams (to cool them as required) is removed by CW in coolers (which are essentially heat exchangers), and then it is transferred from warm CW to ambient air in the cooling tower. The energy required for this heat transfer (i.e., cooling of process streams) is mostly for the cooling tower fan (to push or pull air through the cooling tower) and CW pumps (to overcome pressure drops in piping, control valves and coolers as well as static heads involved). Note that CW return must flow to the top of the cooling tower (5–10 m from the ground level). The cooling tower fan and CW pumps are usually driven by motors, and hence they require electric power.

Electrical energy required for the cooling tower fan and CW pumps as well as operating costs involved for recirculating CW (i.e., for electric power, makeup water and chemicals) are estimated in Example 8.3 in Turton *et al.*[4] According to this example, electric power required for removing 1 GJ/h of thermal energy by recirculating CW is given as follows:

Power required for CW pumps (for assumed pressure drop of 266.7 kPa)

$$= 2.36 \text{ kW/(GJ/h)} \tag{7.9a}$$

Power required for cooling tower fan (for required tower cross-sectional area)

$$= 1.61 \text{ kW/(GJ/h)} \tag{7.9b}$$

Total electric power required

$$= 2.36 + 1.61 = 3.97 \text{ kW/(GJ/h)} \tag{7.9c}$$

Since 1 GJ/h = 0.2778 MW, total electrical energy required for removing 1 MJ of thermal energy from process streams

$$= \frac{3.97}{0.2778} = 14.29 \text{ kJ} \tag{7.9d}$$

Or, total electrical energy required for removing 1 GJ of thermal energy from process streams

$$= 0.01429 \text{ GJ} \quad (7.9e)$$

Operating cost calculations in Example 8.3 in Turton *et al.*[4] indicate that electrical energy, makeup water and chemicals costs are respectively ~67%, ~23% and ~10% of CW cost (i.e., operating cost to cool warm CW in the cooling tower and recirculate CW). These do not include capital cost of the cooling tower, CW pumps and piping. Coolers are essentially process equipment/units in the main plant and hence their capital cost is considered together with that of other process units.

Existing CW systems such as that in Fig. 7.4 can be retrofitted by changing some of the coolers from parallel to series arrangement, to minimize cost. This topic is covered in Chapter 15.

7.3.4 *Chilled Water Production*

Like the CW circuit, the ChW circuit is also closed loop. It consists of a refrigeration cycle/system (instead of the cooling tower in the CW circuit), ChW pump and coolers using ChW in parallel and/or series (Fig. 7.5). The refrigeration cycle consists of an evaporator (wherein ChW is cooled to ~5°C), compressor, condenser (wherein CW is used to remove thermal energy) and expansion valve (to reduce pressure). It is another closed loop through which a refrigerant such as R-152a (1,1 difluoroethane) circulates. The refrigerant goes through vaporization in the evaporator (thus removing thermal energy from ChW return and cooling ChW to ~5°C), gas compression in the compressor, condensation at a higher temperature in the condenser (thus transferring thermal energy from refrigerant to CW) and partial vaporization during pressure reduction in the throttle/ expansion valve.

The refrigeration system of a ChW circuit (Fig. 7.5) is usually located near the process unit(s) requiring ChW. This is to avoid heat gain from the ambient, water vapor condensation and hence corrosion on the piping carrying cold streams. On the other hand, the cooling tower and its

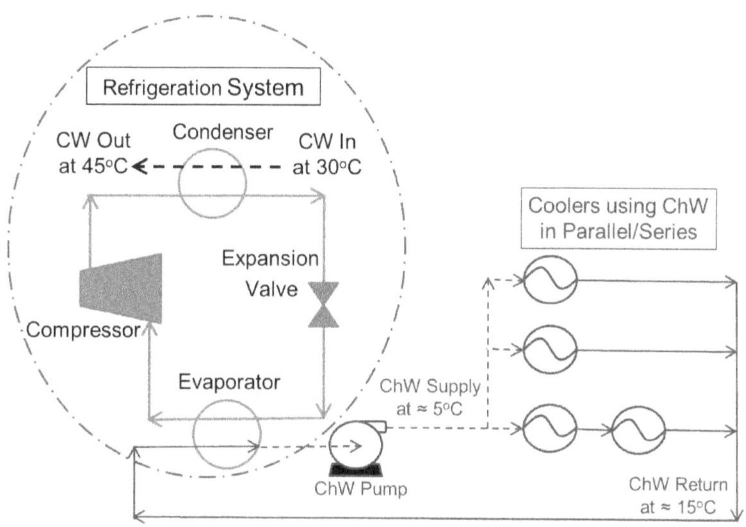

Fig. 7.5: Schematic showing ChW circuit and associated refrigeration system; there may be many coolers but only four are shown for simplicity.

peripherals of CW circuit are generally in the utilities (offsites) section of the plant.

For operating ChW circuit, electric power is required for motors driving the compressor and ChW pump. This can be calculated as follows for removing 1 GJ/h of heat by ChW in coolers (from example 8.5 in Turton *et al.*[4] and compressor efficiency of 75%).

Compression power (shaft work) for 1 GJ/h evaporator duty

$$= 66.5 \text{ kW} \tag{7.10a}$$

Coefficient of performance (COP[g])

$$= \frac{277.8}{66.5} = 4.18 \tag{7.10b}$$

[g] Defined as the ratio of heat removed in the evaporator to the work done (i.e., compressor shaft power).

Assuming motor efficiency of 95%, electric power required for the compressor

$$= \frac{66.5}{0.95} = 70.0 \text{ kW} \tag{7.10c}$$

Electric power required for ChW pump (see Example 7.1)

$$= 1.857 \text{ kW} \tag{7.10d}$$

Total electric power required for 1 GJ/h of cooling by ChW

$$= 70.0 + 1.857 = 71.857 \text{ kW} \tag{7.11a}$$

Since 1 GJ/h = 277.8 kW, total electric power required for 1 kJ of cooling by ChW

$$= \frac{71.857}{277.8} = 0.2587 \text{ kJ} \tag{7.11b}$$

Or, total electric power required for 1 GJ of cooling by ChW

$$= 0.2587 \text{ GJ} \tag{7.11c}$$

Example 7.1: Estimate electric power required for ChW pump for removing 1 GJ/h of energy from process streams. Assume pump efficiency of 75% and motor efficiency of 95%. Total pressure drop of ChW in the closed circuit (i.e., in the piping, valves and coolers) is 200 kPa. State other assumptions, if any, required for calculations.

Solution: Assume ChW is supplied at 5°C and returned at 15°C, density of ChW is 1000 kg/m^3 and heat capacity of ChW is 4.2 kJ/kg.K.

ChW flow rate for removing one GJ/h $= \frac{10^6 \text{ kJ/h}}{4.2\left(\frac{\text{kJ}}{\text{kgK}}\right)\times(15-5)°\text{C}} = 23{,}809 \text{ kg/h}$

$= 6.614 \text{ kg/s} = 0.006614 \text{ m}^3\text{/s}$

Pump shaft power $= \frac{0.006614 \frac{\text{m}^3}{\text{s}} \times 200 \text{kPa}}{0.75} = 1.764 \text{kW}$

Electric power $= \frac{1.764}{0.95} = 1.857 \text{kW}$

As shown in Fig. 7.5, CW is also required for ChW production. The total electric power requirements including that for CW can be calculated as follows.

For CW System:

From Example 8.5 in Turton *et al.*[4] CW (condenser) duty required for 1 GJ/h of ChW duty

$$= 1.24 \text{ GJ/h} \tag{7.12a}$$

From Eq. 7.9, electric power required for 1 GJ/h of CW duty

$$= 0.01429 \text{ GJ/h} \tag{7.12b}$$

Hence, electric power required for 1 GJ/h of ChW duty

$$\begin{aligned} &= 1.24 \times 0.01429 \text{ GJ/h} \\ &= 0.01772 \text{ GJ/h} \end{aligned} \tag{7.12c}$$

From Eq. 7.11, electric power required for 1 GJ/h of ChW duty is 0.2587 GJ/h. Hence, total electrical power required (for the ChW compressor, ChW pump and CW recirculation) for 1 GJ of cooling via ChW

$$= 0.2587 + 0.01772 = 0.2764 \text{ GJ} \tag{7.13}$$

More than 90% of this energy is for the compressor in the refrigeration system. Hence, compressor power and efficiency are important in ChW production/recirculation. In applications, ChW supply and return temperatures may be slightly different from 5°C and 15°C, respectively, assumed in the above calculations. Although these differences will affect the electrical energy required for the compressor, the estimate of total electrical energy (namely, 0.2764 GJ per GJ of cooling by ChW) can be used as a reasonable approximation.

7.4 Energy for Producing/Supplying Utilities

This section presents the estimation of energy required for supplying common utilities (namely, electricity, steam, CW and ChW) to process

Table 7.5: Primary and secondary energy required for producing/supplying 1 GJ of common utilities in plants

Utility	Primary and Secondary Energy (in GJ)	Total Primary Energy (GJ) as LHV from	
		NG	Coal
Electricity	All primary energy	2.387 (Eq. 7.4)	3.257 (Eq. 7.3)
Steam (saturated or superheated)	Fuel and electrical energy	1.2602 (Eq. 7.5d)	1.3425 (Eq. 7.6d)
CW from 30 to 45°C	Electrical energy (0.01429 GJ from Eq. 7.9e)	0.0341 (= 0.01429 × 2.387)	0.0465 (= 0.01429 × 3.257)
ChW from 5 to 15°C	Electrical energy (0.2764 GJ from Eq. 7.13)	0.660 (= 0.2764 × 2.387)	0.900 (= 0.2764 × 3.257)

operations consuming them. This estimation includes expected losses in both production and distribution of utilities. Based on the calculations in Section 7.3, primary and secondary energy required for producing/supplying 1 GJ of energy in the form of electricity and steam, and for removing 1 GJ of energy using CW and ChW are summarized in Table 7.5 for two common fuels: coal and NG. As stated in the beginning of Section 7.3, these values are based on energy requirement for the production/supply of utilities, and do not include energy involved for setting up the production facilities and for manufacturing other materials (e.g., chemicals for water treatment in steam and CW production) used in producing the utilities. To include these contributions, energy for producing utilities in Table 7.5 should be increased by ≈5%.

As indicated for Table 7.4 in Section 7.3.2, readers may want to update the values in Table 7.5, as per their plant conditions and experience. For example, the value of 0.01429 for electrical energy required for CW is based on a total pressure drop of 266.7 kPa in the CW circuit. If this is significantly different in the plant in question, the reader may want to re-calculate the electrical energy required and then the primary energy required.

Typical processes require more than one type of utility, and total primary energy for supplying several utilities can be calculated using the values in Table 7.5. This is illustrated in the following examples.

Example 7.2: A separation process (i.e., distillation with mechanical vapor recompression) requires 2.71 MW of steam, 4.23 MW of electricity and 4.69 MW of CW. Estimate the total primary energy using the values in Table 7.5. Make and state any other required assumptions.

Solution: Primary energy source is not given, and hence assume NG is the fuel. Total primary energy values in Table 7.5 are given as GJ per GJ of utility. They can also be used as MW of primary energy per MW of utility.

Primary energy required for 2.71 MW of steam = 2.71 × 1.2602 = 3.415 MW

Primary energy required for 4.23 MW of electricity = 4.23 × 2.387 = 10.097 MW

Primary energy required for 4.69 MW of CW = 4.69 × 0.0341 = 0.160 MW

Total primary energy required from NG = 3.415 + 10.097 + 0.160 = 13.672 MW. The main contributors to this total primary energy are steam and electricity.

Note the following pointers in connection with the above example and its solution. **First**, total primary energy is not equal to the sum of energy supplied/removed by utilities (i.e., ≠ 2.71 + 4.23 + 4.69 = 11.63 MW). **Second**, some studies estimate total primary energy assuming efficiency of 0.9 for steam and 0.3 for electric power, and neglect energy required for CW (i.e., $= \frac{2.71}{0.9} + \frac{4.23}{0.3} = 17.11$ MW). Efficiency of 0.9 for steam accounts for steam production only, and it does not account for various losses in steam distribution/network, as discussed in Section 7.3.2. Efficiency of 0.3 for electric power is reasonable if coal is the fuel, but it substantially under-estimates efficiency of electric power generation using NG fuel ($= \frac{1.0}{2.387} = 0.419$). Using efficiency of 0.8 for steam and 0.419 for electric power, and neglecting energy required for CW, total primary energy required for utilities in the above example is

$\frac{2.71}{0.8}+\frac{4.23}{0.419}=13.48\,\text{MW}$, which is closer to 13.672 MW estimated above. **Finally**, neglecting energy required for producing/supplying CW underestimates the total primary energy slightly but this may be acceptable. As calculated in Example 7.2, correct estimate of total primary energy for the case of NG fuel is 13.672 MW.

Example 7.3: The existing vacuum system in Fig. 8.12 (Case 2A in Chapter 8), after optimization of the operating conditions, requires MPS of 3.72 ton/h and CW (2140 ton/h from 30 to 31.5°C, and 271.1 ton/h from 31.5 to 41°C). MPS is at 10.5 barg and 220°C. Estimate the total primary energy required for these utilities, using the values in Table 7.5. Ignore the small amount of electric power required for pumps in Fig. 8.12. State any other assumptions for the calculations.

Solution: Note 10.5 barg ≈ 11.5 bar = 1.15 MPa, and 1 ton = 1000 kg. Assume heat capacity of CW is 4.18 kJ/kg.°C. None of the condensate of MPS used is returned to the steam network. Hence, assume all energy in MPS is used in the vacuum system. This is equal to the difference between the enthalpy of MPS at 10.5 barg and 220°C (= 2868 kJ/kg from steam tables) and that of water at 30°C (= 126 kJ/kg). With these assumptions, the following calculations can be performed:

Steam energy supplied = 3.72 × 1000 × (2868 − 126) = 10,200,240 kJ/h = 10.2 GJ/h

Primary energy (from NG) required for steam supply = 10.2 × 1.2602 = 12.85 GJ/h

Energy removed by CW = 2140 × 1000 × 4.18 × (31.5 − 30) + 271.1 × 1000 × 4.18 × (41 − 31.5) = 24,183,000 kJ/h = 24.18 GJ/h

Primary energy (from NG) required for CW supply = 24.18 × 0.0341 = 0.83 GJ/h

Hence, total primary energy from NG for MPS and CW used in the vacuum system is 12.85 + 0.83 = 13.68 GJ/h. Main contributor to this is steam supply.

Example 7.4: This example is related to the previous one. The revamped vacuum system in Fig. 8.17 (Case 2D in Chapter 8) requires MPS of 1.94 ton/h, electric power of 88 kW for liquid ring vacuum pump, CW (362.7 ton/h from 38 to 41°C, and 15.2 ton/h from 30 to 35°C), and 717.9 ton/h of ChW from 7 to 13°C. MPS is at 10.5 barg and 220°C. Estimate the total primary energy for these utilities, using the values in Table 7.5. Ignore the small amount of electric power required for pumps in Fig. 8.17. State any other assumptions for the calculations.

Solution: In addition to the assumptions in Example 7.3, assume that primary energy value given in Table 7.5 can be used for ChW from 7 to 13°C; this assumption implies the use of motor-driven compressor in the refrigeration system (Fig. 7.5). Further, assume that heat capacity of ChW is 4.18 kJ/kg.°C. The following calculations are like those in Example 7.3.

Steam energy supplied = 1.94 × 1000 × (2868 − 126) = 5,319,480 kJ/h = 5.32 GJ/h

Primary energy (from NG) required for steam supply = 5.32 × 1.2602 = 6.70 GJ/h

Electrical energy supplied = 88 kW = 88 × 3600 kJ/h = 0.3168 GJ/h

Primary energy (from NG) required for electrical energy = 0.3168 × 2.387 = 0.76 GJ/h

Energy removed by CW = 362.7 × 1000 × 4.18 × (41 − 38) + 15.2 × 1000 × 4.18 × (35 − 30) = 4,865,938 kJ/h = 4.87 GJ/h

Primary energy (from NG) required for CW supply = 4.87 × 0.0341 = 0.17 GJ/h

Energy removed by ChW = 717.9 × 1000 × 4.18 × (13 − 7) = 18,004,932 kJ/h = 18.00 GJ/h

Primary energy (from NG) required for ChW supply = 18.00 × 0.66 = 11.88 GJ/h

This calculation uses 0.66 GJ of primary energy (NG) for one GJ of ChW (in Table 7.5); this primary energy is for electrical energy required for compressor, ChW pump and CW for producing/supplying ChW (as described in Section 7.3.4).

Hence, total primary energy from NG for steam, electricity, CW and ChW is 6.70 + 0.76 + 0.17 + 11.88 = 19.51 GJ/h. Main contributors to this are for supplying steam and ChW.

Comparing the total primary energy from NG in Examples 7.3 and 7.4, there is an increase of 19.51 − 13.68 = 5.83 GJ. This comparison is for illustration only; it is not on the same basis for a few reasons. First, the revamped vacuum system in Fig. 8.17 (Example 7.4) generates lower vacuum pressure compared to that by the optimized vacuum system in Fig. 8.12 (Example 7.3). Second, in Section 7.3.4 and for Table 7.5, motor-driven compressor and electrical energy from coal/NG are assumed for ChW production. In case the compressor in the chiller (mechanical refrigeration, Fig. 7.5) is driven by hydro, wind or solar power, or absorption chiller (using WH) is employed, the total primary energy in Example 7.4 will be substantially lower, and it will also be lower than that in Example 7.3. Therefore, fair comparison should employ the correct energy sources used for ChW production in the plant. Finally, analysis in Chapter 8 is assuming the use of absorption chiller for ChW production. Total primary energy for this scenario is estimated in the next example.

Example 7.5: The revamped vacuum system in Fig. 8.17 (Case 2D in Chapter 8) is based on the use of absorption chiller and WH, for ChW production. It requires MPS (at 10.5 barg and 220°C) of 1.94 ton/h, electric power (88 kW for liquid ring vacuum pump and 179 kW for chiller and ChW pumps), CW (1438 ton/h from 30 to 38°C, 362.7 ton/h from 38 to 41°C and 15.2 ton/h from 30 to 35°C), and low-low pressure steam (LLPS) at 1 barg (11.38 ton/h). Estimate the total primary energy for these utilities, using the values in Table 7.5. Ignore the small amount of electric power required for pumps in Fig. 8.17. State any other assumptions for the calculations.

Solution: The assumptions are those in Example 7.3. Primary energy values for steam, electricity and CW are for NG fuel in Table 7.5.

ChW is produced by the absorption chiller and it is used in the vacuum system. This requires electric power of 179 kW for chiller and ChW pumps, LLPS of 11.38 ton/h and CW of 1438 ton/h from 30 to 38°C. Of these, we will assume LLPS is available as WH and the primary energy for its production need not be considered here (i.e., it is accounted elsewhere).

MPS energy supplied = 1.94 × 1000 × (2868 − 126) = 5,319,480 kJ/h = 5.32 GJ/h

Primary energy (from NG) required for MPS supply = 5.32 × 1.2602 = 6.70 GJ/h

Electrical energy supplied = 88 + 179 kW = 267 × 3600 kJ/h = 0.9612 GJ/h

Primary energy (from NG) required for electrical energy = 0.9612 × 2.387 = 2.29 GJ/h

Energy removed by CW = 1438 × 1000 × 4.18 × (38 − 30) + 362.7 × 1000 × 4.18 × (41 − 38) + 15.2 × 1000 × 4.18 × (35 − 30) = 52,952,658 kJ/h = 52.95 GJ/h

Primary energy (from NG) required for CW supply = 52.95 × 0.0341 = 1.81 GJ/h

Hence, total primary energy from NG for the utilities in this example, is 6.70 + 2.29 + 1.81 = 10.80 GJ/h. Here, the main contributor is the primary energy required for MPS. Recall that LLPS is taken to be WH and primary energy required for it is not included in this estimate of total primary energy.

The total primary energy from NG (13.68 GJ/h) in Example 7.3 for the existing/optimized vacuum system decreases to 10.80 GJ/h in Example 7.5 for the revamped vacuum system with absorption chiller and liquid ring vacuum pump. This is despite the lower vacuum generated in the case of Example 7.5. In other words, the total primary energy for the vacuum system decreases by 23%. This is achieved by using WH available as LLPS.

Examples 7.3–7.5 are on primary energy analysis and do not consider economics. Cost analysis of existing/optimized vacuum systems and several revamps is presented and discussed in Chapter 8 (namely, Sections 8.6.2 and 8.6.3 and Table 8.4).

7.5 CO_2 Emissions Due to Utilities Used in the Process

Knowing the (total) primary energy required and fuel used, CO_2 emissions can be estimated using CO_2 emission factor, which is defined as the mass of CO_2 emitted per unit energy produced by combustion of fuel (kg of CO_2/GJ). Thus, emission factor depends on the carbon content and heating value (energy) of the fuel. The following examples calculate CO_2 emission factor for methane as fuel.

Example 7.6: Estimate CO_2 emission factor for methane combustion.

Solution: Molecular weight of CO_2, carbon, hydrogen, methane is, respectively, 44.009, 12.0107, 1.00794 and 16.04246.

Hence, carbon fraction in methane $= \frac{12.0107}{16.04246} = 0.7487$

Mass of CO_2 per kg of methane burned $= 0.7487 \times \frac{44.009}{12.0107} = 2.7433\,\text{kg}$.

HHV (gross calorific value) and LHV (net calorific value) of methane is, respectively, 55.5 and 50.0 MJ/kg (https://www.engineeringtoolbox.com/fuels-higher-calorific-values-d_169.html, accessed on 18 April 2021). Note that one GJ = 1000 MJ.

CO_2 emission factor depends on whether HHV or LHV is used for energy produced.

CO_2 emission factor for CH_4 based on HHV $= \frac{2.7433}{55.5} = 0.04943$ kg/MJ = 49.43 kg/GJ

CO_2 emission factor for CH_4 based on LHV $= \frac{2.7433}{50.0} = 0.05487$ kg/MJ = 54.87 kg/GJ

If NG is assumed to be 100% methane, then CO_2 emission factor for NG will be the same as above.

As can be seen from the above example, the emission factor based on LHV is higher by about 10% compared to that based on HHV. This is due to the difference between HHV and LHV of the fuel. Table 7.3 presents HHV and LHV of selected fuels. It shows that the difference between the two heating values ranges from 4% to 17%; the smallest difference is for bituminous coal, whereas the largest difference is for hydrogen. Generally, HHV/LHV and their difference decrease from hydrogen (with no carbon) to NG (mostly methane), propane, butane, crude oil and coal (i.e., with decreasing hydrogen fraction).

In general, heating value (HHV/LHV) and carbon fraction vary with composition of fuels such as coal, crude oil and NG, which are mixtures of components. Although mostly methane, NG may have other components such as CO_2, nitrogen, H_2S, ethane, propane and butane, and its composition varies with the NG reservoir and purification after its production. All these affect CO_2 emission factor of a particular fuel.

CO_2 emission factors for different fuels can be found in the literature including different websites. Their values from four diverse sources are compiled in Table 7.6. Values for some fuels are not available (NA) in one or more sources. Emission factor values from source 1 are generally lower than those in sources 2 and 3. Possible reasons for this are use of HHV

Table 7.6 CO_2 emission factors for selected fuels

Fuel	CO_2 Emission Factor (kg/GJ) from			
	Source 1	**Source 2**	**Source 3**	**Source 4**
Anthracite coal	98.28	98.3	NA	104.0
Bituminous coal	88.41	NA	NA	79.0
Lignite coal	92.62	101.0	101.2	79.0
Hard coal	NA	NA	94.6	NA
Natural gas	50.29	56.1	56.1	50.0
Ethane	56.49	61.6	NA	NA
Propane	58.25	NA	NA	60.0
Butane	61.39	NA	NA	62.0
Liquefied petroleum gas	58.49	63.1	63.1	61.0
Crude oil	70.65	73.3	73.3	NA

Source 1: https://www.epa.gov/sites/production/files/2015-07/documents/emission-factors_2014.pdf; Source 2: Table 3 at https://www.nea.gov.sg/docs/default-source/our-services/climate-change/m-r-appendix-(ver-14-feb-2018).pdf, which gives CO_2 emission factors for numerous fuels and also uncertainty % in them; Source 3: https://www.volker-quaschning.de/datserv/CO2-spez/index_e.php; Source 4: https://www.engineeringtoolbox.com/co2-emission-fuels-d_1085.html (Access date: March 2021)

(particularly for ethane) and/or composition of fuel (in the case of lignite coal, NG, liquefied petroleum gas and crude oil). Interestingly, CO_2 emission factor for anthracite coal is almost same in sources 1 and 2. Sources 2 and 3 have almost same values for lignite coal, NG, liquefied petroleum gas and crude oil. Some CO_2 emission factors from source 4 are comparable, whereas others are somewhat different from those in sources 1–3.

In summary, the value of CO_2 emission factor for a fuel may vary across websites and other literature sources. Hence, the engineer may have to choose one of them as per his/her (reporting) requirements such as specific website to be used, basis/assumptions and dimensions. Recall that primary energy for providing common utilities in Table 7.5 is LHV of the fuel. Hence, for consistency with using primary energy values in Table 7.5, CO_2 emission factor of the fuel should be based on LHV (and not HHV).

Besides CO_2, there may be emissions of methane and nitrous oxide (N_2O), which cause global warming. Emission factors of these components are also given in Table 3 available at https://www.nea.gov.sg/docs/default-source/our-services/climate-change/m-r-appendix-(ver-14-feb-2018).pdf (accessed on 18 April 2021). For example, emission factors for NG are respectively 0.001 kg of methane/GJ and 0.0001 kg of nitrous oxide/GJ. Although these values are negligible compared to CO_2 emission factor of about 50 (Table 7.6), global warming potential of methane and nitrous oxide are, respectively, about 25 and 300 times that of CO_2 over 100 years (https://www.epa.gov/ghgemissions/overview-greenhouse-gases, accessed on 18th April 2021). If required, emissions of methane and nitrous oxide can be included suitably in the estimation of total CO_2 emissions.

Electricity grid emission factor (i.e., mass of CO_2 emission associated with each unit of electricity provided by an electricity system) values can be found on the internet. For example, its value for 2019 for Singapore is reported as 0.4085 kg CO_2/kWh, at https://www.ema.gov.sg/cmsmedia/18RSU.pdf (accessed on 18th April 2021). This is for electricity production, probably using NG fuel, and it is close to 0.4128 kg of CO_2/kWh for NG-based electricity production for the United States, which is calculated based on the tabulated data at https://www.eia.gov/tools/faqs/faq.php?id=74&t=11 (accessed on 18 April 2021). The following example compares these values with that estimated based on primary energy required for electricity (Table 7.5) and CO_2 emission factors of fuels (Table 7.6).

Example 7.7: Estimate CO_2 emission for electricity production based on Tables 7.5 and 7.6. Then, compare it with the value in the above paragraph. Assume NG is the fuel for electricity generation.

Solution: From Table 7.5, primary energy required for producing one GJ of electricity from NG fuel is 2.387 GJ (of LHV).

As per Sources 2 and 3 in Table 7.6, the CO_2 emission factor for NG is 56.1 kg/GJ of energy (likely to be based on LHV, if we compare with the values estimated in Example 7.6).

Hence, CO_2 emission is $2.387 \times 56.1 = 133.91$ kg CO_2/GJ of electricity (or 0.4821 kg CO_2/kWh of electricity).

If Source 1 in Table 7.6 is used, CO_2 emission factor for NG is 50.29 kg/GJ of energy, and CO_2 emission is $2.387 \times 50.29 = 120.04$ kg CO_2/GJ of electricity (or 0.4321 kg CO_2/kWh).

In the paragraph before Example 7.7, electricity grid emission factor for 2019 and NG as the fuel, from the websites is ≈0.41 kg CO_2/kWh, which is equal to ≈114 kg CO_2/GJ. This is lower than the estimates in Example 7.7. One reason for this is that the estimates based on Tables 7.5 and 7.6 are for electricity delivered to and used by the user after deducting distribution losses of 5% (as stated in Section 7.3.1). Perhaps, the electricity grid emission factors reported at the cited websites do not include distribution losses of electrical energy.

7.5.1 *Life Cycle Assessment of CO_2 Emissions in Electricity Generation*

CO_2 emissions described above are for the use of coal and NG (fossil fuels) as primary energy in the production utilities. Further, they are only for the operation of the utility plant, and do not include CO_2 emissions associated with (i) production of materials for equipment in and construction of the plant and (ii) decommissioning/disposal of the plant (at the end of its life). Both these are addressed for electricity generation using fossil and renewable energy, in this sub-section.

Researchers at the National Renewable Energy Laboratory (NREL) thoroughly reviewed and harmonized (for consistency) the reported values on CO_2 emissions for electricity generation from different energy sources.[6] These emissions are for the entire life cycle of the generation plant starting from the production of materials for equipment in the plant, construction of the plant, plant operation and plant decommissioning/disposal at the end of its life. The energy sources include hydro, wind, solar photovoltaics (PV), concentrating solar power, nuclear, NG and coal. Results on life cycle greenhouse gas (GHG) emissions (grams of CO_2 equivalent per kWh) are available at https://www.nrel.gov/analysis/life-cycle-assessment.html (accessed on 11 July 2021). Some of these are selected and presented below.

Table 7.7 Life cycle GHG emissions for electricity generation from different energy sources

	GHG Emissions (kg of CO_2 Equivalent/kWh)				
Energy Source	**Maximum**	**Third Quartile**	**Median**	**First Quartile**	**Minimum**
Hydropower*	0.1651	0.0149	***0.0066***	0.0043	0.0007
Wind	0.0448	0.0184	***0.0109***	0.0085	0.003
Solar PV	0.1826	0.0501	***0.0441***	0.0372	0.0180
NG*	0.9875	0.5417	***0.4768***	0.4272	0.3068
Coal	1.3664	1.0483	***0.9787***	0.9330	0.7292

* Values for these sources are based on original data (i.e., without harmonization).

Not surprisingly, there are considerable variation in the reported CO_2 emissions (Table 7.7). Reasons for this include different system designs, operating assumptions, technological improvements over time and life cycle assessment methods/assumptions. The systematic harmonization by NREL researchers reduced the variability. However, it did not have much effect on the median of CO_2 emissions.

Table 7.7 presents life cycle GHG emissions as kg of CO_2 equivalent per kWh of electricity generated from hydropower, wind, solar PV, NG and coal (https://openei.org/apps/LCA/, accessed on 11 July 2021). Variability in the values reported in the literature can be seen from the maximum, third quartile, median, first quartile and minimum for each energy source. Values in Table 7.7 are likely for electricity at the generation plant (i.e., do not include 5% loss in the transmission and distribution). In terms of median of CO_2 emissions reported, hydropower has the lowest, followed by wind, solar PV, NG and coal, among the energy sources in Table 7.7. As expected, renewable sources (hydropower, wind and solar PV) have significantly lower (by at least a factor of 10) CO_2 emissions compared to fossil sources (NG and coal). Further, CO_2 emissions in case of NG is about half of that for coal.

In the previous section, electricity grid emission factor for 2019 and NG as the fuel is $\approx$0.41 kg CO_2/kWh, which is $\approx$15% lower than the median of life cycle CO_2 emission for electricity generated from NG. This is probably because the former is for plant operation only (i.e., excluding CO_2 emissions associated with (i) production of materials for equipment in and construction of the plant and (ii) decommissioning/disposal of the plant). For the present, it is reasonable to use the median values in Table 7.7 along with a multiplication factor of 1.05 (to account for loss in the transmission and distribution of electricity), for estimating life cycle GHG emissions due to electricity used in the plant.

Life cycle GHG emissions for using common utilities produced by different energy sources can be estimated using the primary and secondary energy requirements in Table 7.5 along with the median CO_2 emissions in Table 7.7. This is particularly straightforward for CW and ChW, which require only electrical energy; this is illustrated in the following example.

Example 7.8: Estimate life cycle GHG emissions due to the use of one GJ of ChW in a plant. Assume electricity energy source is solar PV and use the data in Tables 7.5 and 7.7.

Solution: From Table 7.5, 1 GJ of ChW requires 0.2764 GJ of electrical energy. Note that total primary energy is not required for the estimation in this example. From Table 7.7, life cycle GHG emissions for electricity from solar PV is 0.0441 kg of CO_2 equivalent per kWh. Using the conversion factor of 1 GJ = 277.778 kWh,

$$\text{Life cycle GHG emissions due to the use of 1 GJ of ChW}$$
$$= 0.2764 \times 277.778 \times 0.0441 = 3.3859 \text{ kg of } CO_2 \text{ equivalent}$$

Currently, steam is mostly generated using a fossil fuel, and then distributed and used in the plant. Hence, steam generation requires fuel energy. If renewable electricity becomes abundant, it is not clear whether the industry will use it for steam generation centrally (followed by its distribution) or directly for heating in the plant. In case steam production continues to be centrally in the utilities section, many heat losses and electrical energy in Table 7.4 will still be applicable but not others such as heat loss in flue gas. Accordingly, electricity required (instead of fuel energy) for generating steam should be estimated. Direct heating using electricity requires major changes in the existing equipment, new infrastructure for electricity distribution, addressing safety issues, etc. In either case, life cycle GHG emissions can be estimated knowing the electricity

Table 7.8 Contribution of three stages of a power plant to life cycle GHG emissions for electricity generation from different energy sources

Energy Source	Contribution to Life Cycle GHG Emissions		
	Extraction/Production of Materials for Equipment and Plant Construction	**Operation of the Plant**	**Decommissioning and Disposal of the Plant**
Wind	≈86%	≈9%	≈5%
Solar PV	60%–70%	21%–26%	5%–20%
Coal	<1%	>98%	<1%

required, renewable energy source, median values of emissions in Table 7.7 and electricity loss in transmission and distribution.

As stated earlier, life cycle GHG emissions of a plant is from three stages: extraction/production of materials for equipment and plant construction, operation of the plant, and decommissioning and disposal of the plant. Contributions of these three stages to life cycle GHG emissions for electricity generation from wind, solar PV and coal are presented in Table 7.8; these are from https://www.nrel.gov/analysis/life-cycle-assessment.html (accessed on 11th July 2021). Clearly, major part in case of wind and solar PV sources is from the extraction/production of materials for equipment and plant construction; in case coal is the energy source, almost all CO_2 emissions is from the plant operation.

7.6 Summary

This chapter is on primary and secondary energy required for providing common utilities (namely, electricity, steam, CW and ChW) in process plants, and CO_2 emissions due to primary energy used for providing these utilities. Main learning points of this chapter are as follows.

- Primary energy for providing common utilities is mainly from coal and NG, which together account for ≈50% of energy consumption and for ≈60% of electricity generation in the world.
- Primary and/or secondary energy are required for providing electricity, steam, CW and ChW to the process users in the plant.

- Efficiency of coal- and NG-based power plants in the year 2020 is, respectively, 32.3% and 44.1%. These values are based on net electricity produced at the generation plant.
- After including 5% losses in the transmission and distribution of electricity, only 30.7% or 41.9% of chemical energy (LHV) in coal or NG, respectively, is delivered as electrical energy to the final user.
- Efficiency of a boiler using NG as the fuel to produce steam is 92%–94%, whereas that of a boiler using coal as the fuel is 88%–90%. These values are based on LHV of the fuel.
- There are many losses in steam generation and distribution (Table 7.4). Electric power is required for pumps and blowers in the steam boiler and network. Accounting for all these, around 80% of primary energy (LHV of the fuel) is supplied as thermal (steam) energy to the process units requiring steam.
- In plants, CW is often cooled in a cooling tower and reused in a closed cycle. Providing CW to process users requires electric power for pumps and cooling tower fan.
- Production of ChW is generally by mechanical refrigeration involving a compressor, which accounts for more than 90% of total primary energy required for ChW production and distribution. Besides electrical energy, ChW production requires CW.
- Table 7.5 summarizes primary and secondary energy required for providing 1 GJ of electricity, steam, CW and ChW to a process plant, using coal and NG as fuel. Total primary energy (LHV of fuel) required for providing CW is low (≈0.04 GJ) compared to that required for providing electricity, which is quite high at 2.3–3.3 GJ.
- Total primary energy is not equal to the sum of energy supplied to or removed from the process units, using utilities. It should be estimated based on total primary energy required to provide the different utilities required (e.g., using values in Table 7.5).
- HHV and LHV of a fuel such as NG and coal depend on fuel composition. HHV of a fuel is 4%–17% more than that of LHV (Table 7.3).
- CO_2 emission factor is the mass of CO_2 emitted per unit of energy. Its value for a fuel can be calculated based carbon content and HHV/LHV of the fuel.

- CO_2 emission factor for a fuel can differ from one reference/source to another. Reasons for this include the heating value basis (HHV or LHV) and fuel composition.
- CO_2 emissions due to use of utilities in the plant can be estimated from primary energy required per unit and quantity of each utility used, and CO_2 emission factor of the fuel providing the primary energy (Tables 7.5 and 7.6).
- Boiler efficiency and CO_2 emission factor values in different sources in the literature may be based on LHV or HHV.
- Life cycle GHG emissions for electricity from renewable sources and fossil fuels are described in Section 7.5.1.

More details on steam, CW and ChW production and distribution are available in Chapter 8 in Turton *et al.*[4] and in Chapters 23 and 24 in Smith.[7]

References

1. Huichao X, Chao Y, Srinivasan D, Rangaiah GP, Poh TK. (2016) "Modelling and Validation of Parameters of Combined Cycle Power Plant," 2016 IEEE International Conference on Power System Technology (POWERCON), Wollongong, NSW, pp. 1–6, doi:10.1109/POWERCON.2016.7753893.
2. Gulen SC. (2010) *Plant Efficiency: Begin with the Right Definitions, Power*. https://www.powermag.com/plant-efficiency-begin-with-the-right-definitions/
3. Paffel K. (March 2020) *Steam System Thermal Cycle Efficiency, Plant Engineering*. https://www.plantengineering.com/articles/502971/
4. Turton R, Bailie RC, Whiting WB, Shaeiwitz JA, Bhattacharyya D. (2013) *Analysis, Synthesis, and Design of Chemical Processes*, 4th ed. Pearson Education International, Singapore.
5. Gadalla MA, Olujic Z, Jansens PJ, Jobson M, Smith R. (2005) Reducing CO_2 emissions and energy consumption of heat-integrated distillation systems. *Environ Sci Technol* **39**: 6860–6870.
6. Heath GA, Mann MK. (2012) Background and reflections on the life cycle assessment harmonization project. *J Ind Ecol* **16**: S8–S11.

7. Smith R. (2005) *Chemical Process Design and Integration.* John Wiley.

Acronyms and Notation

Acronyms

BFW	Boiler Feed Water
ChW	Chilled Water
CW	Cooling Water
DM	Demineralized Water
ESP	Energy in Steam Produced
GHG	Green House Gas
HHV	Higher Heating Value
HP	High Pressure
HPS	High-Pressure Steam
LHV	Lower Heating Value
LLPS	Low-Low Pressure Steam
LP	Low Pressure
LPS	Low-Pressure Steam
MP	Medium Pressure
MPS	Medium-Pressure Steam
NA	Not Available
NG	Natural Gas
NREL	National Renewable Energy Laboratory
PV	Photovoltaics
WHR	Waste Heat Recovery

Notation

A	Heat transfer area in the heat exchanger (m^2)
C_p	Heat capacity at constant pressure (kJ/kg.K = kJ/kg.°C)
h_P	Enthalpy of steam delivered to the process (kJ/kg)
U	Overall heat transfer coefficient ($kW/m^2 \cdot K = kW/m^2 \cdot °C$)

Greek Symbol

λ_P	Latent heat of steam delivered to the process (kJ/kg)

Exercises

7.1 State the primary energy source(s) used for electricity supplied in your city, region or country. Briefly discuss potential changes in these energy sources in the next 5 years, and the reason(s) for them.

7.2 State the primary/secondary energy source(s) used for steam supplied to your plant. Briefly discuss potential changes in these energy sources in the next 5 years, and the reason(s) for them.

7.3 Estimate electric power required for CW pump for removing one GJ/h of energy from process streams. Assume pump efficiency of 75% and motor efficiency of 95%. Total pressure drop of CW in the closed circuit (i.e., in the piping, valves and coolers) is 400 kPa. Assume CW is supplied at 30°C and returned at 45°C; density and heat capacity of CW are 1000 kg/m^3 and 4.18 kJ/kg.K, respectively. State other assumptions, if any, required for calculations.

7.4 Solve each of the Examples 7.2 to 7.4 assuming coal is the fuel. Compare the required total primary energy from coal with that from NG.

7.5 Estimate CO_2 emission factor for combustion of (a) propane (C_3H_8) and (b) bituminous coal (wet basis); compare your results with those given in Table 7.6. Use LHV and HHV values of these fuels given in Table 7.3. Assume carbon fraction of bituminous coal is 0.8.

7.6 Estimate CO_2 emissions due to the use of utilities stated in each of the Examples 7.3, 7.4 and 7.5. What is the change in CO_2 emissions in Examples 7.4 and 7.5 compared to that in Example 7.3? Neglect CO_2 emissions associated with the waste heat used in the absorption chiller. State other assumptions, if any.

7.7 Answer the following with yes (right) or no (wrong), and give brief reasoning, where applicable.
   (A) Primary energy is the same as secondary energy.
   (B) More than 70% of chemical energy in fuel goes into electrical energy.
   (C) Steam boiler efficiency includes all heat losses in the steam distribution network.

(D) More than 90% of energy in fuel goes into steam delivered to the process user.

(E) Steam generation and distribution requires electrical energy for pumps and blower.

(F) No electric power is required for CW production/supply.

(G) Besides electrical energy, cooling (e.g., using CW) is required for producing ChW by mechanical refrigeration.

(H) Required primary energy is more than the energy supplied by electricity.

(I) Required primary energy is less than the energy supplied by steam.

(J) Required primary energy is equal to the energy removed by CW.

(K) Required primary energy is more than the energy removed by ChW.

Chapter 8

Waste Heat Reduction in Vacuum Systems

8.1 Overview

Vacuum systems are very important in process industries, for distillation of heat-sensitive substances, drying and evaporation. They can improve product separation and/or avod product degradation. Power plants use vacuum to maximize power generation by extracting and condensing steam under vacuum conditions. Petroleum refineries utilize vacuum for vacuum distillation of atmospheric residue, internal power generation and for diesel drying. Food industries use vacuum for preventing food degradation. Low-temperature thermal desalination plants employ vacuum for maximizing water evaporation from seawater. Vacuum pumps are used for removing air from process vessels, during plant start-up and for pumping water from underground pits. They are also employed for pneumatic conveying and deaeration of boiler feed water.

Process industries mainly employ steam jet ejectors (SJEs), liquid ring compressors (liquid ring vacuum pumps [LRVPs]) and dry vacuum pumps (DVPs), for vacuum generation. Rough vacuum from 1 mm Hg to 760 mm Hg is common in process industries. Although vacuum is very beneficial, it is highly energy intensive due to low efficiencies of vacuum pumps. SJEs need lot of steam for vacuum generation and eventually condense it using cooling water. Hence, they discard lot of waste heat (WH) to environment. The only way to recover this low-grade WH is by reducing steam requirement as much as possible. Optimization of multi-stage SJE and use of LRVP/DVP are essential for this. Each of vacuum producing equipment has benefits and drawbacks.[1,2] Hence, appropriate equipment selection for and energy optimization of vacuum system are essential to minimize WH generation and hence the need for WHR system.

The main objective of this chapter is to review and develop a systematic methodology for the selection and optimization of vacuum generation systems, for minimum WH generation, capital cost (CC) and operating cost (OC). This chapter begins with description of vacuum pumps, in Section 8.2. Next, benefits and constraints of vacuum generation equipment are discussed in Section 8.3. The subsequent section presents design principles and utility requirements for SJE, LRVP and DVP, with examples. Section 8.5 describes useful strategies for designing a new vacuum system and for retrofitting an existing vacuum system, for minimum WH generation. Several industrial case studies (involving hazardous substances, use of pre-condenser and use of WH-powered chilled water) for design of a new vacuum generation system and for retrofitting an existing vacuum system are presented and analysed in Section 8.6. The chapter ends with summary in Section 8.7. Contents and case studies of this chapter are useful for understanding, selection and design/retrofit of vacuum generation systems for increased sustainability.

Learning outcomes of this chapter are as follows:

1. Describe common vacuum generation equipment (pumps) in industries.
2. Discuss the advantages and disadvantages of SJE, LRVP and DVP.
3. Select and apply sizing methods for vacuum equipment.
4. Estimate energy consumption for common vacuum producing equipment.
5. Explain and apply WH minimization strategies including energy optimization, for design of a new vacuum systems and for retrofitting an existing vacuum system.
6. Discuss the vital role of other WHR methods such as chilled water generation, for reducing WH generation in a multi-stage SJE system.

8.2 Working Principles of Main Vacuum Generation Equipment

SJE, LRVP and DVP are the main equipment used for vacuum generation in process industries and are described in this section. Ryans and Bays[1] presented a comprehensive review of various vacuum generation equipment capacities, their operating ranges and efficiencies.

8.2.1 *Steam Jet Ejectors*

An SJE[a] converts pressure energy of motive steam to kinetic energy by passing it through motive steam nozzle. It creates a low-pressure region, lower than the suction gas pressure, at the motive nozzle's tip. This causes suction effect and hence sucks suction gas into the ejector body and eventually mixes it with motive steam. The fluid mixture then passes through a diffuser, where velocity decreases due to the diverging geometry and hence increases the pressure. SJEs are the workhorse for producing vacuum in process industries for many decades. They are very reliable, are available in many materials of construction, have almost unlimited suction capacity, have no suction pressure restrictions and can handle hazardous vapour safely. Ready availability of steam and low maintenance costs (as there are no moving parts), make SJEs the preferred vacuum generation equipment in process industries. Medium-pressure steam (MPS) or high-pressure steam (HPS) is generally used as motive fluid for compressing suction gas/vapour in SJEs, as illustrated in Fig. 8.1.

SJE are generally designed for a single suction load. Hence, turndown operation (for reduced suction gas/vapour flow rate) is difficult. With some compromise on energy consumption, turndown operation can be achieved by implementing a bypass flow control from SJE discharge to SJE suction pipe (as shown in Fig. 8.1). Alternatively, turndown operation can be provided by installing multiple smaller capacity SJEs in parallel,

---

[a]View a video of SJE operating principle at: https://www.youtube.com/watch?v=0yK4Mzv7ei0, accessed on 03-01-2020. At 0.26 s, observe three tall SJEs and four condensers on the elevated platform.

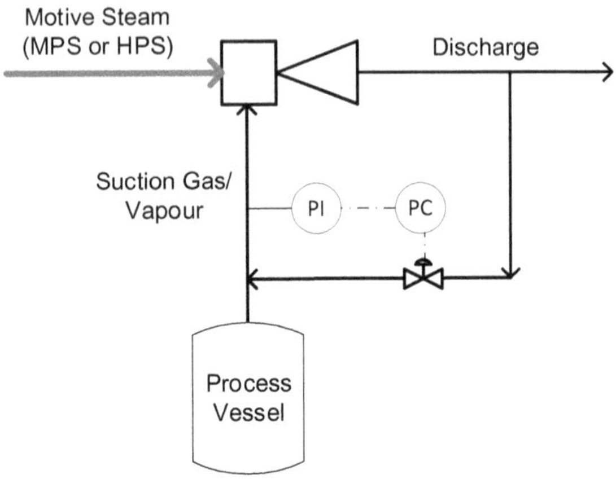

Fig. 8.1: Single-stage SJE with bypass flow control.

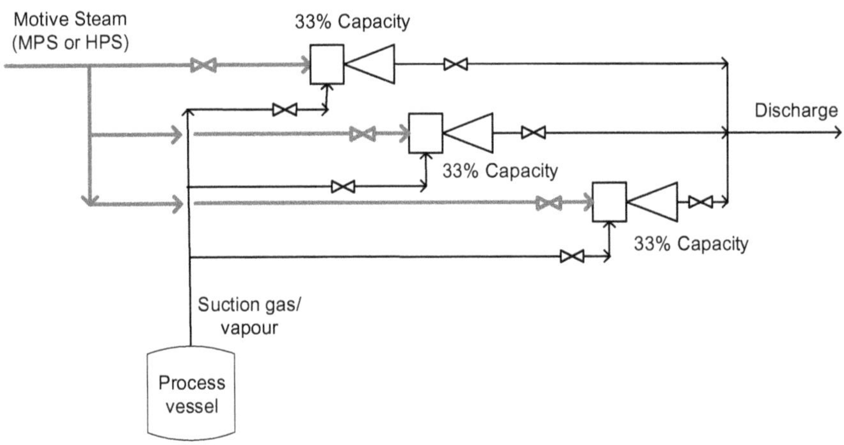

Fig. 8.2: Parallel arrangement of three SJEs (each of one-third capacity) for possible turndown operation.

as shown in Fig. 8.2. Although this increases CC, one or more of SJEs can be isolated for turndown operation.

Depending on the compression ratio (i.e., ratio of discharge pressure to suction gas/vapour pressure) and allowable energy consumption, single-, two- or three-stage SJEs in series are used for optimizing steam and cooling water and/or chilled water consumption. Generally higher the number of stages, lower will be steam and cooling water consumption.

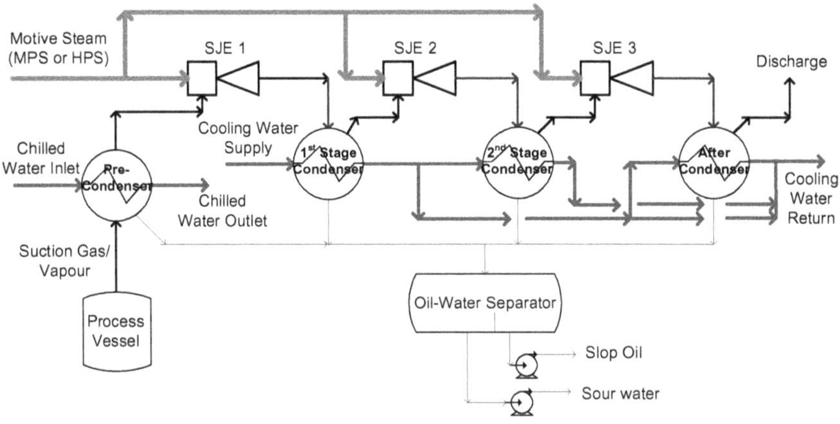

Fig. 8.3: A three-stage (in series) SJE system with pre-, inter-stage and after condensers.

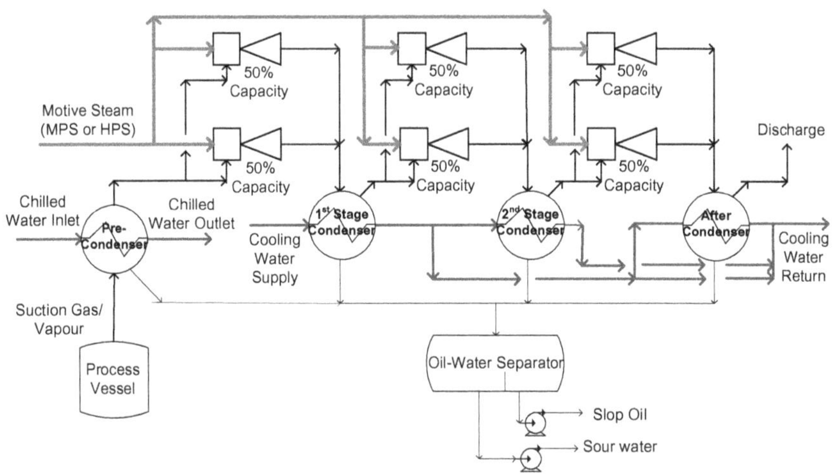

Fig. 8.4: Typical series-parallel arrangement of SJEs with pre-, inter-stage and after condensers.

A three-stage SJE system with a pre-condenser, inter-stage and after-stage condensers is shown in Fig. 8.3. Pre-condenser and inter-stage condensers reduce steam load to the respective downstream SJEs. SJEs can be arranged in series-parallel combination as shown in Fig. 8.4, depending on the level of vacuum required, turndown operation requirement and utility optimization.

Condensers are generally non-contact type (i.e., using cooling/chilled water on tube side of the condenser) to avoid contamination of cooling/chilled water. If there are no chemicals or hydrocarbons in the condensing vapour such as condensation of steam under vacuum in power plants, barometric or direct contact (of cooling water and ejector outlet gas/vapour) type condensers can be used. Chilled water, if available, is often used in pre-condenser as it can reduce the load for the SJE system and hence can reduce OC (by reducing both steam and cooling water consumption).

Although SJEs are not expensive, the total vacuum system installation cost (includes costs for condensers, oil-water separation drum/vessel, slop oil and water removal pumps and piping system) and OC are very significant, especially for the multi-stage SJE system. OC of SJEs is high as they are highly inefficient at < 10%.[1] The inefficiency is mainly attributed to lack of moving parts to efficiently convert fluid velocity to pressure. OC of SJE system can be reduced by using multi-stage SJE systems with cooling/chilled water-cooled condensers. For minimizing utility costs, almost all the motive steam (typically 90% in the first stage and 98% in other stage condensers) should be condensed and separated before it reaches as suction load to the next stage SJE. This has the additional benefit of reduced wastewater generation from the vacuum system.

8.2.2 *Liquid Ring Vacuum Pumps*

LRVPs operate by compressing the suction gas/vapour by a liquid ring[b] formed by the rotating impeller's centrifugal force. The compression process in LRVP is nearly isothermal. Although many liquids can be used as the seal liquid, water is common in process industries since it is inherently safer to use, separate and dispose through treatment in the wastewater plant. Generally, an LRVP system consists of LRVP itself, outlet separation vessel, a pump for circulating seal water and a heat exchanger (HE) to cool the seal fluid (Fig. 8.5). LRVPs are mainly driven by electric motors, equipped with variable speed drives (VSDs) if turndown operation

---

[b] View a video of LRVP operating principle at https://www.gardnerdenver.com/en-bg/nash/technologies-by-nash/technology-of-liquid-ring-vacuum-pumps, accessed on 03-01-2020. In this video, observe liquid ring formation at 01:02 sec and gas compression by liquid seal from1:40 sec to 2:06 sec.

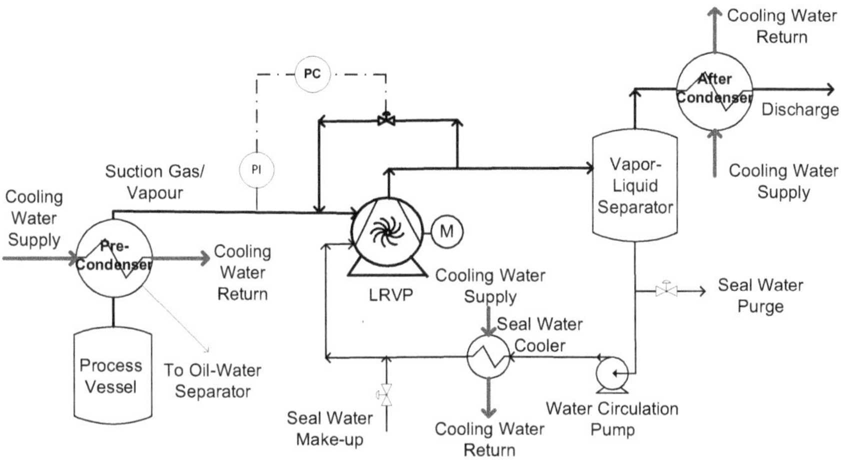

Fig. 8.5: An LRVP system for maintaining vacuum in the process/pressure vessel.

is desired. They can be operated at between 75% and 150% of the nominal speed.[3] So, suction capacity can also be controlled by using VSD. For bigger suction capacity and/or turndown operation, LRVPs can be arranged in parallel as shown in Fig. 8.6. Recycling some LRVP discharge flow to suction side by using pressure control valve is another capacity control method; however, this wastes energy.

For handling large compression ratios, LRVPs can be arranged in series as shown in Fig. 8.7. They can also be used in series operation with other vacuum generation equipment such as SJE and/or DVP. In large vacuum systems like vacuum distillation unit of petroleum refineries, LRVPs are generally used at the backend of (i.e., after) SJEs, mainly due to capacity (generally lower than 18,000 acfm, defined in the footnote[c]) and the suction pressure limitations. Power[2] specified a minimum practical suction pressure limit of 75 torr or 10 kPa (for single-stage LRVP) and 40 torr or 5.33 kPa (for two-stage LRVP). However, at such low suction pressure, density of gas/vapour will be very small (i.e., very large specific volume) and LRVP cannot be used unless suction load is very small. In petroleum refineries, LRVP is a good choice for suction pressures of 150 torr or 20 kPa and above.

[c] This unit refers to actual cubic feet per minute, which is common in the literature on vacuum pumps. One cfm is equal to 1.69901082 m^3/hr.

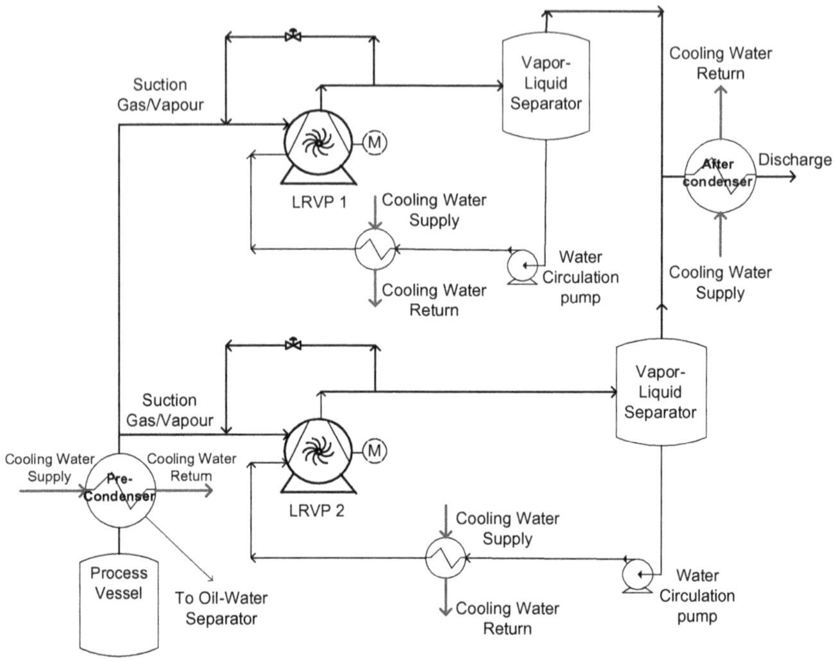

Fig. 8.6: Two LRVPs in parallel for large suction gas/vapour flow rate.

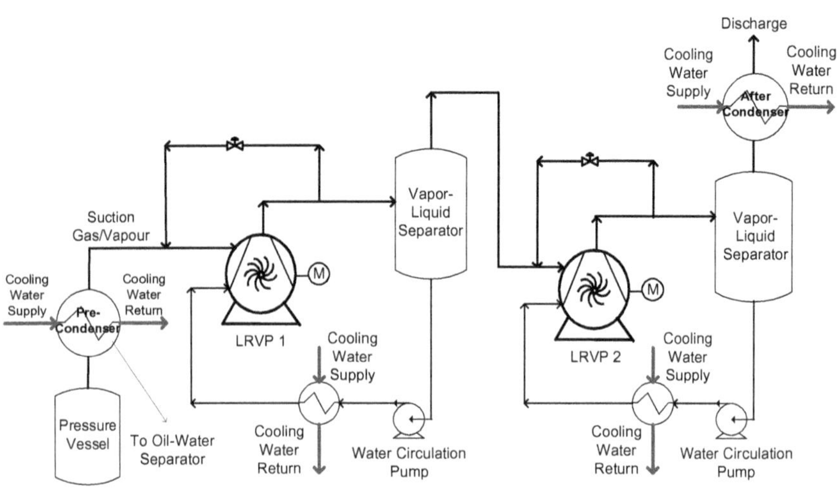

Fig. 8.7: Two LRVPs in series for higher compression ratios.

8.2.3 *Dry Vacuum Pumps*

As the name suggests, DVPs do not use any liquid media or steam for compression. They create vacuum through two electrically driven rotating screws/lobes/claws, arranged in parallel, rotating in opposite directions. The rotating elements enclose/trap the suction vapour/gas on the inlet side and deliver it to the high-pressure/discharge side.[d] During the compression process, rotors do not contact each other. Hence, there is no need for lubrication in the compression chamber. Further, due to compression heat, there will not be any liquid condensation in the DVP.

Generally, DVPs are installed with air- or water-cooling using jackets to control pump temperature. Like SJEs and LRVPs, they can be arranged in parallel, series, series–parallel and/or in combination with other vacuum producing equipment. Efficiency of a DVP (~ 50%) is greater than that of SJE (~ 10%) and LRVP (25%–40%). However, DVPs cannot tolerate any liquid slugs. The capacity of a DVP is generally limited to 1400 acfm or 2379 m^3/hr. Because of this limited size, potential safety (may require flame arrestors in the suction and discharge pipes) and reliability concerns,[4] they are generally used at the backend of SJEs or LRVPs, for vacuum generation in process industries. However, they are used directly at the outlet of condensing steam turbines (STs) and other applications requiring low capacities and involving only non-flammable/non-explosive gas.

8.3 Benefits and Constraints of Vacuum Pumps

For SJE, LRVP and DVP, lower/higher the gas molecular weight (MW) is, higher/lower are the CC and OC. On the other hand, lower/higher gas temperatures require lower/higher CC and OC. Relative merits and constrains of SJE, LRVP and DVP are summarized in Table 8.1. They are useful for the proper selection of vacuum generation equipment.

---

[d] View a video of DVP operating principle at https://www.gardnerdenver.com/en-in/nash/technologies-by-nash/dry-screw-vacuum-pump-technology, accessed on 03-01-2020. In this video, observe animation of gas compression path along the screws in DVP.

Table 8.1: Relative merits and constraints of SJE, LRVP and DVP

Factor	**SJE**	**LRVP**	**DVP**
Efficiency	Least efficient (~ 10%)	Higher efficiency compared to SJE (25%–40%)	Generally higher efficiency compared to LRVP (~ 50%)
Suction capacity	No limit (up to 1,000,000 acfm[2])	Restricted to 18,000 acfm[2] or 22,000 acfm[1]	Generally restricted to 1400 acfm[1]
Suction pressure	They can be designed for any suction pressure; minimum suction pressure for three-stage SJE is 1.5 torr.[2] CC and OC increase with decrease in the suction pressure and with increase in the discharge pressure.	Minimum suction pressure depends on vapour pressure of seal liquid. With water (at 15°C) as seal fluid, minimum suction pressure for two-stage LRVP is 40 torr.[2] Power consumption increases with decrease in the suction pressure and with increase in the discharge pressure.	DVP can be used from suction pressure above 0.1 torr. Power consumption and CC increase with decrease in the suction pressure and with increase in the discharge pressure.
Suction temperature	No limit	Limited by vapour pressure of seal liquid. Generally, 100°C is the limit with water as the seal liquid.	Generally, no limitation
Main driving force	MPS or HPS	Electricity	Electricity
Other utility requirements	Cooling and/or chilled water (instrument air is required only if discharge bypass is used for capacity control)	Make up water, cooling water and instrument air	Nitrogen, air and/or cooling water for DVP cooling, and instrument air

Flammable gas/ vapour stream	No concern	No concern	Generally, not recommended
Occasional liquid slugs in vapour/gas stream	No major concern	No concern	Not suitable. Any liquid slug accumulation and/or formation in suction piping must be avoided.
Turndown operation	Turndown operation is very limited. It can be improved by installing multiple smaller SJEs in parallel (Fig. 8.2) and/or discharge gas/ vapour bypass to suction side using a suction pressure control valve (Fig. 8.1)	Turndown operation is possible within limits imposed by the manufacturer, by using VSD for motors[7] or by installing smaller sized LRVPs or DVPs in parallel arrangement.	
Space requirements	SJEs are compact. But a multi-stage SJE system includes condensers and condensate separator, which require significant space. Also, draining of condensate from pre- and inter-stage condensers under vacuum require them to be elevated sufficiently above the separator vessel/drum. Hence, it requires multi-stage platforms for the system installation.	Generally, requires less space, compared to a multi-stage SJE system. LRVP system can be installed on ground level.	DVP system requires the smallest plot space compared to SJE and LRVP. It can be installed on ground level.

(*Continued*)

Table 8.1: (*Continued*)

Factor	SJE	LRVP	DVP
Effluent handling	If suction vapour contains hydrocarbons, SJE-condenser system generates foul/oily water, which can be treated in foul water stripper or wastewater plant.	Less wastewater is generated in LRVP and it can be treated in foul water stripper or wastewater plant.	DVPs do not produce any effluent and so do not require any effluent handling facilities.
Capital and operating costs	High OC and CC for the entire system including condensers and separators.	Low OC and CC	Generally, DVP incurs lower OC compared to SJE and LRVP but its CC is more than that of SJE and LRVP. However, for smaller capacities, total installation cost of DVP can be lower compared to SJE and LRVP.
Cooling water supply temperature	CC and OC are directly proportional to the cooling water supply temperature.		DVP requires small amount of cooling water in jackets, for pump temperature control. So, its costs are unaffected by cooling water supply temperature.
Reliability, availability and maintainability considerations	They are highly reliable as they do not have any moving parts, their availability is almost 100% and they require minimal maintenance.	Generally, LRVP is reliable; nearly 100% availability can be achieved by installing spare LRVPs in parallel; and require some maintenance due to moving parts and seals.	DVP reliability is lower compared to SJE or LRVP due to close clearances of moving parts and dry operation. Its availability can be increased by installing spare

			DVPs in parallel. It requires more maintenance compared to SJE or LRVP.
Vacuum equipment combinations	SJE can installed in series with SJE first, followed by LRVPs and/or DVPs	LRVP can be used standalone or can be installed with LRVP first, followed by LRVPs or DVPs in series. If suction load is very small, suction gas is non-hazardous and low vacuum level such as < 33 torr is required, SJE or DVP can be installed first, followed with LRVP	DVP is generally installed after SJE and/or LRVP. However, sometimes it can be installed before LRVPs.

8.4 Design Principles and Utility Requirements for SJE, LRVP and DVPs

Operating, maintenance and production loss costs of a vacuum system can be several times its CC. Good understanding of vacuum system design principles is essential for reducing WH generation in a vacuum system. Further, it will help to reduce the life cycle cost (involves CC, OC and maintenance costs) for the vacuum system. This section presents useful design principles and utility requirement estimation methods for SJEs, LRVPs and DVPs.

8.4.1 *Estimation of Suction Flow Rate of a Vacuum System*

The suction flow rate or 'load' to a vacuum system mainly consists of non-condensable gases and condensable process vapour. Non-condensable gases mainly result from air leakage into the vacuum system (through flange joints, valve stems, pumps and agitator shafts), dissolved gases and any nitrogen flush for the instruments and nozzles. Sometimes, non-condensable gases may be produced by chemical reactions.

Process equipment are not perfectly air-tight for the whole life cycle, especially under vacuum conditions. Reasonably air-tight equipment is essential for preventing selection of overly sized vacuum generation equipment. Oversized vacuum equipment wastes energy. Since air leak cannot be accurately estimated in actual equipment operation (as it depends on equipment integrity, maintenance quality etc.), some designers count the number of flanges and leakage paths and assume that all of them will leak. This will result in excessively over-designed vacuum system and hence it may lead to high energy cost. In industrial practice, there are two main methods for estimation of air leakage: system volume method and system component method.[5,6]

System volume method estimates the air leakage by considering volume of all equipment under vacuum that are connected to the vacuum system (e.g., distillation column, condensers and other HEs such as reboilers). Seider *et al.*[8] presented the following equation for estimating air leakage rate.

$$M_a = 5 + \{0.0298 + 0.03088[\ln(P_s)] - 0.0005733[\ln(P_s)^2]\} V_s^{0.66} \quad (8.1)$$

Here, M_a, P_s and V_s are weight of air leakage in lb/hr, system pressure in torr and system volume in ft^3, respectively.

System component method estimates the air leakage by counting and assigning air leakage values for every air leakage location (e.g., flanges, valve seals, pump seals, sight glasses, stuffing boxes of agitators, safety valves and vacuum breakers), without considering any system volume. Coker[5] published average air leakage values for various system components. Air leakage estimated by the system component method is often multiplied by a safety factor of 2 or 3.[5] Some designers use a hybrid approach by combining system volume and component methods. Such a method estimates air leakage using both the methods and uses their summation value with a safety factor of 1.5–2.[5] Hybrid method is more commonly used in industrial practice.

8.4.2 *Steam Jet Ejectors*

Both CC and OC of an SJE system can be reduced by increasing the number of SJE stages in series (usually two to four stages are used) along with pre- and inter-stage condensers. Number and size of SJEs, and size of condensers depend on many factors such as suction and discharge pressures, suction temperature, vapour quality (amount of non-condensable gases and MW), cooling water flow rate, supply and return temperatures, availability and cost of chilled water, cost of steam and/or electricity, project economic requirements, plot space and structural availability. Lower the required compression ratio (i.e., ratio of SJE discharge and suction pressure), lower would be the required number of SJE stages.

One of the most important parameters that limits the minimum attainable vacuum level of an SJE is the cooling water temperature. Lower the cooling water temperature, lower will be the minimum attainable vacuum level. Use of chilled water instead of cooling water in pre-condenser can lower the attainable vacuum level and can also reduce steam/power consumption of the vacuum system (if LRVP or DVP is used in combination with SJE). Although chilled water is expensive compared to cooling water, the former can be very attractive if WH is available for chilled water production or electricity price is low.

For sizing, SJEs suction gas/vapour is generally converted to dry air equivalent (DAE). HEI[9] published curves for this conversion using four ratios:

(i) Molecular weight entrainment ratio for gases other than water vapour based on average MW (MW_{EROG}), equal to the ratio of weight of gas to weight of air.
(ii) Molecular weight entrainment ratio of steam (MW_{ERS}), equal to ratio of weight of steam to weight of air.
(iii) Temperature entrainment ratio for air (T_{ERA}), equal to ratio of weight of air at actual temperature to weight of air at 70°F (21.1°C).
(iv) Temperature entrainment ratio for water vapour (T_{ERS}), equal to the ratio of weight of water vapour at actual temperature to weight of water vapour at 70°F (21.1°C).

Based on curves in HEI,[9] Reddy *et al.*[10] developed the following equations for estimating T_{ERA}, T_{ERS}, MW_{EROG} and MW_{ERS}, which are convenient for using in computer programs. In the following, T is the temperature of air/gas/vapour (in °C) and MW is the molecular weight of steam or gases/vapour other than steam.

$$T_{ERA} = -4 \times 10^{-10}\,T^3 + 3 \times 10^{-7}\,T^2 - 0.0005\,T + 1.0131 \tag{8.2}$$

Error: 0%–1.73%

$$T_{ERS} = -1 \times 10^{-13}\,T^4 - 7 \times 10^{-12}\,T^3 + 8 \times 10^{-8}\,T^2 - 0.0006\,T + 1.015 \tag{8.3}$$

Error: 0% to −0.4%

For MW = 0–60, MW_{EROG} or $MW_{ERS} = 1 \times 10^{-5}\,MW^3 - 0.0013\,MW^2 + 0.0642\,MW + 0.0161$ (8.4a)

Error: −7.2% to 2.4%

For MW= 60–150, $MW_{EROG} = -2 \times 10^{-5}\,MW^2 + 0.0077\,MW + 0.9464$ (8.4b)

Error: +0.17% to 2.18%

Equation (8.4a) can be used for estimating molecular weight entrainment ratio of steam (as steam MW is 18) and other gases with average

MW up to 60. Steam and other components in the suction gas can be converted to DAE using the entrainment ratios from Eq. 8.2–8.4 and the following equation:

$$\text{DAE of suction gas or vapour} = \frac{M_{\text{OG}}}{T_{\text{ERA}} \times \text{MW}_{\text{EROG}}} + \frac{M_S}{T_{\text{ERS}} \times \text{MW}_{\text{ERS}}} \quad (8.5)$$

Here, M_{OG} and M_{S} are the mass flow rate of gases/vapour other than steam and of steam, respectively. Calculated DAE is in the units of these mass flow rates.

Example 8.1: Estimate DAE value of the suction gas at 70 torr and 70°C, and containing 100 kg/hr of air, 50 kg/hr of methane, 5 kg/hr of H_2S and 350 kg/hr of steam.

Solution

Average MW of OG (gases other than water vapour/steam) can be calculated as follows:

	Mass flow rate, kg/hr	MW	Number of moles
Air	100	29	$\frac{100}{29} = 3.448$
Methane	50	16	$\frac{50}{16} = 3.125$
H_2S	5	34	$\frac{5}{34} = 0.147$
Total	155		6.720
Average MW = 155/6.720 = 23.06			

Total suction flow rate = 100 + 50 + 5 + 350 = 505 kg/hr
Mass flow rate of steam (MW = 18) = 350 kg/hr
Mass flow rate of other gases (OG) (MW =23.06) = 155 kg/hr
$T = 70°\text{C}$
From Eq. 8.2,

$T_{ERA} = -4 \times 10^{-10} \times (70^3) + 3 \times 10^{-7} (70^2) - 0.0005 \times 70 + 1.0131 = 0.9794$

From Eq. 8.3,

$T_{ERS} = -1 \times 10^{-13} \times (70^4) - 7 \times 10^{-12} (70^3) + 8 \times 10^{-8} \times (70^2) - 0.0006 \times 70 + 1.015 = 0.9734$

From Eq. 8.4a,

$MW_{EROG} = 1 \times 10^{-5} \times (23.06^3) - 0.0013 \times (23.06^2) + 0.0642 \times (23.06) + 0.0161 = 0.9280$

Also, from Eq. 8.4a,

$MW_{ERS} = 1 \times 10^{-5} \times (18^3) - 0.0013 \times (18^2) + 0.0642 \times (18) + 0.0161 = 0.8088$

From Eq. 8.5, DAE of suction stream $= \frac{155}{0.9794 \times 0.9280} + \frac{350}{0.9734 \times 0.808} =$ 615.1 kg/hr

Once DAE of suction gas/vapour is estimated, required amount of motive steam to compress unit DAE mass of suction gas/vapour in an SJE (defined as R_a) can be found from Fig. 4.8 in the book by Power,[2] using SJE suction and discharge pressures. The curves in this figure cover suction pressures from 0.1 to 1000 torr, discharge pressures from 1 to 1000 torr, and R_a from 0.5 to 20. For better accuracy, various correction factors for this estimated R_a value are also available in Figs. 4.9–4.11 in Power.[2]

Alternately, for rough estimation of SJE's steam consumption, the following equation[11] can be used.

$$R_a = \left[\frac{P_d}{P_s}\left(0.434 - \frac{1.338}{P_s} + 0.475\times10^{-3} P_s\right) - 0.187\right]\left(1.2 - \frac{P_m - 10.2}{20}\right) \quad (8.6)$$

Here, P_s, P_d and P_m are suction and discharge pressures (torr), and motive steam pressure (barg), respectively. For most cases investigated by the first author of this book, R_a estimated by the above equation is ~ 25% higher compared to that from the plots in Power.[2] The use of Eq. 8.6 is recommended only for quick and conservative estimation of steam consumption for SJE and preliminary design and/or revamp of multi-stage SJE systems. For accurate results, obtain and use technical quotations from SJE vendors.

Finally, motive steam requirement for the given mass of suction gas/vapour to the ejectors can be estimated by multiplying DAE with R_a.

For condenser area calculations, overall heat transfer coefficient, U W/(m^{2}K) for condensing air/non-condensable gases and steam/water vapour mixtures, can be estimated using the following equations, which were developed by Reddy *et al.*[10] based on Fig. 5.15 in Power.[2]

For gas–vapour mixture with non-condensable vapour mole percentage, NC from 1%–50%:

$$U = 5.678\ \{220.0417 + 1.6919\ \ln(\mathrm{NC}) - 2.67975[\ln(\mathrm{NC})]^2 - 1.5465[\ln(\mathrm{NC})]^3\} \tag{8.7}$$

Error: 0% to 4.5%

For gas/vapour mixture with NC from 50% to 95%:

$$\mathrm{U} = 5.678\{-245896 + 233845.3\ \ln(\mathrm{NC}) - 83300.5\ [\ln(\mathrm{NC})]^2 + 13183.62\ [\ln(\mathrm{NC})]^3 - 782.58\ [\ln(\mathrm{NC})^4]\} \tag{8.8}$$

Error: −2% to +2%

8.4.3 *Liquid Ring Vacuum Pumps*

LRVP compresses suction gas/vapour by nearly isothermal compression. Seal water gets heated up by absorbing the heat from gas compression, friction of LRVP internals, gas cooling and vapour condensation. Heat from the seal liquid is removed by cooling water in the seal-water cooler (Fig. 8.5). To maintain the quality of seal water, some of it is purged and equivalent amount of fresh water is added. Lower the seal-water temperature, lower will be the suction pressure achieved. Hence, if the cooling water supply temperature is high, chilled water may be used as cooling water. Lower seal-liquid/water temperature can also eliminate the cavitation problem associated with LRVPs, if used at low suction pressures. With cooling water temperature of ~ 30°C, LRVPs are used in industries to replace SJEs operating at suction pressures > 150 torr.

LRVPs are generally more expensive compared to SJEs. However, they can be cheaper to install and operate compared to a multi-stage SJE system. Power required by an LRVP can be estimated using the isothermal compression equation[3]:

$$P = \frac{P_s \times V_s \times \ln \frac{P_d}{P_s}}{27{,}000 \times \eta_e} \tag{8.9}$$

Here, P, P_s, P_d and V_s are the isothermal compression power (kW), suction and discharge pressures of LRVP (torr), and volumetric flow rate at suction pressure and temperature (m^3/hr), respectively. In the denominator, 27,000 is the conversion factor (for obtaining P in kW with P_s in torr, P_d in torr and V_s in m^3/hr), and η_e is the isothermal efficiency as a fraction, typically a value of 0.25–0.4.[3]

Example 8.2: Estimate the power required for a LRVP to compress a suction gas stream to 800 torr, for the following suction gas data: flow rate = 500 kg/hr, pressure = 150 torr, temperature = 70°C, and composition of 30 wt% air and rest steam. Assume LRVP efficiency of 0.25.

Solution

Total suction flow rate = 500 kg/hr
Mass flow rate of air = 500 × 0.3 = 150 kg/hr
Mass flow rate of steam = 500 × 0.7 = 350 kg/hr
Molecular weight of air = 29
Molecular weight of steam = 18
Suction temperature = 70°C = 70 + 273.15 = 343.15 K
Suction pressure = 150 torr = 199.983 mbar
Moles of air per hour = $\frac{150}{29} = 5.17$
Moles of steam per hour = $\frac{350}{18} = 19.44$
Using ideal gas equation with gas constant of 83.14 $\frac{\text{mbar}, m^3}{\text{kmol, K}}$,

$$V_s = \left(\frac{83.14 \times (5.17 + 19.44) \times 343.15}{199.983} \right) = 3511.83 \text{ m}^3\text{/hr}$$

From Eq. 8.9, $P = \dfrac{150 \times 3511.83 \times \ln \frac{800}{150}}{27000 \times 0.25} = 130.64 \text{kW}$

8.4.4 *Dry Vacuum Pumps*

DVPs minimize the WH generation by avoiding the use of steam for gas/vapour compression. They utilize screws, roots, lobes or vanes for compressing suction gas/vapour using electricity, and hence they are energy efficient and compact. DVPs do not produce any wastewater, and so wastewater handling facilities are not required. However, their main drawbacks are lower capacity, safety risk for handling flammable vapour/gases and inability to handle liquid slugs. A spark may be generated by accidental contact between the rotating parts and the casing, and it can potentially lead to fire/explosion. Auto-ignition is another concern if DVP's jacket cooling system fails. DVPs are applicable for suction pressures greater than 0.1 torr.

Purchase cost of DVP is generally higher compared to that of SJE or LRVP of similar capacity. However, DVPs are compact and generally do not require inter-stage condensers to condense the vapour. They do not require closed circulation of seal liquid and its cooler, which are required for LRVP. DVP's OC is minimum compared to LRVP or SJE system. Hence, for non-hydrocarbon applications, involving smaller suction loads (< 1400 acfm), DVPs are highly recommended. Power required by a DVP can be estimated using[1]:

$$P = \frac{M \times \left(\left(\frac{P_d}{P_s} \right)^{0.286} - 1 \right)}{20 \times \eta_d} \tag{8.10}$$

Here, M is suction gas/vapour flow rate (in lb/hr), P_s and P_d are, respectively, suction and discharge pressures (in torr), η_d is the adiabatic thermal efficiency of DVP as a fraction and P is the power required (in hp). For general process applications, typical η_d is between 0.3 and 0.5. Note that Eq. 8.10 gives power (P) in hp and it should be multiplied by 0.7457 to obtain P in kW.

Example 8.3: Estimate the power required for a DVP to compress a gas stream to 800 torr, using the following suction gas data: flow rate = 500 kg/hr, pressure = 150 torr, temperature = 70°C, and composition = 30 wt% air and rest steam. Assume DVP efficiency of 0.5.

Solution

Total suction flow rate = 500 kg/hr = $\frac{500}{0.4536} = 1102.29\,\text{lb/hr}$
Suction pressure = 150 torr
Discharge pressure = 800 torr

From Eq. 8.10, $P = \frac{1102.29 \times \left(\left(\ln \frac{800}{150} \right)^{0.286} - 1 \right)}{20 \times 0.5} = 67.69\,\text{hp}$ = 67.69 × 0.7457 = 50.48 kW

Table 8.2: Typical utility requirements for production of chilled water

Type of Chiller	Driving Force	Typical Steam and Electrical Power Requirements	Typical Cooling Requirements	Typical Coefficient of Performance
Single-stage absorption (LiBr system)	LLP steam (~ 1 barg) or hot water or hot stream > 93°C	Steam: 2.36–2.41 kg and electrical power: 0.0028–0.0114 kW, both per kW of refrigeration	2.5 kW per kW of refrigeration	0.6–0.75
Mechanical compression (Propane)	Motor or engine driven compressor	Electrical power: 4.5 kW/kW of refrigeration	1.283–1.125 kW/kW of refrigeration	4.5

8.4.5 *Chilled Water Generation*

Chilled water can be generated using heat pumps as described in Chapter 5. WH-powered absorption chiller or electricity driven mechanical chillers are widely used in process industries. Typical utility requirements of absorption and mechanical chillers are presented in Table 8.2.[12]

8.5 Strategies for Minimizing Waste Heat Generation in Vacuum Systems

Total life cycle OC of vacuum systems, especially SJE type, is several times their installation cost. Most of the energy used is discarded to atmosphere as WH. As this WH is not economically recoverable, the best strategy is to minimize WH generation. WH generation in LRVP or DVP is insignificant and hence it is not discussed here. Accordingly, this section mainly focuses on minimizing steam usage in SJEs. Further, other WHR methods such as chilled water generation can reduce steam usage and consequently WH generation in SJEs. Often, they may also reduce the CC of the vacuum system.

A strategy for minimizing WH generation for a new or revamp deign of a vacuum system is presented in Fig. 8.8. First, condensing any condensable vapour with cooling water at pre-condenser shall be maximized.

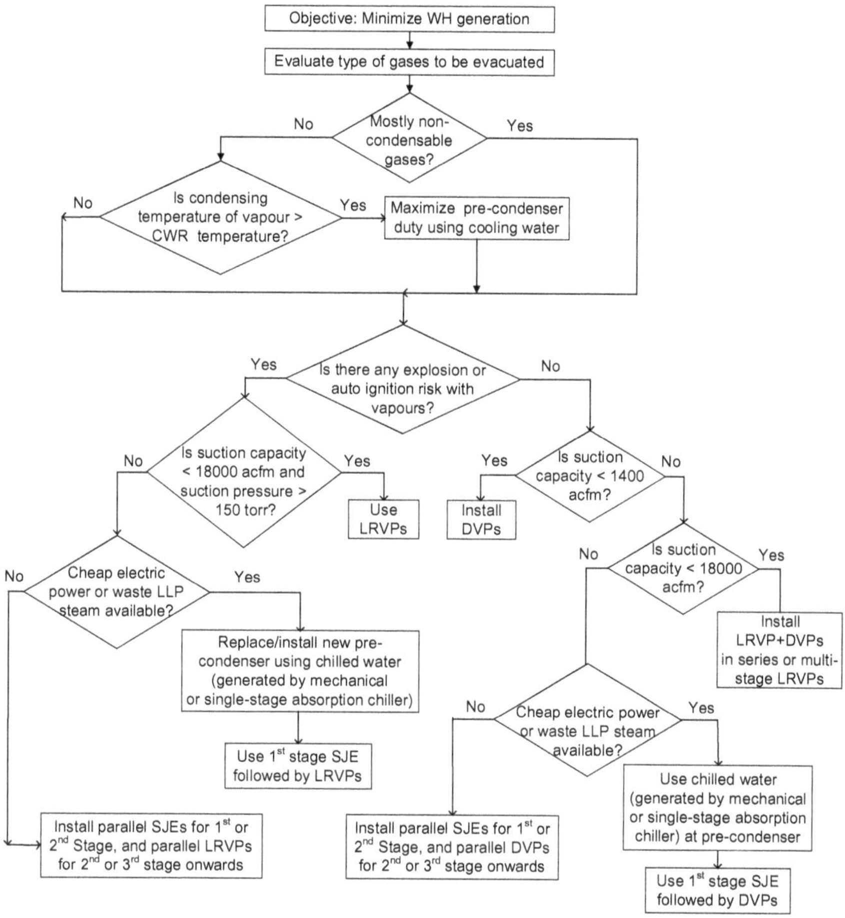

Fig. 8.8: Design strategy for minimizing WH in a vacuum system.

This will reduce the load to the vacuum system. Next, auto-ignition and/or explosion risks of the gas/vapour need to be evaluated. If there are no such concerns, DVPs can be used as long as suction capacity is small (< 1400 acfm). If the capacity is between 1400 and 18000 acfm, LRVPs can be used. Otherwise, the possibility of using chilled water in the pre-condenser can be explored, to reduce the load to the vacuum system by removing more condensable vapour, so that it falls within the range of LRVP's. If the plant site has chilled water readily available or it can be generated economically by WHR, chilled water can be used in the pre-condenser. If the site has cheap electricity, chilled water can be generated by mechanical refrigeration. Then, DVPs or LRVP+DVP combination can be selected for the vacuum system design/revamp, depending on the suction load. If chilled water usage is not economical, install one or two stages of SJEs, followed by DVPs.

If the suction gas/vapour compression involves auto-ignition and/or explosion risks, suction capacity is < 1800 acfm and suction pressure is > 150 torr, LRVPs can be used for minimizing WH generation. Otherwise, economic use of chilled water can be evaluated. If chilled water use is feasible, it should be used in the pre-condenser to minimize the suction load to the vacuum system. Single-stage SJE followed with LRVPs can be used for minimizing WH generation. If chilled water usage is not economical, install one or two stages of SJEs, followed by LRVPs.

For minimizing steam consumption, optimization of the selected vacuum system is required. This determines the optimum discharge pressures and MPS consumption for each stage of the multi-stage SJE system (shown in Fig. 8.9). This can be easily obtained by solving the following optimization problem, for example, using the Solver tool in MS Excel. The objective function is to minimize total steam consumption for a multi-stage SJE system.[10]

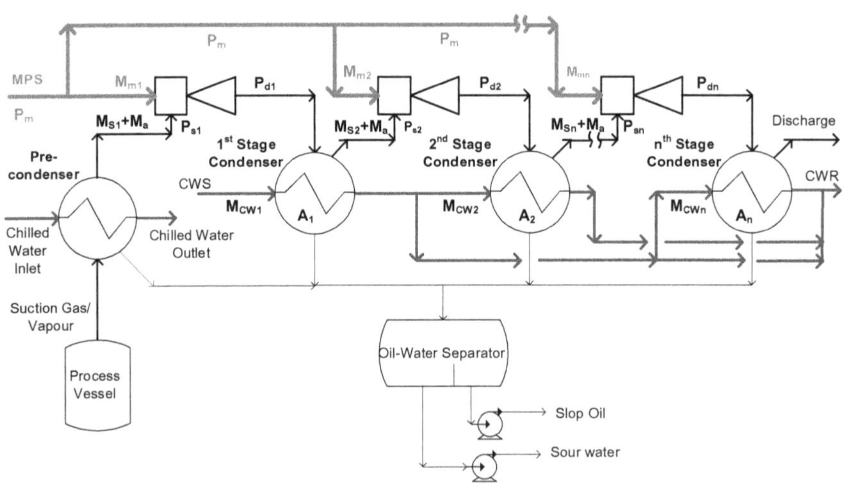

Fig. 8.9: Schematic of a multi-stage SJE system.

$$\text{Minimize}\, M_m = \sum_{i=1}^{n} M_{\text{mi}} = \sum_{i=1}^{n} \left[\frac{P_{\text{di}}}{P_{\text{Si}}} \left(0.434 - \frac{1.338}{P_{\text{si}}} + 0.000475\, P_{\text{si}} \right) - 0.187 \right]$$

$$\left(1.2 - \frac{P_m - 10.2}{20} \right) \left(\frac{M_{OG}}{\text{MW}_{\text{EROG}} \times T_{\text{ERA}}} + \frac{M_{\text{Si}}}{\text{MW}_{\text{ERA}} \times T_{\text{ERS}}} \right) \tag{8.11}$$

Here, P_{di} and P_{si} are the discharge and suction pressures for ith stage SJE, and steam consumption in each stage is calculated using Eqs. 8.5 and 8.6. The quantities in Eq. 8.11 can be calculated as follows.

Considering pressure drop in ith stage condenser (DP_{i}), suction pressure for (i + 1)th stage is:

$$P_{s(\text{i}+1)} = \text{P}_{\text{d(i)}} - \text{DP}_{\text{i}} \tag{8.12a}$$

Steam/water vapour flow rate at the inlet of (i + 1)th stage is:

$$M_{\text{S(i+1)}} = \frac{\text{MW}_{\text{s}} \times \text{Pv}_{\text{i}} \times M_{\text{OG}}}{\text{MW}_{\text{OG}} \times (P_{s(\text{i+1})} - \text{Pv}_{\text{i}})} \tag{8.12b}$$

This is assuming that mole fractions of steam/water vapour, air and non-condensable mixture are proportional to their partial pressures. Saturation pressure of steam, Pv_{i} corresponding to the vent temperature of ith stage condenser (T_{coi}) is a function of this vent temperature:

$$\text{PV}_{\text{i}} = \text{f}(T_{\text{coi}}) \tag{8.12c}$$

Steam properties in Eqs. 8.12c and 8.12h can be calculated using the Excel spreadsheet with inbuilt steam properties.

Considering approach temperature (T_{APi}), vent temperature for the ith stage condenser is:

$$T_{\text{coi}} = T_{\text{CWRi}} + T_{\text{APi}} \tag{8.12d}$$

Here, T_{CWRi} is the cooling water return temperature at the ith stage condenser. T_{AP} values for the condensers are user-defined values. Guidance for choosing them is presented in Section 8.6.2.

Mole fraction of steam/water vapour at the ith stage condenser inlet is

$$mf_i = \left[\frac{M_{si} + M_{mi}}{18}\right] / \left[\frac{M_{OG}}{MW_{OG}} + \frac{M_{si} + M_{mi}}{18}\right] \tag{8.12e}$$

Here, MW_{OG} is the MW of gases other than water vapour/steam, which include air and non-condensable gases. Mole percent of non-condensable gases at the ith stage condenser inlet is:

$$NC_i = (1 - mf_i) \times 100 \tag{8.12f}$$

Partial pressure of steam/water vapour at the ith stage condenser inlet is:

$$PP_{Si} = P_{di} \times mf_i \tag{8.12g}$$

Saturation temperature of water vapour corresponding to its partial pressure at ith stage condenser inlet is a function of its partial pressure at this inlet:

$$T_{cii} = f(PP_{Si}) \tag{8.12h}$$

Log mean temperature difference for ith stage condenser is:

$$LMTD_i = \frac{(T_{cii} - T_{CWR}) - (T_{coi} - T_{CWS})}{\ln\frac{(T_{cii} - T_{CWR})}{(T_{coi} - T_{CWS})}} \tag{8.12i}$$

Overall heat transfer coefficient for ith stage condenser, U_i is given by Eq. 8.7 or 8.8. Area and cooling water required for the ith stage condenser are:

$$A_i = \frac{L_i \times (M_{Si} + M_{mi} - M_{S(i+1)}) \times 1000}{U_i \times LMTD_i \times 3600} \tag{8.12j}$$

$$M_{CWSi} = \frac{L_i \times (M_{Si} + M_{mi} - M_{S(i+1)})}{4.184 \times (T_{CWR} - T_{CWS})} \tag{8.12k}$$

Here, L_i is latent heat of water vapour in kJ/kg and 4.184 is the specific heat of water in kJ/kg.K.

In the above optimization problem, decision variables are P_{di} for i = 1, 2, ..., $n - 1$, and their bounds are $0 < P_{di} < P_{dn}$. Note that P_{dn} is a specified value as part of the SJE system requirements. Other constraints are: $M_{Si} > 0$ for i = 1, 2, ..., $n - 1$ or n. The optimization problem (Eq. 8.11) and the model equations (Eqs. 8.12) are employed in the case studies presented in the next section.

As most of the vacuum systems involve air, non-condensable gases and steam/water vapour only, the above optimization problem is very useful for preliminary design/retrofit and/or verify vendor's proposals. The above model equations ignore any vapour superheat at the ejector inlet and sub-cooling of vapour and condensate in condensers; also, they do not include LMTD correction factor. T_{ERA} and T_{ERS} values are very close to unity; hence, they are considered as constants with a value of 1.0 each. So, area calculations in this method are approximate but sufficient for optimization. For accurate condenser design, one can use commercial design software such as HTRI, ASPEN EDR or HTFS. For vacuum systems involving condensable gases (besides steam), optimization problem can be solved by including simulation data from process simulators such as Aspen Hysys, Aspen Plus and Pro/II.

Reddy *et al.*[10] checked the accuracy of the model in Eqs. 8.11 and 8.12 for an example problem in Power[2] and found that condenser's vent pressures tally well with those given in this book, with error in the range of +14% to −8%. Motive steam flow rates estimated by the above model is higher (up to 25%) for three- and four-stage SJE, and lower (by ~ 20%) for two-stage SJE, compared to the reported values in Power.[2] These are mainly due to differences in R_a values from the graphs in Power[2] and Eq. 8.6 used in the above model. For accurate estimation of steam flow rates, use the optimized vent pressures and motive steam rates from the actual performance graphs supplied by SJE vendor. However, the optimization procedure becomes partly manual, and requires iterations and hence lot of time and effort.

The objective function, Eq. 8.11 is for minimizing total steam required by an SJE system. Another possible objective is total annual cost (including both CC and OC), which requires cost estimates of equipment and utilities described in Chapter 6. It is also possible to consider more than one objective at the same time and perform multi-objective optimization in MS Excel. See Rangaiah *et al.*[13] for a tutorial-cum-review of procedure, techniques, applications, and programs for multi-objective optimization.

8.6 Case Studies

In this section, two case studies with many sub-cases are described to illustrate key design options and their impact on minimizing WH generation in SJEs. Two other case studies are available in Reddy and Rangaiah[14]

8.6.1 *Case Study 1: Vacuum System Design for Condenser of a Condensing Steam Turbine*

Condensing STs maximize power generation by extracting steam to vacuum level. This steam condenses in a surface condenser by exchanging heat with cooling water or air. For maintaining vacuum in the condenser, uncondensed vapour including air and steam should be removed by SJE or LRVP or DVP. As this service does not involve any hydrocarbons, DVP is beneficial for reducing WH generation and OC.

Case study 1 considers vacuum system for a surface condenser for condensing flash steam from a condensing ST, driving a huge centrifugal compressor in a process plant. Operating conditions of this system are shown in Fig. 8.10. ST is using 30,000 kg/hr of HPS and exhausts it in partially wet condition. Steam condensate (3000 kg/hr) is separated from the ST's discharge header by gravity and pumped to the steam condensate drum. Steam (= 27,000 kg/hr) from the discharge header is condensed in a surface condenser using cooling water.

Fig. 8.10 shows the operating conditions of the surface condenser (Case 1A), installed with two SJEs in parallel (of which, one is in operation and the other is in standby mode) and another condenser. Present operation of SJEs is using 923.2 kg/hr of MPS, which is ultimately condensed in a condenser for recovering heat and water. Although this looks reasonable from energy perspective, it wastes pressure energy of MPS that can be used to generate electric power in a backpressure type ST.

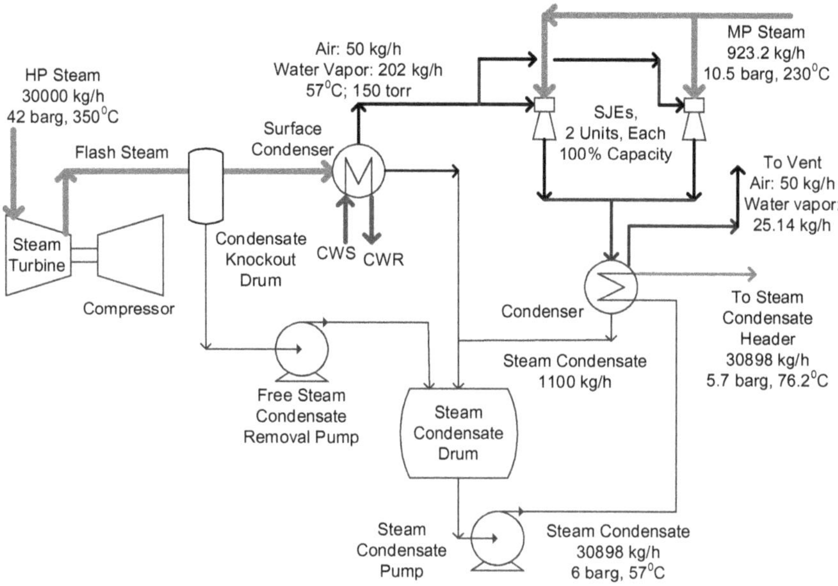

Fig. 8.10: Surface condenser installed with steam jet ejectors (Case 1A).

Moreover, if the condensate header is not well insulated, to prevent the problem of corrosion under insulation, most of the recovered energy will be lost to atmosphere.

Hence, it is beneficial to avoid the use of MPS for SJEs as much as possible. One such design is presented as retrofit case (Case 1B) in Fig. 8.11. This design uses two DVPs (with one in operation and another in standby mode), as they are the most efficient vacuum-generating equipment, occupy less space, do not generate any foul water and suitable for non-hazardous service. Note that LRVPs are less attractive to DVPs for this case due to higher power consumption and more piping requirements for seal water and foul water. These piping costs can be significant if the process unit does not have nearby tie-in points for seal water and foul water. For Case 1B, DVP requires 25.2 kW of electricity and eliminates MPS required for SJE; resulting OC savings due to this modification are US$147075/year (assuming MPS cost of US$22.28/ton, electricity cost of US$0.15/kWh and annual operating hours of 8760). Installed cost of two DVPs (with explosion proof motors), based on vendor quotation, is ~US$300,000. So, payback for this retrofit modification is ~2 years.

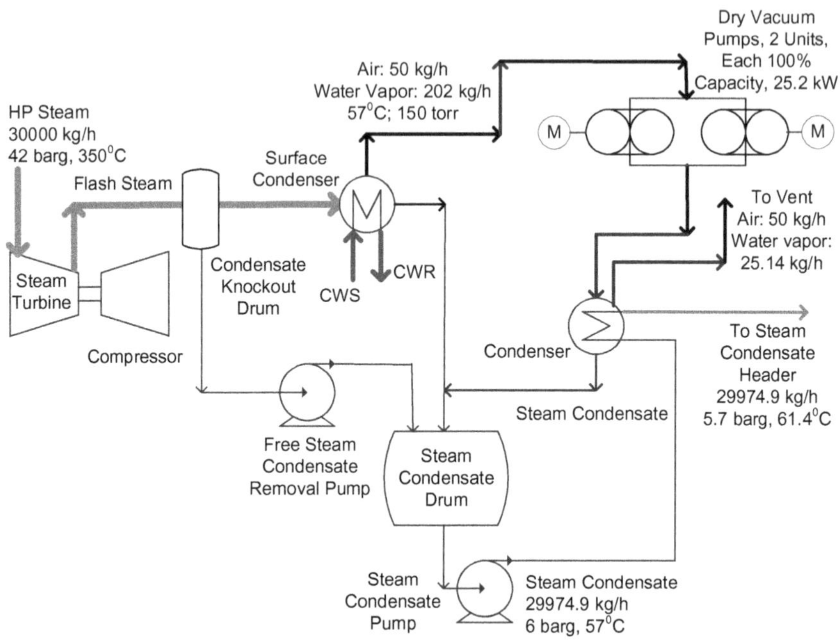

Fig. 8.11: Surface condenser retrofitted with DVPs (Case 1B).

Existing condenser is oversized for the modified case. However, as steam and condensate are clean, there is no concern of fouling.

8.6.2 *Case Study 2: Design of a Large Vacuum System Involving Hydrocarbons*

This case study is on the design of a vacuum system of a crude vacuum distillation column in a typical petroleum refinery. Overhead pressure of the distillation column is 50 torr, vacuum system suction flow rate is 6900 kg/hr (6500 kg/hr of steam/water vapour + 400 kg/hr of air and non-condensable hydrocarbons with average molecular weight of 40), and discharge pressure required is 850 torr. Cooling water supply and maximum return temperatures are 30°C and 41°C, respectively. Optimized design details for Case 2A with a two-stage SJE system, are shown in Fig. 8.12. Here, cooling water exiting pre-condenser is re-used in first stage and after condensers. Case 2A consumes 3.72 ton/hr of MPS and 2140 ton/hr of cooling water. Eqs. 8.11 and 8.12 given earlier are employed to find

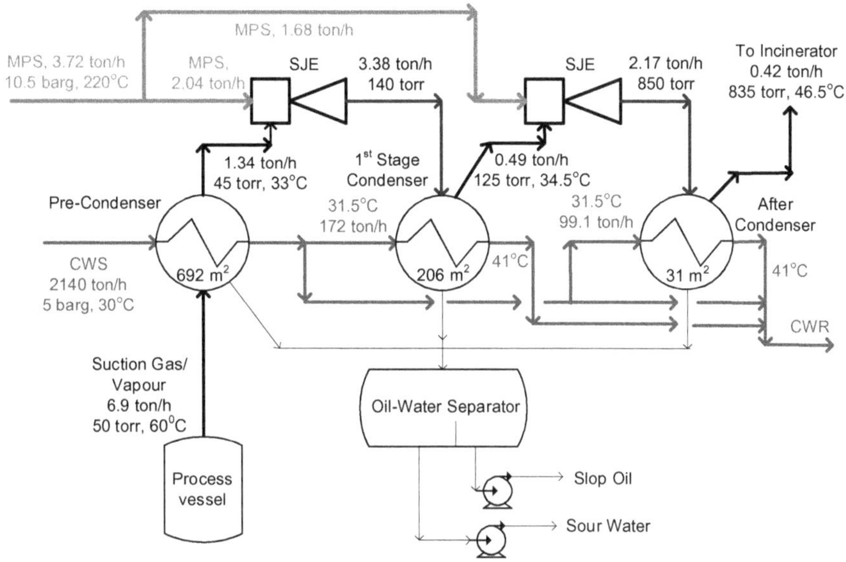

Fig. 8.12: Optimized vacuum system of a crude distillation column (Case 2A).

optimal values of required MPS, cooling water and condenser areas in this case study.

Case study 2A uses 3°C, 3°C and 15°C approach temperatures (i.e., difference in condenser vapour outlet temperature and cooling water inlet temperature to the condenser) for pre-, first stage and after condensers, respectively. Use of minimum possible approach temperature at pre-condenser and first stage condenser is essential to achieve minimum MPS consumption and lower OC of the vacuum system. Effect of different approach temperatures are illustrated by two sub-cases of Case 2A: Alt-1 with approach temperatures: 3°C, 5°C, 15°, for pre-, first stage and after condensers, respectively (Fig. 8.13), and Alt-2 with approach temperatures: 5°C, 5°C, 15°, for pre-, first stage and after condensers (Fig. 8.14).

Heat transfer areas and utility requirement for Case 2A and its sub-cases are summarized in Table 8.3. All these three cases use 5, 15 and 15 torr for pre-, first stage and after condensers, respectively. It can be seen from Table 8.3 that, as the approach temperature for pre-condenser increases from 3°C to 5°C, there is a significant rise in MPS consumption (~ 79% between Alt-1 and Alt-2 sub-cases) for the SJEs, which produces more WH. Increase in first stage approach temperature increases MPS

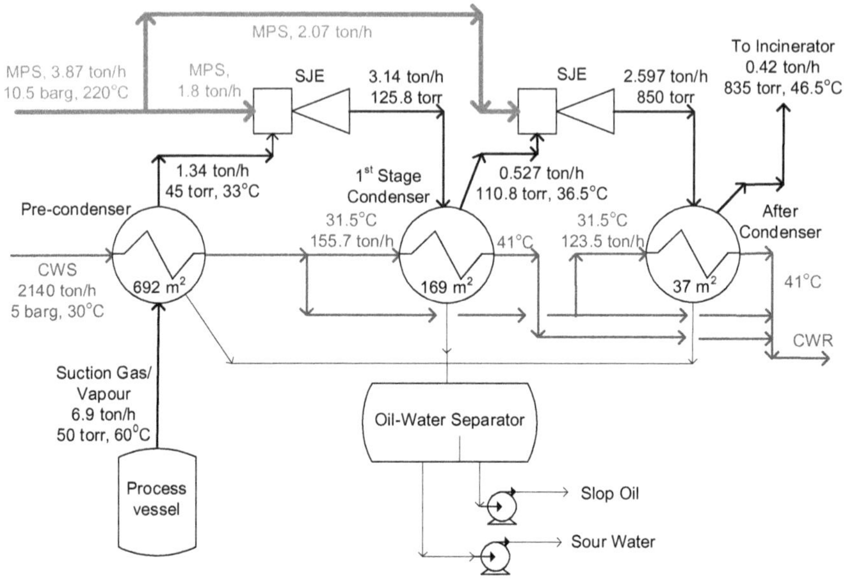

Fig. 8.13: Design details of sub-case Alt-1 of Case 2A.

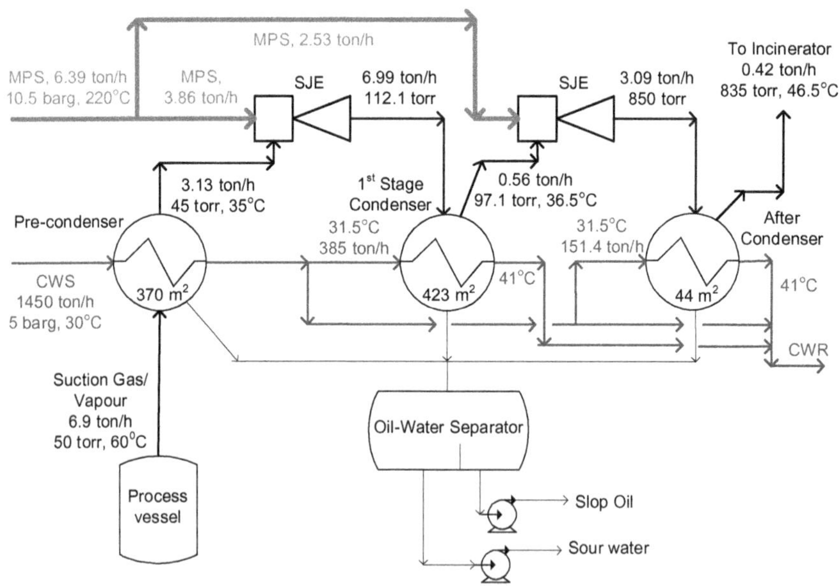

Fig. 8.14: Design details of sub-case Alt-2 of Case 2A.

Table 8.3: Effect of condenser approach temperatures on heat transfer areas and utility requirements

Case	Approach Temperatures of Pre-, First Stage and after Condensers (°C)	MPS Flow Rates to First and Second Stages (ton/hr)	Total MPS Flow Rate (ton/hr)	Heat Transfer Areas of Pre-, First Stage and after Condensers (m^2)	Total Heat Transfer Area (m^2)	Fresh Cooling Water Flow Rate to Pre-Condenser (ton/hr)	Cooling Water Flow Rates to First Stage and After Condensers (ton/hr)
Case 2A	3, 3, 15	2.04, 1.68	3.72	692, 206, 31	929	2140	172, 99.1
Case 2A: Alt 1	3, 5, 15	1.8, 2.07	3.87	692, 169, 37	898	2140	155.7, 123.5
Case 2A: Alt 2	5, 5, 15	3.86, 2.53	6.39	370, 423, 44	837.6	1450	384.8, 151.4

consumption marginally (Cases 2A and Alt 1). Further, lower pre-condenser approach temperature increases both heat transfer area and fresh cooling water flow rate. However, more MPS consumption leads to more WH generation and more OC, and so it is recommended to use as low approach temperature as practical. A good cooling water filtration system and high velocity (2–3 m/sec) of water in condenser tubes help to maintain sustainable heat transfer in condensers with low approach temperatures. Pressure drops for the pre-condenser shall be as minimum as possible (typically 3 torr) to maximize steam/water vapour condensation and hence minimize MPS consumption.

Case 2A with lower approach temperatures can be taken as the final design case. Assume that after a few years of operation, the plant engineer found an opportunity to revamp this vacuum system for better product separation and hence product yield improvement and higher profitability. For this, the new vacuum pressure required at the distillation column overhead is 30 torr (lower than 50 torr in the current operation). Further, new loads are 500 kg/hr of air with non-condensable hydrocarbons and 7500 kg/hr of steam/water vapour. If it is desirable to use SJEs only, then a three-stage SJE system is required for this revamp case (Case 2B), as shown in Fig. 8.15.

In Case 2B, with the use of cooling water (supplied at 30°C), it is not possible to condense any steam/water vapour in the pre-condenser and

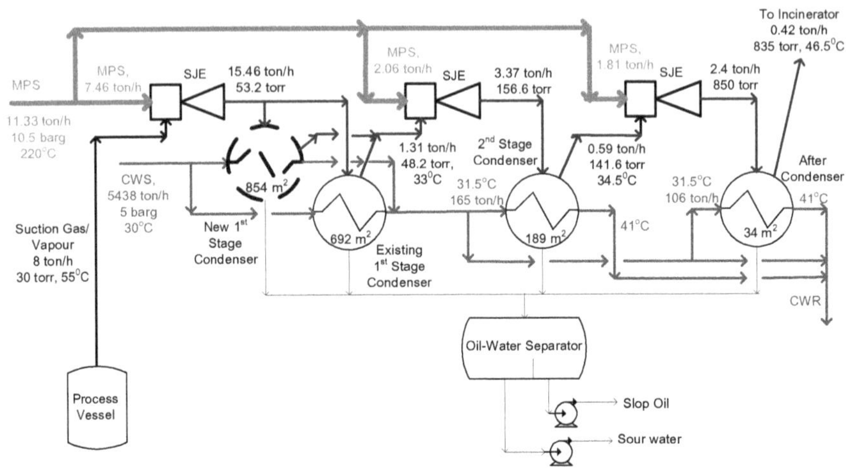

Fig. 8.15: A three-stage SJE system to meet revamp requirements (Case 2B); here, the Existing first stage condenser is the pre-condenser in Case 2A (Fig. 8.12).

hence it is not used in Fig. 8.15. If two-stage SJE design is employed, MPS demand and heat transfer area requirements will be even higher. Revamp design should re-use existing HEs (more expensive than SJEs) as far as possible. Original first stage and after condensers (in Case 2A) can be re-used as second stage and after condenser in the updated system. Pre-condenser in Case 2A does not meet the new area requirement of 1546 m^2 for first stage inter-condenser in Case 2B. Hence, a new HE (labelled as new first stage condenser in Fig. 8.15) of 1546–692 = 854 m^2 can be added in parallel to the existing first stage condenser, to meet the new area requirement for the first stage condenser. All other equipment such as oil–water separation tank, oil and water pumps are re-used. Generally, SJEs cannot be re-used for the new design, as they cannot operate stably under conditions different from the original design. Hence, the proposed revamp design uses three new SJEs.

Can MPS consumption for Case 2B be reduced by avoiding the third/last SJE stage? The answer is yes. As this service involves hydrocarbons, DVP is not recommended. However, LRVP can be used if the suction pressure is $\geq$ 150 torr and suction volume is < 18,000 acfm. In Fig. 8.15, second condenser vent pressure is 141.6 torr and suction flow is 0.59 ton/hr (= 1404 acfm < 18,000 acfm). Although vent pressure is slightly lower than 150 torr, LRVP can be used. This will reduce 1.81 ton/hr of MPS but will require some electrical power for the LRVP. Economic analysis of this option is given in the next sub-section.

Is it possible to further improve the revamp design to reduce number of SJE stages, MPS consumption and cooling water consumption as well as avoid adding a new HE at the first stage condenser? The answer is yes. The key idea for achieving this is to use chilled water, possibly generated by WHR (wherein thermal energy cost can be considered as zero for the operation of absorption chiller) at pre-condenser. WH for chilled water generation is often available in petroleum refineries and other process plants. Next two cases present the concept of chilled water usage at pre-condenser and replacing last stage with LRVP.

Case 2C, shown in Fig. 8.16, uses chilled water, generated by a single-stage LiBr chiller (whose details are available in Chapter 5), in the pre-condenser. With the use of chilled water supplied at 7°C, most of the steam/water vapour is condensed in the pre-condenser and hence it would require just two SJE stages. As can be seen in Fig. 8.16, MPS

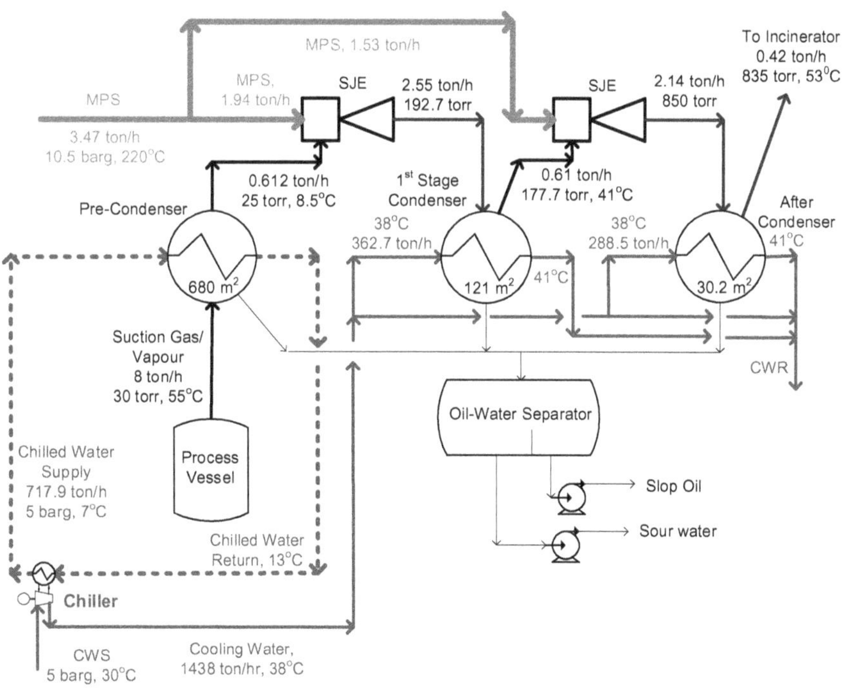

Fig. 8.16: A two-stage SJE with pre-condenser cooled by chilled water (Case 2C).

consumption dropped from 11.33 ton/hr (in Case 2B) to 3.47 ton/hr (by 7.86 ton/hr). Pre-condenser area requirement now is just 680 m^2 (less than 692 m^2 in Case 2A). Hence, no new HE is required at first stage condenser since this scenario requires first stage condenser area of 121 m^2, which is less than the area (206 m^2) of the first stage condenser, already available from Case 2A. Cooling water consumption is significantly dropped from 5438 ton/hr in Case 2B to 1438 ton/hr in Case 2C.

Is it possible to decrease MPS consumption even further? It is indeed possible. As the first stage condenser vent pressure is 177.7 torr and suction flow rate < 18,000 acfm, second stage SJE can be replaced with LRVP as illustrated in Fig. 8.17 as Case 2D. Then, MPS consumption reduces to 1.94 ton/hr at the expense of 88 kW of electricity and 15.2 ton/hr of additional cooling water consumption. Also, this case requires only two of three existing HEs in Case 2A; HE of area 37 m^2 is not required and can be employed elsewhere.

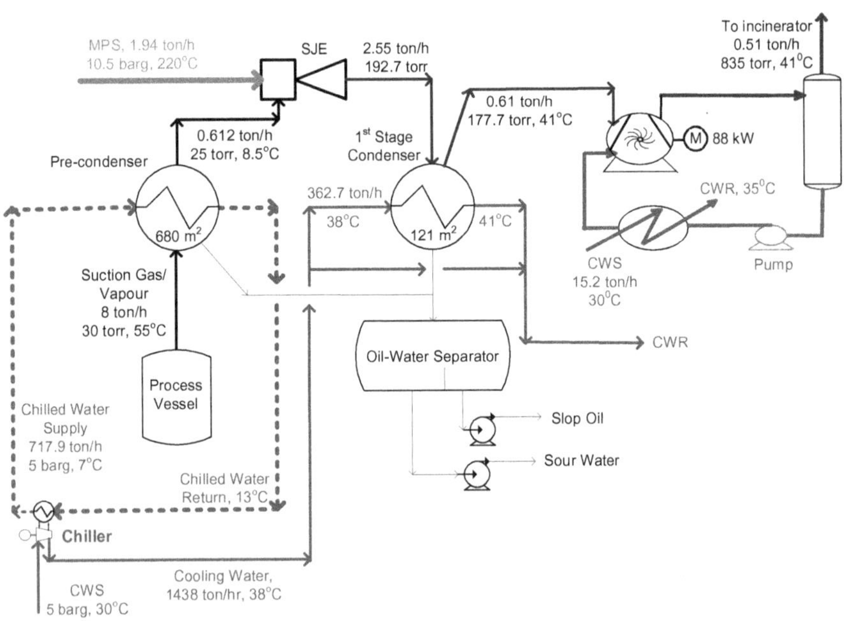

Fig. 8.17: A single-stage SJE with pre-condenser cooled by chilled water and LRVP: Case 2D.

8.6.3 *Analysis of Revamp Alternatives: Cases 2B to 2D*

Economics of revamp alternatives for the vacuum system of a crude distillation column (Cases 2B to 2D developed above) are analysed in this sub-section. For this, CCs of SJEs are estimated from Fig. 10.2 in Power,[2] and they are adjusted for inflation using the Chemical Engineering Plant Cost Index of 598.3 for July 2020. Purchase costs of absorption chiller (single stage using LiBr solution) and LRVP are based on vendor quotations. Installed cost of LRVP is taken as twice its purchase cost. Absorption chiller is a packaged unit and so involves less cost for installation; hence, its installed cost is taken as 1.5 times its purchase cost. Utility requirements for foul water treatment are 0.345 MM BTU of LLP steam, 2.73 kWh of electricity and 135 Imperial Gallons of cooling water, all per ton of foul water, from Prakash.[15] Unit utility costs from an industrial facility are: MP steam = US$23.56/ton, LLP steam = US$22.64/ton, cooling water = US$0.11/ton and electric power = US$0.15/kWh.

CC, OC and payback period for Cases 2B to 2D are summarized in Table 8.4. Payback period is defined as the ratio of incremental CC for vacuum system revamp to benefit from revamp. Here, the benefit is the sum of utility cost savings from revamp and cost savings (estimated as US$5 millions) from the crude distillation column operation at lower vacuum after revamp. Utility cost savings from revamp is the difference in OC for Case 2A and OC for revamp case (Case 2B, 2C or 2D). As can be seen in Table 8.4, annual OC of a vacuum system is very significant compared to the installed cost. This is true especially for vacuum system revamp projects involving re-use of equipment as much as possible.

In all cases in Table 8.4, first stage SJE consumes more MPS and generates more WH, followed by second stage and later stages, if present, in decreasing order. The main reason for this is decreasing amount of water vapour handled by SJE stages as we move from first stage to last stage. Hence, by reducing the suction load to first stage, the number of stages, cooling water consumption, MPS consumption and hence WH generation can be reduced.

The base case (Case 2A) represents the optimized design for achieving 50 torr pressure at distillation column overhead. When this system requires to be revamped for achieving 30 torr vacuum pressure, with only

Table 8.4: Analysis of alternatives (cases 2B to 2D) for retrofitting vacuum system (Case 2A) of a distillation column in a petroleum refinery

Details of Vacuum system revamp		**Case 2A: Base case with pre-condenser, cooled with cooling water, and two-stage SJE**	**Case 2B: Revamp case with three-stage SJE**	**Case 2C: Revamp case with pre-condenser, cooled with chilled water, and two-stage SJE**	**Case 2D: Revamp case with pre-condenser, cooled with chilled water, single-stage SJE and LRVP**
Condenser area, m^2	Pre-condenser	692	0	680	680
	Condenser-1	306	1546	121	121
	Condenser-2	31	189	30.2	0
	Condenser-3	0	34	0	0
	Total area	**1029**	**1769**	**831**	**801**
Cooling water flow rate for condensers (ton/h)	Pre-condenser	2140	0	0	0
	Condenser-1	206	5438	362.7	362.7
	Condenser-2	31	165	288.5	0
	Condenser-3	0	106	0	0
Chilled water flow rate	Pre-condenser	0	0	717.9	717.9
MPS flow rate for SJEs (ton/h)	1st stage ejector	2.04	7.46	1.94	1.94
	2nd stage ejector	1.68	2.06	1.53	0.00
	3rd stage ejector	0	1.81	0.00	0.00
Single-stage absorption chiller	Cooling water flow rate (ton/h)	0	0	1438	1438
	LLP Steam flow rate (ton/h)	0	0	11.83	11.83

	Power for chiller and chilled water pumps (kW)	0	0	179	179
Total utility requirements	MPS (ton/h)	3.72	11.33	3.47	1.94
	LLP steam (ton/h)	0	0	11.83	11.83
	Chilled water (ton/h)	0	0	717.9	717.9
	Cooling water(ton/h)	2140	5438	1438	1438
	Power requirement (kW)	0	0	179	267
Savings at foul water stripper	LLP steam saved (ton/h)	0	0.00	0.04	0.31
	Power saved (kW)	0	0.00	0.68	4.86
	Cooling water saved (ton/h)	0	0.00	0.15	1.09
Net savings	Total MPS saved (ton/h)	0	0.00	0.25	1.78
	Total LLP steam saved (ton/h)	0	0.00	–11.78	–11.52
	Total cooling water saved (ton/h)	0	0	702	703
	Total power saved (kW)	0	0	–178	–262
Operating profit, Million US$/year	Due to better separation and yield at the column	0	5	5	5

(Continued)

Table 8.4: *(Continued)*

Details of Vacuum system revamp		Case 2A: Base case with pre-condenser, cooled with cooling water, and two-stage SJE	Case 2B: Revamp case with three-stage SJE	Case 2C: Revamp case with pre-condenser, cooled with chilled water, and two-stage SJE	Case 2D: Revamp case with pre-condenser, cooled with chilled water, single-stage SJE and LRVP
Operating cost, Million US$/year	If LLP steam for absorption chiller costs US$22.64/ton	**2.83**	**7.58**	**4.67**	**4.41**
	If LLP steam for absorption chiller is available at zero cost	**2.83**	**7.58**	**2.33**	**2.07**
Capital cost for revamp, Million US$	Steam ejectors	—	0.42	0.07	0.06
	LRVP	0	0	0	0.32
	Single-stage absorption chiller	0	0	3.35	3.35
	Additional 1st stage condenser	0	0.345	0.00	0.00
	Total installation cost	**0**	**0.769**	**3.42**	**3.73**
Payback period for revamp, years	If LLP steam for absorption chiller costs US$22.64/ton	—	3.06	1.08	1.09
	If LLP steam for absorption chiller is available at zero cost	—	3.06	0.62	0.65

SJEs, a three-stage SJE system (Case 2B) is required. As pre-condenser with cooling water is not feasible for this case, both MPS and cooling water consumption are significantly increased to 11.33 ton/hr and 5438 ton/hr, respectively, compared to 3.72 ton/hr and 2140 ton/hr, respectively, in Case 2A. Accordingly, OC of Case 2B increases to US7.58 million from US$2.83 million for Case 2A. Also, it requires an additional HE with 854 m^2 area, at first stage condenser (even after re-using the pre-condenser from Case 2A). Payback period for Case B is still reasonable at 3.06 years.

With the use of chilled water, revamp design in Case 2B can be improved to Case 2C, which uses a pre-condenser (cooled with chilled water) and two-stage SJE system. Both MPS and cooling water consumption are significantly reduced from 11.33 ton/hr and 5438 ton/hr in Case 2B to 3.47 ton/hr and 1438 ton/hr in Case 2C, respectively. This improved case requires 11.83 ton/hr of LLP steam for the absorption chiller to produce 717.9 ton/hr of chilled water for pre-condenser. Cooling water used in the chiller is re-used in SJE system (Figs. 8.16 and 8.17). This method re-uses all the condensers in Case 2A and hence no new condensers are required.

Two OC scenarios for chilled water generation are considered in Table 8.4. The first scenario uses LLP steam cost of US$22.64/ton based on steam enthalpy value, and the second scenario assumes zero cost of LLP steam (by using WHR of sufficient vented flash steam in the facility). Payback period for the first scenario is still attractive at only 1.08 years for revamping as in Case 2C. Obviously, for WH-powered chilled water scenario, it is even more attractive at 0.62 years. Availability of plot space and maintenance cost are other critical issues for chiller installation. Mechanical chillers powered by electricity can also be used for chilled water generation. However, this method will be attractive only if electricity is very cheap and hence is not discussed here. See Chapter 5 for more details on mechanical chillers.

MPS consumption of 3.47 ton/hr in Case 2C is further reduced to 1.94 ton/hr in Case 2D by replacing the last SJE stage with LRVP, which requires 0.32 million US$ for its purchase and installation. Even then, payback period for this case is between 1.09 and 0.65 years, depending on chilled water cost, which is similar to that in Case 2C (Table 8.4).

8.7 Summary

This chapter presents design principles and applications of main vacuum generation equipment (pumps) in process industries.

- SJEs are most widely used as they have almost unlimited capacity, safer for handling hazardous vapours and do not have any suction pressure limitation. However, they are highly energy inefficient and generate large quantity of WH, which is eventually discarded to atmosphere via cooling water.
- Steam consumption for SJE system can be reduced by condensing maximum vapour in a pre-condenser, increasing number of stages and/or replacing SJEs with LRVP and/or DVP.
- After SJEs, LRVP and DVP are more common and efficient but they have capacity constraints.
- LRVP is not suitable for applications, where suction pressure is lower than the vapour pressure of seal liquid.
- DVPs are not suitable for handling hazardous substances and cannot handle any liquid slugs.
- For minimizing cooling water consumption, water used in pre-condenser can be re-used in inter-stage and after condensers.
- The best strategy for reducing steam consumption is to condense and separate as much water vapour and other condensable components in the suction gas as possible in a pre-condenser. Use of chilled water, powered by flash steam (generated from WH, if available) is beneficial for this purpose.
- Low approach temperature and pressure drop in the pre-condenser reduce motive steam and cooling water required for a vacuum system.
- SJE optimization illustrated in this chapter is useful for conceptual design of new as well as revamp design of vacuum systems. It minimizes WH generation, motive steam consumption, cooling water and CC.

Power[2] is a holistic book on SJE, DVP and LRVPs, whereas Bannwarth[3] is a comprehensive book on LRVP. They are recommended for further reading on the topics of this chapter. The book by Coker[5] is an excellent resource for estimation of air leakage and design calculations of SJEs.

References

1. Ryans J, Bays J. (October 2001) Run clean with dry vacuum pumps. *Chem Eng Prog* **97**:32–41.
2. Power RB. (2005) *Steam Jet Ejectors for the Process Industries*, 2nd ed. McGraw-Hill, Charleston.
3. Bannwarth H. (2005) *Liquid Ring Vacuum Pumps, Compressors and Systems*. Wiley-VCH Verlag GmbH & Co. KGaA, Weinheim.
4. Tunna C. (May 2005) *Pumping Potentially Explosive Atmospheres*. *Chem Eng* 30–31.
5. Coker AK. (2007) *Ludwig's Applied Process Design for Chemical and Petrochemical Plants*, Vol. 1, 4th ed. Burlington.
6. Couper JR, Penney WR, Fair JR, Walas SM. (2012) *Chemical Process Equipment-Selection and Design*, 3rd ed. Butterworth-Heinemann, Waltham.
7. Skelton K. (January 1998) Variable frequency drives for a vacuum pump system, *Pumps and Systems Magazine*. http://www.graham-mfg.com/usr/pdf/TechLibVacuum/219.PDF
8. Seider WD, Seader JD, Lewin DR. (2009) *Product and Process Design Principles Synthesis, Analysis and Evaluation*, 3rd ed. Wiley.
9. HEI. (2012) *Standards for Steam Jet Vacuum Systems, Heat Exchange Institute Inc.*, 7th ed. Cleveland, Ohio.
10. Reddy CCS, Naidu SV, Rangaiah GP. (2013) *Energy Efficient Vacuum Systems*. PTQ 2Q, pp. 125–135.
11. Trambouze B. (1999) *Petroleum Refining, Vol. 4, Materials and Equipment*. Editions Technip, Paris.
12. Reddy CCS, Rangaiah GP, Naidu SV. (January 2013) Waste heat recovery methods and technologies. *Chem Eng* 28–38.
13. Rangaiah GP, Feng Z, Hoadley AF. (2020) Multi-objective optimization applications in chemical process engineering: Tutorial and review. *Processes* 8: 508. doi:10.3390/pr8050508 (2020).
15. Reddy CCS, Rangaiah GP. (2016) Retrofit of vacuum systems in process industries. In: Rangaiah GP (ed), *Chemical Process Retrofitting and Revamping: Techniques and Applications.* Wiley.

16. Prakash S. (2003) *Refining Processes Handbook*. Gulf Professional Publishing, Burlington.

Acronyms and Notation

Acronyms

CC	Capital Cost
DP	Differential Pressure (torr)
DVP	Dry Vacuum Pump
FS	Flash Steam
HE	Heat Exchanger
HPS	High-Pressure Steam
LMTD	Log-Mean Temperature Driving Force (°C)
LLP	Low-Low Pressure Steam
LRVP	Liquid Ring Vacuum Pump
MPS	Medium-Pressure Steam
MW	Molecular Weight
NC	Non-Condensable Gases
OC	Operating Cost
OP	Operating Pressure
SJE	Steam Jet Ejector
ST	Steam Turbine
VSD	Variable Speed Drive
WH	Waste Heat
WHR	Waste Heat Recovery

Notation

A	Heat transfer area of a heat exchanger (m^2)
i	Index for SJE stage
L	Latent heat of vapour (kJ/kg)
M	Mass flow rate (kg/sec)
mf	Mole fraction
n	Number of SJE stages
P	Pressure (bar) or power (kW)
PP	Partial pressure (torr)

Pv	Saturation pressure of steam corresponding to the vent temperature (torr)
R_a	Amount of motive steam required to compress unit DAE mass of gas/vapour in a SJE
T	Temperature (°C or K)
U	Overall heat transfer coefficient ($W/m^2.K$)
V	Volumetric flow rate (m^3/hr)

Greek Symbols

η_e	Isothermal efficiency
η_d	Adiabatic thermal efficiency
ρ	Density (kg/m^3)

Subscripts

A	Air
APi	Approach temperature for i^{th} stage condenser
cii	i^{th} stage condenser inlet condition
coi	i^{th} stage condenser outlet condition
CWR	Cooling water return
CWS	Cooling water supply
CWSi	Cooling water supply for i^{th} stage condenser
d	Discharge
di	Discharge condition for i^{th} stage SJE
dn	Discharge condition for n^{th} stage SJE
ERA	Entrainment ratio for air
EROG	Entrainment ratio for gases other than steam or water vapour
ERS	Entrainment ratio for steam or water vapour
ie	existing i^{th} stage heat exchanger
m	Motive steam
mi	Motive steam for i^{th} stage SJE
OG	Gases other than steam or water vapour
s	Steam/system/suction
si	Suction condition for i^{th} stage SJE
Si	steam or water vapour for i^{th} stage SJE

Exercises

8.1 Estimate DAE value of suction gas at 40 torr and 60°C. Suction gas contains 200 kg/hr of air, 100 kg/hr of methane and 200 kg/hr of steam.

8.2 Estimate the power required for LRVP and DVP to compress a suction gas stream to 800 torr, using the following data: flow rate = 1500 kg/hr, suction pressure and temperature are 170 torr and 55°C, and composition is 25 wt% air and rest steam. Assume LRVP efficiency of 0.25 and DVP efficiency of 0.4.

8.3 Describe the advantages and disadvantages of SJE, DVP and LRVP.

8.4 Estimate the required utilities and condenser areas for a three-stage SJE system for the following application: suction pressure = 50 torr; suction temperature: 55°C; discharge pressure = 800 torr; water vapour flow rate = 5000 kg/hr; flow rate of non-condensable hydrocarbons (of molecular weight of 38) = 350 kg/hr; air leakage rate = 200 kg/hr; motive MPS pressure = 10.5 barg; cooling water supply and return temperatures = 30°C and 40°C; pressure drop in pre-condenser = 3 torr, first stage condenser = 5 torr; pressure drop in each of other condensers = 10 torr; approach temperature for pre-condenser and first stage condenser = 3°C, and approach temperature for other condensers = 15°C. Use optimization procedure given in this chapter.

8.5 Estimate the required utilities and condenser areas by increasing the approach temperature for pre-condenser and first stage condenser to 6°C; use remaining data given in the previous exercise. What conclusion(s) can be made from the results?

8.6 Use the data in Exercise 8.4 and estimate the required utilities and condenser areas for the following cases and compare the results:
   (A) By applying additional 50% safety factor on air leakage (i.e., new air leakage = 300 kg/hr; use all the remaining data from Exercise 8.4).
   (B) Applying safety factor of 5 torr on suction pressure (i.e., new suction pressure is 45 torr; use all the remaining data from Exercise 8.4).

Which of the above two safety factors will lead to more OC penalty? Which safety factor do you recommend, and why?

8.7 Estimate utilities consumption in Exercise 8.4 if chilled water, available at 7°C and to be returned at 13°C to the chiller, is used in the pre-condenser.

8.8 Use LRVP to replace the last stage of SJE in Exercise 8.7 and re-calculate the utilities required. Assume LRVP and motor efficiency of 0.25 and 0.95, respectively.

8.9 Assuming the non-condensable gas in Exercise 8.7 is non-hazardous, calculate the utility consumption by replacing the last SJE stage in Exercise 8.7 with a DVP. Assume DVP and motor efficiency of 0.4 and 0.95, respectively. Compare the utility consumption in Exercises 8.4, 8.8 and 8.9.

Chapter 9

Waste Heat Recovery in Distillation

9.1 Overview

Distillation is the most important and common process for the separation of liquid mixtures in chemical and related industries. It handles more than 90% of such separations industry-wide and this trend is likely to continue in the future. However, distillation is highly inefficient; it consumes significant amount (~40%) of the energy used in process industries. In addition to heating duty, it requires significant amount of cooling duty at the overhead condenser.

For improving energy efficiency of distillation, it is necessary to reduce both the heating and cooling duties. For this, many waste heat recovery (WHR) methods such as operational changes (e.g., operating at the lowest possible pressure and pre-heating the feed), use of low-pressure drop internals, application of absorption heat pumps (AHPs) and mechanical vapour recompression (MVR), as discussed in Chapter 5, heat integration (described in Chapter 3), dividing-wall column (DWC), power generation methods (e.g., organic Rankine cycle, ORC and Kalina cycle [KC]) can be used. Operational optimization is preferred as the first step for improving distillation efficiency as it does not require any equipment modifications. Next step of improvement involves selection and implementation of a cost-effective solution among the many applicable WHR options, by performing a systematic study.

This chapter outlines WHR methods for distillation systems, and then presents their application to industrial cases. It begins with Section 9.2

with the discussion of WHR opportunities for distillation. Section 9.3 presents operation optimization for minimizing energy consumption for distillation. The subsequent section presents low cost/complexity methods for WHR. Section 9.5 outlines the application of heat pumps (HPs) for distillation WHR and debottlenecking. Sections 9.6 and 9.7 describe power generation options and application of both external and internal heat integration for distillation, respectively. Section 9.8 presents several industrial case studies on the application of WHR techniques to distillation. Finally, Section 9.9 summarizes the main points for reducing heating and/or cooling duties required for distillation.

Learning outcomes of this chapter on the application of WHR techniques for distillation are as follows.

1. Outline WHR opportunities in distillation.
2. Describe low, medium and high cost/complexity WHR solutions for distillation.
3. Discuss the application of HPs for reducing both heating and cooling duties of distillation.
4. Describe power generation opportunities for reducing cooling duty of distillation.
5. Describe external and internal heat integration opportunities in distillation.
6. Apply WHR techniques to industrial distillation examples.

9.2 WHR Methods for Distillation

A typical feed to distillation is a mixture of two or more components, often in liquid and/or vapour phases. Hot utility such as steam is used in the reboiler, for providing vapour phase for the separation of feed into top and bottom products. In the condenser, heat is removed from and to condense the overhead vapour stream. Part of the condensate is returned to the column as the reflux to provide liquid phase for separation, and the rest is withdrawn as the top product.

In a distillation column, the latent heat of condensation of the overhead vapour is usually released/wasted to atmosphere via cooling water or air. This waste heat (WH) is often significant and can be recovered to provide partial/total heating duty at the reboiler using HPs (covered in Chapter 5). It can also be used in ORC or KC, which can work with WH streams at 80°C–200°C, to generate electricity. ORC uses an organic component having a lower boiling point, as the working fluid, whereas KC uses ammonia and water mixture as the working fluid. Efficiency of both ORC and KC is lower than that of conventional Rankine cycle, which uses steam as the working fluid. However, ORC and KC are very beneficial as they operate using WH and hence do not incur any fresh fuel consumption.

Among the several options available for improving energy efficiency of distillation, the simplest is to minimize the energy usage by optimizing the column operation, which does not require any equipment modifications. Further energy reduction is possible by evaluating and implementing modification schemes ranging from low to high cost/complexity, which are described in Sections 9.4–9.6.

9.3 Minimizing Energy Consumption of Distillation by Operation Optimization

In industrial plants, distillation throughput and/or product specifications may change over time; hence, there are the following opportunities for energy minimization.

- <u>Operating the column at the lowest possible pressure</u>: Higher column pressure increases the difficulty of separation of components, and also reduces vaporization, requiring the column to operate at a higher temperature, and hence a higher hot utility consumption at the reboiler; high-pressure operation increases cooling duty at the condenser as well. Opportunities for reducing column pressure can be identified by a careful look at the separation and product purity requirements. Seasonal fluctuations of temperature of cooling medium used in the condenser (e.g., lower temperature of cooling water during winter) can allow the column to operate at a lower temperature/pressure.
- <u>Selecting optimal feed location</u>: Some distillation columns are provided with feed nozzles at multiple locations, to cater for changes in feed and/or product compositions. Introducing the feed at the wrong

location means the trays around the feed entrance operate inefficiently and hence require more energy, to achieve the required separation. In such cases, selecting the correct feed location enables lower duties at both reboiler and condenser. A process simulation package such as Aspen Hysys or Pro/II can be used for quick selection of the feed location. Significant effect of feed location on energy consumption and costs, and the need for rigorous (instead of shortcut) simulation of distillation for determining optimal feed location are illustrated for the separation of seven binary and six multicomponent mixtures, by Lek *et al.*[1]

- Avoid over-purifying the products: Operators often operate the distillation column with over purification of products due to concerns of producing off-spec products. This increases both reboiler and condenser duties, and wastes energy. With the use of advanced process control schemes, distillation column can be operated 'tighter' near the desired product specifications and hence save energy. Successful realization of this requires education and training of operators on advanced process control, to change their mindset.
- Use of variable reflux rate: High reflux rate can achieve high product purities. However, it increases the energy consumption for the column operation. Feed rate to distillation column may change from time to time, and hence adjusting the reflux rate according to feed flow rate using automatic control, saves energy.

9.4 Low-Cost/Complexity Modifications for Reducing Energy Consumption

Low-cost/complexity solutions for reducing energy and/or increasing WHR are generally easy to implement and often provide short payback periods (<2 years). They are summarized in Table 9.1.

9.5 WHR in Distillation Columns using Heat Pumps (Medium Cost/Complexity)

WHR in distillation columns can be achieved by using HPs such as MVR, AHP and absorption heat transformer (AHT). There are three variations of MVRs, namely, MVR, bottom flash (BF) and closed cycle compression

Table 9.1: Low-cost solutions to reduce energy consumption in distillation columns

Solution	Option(s)
Use of internals having low-pressure drop	• Use low-pressure drop and high-capacity trays • Replace trays with structured packing
Clean (i.e., free from fouling) operation of the condenser	• Design the condenser with cooling water velocity of 2–3 m/sec in the tubes. • Avoid high overdesign of condenser. • Use properly sized filters for cooling water supply • Clean condensers on time
Reduce fouling in the reboiler	• Design the reboiler to operate under nucleate boiling range. For this, avoid excessive temperature driving force. • Use special tubes such as high flux tubes for decreasing the temperature driving force for heat transfer in the reboiler. • Avoid high overdesign of Reboiler.
Use correct pressure level of steam to the reboiler	• Use minimum allowable pressure of steam in the reboiler • If required steam pressure is not available, use steam turbine to reduce the steam pressure to a level suitable for use at the reboiler. This is described in Section 16.5.2 of Chapter 16.
Feed-bottom heat exchanger (HE)	• Pre-heating of feed with column bottom stream in this feed-bottom HE reduces both reboiler and condenser duties. Such an arrangement is shown in Fig. 9.1.
Feed pre-heating using low-low pressure steam (LLPS) at 1–2 bar	• Pre-heating feed with LLPS can reduce the reboiler duty and is beneficial if the LLPS pressure is lower than the steam pressure used at the reboiler[2]
Use of pre-flash vessel	• Installing a pre-flash vessel can reduce the energy consumption for distillation of hot-pressurized feed. This arrangement (Fig. 9.2) is frequently used for crude distillation units in petroleum refineries.

(CCC) shown in Figs. 9.3–9.5, respectively. All of them use the same working principle of varying dew and bubble points via pressure change, caused by a compressor. Centrifugal compressors driven by electric motors, are generally used for MVR applications.

In MVR-assisted distillation (Fig. 9.3), the overhead vapour is compressed to raise its dew point temperature above the reboiler temperature and then used as the heating medium in the reboiler. This transfers the latent heat of the compressed overhead stream to the reboiler. The cooled/condensed overhead stream from the reboiler is then flashed and sent to

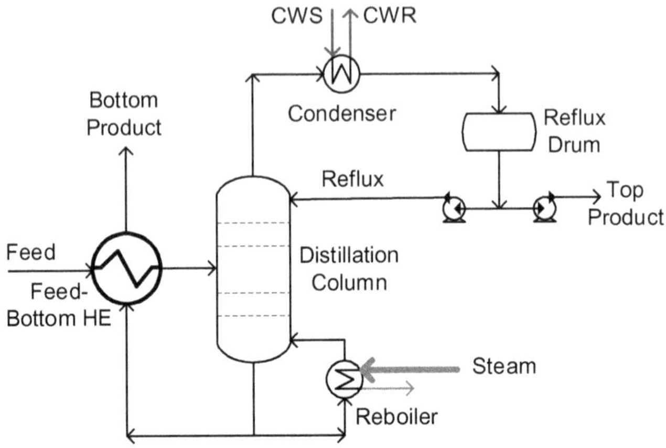

Fig. 9.1: Distillation column with a feed-bottoms HE.

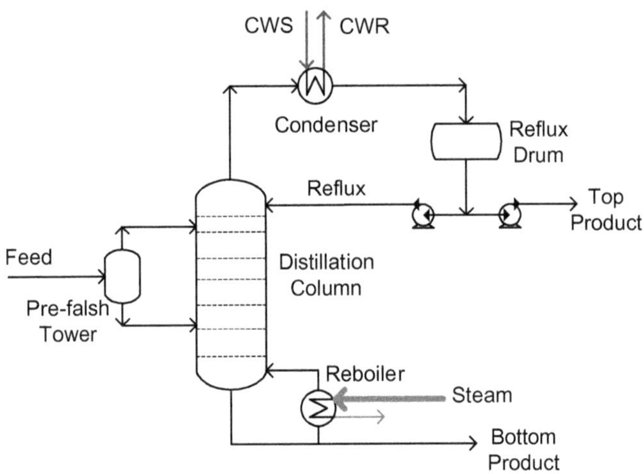

Fig. 9.2: Distillation column with a pre-flash drum.

the reflux drum. Sometimes, a trim condenser (using cooling water or air cooling) may be required before sending the overhead stream to the reflux drum. In cases where the pressure ratio required for a single stage exceeds 3.5 or compressor discharge temperature exceeds 240°C, multi-stage compressor is used.

The MVR configuration (Fig. 9.3) is usually suitable if the overhead stream flow rate is very high such as in light end hydrocarbon separations

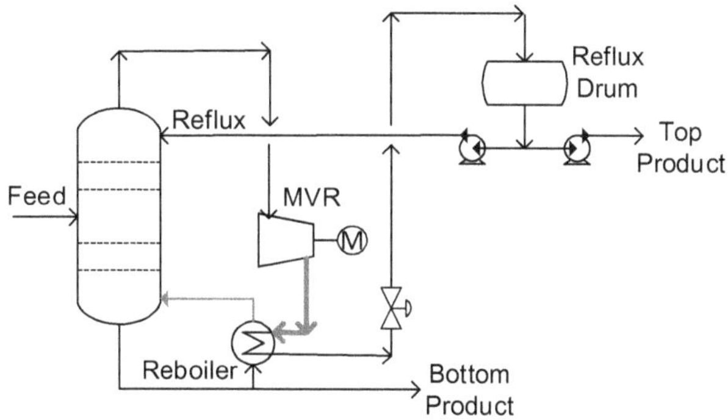

Fig. 9.3: Typical distillation column with MVR; an auxiliary condenser or reboiler may be required in some applications.

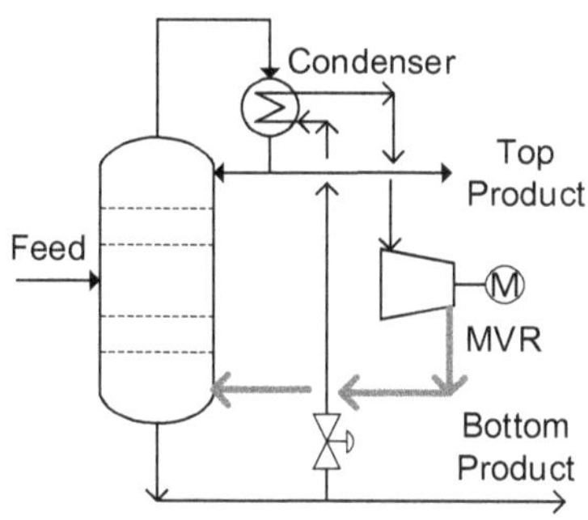

Fig. 9.4: Distillation with bottom flash; an auxiliary condenser or reboiler may be required in some applications.

(e.g., ethylene/ethane, propylene/propane, isobutane/n-butane separation) and difference between the column bottom and overhead temperatures (i.e., temperature lift, T_{lift}) is <30°C. Temperature driving force for heat transfer in reboiler and condenser are not considered in this definition of T_{lift}, which is employed in industries for assessing suitability of MVR for distillation applications. Instead of T_{lift}, the recent literature is employing the coefficient of performance.[3] If the overhead stream is relatively pure, it can condense over a narrow temperature and pressure range in the reboiler, and hence avoids the need for very high-pressure compressor. As compressor is used to raise the pressure, MVR is highly flexible for WHR in distillation columns and hence has wider application range.

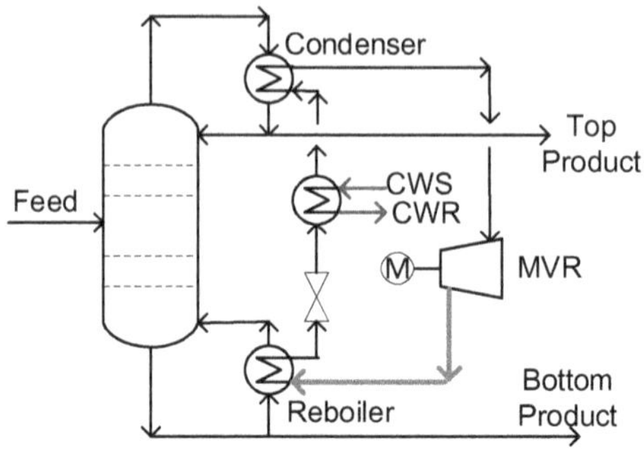

Fig. 9.5: Distillation with CCC; an auxiliary condenser or reboiler may be required in some applications.

In a BF configuration (Fig. 9.4), the high temperature (HT) bottom stream from the distillation column is flashed to a lower pressure, where its bubble point temperature is below the condenser temperature. The flashed bottom stream is then used to condense the overhead stream in the condenser. In this step, the flashed bottom stream vaporizes fully, which is subsequently compressed by MVR and sent directly into bottom of the distillation column. This provides the necessary vapour/heat for distillation. Flashing of bottom streams to vacuum should be avoided as it would require costly combination of dry vacuum pump and centrifugal compressor, in series arrangement for gas compression. Hence, BF is more suitable for columns operating at high pressure (typically above 10 bar).

In CCC configuration (Fig. 9.5), an external working/heat transfer fluid such as water is used for WHR and transfer from the column overhead stream to the reboiler. Hence, column overhead stream (as in MVR) or bottom stream (as in BF) are not subjected to any changes in pressure by compression. The CCC configuration is more appropriate if the process streams are sensitive to pressure changes, such as undesirable reaction or chemical degradation at high pressures and/or temperatures. The working fluid operates in the CCC loop between two pressure levels, determined by the temperatures of the column condenser and reboiler.

Working fluid at low pressure recovers WH from the overhead stream and fully vaporizes (Fig. 9.5). It is subsequently compressed by a compressor to a pressure, which is suitable for transferring mainly latent heat at the reboiler. This condenses the working fluid at high pressure. In some cases, a cooler cooled by cooling water is used to cool the working fluid, after heat exchange at the reboiler. High-pressure working fluid is depressurized across a throttling valve to a lower pressure, suitable for providing cooling duty by vaporization in the condenser. Compared to one-time heat transfer between overhead and bottom streams in MVR and BF, CCC requires heat exchange twice between process fluids (Fig. 9.5) and the working fluid. Due to required temperature driving forces in condenser and reboiler, less WHR is achieved in CCC compared to BF and MVR. Similar to BF configuration, vacuum operation should be avoided in CCC; hence, with water as the working fluid, CCC is suitable for distillation with condenser temperature more than 100°C, which is the dew point temperature of steam at atmospheric pressure.

AHP and AHT working principles are presented in Chapter 5. AHP uses medium-pressure steam (MPS) or high-pressure steam (HPS) to amplify WH from the condenser to a temperature, suitable for heat exchange at the reboiler. However, it does not require any cooling utility. AHP-assisted distillation column is presented in Fig. 9.6. See Section 5.4 for more details on AHPs.

AHT upgrades nearly 50% of WH from the condenser to a higher temperature by a T_{lift} of 30°C to 40°C; higher T_{lift} is possible with chemical HPs (described in Chapter 5). It discards the remaining ~50% of WH to cooling water in AHT's condenser. Typical application of AHT in distillation column is illustrated in Fig. 9.7. Auxiliary reboiler and condenser in Figs. 9.6 and 9.7, respectively, are useful for operational flexibility. See Section 5.4 for more details on AHTs.

9.5.1 *Use of WHR for Reducing Operating Pressure of Distillation*

Lower operating pressure leads to lower energy consumption in distillation columns. Water as coolant in condenser limits the overhead stream temperature to 40°C (assuming cooling water supply temperature of 30°C and a temperature approach of 10°C). This poses a problem for lowering operating pressure of distillation column. Chilled water generated using WH (from other WHR applications in the plant or steam condensate from the reboiler) can be used to reduce the column's overhead pressure and

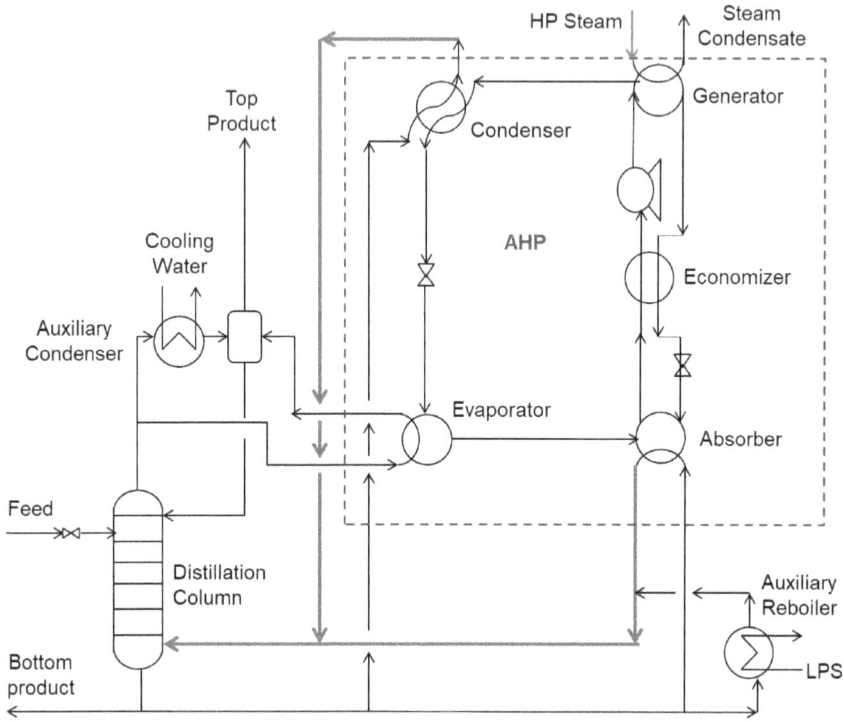

Fig. 9.6: Distillation column with AHP in the dashed area.

hence reduce the reboiler duty. This option is very useful to debottleneck a distillation column involving significant quantities of lighter material (e.g., pentane and lighter gases), which generally uses a compressor as shown in Fig. 9.8. With the installation of a trim-cooler using chilled water produced by the absorption chiller utilizing MPS condensate from the reboiler (Fig. 9.9), the overhead stream can be further cooled to 10°C–12°C (assuming chilled water supply temperature of 7°C), thus condensing more vapour. This reduces vapour load and power consumption of the compressor.

9.6 WHR in Distillation for Power Generation (Medium Cost/Complexity)

WH from condenser can be utilized to generate electric power using ORC (Fig. 9.10). It uses an organic working fluid with a boiling point much

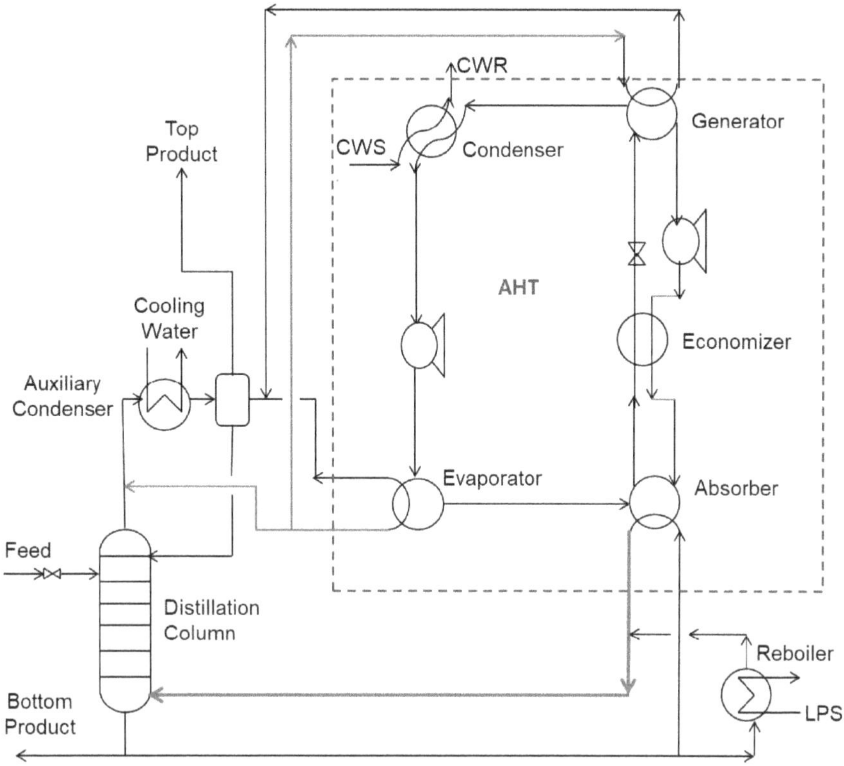

Fig. 9.7: Distillation column with AHT in dashed area.

lower than that of overhead stream temperature. Working fluid at high pressure vaporizes by heat gain from overhead stream at the condenser and generates high-pressure vapour, which expands subsequently in a gas expander to generate electricity. Then, low-pressure vapour from the expander is cooled in the economizer and condensed often using cooling water in the ORC condenser. Finally, condensed working fluid is pumped back to its original high pressure to complete the cycle. Efficiency of the ORC is improved by installing an economizer to pre-heat the high-pressure working fluid with the hot gas leaving the expander (Fig. 9.10).

A rough estimate of the efficiency of ORC cycle (η_o) for an application can be estimated by:

$$\eta_o = \eta_e \left(1 - \frac{T_c}{T_h}\right) \tag{9.1}$$

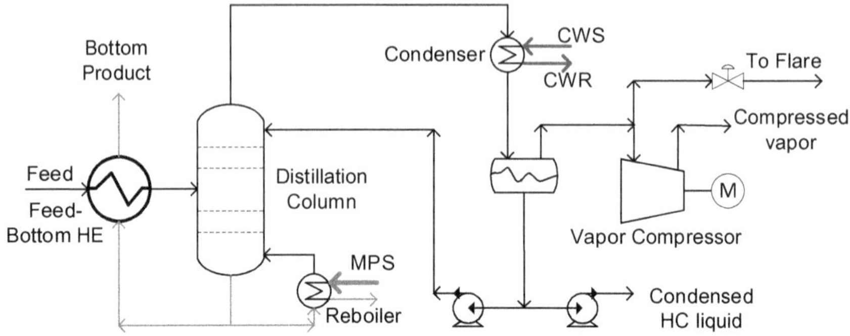

Fig. 9.8: Typical distillation column with overhead vapour compressor.

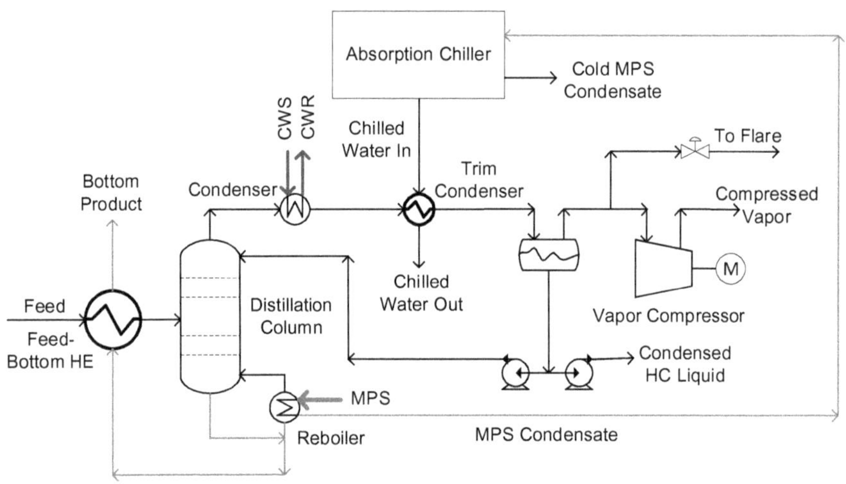

Fig. 9.9: Distillation column with condenser and trim/auxiliary condenser in series.

Here, η_e is the exergy efficiency, which can be taken as 0.5.[4] T_c (K) and T_h (K) are the temperatures (in absolute units) of cold utility (e.g., cooling water) and WH source, respectively. Thus, ORC system efficiency depends mainly on the hot stream temperature. As shown in Fig. 9.11, it increases from 0.094 to 0.180 with increase in WH source temperature from 100°C to 200°C; here T_c is assumed to be 30°C.

Kalina Cycle: Working principle of KC (Fig. 9.12) is the same as that of ORC, except that the working fluid is ammonia and water mixture. As the working fluid is a mixture, it evaporates only partially in the evaporator.

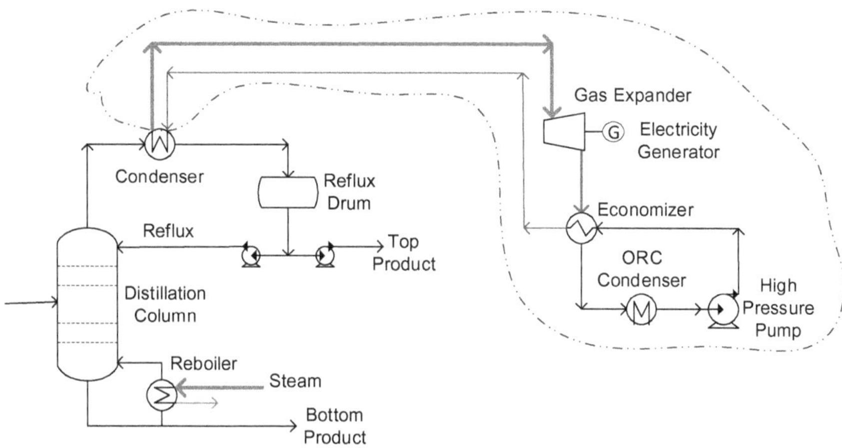

Fig. 9.10: Distillation column with ORC inside the dashed area.

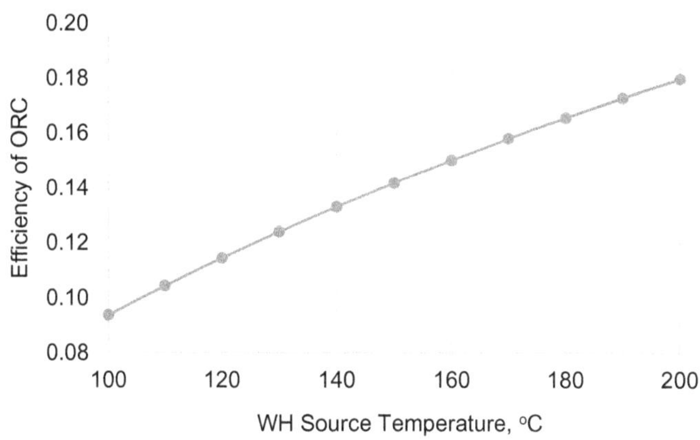

Fig. 9.11: Variation of efficiency of ORC with WH source temperature for cold utility temperature of 30°C.

Vapour and liquid portions are separated in a knockout (KO) drum. Vapour stream at high pressure from the top of the KO drum expands in an expander and hence generates electric power. Low-pressure vapour from the expander is mixed with the liquid stream from KO drum and cooled in the low temperature (LT) recuperator and subsequently condensed in a (KC) condenser, cooled by cooling water. The condensed stream from this condenser is pressurized by a pump and heated in the LT

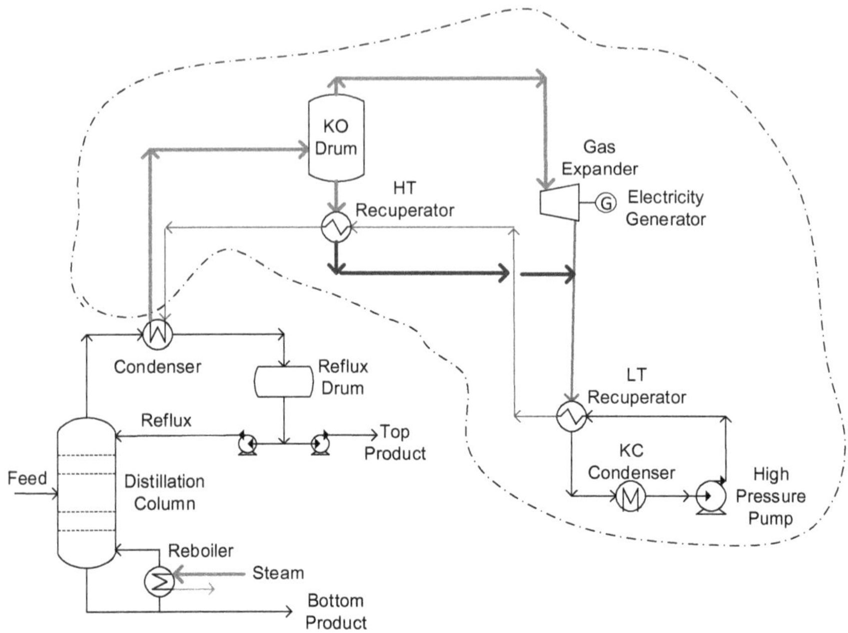

Fig. 9.12: Details of distillation column with KC system inside the dashed area.

recuperator, followed by the HT recuperator, where the hot fluid from KO drum is used to heat the stream from LT recuperator. The high-pressure working fluid from HT recuperator is sent to the condenser of the distillation column for further heating and evaporation. This completes the cycle. KC has 2 degrees of freedom: working pressure, which is high, and composition of ammonia in the working fluid mixture. Typically, ammonia concentration between 70 and 80 mole % is used. As ammonia is very volatile, working pressure up to 80 bar is required.

9.7 Other Heat Integration Methods for Distillation Columns

Many heat integration methods/techniques are available for reducing energy required for distillation; they range from external heat integration (i.e., heat integration outside the distillation column) to internal heat integration (i.e., heat integration within the distillation column). Main benefit of these heat integrations outlined in this section is reduced reboiler/condenser duties, thus resulting in energy savings. Although some capital investment is required, reduced energy cost often leads to attractive payback period (PBP) for this investment.

External Heat Integration Options: In these, it is important to ensure sufficient driving force for heat transfer in the reboiler using WH.

- WH from another process can be used in the reboiler of a distillation column as shown in **Fig. 9.13**: An optional, additional reboiler (not shown in Fig. 9.13) may be required if the distillation column is required to operate when WH is not available (due to shutdown of the process, supplying WH stream).
- In process plants with multiple distillation columns, there may be opportunities to integrate WH among them. In one such arrangement (Fig. 9.14), the hot overhead vapour stream from one column is thermally coupled with a reboiler in another column. Generally, such an integration is attractive if both the distillation columns are in close proximity and in the same process unit. In some cases, MVR may be required to raise the temperature of the vapour stream for heat exchange in the reboiler as shown in Fig. 9.15. Instead of MVR, AHT (either lithium bromide–water or chemical type described in Chapter 5) can be used if higher T_{lift} is required. In both Figs. 9.14 and 9.15,

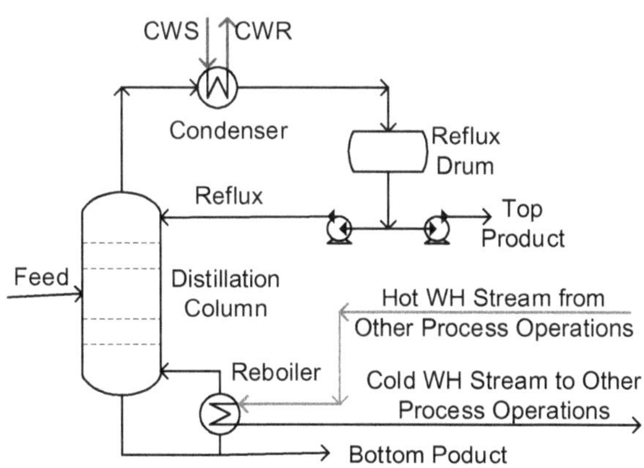

Fig. 9.13: Heat integration of a distillation column with WH stream from another process.

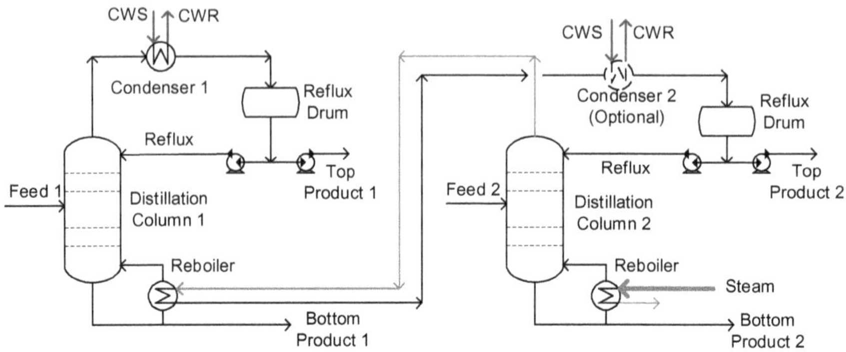

Fig. 9.14: Heat integration between two distillation columns; optional reboiler may be required for the first column if the duty from the second column vapour stream is not sufficient.

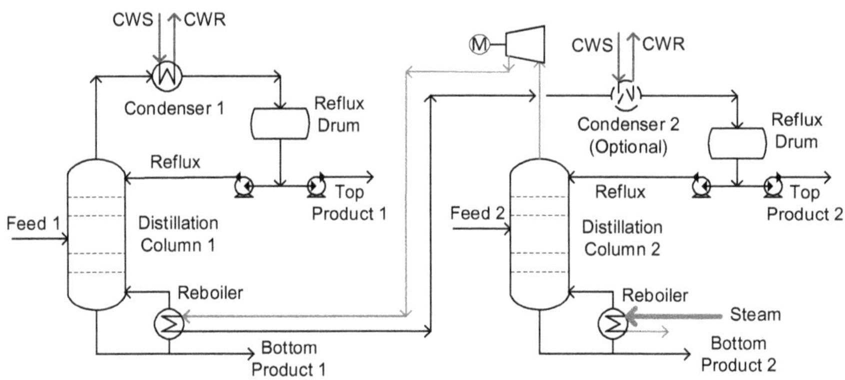

Fig. 9.15: Heat integration between two distillation columns, with an MVR.

optional coolers are shown for operational flexibility. Compared to the arrangement in Fig. 9.14, HP-assisted arrangement shown in Fig. 9.15 is more common in industry as it maximizes WH utilization and avoids the need for auxiliary reboiler.

- Side/intermediate reboiler (IR) and/or condenser can be installed if they are cost effective to reduce reboiler and condenser duties, respectively. In some cases, overhead vapour stream can be compressed by MVR to provide heat to an IR (instead of providing heat to the main reboiler at the bottom of the column).

Internal Heat Integration Options: Main techniques for internal heat integration are outlined below.

- The rectifying and stripping sections in a distillation column can be internally heat integrated. One such arrangement known as heat-integrated distillation column (HIDiC in Fig. 9.16), maximizes energy efficiency and hence can save energy up to 70%,[5] compared to conventional distillation with reboiler and condenser. In MVR-assisted distillation (Fig. 9.3), overhead vapour is compressed and sent to the reboiler for heat exchange. However, in HIDiC column (Fig. 9.16), stripper section is at a lower pressure compared to the rectifier section, and so vapour stream from the top of stripper is compressed and supplied to the bottom of rectifier. Lighter product is collected at the top of the rectifier. High-pressure liquid stream from the rectifier is depressurized, mixed with feed and then sent to the top of the stripper. Heavy product is withdrawn from the bottom of the stripper. Heat is transferred from the rectifier to stripper by direct contact of these two sections (Fig. 9.16). In some cases, optional reboiler and condenser may be used at stripper and rectifier, respectively. HIDiC is very

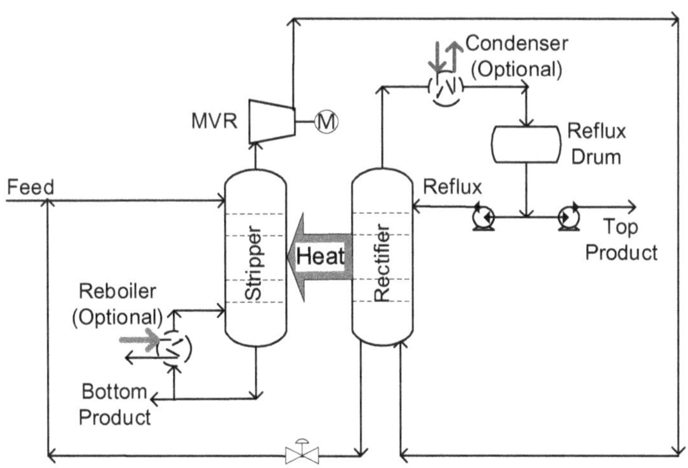

Fig. 9.16: Details of HIDiC; stripper pressure/temperature is lower than rectifier pressure/temperature for heat transfer from rectifier to stripper sections.

complex as it involves complex internals inside a single column, coupled with an MVR; see Kiss[2] for details on HIDiC internals and their arrangements. Industrial installations of HIDiC are not known.

- For separation of a ternary mixture by conventional distillation, two or three columns in series are required; each of them is a conventional column with a reboiler and a condenser. Petyluk configuration in Fig. 9.17 is an improvement for such separation of mixtures with three or more components. Although it has two columns, it requires only one reboiler and condenser. Moreover, the two columns in Petyluk configuration can be realized in a single column (known as DWC) described below.
- A DWC can be used to separate a ternary mixture into three product streams of desired purity. It has only one distillation column, one reboiler and one condenser, and often requires lower reboiler/condenser duties. DWC integrates the two columns of a Petyluk system into one column shell (with a vertical partitioning wall), as shown in Fig. 9.18. Although trays are shown in this figure, packing can be used on either or both sides of the dividing wall. Compared to conventional sequence of columns for separating mixtures of three or more components, DWC can save both capital and operating costs

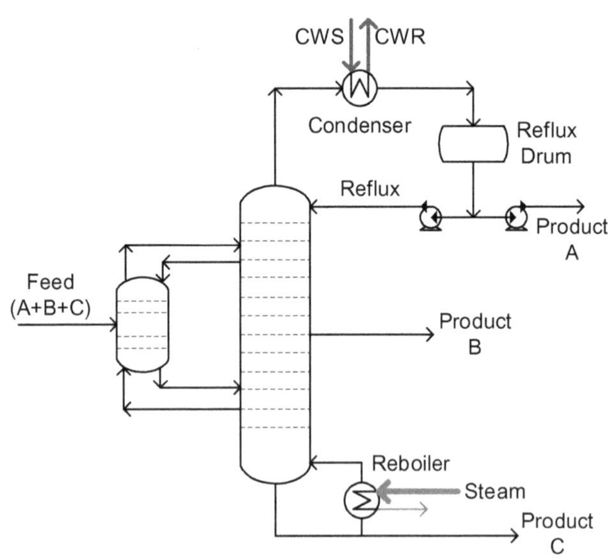

Fig. 9.17: Details of Petyluk column for separating a ternary mixture.

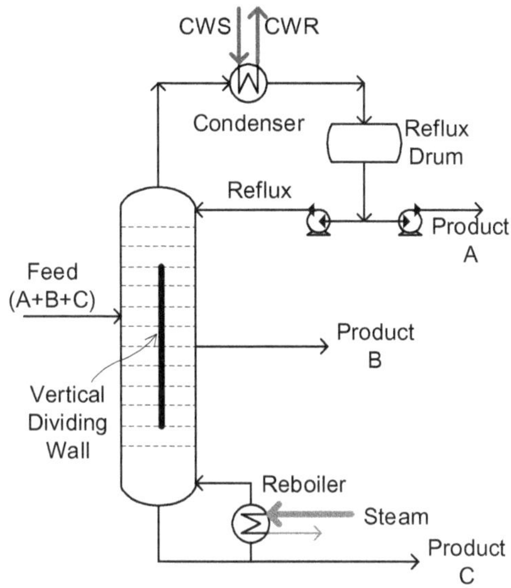

Fig. 9.18: Details of DWC for separating a ternary mixture.

(up to 30%) and significantly reduce the plot space requirement (up to 40%). These benefits are possible for both new systems and revamping existing systems. Premkumar and Rangaiah[6] analysed revamping of existing conventional two-column system to a DWC for six industrial applications; their results show that PBP for this revamping is less than 3 years for five applications and about 4 years for one application. DWCs have been implemented industrially for several decades; Olujic *et al.*[7] stated that more than 70 packed DWCs are in operation by BASF alone in various regions of the world. There must be more applications of DWC in industries of other companies in the past as well as since 2009.

Application Guidelines for WHR configurations for Distillation are summarized in Table 9.2.

9.8 Industrial Case Studies

Several industrial case studies are presented in this section to demonstrate the attractiveness of some of the techniques described in the previous sections.

Table 9.2: WHR configuration guidelines for distillation columns

Consideration	Guidelines
Process conditions	• Distillation applications with column overhead temperature (T_{OH}) > 50°C and $T_{lift} \leq$ 100°C will have at least one technically feasible WHR, either MVR, BF, AHT or IR. • MVR is very attractive if T_{lift} is < 30°C. • Distillation applications with T_{OH} > 150°C usually have favourable ORC and KC configurations.
Electricity price and requirement	• If local electricity price is low, BF or MVR configuration is usually economical. • If local electricity is expensive, AHP, AHT, ORC or KC configuration may be attractive. • If compressor power requirement is more than three times reboiler duty, VR and BF configurations are unlikely to be economical.
Sustainability	• AHP, AHT, ORC and KC configurations are more sustainable than other WHR configurations, mainly due to minimum or no net requirement of electric energy.
Selection between feasible configurations	• BF is typically more economical and environmentally favourable than MVR. • Generally, KC is slightly more favourable than ORC.

9.8.1 *Propylene–Propane Separation*

Propylene–propane separation by the conventional and MVR-assisted distillation (shown in Figs. 9.18 and 9.19, respectively) was studied and compared by Chaudhari *et al.*[8]; one of these authors is from the Reliance Industries. This separation is important in petroleum refining and petrochemical plants. Feed, products and utility specifications for this case study are given in Table 9.3.

Conventional distillation for propylene–propane separation requires 180 ideal trays (Table 9.4) and uses two columns arranged in series, with an inter-column transfer pump, as shown in Fig. 9.19. This arrangement is mainly to achieve lower capital cost by avoiding a single very tall column. Due to the use of cooling water as coolant in the condenser, distillation requires high operating pressure of 17.684 bar and 18.554 bar at the

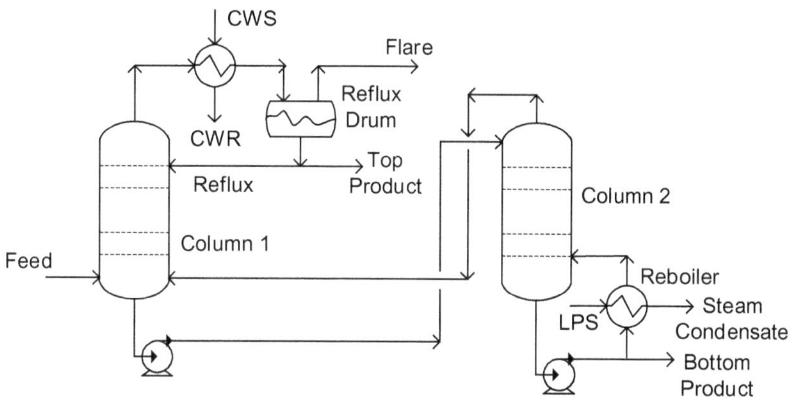

Fig. 9.19: Conventional two distillation columns for propylene–propane separation.

Table 9.3: Feed, products and utility specifications for propylene–propane separation[8]

Feed details	50,000 kg/hr having 25 wt.% propane and 75 wt.% propylene, at 40°C and 24.52 bar
Product requirements	Propylene recovery of 95% and purity of 99.7 wt.%
Available utilities	Low-pressure steam (LPS, saturated at 4.445 bar) and cooling water (32°C supply and 38°C return)

overhead of the first column and bottom of the second column, respectively. This design requires significant utilities (79.69 ton/hr of LPS at reboiler, 8725 m^3/hr of cooling water at condenser) and 687.3 kW of power (for inter-column transfer and reflux pumps).

An MVR-assisted distillation design (Fig. 9.3) generally uses a single column as it requires lower number of ideal trays (160 for this case). In this arrangement, there is no need for a dedicated condenser using cooling water as the column overhead vapour is compressed by MVR and then condensed in the reboiler. (However, an auxiliary condenser using cooling water, especially during unit start-up and shut down operations, is generally required to balance the condensation and vaporization duties, and for operational flexibility during turndown operations.) Hence, reboiler acts as both reboiler and condenser. The condensed vapour in reboiler is sent to the reflux drum, which provides reflux flow and propylene product.

Table 9.4: Comparison of results for conventional and MVR-assisted distillation designs for propylene–propane separation[8]

Quantity	Conventional Distillation with Two Columns	MVR-Assisted Single Distillation Column
Reflux ratio	16.4	14.3
Number of ideal trays	180	160
Feed tray	142	127
Reboiler duty, MW	52	49.97
Condenser duty, MW	51.8	7.72 (auxiliary)
Column diameter, m	6.25	5.8
Overhead product rate, kg/hr	35,733	35,733
Inter-column transfer pump capacity, m^3/hr	1400	NA
Column pressure at top/bottom, bar	17.684/18.554	12.778/13.67
Column temperature at top/bottom, °C	42.5/51.5	28.5/38
Reboiler steam, ton/hr	79.69	NA
Condenser cooling water, m^3/hr	8,725	1295 (auxiliary)
Reflux pump duty, kW	287.3	201
Inter-column transfer pump, kW	400	NA
Compressor power, kW	NA	7,045

In this case, column overhead pressure is not governed by cooling water temperature and hence it can be lowered to 12.778 bar. This lowers both the required energy and number of trays. Column diameter for MVR-assisted operation is lower at 5.8 m compared to 6.25 m for conventional distillation. Energy requirements for both conventional distillation and MVR-assisted distillation columns are summarized in Table 9.4. Chaudhari *et al.*[8] reported an operating cost saving of ≈$7 million/year for MVR-assisted distillation, which is around 70% decrease compared to the conventional distillation.

It is evident from Table 9.4 that, for separation of close boiling mixtures such as propylene and propane, MVR is very attractive. Although MVR requires capital investment, it results in lower system pressure and column dimensions, and avoids the need for significant amount of LPS piping, cooling water piping, pumps and cooling tower. Hence, overall capital cost of MVR-assisted distillation for this case is comparable to that for the conventional distillation column.

9.8.2 *Foul Water Stripper*

Typical details of an operating foul water stripper (FWS), also known as sour water stripper, from Reddy *et al.*[9] are shown in Fig. 9.20. This process is commonly used for the purification of wastewater (having ammonia and hydrogen sulphide impurities) in process plants. Here, foul/sour water feed is heated by hot bottom stream in the feed pre-heater, and then sent to FWS for removing dissolved impurities, using LPS at the reboiler. Overhead vapour from FWS is condensed in the condenser using cooling water.

In the BF configuration (Fig. 9.4), the bottom stream is usually flashed to the minimum allowable pressure of 1 atm (to avoid vacuum operation) to maximize flashing and heat recovery from the overhead

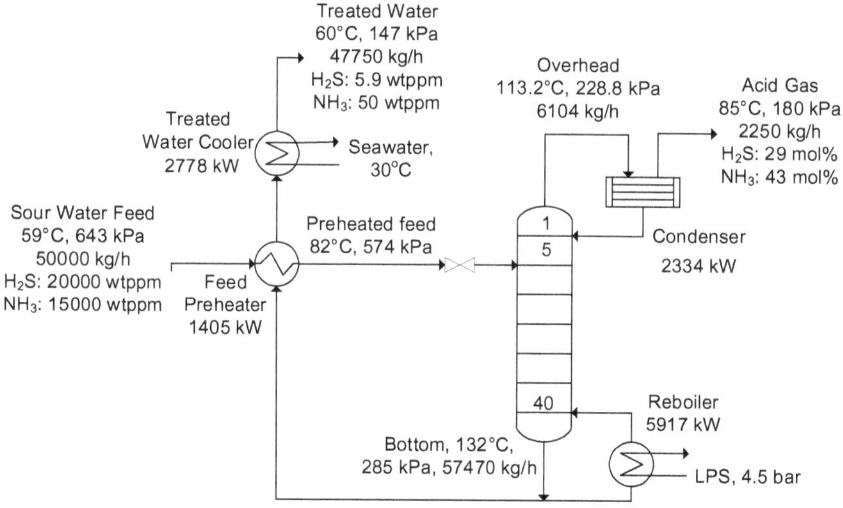

Fig. 9.20: Process flow diagram of FWS with the current operating conditions[9]; numbers in the column refer to actual trays (excluding condenser and reboiler); a tray efficiency of 0.2 is used.

stream in the condenser. In the present case, with bottom stream flashing to 1 atm, only 6.7 kW of heat can be recovered from the overhead stream before reaching the minimum temperature approach (of 10°C) in the overhead/bottom heat exchanger (i.e., condenser). Hence, conventional BF is not economical for this case.

However, it was found that bottom stream flashing to 1 atm generates 3607 kg/hr of flash stream, which can be compressed using an MVR and supplied as substantial portion of the heat input to the FWS, by direct injection of compressed flashed steam/vapour along with LPS. Power of 267.4 kW is required for compressing flashed vapour to 285 kPa in a compressor with adiabatic efficiency of 0.75. The remaining heat required for the reboiler of FWS is supplied by 6630 kg/hr LPS, as shown in Fig. 9.21. LPS condensate will leave with treated water. This generates more treated water for recycle into process operations (although LPS condensate will not be returned to steam generation). Overall, 1940 kW of heating duty can be saved at an expense of 267.4 kW of electric power. Based on economic analysis by Reddy *et al.*[9] the modified BF scheme in Fig. 9.21 has PBP of 0.9 years for capital investment of US$ 623,285. This

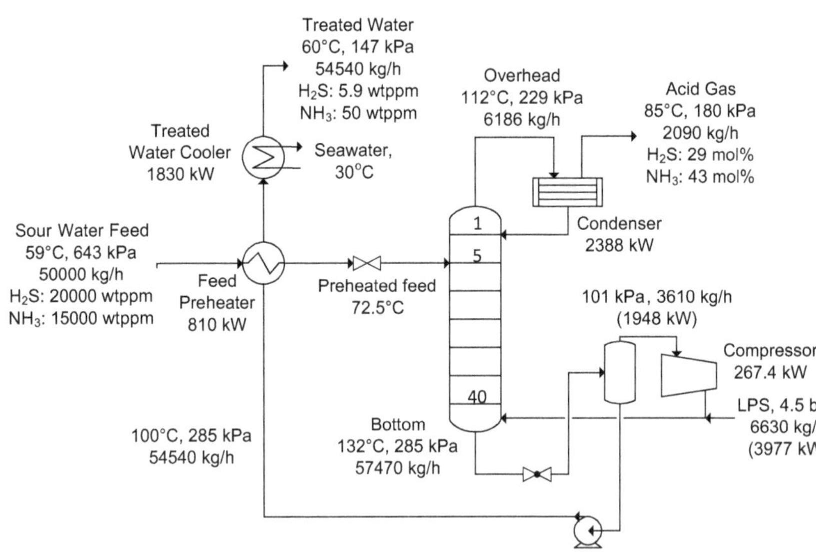

Fig. 9.21: Process flow diagram of FWS with modified BF[9]; numbers in the column refer to actual stages (excluding condenser and reboiler).

scheme of flashing the column bottom stream and then increasing its pressure by MVR may be economical for all distillation columns; hence, it should be assessed for improving energy efficiency of distillation. See Reddy *et al.*[9] for design and economic evaluation of several other WHR options for FWS.

9.8.3 *C4 Separation*

C4 separation is required in the alkylation unit of a typical petroleum refinery. Alkylation product (known as alkylate) and unconverted reactants (n-butane and iso-butane) are separated in a series of two distillation columns: deisobutanizer (DIB) and debutanizer (DeC4). Design/operational details of these columns are presented in Fig. 9.22, and specifications of feed and product streams are given in Table 9.5. The top product of DIB is iso-butane and the bottom stream of DIB is fed to DeC4, which separates its feed into n-butane as the top product and alkylate as the bottom product.

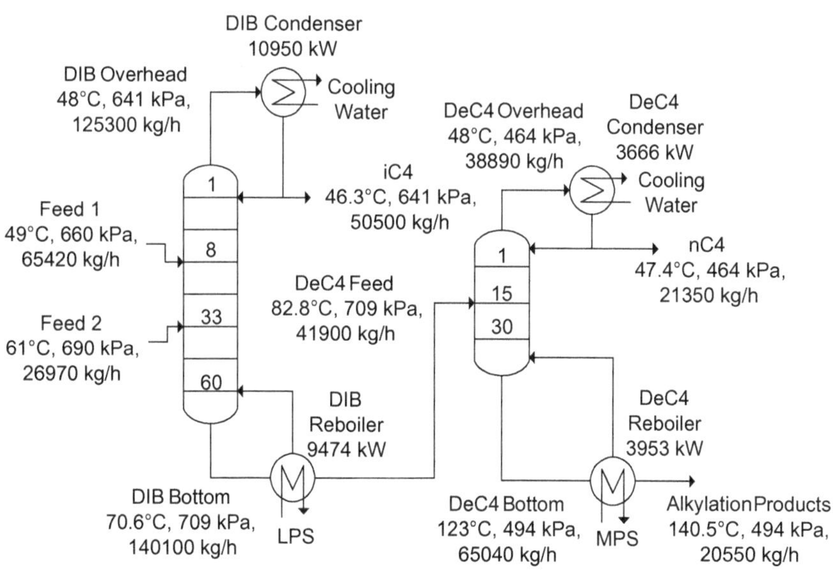

Fig. 9.22: Process flow diagram of C4 separations unit, showing current operating conditions[9]; numbers in each column refer to actual trays (excluding condenser and reboiler); a tray efficiency of 0.9 is used.

Table 9.5: Stream data for C4 separation in Fig. 9.22

Quantity	Feed 1	Feed 2	DIB Distillate (iC4)	DIB Bottom	DeC4 Distillate (nC4)	DeC4 Bottom (Alkylation Products)
Temperature (°C)	49.0	61.0	46.3	82.8	47.4	140.5
Pressure (kPa)	660.3	689.7	640.7	709.3	464.2	493.6
Mass flow rate (kg/hr)	65420	26970	50500	41900	21350	20550
Composition (wt%)						
C3	1.90	2.6	3.85	0.00	0.00	0.00
nC4	12.81	70.01	13.03	49.36	95.98	0.93
iC4	55.59	23.29	83.11	1.63	3.20	0.00
C5	0.55	4.09	0.01	3.48	0.82	6.25
C_7H_{16}	13.62	0.00	0.00	21.27	0.00	43.37
C_8H_{18}	15.53	0.00	0.00	24.25	0.00	49.45

MVR-assisted design for DIB in Fig. 9.23, saves 9300 kW of hot utility at reboiler and 7580 kW of cold utility at condenser but consumes 1,730 kW of electric power. Based on detailed assessment, Reddy *et al.*[9] have reported PBP of 1.24 years and hence it is an attractive WHR option. As stated earlier, MVR is attractive if T_{lift} is low. For DIB in this application, T_{lift} is 82.8 − 48 = 34.8°C. On the other hand, for DeC4, T_{lift} is very high (i.e., 123 − 48 = 75°C). Hence, MVR is likely to be uneconomical. So, for such applications, IR is an attractive solution to replace part of MPS duty in the bottom reboiler with less expensive LPS. See Reddy *et al.*[9] for design and economics of several other WHR options for DIB.

Based on the column temperature profile in DeC4 column (Fig. 9.24), obtained from Aspen Hysys simulation, the biggest temperature rise occurs from tray 28 onwards. Hence, placing IR at or above tray 28 can be beneficial for improving energy efficiency of the column. Trial and error on several positions gave the optimal IR at tray 28. Accordingly, liquid is withdrawn from tray 28 and sent to IR for some vaporization; arbitrary vapour fraction of 0.2 is chosen for the outlet stream from IR, which is returned to tray 28 itself. These and other details of IR installation are presented in Fig. 9.25.

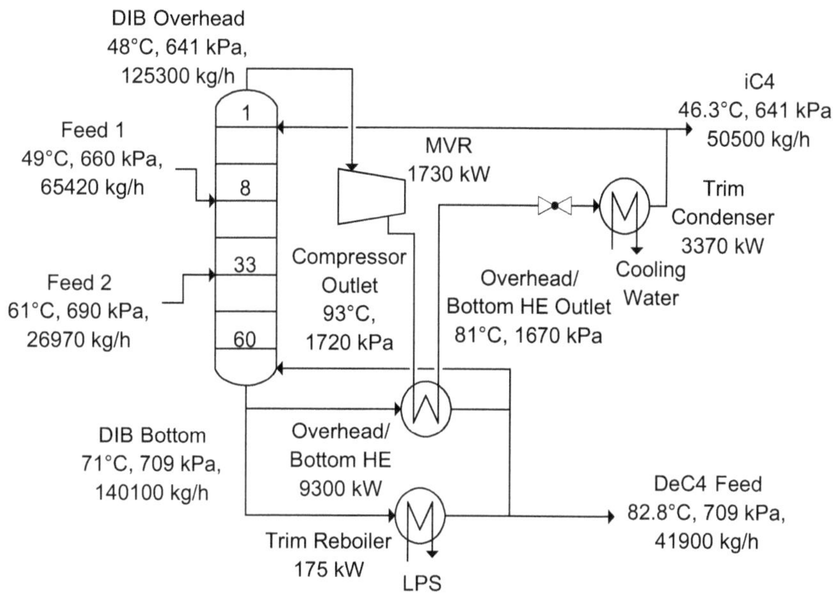

Fig. 9.23: Process flow diagram for DIB with MVR, showing optimal design conditions.[9]

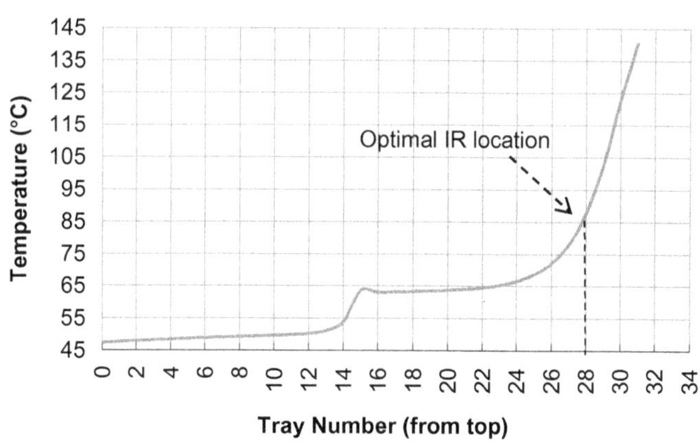

Fig. 9.24: Temperature profile in DeC4 column, for choosing IR location.[9]

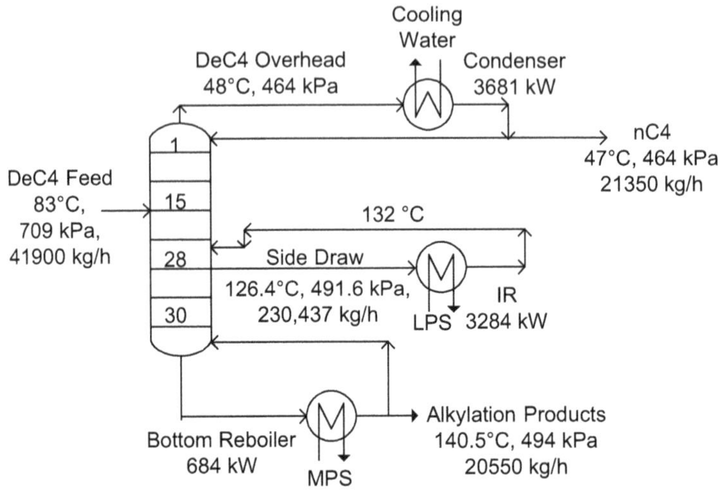

Fig. 9.25: Process flow diagram of debutanizer with IR Configuration.[9]

For this case of DeC4 column, IR uses 3284 kW of LPS and reduces nearly the same amount of MPS, at the bottom reboiler (i.e., 3953 at the bottom reboiler from Fig. 9.22 − 684 = 3,269 kW). The saved MPS can be used at a steam turbine to generate electric power and LPS that can be used at the IR. For IR alone option, Reddy *et al.*[9] reported capital investment of US$143,228 and PBP of 2.05 years. They also analysed design and economic evaluation of several WHR options for DeC4 column.

9.8.4 *Application of WHR Methods to DWC*

Chew *et al.*[4] analysed the feasibility of implementing several WHR methods for six DWCs for separation of (a) alkanes; (b) benzene, toluene and ethylbenzene (BTE); (c) benzene, toluene and p-xylene (BTX); (d) depropanizer/debutanizer (DeC_3/C_4); (e) ehanol, 1-propanol and 1-butanol (EPB) and (f) ethanol, water and ethyl glycol (EWE). Feasible WHR methods for these six DWC columns are summarized in Table 9.6. VR and BF are feasible for five and four applications, respectively; thus, they are more feasible compared to AHP, ORC and KC.

Capital investment, operational cost savings, PBP and net present value (NPV) for the most economical WHR method for each of the six

Table 9.6: Summary of feasible WHR methods and economics of the most economical WHR method for six DWC applications

Application	Alkanes	BTE	BTX	DeC3C4	EPB	EWE
Feed rate (kmol/hr)	500	500	500	1600	500	500
Pressure (kPa)	510	175	1050	1490	101	150
T_{OH} (°C)	93	97.3	179	36.5	75.5	42.3
T_{bot} (°C)	161.8	162.0	251.2	138.2	116.1	105.1
T_{lift} (°C)	68.8	64.7	72.2	101.7	40.6	62.8
Condenser duty (kW)	4498	6654	6538	10,470	8798	5250
Reboiler duty (kW)	4849	6974	6784	10,740	10,640	4880
Feasible WHR methods	VR, BF	VR, BF	VR, BF, ORC, KC	BF	VR, AHP	VR, AHP
Cost Analysis						
Most Economical WHR Method	**BF**	**BF**	**BF**	**BF**	**VR**	**VR**
Capital investment (US$)	$2,660,842	$3,422,498	$3,426,782	$4,634,062	$3,295,542	$2,887,889
Operating cost savings (US$/year)	$748,898	$2,036,613	$2,189,828	$830,381	$2,704,826	$1,333,269
PBP (years)	3.55	1.68	1.56	5.58	1.22	2.17
NPV (US$)	$1,940,814	$9,091,608	$10,028,766	$468,270	$13,324,440	$5,304,473

DWCs are given in the later part of Table 9.6. Capital investment for the new equipment is estimated assuming Chemical Engineering Plant Cost Index of 585.6 in 2012, 80% ammonia and 20% water mixture for KC and cost of 2000 EUR/kW for KC plants. Operating cost savings are assuming the following utility prices (all in US$/kWh): electricity = 0.0605, HPS = 0.0637, MPS = 0.0534, LPS = 0.506 and cooling water (30°C–45°C) = 0.0013. Results in Table 9.6 show that there is at least one feasible and attractive WHR option for each DWC application. PBP of 1.2–2.2 years for recovering the capital investment for BF/VR is very attractive for BTE, BTX, EPB and EWE applications. BF for alkanes and DeC3C4 applications has PBP of 3.55 and 5.58 years, and it may be acceptable for reducing energy consumption (and consequently CO_2 emissions). Chew *et al.*[4] also presented sensitivity analysis of results for increased cost of electricity.

9.8.5 Application of WHR Methods to Reactive Distillation

Reactive distillation (RD) is an intensified process, wherein both reaction and separation occur in the same column (instead of a reactor and a separator individually). It can reduce total energy requirement for reversible equilibrium-limited reactions, whose reacting conditions (i.e., temperature and pressure) are like those required for separation. RD can achieve 100% conversion due to continuous removal of products. Industrial applications of RD include production of methyl tertiary-butyl ether in petroleum refineries and methyl/other acetates in petrochemical plants.

Energy requirements of some RD can be further reduced by implementing WHR methods. For example, Feng *et al.*[10] investigated the benefits of reactive dividing-wall column (RDWC that integrates RD with another, adjacent distillation column), heat integration between feed and product streams and MVR, for producing n-propyl acetate by esterification of n-propanol with acetic acid. The conventional RD process consists of an RD and a product column (PC) for producing pure n-propyl acetate. Reaction of n-propanol and acetic acid occurs on some stages of RD having a solid catalyst. PC is required because of azeotropic mixture in the overhead stream of RD. Feng *et al.*[10] integrated RD and PC into RDWC for producing n-propyl acetate. Then, they explored heat recovery from RD overhead vapour stream to pre-heat acetic acid feed and also from n-propyl acetate product stream to pre-heat n-propanol feed. Finally, they evaluated MVR for compressing RD overhead stream (after pre-heating acetic acid feed) to supply all the duty of the main reboiler of RDWC.

For n-propyl acetate production, total capital cost of conventional RD and PC, RDWC and feed pre-heated RDWC is comparable, whereas that of RDWC with both feed pre-heating and MVR is nearly 68% more than that of conventional RD and PC, mainly because of compressor and larger reboiler (Table 9.7). On the other hand, operating cost decreases by around 10%, 20% and 75%, respectively, by use of RDWC, feed pre-heating and feed pre-heating together with MVR, compared to conventional RD and PC process. Overall, total annual cost decreases by about 7%, 13% and 26% when conventional RD and PC are integrated into RDWC, addition of feed pre-heating to RDWC, and implementation of MVR to pre-heated RDWC, respectively (Table 9.7). Feng *et al.*[10] demonstrated that control of RDWC with both feed pre-heating and MVR is

Table 9.7: Comparison of conventional RD and PC, RDWC, RDWC with feed pre-heating and RDWC with both feed pre-heating and MVR, for producing 40,500 ton/year of n-propyl acetate; percent change compared to conventional RD and PC is given in brackets

Quantity	Conventional RD and PC	RDWC	Feed Pre-Heated RDWC	Feed Pre-Heated RDWC with MVR
Total capital cost (US$)	1,906,390	1,915,750 (+ 0.49%)	1,908,410 (+ 0.11%)	3,201,270 (+ 67.92%)
Operating cost (US$/year)	1,214,000	1,087,290 (−10.44%)	978,490 (−19.40%)	309,140 (−74.54%)
Total annual cost (US$/year)	1,849,510	1,725,880 (−6.68%)	1,614,620 (−12.70%)	1,376,230 (−25.59%)

Note: Total annual cost is the sum of annual operating cost and total capital cost divided by PBP of 3 years.

possible using proportional–integral controllers. See Feng *et al.*[10] for more details on the design and control of RD/RDWC processes for n-propyl acetate production.

9.9 Summary

This chapter describes the application of WHR methods/techniques for reducing heating and/or cooling duties required for distillation, which is the most common separation method in the process industry. Main points of this chapter are as follows.

- Many WHR options ranging from operation optimization to modifications of low-, medium- and high-cost/complexity are available for reducing energy requirement of distillation.
- Operation optimization is the simplest and easiest option for maximizing energy efficiency of distillation columns.
- Next, low-cost/complexity options such as replacing trays with packing, designing condenser and reboiler for low fouling, feed-bottom stream heat exchange and use of pre-flash drum can be considered.
- Application of HPs is a medium-cost/complexity option. MVR is very attractive if T_{lift} required is low (<30°C). Lower the required T_{lift}, better will be the economic attractiveness of MVR.

- BF is very attractive for WHR from distillation columns operating at above atmospheric pressure. Often, it can be more attractive compared to MVR.
- Application of IR without/with MVR can be economical for high T_{lift} applications, with steep increase in temperature at the bottom section of the distillation column.
- DWC and HIDiC are high complexity options; however, usually they achieve higher energy savings. Of these, DWC has been employed for several hundred industrial applications.
- WHR methods can further reduce energy requirements of DWCs and RDs, and hence they should be considered.

Kiss[2] is a comprehensive book on advanced distillation technologies including DWC and HIDiC. More details on application of WHR techniques in conventional distillation and DWC are available in papers by Reddy *et al.*[9] and Chew *et al.*,[4] respectively. The book by Long and Lee[11] describes the recent technologies and their practical applications to retrofitting/revamping existing distillation columns.

References

1. Lek CM, Rangaiah GP, Hidajat, K. (September 2004) *Distillation: Revisiting some rules of thumb. Chem Eng* 50–55.
2. Kiss AA. (2013) *Advanced Distillation Technologies: Design, Control and Applications*. Wiley.
3. Plesu V, Bonet Ruiza AE, Bonet J, Llorens J. (June 15–18, 2014) Simple equation for suitability of heat pump use in distillation. In: Klemes JJ, Varbanov PS, Liew PY, (eds), *Proceedings of the 24th European Symposium on Computer Aided Process Engineering (ESCAPE 24)*, Budapest, Hungary, pp. 1328–1332.
4. Chew JM, Reddy CCS, Rangaiah GP. (2014) Improving energy efficiency of dividing-wall columns using heat pumps, organic Rankine cycle and Kalina cycle. *Chem Eng Process: Process Intensification* **76:** 45–59.
5. Kiss AA, Olujic Z. (December 2014) A review on process intensification in internally heat-integrated distillation columns. *Chem Eng Process: Process Intensification* **86:** 125–144.
6. Premkumar R., Rangaiah GP (2009), Retrofitting conventional column systems to dividing-wall columns, *Chem Eng Research & Design*, **87**: 47–60.

7. Olujic O, Jodecke M, Shilkin A, *et al*. (2009) Equipment improvement trends in distillation. *Chem Eng Process: Process Intensification* **48:** 1089–1104.
8. Chaudhari A, Gune P, Divey J. (March 2012) Distillation optimization by vapor recompression. *Chem Eng* 43–47.
9. Reddy CCS, Fang Y, Rangaiah GP (2014). Improving energy efficiency of distillation using heat pump assisted columns. *Asia Pac J Chem Eng* **9**: 905–908.
10. Feng ZM, Shen WF, Rangaiah GP, *et al*. (2019) Process development, assessment and control of reactive dividing-wall column with vapor recompression for producing n-propyl acetate. *Ind Eng Chem Res* **58:** 276–295.
11. Long NVD, Lee MY. (2017) *Advances in Distillation Retrofit*. Springer.

Acronyms and Notation

Acronyms

AHP	Absorption Heat Pump
AHT	Absorption Heat Transformer
BF	Bottom Flash
CCC	Closed Cycle Compression
CWS	Cooling Water Supply
CWR	Cooling Water Return
DeC4	Debutanizer
DIB	Deisobutanizer
DWC	Dividing-Wall Column
FWS	Foul Water Stripper
HIDiC	Heat-Integrated Distillation Column
HP	Heat Pump
HPS	High-Pressure Steam
HT	High Temperature
IR	Intermediate Reboiler
KO	Knockout
LLPS	Low-Low Pressure Steam
LPS	Low-Pressure Steam
LT	Low Temperature
MPS	Medium-Pressure Steam
MVR	Mechanical Vapour Recompression

NPV	Net Present Value
PBP	Payback Period
RD	Reactive Distillation
RDWC	Reactive Dividing-Wall Column
WH	Waste Heat
WHR	Waste Heat Recovery

Notation

T	Temperature, K

Greek Symbols

η_e	Exergy efficiency
η_O	Efficiency of ORC
ΔT	Temperature difference

Subscripts/Superscripts

bot	Bottom of distillation column
c	Cold utility
h	WH stream
lift	Temperature lift
min	Minimum
OH	Overhead of distillation column

Exercises

9.1 Briefly discuss the potential energy minimizing solutions for distillation, which can be implemented without any need for distillation column shutdown.

9.2 Identify low-, medium- and/or high-temperature lift distillation columns in your plant. Identify applicable WHR methods and arrange them in the order of complexity.

9.3 Answer the following with yes (right) or no (wrong), and give brief reasoning, where applicable.

(A) Generally, there is very little scope for reducing the energy required for distillation operation.
(B) Reducing energy supply to a distillation column results in off-spec products.
(C) Lower the operating pressure, lower are the energy requirements at reboiler and condenser of the distillation column.
(D) Overdesigning of reboiler and condenser areas is beneficial.
(E) Application of MVR for distillation is always the most cost-effective WHR option.
(F) BF is often an attractive WHR option.
(G) Power generation from distillation operation is not generally feasible.
(H) Use of steam pressure at the highest possible value is preferred as it provides the maximum temperature approach and hence leads to smaller reboiler.
(I) Application of WHR methods on DWC is not generally attractive.

Chapter 10

Waste Heat Recovery from Electric Power Generation

10.1 Overview

Process plants use significant amount of electrical energy. They may generate electricity using inhouse power generation plant or purchase electricity from an external utility supplier. Efficiency of electric power generation equipment such as condensing steam turbine (CST), back pressure type steam turbine (BST), gas turbine (GT), micro turbine (MT), reciprocating/diesel/gas engine (RE) alone, is low at 25 to 50%. Standalone CST-based power generation rejects lot of waste heat (WH) by condensing flash steam under vacuum conditions. Similarly, GT, MT and RE discard significant high temperature flue gas (FG) to atmosphere. This WH rejection is the main reason for low efficiency of electric power generation.

Co-generation to produce electric power and steam, using GT, RE, BST and MT, can significantly improve the system's energy efficiency to 80% or above. BST enables steam energy to be used for process heating applications. WH from power generation can be utilized to produce chilled water (ChW) using lithium bromide (LiBr) absorption chiller (covered in Section 5.5 of Chapter 5); then, ChW can be used to cool the combustion air intake and hence increase the power generation from GT. It can also be utilized in organic Rankine cycle (ORC) or Kalina cycle (KC) for generating more electric power. In general, recovering WH from power generation will increase the energy efficiency of the overall process.

Tri-generation can produce electric power, steam/hot water and ChW or desalinated water (DW). Quad-generation simultaneously produces electric power, steam/hot water, ChW and CO_2. As post-combustion CO_2 is captured in quad-generation, it can reduce/eliminate CO_2 emissions from power generation. This can be very profitable for process plants, especially plants such as beverage plants, which require CO_2. Application of co-, tri- or quad-generation can substantially increase the energy efficiency and profitability of the process/power plants.

Many waste heat recovery (WHR) methods are available for improving energy efficiency and/or increasing power output from power generation processes. Hence, a detailed understanding of WHR applications is essential for improving the efficiency and profitability of power generation systems. Accordingly, this chapter outlines power generation and WHR methods, and then describes the application of WHR methods for improving economics and efficiency of power generation. It begins with Section 10.2 on power generation methods in process and power plants. Section 10.3 outlines the co-generation or combined heat and power (CHP) methods. Sections 10.4 and 10.5 describe the tri-generation and quad-generation, respectively, to further improve the efficiency of power generation. Section 10.6 presents WHR methods to maximize power generation. Finally, Section 10.7 summarizes the main points for utilizing WH to maximize electric power generation.

Learning outcomes of this chapter on the use of WHR for maximizing electric power generation are as follows.

1. Describe the power generation methods and equipment.
2. Elaborate CHP systems and their advantages and disadvantages.
3. Discuss tri- and quad-generation processes for improving power generation systems.
4. Explain the WHR methods to increase the output from the power generation system.

10.2 Industrial Power Generation Methods

There are many methods to produce electric power in process and power plants. Common methods use GT, CST, BST, RE, MT, ORC and KC.

Simple CST- or BST-based power generation use (steam) Rankine cycle, and it involves four main steps.

1. Pressurized boiler feed water (BFW) is supplied to a boiler.
2. BFW is heated and vaporized in the boiler tubes using energy from fuel combustion. This produces saturated, high-pressure steam (HPS), which is separated from BFW in the steam drum and then superheated by heat from hot FG to increase its enthalpy and eliminate moisture.
3. Steam is expanded in the CST/BST to a vacuum pressure or above atmospheric pressure (medium-pressure steam, MPS, and/or low-pressure steam, LPS). This results in conversion of part of steam's thermal energy to kinetic energy of the turbine shaft, which can be used to generate electric power at a generator (or drive a compressor or pump).
4. Extracted/outlet steam from CST is condensed by a cooling medium, usually cooling water, in the condenser. In BST, extracted steam (MPS and/or LPS) is supplied to the process for heating, where condensation occurs. Condensed steam from CST or process heat exchangers, is returned to the deaerator for producing BFW (for recycle to the boiler).

Power Generation Using CST: HPS from a boiler (using fuel), process WH boiler or heat recovery steam generator (HRSG), is extracted from CST to a vacuum pressure and then condensed by the cooling medium, generally cooling water. Because of the higher pressure difference of steam across CST, it produces more power than BST. However, CST discards all the latent heat in the extracted steam to cooling water. In a boiler-based power plant installed with a CST (Fig. 10.1), only ≈33% of fuel's (chemical) energy is converted into electrical energy. Typical energy distribution in such a plant is shown in Fig. 10.2.

Power Generation Using BST: Typical details of a boiler-based power plant with a BST are presented in Fig. 10.3. Superheated HPS produced by a boiler, WH boiler or HRSG is supplied to a BST, where part of the steam energy is used to generate electric power and the remaining energy is supplied as MPS and/or LPS to the process. Steam condensate,

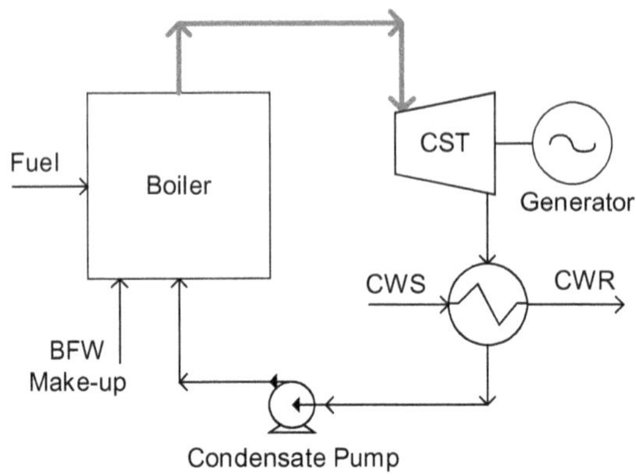

Fig. 10.1: Schematic of a boiler-based power plant with a CST.

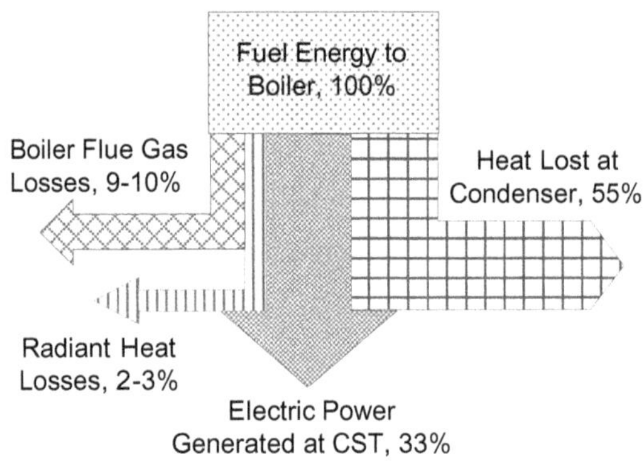

Fig. 10.2: Energy distribution for a boiler-based power generation system, installed with a CST.

generated by the use of MPS and LPS in process heating applications, returns to WH boiler or HRSG for HPS generation again. The main advantage of this system is high energy efficiency (of 80% or more). However, electric power generation is dictated by the steam demand. BST is preferred for small- to medium-scale electric or driver power in process plants, where steam generation and distribution network is available.

Power systems using BST or CST allow the use of low-pressure gaseous/liquid fuel in the boilers. This is an important advantage of steam turbine (ST)–based electric power generation compared to GT-based

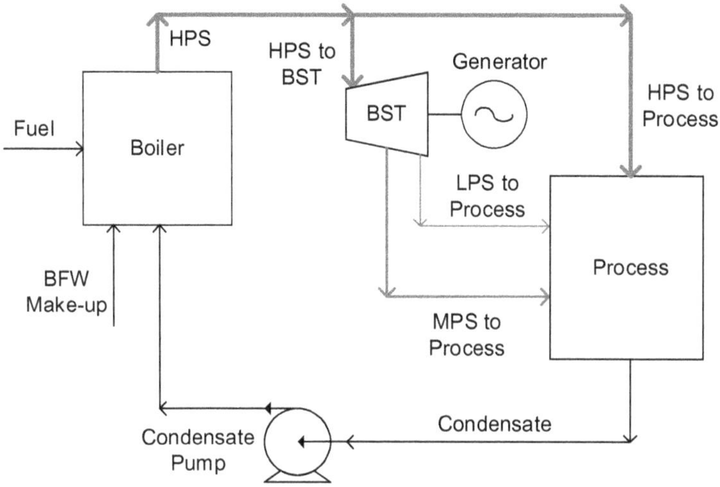

Fig. 10.3: Typical details of a boiler power plant with a BST.

power generation, which requires fuel at a high pressure (of 20 bar or more). Compared to CST, BST generates lower electric power but produces MPS and/or LPS, which can be used for process heating and hence increases the system energy efficiency. Process plants such as refinery, petrochemical and chemical plants usually require both steam and electric power. Hence, the use of BST for co-generation of electric power and steam is attractive. Use of BST in a power cycle is termed a co-generation system as it generates both electric power and steam. This is explained further in the next section.

Power Generation Using GT: A typical GT power generation system is shown in Fig. 10.4. In this, atmospheric air is compressed by an air compressor. The compressed air and high-pressure fuel enter the combustion chamber, where fuel is combusted. Combustion gases at ≈3 bar (for smaller units) to ≈40 bar (for larger complex units) and 1200 to 1400°C (depending on metallurgy of blades of GT) enter the turbine for expansion, which is connected to an electric generator. Rotation of the common shaft of turbine and generator, produces electric power at the generator. Standalone GT efficiency is low at ≈35%. Integrating HRSG at the outlet of GT increases the overall system efficiency to 80% or higher.

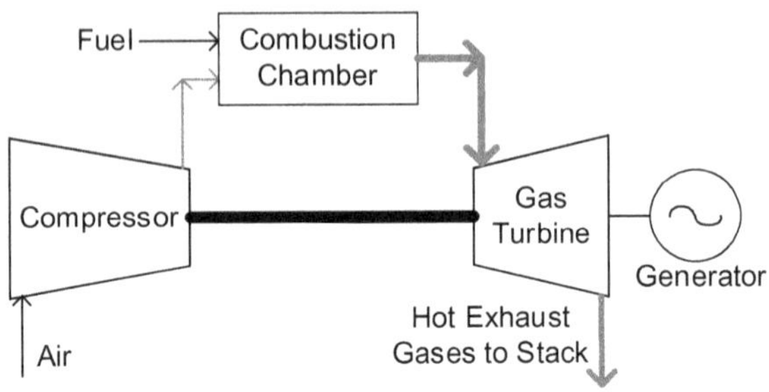

Fig. 10.4: Schematic of a power generation system using GT without HRSG.

Power Generation Using MT: MTs are essentially smaller scale GT generators and are available in the capacity range of 25 to 250 kW. They can use natural gas (NG), biogas or any hydrocarbon gases for combustion, at electric power generation efficiency of 20 to 30%. WHR system can be installed for recovering WH from high-temperature FG from MT. This can increase the overall system energy efficiency to 80% and above.

Power Generation Using RE: There are mainly two types of REs, which are essentially internal combustion engines common in vehicles. They are spark-ignited engines (that operate on Otto cycle and use gaseous or readily vaporizable liquid fuels such as NG, gasoline, propane, biogas and other combustible gases) and self-ignited or diesel engines (that use liquid fuels and achieve ignition through the heat of compression). Combustion of air and fuel mixture followed by expansion moves the pistons in RE. Linear movement of pistons is translated into rotational movement of a crankshaft, which generates electricity via a generator. RE generates electric power with thermal efficiencies of 25 to 45% (based on lower heating value), over a wide range of capacities. Detailed description of various types of REs is available in Chapter 9 of Petchers.[1]

Typical details of RE for generating electric power are shown in Fig. 10.5; for simplicity, lube oil and other smaller coolers are not shown in this figure. RE has higher thermal efficiency and greater power output per unit of fuel energy input, compared to GT. It is advantageous compared to GT, for smaller electricity demand with LPS or hot water demand and when both these demands vary widely. RE is also preferred if there is more demand for LPS or hot water and less demand for HPS. However, it

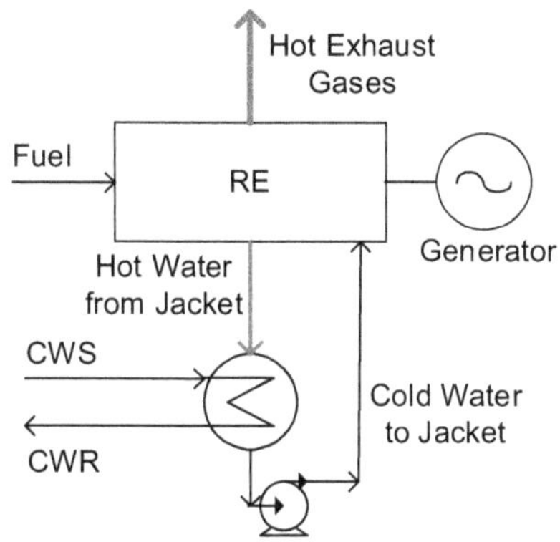

Fig. 10.5: Schematic of an RE-based power generation system.

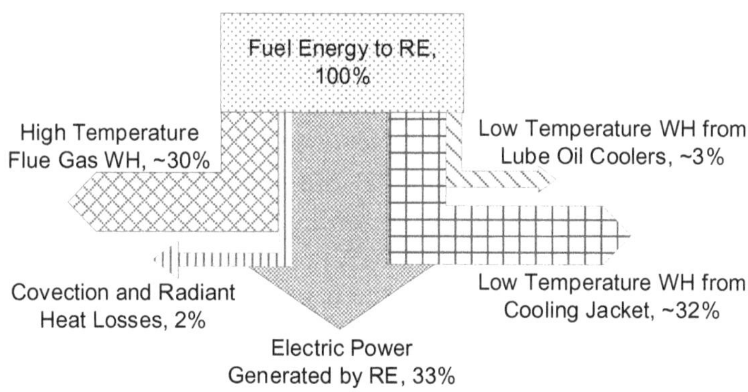

Fig. 10.6: Typical energy distribution in RE.

is relatively large and heavy compared to GT and ST, for a given electric power generation capacity. RE generates significant amount of WH, and hot FG from it has high-temperature WH. RE jacket cooling, lube oil cooling and other miscellaneous coolers generate low-temperature WH. Fig. 10.6 presents the typical distribution of energy output from RE. Most (≈62%) of WH can be recovered by installing good WHR system, and only very small amount (≈2%) of WH is lost to the atmosphere through radiation and convection. RE is a packaged unit (requiring minimal installation cost) with optional WHR unit, and is available in the capacity range of 50 to 10,000 kW.

Power Generation Using ORC and KC: ORC and KC use WH at 100°C to 300°C, to generate electricity at 10 to 20% efficiency, which mainly depends on WH temperature. Higher the WH temperature, higher is the power generation efficiency. Working fluid of ORC and KC is an organic component and a mixture of ammonia and water, respectively. Both ORC and KC, described in Section 9.6 of Chapter 9, are generally expensive compared to ST operating on Rankine cycle. However, they do not utilize any fuel; hence, they are very useful to recover WH and generate additional electricity. ORC and KC are packaged units, available in sizes from 30 to 10,000 kW.

10.3 Co-Generation System

Co-generation (also known as CHP) is a process that produces electricity and heat simultaneously. Some examples of CHP are boiler-BST/CST, GT-HRSG, GT-HRSG-BST/CST, MT-WHR and RE-WHR, which are outlined in this section. Compared to low energy efficiency (≈33%) of simple thermal power plants (i.e., boiler and CST), the efficiency of CHP plant is very high (typically 80 to 95%). Main details of a boiler-CST and boiler-BST–type CHP system are shown in Figs. 10.1 and 10.3, respectively. Large-size CHP systems (>5 MW) use GT, HRSG and BST/CST. Medium-size CHP systems (0.25 to 5 MW) use RE and WHR. Small-size CHP systems (<0.25 MW) use MT and WHR. Benefits of a CHP system are high thermodynamic efficiency and hence lower fuel cost, high reliability for power and steam supply, lower CO_2 emissions, lower emission treatment costs and lower FG stack height.

CHP Systems using RE: RE generates electric power and WH at both high and low temperatures. High-temperature exhaust gases from RE, at 300 to 540°C, can be used for generating HPS and/or MPS. Low-temperature WH at 38 to 127°C is available from RE's water jacket, lubricating oil, valve cage and charge air cooling systems. It can be used to produce LPS or hot water as required by the process plant. Typical RE-based CHP system is shown in Fig. 10.7. It can achieve an overall energy efficiency of up to 95%.[a] Many examples of RE-based co-generation systems are presented in Chapter 9 of Petchers.[1]

[a] https://www.powermag.com/benefits-of-reciprocating-engines-in-power-generation/, accessed in September 2021.

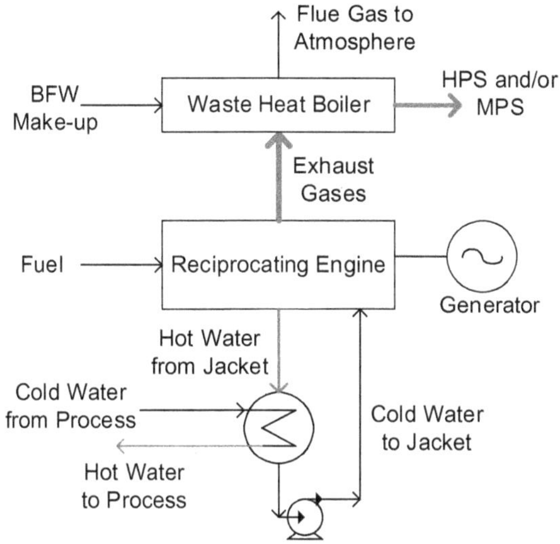

Fig. 10.7: Schematic of a CHP process using RE.

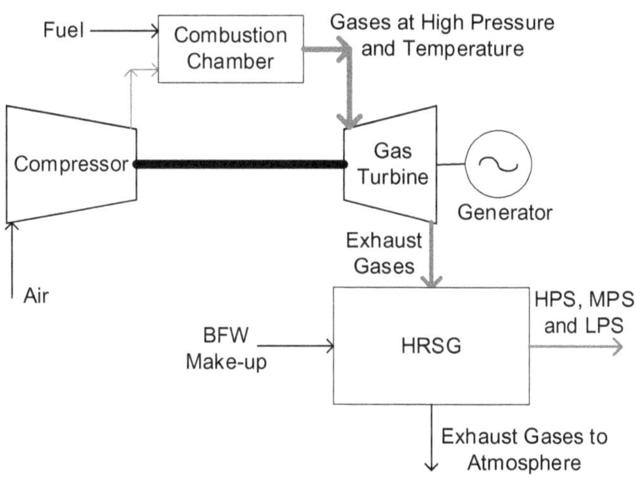

Fig. 10.8: Schematic of a CHP process using GT and HRSG.

CHP Systems Using GT: Besides electric power, GT generates hot exhaust gases, which can be used to generate steam and/or hot water. Typical GT-HRSG CHP system is shown in Fig. 10.8. Exhaust gas from GT is at a very high temperature (≈600°C); hence, it is useful for the generation of HPS, MPS and LPS in HRSG. It can also be used as combustion air in fired heaters as it has high oxygen content and is at a high temperature. The high oxygen content (≈15% by volume) of GT exhaust gases also allows for supplementary fuel firing (known as duct firing), before HRSG, providing additional steam production in HRSG at very high efficiency. This helps to meet any peak demand for steam consumption. Heat recovery at HRSG is limited mainly by steam pressure/temperature requirement and the practical limitation to keep FG above its dew point temperature (covered in Section 18.4 of Chapter 18). GTs are available in capacities of up to several hundred MWs.

GT-based CHP is generally favourable when thermal load requirements are high compared to electric loads and/or when higher temperature steam is required. Its economic performance increases with system size (i.e., increase in demand for power and steam). In applications where most of a plant's thermal energy requirements is in the form of HPS, with limited requirements for LPS, GT is the most suitable choice. If required, more electricity can be generated in the combined cycle process by using part of HPS (from HRSG) in BST or CST. Typical details of a GT-based combined cycle CHP process is shown in Fig. 10.9. This scheme employs BST to generate electric power and also MPS/LPS for process heating applications.

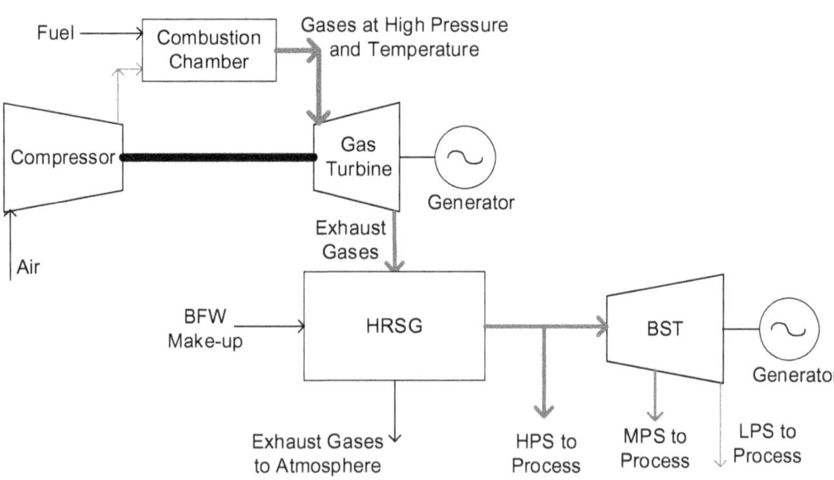

Fig. 10.9: Schematic of a GT-based CHP process with GT, HRSG and BST to produce electric power and HPS/MPS/LPS.

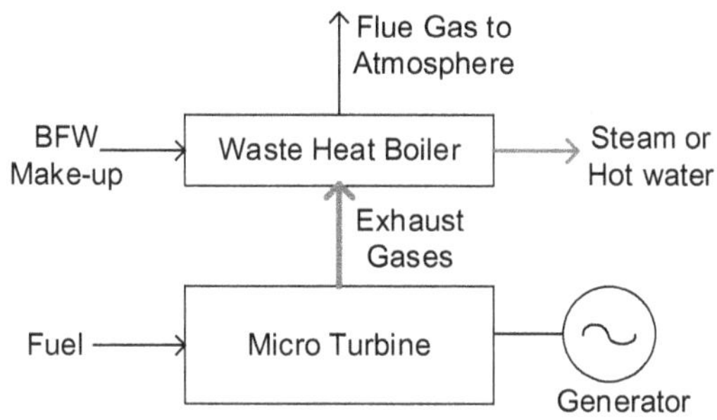

Fig. 10.10: Typical details of a CHP system using MT.

CHP Systems using MT: Schematic of an MT-based co-generation process is presented in Fig. 10.10. MT systems are compact and very suitable for small-scale CHP.

Above described co-generation technologies are compared in Table 10.1.

10.4 Tri-Generation

Tri-generation process produces three utilities (usually, electricity, steam/hot water and ChW/DW/CO_2) simultaneously, using the combustion of fuel. It generates ChW (>5°C), recovers CO_2 or produces DW, in addition to steam/hot water and electric power, generated in a CHP process. Tri-generation is very useful if a site requires electricity, steam/hot water, and ChW/CO_2/DW, and can result in significant operating cost savings. FG discharged to atmosphere (through FG stack) in co-generation plants, is commonly at 180 to 200°C. Hence, it is suitable for use in LiBr absorption chiller for producing ChW. Note that absorption chiller can generate ChW using MPS, LPS or hot water produced in HRSG. Tri-generation system can also be used for producing DW (using thermal desalination, covered in Section 13.6 of Chapter 13).

A tri-generation system to produce electricity, steam and ChW is shown in Fig. 10.11. In this system, combined cycle is used for power generation and HRSG produces HPS. WH from FG leaving HRSG, is extracted in WHR heat exchanger to produce saturated low-low pressure steam (LLPS) at 1.2 bar. After heat exchange in the single-stage LiBr

Table 10.1: Comparison of co-generation technologies: MT, RE, GT and ST

Quantity	MT	RE	GT	BST and CST
Electric power capacity range, kW	30 to 250	50 to 10,000	1000 to >300,000	80 to >500,000
Electric power generation efficiency	20 to 30%*	30 to 50%*	Up to 40%*	25 to 40%*
Fuel	NG, propane, other fuel gases	NG, diesel, fuel oil	NG, propane, other fuel gases	NA
Fuel pressure, bar	2.5 to 7	0.3 to 3	10 to 30	NA
FG temperature, °C	200 to 340°C	Up to 540°C	≈ 600°C	NA
Advantages	• High reliability • Light weight and compact design • No cooling requirement • Low emissions • High-temperature WH is available for recovery • Low cost	• High electrical efficiency even at part load • Both high- and low-temperature WH is available for recovery • Can work with low-pressure fuel gas	• High reliability • Low emissions • High-temperature WH is available for recovery • Minimal cooling requirement. • Operational flexibility in heat to power ratio using supplementary firing • High power-to-weight ratio	• High thermal efficiency for larger sizes • High reliability • Operates using steam and does not use any fossil fuel • Can support variable load operation efficiently

Disadvantages	• Requirement of low- to medium-pressure fuel gas • Higher capital cost per unit electric power output • Lower electric power efficiency • Produces lower grade WH • High maintenance costs	• Requires engine cooling • High emissions • High maintenance costs due to combustion and moving parts	• High capital cost • Requirement of medium- to high-pressure (≈20 to 40 bar) fuel gas • Poor efficiency at part load • Electric power output reduces with increase in ambient temperature • High maintenance costs • Lower electric power efficiency than RE • High noise level requiring acoustic enclosure • Large volume of hot gas to be handled from the exhaust	• Low thermal efficiency and high capital cost per unit electric power, for smaller STs • Requires steam at high pressure (usually HPS or MPS) • Low cycle efficiency for systems using CSTs. • Steam leakage losses

*However, overall efficiency is 80% or higher for co-generation mode producing electric power and steam.

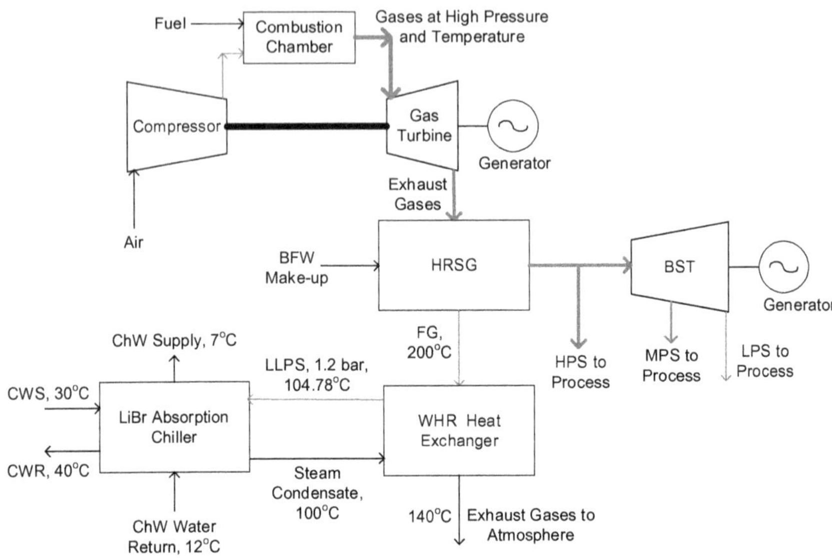

Fig. 10.11: Schematic of a tri-generation process producing electricity, steam and ChW.

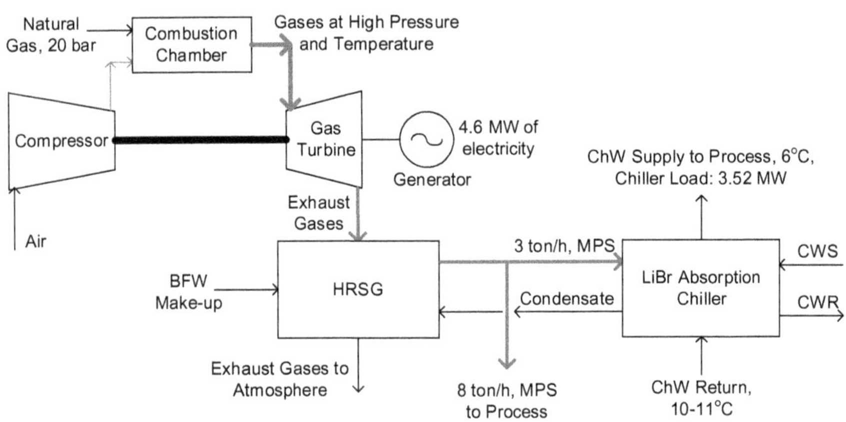

Fig. 10.12: Main units of a tri-generation system in an industrial plant.[b]

absorption chiller, condensed steam is returned to WHR heat exchanger for LLPS generation. An industrial tri-generation system with a GT, HRSG and two-stage LiBr absorption chiller is shown in Fig. 10.12.

[b]http://www.e2singapore.gov.sg/data/0/docs/Case_Study_of_Tri-generation_Project_in_Pfizer.pdf. (accessed on 28 August 2021).

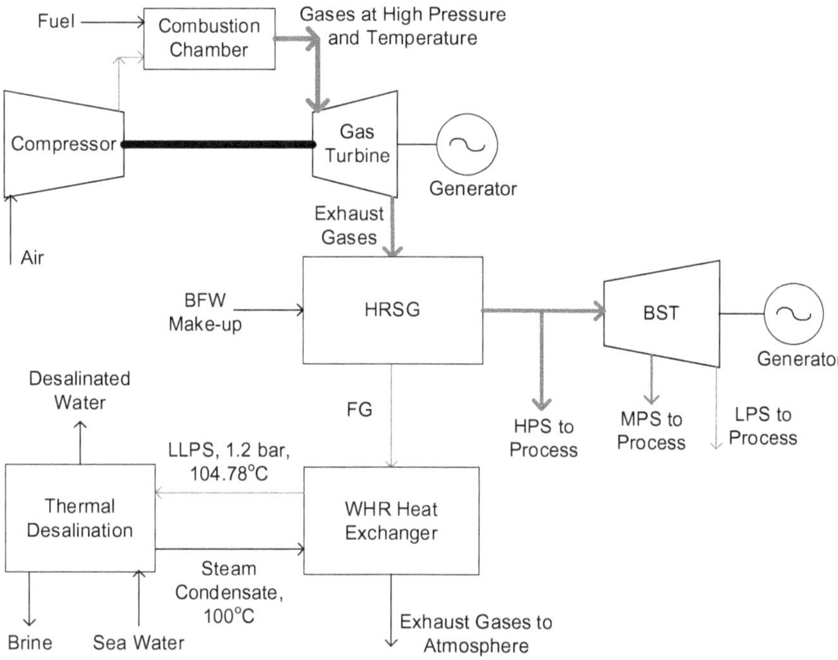

Fig. 10.13: Typical tri-generation process to generate electricity, steam and DW.

HRSG in this system produces 11 ton/hr of MPS, of which 3 ton/hr supplies the required heat to the absorption chiller to produce 1000 ton of refrigeration (= 3.52 MW) in the form of ChW at 6°C.

Tri-generation system can also be used to produce DW instead of ChW. Such a system to generate electricity, steam and DW (generated by thermal desalination) is shown in Fig. 10.13. Such systems are popular in the Middle East.

In the systems in Figs. 10.11 and 10.13, GT is used. However, RE (instead of GT) can be used in a tri-generation system. RE coolant system can generate hot water or LPS that can be used for single-stage LiBr absorption chiller. Exhaust gas of RE or MPS can be used for high-efficiency two-stage LiBr absorption chiller. A tri-generation system using RE for producing electricity, hot water, and CO_2 that can be used at a greenhouse, is shown in Fig. 10.14. Clean fuel such as NG is used in RE for the generation of electricity. Exhaust gas from RE is treated in a catalytic process to remove NOx and convert carbon monoxide to CO_2. Treated exhaust gas, after heat transfer with circulating water at HE-2 (Fig. 10.14), is

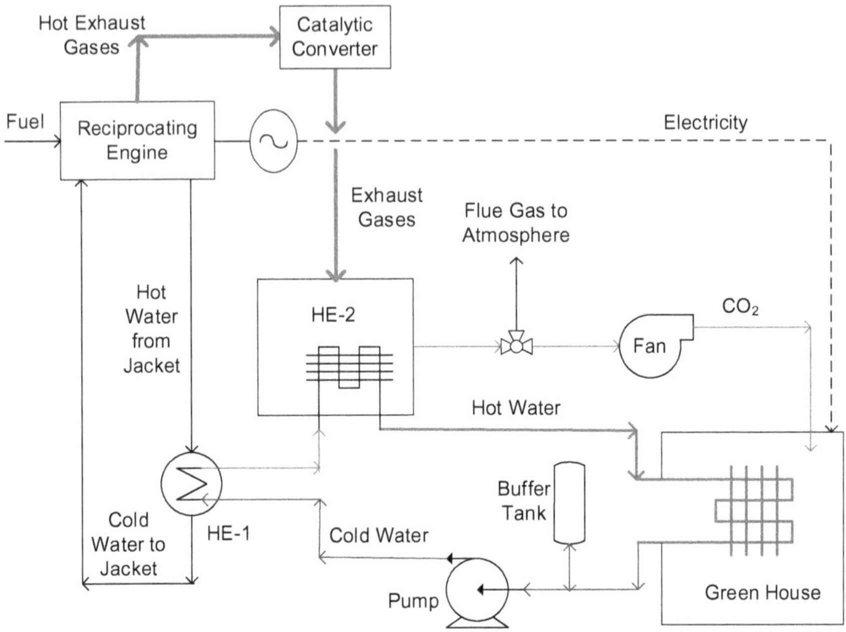

Fig. 10.14: Typical tri-generation system to produce electricity, hot water and CO_2.[c]

compressed by a fan (to typically 1.04 bar) and supplied as CO_2 source for the green house. Heat from cooling jacket of RE is used for heating circulating water at HE-1. Electricity and hot water produced are used for providing light and heat energy, respectively, in the greenhouse. Tri-generation system to produce electricity, steam/hot water and CO_2 is also useful in food and beverage industry, where CO_2 is required for process operations.

10.5 Quad-Generation

Quad-generation produces electricity, heat (steam) and ChW as well as recovers CO_2, for example, from the FG of RE.[d] It is economical and

[c]https://www.clarke-energy.com/wp-content/uploads/Greenhouse-Power1.pdf (accessed on 29 August 2021).

[d]RE is commonly used in industrial installations of quad-generation as it provides sufficient CO_2 for beverage industry. CO_2 from GT will be too much and it will not be economical unless there is use for CO_2 such as carbon sequestration. CO_2 from MT is too small, requires many MTs to meet CO_2 demand, and hence it is not economical.

attractive if there is sufficient demand for the captured CO_2 in the process plant. The exhaust gases from RE can be used to produce HPS and/or MPS. Then, the residual heat in the exhaust gases can be used to produce LLPS at ≈1 bar or hot water at ≈100°C, which can be used in a single-stage LiBr chiller to produce ChW. FG is treated for the removal of NOx and conversion of CO to CO_2, and then CO_2 in it is absorbed using a suitable solvent and released as pure gas from the solvent regenerator.

Quad-generation is now a reality. One example is such an installation in the beverage industry.[e] Benefits of quad-generation are as follows.

- Potential for low or zero carbon emissions.
- Operational cost savings compared to separate purchase of electricity, heat, ChW and CO_2.
- Separated CO_2 can be used for several purposes: as a growth enhancer in organic farming, food and beverage industry, industrial processes and so on.

10.6 Increasing Power Generation using WHR Methods

WHR from a CHP system can be used to maximize power recovery. Three useful methods are steam re-injection, cooling inlet air to air compressor of GT and addition of ORC to the CHP process. First two methods are applicable for GT only. The third method (i.e., power generation by ORC using WHR) can be applied to BST, if LPS generation at BST is more than the process demand. However, this method is generally uneconomical for application on systems such as RE or MT, due to their smaller capacities. The three methods for power augmentation are elaborated in this section.

Steam re-injection: If steam demand is low and power demand is high, additional electric power can be generated at GT by injecting some of the

---

[e] https://www.powerengineeringint.com/world-regions/africa/quad-generation-puts-the-fizz-into-coca-cola/, accessed in September 2021.

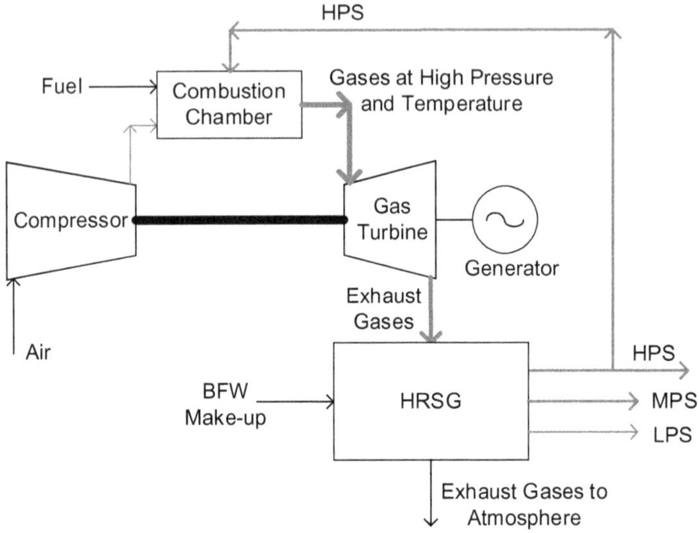

Fig. 10.15: GT power augmentation with HPS injection.

HPS from HRSG back into the GT inlet, as shown in Fig. 10.15. Injecting, usually the superheated HPS (or MPS in some GT models), back to the GT, increases the mass flow through it and hence increases the power output by as much as 50%.[1] This steam injection cycle system is less complex, cheaper and requires smaller space (compared to the combined cycle plant that involves CST or BST). However, it should be provided at the system design stage itself. For any retrofit application, GT manufacturer should be consulted. Steam re-injection requires higher water consumption as the injected steam after expansion in the GT is lost to atmosphere along with the exhaust FG. On the other hand, steam injection helps to reduce NO_X emissions.

GT's Inlet Air Cooling: Electric power generation capacity of GT is generally estimated by its manufacturer assuming intake air at 15°C, 1.013 bar and 60% relative humidity of International Standards Organization (ISO). However, at site conditions, GT may operate at a higher intake air temperature such as 30°C and actual power output can be lower than that achieved under ISO conditions. Hence, cooling the air intake to 15°C increases the electric power output.

A typical application of LiBr absorption chiller is for WHR from FG of HRSG to produce ChW for subsequent use to cool the inlet air to the compressor of GT. ChW is sent through a finned coil heat exchanger, placed inside the intake air filter housing of GT/compressor. This cools the air passing through the filter housing. By this, air density at the compressor inlet increases and hence increases mass flow rate through the compressor. This is the most cost-effective method for increasing power output of GT. Cooling inlet air to the air compressor increases net incremental power output more than incremental fuel input to GT, resulting in improved overall fuel efficiency of GT. Its effect on power output of a typical GT is shown in Fig. 10.16, and details of inlet air cooling of a GT are shown in Fig. 10.17. For retrofit projects, care must be taken to ensure that the air filter housing is made of stainless steel (to prevent corrosion) and separated water from intake air is drained out properly from the filter housing. This prevents any potential damage to the air compressor due to the entry of free water into the compressor.

Another GT-based CHP process that can increase power generation is by using ORC, as shown in Fig. 10.18. This method recovers WH from FG at the outlet of HRSG for generation of additional electric power via ORC.

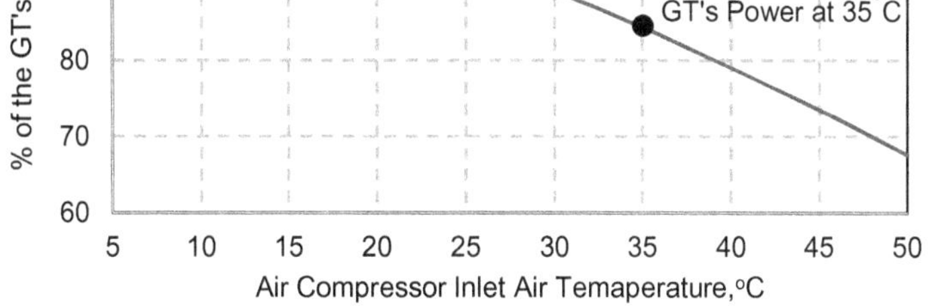

Fig. 10.16: Effect of inlet air temperature on power output of a typical GT.[2]

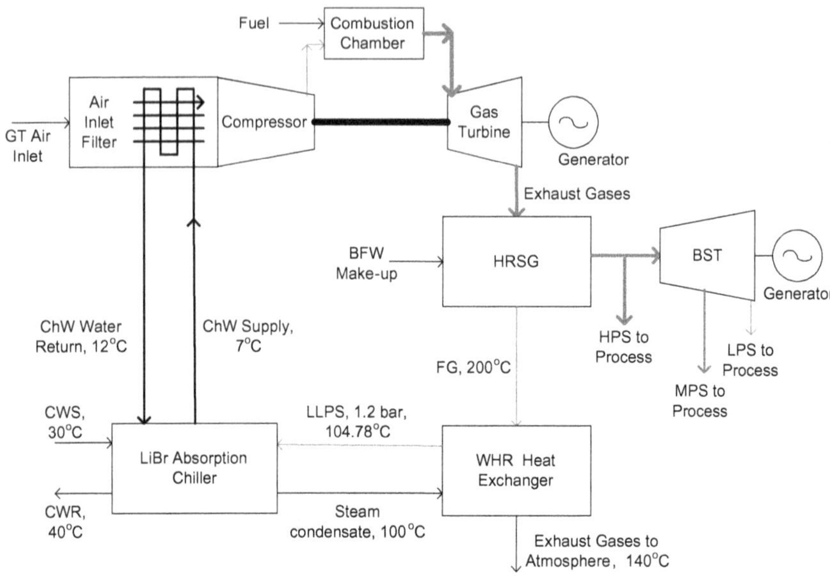

Fig. 10.17: GT with inlet air cooling system.

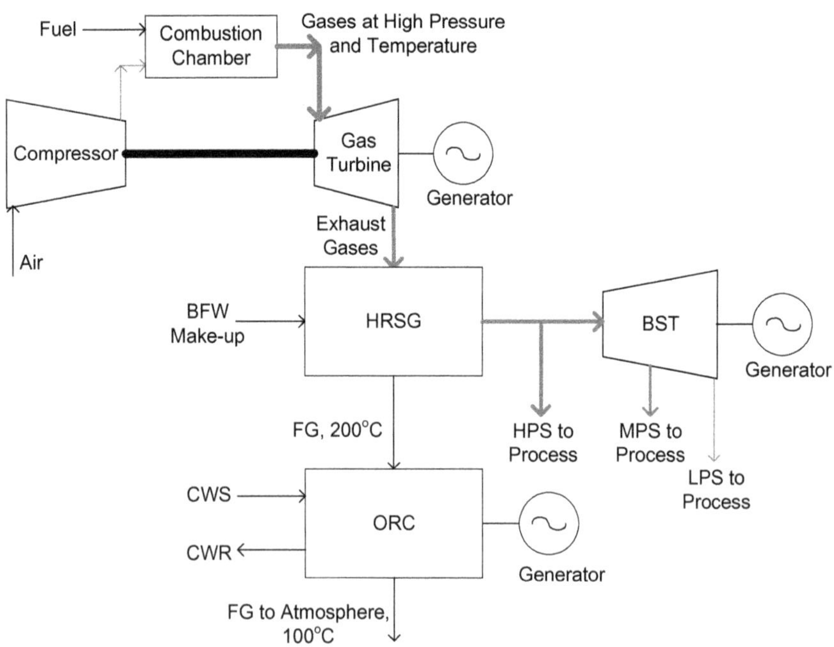

Fig. 10.18: Schematic of a GT-based CHP process with GT, HRSG, BST and ORC.

10.7 Summary

This chapter describes the use of WHR for maximizing electric power generation from BST, CST, GT, RE and MT. Main points of this chapter are as follows.

- Electric power generation equipment common in process and power plants are BST, CST, GT, RE and MT.
- A CHP system allows the recovery of WH from FG and hence significantly increases energy efficiency and fuel savings compared to separate production of electric power and steam.
- Use of BST instead of CST improves the energy efficiency of a CHP system.
- Tri-generation is very useful if there is demand for ChW, DW or CO_2 in process operations. It is also useful for increasing electric power output from GT. As ChW is generated by WH, tri-generation is more beneficial compared to CHP system.
- Quad-generation is very useful if there is demand for CO_2 in the process operations such as in beverage plants. It is also very useful for CO_2 sequestration.

Petchers[1] is a comprehensive book on CHP systems. Chapters 9 to 12 in this book are recommended for detailed applications of RE, GT, ST and CHP systems.

References

1. Petchers N. (2020) *Combined Heating, Cooling & Power Handbook: Technologies & Applications: An Integrated Approach to Energy Resource Optimization*. River Publishers.
2. Green DW, Perry RH. (2008) *Perry's Chemical Engineers' Handbook*, 8th ed. McGraw- Hill.

Acronyms

BFW Boiler Feed Water
BST Back Pressure Type Steam Turbine
CHP Combined Heat and Power

ChW Chilled Water
CST Condensate Type Steam Turbine
CWS Cooling Water Supply
CWR Cooling Water Return
DW Desalinated Water
FG Flue Gas
GT Gas Turbine
HPS High-Pressure Steam
HRSG Heat Recovery Steam Generator
ISO International Standards Organization
KC Kalina Cycle
LiBr Lithium Bromide
LLPS Low-Low Pressure Steam
LPS Low-Pressure Steam
MPS Medium-Pressure Steam
MT Micro Turbine
NG Natural Gas
ORC Organic Rankine Cycle
RE Reciprocating/diesel/gas Engine
ST Steam Turbine
WH Waste Heat
WHR Waste Heat Recovery

Exercises

10.1 Briefly discuss the advantages and disadvantages of CST compared to BST.

10.2 List the power generation equipment commonly used in power and process plants.

10.3 State the advantages of tri-generation systems compared to CHP systems.

10.4 Answer the following with yes (right) or no (wrong), and give brief reasoning, where applicable.
(A) Use of CST for electric power generation is more beneficial compared to the use of BST.

(B) ChW can be produced using WHR from power generation systems.
(C) DW can be produced from WH from power generation systems.
(D) ORC is not recommended for WHR from a power plant as its efficiency is low.
(E) Tri-generation is always more beneficial compared to CHP process.
(F) Quad-generation is not always suitable for power plants.
(G) Steam re-injection is generally cheaper to install compared to a combined cycle system.

Chapter 11

Waste Heat Recovery from Flue Gas Systems

11.1 Overview

Process industries require combustion systems (e.g., boilers, fired heaters/furnaces and cogen plants) for the generation of steam, thermal duty and/or electricity for use in various unit operations. Flue gas (FG) produced in such combustion systems contains significant amount of waste heat (WH). Recovering this WH as much as possible, re-using it in the same equipment and/or exporting it for use in other equipment, is essential for increasing the energy efficiency of the process. This also helps to reduce operating and maintenance costs as well as air and/or water pollution. Hence, flue gas waste heat recovery (FGWHR) is necessary for sustainability of process plants. Many types of FGWHR equipment are available commercially. Proper selection of one of them requires thorough understanding of WH uses (such as direct heating and indirect use via steam, hot water or oil circuits) and constraints such as acid dew point (ADP) corrosion, cost, payback period, space and reliability. Systematic review of FGWHR options is necessary before making the final selection.

Two main purposes of this chapter are to provide a comprehensive review of FGWHR equipment, methods and to present illustrative applications for selecting one of them. This chapter begins with detailed description of FGWHR methods, highlighting their advantages and disadvantages in Section 11.2. Subsequent section presents FGWHR in boilers, with examples. Section 11.4 discusses application of condensing waste heat recovery (WHR) with an example. FGWHR applications in boilers and fired heaters/furnaces, using air preheater (APH), are presented in Sections 11.5 and 11.6, respectively. WHR using heat pipes is presented in Section 11.7. Section 11.8 presents limitations and potential solutions for APH applications. FGWHR in steam generation is discussed in Section 11.9. Next, synergy of FGWHR with pollution reduction and carbon capture technologies is outlined in Section 11.10. This chapter ends with summary in Section 11.11.

Learning outcomes of this chapter are as follows.

1. Identify various types of FGWHR equipment.
2. Discuss the advantages and disadvantages of FGWHR equipment.
3. Explain sizing method for APH and economizers, with considerations to minimize ADP corrosion.
4. Select proper working medium for heat pipes for FGWHR.
5. Discuss indirect WHR using water or oil loops, and steam generation.
6. Discuss FGWHR equipment applications via illustrative examples.
7. Discuss strategies for maximizing FGWHR.

11.2 Advantages and Disadvantages of Flue Gas WHR Methods

FGWHR is essential for increasing energy efficiency of the combustion processes. There are four main types of FGWHR methods and their relative merits are presented in this section. In general, all FGWHR equipment is subject to ADP corrosion if proper care is not taken in their design and throughout their operation to mitigate this damage mechanism.

1. **Direct Heat Exchange:** In boilers, direct heat exchange is carried out between hot FG and boiler feed water (BFW), in economizers. In fired heaters/furnaces, direct heat exchange between hot FG and cold combustion air (CCA) is carried out using recuperative type APH (e.g., tubular, plate type or heat pipe) or regenerative type heat

wheels.[a] Advantage of this method is that recovered WH is recycled back to the WH generating equipment and it will reduce the fuel consumption. Economizers and heat pipes are compact equipment compared to APH. In fact, heat pipe is the most compact and efficient FGWHR equipment, mainly due to condensation and evaporation of working fluid. APHs are generally very big in size mainly due to very low heat transfer coefficients (i.e., gas-gas heat transfer and involves only sensible heat transfer). Hence, they require bigger plot size. Regenerative APHs are generally compact compared to tubular type recuperative APH due to high heat transfer coefficients. It should be noted that plate type and heat-pipe type recuperative APHs are also compact. However, recuperative APH can have significant air leakage (>10%) from CCA to hot FG side. Further, due to rotating parts, they require more maintenance compared to recuperative APHs. Economizers and APH are explained in detail in Sections 11.3 and 11.4/11.5, respectively.

2. **Indirect heat exchange** between hot FG and CCA using a secondary liquid stream such as pressurized water or hot oil: This method is especially useful when there are space constraints for installing APH for exchanging heat directly between FG and CCA. Disadvantage of this method is the requirement of more capital cost for installing WHR heat exchanger, water circulation pump and holding vessel/ drum. This method of indirect heat exchange is described in detail in Section 11.5.
3. **Combinations of FGWHR equipment** include recuperative or regenerative APH, heat pipe, condensing economizer (CE), which are discussed in Section 11.4. Combination of metal and polymer type APH is used in fired heaters/furnaces for recovering WH from FG by

[a]A recuperative heat exchanger has separate paths for flow of cold and hot streams simultaneously; heat transfer from hot to cold stream occurs through a wall. Double pipe heat exchanger (outlined in Chapter 2) is a simple example of recuperative type. On other hand, in a regenerative heat exchanger, cold and hot streams may flow in the same path alternately (static type) or may have separate flow paths (dynamic type); the dynamic type has a rotating wheel with one part being heated by the hot stream and the remaining part being cooled by the cold stream at the same time.

reducing its temperature below acid gas and water dew points. This is explained further in Section 11.7. Similarly, in large power plants, tubular or plate type APH and regenerative APH (heat wheels) are used in series, mainly for optimizing WHR and equipment sizes. The only disadvantage of this method is that it requires more FGWHR equipment.

4. **Generation of steam** from FGWHR (e.g., FG from incinerators): Generated steam can be used for heating applications in other process units, heating CCA in fired heaters/furnaces or utility applications. This option generally requires more capital cost, especially if plant's main steam header is far away from the FGWHR equipment.

11.3 Flue Gas WHR in Boilers with Non-Condensing Economizers

FG generated at a boiler carries a lot of WH. Economizer is a heat exchanger, used in FG ducting of a boiler, to recover some of this WH from FG and utilize it to heat BFW. Typical economizer is shown in Fig. 11.1. Economizer generally uses finned (on external surface) coil made up of carbon steel (CS) or stainless steel (SS), for non-fouling applications such as FG generated with the use of natural gas (NG) or gaseous fuels. For fouling applications such as coal-fired boilers, usually plain tube economizers are used to avoid excessive fouling and for easy cleaning. In an economizer, BFW and FG flow inside and outside the tubes, respectively.

The economizer is installed in FG ducting, between superheater and APH (if used), on its way to the FG stack. Typical arrangement of an economizer is shown in Fig. 11.2 Although flow of water and FG in economizers is a mix of cross and counter flow, they are normally evaluated as counter flow.[1] As a rule of thumb, for everyone 1°C increase of BFW outlet temperature from the economizer, FG temperature decreases by 3–3.5°C. An economizer typically increases BFW temperature by 50–55°C, while lowering FG temperature by 150–170°C. For a given heat transfer area, it recovers more WH from FG due to gas-liquid heat transfer compared to an APH involving gas-gas heat transfer.

To avoid steam generation, which can damage the economizer, water outlet temperature is maintained 15–20°C below the saturation

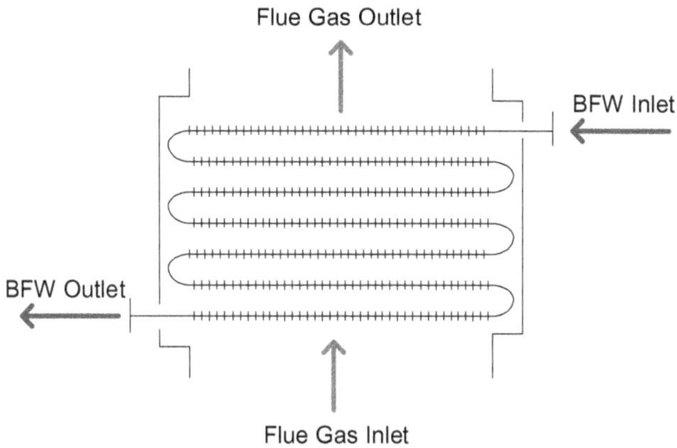

Fig. 11.1: Schematic of a typical economizer.

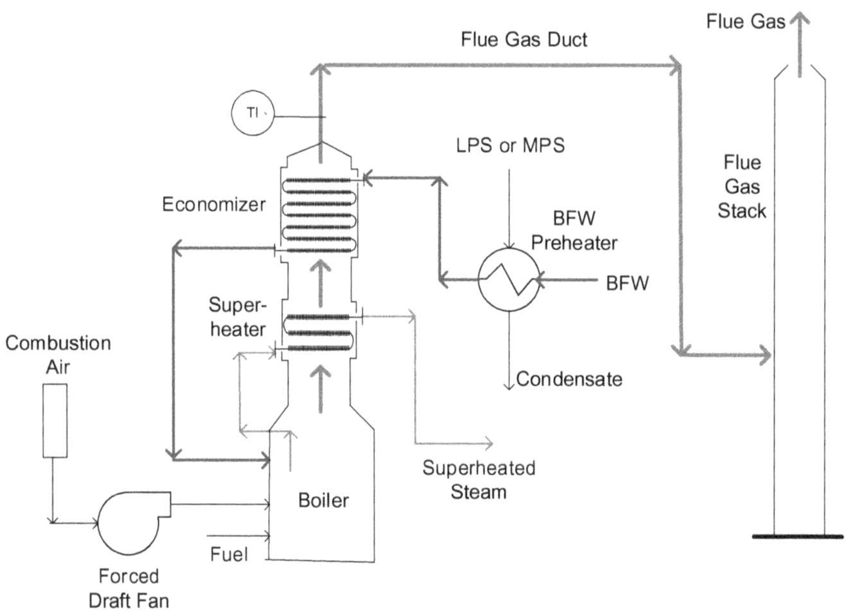

Fig. 11.2: Typical economizer usage in a boiler's FG ducting, without APH.

temperature of boiler's steam drum. Skin/outer temperature of heating coil at BFW inlet (i.e., cold end of the economizer) shall be at least 10°C above ADP temperature (ADPT). If skin temperature indication is not readily available, FG temperature at economizer outlet (TI shown in

Fig. 11.2) may need to be maintained at 28–56°C (50–100°F) above the calculated SO_2 dew point temperature.[2] With NG as fuel, ADP is not a concern. However, if high sulfur fuel oil or other corrosive substances are used as boiler fuel, ADP shall be the most important criterion.

ADPT estimation is illustrated in Chapter 18. In economizers, skin/tube wall temperature of metal is very close to BFW inlet temperature. Ganapathy[3] has presented the following equation for estimation of tube wall temperature.

$$T_w = 0.5 \times \left(T_{bfw\text{-}i} + T_{fg\text{-}o} - U \times \left(T_{fg\text{-}o} - T_{bfw\text{-}i} \right) \times \left(\frac{1}{h_{fg}} - \frac{1}{h_{bfw}} \right) \right) \quad (11.1)$$

Here, T_w, $T_{fg\text{-}o}$ and $T_{bfw\text{-}i}$ are the tube wall, FG (at economizer outlet) and BFW (at economizer inlet) temperatures (all in °F), respectively, U is the overall heat transfer coefficient (Btu/ft^{2}h°F), and h_{fg} and h_{bfw} are the heat transfer coefficients (Btu/ft^{2}h°F) of FG and BFW, respectively. After estimating ADPT, one must ensure that BFW temperature entering economizer shall be at or above ADPT.

11.3.1 *Avoiding Cold-End Corrosion in Economizers*

Several methods to avoid cold-end corrosion in economizers are outlined below.

1. Increase deaerator operating pressure to ensure that BFW temperature entering the economizer is greater than ADP. This is illustrated in Section 16.5.1.1.
2. Use low-pressure steam (LPS)/medium-pressure steam (MPS) heated BFW preheater before the economizer, as shown in Fig. 11.2. In some applications, a portion of high-pressure steam (HPS) generated in the boiler is used for heating BFW. However, this will waste recoverable pressure energy from HPS by passing it through a steam turbine.
3. Hot BFW from economizer outlet can be used to preheat BFW in BFW preheater as shown in Fig. 11.3. This option increases BFW preheater cost as it would require high design pressure for BFW preheater shell.

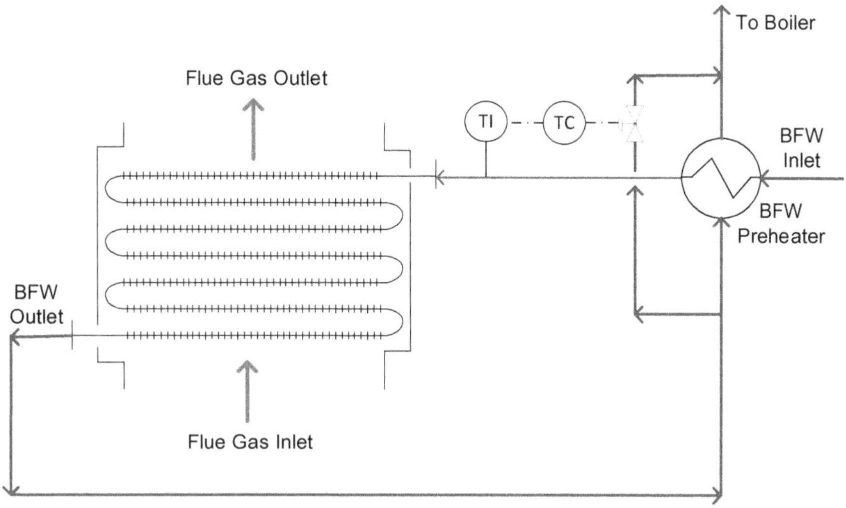

Fig. 11.3: Use of BFW leaving the economizer to preheat BFW at BFW preheater.

4. Some of the BFW from the economizer can be pumped back and mixed with BFW make-up as shown in Fig. 11.4, to preheat BFW by direct contact heating. This option requires two pumps (one in operation and another in standby with auto start function) for recycling hot BFW from economizer's outlet.
5. A coil located inside the steam drum may be used to preheat BFW to near ADPT before BFW enters the economizer.[4]
6. Use of CE (wherein water in FG may condense) for heating BFW and sending the hot BFW above ADPT to non-CE as illustrated in Fig. 11.5.

The last method is the most beneficial method for FGWHR if corrosion issues can be mitigated. First method is the next beneficial method for WHR as it can use LPS recovered from WHR in deaerator to increase the BFW to a temperature equal to ADPT.

Example 11.1: Size an economizer for heating BFW make-up, supplied at 90 ton/hr, 140°C and 60 barg to 185°C. FG flow rate and temperature at the economizer inlet are 96 ton/hr and 350°C, respectively. Assume an overall heat transfer coefficient of 40 $W/m^2 \cdot °C$ for the economizer using

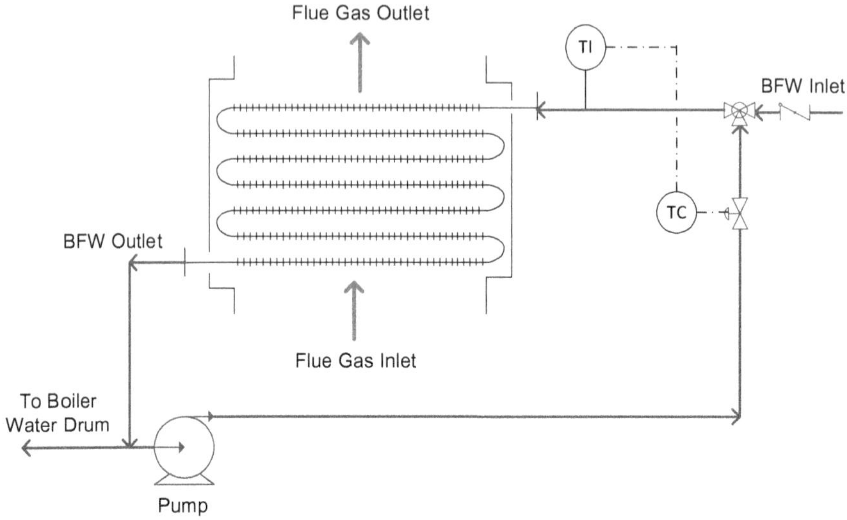

Fig. 11.4: Pumping some of the BFW Outlet from the economizer to heat BFW using a mixing nozzle.

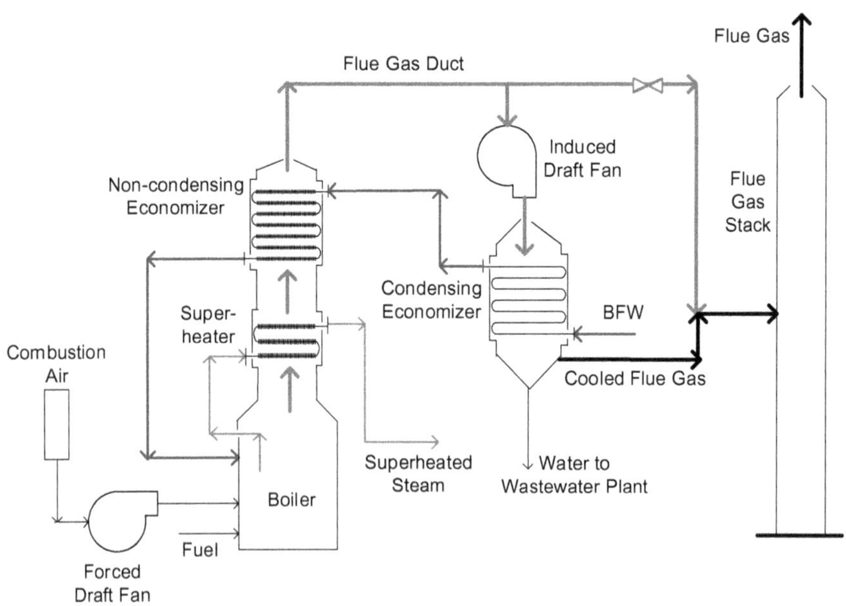

Fig. 11.5: Typical arrangement of non-CE and non-contact type CE.

finned tubes. Average heat capacity of BFW is 4.35 kJ/kg·°C and that of FG is 1.22 kJ/kg·°C

Solution:

BFW temperature at the inlet of the economizer = 140°C
BFW temperature at the outlet of the economizer = 185°C
BFW flow rate = 90 ton/hr
FG temperature at the inlet of the economizer = 350°C
FG flow rate = 96 ton/hr

Using energy balance across economizer, heat gained by BFW = heat lost by FG 90,000 × 4.35 × (185 − 140) = 96,000 × 1.22 × (350 − $T_{fg\text{-}o}$),

By solving the above equation, $T_{fg\text{-}o}$ = 197.79°C. Thus, FG temperature change of 152.21°C is about thrice that in BFW temperature of 45°C.

Log mean temperature difference (LMTD) = $\frac{(350-185)-(197.79-140)}{\ln\frac{(350-185)}{(197.79-140)}}$ = 102.19°C

Heat transfer area required in the economizer = $\frac{90000\times 4.35\times(185-140)\times 1000}{40\times 102.19\times 3600}$ = 1197.22m^2

Example 11.2: Estimate tube wall temperature in economizer for Cases A and B, using the following conditions: BFW inlet temperature to economizer = 284°F (140°C), and FG temperature at the outlet of economizer = 707°F (375°C) for Case A and 347°F (175°C) for Case B. Assume tube inside (h_{bfw}), outside (h_{fg}) and overall heat transfer coefficients of 980 Btu/ft^{2}hr°F, 13 Btu/ft^{2}hr°F and 12.83 Btu/ft^{2}hr°F, respectively.

Solution:

Case A: From Eq. 11.1,

$$T_w = 0.5\times\left(284+707-12.83\times(707-284)\times\left(\frac{1}{13}-\frac{1}{980}\right)\right)$$ = 289.5°F or 143.1°C

Case B: From Eq. 11.1,

$$T_w = 0.5\times\left(284+347-12.83\times(347-284)\times\left(\frac{1}{13}-\frac{1}{980}\right)\right)$$ = 284.8°F or 140.5°C

From these two cases, one can see that FG temperature at BFW inlet section of the economizer has very little influence on the economizer's tube wall temperature at the cold end. For this case study, 360°F (180°C) increase in FG temperature increases economizer's cold-end tube skin temperature by only 4.7°F (2.6°C). Hence, maintaining BFW inlet temperature above ADPT is essential to minimize any chances of ADP corrosion.

11.4 Flue Gas WHR in Boilers with Condensing Economizer

FG can have significant amount of water vapor, as lot of water vapor is generated by the combustion process. For example, for combustion of methane, the stoichiometric equation $CH_4 + 2O_2 \rightarrow CO_2 + 2H_2O$ shows that one kmole (16 kg) of methane combustion generates two kmoles ($2 \times 18 = 36$ kg) of water vapor. So, methane generates 2.25 $\left(=\frac{36}{16}\right)$ times its mass, of water vapor.

Water vapor in FG contains not only sensible heat but also latent heat. Non-contact type (i.e., BFW and FG do not contact with each other) CE (shown in Fig. 11.5) is very useful for extracting large portion of both sensible and latent heat of water vapor from FG. If contact type CE[b] (illustrated in Fig. 11.6) is used, WH is exchanged with an external water circuit to heat water to 40–50°C only. Hence, it is not useful in process plants. Further, before water condenses, all the sulfuric acid in FG will condense. This may cause severe corrosion damage in CE, especially if high sulfur fuel is used in the boiler. This problem will be minimal with NG firing and hence condensing heat transfer can be attractive in that case. Further, corrosion resistant material of construction and/or corrosion resistant coatings can be used effectively.

Bare tubes are used for CE and hence its size will be quite substantial. Some materials of construction for CE tubes are borosilicate glass (suitable up to 232°C and ~7 barg[3]), Teflon or glass-coated CS tubes. For reducing the capital cost, CEs are generally installed after non-CE, as shown in Figs. 11.5 and 11.6. In CE, BFW enters at the bottom and flows

---

[b] https://www.youtube.com/watch?v=exJ3MAEQYKg, accessed on 5th January 2020.

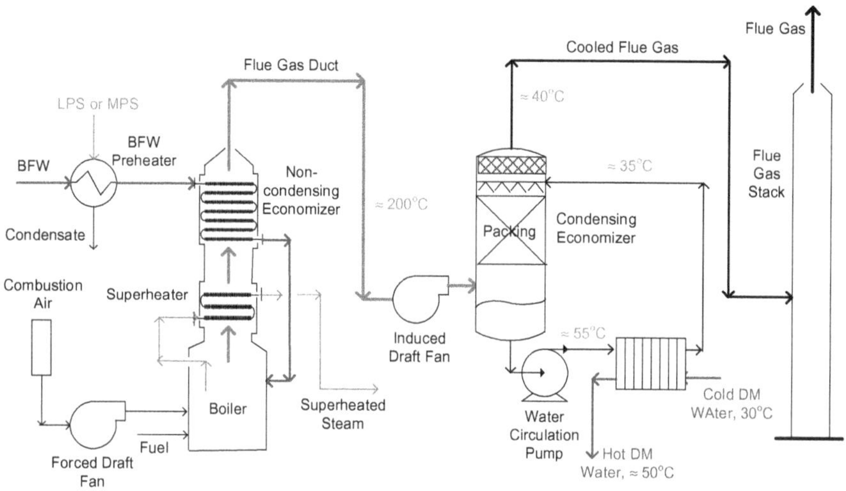

Fig. 11.6: Typical arrangement of non-CE and contact type CE.

upwards inside plain tubes, while FG enters from the top and flows down. After passing through CE, FG flows up to the stack. This arrangement helps in free draining of condensate and acid formed on tubes, to the bottom collection drum. Also, this minimizes any liquid carryover to the FG stack, which can cause severe damage in the stack. Water vapor condensed in CE cannot be directly used for process applications unless it is properly treated. Usual economic treatment route for this water re-use is processing it in the wastewater plant and then through reverse osmosis plant. CEs are generally economical only for larger systems.

Example 11.3: Estimate BFW (DM water) temperature at the outlet of a non-contact type CE using the following conditions. DM water flow rate, supply pressure and temperature to CE are 120 ton/hr, 10 barg and 30°C, respectively. FG pressure, inlet and outlet temperatures at CE are 104.3 kPa, 350°C and 40°C, respectively. FG flow rate is 142 ton/hr (CO_2: 19387.4 kg/hr, H_2O: 18756.1 kg/hr, N_2:101033.8 kg/hr and O_2: 2822.7 kg/hr). Assume pressure drops of 1.5 kPa and 0.5 bar for FG and DM water, respectively, in CE.

Solution: Hysys simulation is performed for this example, and its output is shown in Fig. 11.7. As can be seen DM water can be heated to a

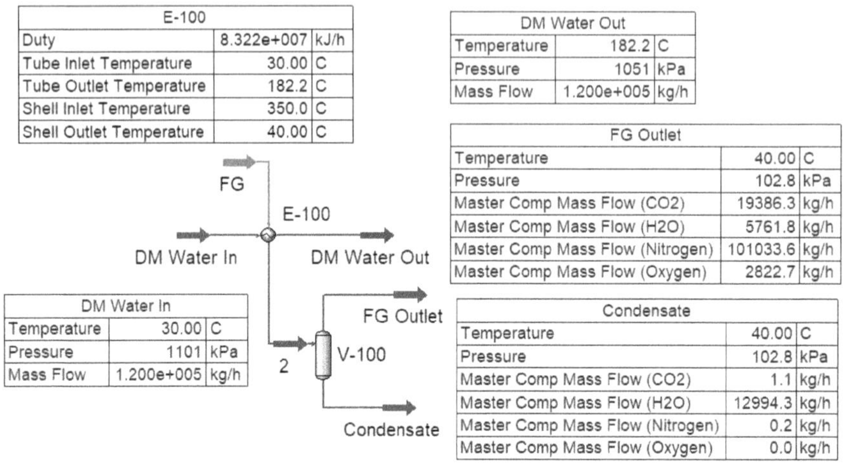

Fig. 11.7: Hysys simulation for exercise 11.3.

temperature of 182.2°C, which maximizes FGWHR by heating feed water. The same result can be obtained without Hysys simulation, by using steam tables, and the procedure for this is given below.

FG inlet temperature = 350°C

FG outlet temperature = 40°C

FG pressure = 104.3 kPa

FG flow rate = 142,000 kg/hr and its component flow rates are: CO_2 (mol. wt. of 44) = 19387.4 kg/hr; H_2O (mol wt. of 18) = 18,756.1 kg/hr; N_2 (mol. wt. of 28) = 101,033.8 kg/hr; O_2 (mol. wt. of 32) = 2822.7 kg/hr

Mass flow rate of gases other than water vapor = 19,387.4 + 101,033.8 + 2822.7 = 123,243.9 kg/hr

DM water inlet temperature = 30°C

DM water flow rate = 120,000 kg/hr

Pressure drop on CE's shell side = 1.5 kPa

Approach temperature at cold-end of CE = 40 − 30 = 10°C

FG pressure at CE outlet (P_{CE-out}) = 104.3 − 1.5 = 102.8 kPa

From steam tables, vapor pressure of water vapor at FG outlet temperature (i.e., 40°C), Pv_{CE-out} = 7.361 kPa

Moles of CO_2 in FG at CE inlet $= \frac{19387.4}{44} = 440.62$

Moles of N_2 in FG at CE inlet $= \frac{101033.8}{28} = 3608.35$

Moles of O_2 in FG at CE inlet $= \frac{2822.7}{32} = 88.21$

Moles of gases other than water vapor in FG at CE inlet = 440.62 + 3608.35 + 88.21 = 4137.2

Average molecular weight of gases other than water vapor (MW_{OG}) in FG at CE inlet = $\frac{440.62}{4137.2} \times 44 + \frac{3608.35}{4137.2} \times 28 + \frac{88.21}{4137.2} \times 32 = 29.79$

Mole fraction of water vapor in FG at CE inlet =

$\dfrac{\frac{18756.1}{18}}{\frac{19387.4}{44} + \frac{18756.1}{18} + \frac{101033.8}{28} + \frac{2822.7}{32}} = 0.201$, and so mole percent of non-condensable components = (1 − 0.201) × 100 = 79.9%

Partial pressure of water vapor in FG at CE inlet = Inlet pressure × Mole fraction = 104.322 kPa × 0.201 = 20.97 kPa

Amount of water vapor leaving the CE is given by:

$$M_{S(CE-out)} = \frac{MW_S \times Pv_{CE-out} \times M_{OG}}{MW_{OG} \times \left(P_{CE-out} - Pv_{CE-out}\right)}$$

$$= \frac{18 \times 7.361 \times 123243.9}{29.79 \times \left(102.8 - 7.361\right)} = 5743.52 \text{ kg/h}$$

Here, MW_s and MW_{OG} are the molecular weights of water vapor and average molecular weight of gases other than water vapor (i.e., CO_2, N_2 and O_2). M_{OG} is the mass flow rate (kg/hr) of the gases other than water vapor. P_{CE-out} and Pv_{CE-out} are the pressure (kPa) of FG at CE outlet and vapor pressure (kPa) of water vapor at CE inlet, respectively.

Small difference of 0.31% in values, estimated by Hysys simulation (5761.78 kg/hr) and using steam tables (5743.52 kg/hr), is due to differences in steam property methods used.

Approximate value of BFW (DM water) temperature at the outlet of a non-contact type CE can be calculated as follows.

Saturation temperature of vapor at CE inlet = 61.2°C (from steam tables, corresponding to partial pressure of water vapor of 20.97 kPa)

Average heat capacity of FG from 350°C to 61.2°C = 1.178 kJ/kg°C (from Hysys)

Average heat capacity of FG from 61.2°C to 40°C = 1.1 kJ/kg°C (from Hysys)

Latent heat of water vapor at CE outlet pressure = 2255.2 kJ/kg (from steam tables)

Amount of water vapor condensed = 18,756.1 − 5743.52 = 13,014.5 kg/hr

For approximate calculation, it is assumed that all 13,014.5 kg/hr of water vapor condenses at 61.2°C. In reality, water vapor progressively condenses as FG is cooled below 61.2°C. For accurate calculation, one should use a process simulator such as Hysys.

From heat balance, sensible heat gained by DM water = sensible heat lost by FG + latent heat lost by condensed water vapor + sensible heat lost by condensed water

Sensible heat lost by FG = Heat lost by 142,000 kg/hr of FG from 350°C to 61.2°C + Heat lost by 128985.5 (= 142,000 − 13,014.5) kg/hr of FG from 61.2°C to 40°C = (142,000 × 1.178 × (350 − 61.2)) + (128,985.5 × (61.2 − 40)) = 48,309,308.8 + 3,007,941.86 = 51,317,250.7 kJ/hr

Latent heat lost by the condensed water = 13,014.5 × 2255.2 = 29,350,300.4 kJ/hr

Sensible heat lost by the condensed water = 13,0145 × 4.18 × (61.2 – 40) = 1,153,292.93 kJ/hr

Here, 4.18 is the average heat capacity of water, between 61.2 °C and 40 °C.

Total WH recovered to DM water = 51,317,250.7 + 29,350,300.4 + 1,153,292.93 = 81,820,844 kJ/hr

Since outlet temperature of DM water is not known, as a first trial, let us use heat capacity of DM water at 30°C = 4.18 kJ/kg °C (from steam tables).

Expected temperature rise of DM water $= \frac{81{,}820{,}844}{120{,}000 \times 4.18} = 163.1°C$

Outlet temperature of DM water at the non-contact type CE = 163.1 + 30 = 193.1°C

For 2nd trial, average temperature of DM water $\left(= \frac{30 + 191.09}{2} = 111.6°C\right)$ can be used.

Heat capacity of DM water at 111.6°C = 4.2327 kJ/kg °C (from steam tables)

Temperature rise of DM water $= \frac{81{,}820{,}844}{120{,}000 \times 4.2312} = 161.145°C$

Outlet temperature of DM water at the non-contact type CE = 161.09 + 30 = 191.09°C

For 3rd trial, average temperature of DM water $\left(= \frac{30+191.09}{2} = 110.545°C\right)$ can be used.

Heat capacity of DM water at 110.545°C = 4.2312 kJ/kg°C (from steam tables)

Temperature rise of DM water $= \frac{81{,}820{,}844}{120{,}000 \times 4.2312} = 161.145°C$

Since temperature raise of DM water from 2nd and 3rd trials is only 0.055°C, further trials are not required. Hence, outlet temperature of DM water at the non-contact type CE = 161.1 + 30 = 191.1°C

Corresponding value by Hysys simulation is 182.2°C. The estimated value by approximate procedure is 4.91% higher than that obtained by Hysys simulation. This error is mainly due to the use of simplified condensation assumption and average heat capacity values.

11.5 Flue Gas WHR in Boilers with Air Preheater

A boiler produces utility steam, mainly for process heating purposes. With small to medium size NG or oil-fired boiler applications, economizer installation is more beneficial compared to APH installation. Further, if there are concerns on increase in NO_x emissions, APHs cannot be used. This is because APH increases the combustion air temperature in the boiler/fired heaters and hence elevates the flame temperature, which leads to increases in NO_x levels. This may require NO_x reduction/control equipment, if very tight NO_x regulations are applicable at the site. For example, with NG as fuel, at 15% excess air, flame temperature increases by about 36°C for 55°C increase in combustion air temperature.[4] Additionally, APH incurs higher pressure drop (100–200 mm H_2O) compared to the economizer (25–50 mm H_2O). Hence, they incur higher operating cost due to higher fan power requirement. However, in some cases, CCA preheating using APH is used, especially in boiler applications involving burning low heating value or solid fuels (where flame stability is improved by using hot air) and for fired heaters (where economizer is not generally applicable). Normally, ambient air is heated to 300–350°C in APH. For each degree increase in combustion air temperature, roughly 0.8 degree drop in FG temperature is achieved. Boiler's thermal efficiency increases

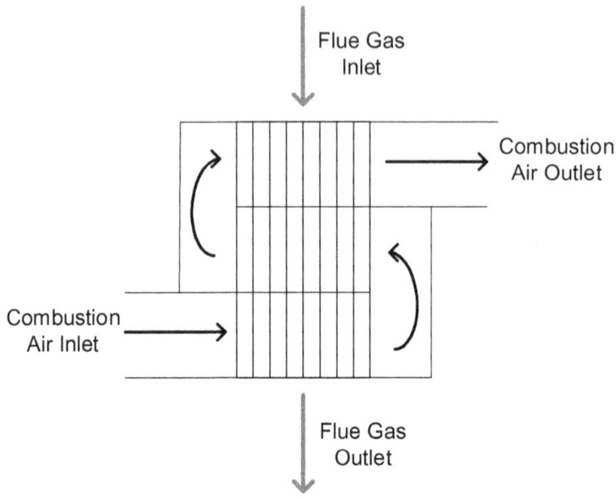

Fig. 11.8: APH with three passes on combustion air side and single pass on FG side.

by 1% for every ~20°C rise in combustion air temperature.[5] Typical APH installation arrangement is shown in Fig. 11.8.

Three types of APH are in common use for FGWHR:

- Direct WHR using recuperative APH
- Direct WHR using regenerative type such as heat wheel, packed with metal baskets (Ljungstrom type)
- Indirect WHR using intermediate circulating fluid such as water or heat transfer oils

Recuperative APHs are provided in FG stacks of small to medium capacity boilers. They can be tubular or plate type. FG and CCA are prevented from mixing, without need of any seals. APH can be designed and installed in many arrangements. APH, heat pipe or heat wheels may use materials of construction such as CS, corten steel,[c] SS, glass, glass-coated CS and polymers. Some of the arrangements of APH are shown in Figs. 11.8–11.10.

[c] Coten steel is a high-strength low alloy steel. This material has high thermal conductivity and high resistance to corrosion and erosion.

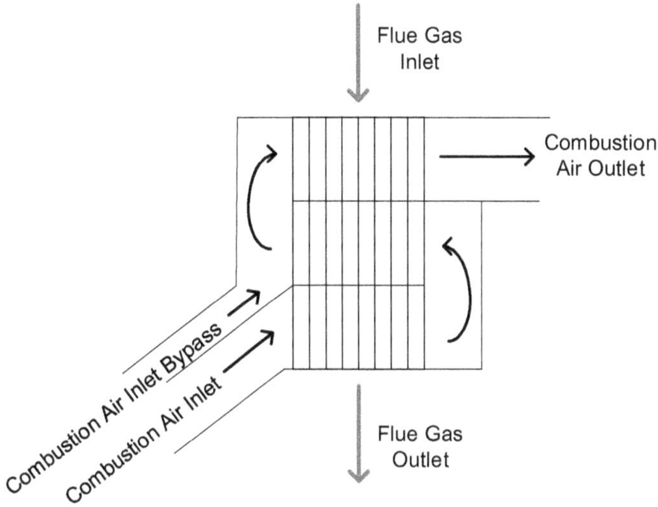

Fig. 11.9: APH with three passes on combustion air side and single pass on FG side as well as CCA bypass for corrosion prevention.

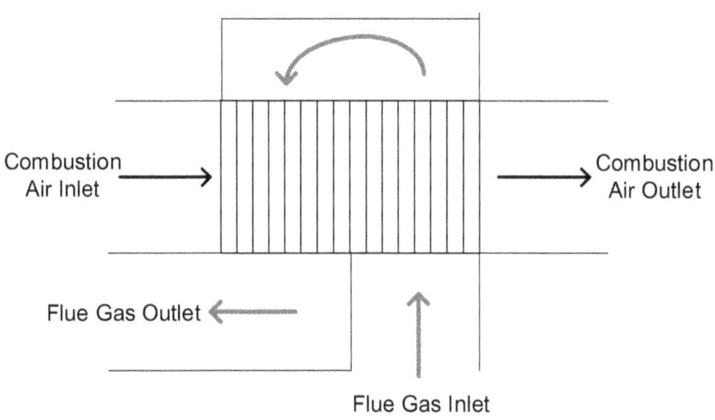

Fig. 11.10: APH with two passes on FG side and single pass on combustion air side.

In Fig. 11.8, APH uses three passes on CCA side and one pass on FG side. FG direction is top to bottom, which is very useful to minimize the ducting work between APH and induced draft fan (IDF), installed at the bottom of APH. APH arrangement in Fig. 11.9 is similar to that in Fig. 11.8, except a CCA bypass is provided. This is useful if there is a concern on cold-end corrosion due to acid condensation. In this arrangement, some of the CCA bypasses the 1st pass (bottom section) and mixes directly with hot air from 2nd pass. With proper design of bypass flow rate, skin temperature of metal in 1st pass can be maintained above ADPT. Thus, corrosion-erosion damage and fouling can be minimized. Number of CCA passes in both Figs. 11.8 and 11.9 can be adjusted based on the heat exchange requirement and available pressure drop. Figure 11.10 shows another possible arrangement, where FG passes through two passes whereas CCA passes through a single pass. In this arrangement, FG flows up and then down and so it is very useful for handling FG with solid particles such as in coal-fired boilers.

For very large capacity boilers, such as producing >500 ton/hr of steam, recuperative APH size becomes very big and hence regenerative APHs are used. Regenerative type APH generally includes two or more heat storage segments. FG and combustion air take turns to pass through each regenerator segment. By this, hot FG heats the storage segment first and then CCA extracts this heat when it passes through it later. Sometimes, a combination of recuperative and rotary regenerative APHs (with tubular recuperative type for primary air heating and regenerative type for the secondary air heating) is employed. Regenerative APHs are compact and can be used in either stationary or rotating arrangements.

The regenerative type APH[d] generally has a major problem of air leaks across mechanical seals. In many of the older plants installed with this type of APH, it is being replaced with recuperative APH, mainly due to high leakage (>10%) of CCA to FG and high maintenance costs. CCA leak to FG in APH creates many problems such as (i) reduces WHR, (ii) reduces CCA supply to fired heater or boiler resulting in oxygen deficient combustion and hence creating destructive hazard of after-burn and (iii) wastes forced draft fan (FDF) power as some CCA goes to stack directly with FG. Air leakage in any APH may be easily identified by taking FG samples at APH inlet and outlet, and analysing them for oxygen content. If FG oxygen concentration at APH outlet is more than that at APH inlet, it indicates air leak; the more the difference in these oxygen concentration values, the higher is the air leakage.

---

[d] https://www.youtube.com/watch?v=XuLfASz0YkU, accessed on 5th January 2021.

For the above reasons, regenerative type APH are not covered further in this chapter.

Typical APH arrangement in a boiler's FG stack is shown in Fig. 11.11. For large boilers, both economizer and APH may be used (shown in Fig. 11.12) if they can be economically justifiable. Advantages of use of APH and economizers in boilers are: (i) increases WHR and

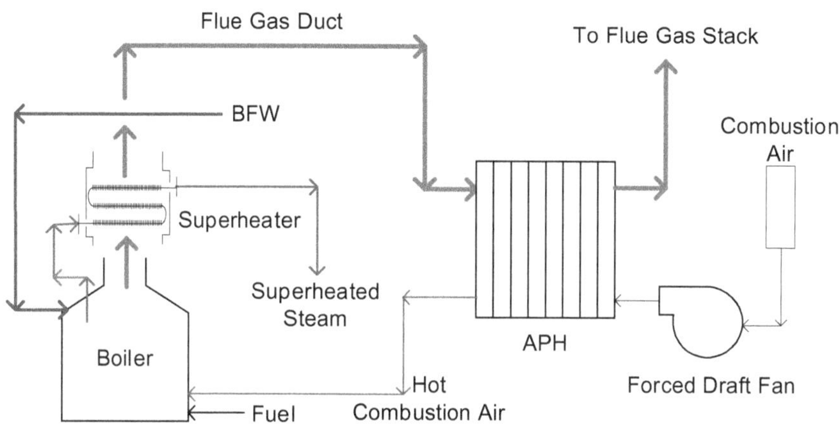

Fig. 11.11: Typical APH use in a boiler FG ducting.

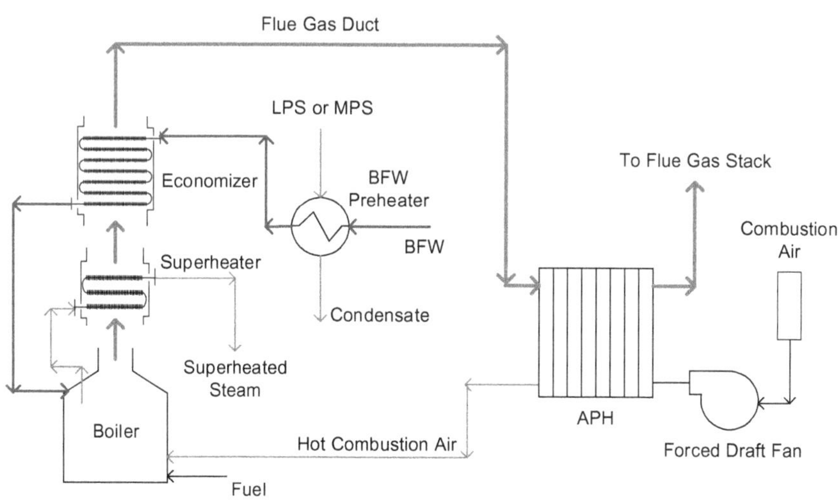

Fig. 11.12: Typical arrangement of economizer and APH in a boiler FG ducting.

hence reduces the fuel consumption, (ii) reduces cooling water requirement for FG desulphurization and/or CO_2 capture and (iii) may reduce FG stack height requirements.

11.6 Flue Gas WHR in Fired Heaters/Furnaces Using Air Preheater

Fired heater/furnaces in process industries heat the process fluid. Boilers may handle water inside or outsides the tubes. However, fired heaters/furnaces handle process fluid only inside the tubes. Common features of boilers and fired heaters are that both involve combustion process and generate FG. Hence, FGWHR is essential for increasing the efficiency of these equipment.

For fired heaters, economizers for heating BFW are not generally applicable as convection section recovers as much FG heat as economically possible by proper use of approach temperature between cold process fluid and FG at convection section outlet. APH is the most commonly used solution for further FGWHR, unless there are NO_x emission constraints. Although fired heaters do not require any BFW heating, CE discussed in Section 11.4 can still be used for suppling hot water to other processes and/or utility WHR applications such as absorption chillers (described in Chapters 5 and 9), organic Rankine cycle (covered in Chapter 10) and thermal desalination (presented in Chapter 13). If employing APH exceeds NO_x emission limits, other WHR methods such as steam generation, water or oil system can be utilized to recover WH in the convection section of fired heaters, for use in other process equipment.

Before applying APH, WH in FG shall be minimized as much as possible; this can reduce the size and hence cost of WHR equipment. For this, first, maintaining the recommended air-fuel ratio is very important. Low ratio leads to incomplete combustion, low efficiency and unsafe operating conditions; but also, maintaining low excess air will reduce NO_x generation.[6] On the other hand, high ratio leads to reduced efficiency. Typical thermal efficiency variation of a fired heater with excess air and stack temperature is shown in Table 11.1. The values in this table are extracted from Figure 2 in Platvoet.[7] Optimal excess air levels are 15% for natural draft (ND) and 10% forced/balanced draft (FD/BD) operation of fuel-gas/

Table 11.1: Variation of thermal efficiency of a fired heater with excess air and FG temperature

Excess Air, %	O_2 in FG, %	Temperature of Flue Gas, °C							
		149	**204**	**260**	**316**	**371**	**427**	**482**	**538**
0	0	92.5	90.1	87.5	85.1	82.5	80.1	78.0	75.1
5	1	92.3	89.9	87.3	84.7	82.0	79.5	77.1	74.2
10	2	92.0	89.4	87.0	84.0	81.5	78.8	76.2	73.2
15	3	91.8	89.1	86.4	83.6	80.6	78.1	75.3	72.4
20	3.8	91.5	88.8	86.0	83.2	80.3	77.4	74.4	71.4
25	4.6	91.3	88.4	85.6	82.6	79.6	76.6	73.6	70.4
30	5.2	91.1	88.1	85.1	82.1	79.0	75.9	72.7	69.5

NG fired heater, and 20% for ND and 15% for FD/BD operation of fuel-oil fired heater.

Excess air can be monitored and optimized using an oxygen analyser, installed at the inlet of convection section of the fired heater. This control is very useful for APH installations. In addition, the use of carbon monoxide equivalent (COe) analyzer[e] provides assurance that there is sufficient oxygen for the combustion process, as it indicates COe reading if oxygen is insufficient for combustion. It is recommended to:

— Operate at ~2% excess oxygen (and not air) in FG, for NG or fuel-gas firing.
— Avoid turndown operation of burners as much as possible by shutting down some burners.
— Maximize primary air to burners if they have both primary and secondary air.

Second, it is important to minimize any air leakage through damaged ducting, openings in convection section and FG ducting. Due to vacuum pressure in these sections, cold ambient air can enter FG ducting and hence can increase FG flow rate and reduce FG temperature. This reduces

[e] A COe analyzer detects and indicates unburnt fuel in FG as carbon monoxide equivalent value.

FGWHR at APH. Hence, convection section and FG system shall be maintained with good mechanical integrity to prevent ingress of atmospheric air. According to Liebermann and Lieberman,[2] percent fuel loss due to air ingress into FG system can be calculated by:

$$\%\text{fuel loss} = \frac{\left(T_{fg} - T_a\right) \times \left(O_{2fg} - O_{2ic}\right)}{500} \tag{11.2}$$

Here, T_{fg} and T_a are FG (at stack) and ambient temperatures in °F, respectively, and O_{2fg} and O_{2ic} are oxygen concentration (volume percent) in FG at stack and at the convection section inlet, respectively.

Example 11.4: Estimate percentage of fuel lost, if temperature and oxygen concentration of FG at stack are 300°C and 8%, respectively. Oxygen concentration in FG at the inlet of the convection section is 4% (by volume). Temperature of ambient air is 30°C.

Solution:

FG temperature = 300°C = $\frac{300\times 9}{5}+32$ = 572°F

Ambient temperature = 30°C = $\frac{30 \times 9}{5}+32$ = 86°F

Using Eq. 11.2, $\%\text{fuel loss} = \frac{(572-86)\times(8-4)}{500} = 3.89\%$

It is not possible to use APH in fired heaters/furnaces with ND (Fig. 11.13), mainly due to insufficient pressure for flow of FG and CCA through APH as FDF and IDF are not available. ND heaters/furnaces use stack/sucking effect[f] to pull combustion air through the burner(s) into the combustion zone, to create negative pressure inside the heater.

APHs are also not generally applicable for ID furnaces. If the draft created by the FG stack is not sufficient for safe operation of the fired heater, IDF is used. General arrangement of ID type fired heater is shown in Fig. 11.14.

On the other hand, fired heaters with FD (Fig. 11.15) utilise FDF to push combustion air through the burner(s) and into the combustion zone. FD allows good air-fuel mixing and hence leads to efficient combustion through smaller burners. FDF delivers pressurized (usually ~500 mm H_2O, depending on air control valve, air duct and burner pressure drop)

---

[f]Stack effect is the air movement caused by temperature difference (and consequently density difference) between temperature of FG and ambient air.

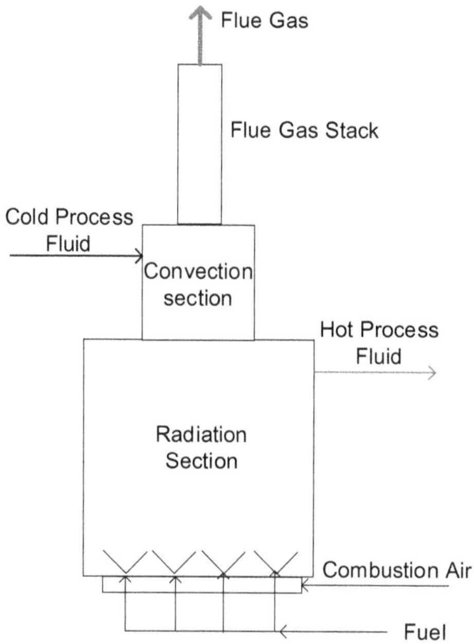

Fig. 11.13: Schematic of a typical ND furnace, wherein APH is not feasible.

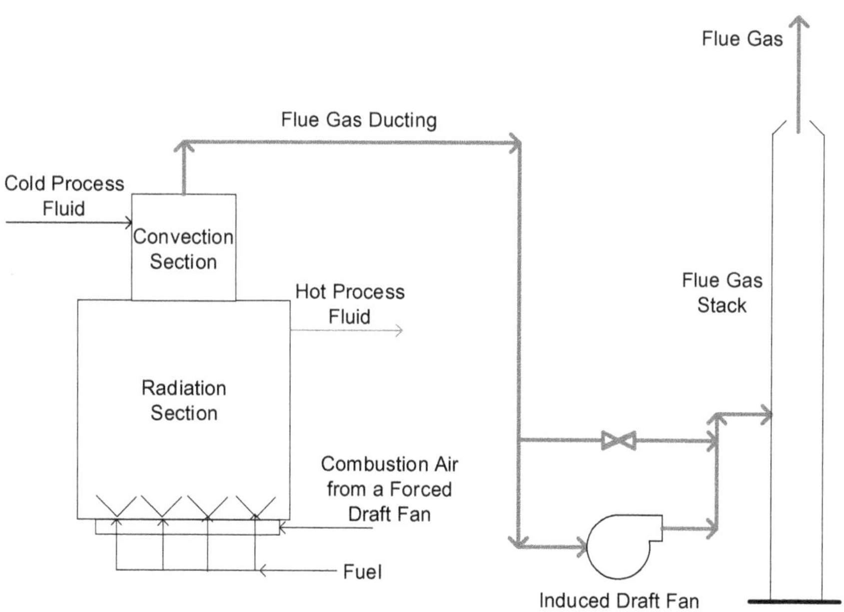

Fig. 11.14: Details of an ID furnace.

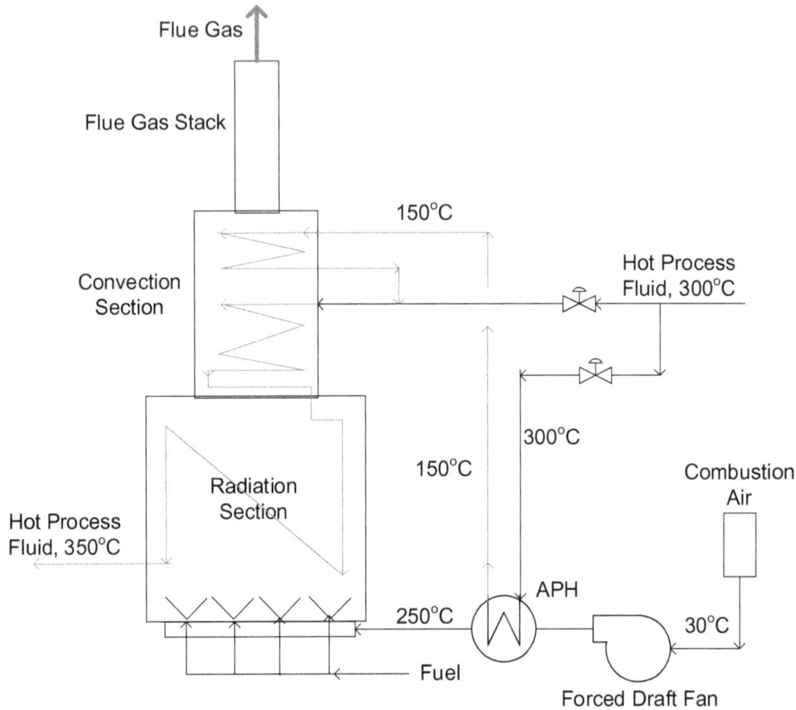

Fig. 11.15: Details of a FD furnace with APH, heated by a part of hot process fluid.

combustion air to fired heater/furnace with FD. This allows installation of APH.

FG stack may be installed directly above ND/ID/FD heater as they are generally used for providing smaller heating duties and hence handle low FG quantity.

In some process heating applications, process fluid temperature at the inlet of the convection section can be significantly high (e.g., heating of atmospheric distillate in the fired heater of a vacuum distillation unit). This allows a part of the process feed to exchange heat with CCA in an APH, as illustrated in Fig. 11.15. Another alternative arrangement for heating CCA is shown in Fig. 11.16; this arrangement uses LPS or an external hot process stream.

BD fired heaters generally serve large heating duties and use both FDF and IDF to push and pull combustion air through the burner(s) into

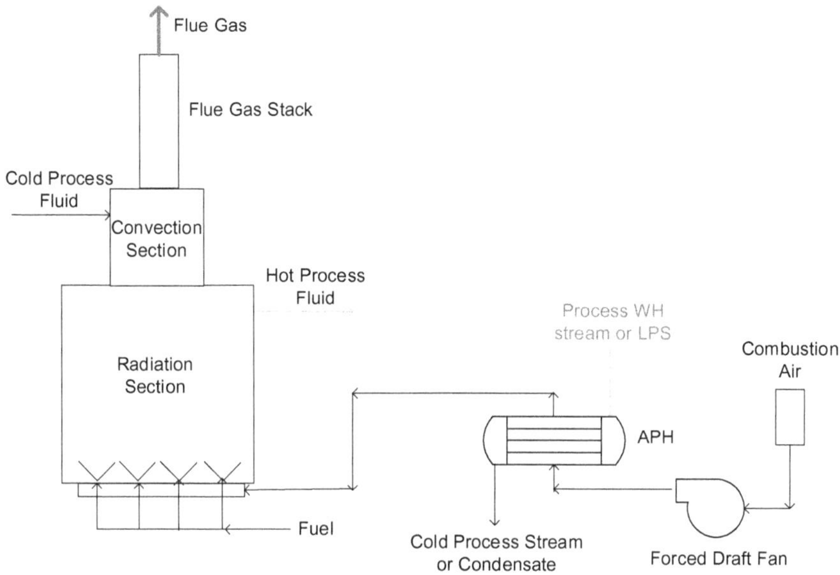

Fig. 11.16: Use of process WH stream or LPS to heat CCA in a FD furnace.

the combustion zone (Fig. 11.17). IDF extracts FG under vacuum condition (usually 250–500 mm H_2O) and compresses it to a pressure above atmosphere pressure, sufficient to overcome frictional pressure drops in FG ducting and stack. BD furnaces generally use an external FG stack. With the use of IDF, APH can be used, as long as IDF can meet pressure drop requirements for APH as well. Typical BD fired heater/furnace with plate type APH is shown in Fig. 11.18. This arrangement uses one LPS heater (to heat CCA to a temperature high enough to prevent ADP corrosion) before APH. LPS generated in other process equipment/WHR can be used for this purpose.

Example 11.5: Size an APH for heating CCA of 90 ton/hr, from 100°C to 300°C. FG flow rate and temperature at APH inlet are 95 ton/hr and 380°C, respectively. Consider an overall heat transfer coefficient of 27 W/m^2°C for APH. Average specific heat of FG and combustion air are 1.135 kJ/kg °C and 1.1 kJ/kg °C, respectively.

Solution:

Amount of heat transferred to combustion air = 90,000 × 1.1 × (300 − 100) = 19,800,000 kJ/hr $= \frac{19800000}{3600} = 5500\,\text{kW}$

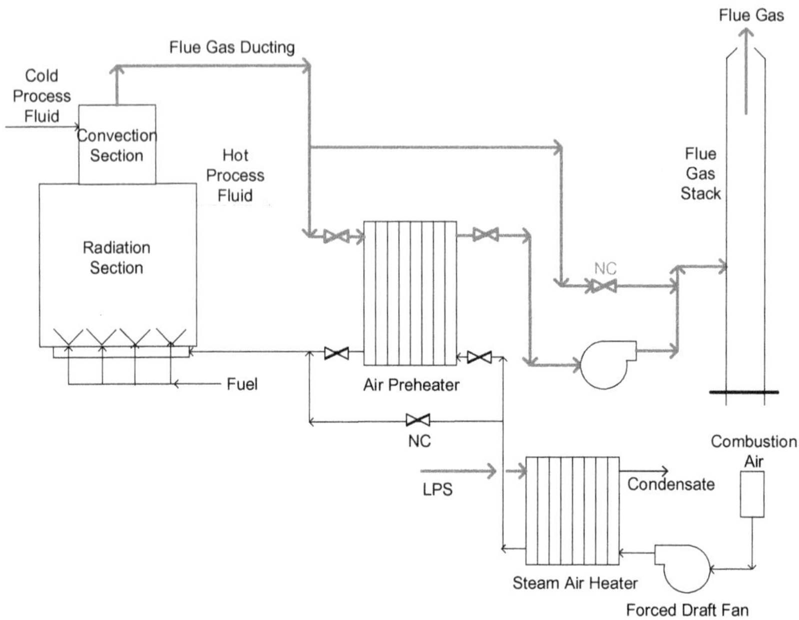

Fig. 11.17: Typical details of a BD furnace with APH and steam APH.

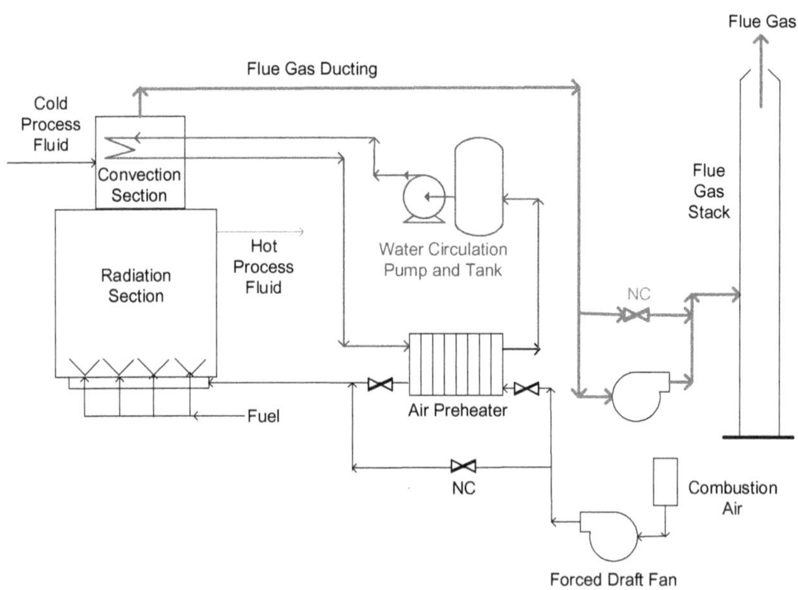

Fig. 11.18: Typical details of a BD furnace with APH and water loop for transferring FG heat to combustion air.

By energy balance, FG temperature at outlet of APH = FG temperature at APH inlet $-\frac{\textit{Heat exchanged to CCA}}{\textit{Mass flow rate of FG} \times \textit{average specific heat of FG}} = 380 - \frac{19800000}{95000 \times 1.135} = 380 - 183.63 = 196.37\,°C$

Approach temperature at hot end of APH = 380 − 300 = 80°C

Approach temperature at cold end of APH = 196.37 − 100 = 96.37°C

$$\text{LMTD} = \frac{(96.37-80)}{ln\frac{96.37}{80}} = 87.93°C$$

Overall heat transfer coefficient, U = 27 W/m^2°C

Heat transfer area required $= \frac{\text{Heat exchanged}}{\text{U} \times \text{LMTD}} = \frac{5500 \times 1000}{27 \times 87.93} = 2316.66\,\text{m}^2.$

There are many types of APHs such as tube type, plate type (includes plain plates or corrugated plates or finned plates), tubes/plates + glass tubes/glass-coated tubes combination. For accurate sizing, one can get free sizing proposal from vendors.

Due to low heat transfer coefficients for air-to-air heat exchange, APH sizes are generally huge and require considerable plot space for installation. Plate type APH is compact compared to tube type, for the same FGWHR duty. Owing to requirements of very low pressure drops, air and FG duct sizes will be very big. Hence, APH installation with ducting connection will require significant amount of space. In some older plants, sufficient plot space may not be available for adding APH. In such cases, arrangement shown in Fig. 11.18, which uses a water or other working fluid circuit to transfer WH from FG to CCA, is possible. This arrangement requires finned coils in convection section and/or FG ducting to extract WH from FG to circulating water/working fluid under pressure (5–15 barg depending on WH temperature, to prevent steam generation), by a centrifugal pump connected to water holding vessel/drum. After heat transfer in APH, water/working fluid returns to holding vessel/drum and the cycle repeats. The arrangement in Fig. 11.18 can reduce APH size due to higher heat transfer coefficient of water compared to FG, and also space required for FG ducting with APH. Additionally, it prevents any mixing of combustion air with FG. However, overall cost for this arrangement is generally high due to the requirement of holding vessel, pumps, piping and instrumentation.

The extent of heat recovery using an APH is usually constrained by the ADPT constraint. FG temperature at APH outlet is generally maintained at >150°C. For maximizing WHR from FG, a metal APH can be

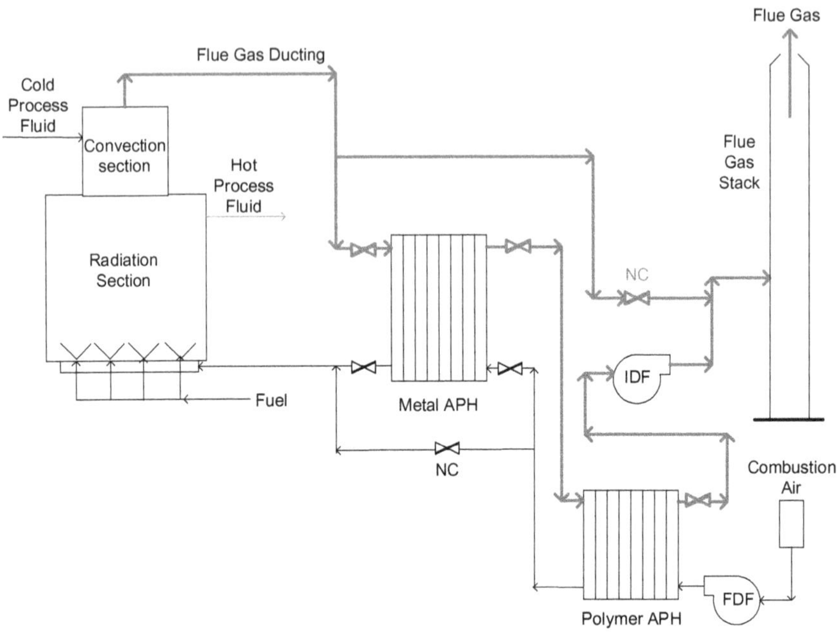

Fig. 11.19: Typical details of a BD furnace with metal and polymer APHs, installed in series.

used in series arrangement with a polymer APH[g] as shown in Fig. 11.19. In this arrangement, FG temperature at the inlet of the polymer APH shall be maintained above the ADPT constraint of 150°C for metal APH but below ~160°C to avoid damaging the polymer heat transfer elements of the polymer type APH. The polymer APH can then cool FG below both the ADPT and water dew point temperature, without concerns of ADP corrosion. Hence, FG temperature to FG stack can be reduced to below 100°C. Note that, due to low heat transfer coefficients, low approach temperatures and low thermal conductivity of polymer material, polymer APH size will be quite large. Further, polymer APH increases combustion air temperature to metal APH, which will increase the size of metal APH also. Owing to use of two APHs in series, pressure drop can be substantial. So, a WHR revamp project to add polymer APH

[g] https://heatmatrixgroup.com/products/air-preheater/, accessed on 5th January 2021.

may require upsizing of IDF and consequently can increase capital cost further.

11.7 Flue Gas WHR Using Heat Pipes

Heat pipe exchanger consists of a bundle of tubes; each tube contains a wick, is partly filled with a working fluid such as ethanol, water or toluene and is sealed. It is generally installed slanted (6°–10° slope from the horizontal) or vertical. Typical heat pipe details are shown in Fig. 11.20. At the bottom of the heat pipe, working fluid is in liquid state and it evaporates by absorbing heat from the hot FG. Vapour generated at the bottom end of the heat pipe rises to top end (due to low vapour density), where it releases

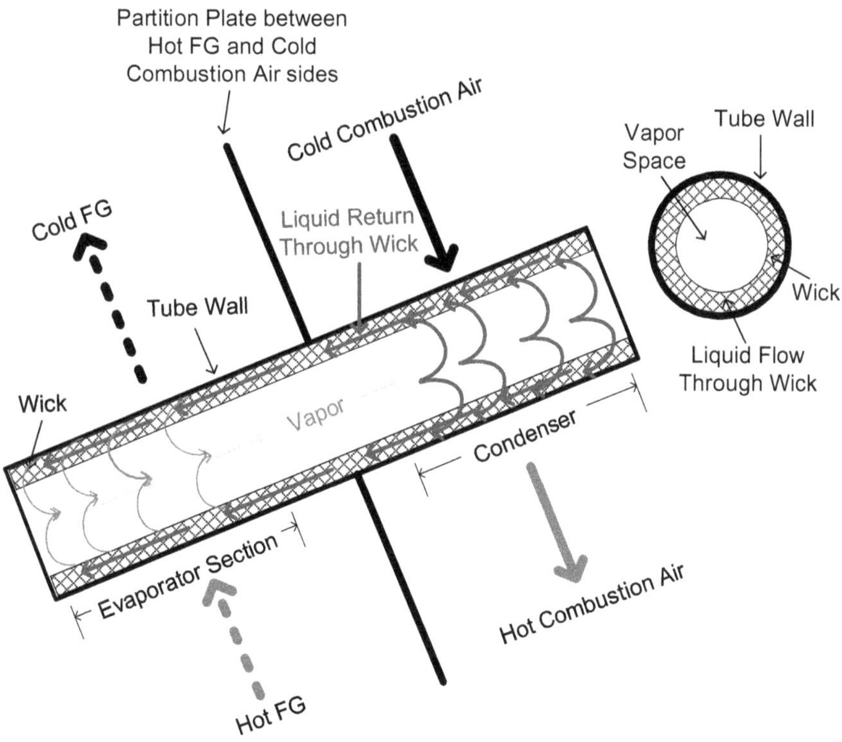

Fig. 11.20: Typical details of a tube in a heat pipe exchanger; cross section of the tube is shown at the top right side.

heat to CCA and eventually condenses. Condensed liquid at top end flows down by gravity (due to high liquid density) through the internal capillary wick to the lower end. This cycle of vaporization and condensation continues repeatedly, and WHR from hot FG is transferred to CCA.

Heat pipe's external surface can be finned (for clean FG services) or without fins (for heavy fouling FG services), to promote heat transfer and hence make the unit compact. A partition plate (usually installed perpendicular to tubes) separates and seals the hot FG and CCA sides. It also provides the intermediate support for tubes. Heat transfer is nearly isothermal across the heat pipe portions in hot and cold sections. Heat pipes can be used over broad range of services from refrigeration to high temperature WHR. They can be used in both boilers and fired heaters/furnaces.

Typical arrangement of heat pipe type APH for FGWHR is shown in Fig. 11.21. Heat pipes have the following advantages over APHs.

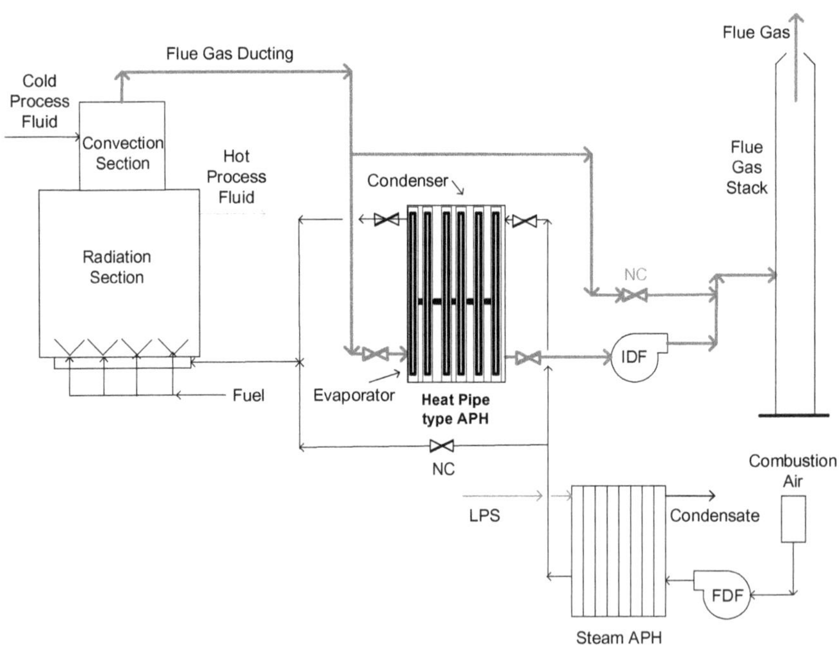

Fig. 11.21: Typical arrangement of heat pipe type APH for FG WHR; Condenser and evaporator of the heat pipe exchanger are separated by a horizontal partition plate.

1. They can operate with low approach temperatures.
2. They prevent leakage of CCA into FG side, which is a major problem in regenerative APHs.
3. They have very high heat transfer coefficients, mainly due to condensation, vaporization of the working fluid and external finned surface. Hence, they are very compact and lighter compared to other type of APHs.
4. There are no rotating parts (which are present in some regenerative type heat exchangers), and hence do not require any auxiliary power.
5. FG and CCA side pressure drops are generally lower compared to other types of APHs.
6. They can be designed to prevent ADP corrosion by proper selection of working fluid.
7. Fin density of FG and CCA side portions of the heat pipe can be selected independently. As CCA is clean service, high fin density can be used. Depending on cleanliness of the FG, fin density on FG side can be selected differently.

Disadvantages of heat pipes are: they are generally expensive, and operating temperature range of cold and hot fluids are fixed depending on the selected working fluid and its pressure. If operating temperatures exceed this temperature range, the heat pipe fails to operate.

Theoretical operating temperature range of a working fluid in a heat pipe exchanger varies between the triple and the critical point temperatures of the fluid.[8] Typical properties and practical range for some working fluids for WHR are listed in Table 11.2. Working fluid selection for heat pipes depends on many parameters such as good thermal stability, maximum and minimum temperature ranges of FG and combustion air, corrosivity, heat transfer capabilities, high latent heat, good wettability of wick, vapour pressure, and compatibility of the working fluid with both tube and wick materials. Copper, aluminium and SS are the most commonly used tube materials. Copper is preferred for low-temperature FG WHR applications between 0 and 200°C. With water as working fluid, SS is not used. For very high temperature applications, metals such as Inconel, nickel, tungsten, molybdenum and niobium are used.

Table 11.2: Properties of working fluids for WHR

Fluid	Melting Point, °C	Boiling Point, °C	Useful Range, °C
Acetone	−95	57	0–120
Methanol	−98	64	10–130
Fluetec PP2	−50	76	10–160
Ethanol	−112	78	0–130
Heptane	−90	98	0–150
Water	0	100	30–200
Toluene	−95	110	50–200
Fluetec PP9 or Dowtherm A	−70	160	0–225
Thermax	12	257	150–350
Mercury	−39	361	250–650
Caesium	29	670	450–900

Table 11.3: Summary of APH applications for fired heaters/furnaces: Applicable (A) and Not Applicable (NA)

Draft Type	Regenerative APH	Recuperative APH	Heat Pipe Exchanger	Indirect Loop Using Water	Steam Heater
ND	NA	NA	NA	NA	NA
FD	NA	A	A	A	A
ID	A	A	A	NA	NA
BD	A	A	A	A	NA

11.8 Limitations and Potential Solutions for Air Preheater Applications

APH applications for fired heaters/furnaces are summarized in Table 11.3. If the fired heater's efficiency is very low, APH installation projects may encounter the following problems.

- Use of APH may increase combustion air temperature, which in turn may increase NO_x emissions above acceptable limits.
- Increase in combustion air temperature increases radiant coil heat flux, which may exceed the acceptable limit.

- Increase in combustion air temperature raises tube-support and/or guide temperatures, which may exceed their capabilities.

Potential solutions for the above problems are as follows.

- Provide combustion air temperature control capability by adding a temperature control loop, which uses CCA bypass flow control across APH. This controls the combustion air temperature by mixing CCA with hot combustion air from APH. This feature increases the heater's turndown capability (without NO_x emissions problems).
- Limit the APH size to maintain the combustion air temperature to achieve acceptable NO_x emissions limit, achieve acceptable radiant average/peak heat flux rates and stay within the recommended tube-support and/or guide temperatures.
- Consider upgrading burners to low or ultra-low NO_x type burners.
- Use alternative WHR method such as steam generation if there is demand for additional steam at the site.
- Consider reducing excess air.
- Consider installation of NO_x reduction systems such as selective catalytic reduction (SCR) reactor in the FG stream downstream of the convection section.

Use of APH can lead to corrosion and erosion at the cold end. Following solutions can be used to prevent this problem.

- Glass tubes or glass-coated tubes at cold end: Borosilicate or glass-coated CS tubes at the cold end of APH can be used for minimizing ADP corrosion and erosion. This will help to maximize WHR.
- Some portion of CCA bypass across APH: This method uses a portion of CCA to bypass APH and mix with hot air from APH. The reduction of CCA flow through APH results in a lower air side heat transfer coefficient, and hence results in higher metal temperature at the cold end of APH, which prevents corrosion and erosion.
- External preheat of cold air: This method uses CCA to be heated by LPS or any hot process fluid to a temperature higher than ADPT. The heated air then enters APH and hence metal temperature at APH's

cold end can be maintained sufficiently high to prevent corrosion and erosion problems.

- Hot air recirculation[9]: This method recycles heated combustion air from APH's outlet to the FDF's suction to obtain mixed-air temperature, which is equal or higher than ADPT. Although this method can avoid corrosion and erosion, it requires bigger FDF and hence results in more operating cost.
- Working fluid temperature control: In indirect FGWHR system such as that shown in Fig. 11.18, APH cold-end metal temperature can be maintained at above ADPT by controlling the temperature of the heat transfer fluid (e.g., pressurized water) at the inlet of APH. Working fluid temperature can be increased by either bypassing a portion of the working fluid around the working fluid heater (heated with FG) or decreasing the working fluid flow rate by varying the speed of the working fluid circulation pump using a variable speed drive.

11.9 Flue Gas WHR for Steam Generation

FGWHR can be utilized for steam generation if use of APH creates problem of high NO_x or other constraint such as need for replacement of burners. Steam generation from FG can be implemented in natural or forced circulation of water. Typical natural circulation arrangement for FGWHR from a fired heater is shown in Fig. 11.22. In this arrangement, BFW flows through an economizer to be heated to 15°C to 20°C lower than the saturation temperature of water in the steam drum, and then to the steam drum. Hot water from the steam drum flows to steam generating coils by static head. After heat gain in steam generating coils, steam/water mixture is collected in an outlet header, which returns it to the steam drum. In the steam drum, steam is separated from hot saturated water and then sent to the steam header.

Depending on the pressure of steam generated and end use of steam, superheater coils are optional. If steam requires superheating, saturated steam passes through a superheating coil placed before the inlet of the steam generating coil. If saturated steam is required for consumption at a nearby equipment, superheater is not required. However, steam superheater is required if the generated steam is exported to the plant's main steam header for distribution over long distances. Economizer, steam generation coils and superheater use finned serpentine coils (if the fuel is clean such as NG) or studded tube coil (if high sulfur fuel oil is used as the fuel).

Natural circulation arrangement requires steam drum to be sufficiently elevated above the convection section, to ensure sufficient water circulation ratio (CR) through the steam generating coils, to maintain heat flux (heat transfer rate per unit area of tube surface), in nucleate boiling regime. CR is defined as:

$$\mathrm{CR} = \frac{\text{Total steam and water mass flow rate in the steam generating coil}}{\text{Total steam mass flow rate separted in the steam drum}}$$

During operation, steam/water mixture at outlet of the steam generating coil is less dense than water at the coil inlet. Water/steam flow occurs within the circuit at a rate that corresponds to balance the static head (difference in elevation between the water level in the steam drum and the coil inlet) with the sum of pressure drops in the steam generating coil and its outlet pipe to the steam drum. Typically, CR is maintained at 15:1 to 20:1.[8] The arrangement in Fig. 11.22 is frequently used at the outlet of convection section in fired heaters/furnaces.

If CR is low, heat flux can be excessive, which results in rapid steam generation and hence causing a steam film to be formed at the tube wall surface. This phenomenon causes departure from nucleate boiling[h] and hence failure of tubes due to very high tube temperature. If heat flux increases above 1262 kW/m^2,[10] nucleate boiling changes to film boiling. WHR projects shall maintain a CR sufficient to operate under nucleate boiling, for all operating conditions.

The CR depends on the static head differences (between the steam/water drum level and steam generating coil inlet), flow resistances in the

[h]Nucleate boiling occurs when the coil's surface temperature is greater than the saturated fluid temperature (flowing either inside or outside the coil) by a certain temperature range, where the heat flux is below the critical heat flux.

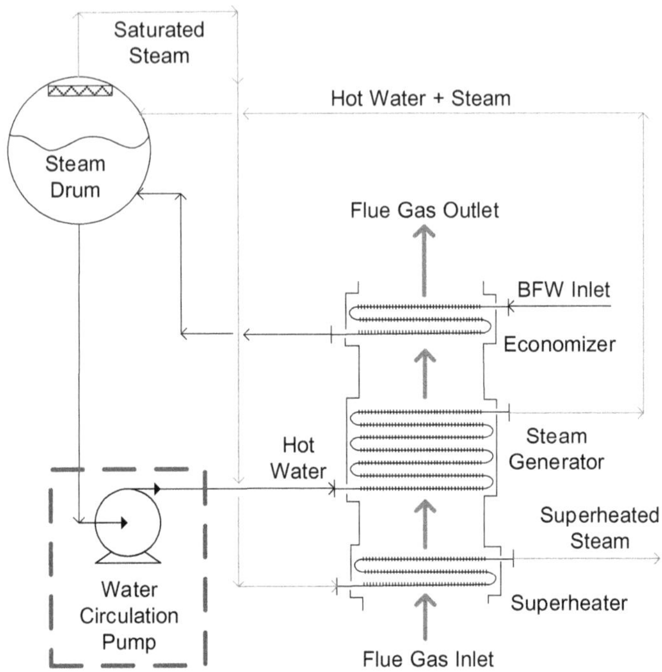

Fig. 11.22: Steam generation by FGWHR using natural/forced circulation configuration; water circulation pump shown in dashed box is for forced circulation configuration only.

circuit (i.e., steam generating coil, piping system from steam generating coils and steam drum), system pressure, and quantity of steam generated. CR can be increased by raising the height of the steam drum and/or by reducing flow resistance (i.e., using bigger steam generating coil, piping and/or reducing piping bends).

Due to structural and/or space limitations, if the steam drum cannot be placed sufficiently above steam generating coils (i.e., sufficient natural circulation cannot be achieved), a forced circulation steam generation arrangement, shown in Fig. 11.22, is used. In this arrangement, steam drum elevation is not important as CR is controlled by a water circulation pump. A minimum CR of 10:1 is used.[10] Advantages and disadvantages of natural and forced circulation steam generation arrangements are summarized in Table 11.4.

Table 11.4: Advantages and disadvantages of natural and forced circulation WHR arrangements

Circulation Type	Advantages	Disadvantages
Natural circulation	• No pump is required. • Once adequately designed, incurs lower maintenance cost. • More reliable	• Steam drum is required to be elevated. • Requires more plot space and structural work. • Costly to revamp for higher capacity as it would require increasing steam drum elevation and/or upsizing the piping.
Forced Circulation	• Requires lower plot space and structural work. • Steam drum can be installed at any elevation. • Relatively easier to revamp for higher duty.	• Requires water circulation pumps (one in operation and one spare, with auto start provision). • Higher maintenance cost, mainly due to pumps

Example 11.6: Calculate the amount of saturated steam (at 3.5 bar pressure) that can be generated from FGWHR, for the following conditions: FG flow rate of 100 ton/hr, supply temperature of 350°C, target FG temperature to stack of 170°C. BFW is available at 10 barg and 120°C. Average specific heat of FG is 1.22 kJ/kg°C.

Solution:

Heat transfer from FG, $H_{fg} = 100{,}000 \times 1.22 \times (350 - 170) = 21{,}960{,}000$ kJ/hr

From steam tables, saturation temperature of water/steam at 3.5 bar = 138.9°C

Specific enthalpy change for heating water from 120°C to 138.9°C, $\Delta H_{DM\text{-}sh}$ = Specific enthalpy of water at 138.9°C − Specific enthalpy of water at 120° = 584.311 − 503.785 = 80.526 kJ/kg

Specific latent heat of vaporization of water at 3.5 bar and 138.9°C, $\Delta H_{DM\text{-}lh}$ = 2147.65 kJ/kg

Amount of steam generated $= \frac{H_{fg}}{(\Delta H_{DM-sh} + \Delta H_{DM-lh})} = \frac{21960000}{(80.526+2147.65)}$ = 9855.6kg/hr

11.10 Synergy of Flue Gas WHR with Emissions Reduction and Carbon Capture Technologies

Combustion systems generate NO_x, SO_x, CO, CO_2, particulate matter (e.g., produced by fuel combustion such as soot, eroded refractory particles, from ducting and tube scales) and volatile organic components (VOC). Air pollutants generated from combustion process, their mitigation methods and synergy with FGWHR are summarized in Table 11.5.

Table 11.5: Synergy of air pollution control methods with FGWHR

Pollutants	Control Methods for Reducing Pollutants Release to Atmosphere	Impact on FGWHR	Synergy of WHR Method with Pollution Control Method
NO_x emissions (NO and NO_2)	• Minimize excess air. • Substitute fuel oil with NG. • Use oxygen enrichment of CCA. • Use NO_x additives. • Reduce air preheat temperature (e.g., by CCA bypass across APH). • Use staged air burners such as low or ultra-low NO_x. • FG recirculation to burners • Use SCR • Selective non-catalytic reduction	• High NOx generation generally reduces FGWHR using APH. However, FGWHR can be increased by using other methods (e.g., steam generation). • If SCR is used along with its downstream APH, a water wash or other salt cleaning at APH shall be provided[11]	• Fuel oil change to NG can increase FGWHR by adding bigger APH due to minimal concern on ADP corrosion.
CO (carbon monoxide) and VOC emissions	• Use correct amount of excess air/oxygen. • Use correct type of burner to ensure good mixing of air and fuel.	• No Impact.	• Maintaining correct air ratio for FGWHR reduces CO and VOC emissions.

Table 11.5: (*Continued*)

Pollutants	Control Methods for Reducing Pollutants Release to Atmosphere	Impact on FGWHR	Synergy of WHR Method with Pollution Control Method
Particulates emissions	• Use mechanical dust separators such as sonic horns and/or soot blowers • Maintaining good condition of insulation, electrostatic separators and scrubbers	• Results in erosion and fouling. • Generally, reduces FGWHR	• FGWHR before scrubber reduces substantial amount of water requirement for scrubbing.
SO_2 and SO_3	• FG desulfurization (FGD)	• Results in corrosion-erosion and fouling. • Generally, reduces WHR.	• FGWHR is required before FGD. It saves significant amount of water required for FGD.
CO_2 emissions	• CO_2 absorption/ scrubbing by amine- or ammonia-based solvents	• No impact	• FGWHR is required before carbon capture unit as FG temperature shall be reduced to below 100°C for treatment with amine-based solvents.

11.11 Summary

FGWHR concepts and applications presented in this chapter are very useful for practicing engineers and researchers. They provide deeper insight to select the appropriate FGWHR equipment. Following are the salient points of this chapter.

- Use of economizer is very beneficial for FGWHR in boilers.
- CEs can maximize FGWHR in boilers. Although fired heaters do not require BFW heating, they can still use condensing WHR for suppling hot water to other process and/or utility applications.
- APH is essential for increasing FGWHR mainly in fired heaters/ furnaces.
- Understanding various APH configurations presented in this chapter provides sound guidance for selecting the right configuration for FGWHR opportunity.
- Heat pipes are the most compact FGWHR equipment, and they can be custom designed to maximize WHR by choosing suitable working fluid, design and operating conditions.
- Steam generation by WHR is an alternative option for FGWHR; it can utilize excess WH in the fired heater for generating steam, which can be used in some other equipment/system.
- Limitations and potential solutions for APH applications, highlighted in this chapter, are very useful for proper design of APH and sustaining FGWHR.
- Lastly, FGWHR is an enabler for reducing capital and operating costs of many air pollution control systems.

Ganapathy[4] is a holistic book on economizers, WH boilers and steam generators. It is recommended for further reading on the topics of this chapter.

References

1. Kakaç S. (1991) *Boilers, Evaporators, and Condensers*. John Wiley & Sons.
2. Lieberman NP, Lieberman ET. (2014) *Working Guide to Process Equipment*. McGraw-Hill Education.
3. Ganapathy V. (1991) *Waste Heat Boiler Deskbook*. The Fairmont Press Inc.
4. Ganapathy V. (2015) *Steam Generators and Waste Heat Boilers: For Process and Plant Engineers*. CRC Press.
5. Raju KS. (2011) *Fluid Mechanics, Heat Transfer, and Mass Transfer: Chemical Engineering Practice*. John Wiley & Sons.
6. *API RP 538, Industrial Fired Boilers for General Refinery and Petrochemical Service*, 1st ed. (October 2015).

7. Platvoet E. (2020) When excess air becomes too much. *PTQ* **Q2**: 65–68.
8. Jouhara H, Chauhan A, Nannou T, *et al.* (2017) Heat pipe-based systems — Advances and applications. *Energy* **128**: 729–754.
9. *API Std. 560, Fired Heaters for General Refinery Service*, 5th ed. (February 2016).
10. *API RP 534, Heat Recovery Steam Generators*, 2nd ed. (October 2013).
11. Malhotra K. (November 2019) Rethink fired heater design for emissions. *Hydrocarbon Processing* 47–51.

Acronyms and Notation

Acronyms

ADP	Acid Dew Point
ADPT	Acid Dew Point Temperature
APH	Air Preheater
BD	Balanced Draft
BFW	Boiler Feed Water
CCA	Cold Combustion Air
CE	Condensing Economizer
CR	Circulation Ratio
CS	Carbon Steel
DM	Demineralized
FD	Forced Draft
FDF	Forced Draft Fan
FG	Flue Gas
FGD	Flue Gas Desulfurization
FGWHR	Flue Gas Waste Heat Recovery
HPS	High-Pressure Steam
ID	Induced Draft
IDF	Induced Draft Fan
LMTD	Log Mean Temperature Difference
LPS	Low-Pressure Steam
MPS	Medium-Pressure Steam
NG	Natural Gas
SCR	Selective catalytic reduction
SS	Stainless Steel

VOC	Volatile Organic Components
WHR	Waste Heat Recovery

Notation

A	Heat transfer area in the heat exchanger (m^2)
C_p	Heat capacity at constant pressure (kJ/kg K = kJ/kg°C)
h	Heat transfer coefficient (kW/m^2 K = kW/m^2°C)
M	Mass flow rate (kg/h)
MW	Molecular weight (kg/kmole)
O_{2fg}	Oxygen concentration in FG at stack (%)
O_{2ic}	Oxygen concentration in FG at the inlet of convection section (%)
P	Pressure (kPa)
Pv	Vapor pressure (kPa)
T	Temperature (°C)
U	Overall heat transfer coefficient (kW/m^2.K = kW/m^2°C)

Greek Symbols

ΔH	Enthalpy change (kW)

Subscripts/Superscripts

a	Ambient
bfw	Boiler feed water
bfw-i	Boiler feed water at economizer inlet
CE-out	Condensing economizer outlet
DM-sh	Sensible heat for DM water
DM-lh	Latent heat for DM water
fg	Flue gas
fg-o	Flue gas temperature at the economizer outlet
i	inside the tube
o	outside the tube
OG	Gases other than water vapor
s	Steam

Exercises

11.1 State different FGWHR methods applicable for boilers and fired heaters. What are their advantages and disadvantages?

11.2 A boiler uses fuel oil with 3 wt% sulfur. Identify the best method for maximizing FGWHR. Comment how the selected method can be further improved if fuel is changed to NG.

11.3 State advantages and disadvantages of CEs. List important considerations for designing a sustainable CE.

11.4 Size an economizer for heating BFW make-up, supplied at 100 ton/hr, 120°C and 60 barg to 175°C. FG flow rate and temperature at the economizer inlet are 110 ton/hr and 340°C, respectively. Consider an overall heat transfer coefficient of 38 $W/m^2°C$ for the economizer with finned tubes. Average heat capacity of BFW and FG are 4.35 kJ/kg°C and 1.22 kJ/kg°C, respectively.

11.5 Calculate the amount of saturated steam (at 5 bar) that can be generated from FGWHR, under the following conditions: FG flow rate of 80 ton/hr; supply temperature of 350°C; target FG temperature to stack of 180°C. BFW is available at 10 barg and 30°C. Average specific heat of FG is 1.23 kJ/kg°C.

Chapter 12

Waste Heat Recovery in Compression Systems

12.1 Overview

Process plants use compressors for increasing vapour/gas stream pressure to supply utilities such as plant air (PA),[a] instrument air (IA), nitrogen and fuel streams (e.g., natural gas), supply high-pressure gas streams as reactor feed/recycle streams, support refrigeration and product storage (e.g., liquified natural gas, carbon dioxide, oxygen, nitrogen, ethylene, propane and propylene). Gas compression generates significant thermal energy (heat), and most of it is discarded to atmosphere through cooling water or air, at the interstage and/or oil coolers. Multi-stage compression with interstage cooling is used for producing higher pressures and reducing power/energy consumption and waste heat (WH) generation. Still significant amount of recoverable WH is generally wasted to cooling water or ambient air. For example, in process industries, nearly 96% of the WH generated from air compression is transferred to atmosphere, through interstage and/or lube oil coolers. With proper design, most of this WH can be recovered and used for heat pumps, desalination, process/space heating and generation of chilled water.

This chapter covers waste heat recovery (WHR) applications from air, process and refrigeration systems, as they are common high-energy-

[a] PA is essentially CA with saturated water (at its operating temperature), whereas IA is CA which is relatively moisture free (i.e., dried to a specified moisture content).

consuming equipment in process plants. It begins with Section 12.2 on descriptions of compressed air (CA) systems, classification of air compressors (ACs) and performance evaluation. Section 12.3 outlines the methods to reduce energy consumption and WH generation in CA systems. Sections 12.4 presents WHR from CA system, for utilization in process equipment and space heating. IA drying options including WH utilization are discussed in Section 12.5. The subsequent section covers WHR from process compressors. Section 12.7 presents WH reduction and WHR methods from refrigeration systems. Finally, Section 12.8 summarizes salient points for WH minimization and WHR options for compression systems.

Learning outcomes of this chapter on WHR in compression systems are as follows.

1. Outline the CA system and its components.
2. Identify options and design practices for reducing WH generation and WHR from CA systems.
3. Assess the opportunities for WHR from CA systems.
4. Identify the opportunity to improve IA dryer efficiency using WHR.
5. Identify WHR opportunities in process compressors.
6. Describe the design practices to reduce WH generation and/or increase WHR in refrigeration systems.

12.2 Compressed Air System

CA is generally one of the most expensive utilities in process industries. It has several industrial uses. CA system includes ACs, IA and PA systems, which are essential for process plant operations involving field instruments and controls. IA is mainly used for controlling valve actuators, whereas PA is used for air cleaning, purging, laboratories, rotating equipment seals, pneumatic equipment (e.g., air-driven pumps and agitators, pneumatic hammers, drills and grinders), paint spraying, powder coating, vessel purging and ventilation and so on.

Typical details of an AC are shown in Fig. 12.1, whereas those of the CA system are shown in Fig. 12.2. Larger process plants are usually designed with multiple ACs and some spare capacity. IA and PA systems

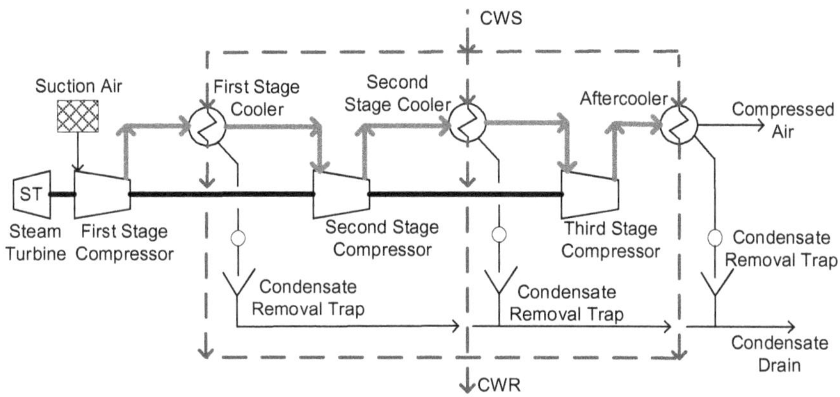

Fig. 12.1: Typical details of an AC; this system has one 3-stage compressor driven by a steam turbine.

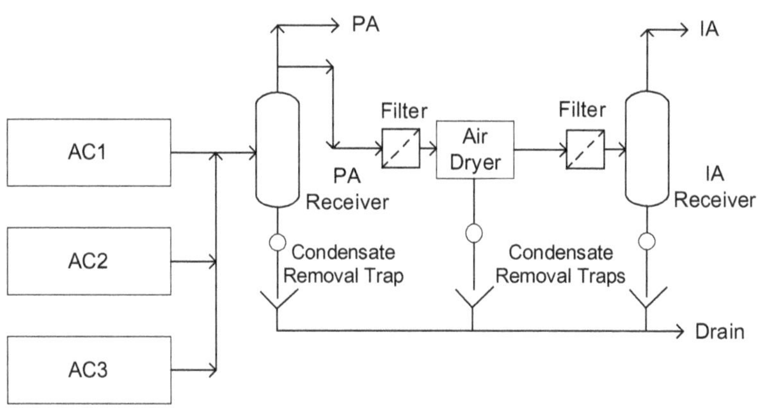

Fig. 12.2: Details of PA and IA system (with three ACs in parallel) in process plants.

are generally designed to provide air at 8–10 bar for the end users. Only a small portion of large control valves, with fast closing requirement, may need such a high pressure whereas many control valves require air at 4–5 bar only. Hence, compressing all the air to high pressure leads to wastage of compression energy.

The main difference between IA and PA is the drying process (Fig. 12.2). Removal of water or drying of PA produces IA. This drying is very important

Table 12.1: Typical operating conditions of PA and IA

Quantity	**PA**	**IA**
Supply pressure, bar	8–10	9–9.5
Supply temperature, °C	Ambient	Ambient
Drying requirement	NA	–40°C dew point
Maximum solid particle size, mm	NA	< 3
Maximum contaminants (oil, liquid)	0.1 mg/m^3	

to prevent fouling, corrosion and possible freezing of water in the IA system and instruments. A −40°C dew point specification is common, but this specification varies according to the climatic conditions of the site. Typical operating conditions of PA and IA are given in Table 12.1.

The main components of CA system (Figs. 12.1 and 12.2) are as follows.

- **Air intake filters:** They prevent dust entering the AC.
- **Interstage coolers:** These reduce the air temperature (close to the temperature of air at the suction of the first stage) before it enters the next stage, to reduce the compression work and increase energy efficiency. This will bring the compression process close to isothermal compression. Interstage cooling increases air density and reduces volumetric flowrate of air going to the next stage. Due to this volumetric flow reduction, compression work and hence required compressor power is reduced. Interstage cooling can be done using water or air. Generally, it uses water coolers because of their compact design and the ability to cool air to lower temperatures, compared to air coolers. In Fig. 12.1, first-stage and second-stage coolers are the interstage coolers.
- **Aftercoolers:** They remove most of the water in the CA by heat exchange with water or air. Again, water-cooled aftercoolers are mostly used. Aftercooling is very important to prevent the high condensation of water and hence reduce corrosion, pressure drops and leakages in air receivers and air distribution network. In Fig. 12.1, the aftercooler is after the third-stage compressor.

- **IA dryers:** Air used in instruments and pneumatic equipment must be relatively free of moisture. IA dryers (Fig. 12.2) remove traces of moisture left in air, after condensation and removal of moisture in the aftercooler. There are many types of air dryers, which are described in Section 12.5.
- **Moisture/condensate drain traps:** Moisture/condensate drain traps are used for the removal of condensed water from the PA/IA equipment. They work on the same working principle as steam traps.
- **PA/IA Receivers:** These store air and provide buffer volume for sudden variations in demand/supply. In large facilities, PA and IA receivers are instaled at utility area and/or multiple locations in the plant. They help to ensure sufficient air volume and pressure for safe operation of large air users, even during minor upsets in PA/IA supply.

Industrial ACs are classified in Fig. 12.3. They can be dynamic or positive displacement type. Centrifugal compressors are well suited for high pressure (≥8 bar) and high capacity (10–30 ton/hr). They are oil free by design. Multi-stage centrifugal compressors have a lower specific power consumption compared to single-stage compressor, operating over the same pressure differential. Axial type AC is used for very large

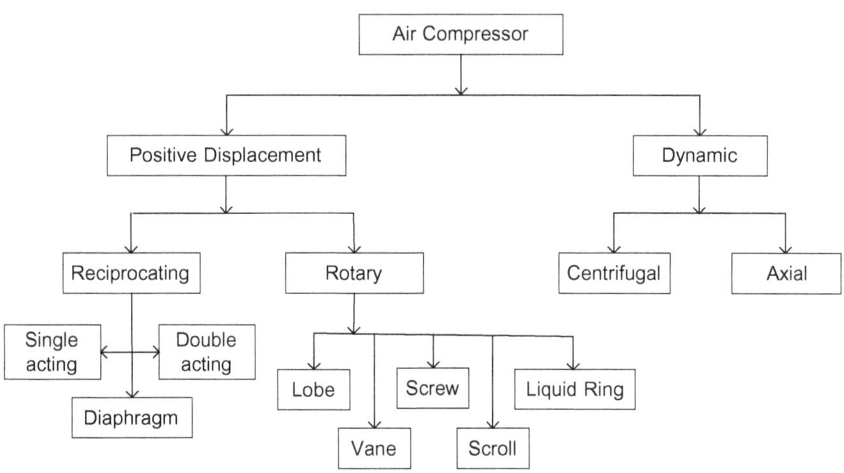

Fig. 12.3: Classification of ACs employed in process industries.

capacity (>100 ton/hr) and medium pressure (4–5 bar). For lower capacities, rotary and reciprocating compressors are generally used. Reciprocating compressors can be single acting or double acting type[b]; they use a single cylinder for low capacity and head applications. For higher discharge pressure, two-stage reciprocating compressors or diaphragm-type reciprocating compressors are used.

Lobe blowers are used for low discharge pressures (~2 bar in single-stage design). The most common rotary AC is screw compressor (SC), which can be single or two staged. SC is available with both lube oil-flooded and dry/oil free designs. It usually consists of two or more rotors within a casing, where the rotors compress the air internally. There are no valves. Cooling of the compressor is provided by oil coolers, interstage and aftercoolers. In oil-cooled units, the oil seals the internal clearances; since oil removes the heat of compression, cooling takes place right inside the compressor. An external oil cooler is used to cool the lube oil before returning it back to the SC. In oil-cooled ACs, oil must be separated from AC discharge air.

The oil-free rotary SC uses specially designed air ends to compress air without oil in the compression chamber. It produces oil-free air. Oil-free SCs are very useful for providing oil-free IA for process instrumentation. However, they have higher specific power consumption (kW/(m^3/hr)) as compared to lubricated types. Oil-free SCs deliver oil-free air and are available in sizes up to 34,000 m^3/hr and pressure up to 15 bar. Oil-lubricated types are available in smaller sizes ranging from 170 to 1700 m^3/hr, with discharge pressure up to 10 bar. Vane/lobe and scroll-type rotary ACs are also used in process plants.

Liquid ring compressors, which use liquid seal to perform air compression, are not commonly used in process industries due to low efficiency (25%–40%), higher instalation and operating costs (because of high energy, water and waste disposal of the purged liquid, which is frequently water). However, they are very useful in vacuum applications (covered in detail in Chapter 8).

Among the various ACs in Fig. 12.3, integrally geared centrifugal compressors, multi-stage reciprocating compressors and SCs (both oil-flooded and oil-free) are widely used for producing PA/IA.

---

[b] In single-acting-type compressor, piston compresses air only in up-stroke compared to both up- and down-stroke compression in double acting type.

12.2.1 *Performance Equations for Air Compressors*

Isothermal compression obeys PV = constant and requires least amount of power compared to adiabatic or polytropic compression. Isothermal power consumption (W_i kW) of an AC can be estimated by

$$W_i = \left(\frac{mRT_S}{M_i}\right) \times \ln \frac{P_D}{P_S} \tag{12.1}$$

Here, R is the universal gas constant (8.314 J/K.mol), m and M are the mass flow rate (kg/sec) and molecular weight of air, respectively, η_i is the isothermal efficiency, P_D and P_S are the discharge and suction pressure (bar), respectively, and T_S is the suction temperature (K) of the AC. Oil-flooded ACs closely follow isothermal compression.

Adiabatic compression obeys PV^k = constant and requires more power compared to both isothermal and polytropic compression. Power for (W_a kW) and approximate discharge temperature (T_D *K*) of adiabatic compression can be estimated by:

$$W_a = \left(\frac{mZRT_S}{M\eta_a}\right) \times \left(\frac{k}{k-1}\right) \times \left(\left(\frac{P_D}{P_S}\right)^{\frac{(k-1)}{k}} - 1\right) \tag{12.2}$$

$$T_D = T_S \times \left(\frac{P_D}{P_S}\right)^{\frac{(k-1)}{k}} \tag{12.3}$$

Here, Z is the compressibility factor (in PV = ZRT), η_a is the adiabatic efficiency and k is the isentropic (adiabatic) exponent (also known as the ratio of specific heats, C_p/C_v).

Similarly, polytropic compression obeys PV^n = constant. Power consumption (W_p) of and discharge temperature (T_D) of polytropic compression can be estimated approximately by:

$$W_p = \left(\frac{mZRT_S}{M\eta_p}\right) \times \left(\frac{n}{n-1}\right) \times \left(\left(\frac{P_D}{P_S}\right)^{\frac{(n-1)}{n}} - 1\right) \tag{12.4}$$

$$T_D = T_S \times \left(\frac{P_D}{P_S}\right)^{\frac{(n-1)}{n}} \tag{12.5}$$

Here, η_p is the polytropic efficiency and n is the polytropic exponent (in PV^n = constant), respectively. Isentropic and polytropic exponents are related by:

$$\frac{n}{(n-1)} = \frac{k}{(k-1)} \times \eta_p \tag{12.6}$$

Exercise 12.1: Estimate the isothermal, adiabatic and polytropic power consumptions for an AC using the following data: suction pressure and temperature = 1 bar and 30°C; discharge pressure = 8 bar; polytropic exponent = 1.3; adiabatic exponent = 1.4; mass flow rate of air at the AC suction = 1 kg/sec; molecular weight of air = 28.97; compressor isothermal/adiabatic/polytropic efficiency = 0.75 and compressibility factor = 1.

Solution

Using Eqs. 12.1, 12.2 and 12.4, compression power for isothermal, adiabatic and polytropic compression are respectively:

$$W_i = \left(\frac{1 \times 8.314 \times (273.15 + 30)}{28.97 \times 0.75}\right) \times \ln\left(\frac{8}{1}\right) = 241.22\,\text{kW}$$

$$W_a = \left(\frac{1 \times 1 \times 8.314 \times (273.15 + 30)}{28.97 \times 0.75}\right) \times \left(\frac{1.4}{1.4 - 1}\right) \times \left(\left(\frac{8}{1}\right)^{\frac{(1.4-1)}{1.4}} - 1\right)$$

$$= 329.45\ \text{kW}$$

$$W_p = \left(\frac{1 \times 1 \times 8.314 \times (273.15 + 30)}{28.97 \times 0.75}\right) \times \left(\frac{1.3}{1.3 - 1}\right) \times \left(\left(\frac{8}{1}\right)^{\frac{(1.3-1)}{1.3}} - 1\right)$$

$$= 309.58\,\text{kW}$$

The above results confirm that isothermal compression requires the lowest power followed by polytropic and adiabatic compression.

Except liquid ring and oil-injected ACs (which are closer to isothermal compression), all other industrial compressors closely follow polytropic compression.

To minimize the power requirement, often multi-stage compression with interstage coolers is used. Minimum compression power is achieved when the following equation is satisfied:

$$r_S = \left(\frac{P_D}{P_S}\right)^{(1/s)} \tag{12.7}$$

Here, r_s is the compression ratio per stage and s is the number of compression stages.

Exercise 12.2: Estimate the power of compression for single-, two- and three-stage compression, for compressing air from 1 to 8 bar. Assume the following data for the calculations: suction temperature = 30°C; polytropic exponent = 1.3; mass flow rate of air at the AC suction = 1 kg/sec; molecular weight of air = 28.97; polytropic efficiency = 0.75 and compressibility factor = 1. Neglect the pressure drop in interstage and aftercoolers. Assume interstage cooling achieves same temperature as suction temperature.

Solution

Using Eq. 12.4, power requirement for singe stage polytropic compression

$$= \left(\frac{1\times1\times8.314\times(273.15+30)}{28.97\times0.75}\right)\times\left(\frac{1.3}{1.3-1}\right)\times\left(\left(\frac{8}{1}\right)^{\frac{(1.3-1)}{1.3}}-1\right) = 309.58\,\text{kW}$$

For a two-stage compression, compression ratio for minimum compression power using Eq. 12.7 is $(8)^{(1/2)} = 2.8284$.

Using Eq. 12.4, power requirement for first stage

$$= \left(\frac{1\times1\times8.314\times(273.15+30)}{28.97\times0.75}\right)\times\left(\frac{1.3}{1.3-1}\right)\times\left(\left(\frac{2.8284}{1}\right)^{\frac{(1.3-1)}{1.3}}-1\right)$$

$$= 136.31\,\text{kW}$$

Using Eq. 12.4, power requirement for second stage

$$= \left(\frac{1\times1\times8.314\times(273.15+30)}{28.97\times0.75}\right)\times\left(\frac{1.3}{1.3-1}\right)\times\left(\left(\frac{8}{2.8284}\right)^{\frac{(1.3-1)}{1.3}}-1\right)$$

$$= 136.31\,\text{kW}$$

So, total power consumption for a two-stage compression = 2 × 136.31 = 272.62 kW. Compared to single-stage compression, power required is reduced by 11.9%.

For a three-stage compression, compression ratio for minimum compression power using Eq. 12.7 = $(8)^{(1/3)} = 2$.

Using Eq. 12.4, power requirement for each of the stages

$$= \left(\frac{1\times1\times8.314\times(273.15+30)}{28.97\times0.75}\right)\times\left(\frac{1.3}{1.3-1}\right)\times\left(\left(\frac{2}{1}\right)^{\frac{(1.3-1)}{1.3}}-1\right)$$

$$= 87.19\,\text{kW}$$

So, total power consumption for a three-stage compression = 3 × 87.19 = 261.57 kW. This is less (by 4.05%) than that for two-stage compression. Compared to single-stage compression, power required is reduced by 15.5%.

Exercise 12.3: Estimate the power of compression for AC using suction temperatures of 20°C, 30°C and 40°C. Assume the following data for the calculations: suction and discharge pressure = 1 bar and 10 bar, respectively; mass flow rate of air at the AC suction = 1 kg/sec; molecular weight of air = 28.97; polytropic exponent = 1.3; polytropic efficiency = 0.75; and compressibility factor = 1. Neglect the pressure drop in interstage and aftercoolers.

Solution

Using Eq. 12.4, power requirement for suction temperature of 20°C, 30°C and 40°C are respectively:

$$\left(\frac{1\times1\times8.314\times(273.15+20)}{28.97\times0.75}\right)\times\left(\frac{1.3}{1.3-1}\right)\times\left(\left(\frac{10}{1}\right)^{\frac{(1.3-1)}{1.3}}-1\right)=304.87\,\text{kW}$$

$$\left(\frac{1\times1\times8.314\times(273.15+30)}{28.97\times0.75}\right)\times\left(\frac{1.3}{1.3-1}\right)\times\left(\left(\frac{10}{1}\right)^{\frac{(1.3-1)}{1.3}}-1\right)=352.50\,\text{kW}$$

$$\left(\frac{1\times1\times8.314\times(273.15+40)}{28.97\times0.75}\right)\times\left(\frac{1.3}{1.3-1}\right)\times\left(\left(\frac{10}{1}\right)^{\frac{(1.3-1)}{1.3}}-1\right)=364.13\,\text{kW}$$

So, an increase of 10°C in suction temperature from 20°C to 30°C and from 30°C to 40°C increases power consumption by 3.4% and 3.3%, respectively. So, for every 10°C, rise in suction temperature, power consumption typically increases by ~3.3%.

Exercise 12.4: Estimate the power of compression for AC using discharge pressures of 10, 8 and 6 bar. Assume the following data for the calculations: suction pressure and temperature = 1 bar and 30°C, respectively; mass flow rate of air at the AC suction = 1 kg/sec; molecular weight of air = 28.97; polytropic exponent = 1.3; polytropic efficiency = 0.75 and compressibility factor = 1.

Solution

Using Eq. 12.4, power requirement for a discharge pressure of 10, 8 and 6 bar are respectively:

$$\left(\frac{1\times1\times8.314\times(273.15+30)}{28.97\times0.75}\right)\times\left(\frac{1.3}{1.3-1}\right)\times\left(\left(\frac{10}{1}\right)^{\frac{(1.3-1)}{1.3}}-1\right)=352.50\,\text{kW}$$

$$\left(\frac{1\times1\times8.314\times(273.15+30)}{28.97\times0.75}\right)\times\left(\frac{1.3}{1.3-1}\right)\times\left(\left(\frac{8}{1}\right)^{\frac{(1.3-1)}{1.3}}-1\right)=309.56\,\text{kW}$$

$$\left(\frac{1\times1\times8.314\times(273.15+30)}{28.97\times0.75}\right)\times\left(\frac{1.3}{1.3-1}\right)\times\left(\left(\frac{6}{1}\right)^{\frac{(1.3-1)}{1.3}}-1\right)=257.4\,\text{kW}$$

So, a decrease of 2 bar in discharge pressure from 10 to 8 bar and from 8 to 6 bar decreases power consumption by 12.2% and 16.9%, respectively. Thus, decrease in power consumption for discharge pressure change from 8 and 6 bar is more compared to that from 10 and 8 bar. Hence, power savings are higher with lower AC discharge pressure.

Above exercises did not consider the effect of interstage and after coolers on power consumption of ACs. Consideration of these HEs requires solution of the following non-linear optimization problem, to find the optimal discharge pressure of each stage.

Objective Function: For a three-stage AC, minimize the total compression power required,

$$W_\mathrm{p} = \left(\frac{n}{n-1}\right) \times \left(\frac{mZR}{M\eta_\mathrm{p}}\right) \times \sum_{j=1}^{3} \left\{ T_\mathrm{Sj} \times \left[\left(\frac{P_\mathrm{Dj}}{P_\mathrm{Sj}}\right)^{\frac{(n-1)}{n}} - 1 \right] \right\} \tag{12.8}$$

$$P_{\mathrm{S}(j+1)} = P_{\mathrm{D}(j)} - \mathrm{DP}_j \qquad \text{for } j = 1 \text{ and } 2 \tag{12.9}$$

Here, $\mathrm{DP_j}$ is the pressure drop in j^{th} interstage cooler.

$$\textbf{Constraints: } P_{\mathrm{D}j} > 0 \tag{12.10}$$

$$\textbf{Decision variables: } P_{\mathrm{D1}} \text{ and } P_{\mathrm{D2}} \tag{12.11}$$

This non-linear optimization model can be readily solved using Solver tool in Microsoft Excel. This is illustrated with the next exercise.

Exercise 12.5: Estimate the power of compression for single-, two- and three-stage compression, for compressing air from 1 to 8 bar (at the outlet of the aftercooler). Assume the following data for the calculations: suction temperature for first stage = 30°C; suction temperatures for subsequent stages = 40°C; polytropic exponent = 1.3; mass flow rate of air at the AC suction = 1 kg/sec; molecular weight of air = 28.97; polytropic efficiency = 0.75; compressibility factor = 1 and pressure drop in each of interstage and after-coolers = 0.2 bar.

Solution: For a single-stage AC, $P_\mathrm{S} = 1$ bar and $P_\mathrm{D} = 8.2$ bar (= pressure at the outlet of the aftercooler + 0.2 bar DP across the aftercooler). Using Eq. 12.4, (as no optimization is required),

Power required for a single-stage AC

$$= \left(\frac{1\times1\times8.314\times(273.15+30)}{28.97\times0.75}\right)\times\left(\frac{1.3}{1.3-1}\right) \times\left(\left(\frac{8.2}{1}\right)^{\frac{(1.3-1)}{1.3}}-1\right)=314.22\,\text{kW}$$

For a two-stage AC, $P_{S1} = 1$ bar, $P_{D2} = 8.2$ bar, pressure drop for first stage intercooler (DP_1) = 0.2 bar, $T_{S1} = 30°C$ and $T_{S2} = 40°C$. With these values, solution of Eqs. 12.8–12.11 (simplified for two stages) using Excel's Solver gives 3.581 bar for P_{D1}. Hence, suction pressure for second-stage compression (P_{S2}) = 3.581 – 0.2 = 3.381 bar.

Using Eq. 12.4, power consumptions for first and second stages of compression are as follows:

$$W_{p1} = \left(\frac{1\times1\times8.314\times(273.15+30)}{28.97\times0.75}\right)\times\left(\frac{1.3}{1.3-1}\right)\times\left(\left(\frac{3.581}{1}\right)^{\frac{(1.3-1)}{1.3}}-1\right)$$
$$=172.06\,\text{kW}$$

$$W_{P2} = \left(\frac{1\times1\times8.314\times(273.15+40)}{28.97\times0.75}\right)\times\left(\frac{1.3}{1.3-1}\right)\times\left(\left(\frac{8.2}{3.381}\right)^{\frac{(1.3-1)}{1.3}}-1\right)$$
$$=117.79\,\text{kW}$$

Hence, total power consumption $W_P = W_{P1} + W_{P2} = 172.06 + 117.79 = 289.85$ kW, which is 7.8% less compared to single-stage compression. This saving is less than 11.9%, obtained for the second-stage case in Exercise 12.2. This is due to pressure drops in interstage and aftercoolers, in the current exercise.

For a three-stage AC, $P_{S1} = 1$ bar, $P_{D3} = 8.2$ bar, $DP_1 = 0.2$ bar, $DP_2 = 0.2$ bar, $T_{S1} = 30°C$, T_{S2} and $T_{S3} = 40°C$. With these values, solution of Eqs. 12.8–12.11 using Excel's Solver gives $P_{D1} = 2.970$ bar and $P_{D2} =$ 5.282 bar. Hence, $P_{S2} = 2.970 – 0.2 = 2.770$ bar and $P_{S3} = 5.282 - 0.2 =$ 5.082 bar.

Using Eq. 12.4, power consumptions for first, second and third stages of compression are as follows:

$$W_{P1} = \left(\frac{1\times1\times8.314\times(273.15+30)}{28.97\times0.75}\right)\times\left(\frac{1.3}{1.3-1}\right)\times\left(\left(\frac{2.970}{1}\right)^{\frac{(1.3-1)}{1.3}}-1\right)$$

$$= 143.53\,\text{kW}$$

$$W_{P2} = \left(\frac{1\times1\times8.314\times(273.15+40)}{28.97\times0.75}\right)\times\left(\frac{1.3}{1.3-1}\right)\times\left(\left(\frac{5.282}{2.770}\right)^{\frac{(1.3-1)}{1.3}}-1\right)$$

$$= 83.43\,\text{kW}$$

$$W_{P3} = \left(\frac{1\times1\times8.314\times(273.15+40)}{28.97\times0.75}\right)\times\left(\frac{1.3}{1.3-1}\right)\times\left(\left(\frac{8.2}{5.082}\right)^{\frac{(1.3-1)}{1.3}}-1\right)$$

$$= 60.61\,\text{kW}$$

Hence, total power consumption $W_P = W_{P1} + W_{P2} + W_{P2} = 143.53 + 83.43 + 60.61 = 287.56$ kW, which is 8.5% less compared to single-stage compression. This saving is lower than the 15.5% obtained for the third-stage case in Exercise 15.2, again because of the pressure drops.

In summary, power saving in the realistic scenario of pressure drop in interstage and aftercoolers is lower compared to simple calculations that ignore these. So, accounting for pressure drop in interstage and aftercoolers is important for correct estimation of power savings. For example, in the earlier exercise, power reduction for a three-stage AC compared to a two-stage AC is only 2.29 kW (=289.85 − 287.56), which may not justify the selection of three stages over two stages. On the other hand, this power reduction without considering pressure drop in interstage and aftercoolers (Exercise 12.2), is 272.61 − 261.57 = 11.04 kW, which appears to be attractive.

12.3 Methods to Reduce Energy Consumption and WH Generation in Air Compressors

CA is the most inefficient utility in process plants. About 4% of electricity supplied to the AC is used for the production of CA and the rest is discarded as WH to cooling water or ambient air. Minimizing the power requirement for AC reduces the WH generated and hence the cost of WHR equipment. Following are the options for reducing the power requirement for ACs.

- Air intake for AC should be taken from well-ventilated areas. Avoid air suction near cooling towers, which increase the humidity. Every 10°C rise/drop in air inlet temperature can increase/decrease the AC power consumption by ~3.3%. Installing inlet air chillers (using chilled water, preferably generated by a WHR chiller) can increase the AC capacity and/or decrease the power consumption for a given capacity.
- Maintain air inlet filters clean by installing automatic differential pressure (DP) reading and high DP alarm in DCS. Clean the filters timely. AC efficiency reduces by ~2% for every 250 mm H_2O DP across the air intake filter.
- If reciprocating AC is used, regularly monitor the suction and discharge valves of the compressor. Worn-out/damaged valves can reduce AC efficiency by as much as 50%. Hence, replace damaged valves with timely inspections, at least once in 6 months.
- Use heat of compression (HOC) air dryers between pre- and after filters (in Fig. 12.2) for removing moisture from PA.
- Maintain interstage and aftercoolers clean. This helps to reduce pressure drop, AC power consumption and corrosion.
- Use AC of low capacity or with variable speed drive (VSD), during low demand.
- Use multi-stage ACs with interstage coolers. This helps to minimize power consumption and to achieve higher compression ratios. The compression ratio and recommended number of stages are shown in Table 12.2. The compression ratios in Table 12.2 are extracted from Table 5 in Kenkre[2]. Approximate maximum compression ratio per

Table 12.2: Choosing number of stages based on compression ratio

Overall Compression Ratio (P_D/P_S)	Number of Stages
1–4	One stage; sometimes two stages
4–20	Two stages; sometimes three stages
20+	Three or more stages

stage for reciprocating, centrifugal, rotary, and axial flow compressors are 10, 3–4.5, 4 and 1.2–1.5, respectively.[1]

- When using multiple ACs, smallest AC should be used for meeting demand fluctuations, while the bigger ACs operate at full load.
- Use AC control methods such as automatic on/off control, load/unload control (for reciprocating AC), multi-step control and throttling control.
- For multi-level air pressure demand (e.g., 3 and 8 bar), instal separate CA systems, which can meet the air pressure demand closely. This helps to reduce the total power consumption for AC systems.
- Reduce compressor delivery pressure by reducing pressure drops in the CA system, wherever possible.
- IA pressure is generally maintained high to prevent any malfunctioning of pressure-critical instrumentation. By detailed hydraulic assessments, if required, upsizing of IA piping and connection manifolds for pressure-critical valves and instruments, one can lower the IA supply pressure requirement. Accordingly, lower discharge pressure requirement for AC results in lower power consumptions and WH generation.
- Operators always wish to keep IA header pressure as high as possible due to the fear of insufficient air pressure to critical valves/equipment, during fluctuations in IA header pressure. With proper combination of proactive methods (e.g., use of air receivers at strategic locations, advanced process controls) and reactive maintenance methods (e.g., timely repair of piping/tubing air leakages and passing condensate removal traps), IA pressure can be maintained stable and hence change the mindset of plant operators to accept lower IA header pressure.
- There may be occasions of transient high IA flow demands in the plant, due to some batch operations. Instaling buffer IA storage vessels/receivers can easily meet these transient demands, without requiring the maintenance of IA header at high pressure.
- Use properly designed air storage vessels/receivers (instead of operating AC at higher pressures) or adding new ACs, for meeting any cyclic high demand for CA. Place storage capacities next to IA/PA consumers that have high variation in the air demand. The IA/PA storage vessel is sized such that it supplies CA demand for an amount of

time required for the air pressure to drop from compressor discharge pressure to the minimum pressure required at the air consumer end. The size of an air receiver can be calculated by the formula:

$$V = \frac{t \times C \times P_a}{(P_1 - P_2)} \tag{12.12}$$

Here, V = IA/PA receiver volume (m^3), P_1 is the initial pressure (bar), P_2 is the minimum allowed pressure (bar), t is the time for pressure in the receiver to drop from P_1 to P_2 to occur (min), C is the free air delivered[c] at compressor discharge (Nm^3/min) and P_a is the atmospheric pressure (bar).

- A smaller dedicated compressor may be instaled at the process unit requiring critical high pressure and low flow rate of air. This can help to reduce the piping costs and maintain low compressor discharge pressure for the main ACs at the utility area, from where the piping network is usually very long.
- Use reliable condensate traps, at condensate vessels, instaled with level transmitter. This enables continuous monitoring of water removal and efficient operation of condensate removal trap.
- A properly designed CA distribution system should have a pressure drop (from the compressor discharge to the point of use) of less than 10% of AC's discharge pressure. The IA/PA distribution piping should be designed for a low DP (typically 0.1 bar/100 m). Although IA piping is correctly sized initially, over a period of plant operation, IA consumption and pressure drop in the IA piping increase for the following reasons.

  i) Plant expansions: Plant expansions will increase the IA consumption and piping pressure drops requiring high pressure to be maintained at AC outlet. One should evaluate the piping hydraulics and debottleneck piping frictional pressure drops by upsizing the necessary piping and fittings, or instaling new IA headers in parallel to the original IA headers. This allows the AC discharge pressure to be maintained as low as possible.

---

[c] Free air delivered is the volumetric flow rate of air drawn in at the AC suction, under atmospheric conditions.

ii) Chocked filters: Air filters may plug. Hence, these should be checked periodically and cleaned. For large consumer of IA, DP indicators with DCS display of the reading can be considered, to alert maintenance staff to clean the filters.
iii) Piping and fitting leakages: CA leakage of 40–50% is not uncommon (compared to 5–10% in well-maintained plants). Hence, periodic survey of the PA/IA system should be conducted to identify air leaks and prompt maintenance should be conducted to arrest the leakage. Replace leaking parts (e.g., hoses). Use welded piping or flanged joints as much as possible, compared to threaded joints, which can potentially leak.
iv) Undersized air dryers: These increase IA header DP. It can be corrected by instaling a new IA dryer in parallel or replacing the undersized dryer with a bigger unit.

- Instal control equipment such as flowmeters and pressure transmitters to monitor the IA/PA consumption.
- Replace old inefficient ACs with new ACs.
- Use high-efficiency electric motors or other types of drivers such as back pressure type steam turbine and gas/diesel engine.
- Avoid the misuse/inefficient use of PA using some of the energy-efficient solutions suggested in Table 12.3.

Table 12.3: Inappropriate use of PA and alternative energy-efficient solutions

Avoidable PA Usage	Energy-Efficient Solution
Cleaning, blowing and drying process	Electric fans, brushes, low-pressure blowers
Aspiration, atomization	Low-pressure blowers
Agitation, sparging	Electrical stirrers, low-pressure blowers
Vacuum generation	Dry vacuum pumps
Diaphragm pumps	Motor driven pumps
Pneumatic conveying of solids	Low-pressure blowers, belt/screw conveyers
Personal cooling to workers in confined spaces	Electric fans
Atomization	Low-pressure blower
Air-driven motor	Electric motor
Pneumatically operated air cylinders	Hydraulic power packs

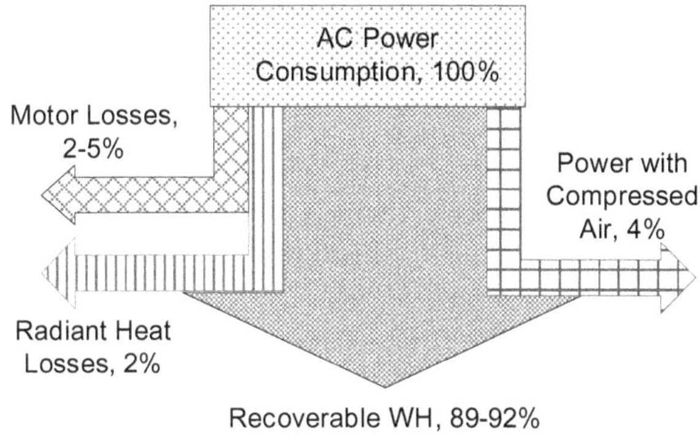

Fig. 12.4: Energy distribution in AC.

12.4 WHR in Air Compressors

Air compression generates significant WH due to compression power, which is removed using cooling water or air-cooled heat exchangers (HEs). In fact, nearly all (~96%) of the electric energy used by an industrial AC is released to atmosphere as illustrated in Fig. 12.4.

For minimizing the compression power, centrifugal compressors usually operate at lower compression temperature or lower pressure ratio per compression stage by using multiple stages (usually two to three stages) with interstage cooling. Hence, WH is available at relatively lower temperature compared to oil-free SCs. Centrifugal compressors run dry and produce oil-free air. The main drawback of centrifugal compressors is surging problems, if operated at low turndown capacity. This requires the use of a recirculation loop, which wastes the compression energy.

Two-stage reciprocating compressors are commonly used for air compression. Like centrifugal compressors, these also generally require recirculation loop for handling turndown operation, which wastes compression energy. Reciprocating compressors have high efficiency. SCs can operate over a wide range of suction and discharge pressures. They can be typically used for compression ratios up to 10 and have comparable efficiency compared to reciprocating compressors, for the same compression range. More percentage (~95%) of compression power can be recovered from oil-free SC than any other type of AC because the former operates at a much higher temperature. Oil-injected SCs use oil as coolant, are more efficient and hence generate less WH.

Table 12.4: AC types, their main features and WHR opportunities

		WHR Method	
Type of AC	**Main Features**	**Air Cooling**	**Water-Cooled Interstage and Aftercoolers**
Oil-injected rotary compressors	Injected oil cools the AC and WH is eventually removed from oil in an oil cooler. Approximately 75% of WH can be removed in oil cooler and remaining in interstage and aftercoolers	WHR from oil, interstage and aftercoolers	*Direct cooling method*: DM water runs once-through oil, interstage and aftercoolers to produce hot DM water *Indirect cooling method*: DM water circulates in closed loop to capture WH (from oil, interstage and aftercoolers) and transfer the recovered heat to process streams in other HEs
Oil-free rotary compressor	Oil is not used to lubricate the screws. Hence, WH can be recovered from interstage and aftercoolers.	WHR from interstage and aftercoolers	Same as for oil-injected compressor except that WHR is from interstage and aftercoolers
Reciprocating compressors	Water or air is used as coolant in cylinder jackets, interstage and aftercoolers	WHR from interstage and aftercoolers	Same as for oil-injected compressor except that WHR is from interstage and aftercoolers
Centrifugal compressors	Air is compressed in two to four stages, with interstage and aftercoolers	WHR from interstage and aftercoolers	Same as for oil-injected compressor except that WHR is from interstage and aftercoolers

Implementation of WHR depends on the type and size of the AC system. Summary of industrial ACs and applicable WHR methods are summarized in Table 12.4. In general, WHR may not be cost-effective in ACs with smaller power (<100 kW) consumption.

Atmospheric air contains some amount of water vapour, which is compressed and heated along with air in the AC. Then, water vapour condenses in the interstage and aftercoolers for separation. For accurate estimation of sensible heat transfer, sensible and latent heat transfer of water vapour in interstage and aftercoolers, process simulation is required. For example, Fig. 12.5 shows Aspen Hysys simulation of an oil-free

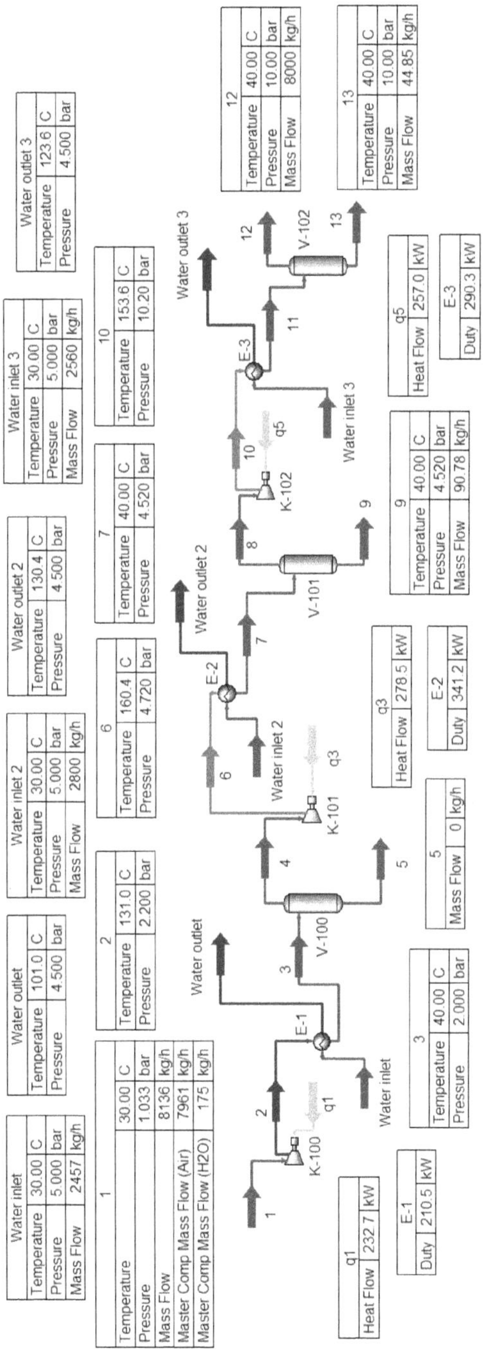

Fig. 12.5: Hysys simulation for an AC.

centrifugal AC, with three stages of compression. Cold- and hot-end approach temperatures of 10°C and 30°C, respectively, for each of coolers: E-1, E-2 and E-3, and polytropic efficiency of 75% for each stage of compression, are used in this simulation. Power requirement for first, second and third-stage compression are 232.7, 278.5 and 257 kW, respectively. So, a total of 768.2 kW of power is required. Assuming a motor efficiency of 95%, 808.6 kW of power is required.

In industrial ACs, cooling water is used in interstage and aftercoolers, with typical supply and return temperatures of 30°C and 45°C, respectively. Increasing the cooling water return temperature beyond 50°C leads to excessive fouling in the interstage and aftercoolers. However, clean water such as pressurized demineralized (DM) water can be used for WHR to achieve high water return temperatures. In Fig. 12.5, water return temperatures from E-1, E-2 (interstage coolers) and E-3 (aftercooler) are 101°C,130.4°C and 123.6°C, respectively, and heat duties of E-1, E-2 and E-3 are 210.5 kW, 341.2 kW and 290.3 kW, respectively. So, a total of 842 kW of WH can be recovered. For this, the combined water outlet temperature from the interstage and aftercoolers is $\frac{((2457\times 101)+(2800\times 130.4)+(2560\times 123.6))}{(2457+2800+2560)}=118.9°\text{C}$. The hot water flow rate in total is 2457 + 2800 + 2560 = 7817 kg/h (Fig. 12.5).

Note that the recovered WH (842 kW) is greater than the compression energy input (768.2 kW) to the AC. This is because of the latent heat of water vapour, present in the suction air, which is later recovered at the interstage and after condensers. This analysis assumes zero heat loss in the compressor and piping. In reality, 2%–3% of WH is lost by radiation and natural convection. Hence, new oil-free ACs can be designed to recover ~95% of the compression energy. WHR from AC may be reduced (to ~75% of WH) if WH from aftercooler is used for IA drying in HOC dryer. If VSDs are used, WHR may reduce when AC operates at part load.

For designing WHR system in the old/operating AC system, it is advisable to contact the compressor manufacturer, to ensure that the proposed modifications do not create any compressor-related problems. WHR from AC system can be used to save fuel/electricity cost for water/process/space/air heating or production of chilled water, as shown in Fig. 12.6. If WHR is used to replace electric heaters for water heating or

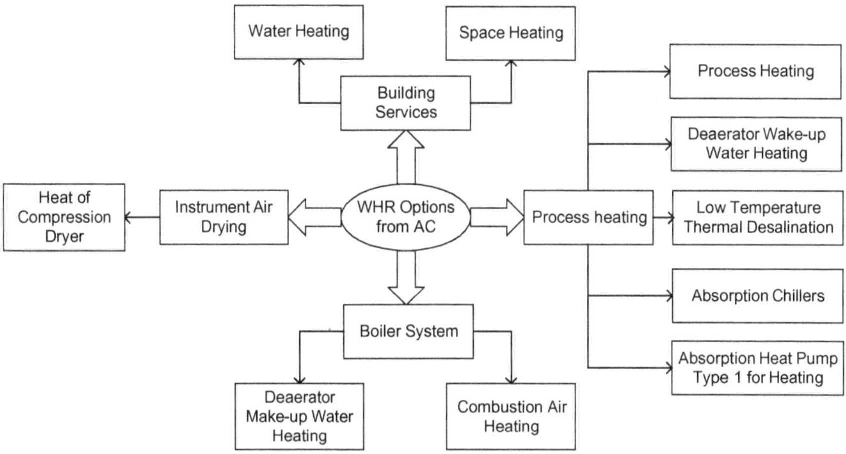

Fig. 12.6: WHR options from AC.

mechanical chillers for production of chilled water, operating cost savings will be even greater.

Air-cooled AC can provide hot air at ~80°C and water-cooled AC can produce hot water at up to ~100°C. WHR from ACs can be used for space heating by instaling ducting from AC to the area that requires heating. The hot air is produced by passing cold air through the interstage, aftercooler and lubricant coolers of the AC. This extracts heat from the AC as well as the lubricating oil (if oil-injected) and provides necessary cooling for the AC system. Hot air produced can also be used as combustion air to boilers and fired heaters to improve the energy efficiency. Hot DM water produced from AC's WHR can be used for use in deaerator (covered in Section 16.5.1.1 of Chapter 16), process heating, in absorption heat pump Type 1 (for heating or production of chilled water, covered in Section 5.4.1 of Chapter 5), low temperature thermal desalination (covered in Section 13.4 of Chapter 13).

Example 12.5: An indirect WHR system is used for an oil-injected SC with a closed-loop water circuit. SC consumes 500 kW of electric power. WHR system recovers 75% of the power consumed by the SC, from the lube oil cooler. Lube oil at 90°C is available at lube oil cooler. Use

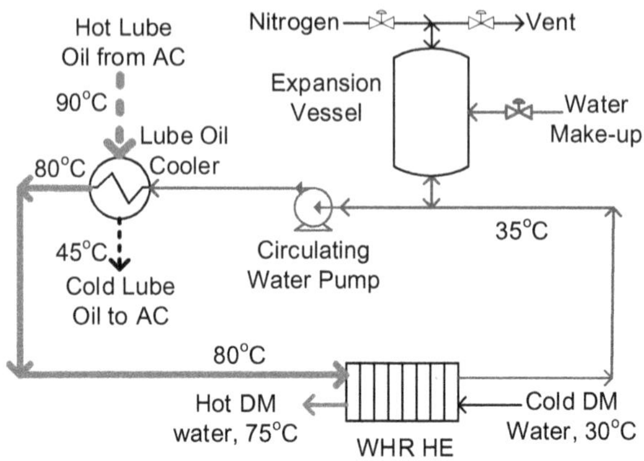

Fig. 12.7: Details of the indirect AC WHR system.

approach temperature of 10°C at both hot and cold ends of the lube oil cooler, and 5°C at both hot and cold ends of the WHR HE, as shown Fig. 12.7. Circulating hot water exchanges heat with a cold DM stream at 30°C in WHR HE. Estimate the flow rates of circulating water and DM water streams. Also, estimate the outlet temperature of DM water at the WHR HE.

Solution: Details of the WHR system are presented in Fig. 12.7.

Hydraulic power consumption at the AC = 500 kW

Amount of WH recovered at the lube oil cooler = 500 × 0.75 = 375 kW

Approach temperature at both hot and cold ends of the lube oil cooler = 10°C

Energy balance across the oil cooler is given by:

$375 = m_w \times 4.184 \times (80 - 35)$

Here, m_w is the mass flow rate in the circulating water loop and 4.184 kJ/kg°C is the heat capacity of water. Hence, $m_W = \frac{375}{4.184 \times (80-35)} =$ 1.992 kg/sec = 7170.2 kg/hr

Approach temperature at both hot and cold ends of the WHR HE = 5°C

Hot DM water temperature at the outlet of WHR HE = 80 − 5 = 75°C

From energy balance across the WHR HE, amount of DM water flow rate is:

$m_{DM} = \frac{375}{4.184 \times (75-30)} = 1.992$ kg/sec = 7170.2 kg/hr

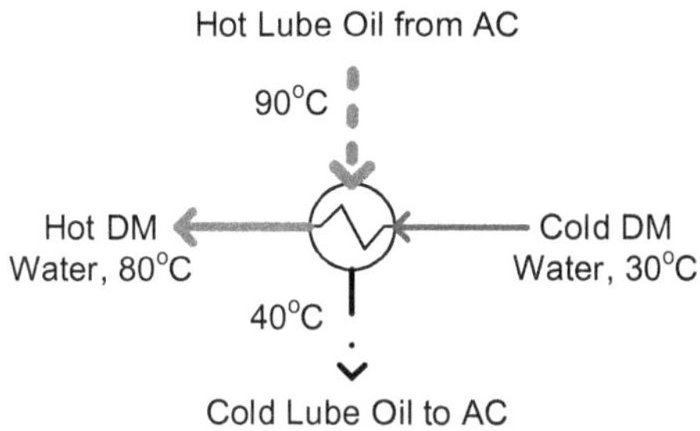

Fig. 12.8: Details of the direct WHR for the AC system.

Indirect cooling method (illustrated in the above example) is used if there is any concern on potential contamination of the DM water with lube oil due to the leakage of HE tubes. If direct WHR method, which involves just once-through flow of DM water through lube oil cooler (as presented in Fig. 12.8), is acceptable, higher amount of WH (i.e., 416.7 kW) would be recovered, at a higher DM water temperature of 80°C. As a result, it can cool the lubricating oil to 40°C (5°C lower compared to indirect method), which improves the efficiency of the AC. Also, direct WHR method requires lower capital cost compared to indirect WHR method. This is mainly due to the absence of circulating water pumps, WHR HE, water buffer tank and associated control systems (Fig. 12.7). However, direct WHR method is suitable only where hot DM water can be directly consumed in the process (e.g. deaerator).

12.4.1 *Heat Recovery from Air Compressor Drivers*

In addition to the WHR from AC, other WHR options are available for ACs driven by diesel/gas engine or steam turbine. Diesel/gas engine drives offer higher temperature WHR option from hot flue gases and hot water generation from casing cooling. Depending on the size of the engine, it may be possible to even generate low pressure steam (LPS) by WHR from the hot flue gas. Engine jacket cooling also produces hot water at 70–110°C. WHR from gas/diesel engines and back pressure–type steam turbines are covered in Section 5.3 of Chapter 5 and Section 16.5.3.4 of Chapter 16, respectively.

Back pressure–type steam turbines as drivers provide cogeneration opportunities. By using high-pressure steam, they provide the shaft energy for the AC and useful LPS or medium-pressure steam for heating process streams. Hence, power required for AC is almost free. WHR from steam turbine–driven AC provides a highly efficient AC system.

12.5 Instrument Air Drying

IA is produced by passing CA or PA through an air dryer. IA should be free of moisture for use in valve actuators. Its drying is usually accomplished in fixed beds, filled with deliquescent salts or adsorbents (such as activated alumina or molecular sieve), by refrigeration or membrane separation. Various IA drying methods are compared in detail, in Table 12.5.

In deliquescent salt bed dryer, moisture is absorbed by the salt and forms brine solution. Hence, salt will get consumed and requires frequent top-ups. Multiple deliquescent salt beds are used to enable continuous operation. In adsorbent or desiccant type dryer, adsorbent adsorbs water from air and gets saturated with water. Saturated adsorbent bed is regenerated by purging it with dry air or by heating it with hot dry air, electric coils or steam coils. There are two types of adsorption dryers, namely, "heatless" and HOC dryers. Less energy efficient type of regeneration is the "heatless" regeneration, which uses nearly ~20% of the cold dry air (from after cooler) to purge moisture away to atmosphere, thus regenerating the adsorbent bed. This wastes the valuable energy used for producing IA. Membrane dryer uses membrane elements to remove the water from CA/PA. It also wastes up to 20% of air in the reject stream.[2]

Comparatively energy-efficient drying methods use refrigeration, WH generated LPS heaters or electric heating coils inside the adsorbent/ desiccant beds. The refrigeration method cools CA/PA to very low temperature and hence condenses out the moisture in air. Resulting cold and dry air is then reheated to ambient temperature. The most energy-efficient dryer is HOC dryer, which uses high-temperature air from the AC discharge (i.e., before aftercooler) for removing the moisture from the adsorbent bed. This method maximizes WHR from AC system.

Table 12.5: Details of various IA dryers

			Desiccant		
Detail	**Chemical Deliquescent**	**Refrigerant**	**Heatless/Heated**	**Heat of Compression**	**Membrane**
Typical configuration	Dual vessels with salt bed	Air to air cooling, followed by refrigerant to cool air and finally heat cold air to ambient temperature	Twin vessels with desiccant packed beds	Generally twin vessels, with desiccant packed beds	Membrane unit
Drying medium	Salt bed of sodium, potassium or calcium	Refrigerant such as R134A and R407C	Desiccant such as silica gel, alumina and molecular sieves/Heating with electric coil or LPS	Hot air from compressor discharge	Membrane unit
Drying principle	Moisture in air is absorbed by the salt bed. Salt forms brine after capturing moisture, which will be periodically drained	AC discharge air is cooled by heat exchange with ambient air, and then cooled by heat exchange with refrigerant. Water gets separated after cooling. Cold air is heated to ambient temperature.	Moisture adsorption in desiccant	Moisture adsorption in desiccant.	Water permeates through membrane walls, while air is dried

(Continued)

Table 12.5: *(Continued)*

Detail	Chemical Deliquescent	Refrigerant	Desiccant		Membrane
			Heatless/Heated	Heat of Compression	
Heat utilization from CA	NA	NA	NA	Uses heat from CA	NA
Drying medium regeneration	Not possible. Salt consumed needs to be topped up (e.g., two to three times in a year)	NA	Possible (need to replace desiccant once in 3–5 years)	Possible (need to replace desiccant once in 3–5 years)	Membranes need replacement, possibly once in 2–3 years
Dew point attained	11°C–26°C lower than air temperature at dryer inlet	2°C–4°C	–73°C to –40°C	–73°C to –40°C	–40°C to 4°C
Purge air loss	0%	0%	Up to 20% for heatless; ~5% for internally or externally heated	0%	Up to 20% of inlet air
Relative capacity	No limit	No limit	No limit	No limit	Low
Relative capital cost	Low	Low	Low for heatless, and medium for heated	High	High

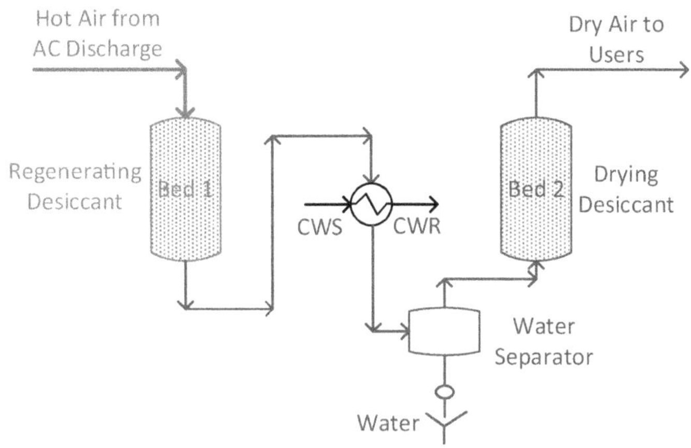

Fig. 12.9: Details of HOC dryer having two beds.

In HOC air dryers (Fig. 12.9), hot air from the AC with discharge temperature higher than 120°C is used directly for the regeneration of the desiccant bed. After regeneration, hot air is cooled to 40°C in a cooler, water is separated and then removed by water trap. Cold air is dried in a second adsorber bed. Thus, both heaters for adsorption bed and purge loss of CA are eliminated. The adsorption beds (generally two) operate repeatedly in batch mode for adsorption and desorption functions. While the first bed is under adsorption mode, the second bed will be in desorption mode and vice versa.

12.6 WHR from Process Compressors

Similar to WHR from ACs, WH can be recovered from process compressors. For example, a gaseous mixture compression system with WHR is shown in Fig. 12.10. Here, a gas stream containing hydrogen and methane is compressed from 15 to 100 bar, in a two-stage compressor; total power consumption is 3535 kW (i.e., 2337 kW and 1198 kW for the first and second stage, respectively). Pressurized DM water at 15 bar is used to recover almost all the compression heat in HEs (E-100 and E-101). Hence, with proper economic design, most of the compression heat from

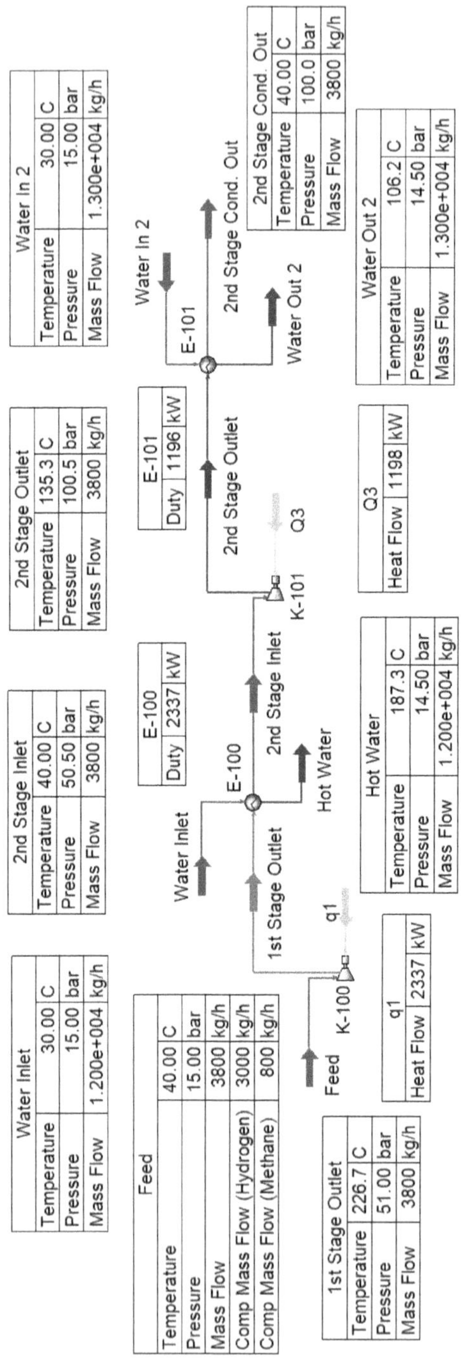

Fig. 12.10: Aspen Hysys simulation of WHR from a process compressor system.

an industrial compressor can be recovered by pressurized DM water. WHR from process compressors is only feasible and beneficial if there is demand for hot water in process and/or other WHR applications in Fig. 12.6. Modifying the operating process compressors for WHR, may require extensive changes. Economical attractiveness of this modification varies from plant to plant due to many factors such as proximity of DM water sources and end user of hot DM water.

12.7 WHR from Refrigeration Compressors

Refrigeration compressors are widely used in process industries. Details of typical refrigeration systems are given in Table 12.6. Refrigeration operations generate considerable quantities of WH, which needs to be removed at the condenser. Refrigeration WH is composed of two components: "superheat" and "condensing" energy. The "superheat" typically accounts for 15% of the total heat of rejection from the refrigeration system, whereas the "condensing" energy accounts for the remaining 85%. As shown in Fig. 12.11, a desuperheater, which is a supplementary HE, can be instaled between refrigeration compressor and condenser. It can recover WH at temperatures, substantially above the condensing temperature, and helps to maximize WHR by generating hot water at 60°C–80°C.

Use of desuperheater in industrial refrigeration systems has several other advantages.

Table 12.6: Details of typical refrigeration systems

System	Temperature Range, °C	Refrigerant
Steam jet	2–21	Water
Lithium bromide–water absorption chiller	4–21	Water
Ammonia–water absorption chiller	–40 to –1	Ammonia
Mechanical compression (centrifugal, reciprocating or SC)	–130 to 4	Ammonia, propane, ethylene, propylene and other commercial refrigerants
Cryogenics	Below –130	Liquified gases

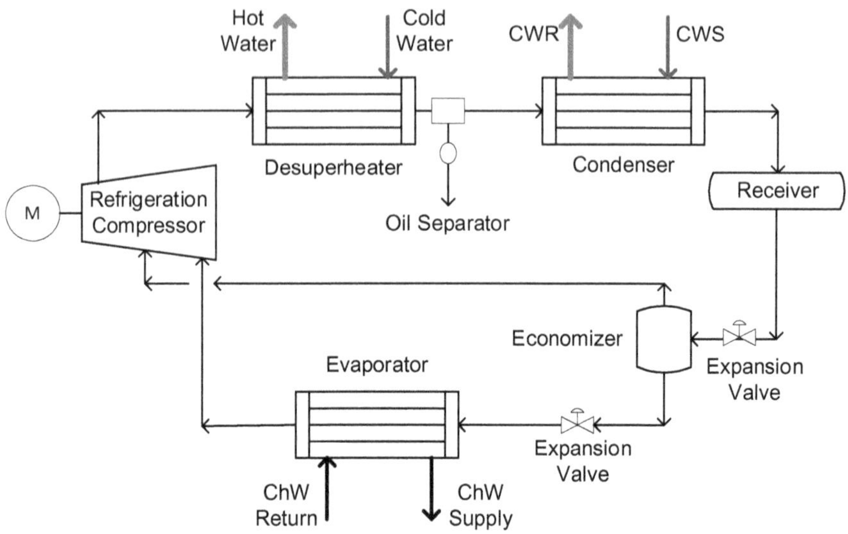

Fig. 12.11: Refrigeration chiller instaled with desuperheater.

1. Condenser duty is reduced when a desuperheater is installed, as in Fig. 12.11. This reduces the condensing pressure, which in turn reduces the discharge pressure and power requirement of the compressor, and/or condenser cooling duty.
2. Use of desuperheater reduces cooling water requirement and fouling at the condenser.
3. Most of the oil that is normally in vapour state will condense in the desuperheater, and condensed oil can be separated. This enables good heat transfer at evaporator, and hence higher evaporator pressure and system energy efficiency. Higher heat transfer at evaporator results in higher evaporator pressure and temperature. As a "thumb rule", a 2°C increase in evaporator temperature decreases refrigeration compressor energy requirement by approximately 6%.

The utilization of refrigeration WH is profitable wherever heating and refrigeration are required simultaneously. Some of the low-temperature process heat sink applications are shown in Fig. 12.6.

Economizer in Fig. 12.11 is a refrigerant cooling vessel, which cools the refrigerant by flashing some of the condensed refrigerant. It is placed between the condenser and evaporator. Economizer is very useful to reduce the refrigeration power consumption. The liquid from the condenser/receiver vaporizes through the expansion valve to the economizer vessel. This cools the liquid inside the economizer. The cold flashed vapour from the economizer enters the refrigeration compressor's intermediate stage, cools the compressed gases and also reduces the suction volumetric flow rate to its first stage. This lowers the total compression power required by the refrigeration compressor. The cooled, liquid refrigerant from the economizer flows through another control valve into evaporator for final vaporization. Use of economizer lowers both capital and operating costs of the refrigeration system. It can save compressor power by 3%–15%.

Options for reducing power consumption and/or increasing WHR from a refrigeration compressor are as follows.

- Increasing compressor suction pressure and/or decreasing condenser pressure reduces the compressor pressure lift, and hence saves significant amount of compressor power.
- Water-cooled condensers can reduce the condensing temperature lower than that achievable by air-cooled condensers.
- Use plate HEs as evaporators and condensers to minimize temperature difference between refrigerant and process fluids.
- Use proper cold insulation to minimize the heat gain in refrigerated pipes. The rule of thumb is to provide sufficient insulation to limit heat gain of 25–30 W/m^2 of pipe surface area to the refrigerated process equipment.
- Minimize mechanical refrigeration for applications above 5°C. In such cases, alternative systems such as chilled water produced by lithium bromide absorption chiller (covered in Section 5.5 of Chapter 5) can be very beneficial.
- Use VSDs on the compressors, for variable load applications.
- Use multiple compressors that can be switched on and off in response to varying loads.

12.8 Summary

This chapter is on WHR in compression systems, mainly focussed on CA system, which is common in process industries. WHR from process and refrigeration compressors are also briefly discussed. Main points of this chapter are as follows.

- Industrial CA systems are highly energy inefficient; hence, it is important to minimize the power of compression of AC as well as maximize WHR from AC.
- Many design options are available for minimizing the power of AC such as the use of multi-stage compression with interstage cooling, decreasing suction temperature and/or decreasing discharge pressure.
- CA system efficiency can be increased by WHR from interstage, after and/or oil coolers.
- WHR from AC is very useful to generate hot water, which can be used in applications such as heat pumps, desalination, evaporation, process and space heating. Non-oil-injected rotary ACs provide the biggest WHR opportunity at interstage and aftercoolers.
- IA drying is also highly inefficient operation if heatless desiccant or membrane type dryers are used because the air purge losses can be as high as 20%. HOC dryer is the most energy-efficient option for IA drying as it does not generally waste CA.
- With proper design, almost all the compression power can be recovered from process compressors if demand for hot water and/or low-temperature heat exists.
- Use of desuperheater enables WHR from industrial refrigeration compressors. Use of economizer reduces the power requirement for the refrigeration compressors.
- Use of lithium bromide absorption chillers should be maximized for cold services with temperatures more than 5°C.

Chapters 11 and 12 in the book by Coker[1] are suggested for further reading on comprehensive coverage of refrigeration and compression equipment.

References

1. Coker AK. (2015) *Ludwig's Applied Process Design for Chemical and Petrochemical Plants*, Vol. 3. Elsevier.
2. Kenkre PD. (January 2013) Design and specification of a compressed air system. *Chemical Engineering* 40–48.

Acronyms and Notation

Acronyms

AC	Air Compressor
CA	Compressed Air
DM	Demineralized
DP	Differential Pressure
HE	Heat Exchanger
HOC	Heat of Compression
IA	Instrument Air
LPS	Low-Pressure Steam
PA	Plant Air
SC	Screw Compressor
VSD	Variable Speed Drive
WH	Waste Heat
WHR	Waste Heat Recovery

Notation

C	Free air delivered at compressor discharge (Nm^3/min)
k	Isentropic (adiabatic) exponent
m	Mass flow rate (kg/sec)
M	Molecular weight (kg/mole)
n	Polytropic exponent
P	Pressure (bar)
r	Compression ratio
R	Universal gas constant (8.314 J/K.mol)
t	Time (min)
T	Temperature (°C or K)
V	Volume (m^3)
W	Compressor power (kW)

Greek Symbols

η	Efficiency

Subscripts/Superscripts

a	Adiabatic or atmospheric
D	Discharge
i	Isothermal
j	Index for compressor stage
p	Polytropic
s	Number of stages
S	Suction

Exercises

12.1 What are the differences between IA and PA?

12.2 List the options for minimizing inefficient consumption of PA/IA in process plants.

12.3 Estimate and compare isothermal and polytropic power consumptions for an AC, for compressing 2 ton/hr of air from a suction pressure of 1 bar to a discharge pressure of 10 bar. Suction temperature is 30°C. Assume both polytropic and isothermal efficiencies to be 0.7. Assume polytropic exponent and compressibility factor of 1.3 and 1, respectively.

12.4 State the advantages and disadvantages of different types of IA dryers. Explain how HOC dryers are the most energy efficient.

12.5 Estimate and compare the required power for single-, two- and three-stage compression, for compressing air from 1 to 8 bar. Assume the following data for the calculations: suction temperature for first stage suction = 35°C; polytropic exponent = 1.3; mass flow rate of air at the AC suction = 3 kg/sec; molecular weight of air = 28.97; polytropic efficiency = 0.75; compressibility factor = 1; pressure drop in each of interstage and aftercoolers = 0.2 bar and suction temperature for each of second and third compression stages is 40°C.

12.6 Repeat the exercise 12.5 with first-stage suction temperature of 25°C and keeping all other data same. What conclusion can be made on the effect of suction temperature on the power required?

Chapter 13

Desalination and Water Recovery Using Waste Heat

13.1 Overview

Water is an important resource for (process) industries. It has many important uses such as raw material in chemical reaction, coolant in process heat exchangers (HEs), dilution of process streams, drinking water, washing and cleaning medium. Presently, over one-third of the world's population lives in water-stressed countries, and this figure is expected to rise in the future. Most (97%) of the available water in the world is seawater (SW). Only the remaining 3% is fresh water, suitable for humans, plants, animals and industrial use. Of this fraction, nearly 2.5% is present in the form of polar ice caps, glaciers and atmosphere. Hence, only about 0.5% of water is available for direct human and industrial consumption.[1]

Water scarcity poses a serious challenge for sustainability of (process) industries in water-stressed areas. Desalination is a key solution for sustainable water supply for industrial and domestic sectors. Technological improvements over the past several decades enabled a significant increase in desalinated water (DW) production. There are many desalination processes available for the production of DW. Two principles used for desalination are thermal evaporation and membrane separation. Common desalination processes are reverse osmosis (RO), multi-stage flash (MSF) and multi-effect distillation (MED).

Thermal desalination processes use thermal energy for the production of DW. Generally, there are many opportunities in process/power plants, to generate low-pressure steam (LPS) from low-temperature waste heat (WH) streams. Hence, waste heat recovery (WHR) in the plants can be used for thermal desalination, which makes it even more attractive for DW in water-stressed areas. Evaporation is commonly used in process industries such as dairy and food industry for concentration; it can recover and recycle water from wastewater (WW). WH, low-low pressure steam (LLPS), LPS, mechanical vapor compression (MVC) or thermal vapor compression (TVC) can be used for this purpose.

This chapter outlines the production of DW using desalination, and recovery and recycle of water from WW. It begins with Section 13.2 on the classification of desalination processes. Section 13.3 outlines the thermal desalination processes. Sections 13.4 and 13.5 discuss major membrane desalination processes and their comparison, respectively. Integration of thermal desalination processes with industrial WH is discussed in Section 13.6. The subsequent section discusses water recovery from WW. Finally, Section 13.8 summarizes the main points of producing DW using desalination processes, especially utilizing WH.

Learning outcomes of this chapter on water production from SW and WW are as follows.

1. Describe the major desalination processes.
2. State the advantages and disadvantages of major desalination processes.
3. Discuss the integration of WH and thermal desalination processes, with few industrial case studies.
4. Outline the evaporation processes for water recovery from WW.

13.2 Classification of Major Desalination Processes

Major desalination processes are presented in Fig. 13.1; they can be broadly classified into thermal and membrane processes. As the name suggests, thermal desalination processes use thermal energy for the separation of salts from water; they do not use any membranes. SW is heated to produce water vapor, which is subsequently condensed to produce DW. The thermal energy may be obtained from WH, conventional fossil fuels or renewable energy sources (e.g., solar, wind and geothermal). Thermal desalination processes include MSF, MED, MVC and TVC. Membrane

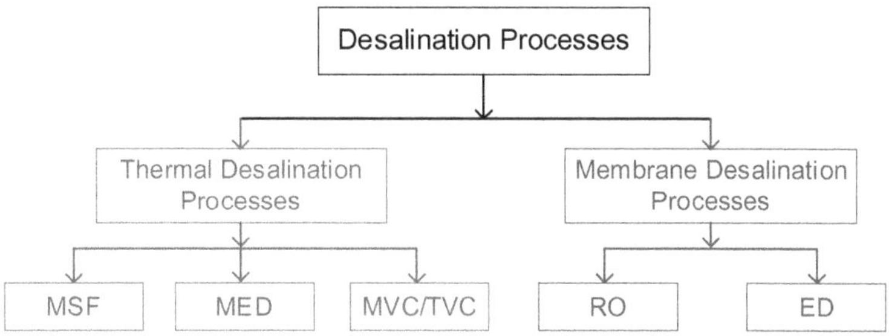

Fig. 13.1: Classification of major desalination processes.

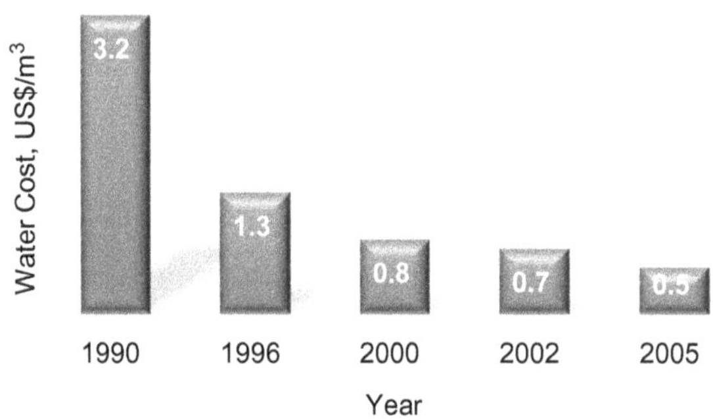

Fig. 13.2: Historical trend of reduction in the cost of DW using SWRO process.[2]

desalination processes include RO and electrodialysis (ED); these require electricity that may be obtained from local electricity network or from renewable power sources such as solar and wind energy.

Currently, RO holds majority market share of ≈60%[a] among desalination processes used in the world. Thermal processes: MSF and MED hold a market share of ≈27% and ≈8%, respectively. They are dominant in the Middle East, where energy costs are lower. RO-based desalination systems are the most used in other regions of the world. Figure 13.2 shows the reduction in price of DW using seawater reverse osmosis (SWRO) process, over the past few decades. Recent cost of DW reported[b] is ~\$0.4/m^3. Cost of desalination plant depends on the type of feed water

---

[a] Waste Heat Recovery for Desalination from Steam Power Plants (wateronline.com) (accessed on 20th August 2021).

[b] https://www.globalwaterintel.com/news/2020/22/we-have-0-40-desal-where-do-we-go-from-here (accessed on 28th August 2021).

(i.e., SW or brackish water with low total dissolved solids [TDS]), plant size, energy and site-specific factors. For example, economies of scale favours larger plants for lower water cost. Hence, the lower end of DW cost is generally valid for larger plants with brackish water as feed. For example, Singapore has several desalination plants based on RO, electrical energy required for these plants is about 3.5 kWh/m^3 of DW and the goal is to reduce the energy required to 1.5 kWh/m^3 and then to 1.0 kWh/m^3 in the long run.[c]

13.3 Thermal Desalination Processes

The TDS content of SW varies considerably from 15,000 to 50,000 ppm, depending upon the location. Thermal desalination processes are generally excellent for removing dissolved minerals and can achieve TDS of 10 ppm or lower in the product water. They are insensitive to changes in salinity and organic activity, which can be troublesome for membrane desalination processes. Further, they can handle higher SW feed temperature, which is a problem for SWRO if it exceeds 45°C. Although capital cost for thermal desalination is higher than that for membrane processes, total cost of water produced by the former can be very attractive if it is powered by WH.

Thermal processes can readily accommodate the processing of highly saline SW; this and low cost of fuel/thermal energy are important reasons for the significant market share of thermal desalination in the Middle East. In these processes, pre-treatment of SW is required to remove debris, particulates, silt and microorganisms, which can pose operational (e.g., blockages and excessive fouling) and equipment integrity problems (e.g., erosion and corrosion). However, thermal processes often require lesser pre-treatment compared to SWRO. The extent of pre-treatment required for them varies with the design of the system. For higher temperature operation, more complex pre-treatment is required to prevent scaling and minimize the corrosive effects. Typical operating temperatures and pre-treatment requirements for thermal desalination are summarized in Table 13.1.

---

[c] https://www.pub.gov.sg/watersupply/fournationaltaps/desalinatedwater (accessed on 27th August 2021).

Table 13.1: Typical operating temperature and pre-treatment requirements for thermal desalination processes

Process	Operating Temperature (°C)	Pre-treatment Requirements
MSF	88	Polyphosphate, chlorination
	113	Polyelectrolyte, degasification, chlorination
MED	71	Polyphosphate, chlorination
	113	Polyelectrolyte, degasification, chlorination
MVC/TVC	55	Degasification, chlorination
	88	Polyphosphate, chlorination

13.3.1 *Multi-Effect Distillation*

MED is the oldest thermal desalination technology. It occurs in a series of evaporator vessels, called effects or stages. The operating pressure reduces from the first effect to the last effect; this permits SW feed to undergo boiling many times without supplying additional heat after the first effect. In the first effect, pre-heated SW feed is sprayed onto the outer surface of a bank of tubes, inside which LLPS or LPS is condensing (Fig. 13.3). This vaporizes a portion of the SW, and the unvaporized SW collects at the bottom of the first effect. The hot water vapor from the first effect flows and condenses inside a bank of tubes in the second effect; thus, it acts as the heat source for the second effect.

Hot unvaporized SW from the first effect is sent to the second effect. Fresh SW feed is sprayed over the tubes, on which the vapor from the first effect is condensing, and this condensate is withdrawn as DW product. Some vaporization occurs on the hot tube surfaces. Unvaporized SW settles at the bottom of the second effect and then flows to the third effect. Vapor produced in the second effect is likewise routed to the third effect, where it is used to vaporize some SW feed. Condensed water vapor inside the tubes of the third stage is withdrawn as DW product. This process continues until the last stage in MED. Unvaporized SW from the last stage is returned to sea as brine. Vapor from the last stage is condensed by heat exchange with fresh SW feed and withdrawn as DW product.

Although Fig. 13.3 shows only four stages, there can be up to 20 effects in industrial MED plants, which are typically built to produce

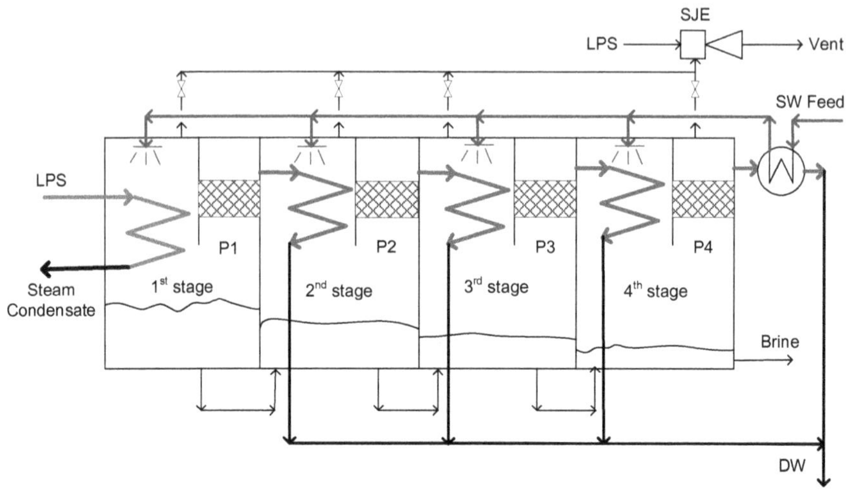

Fig. 13.3: Schematic of a typical MED with four stages, for desalination; note that $P_1 > P_2 > P_3 > P_4$ and steam jet ejector (SJE) for maintaining the required vacuum.

2000 to 20,000 m^3/day, with the highest temperature (in the first effect) of about 70°C. This enables MED to use WH stream such as flash steam (≈1 bar or under vacuum) instead of LPS (in Fig. 13.3) as the heat source.

MED is more efficient compared to MSF in terms of thermal energy consumption, electricity consumption and capital cost. It is usually operated at lower temperatures than MSF, for reducing scale formation but its HE configurations make cleaning more complicated. Hence, they are not as widely used as MSF. Thermal efficiency of MED can be improved by coupling TVC with MED as shown in Fig. 13.4. In this arrangement, LPS required for the first effect flows through an SJE (covered in detail in Section 8.2.1 of Chapter 8), which compresses water vapor generated in the last effect. This reduces the amount of fresh LPS required for MED. MVC can be used instead of TVC, if electricity is very cheap at the plant location.

Vertical- or horizontal-tube falling-film evaporators are generally used in modern MED plants, for achieving higher heat transfer coefficients and hence requiring lower heat transfer area. The main drawbacks of the current MED technology are complex design and lower production capacities. Also, brine temperature needs to be maintained at <70°C, mainly due to scale-forming issues. However, if calcium, which is the

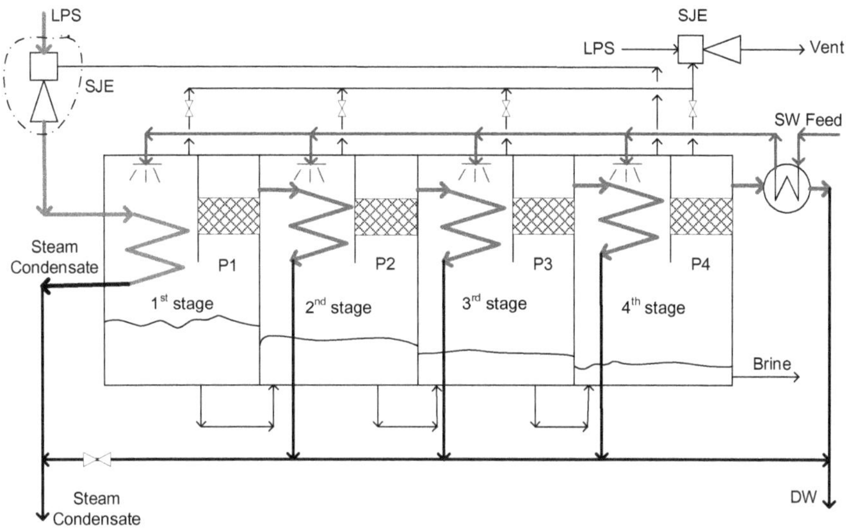

Fig. 13.4: Details of typical MED–TVC desalination; TVC is highlighted with dashed oval at the top left corner.

main contributor of fouling, can be removed using membrane pretreatment, MED can operate at higher temperatures comparable to MSF.

13.3.2 *Multi-Stage Flash*

Because of scaling and low heat transfer rate problems with MED, MSF process was developed in the 1960s. Plant capacity of MSF can be very high (up to 70,000 m^3/day) compared to MED, and so MSF is more popular for desalination. In MSF process (Fig. 13.5), each stage of the process operates at a higher pressure compared to the next stage. When preheated, pressurized SW feed enters a stage, its pressure decreases due to pressure difference between consecutive stages. This causes a portion of SW to boil off or 'flash' and hence cools the remaining water, as heat of vaporization is obtained from the remaining liquid. Since evaporation occurs from the bulk fluid in MSF (rather than at hot heat exchange surface in MED), chances for scaling in MSF are reduced.

SW feed is heated progressively as it flows though coils in the lowest pressure (fourth stage in Fig. 13.5) to the highest pressure stage (i.e., first stage in Fig. 13.5) of MSF unit. It is finally heated with LLPS or LPS, in

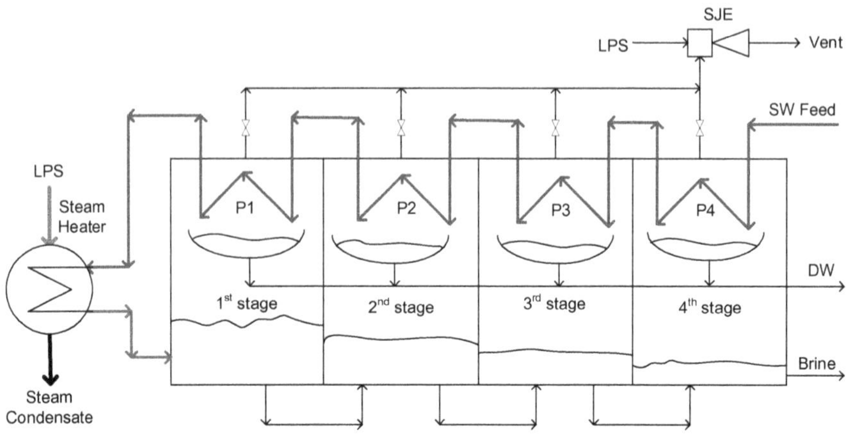

Fig. 13.5: Schematic of a typical MSF desalination; note that $P_1 > P_2 > P_3 > P_4$ and SJE for maintaining the required vacuum.

the steam heater to 110 to 120°C. LLPS or LPS required for MSF can be generated from the extraction steam turbines (described in Section 16.5.3.4 of Chapter 16) or using WH in process plants. The hot SW enters the first stage, which is maintained at a relatively lower pressure; this causes sudden evaporation, called flashing. During flashing, some water vapor is separated from SW. The flashed vapor is then condensed on the outside of tubes carrying relatively colder SW feed. The condensate is withdrawn as DW product.

The unvaporized portion of SW from the first stage flows to the second stage at a pressure lower than that in the first stage, which lowers the boiling point of SW and causes further flash. The flashed vapor is again condensed using heat exchange with SW feed, and the condensate is collected as DW product. This process continues until the last stage (i.e., fourth stage in Fig. 13.5). Flashing in each stage is carried out under vacuum, maintained by using SJE. The concentrated SW or brine is removed from the final stage. MSF plants are built with capacities ranging from 4000 to 70,000 m^3/day. A typical MSF plant may contain up to 25 stages, with each successive stage operating at a lower pressure and temperature than the previous one.

The reliability of MED and MSF can be improved by the use of new high-temperature plastics, coatings for steel surfaces and tube materials. These improvements can reduce the corrosion problems. Higher operating

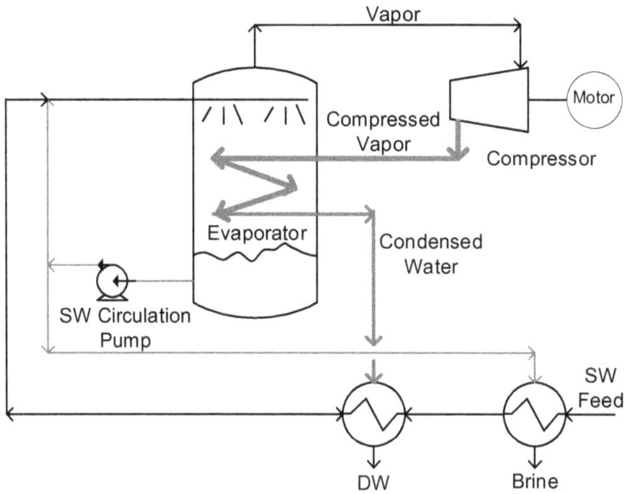

Fig. 13.6: Single-effect MVC desalination process.

temperature of MSF can be accomplished by removing scaling components from SW by nanofiltration and/or using better scale inhibitors.

13.3.3 *Mechanical/Thermal Vapor Compression*

The MVC desalination is generally used for small- and medium-scale production of DW (of 2500 m^3/day to 20,000 m^3/day). In this process (Fig. 13.6), the major energy consumer is the vapor compressor, which accounts for over 80% of the total energy required for DW production. Instead of a mechanical compressor, TVC uses SJE, powered by LPS. MVC and TVC compress water vapor, which then flows inside the tube coils on which SW is sprayed. Sensible and latent heat of compressed vapor is transferred to vaporize part of SW. This generates water vapor, which is sent to the compressor/SJE. The unvaporized SW falls down into the evaporator bottom sump. Some portion of this water is withdrawn as brine, for controlling the TDS. Remaining SW is recirculated to the evaporator along with SW feed, for further vaporization. For minimizing energy consumption, SW feed is heated with hot brine and DW product, before it enters the evaporator.

The MVC desalination plants are ideal for DW production, where low grade WH is not available and electricity is cheap. Compared to smaller capacity plants, large capacity MVC plants have lower specific energy consumption (i.e., economies of scale), making them an attractive desalination option. Use of compressor with higher volumetric flow and head enables the use of more MVC effects, arranged in series and with decreasing pressure. Use of better spray nozzles (for more uniform wetting of the heat transfer coils) and groovy fins on heat transfer coils can increase DW production rate. Desalination plants using TVC can use LPS (e.g., generated by WHR) as motive steam in SJE.

For thermal desalination processes, higher the number of stages/effects, lower will be the thermal energy requirement. However, capital cost increases with increase in the number of stages/effects. Hence, for every application, there will be an optimum number of stages, which can be estimated by performing cost–benefit analysis.

13.4 Membrane Desalination Processes

The main membrane desalination process is SWRO. Another membrane process, namely, ED is applicable and economical for treating brackish water (with low TDS ≈5000 ppm) compared to SW (with TDS up to 50,000 ppm).

13.4.1 *Seawater Reverse Osmosis*

Main steps in the SWRO process are shown in Fig. 13.7. Compared to thermal desalination, more stringent SW pre-treatment steps are required for RO plants, and pre-treatment costs may contribute up to 40% to the overall cost of DW production. Hence, it is an important step in SWRO process. SW is pre-treated to achieve:

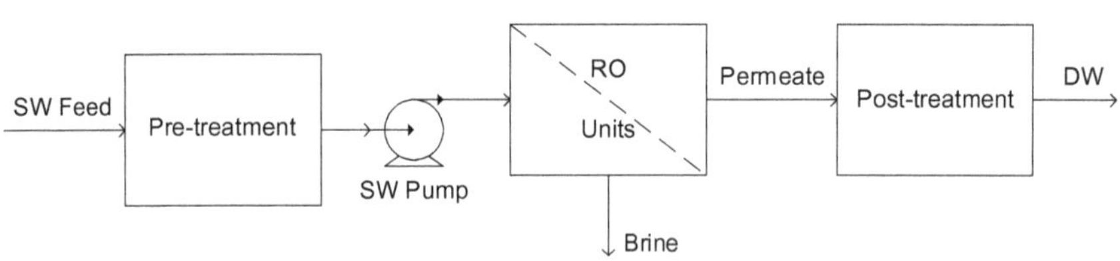

Fig. 13.7: Main steps of SWRO process.

- Removal or minimization of fouling constituents, which can quickly foul the membrane
- Acid or threshold inhibitor addition for scaling control
- Chlorination for micro-organism control (and de-chlorination if chlorine-sensitive membranes are used)
- Fine filtration to remove any free particles

Pre-treated SW is pressurized to about 60–70 bar, by high-pressure pumps, and then sent through RO units. Once the osmotic pressure is overcome, water molecules move across the membrane to the side with lower salt concentration. The permeate stream is sent for post-treatment (generally addition of chemicals such as lime) for producing suitable grade of DW. Reject SW stream (brine) from RO is discharged back to the sea, at a location far away from SW intake station, mainly to prevent mixing of brine with fresh SW intake; details of SW intake station are presented in Section 15.3 of Chapter 15). RO plants are most economical for cases where fuel cost is high and large-scale desalination (>100,000 m^3/day), is required. Lower operating and capital costs compared to thermal desalination make SWRO the most attractive solution.

The amount of DW that could be obtained from SWRO ranges between 30% and 60% of feed water, depending on SW quality, required quality of DW, the technology and RO membranes employed. An energy recovery device (such as Pelton wheel, Francis turbine, hydraulic turbocharger and work exchanger) can transfer pressure energy from the high-pressure RO reject (brine) stream to the low-pressure SW feed, to reduce energy consumption of RO plant by 30% to 50%.[3]

Co-location of SWRO plant at a coastal power or process plant may provide the opportunity to use WH to increase the temperature of SW feed. Standard ROs used for SWRO can handle feed water up to a temperature of 45°C. Increasing the temperature of SW feed increases DW production rate at the expense of a small increase in salt content. The savings in total water cost by elevating temperature from 15°C to 30°C is ≈3%. This option is very useful in cold countries. Further, increasing the temperature of feed SW reduces the operating pressure and hence reduces both capital and operating costs. Raising temperature of SW feed must be done carefully in consultation with RO manufacturer, since it increases the salinity of product water, membrane fouling (thus, requiring more frequent cleaning) and may reduce membrane life.

13.4.2 *Electrodialysis*

ED is a voltage-driven membrane desalination process. An electric potential is used to move salts through a membrane, leaving DW behind as product water. ED utilizes ion exchange membranes and an electrical potential difference for the separation of ionic species from an aqueous solution. ED is mainly suitable and applied for the production of DW from brackish water. It is not economical for desalination of SW.

13.5 Comparison of Major Desalination Processes

Important features of thermal and membrane desalination processes are summarized in Table 13.2, whereas their relative merits are presented in Table 13.3. SWRO is often the least cost method for desalting SW in a

Table 13.2: Important features of major desalination processes

Feature	MSF	MED	TVC	MVC	SWRO	ED
Main energy source	LLPS	LLPS, LPS	LPS	Electricity	Electricity	Electricity
Steam pressure, bar[4]	0.2–0.4	2.5–3.5	NA	NA	NA	NA
Unit size, m^3/day	Up to 70,000	Up to 20,000	Up to 20,000	100–2500	100,000	Up to 60,000
Electrical energy, kWh/m^3 of DW[5]	4–6	1.5–2.5	1.5–2.5	6–12	3–6	2.64–5.50
Thermal energy, kJ/kg of DW[5]	190–390	230–390	145–390	NA	NA	NA
Water quality (TDS), ppm	<10	<10	<10	<10	200–500	150–500
Water cost, \$/m^3 of DW	0.56–1.75	0.52–1.5	0.87–0.95	2.0–2.6	0.45–1.72	0.6–1.05
Plant capital cost, US\$/(m^3 of DW/day)[6]	1500–2000	900–1500	900–1500	1000–1500	900–1200	Not Available

Table 13.3: Relative merits of major desalination processes

Process, Recovery and TDS	Advantages	Limitations
MED Recovery: up to 65% TDS: <10 ppm	• Low energy consumption if WH is used • Minimal pre-treatment • Tolerates changes in salinity, normal levels of suspended and biological matter • Can use WH to minimize water production cost • High-quality product water compared to SWRO	• High energy consumption if thermal energy is provided by fossil fuel • Higher capital cost compared to SWRO • Susceptible to scaling and corrosion • Product water may require cooling before use
MSF Recovery: 25%–50% TDS: <10 ppm	• Economical for larger capacities • Minimal pre-treatment • Very reliable technology with long operating life • Flashing (rather than boiling) reduces scaling • High-quality product water compared to SWRO • Plant process and cost independent of salinity level • Reduces scaling problems associated with MED • Can use WH to minimize water production cost	• Energy intensive process • Higher capital cost compared to SWRO • Susceptible to corrosion • Larger footprint of land and material • Maintenance requires shutdown of the entire plant • Slow start-up. • Product water may require cooling before use
MVC/TVC Recovery: Up to 50% TDS: <10 ppm	• Economical for high-salinity (>50,000 mg/L) feed water • Economical for capacities up to 25,000 m^3/day • Relatively low energy demand • TVC can use WH • Lower temperature requirements reduce the potential of scale and corrosion	• Smaller sized units • In case of MVC, start-up requires auxiliary heating source to generate vapor • High operating and maintenance costs if MVC is used • LPS is required if TVC is used

(*Continued*)

Table 13.3: (*Continued*)

Process, Recovery and TDS	Advantages	Limitations
RO Recovery: 30%–60% TDS: 200–500 ppm	• Lower capital and operating costs • No need of thermal energy and cooling duty • Faster start-up. • Highly modular design and can achieve very high production capacity. • Partial shutdown for maintenance of the plant is possible • Removal of contaminants other than salts is achieved • Longer membrane life of 5–7 years	• High-pressure operation and so potential for mechanical failures • Higher costs for membrane cleaning, chemicals and membrane replacement • Salinity of feed water affects the production rate • More pre-treatment is required compared to thermal processes • Prone to biofouling
ED Recovery: 85%–90% TDS: 150–500 ppm	• Longer membrane life of 7–10 years • Operation at low to moderate pressures	• Attractive for brackish water with up to 5000 ppm TDS • Higher costs for membrane cleaning • Prone to leakages in membrane stacks • Bacterial contaminants not removed by system and post-treatment required for producing potable water

stand-alone process, especially for low salinity waters. However, if thermal energy required for MSF, MED and TVC is supplied by WHR, thermal processes can be operated at a lower cost compared to SWRO.

Typical carbon footprint and airborne emissions for MSF, MED and RO are available in Table 1.7 in Jane.[7] These reported values confirm that emissions of CO_2, NO_x and SO_x from thermal desalination powered by fossil fuels are much higher than those for RO; however, emissions from thermal processes are significantly reduced and become comparable to those of RO, if fossil fuel is replaced by WH or renewable energy source.

13.6 Integration of Thermal Desalination Processes with WH

Thermal desalination processes can be integrated with WH. Refineries, power plants, petrochemical facilities and other process plants often require significant quantities of DW (with stringent water quality), and generally have large quantities of low temperature WH. This provides opportunities to integrate WH with thermal desalination. Multiple benefits such as free energy for desalination, reduction in cooling duty and lower emissions can be obtained.

Some exothermic reactions such as those involved in the production of sulphuric acid, acetic acid and polymers release substantial reaction heat, which can be used for thermal desalination. The produced high-quality water (with TDS <10 ppm) can be used to blend and improve other sources of water such as from SWRO. Integrating thermal and membrane processes in a hybrid design also yields operational efficiency improvements.

Large MSF or MED units are often coupled with steam or gas turbines in power plants for the better utilization of thermal energy and improve the overall process efficiency. Integration of a conventional condensing steam turbine (Fig. 13.8) with a desalination process is shown in Fig. 13.9. Steam generator (either boiler or heat recovery steam generator) in the power plant usually generates high-pressure steam (HPS), which expands through a steam turbine to produce electricity (Fig. 13.8). This generates low (vacuum) to moderate pressure (up to 3 bar) extraction steam exiting the steam turbine. In conventional power plants, steam is extracted until a pressure (≈72 torr;[d] saturation temperature = 45°C) that can be condensed by cooling water in a condenser. This wastes significant amount of latent heat of steam to cooling water.

As shown in Fig. 13.9, the condenser can be used to transfer the latent heat of steam to a MED/MSF plant; this can effectively utilize WH to produce DW required for the power plant and/or export to other plants. However, pressure of extraction steam needs to suit the temperature requirements of MED (≈370 torr or 0.493 bar and saturation temperature

[d] Torr is 1 mm Hg and is equal to 0.00133322 bar.

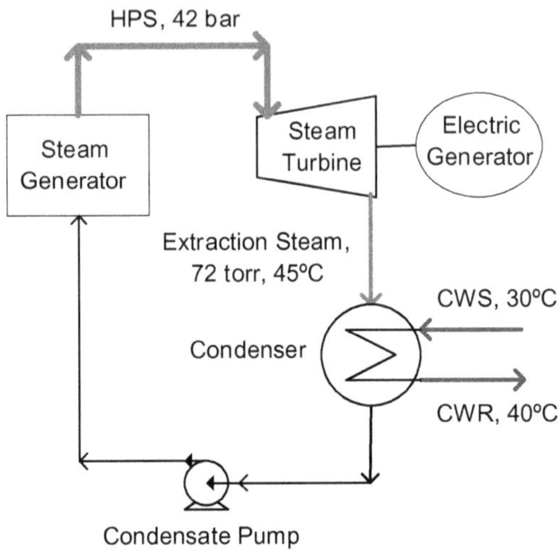

Fig. 13.8: Typical arrangement of power generation using a condensing steam turbine, in a power plant.

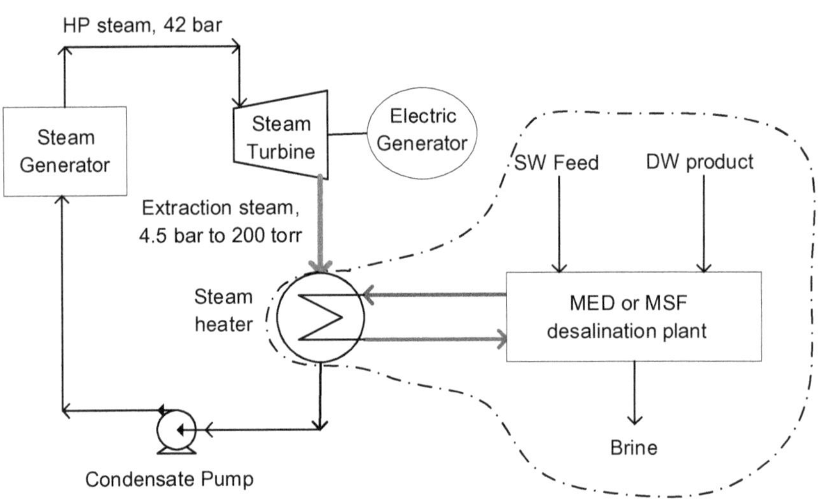

Fig. 13.9: Integration of MED/MSF plant (in dashed envelope) with a power plant.

= 80°C) or MSF (≈2.5 bar). This reduces electricity generation from the steam turbine, for revamp projects. Reduction of power generation from a steam turbine with increase in condenser temperature (equivalent to increasing pressure of extraction steam) is illustrated in Fig. 13.10;

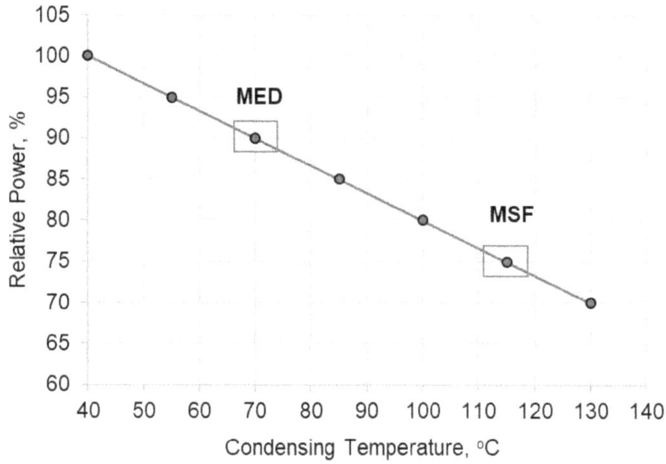

Fig. 13.10: Typical reduction in power generation of a steam turbine with increase in condenser temperature; application range of MED and MSF are indicated by rectangles.[6]

this reduction due to integration of a power plant with MED and MSF is about 10% and 25%, respectively (Fig. 13.9).

Hybrid desalination plants having WH-powered MED/MSF and SWRO units can help to balance fluctuations in electricity and water demands. In such a facility, RO production can be increased/decreased at times of low/high demand for grid electricity. A hybrid desalination plant, shown in Fig. 13.11, can improve the overall profit for the operating company. In this plant, electric power is produced by both gas turbine and steam turbine (i.e., combined cycle power plant), and DW is produced by both SWRO and thermal desalination. Fig. 13.11 is a simplified schematic of a plant in the Middle East.

13.7 Recovery of Water from Wastewater

Several examples of water production from WW are outlined in this section. One is foul water stripping described in Section 9.8.2 of Chapter 9, which uses LPS to purify WW and recover water for re-use in the plant. LPS used for it can be generated from WH in the plant.

Another example is evaporators using MVC (Fig. 13.12); they are widely used in food, pharma, chemical and other process industries for

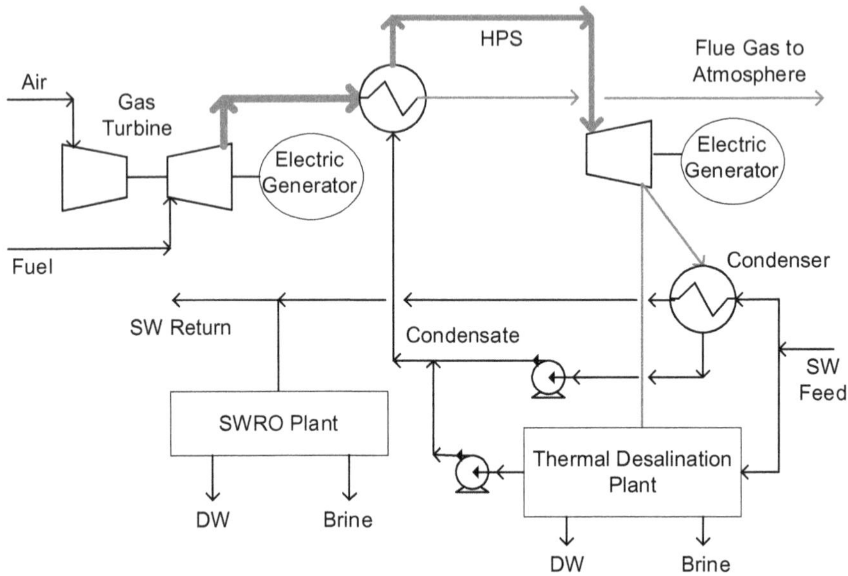

Fig. 13.11: Cogeneration of power and water using combined power cycles and hybrid desalination.

concentrating juices, solutions and WW, and can produce clean water, which can be re-used in the process. WW feed pH is typically adjusted between 5.5 and 6.0 and sent for heat exchange with hot concentrated WW and clean water product, which raises WW temperature to the boiling point. After heat exchange, WW goes to a deaerator, for the removal of non-condensable gases such as carbon dioxide and oxygen. Hot deaerated WW then enters the evaporator's sump, where it mixes with hot recirculating concentrated WW. The liquid from the evaporator's sump is pumped to the top of a heat exchanger (typically with 2-inch diameter tubes), where it flows by gravity in a thin film down the inside of tubes. While the concentrated WW falls down, a small portion of it evaporates and the rest falls into the evaporator's bottom sump, for re-circulation. This heat exchanger is essentially a falling-film evaporator. The water vapor from the sump is drawn into a single stage centrifugal fan, which compresses it to 1.2 to 1.4 bar, and sends it to the shell side of the heat exchanger. A compression ratio of 1.2–1.4 is generally sufficient.

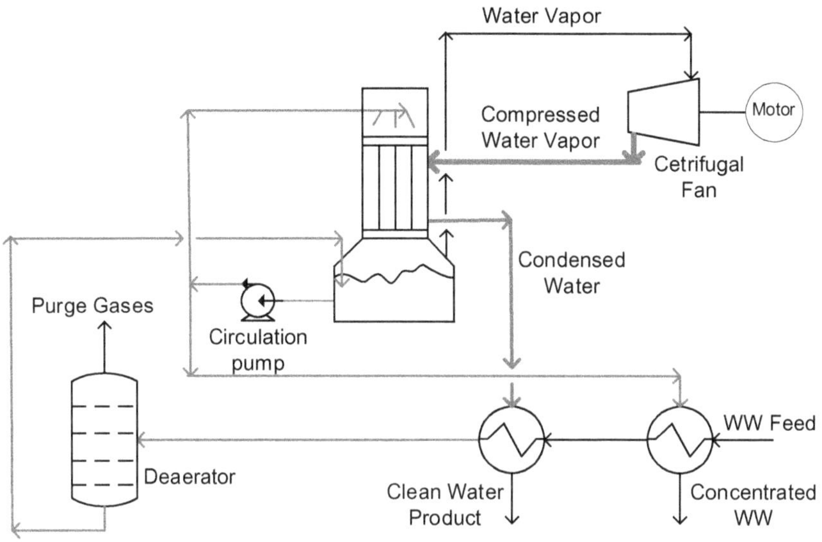

Fig. 13.12: Typical details of a falling-film evaporator with MVC.

As a result of condensation of steam on the shell side of the HE, latent heat of steam is transferred to concentrated WW flowing inside the tubes, and condensate is withdrawn, exchanges heat with WW feed and becomes clean water product (Fig. 13.12). A small amount of the concentrated WW is continuously removed from the discharge of the circulation pump to control density. Typically, 95% of the WW feed can be converted to water (with TDS <10 ppm), for re-use in the plant. The process in Fig. 13.12 requires steam (not shown) to heat WW feed, during the start-up only.

Multi-effect evaporation, similar to MED, can be used for recovering clean water from WW. It is more favourable when sufficient WH is available at or above 80°C. Table 13.4 compares the steam economy (ratio of kilogram of water produced to kilogram of steam used) for single- and multi-effect evaporation, single effect with MVR and triple-effect evaporator with TVR. Energy efficiency increases with the increase in number of effects. Addition of TVR to triple-effect almost doubles the clean water produced for the same amount of steam consumed. Single-effect with

Table 13.4: Comparison of steam economy of different types of evaporators[8]

Type of Evaporator	Steam Economy, Kilogram of Water Product/kg of Utility Steam
Single effect	0.9–0.98
Double effect	1.7–2.0
Triple effect	2.4–2.8
Six effect	4.6–4.9
Triple effect with TVR	4–8
Single effect with MVR	10–30

MVR is most attractive among all the options; hence, it is widely used in the process industry.

13.8 Summary

This chapter is on the production of DW using major desalination processes and recovery of water from WW. It focuses on utilizing WH for improving energy efficiency of these processes. Main points of this chapter are as follows:

- Advancements in desalination processes provide attractive options for production of DW, especially for industries in water-stressed coastal areas.
- Thermal desalination processes, namely, MED, MSF and TVC are well suited for utilization of WH from power and process plants.
- MED can use WH source with temperatures as low as 80°C (assuming 10°C approach temperature for heat transfer).
- TVC can be operated with LPS, generated from WH.
- Combining MED or MSF with TVC (if WH-generated LPS is available) or MVC (if cheap electricity is available) can lower the energy requirements for producing DW.
- MVC-assisted evaporation can be utilized to recover clean water from WW.

More details on desalination processes are available in Chapters 1, 14 and 15 in Voutchkov.[4]

References

1. Gude VG. (2017) Desalination and water reuse to address global water scarcity. *Rev Environ Sci Biotechnol* **16**: 591–560.
2. Voutchkov N. (September 2007) *Advances in Seawater Desalination Technology, Water Conditioning & Purification.* https://wcponline.com/2007/09/16/advances-seawater-desalination-technology-2/
3. Schorr M. (2011) *Desalination: Trends and Technologies*. InTech.
4. Voutchkov N. (2012) *Desalination Engineering: Planning and Design.* McGraw Hill Professional.
5. Curto D, Franzitta V, Guercio A. (2021) A review of the water desalination technologies. *Applied Sciences* **11**: 670.
6. El-Ghonemy AMK. (2012) Future sustainable water desalination technologies for the Saudi Arabia: A review. *Renew Sustain Energy Rev* **16**: 6566–6597.
7. Jane K. (2019) Introduction to desalination. In: K Jane (ed), *Desalination: Water from Water*. Wiley.
8. Hackett BW. (May 2018) The essentials of continuous evaporation. *CEP* 24–28.

Acronyms and Notation

Acronyms

ED	Electro Dialysis
DW	Desalinated Water
HPS	High-Pressure Steam
LLPS	Low-Low Pressure Steam
LPS	Low-Pressure Steam
MED	Multi-Effect Distillation
MSF	Multi-Stage Flash
MVC	Mechanical Vapor Compression
RO	Reverse Osmosis

SJE	Steam Jet Ejector
SW	Seawater
SWRO	Seawater Reverse Osmosis
TVC	Thermal Vapor Compression
WH	Waste Heat
WHR	Waste Heat Recovery
WW	Wastewater

Exercises

13.1 State the advantages and disadvantages of thermal over membrane desalination processes.

13.2 State the advantages of using WH for thermal desalination processes.

13.3 State the energy requirements for major thermal desalination processes. Assess the possibility of DW production, using WH sources in your process plant.

13.4 What are the considerations for retrofitting (i.e., integrating) existing condensing steam turbines with a new thermal desalination plant?

13.5 Specify the methods for recovering water from WW. Discuss their advantages and disadvantages.

Chapter 14

Waste Heat Recovery Using a Heat Transfer Fluid

14.1 Overview

Energy cost significantly influences the profitability of process industries. Hence, reducing energy consumption by waste heat recovery (WHR) is the key to improve process profitability and sustainability. Typically, 20% to 50% of the energy used in the process is rejected as waste heat (WH). A significant part of WH is not recovered because of economic and other reasons; instead, it is discarded to atmosphere via cooling towers, sea water cooling systems, air-cooled heat exchangers (HEs) (air coolers) and/or flue gas (FG) stacks. For example, low-temperature waste heat (LTWH) sources have temperatures lower than 150°C. Hot process streams with temperatures between 100°C and 150°C usually reject WH through air coolers; for cooling process streams below 100°C, fresh water and/or sea water cooling systems are widely used. In general, process streams requiring cooling and those requiring heating are known as heat (or WH) sources and heat sinks, respectively.

Often, there may not be an incentive to recover and utilize WH including LTWH within the same process section/unit.[a] However, it may be possible to recover WH from one section and utilize it in another section within the process plant. Such WHR often involves heat recovery

[a]A process section, also known as unit, is a part of a process plant. Each of these sections/ units plays a role in the overall process of converting raw materials to finished products.

from WH sources (in one process section) into a heat transfer fluid (HTF) and transferring the recovered heat to heat sinks (in another process section). This requires an HTF such as pressurized water or hot oil stream to circulate in a heat transfer loop, for WHR. With careful screening of WH sources and heat sink streams, one may find many possible ways of WHR including LTWH.

Depending on the opportunities in a process plant, there can be one or more WH sources and heat sinks, connected by a simple heat transfer loop or a complex heat exchanger network (HEN[b]), which integrates heat between many heat sources and heat sinks using an HTF. HEN involving an HTF requires additional capital investment for purchasing and installing new HEs, pumps, piping[c] and instrumentation. However, it has large potential to simultaneously reduce both heating and cooling demands of a process plant. Hence, it can be an attractive option; for example, Lai *et al.*[1] stated that 10%–15% of the energy requirement in a process plant can be reduced with a good design of low-temperature heat exchanger network (LTHEN).

This chapter begins with an introduction of closed-loop WHR systems in Section 14.2. Then, Section 14.3 outlines the key considerations for the selection of an HTF for WHR. It also discusses the advantages and disadvantages of HTFs compared to the use of water and lists suitable HTFs for WHR. Section 14.4 outlines hot oil HTF system design; it also discusses options available for low- and medium-temperature (150°C–400°C) WHR and improving thermal efficiency of a fired heater present in an HTF system.[d] The next section presents LTHEN design using pressurized water loop. Use of pressurized water as HTF and inter-unit heat integration using low pressure steam (LPS) are illustrated in Section 14.6. This chapter ends with summary in Section 14.7.

---

[b] See Chapter 3 for pinch analysis and HEN design.

[c] Piping for HEN with heat sources/sinks in two sections of a plant may or may not be feasible, and its cost can be substantial depending on the distance/route between the two sections. These should be assessed carefully.

[d] Heat transfer using HTF involves many equipment such as fired heater, process-HTF HEs, optional coolers and heaters, HTF storage, buffer, drain vessels and pumps. Hence, it can be considered as an HTF system.

Learning outcomes of this chapter are as follows:

1. Describe closed-loop WHR systems using an HTF.
2. Identify the key parameters for selection of an HTF for WHR.
3. Outline the design strategies for developing efficient closed-loop HTF systems, using hot oils and pressurized water.
4. Describe the pinch procedure to establish minimum water flow rate in an LTHEN.
5. Elaborate the design procedure for an LTHEN using pressurized water.
6. Discuss low-temperature waste heat recovery (LTWHR) using pressurized water and subsequent inter-unit heat integration using LPS.

14.2 Closed-Loop WHR Systems

Closed-loop heat transfer systems are widely used in the industry for heat transfer applications such as cooling and heating. Some of the examples are closed-loop cooling water system (Fig. 7.3), chilled water/refrigeration systems (Fig. 7.4), steam systems (Fig. 7.2) and hot oil system. These

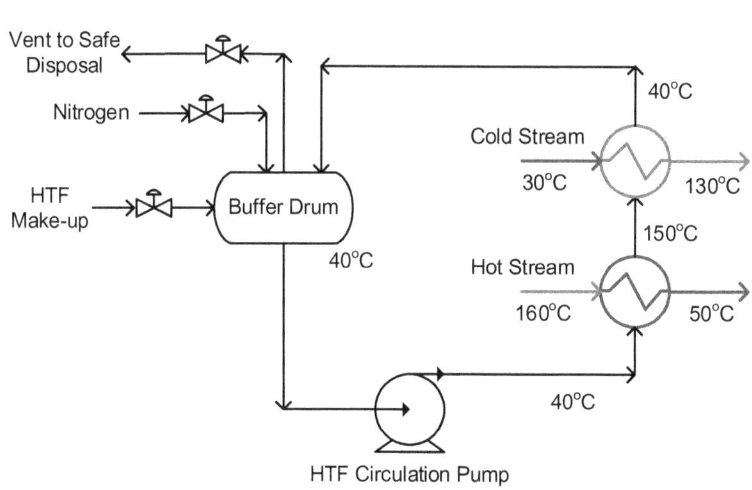

Fig. 14.1: Simple closed-loop WHR configuration providing both process heating and cooling duties; here, hot and cold streams are assumed to be in different sections of the plant.

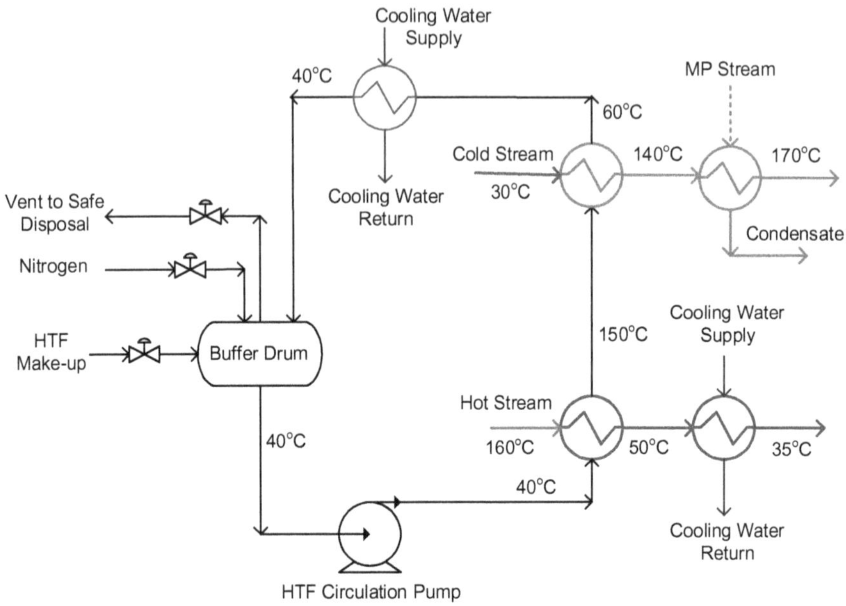

Fig. 14.2: Typical closed-loop LTWHR configuration with additional utility heater and cooler.

closed-loop heat transfer systems provide either process heating or cooling only. On the other hand, a closed-loop WHR system (Fig. 14.1) uses a circulating HTF for providing both cooling and heating duties. When many heaters and coolers are connected by a closed-loop heat transfer system in a low-temperature setting, it is called LTHEN. Often, heating duty available from WH sources may not be equal to the heating duty requirements of heat sinks. Hence, some utility heaters and/or coolers are generally required in closed-loop WHR systems. This is illustrated in Fig. 14.2.

14.3 Heat Transfer Fluid Selection

Many HTFs can be used for WHR; some examples are water, organic and synthetic fluids. Water is safer and generally economical HTF for WHR from WH sources below 150°C. Water may contain excessive hardness or chlorides that can result in high corrosion rates. Hence, if water is used as

HTF, demineralized and deaerated water or cold steam condensate make-up is recommended. In general, use the following guidelines for selecting a suitable HTF.

- **Fluid type:** HTF can be water, water-based chemicals, refined petroleum products, synthetic organic compounds and silicones.
- **Fluid compatibility:** HTF should not pose product contamination or process safety risks if a leak occurs in the HE.
- **Thermal properties:** HTF should have high heat capacity and thermal conductivity but low viscosity for ensuring a high heat transfer coefficient.
  - ***Viscosity*:** HTF viscosity affects the pumping rate, pumping power and heat transfer coefficients. HTFs with a dynamic viscosity lower than 10 mPa-s at the lowest operating temperature of the HTF are preferred.
  - ***Vapour pressure*:** HTFs with vapour pressure of below 600 kPa are generally preferred for economic reasons.[2]
  - ***Flash point*:** HTFs with high flash point (>93°C) are preferred, in order to avoid the requirement of expensive electrical area classification.
- **Temperature range:** HTF should not pose any operational or process safety problems, if operated within the desired temperature range. It should have a high boiling point and low solidification temperature. Viscosity and vapour pressure of the HTF are important for evaluating the suitability of an HTF for a given temperature range.
- **Thermal duty:** Selection of HTF with high heat capacity and heat transfer coefficient will result in lower HTF circulation rate and hence lower operating and capital costs.
- **Toxicity and Flammability:** Non-toxic, odourless and non-flammable HTF is preferred.
- **Degradation:** Synthetic HTF can degrade, especially at high temperatures above 150°C, and the degradation rate roughly doubles with each 10°C increase. Silicone type HTFs do not undergo thermal degradation.
- **System volume:** Total quantity of HTF required for filling the system including the expansion tank should be as small as possible.

- **Cost:** HTF cost should be as low as possible, because there are losses, and make-up is required.

14.3.1 *Use of Hot Oil Heat Transfer Fluids*

A hot oil HTF system can achieve long satisfactory service, without significant physical or chemical change if it is operated within vendor-specified operating limits. However, the hot oil degrades when it is subjected to contamination (e.g., moisture, air and process materials) and thermal stress beyond the acceptable limits set by the vendor. Also, cracking occurs when the hot oil undergoes overheating or is at a high temperature for a long period of time, which destroys the HTF by breaking the molecules into smaller molecules.

The main advantages of hot oil HTFs are the following.

- Low vapour pressure at ambient temperature
- Higher heat capacity compared to normally occurring hydrocarbons
- No corrosion
- Minimal leakage and maintenance
- No requirement of blow down system, steam traps and/or condensate collection and return systems, as well as no water treatment
- High operating temperatures can be achieved with minimal system pressure; only pressure required is to overcome the system pressure drop. Unlike steam/water, hot oil HTFs do not require high system pressure for high-temperature operation. This is mainly due to their low vapour pressure and high boiling point.
- No phase change is involved. Since HTF is always in liquid state, it is easy to handle and pipe sizes are smaller compared to those required for vapour service such as steam.
- Hot oil HTFs can be blended for any desired temperature range. If steam is used, it is generally available at very high pressure (≈101 bar), high pressure (≈43 bar), medium pressure (≈12 bar) and low pressure (≈4.5 bar) (Fig. 14.3). Steam use frequently requires steam headers, steam traps, additional equipment such as de-pressurization valves and desuperheaters, to closely match the heating temperature requirements as well as condensate recovery system.

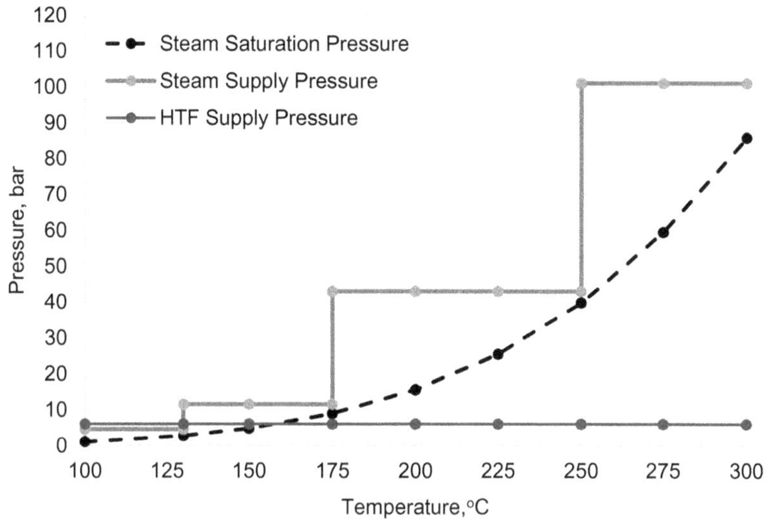

Fig. 14.3: Typical operating pressure requirements of steam and hot oil HTF system for process heating.

The disadvantages of hot oil HTFs are the following.

- Requires safe relief system for any vented vapours.
- If a fired heater is used to heat HTF, overheating of HTF inside the fired heater's coils can cause oxidation and coking. It may degrade the HTF and cause hot spots on the heater coils.
- Use of hot oil HTF may lead to fire hazards, in case of loss of confinement.
- Some hot oil HTFs are expensive.

As the HTF operates under closed-loop, losses are negligible. However, there may be a requirement of partial removal of HTF if degradation happens. In this case, some HTF top-up is required to compensate for the loss. Good design and operation of HTF system is crucial for satisfactory performance, with minimum HTF top-up. Commonly used hot oil HTFs and their important properties are presented in Table 14.1. Hall[2] presented a good compilation of properties and operating temperature range of many organic and synthetic HTFs.

Table 14.1: Commonly used hot oil HTFs and their important properties

Name	Recommended Temperature Range, °C	Density, kg/m^3	Flash Point, °C	Auto Ignition Temperature, °C	Type
BP Transcal N	−10 to 320	875 @ 15°C	221	350	Mineral oil
Dow Syltherm	−100 to 260	852 @ 25°C	47	350	Silicone polymer
DOWTHERM A	Liquid phase: 15–257; Vapour phase: 257–400	1056 @ 25°C	113	599	Synthetic organic fluid
Paratherm HE	53–310	866 @ 15.5°C	210		Mineral oil
CALFO AF	−18 to 316	867 @ 25°C	217	343	Mineral oil
Thermia oil B	−10 to 320	868 @ 15°C	220	360	Mineral oil
Therminol 55	−28 to 300	868 @ 25°C	177	343	Synthetic oil

14.4 Design Strategies for Developing Heat Transfer Fluid Systems Using Hot Oils

An HTF system contains a fired heater or electrical heater or waste heat recovery heat exchanger (WHRHE), which provides the necessary heat required for the process heating demand. Fig. 14.4 shows a typical HTF system using a fired heater, which mainly consists of convection and radiation sections. HTF is first heated in the convection section; it is further heated to a higher temperature in the radiation section.

Fuel is burned inside the fired heater, with a typical thermal efficiency of 80%–85%. This efficiency can be improved to ~93% by installing an air preheater (APH) or economizer. Economizer is an FG to HTF or water HE for heat transfer from FG to HTF or pressurized water, which can be subsequently used to heat combustion air in the APH. Recovered WH from FG saves the equivalent amount of fuel in the fired heater. APH can also use WH of FG to heat the cold combustion air directly in an air to FG HE.

Hot oil HTF system, installed with economizer and APH, is shown in Fig. 14.5. The use of APH for FG to air heat exchange directly (Fig. 14.6) or FG-pressurized water–air heat exchange (Fig. 14.7) is also possible, for

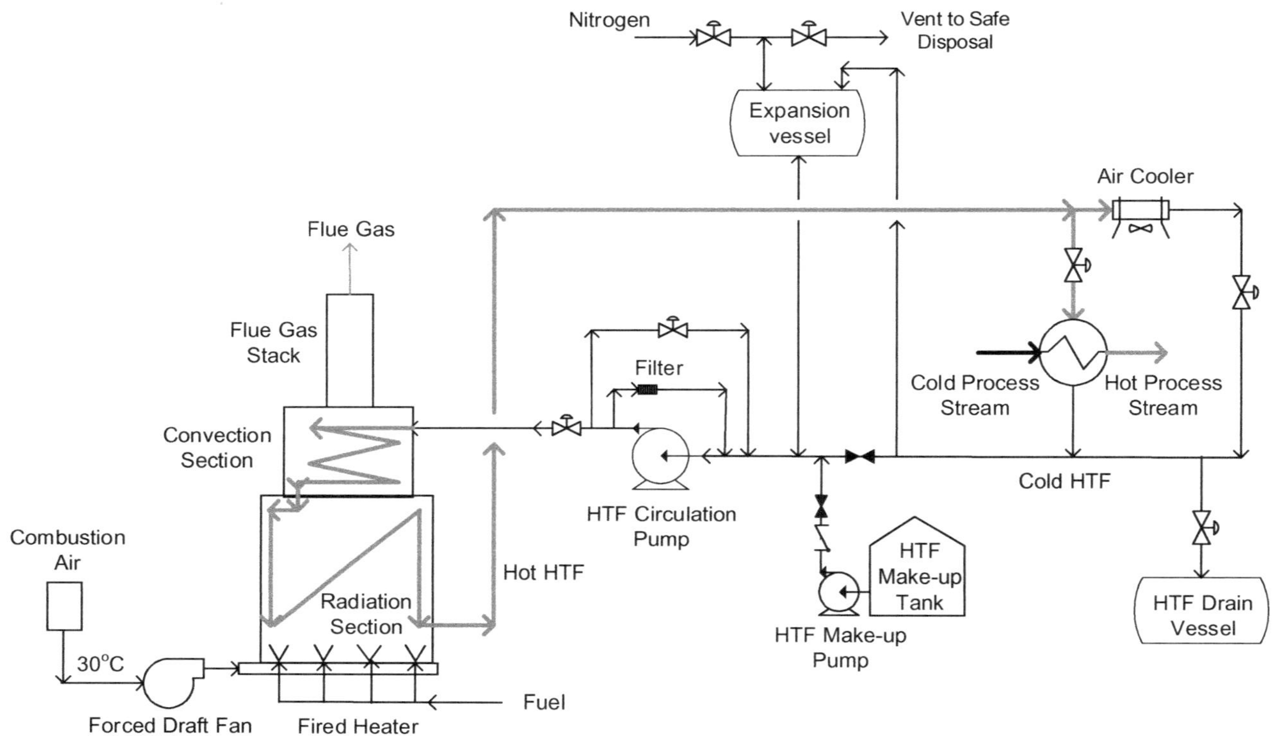

Fig. 14.4: Typical closed-loop hot oil HTF system using a fired heater; bypass air cooler is for operational flexibility for use during turndown or emptying of HTF (described in Section 14.4).

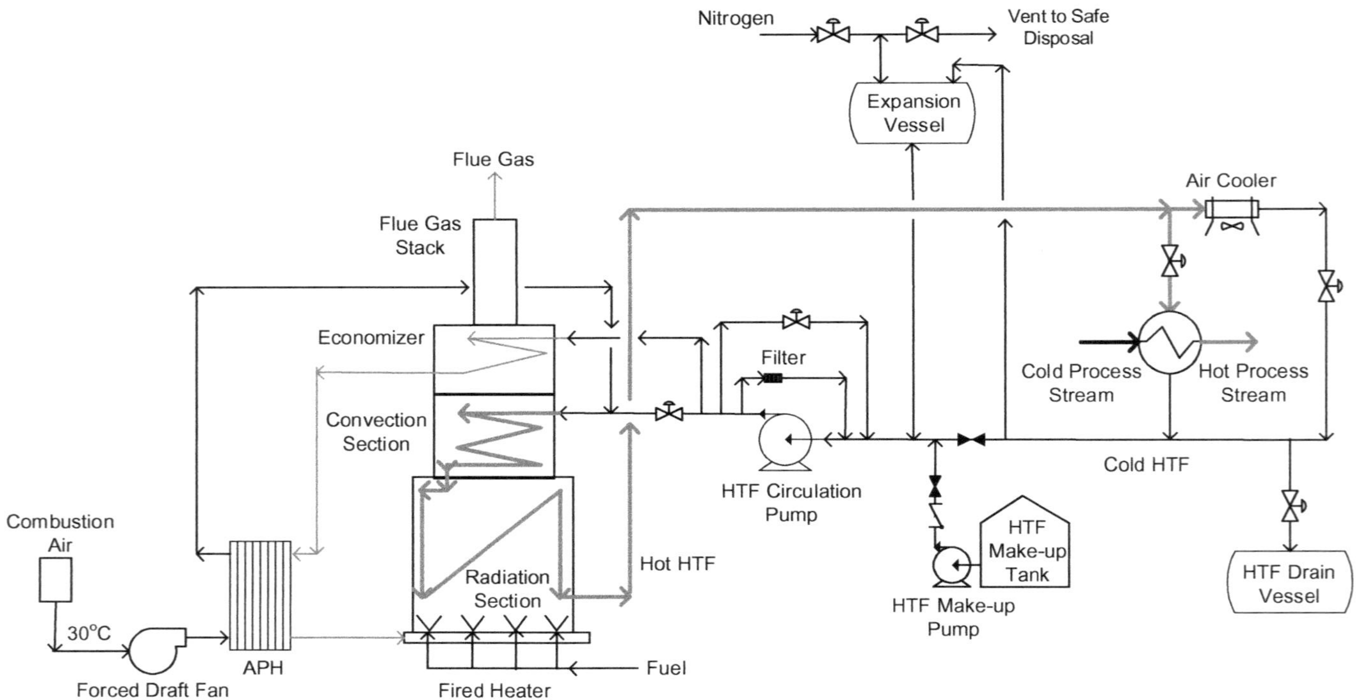

Fig. 14.5: Typical closed-loop HTF system using a fired heater installed with economizer and APH (heated by HTF).

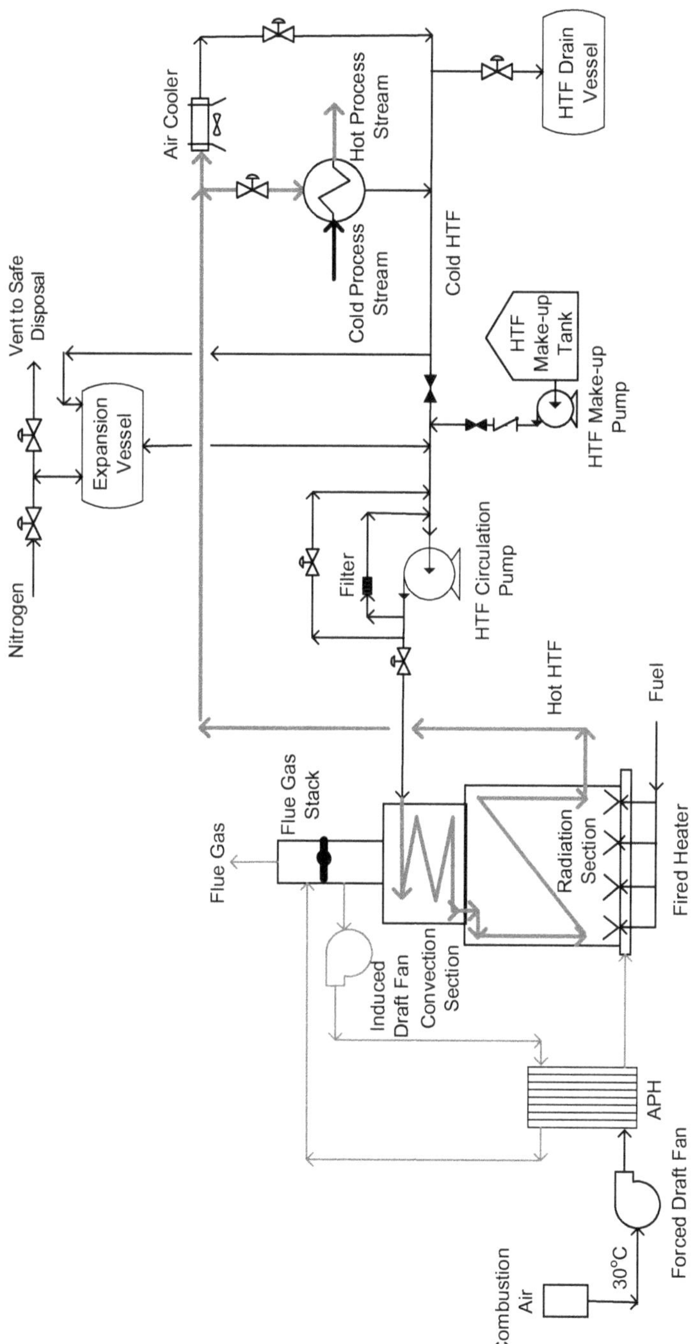

Fig. 14.6: Typical closed-loop HTF system using a fired heater and APH (for heat transfer directly from FG to combustion air).

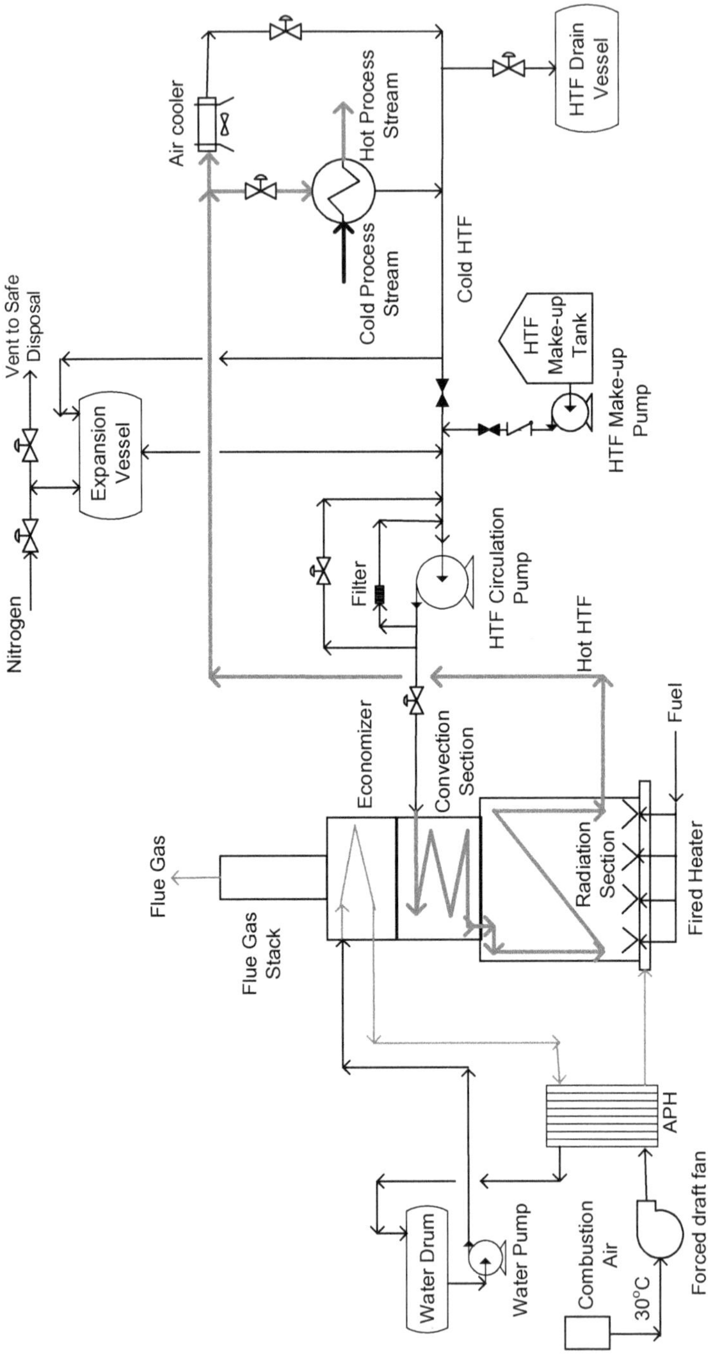

Fig. 14.7: Typical closed-loop HTF system using a fired heater installed with economizer and APH (heated by pressurized water).

WHR from FG to cold combustion air. As shown in Fig. 14.5, a portion of the high pressure HTF, withdrawn from the inlet of the HTF feed flow control valve, enters the economizer for extracting WH of FGH and subsequently transfers the recovered heat to cold combustion air in the APH. Cold HTF from APH is finally returned to the low-pressure side or downstream of HTF feed control valve, to mix with HTF feed. Then, HTF enters the convection section and subsequently the radiation section for the required heating. Balanced draft arrangement, shown in Fig. 14.6, uses an induced draft fan, after closing FG stack damper; the fan extracts FG and sends it to APH for heating cold combustion air. Cold FG from the APH outlet returns to the FG stack (at the outlet portion of the FG damper).

FG WHR system using pressurized water is shown in Fig. 14.7; it uses a water drum (to collect and top up water) to supply water to a pump, which increases the water pressure to the required pressure (i.e., pressure sufficient to suppress steam generation in any part of the water loop) and sends it to the economizer. Water recovers part of WH from FG and later transfers it to the cold combustion air in the APH. Cold water from the APH outlet returns to the water drum.

A detailed discussion of economizers and APH is presented in Chapter 11.

Example 14.1: A fired heater is used to heat 189,080 kg/hr of HTF (STYLTHERM™) from 50°C to 250°C, to supply a heat load of 21.406 MW for process heating. The thermal efficiency of the HTF/fired heater is 85%; efficiency loss through the FG leaving at 308.8°C is 13%, and other heat losses are 2% (total of 15%). The material and energy balances are shown in Fig. 14.8. Estimate the improvement in thermal efficiency of the fired heater by cooling the FG to 150°C, for the following cases.

(A) Use of HTF economizer and APH (illustrated in Fig. 14.5).

(B) Use of balanced draft arrangement for combustion air pre-heat (illustrated in Fig. 14.6). Balanced draft arrangement is explained in Chapter 11.

(C) Use of pressurized water for FG WHR (illustrated in Fig. 14.7).

Average specific heat (heat capacity) of HTF, FG and combustion air are 2.0378 kJ/kg.K, 1.155 kJ/kg.K and 1.015 kJ/kg.K, respectively.

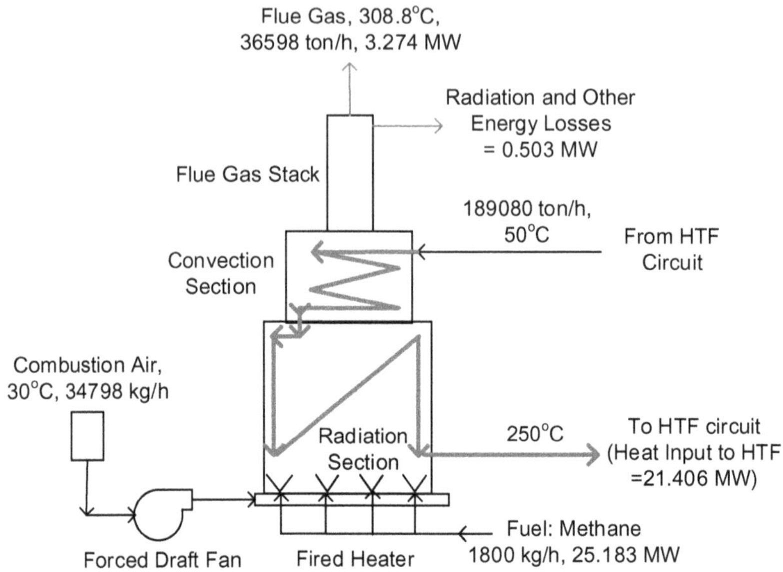

Fig. 14.8: Material and energy balances for Example 14.1.

Assume minimum approach temperature for liquid–gas and gas–gas heat transfer are 20°C and 50°C, respectively.

Solution: The following solution uses many quantities given in Fig. 14.8 as well.

Case A: Use of HTF Economizer and APH

Heat recovered by HTF in economizer = M_{FG} × average specific heat of FG × (FG temperature at convection section outlet – target FG temperature) = 36,598 × 1.155 × (308.8 – 150) = 6,713,385 kJ/hr = $\frac{6713385}{1000 \times 3600}$ = 1.8648 MW

HTF inlet and target outlet temperature at the outlet of economizer = 50°C and 250°C, respectively. Note that this HTF is not that portion going through the convection and radiation zones of the fired heater (Fig. 14.5).

Amount of HTF flow required to flow through the economizer = $\frac{6713385}{2.0378 \times (250-50)} = 16,472.14\,\text{kg/hr}$

Amount of heat that can be transferred to cold combustion air = 6,713,385 kJ/hr

Using energy balance across the APH:

6713385 = 34798 × (Combustion air temperature at the APH outlet – 30) × Specific heat of air

Solution of this equation gives combustion air temperature at APH outlet $= \frac{6713385}{34798 \times 1.015} + 30 = 220.07°C$

With flue gas waste heat recovery (FGWHR) using HTF economizer, fired heater efficiency $= \frac{21.406 + 1.8648}{25.183} \times 100 = 92.4\%$. Hence, there is an increase of 7.4% in the thermal efficiency by using HTF economizer to recover WH from FG.

Economizer can be installed directly on top of the convection section or at any location above the fired heater. Due to FG to HTF and HTF to air heat transfer, overall heat transfer coefficient (~80 W/m^{2}K) is higher compared to that for FG to air heat transfer (~40 W/m^{2}K), used in balanced draft arrangement. Hence, economizer and APH sizes are relatively compact. The only disadvantage for this case is that it requires two HEs (i.e., economizer and APH).

Case B: Use of balanced draft arrangement for FGWHR to cold combustion air

This option requires a new induced draft fan to extract FG from the stack and send it to APH (Fig. 14.6). After exchanging heat at APH, FG returns to the stack.

From Case A, heat recoverable from FG = 6,713,385 kJ/hr = $\frac{6,713,385}{1000 \times 3600} = 1.8648\,\text{MW}$

Combustion air temperature at the outlet of APH $= \frac{6,713,385}{34,798 \times 1.015} + 30 =$ 220.07 °C

As in Case A, thermal efficiency of the fired heater is 92.4%. As the FG to air overall heat transfer coefficient is very low (~40 W/m^{2}K), APH size is relatively bigger. Also, it requires induced draft fan and very large diameter FG ducting, installed with refractory lining. However, it does not require the economizer as required for case A.

Case C: Use of pressurized water for FG WHR

From Case A, heat recoverable from FG = 6,713,385 kJ/hr $= \frac{6,713,385}{1000 \times 3600} =$ 1.8648 MW

Air temperature at the outlet of APH = 220.07°C

Approach temperature for heat exchange between air and water = 20°C

So, water temperature at the inlet and outlet of APH are 220.07 + 20 = 240.07°C and 30 + 20 = 50°C, respectively.

From steam tables, average specific heat of water between 50°C and 240.07°C = 4.474 kJ/kgK and saturation pressure of water at 240.07°C = 33.5 bar.

Rearranging heat balance on water side across the APH, water flow rate required in water loop $= \frac{6{,}713{,}385}{4.474 \times (240.07 - 50)} = 7{,}895.3\,\text{kg/hr}$.

Assume pressure drop in piping, control valves and water coil in the economizer section is 10 bar.

So, required operating pressure at the water pump discharge = 33.5 + 10 = 43.5 bar.

Case C may not be economical because of the requirement of two expensive HEs and piping (due to high operating temperature and design pressure), water storage drum and water pumps (one in operation and another on standby mode). Water-loop heat transfer is generally attractive for low-temperature heat transfer with WH source temperatures <150°C. This can be observed from the low slope of water saturation curve (in Fig. 14.3), where saturation pressure of water rises with temperature linearly but slowly.

Can steam be generated from FGWHR and then use it for combustion air preheating (as shown in Fig. 14.9)? The answer is yes. For combustion air temperature of 220.07°C (as in the earlier example), steam pressure will be at least 35.5 bar (= saturation pressure of 33.5 bar at 240.07°C + 2 bar allowance for pressure drop in steam piping and APH). This solution requires high-pressure steam drum, economizer and APH (Fig. 14.9), and hence it may not be economical.

If the plant has sufficient demand for LPS or MPS or intermediate pressure steam (i.e. between HPS and MPS pressures), steam can be generated at the required pressure level and sent to the equipment through the respective steam header of the plant. When LPS or MPS is generated, it is beneficial to send it directly to plant equipment, instead of its use in APH. This maximizes the FGWHR. This option can be attractive as design involves only a steam drum and economizer. Existing plant LPS/MPS header, if available with sufficient spare capacity, can be used for sending LPS/MPS to its consumers in the plant. LPS generation solution is shown in Fig. 14.10. Assuming BFW temperature of 121°C, 3.17 ton/hr of the LPS can be generated from FGWHR, by cooling FG to 150°C.

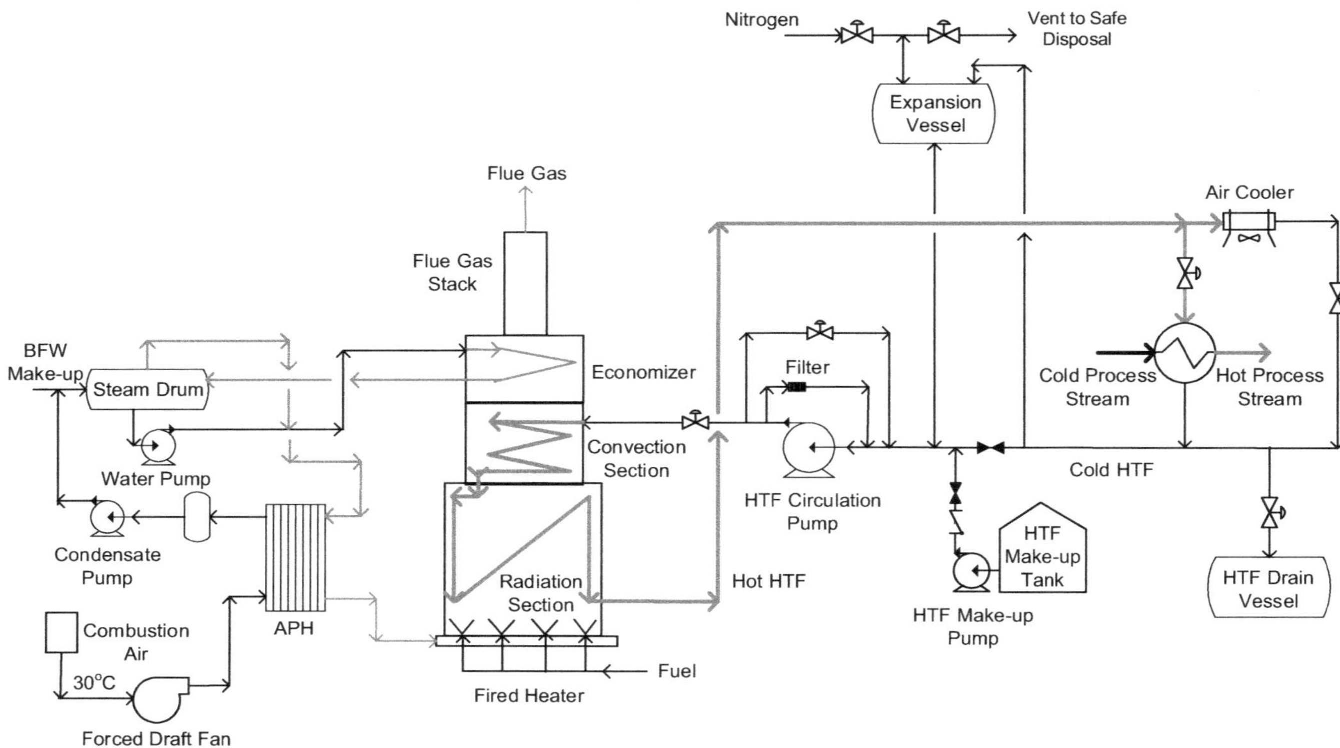

Fig. 14.9: Typical closed-loop HTF system using a fired heater installed with economizer and APH (heated by high-pressure steam).

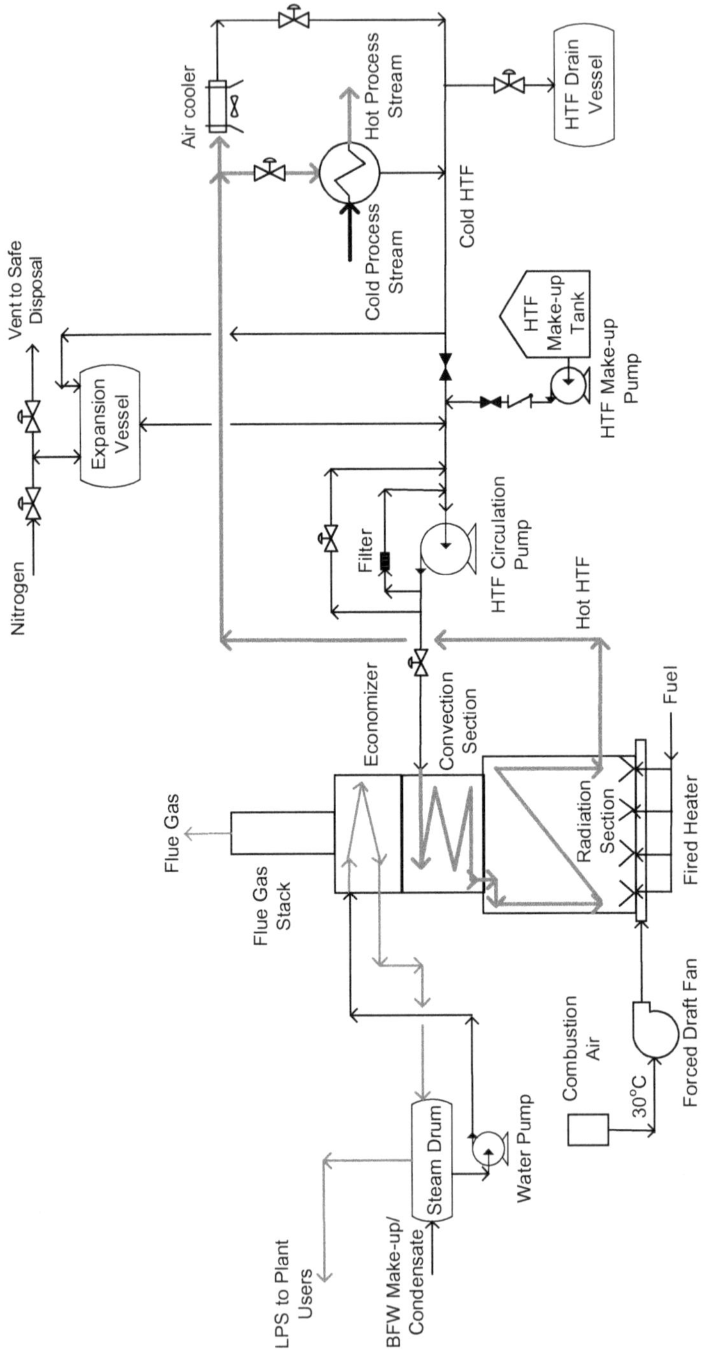

Fig. 14.10: Typical closed-loop HTF system using a fired heater installed with economizer and LPS export for use in the plant itself.

HTF system uses an expansion vessel, which is very useful for the following.

- Providing sufficient net positive suction head for HTF pumps under all operating conditions.
- Serves as a reservoir for the thermal expansion of the HTF when it heats from minimum to maximum operating temperature.
- Venting any non-condensable components, residual air or water vapour during start-up.
- Ensure closed-loop circuit is filled with HTF fluid during start-up after shutdown for equipment maintenance.
- Providing nitrogen blanket for preventing oxidation of the HTF.

The HTF expansion vessel is installed at a high elevation (Figs. 14.5–14.7, 14.9 and 14.10), above all the equipment in the closed-loop, including fired heater/WHRHE and process HEs. It should be sized by careful review of the entire volume of the HTF system (including HEs, piping and fired heater's heating coils) and allowing at least 30% expansion.[2] The vessel should be sized to be approximately 25% full at ambient temperature and 75% full at hot operating temperature. HTF is pumped from this vessel by circulation pumps, to the fired heater/WHRHE for heating. 1%–2% of the HTF flow is generally passed through a side stream filter to remove any solids.[3] Heated HTF flows through the process HEs for providing process heating. Cold HTF returns to the expansion vessel again. A bypass line across the process HEs with air cooler and control valve is used to partially or fully bypass the process HEs, if they are required to operate under low turndown operation or undergo maintenance.

Example 14. 2: Size an HTF (SYLTHERM™[e]) expansion vessel, using the following data. HTF flows inside the tubes of the process HEs. Piping volume based on piping isometric drawings = 30 m^3. Fired heater coils and HEs tube volume = 8 m^3. Density of HTF at ambient (30°C) and maximum operating temperature (250°C) = 844 kg/m^3 and 619 kg/m^3, respectively. Assume a safety factor of 10% for HTF system volume.

[e] SYLTHERM™ XLT Heat Transfer Fluid | Dow Inc.

Solution: Total volume of HTF piping and equipment (including HEs and fired heater/WHRHE tubes) = 30 + 8 = 38 m^3

Total volume with safety factor = 38 × 1.1 = 41.8 m^3

Volume expansion when HTF temperature changes from ambient to maximum operating temperature = Total system volume × $\frac{\text{Density of HTF at ambient temperature}}{\text{Density of HTF at maximum operating temperature}} = 41.8 \times \frac{844}{619} = 56.99\,\text{m}^3$

So, volume expansion = 56.99 − 41.8 = 15.194 m^3

Using an online calculator,[f] a horizontal vessel of 2 m dimeter and cylindrical length of 7 m, with 2:1 elliptical heads is sufficient to provide 15.2 m^3 between 25% and 75% levels.

The vapour space of the expansion vessel is filled with nitrogen to prevent moisture and/or air entry (especially when the system is cooled during shutdown), which can lead to oxidation of HTF, corrosion and safety hazard such as explosion. A nitrogen pressure of 200 to 300 mm of water (1.96 to 2.94 kPa) is generally sufficient for blanketing purposes. If HTF is required to operate near its boiling point, a positive pressure of at least 2 bar above the vapour pressure is maintained in the expansion vessel. This minimizes vapourization of HTF and reduces the oxidation of HTF. In general, HTF design pressure should be at least 1 bar above the highest anticipated fluid vapour pressure.[2]

An HTF make-up/hold tank (with a capacity sufficient to contain all the oil in the system) and HTF drain tank are generally provided for top-up and draining of HTF, respectively (Figs. 14.5–14.7). HTF make-up pump's discharge generally connects to the suction piping of HTF circulation pumps. Any HTF drained from the closed heat transfer loop is diverted to the drain tank, using a drainpipe, cooled by an air cooler (Figs. 14.5–14.7). An HTF velocity of 1.2 to 3 m/s is maintained in the fired heater coil to avoid excessive film temperature on the heater tubes. The average and maximum heat flux on the fired heater coil surface should not exceed 17.35 kW/m^2 and 23.66 kW/m^2, respectively.[4]

Figure 14.11 shows the integration of a process WHRHE into HTF loop. Recovered WH from the process is transferred to cold HTF, and hence it will save fuel in the fired heater. If sufficient WH is available,

[f] Vessel Volume & Level Calculation (checalc.com), accessed on 20 March 2021.

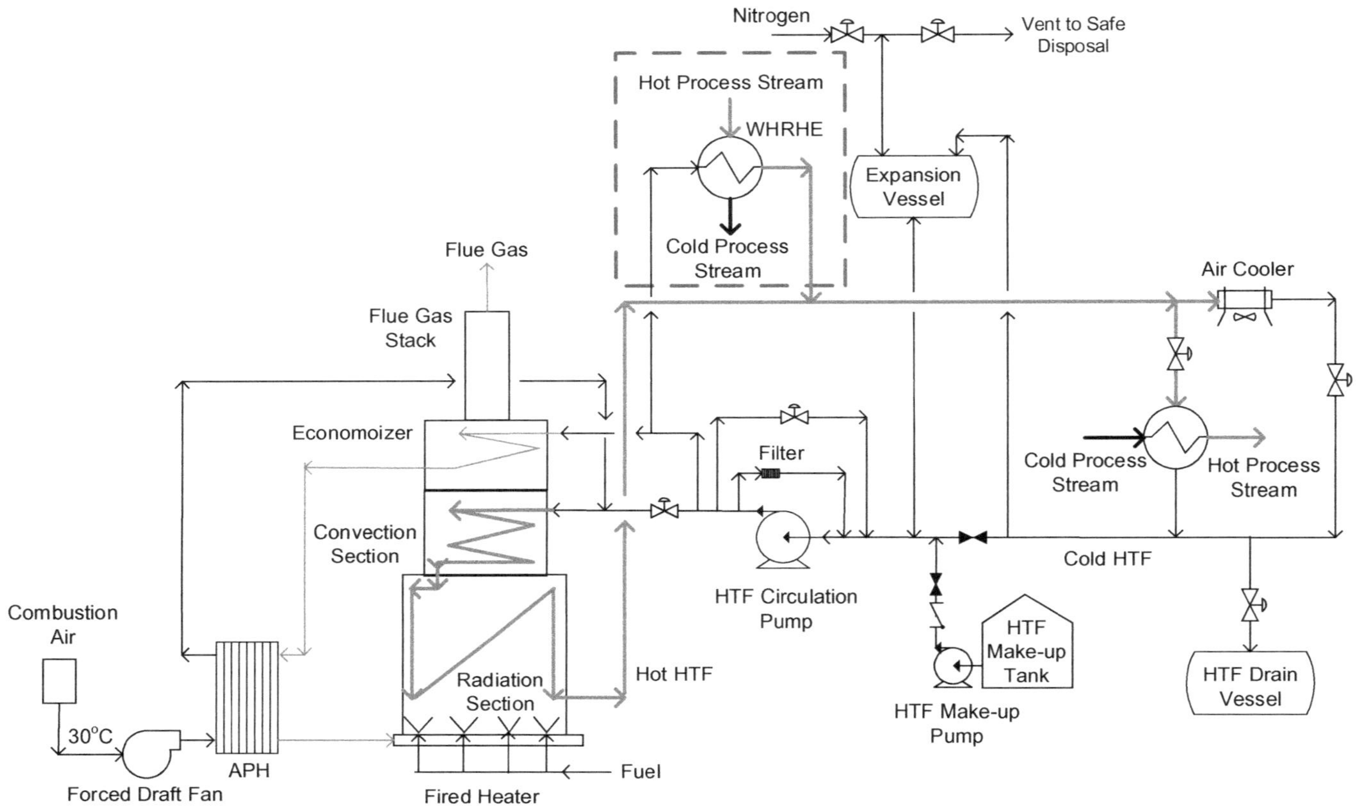

Fig. 14.11: Typical closed-loop HTF system using a fired heater, APH and process WHR (in dashed rectangle at the top).

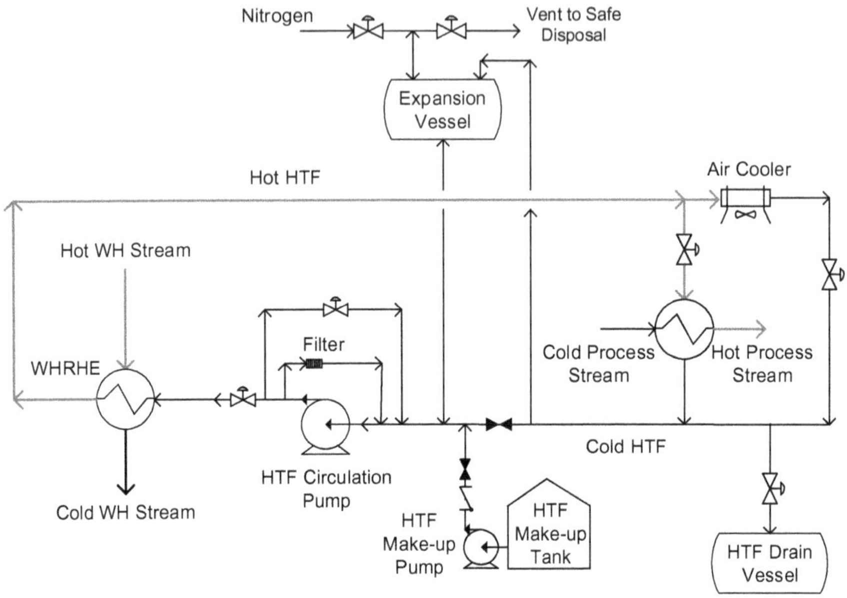

Fig. 14.12: Typical closed-loop HTF system using process WHR; all the heat required for the HTF loop is provided by WHRHE.

fired heater may be totally replaced with a WHRHE (as shown in Fig. 14.12). Efficiency of such an HTF loop can be ~98% (assuming 2% heat losses through insulation in the heat transfer loop).

14.5 Low-Temperature Heat Exchanger Network Using Pressurized Water

Water is a very attractive HTF for temperatures of 150°C and lower. Fig. 14.13 depicts a superstructure for LTHEN using water. HEs are used to exchange process heat to or from circulating pressurized water in heat source or sink sections, respectively. Utility heaters (heated by steam) and coolers (cooled by cooling water) are used, if required, for heating or cooling of heat sink (cold) or source (hot) streams, whose targets are not fulfilled by the LTHEN, respectively. Utility heater and/or cooler may also be used for heating and/or cooling the circulating water as per the heat balance. These utility HEs can be useful if some HEs with heat source or sink streams require

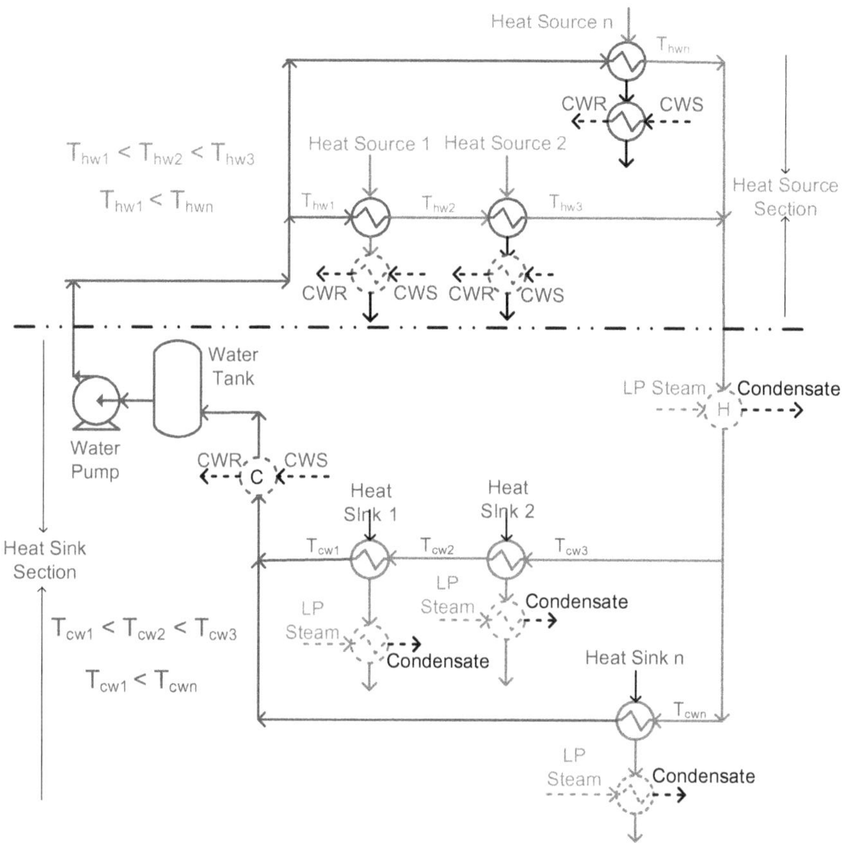

Fig. 14.13: Typical superstructure of LTHEN using the water loop.

maintenance or shutdown. The HEs in heat source and heat sink sections may be arranged in series and/or parallel arrangement. Series arrangement generally requires more HEs, mainly utility heaters and coolers, compared to parallel or series-parallel arrangements, due to restriction on minimum temperature approach between the water and heat source/sink streams.

14.5.1 *A Case Study*

Heat source and heat sink data in a chemical plant are given in Tables 14.2 and 14.3, where heat capacity flow rate is the product of mass flow rate and specific heat capacity. Fig. 14.14 shows heat source and sink

Table 14.2: Heat source (Hot) stream data

Heat Source (Hot) Streams	Heat Available, kW	T_{in}, °C	T_{out}, °C	Heat Capacity Flow Rate, kW/°C
H1	1495	175	60	13
H2	5225	150	55	55
Total cooling duty	6720			

Table 14.3: Heat sink (Cold) stream data

Heat Sink (Cold) Streams	Heat Available, kW	T_{in}, °C	T_{out}, °C	Heat Capacity Flow Rate, kW/°C
C1	3600	30	75	80
C2	2550	35	120	30
Total heating duty	6150			

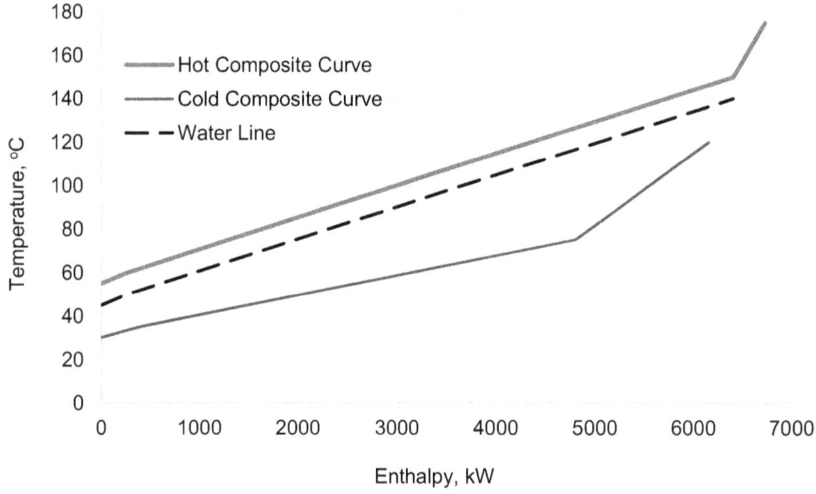

Fig. 14.14: Heat source and sink composite curves.

composite curves (described in Chapter 3). In this figure, water line is drawn with a minimum approach temperature of 10°C from the hot composite curve. Although not optimal, acceptable heat capacity water flow rate can be obtained by taking the reciprocal of slope of the water line (1/0.014 = 67.57 kW/°C). This is good enough for developing quick/

approximate design of LTHEN without optimization that requires more effort.

Example 14.3: Design an LTHEN for the data shown in Tables 14.2 and 14.3 using series and parallel arrangement of HEs in heat source and sink sections. Assume a 10°C approach temperature between water and process streams.

Solution: Let us begin working on the water loop, with the lowest temperature of circulating cooling water = lowest heat sink inlet temperature + approach temperature = 30 + 10 = 40°C.

From Fig. 14.13, heat capacity flow rate of circulating water = 67.57 kW/°C

In series solution, pressurized water flows in series to exchange heat with heat sources and sinks. This arrangement and solution are shown in Fig. 14.15.

So, pressurized water (typically at 10 bar) can be sent to remove heat from heat source stream (H1). The main reason for selecting H1 first, instead of H2, is due to low amount of heat that can be extracted from H1.

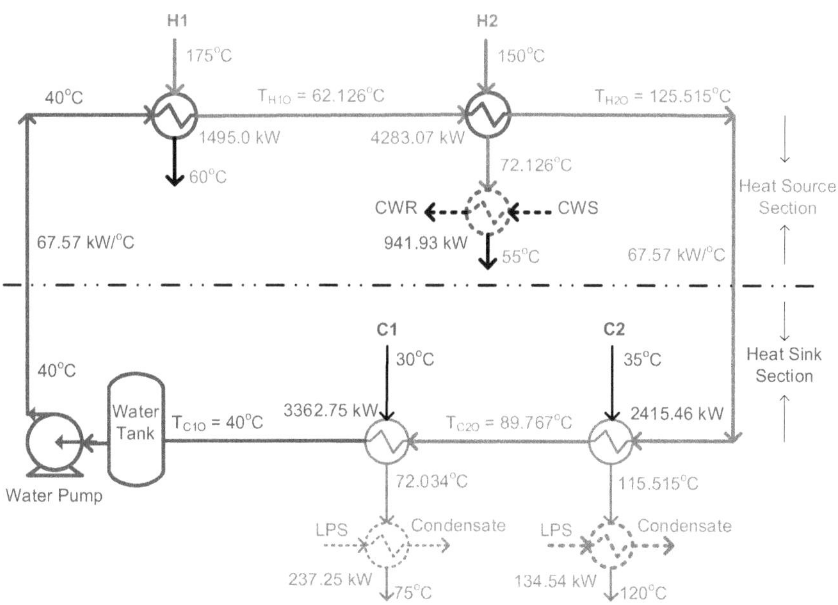

Fig. 14.15: LTHEN solution with series arrangement of HEs.

This results in a smaller increase in the water temperature and hence helps to maximize the heat recovery from next heat source stream (i.e., H2).

Heat balance across hot stream *H*1 and water HE is given by

$$67.57 \times (T_{\mathrm{H1O}} - 40) = 1495$$

Here, T_{H1O} is the outlet temperature of water at H1-water HE. Solving the above equation gives $T_{\mathrm{H1O}} = 62.126$°C.

Using the approach temperature of 10°C, minimum outlet temperature that can be allowed for hot stream H2 = 62.126 + 10 = 72.126°C. Hence, amount of heat that can be transferred from H2 to water stream = Heat capacity flow rate of H2 × (process side inlet temperature of H2 – 72.26) = 55 × (150−72.156) = 4283.07 kW.

Heat balance across H2-water HE is given by

$$67.57 \times (T_{\mathrm{H2O}} - 62.126) = 4283.07$$

Here, T_{H2O} is the outlet temperature of water at H2-water HE. Solving the above equation gives $T_{\mathrm{H2O}} = 125.515$°.

As the H2 stream cannot achieve the target cooling temperature of 55°C, a utility cooler is required to cool H2 from 72.126°C to 55°C. Cooling duty of this utility cooler for H2 is

$$55 \times (72.126 - 55) = 941.93 \text{ kW}$$

This cooling duty can be provided using cooling water.

Now water leaving the H2-water HE at 125.515°C enters the heat sink section. Heat sink stream requiring minimum heat duty (C2) is selected first. This will maximize the temperature driving force for heat exchange for the next heat sink stream (C1).

Using the approach temperature of 10°C, maximum outlet temperature that can be allowed for cold stream C2 = 125.515 − 10 = 115.515°C. Hence, amount of heat that can be transferred from hot water to C2 stream = Heat capacity flow rate of C2 × (115.515 – process side inlet temperature of C2) = 30 × (115.515−35) = 2415.46 kW.

Hence, a steam heater is required to heat stream C2 from 115.515°C to 120°C. Heating duty required for this = 2550 − 2415.46 = 134.54 kW.

Heat balance across cold stream C2 and water HE is given by

$$67.57 \times (125.515 - T_{C2O}) = 2415.46$$

Here, T_{C2O} is the outlet temperature of water at C2-water HE. Solving the above equation gives $T_{C2O} = 89.767$°C.

As T_{C1O} target temperature is 40°C, the maximum amount of heat that can be transferred from water loop to C1 stream = 67.57 × (89.767 − 40) = 3362.75 kW.

C1 stream temperature at the outlet of C1-water HE $= \frac{3362.75}{80} + 30 =$ 72.034°C. Hence, a process heater is required to heat stream C1 from 72.034°C to 75°C, and its heating duty is 3600 − 3362.75 = 237.25 kW.

Due to LTHEN, total steam heating load for streams C1 and C2 is 371.79 kW (= 134.54 + 237.25 kW). Hence, this LTHEN saves 5778.21 kW (or $\frac{(6150 - 371.79) \times 100}{6150} = 93.95\%$) of the steam duty for heating sink (cold) streams in Table 14.3. Total cooling duty with LTHEN is 941.93 kW, Hence, this LTHEN also reduces the cooling duty for heat source (hot) streams in Table 14.2 by 5778.07 (=6720 − 941.93) kW or $\frac{(6720 - 941.93) \times 100}{6720} =$ 85.98%.

In summary, the LTHEN solution in Fig. 14.15 achieves good heat integration between the heat sources and sinks using the water loop. The only disadvantage of series arrangement is that it generally requires more utility coolers and heaters (in this case, one utility cooler and two utility heaters), compared to parallel arrangement, discussed later. The number of heaters for the solution in Fig. 14.15 can be reduced to one if a steam heater is placed on water stream before entering the heat sink section. This is given as Exercise 14.5 at the end of this chapter.

If HEs are present in different sections/units, they may require different shutdown periods. Hence, bypass pipes with isolation valves should be installed across the heat source/sink and water HEs. These bypass valves (not shown in Figure 14.15) should be normally kept closed, and they are open during any HE maintenance and/or shutdown.

Next, consider the parallel arrangement, shown in Fig. 14.16. In this case, water flows in parallel arrangement across HEs in each of heat

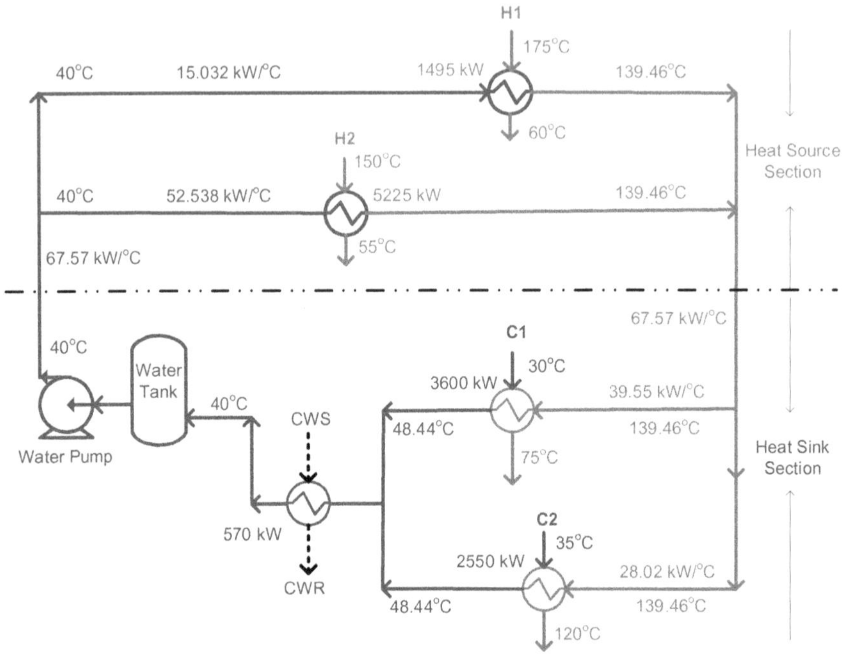

Fig. 14.16: LTHEN solution with parallel arrangement of HEs.

source and sink sections. Again, design starts at water temperature of 40°C in the heat source section. Since this arrangement requires heat source streams (H1 and H2) to heat circulating water in parallel arrangement, water flow rate needs to be split across the (two) HEs. It can be done by splitting water flowrate proportional to the heat that should be transferred from H1 and H2 streams.

Heat capacity flow rate of water through HE for cooling H1 stream = $\frac{1495}{6720} \times 67.57 = 15.032\,\text{kW/°C}$

Heat capacity flow rate of water through HE for cooling H2 stream = $67.57 - 15.032 = 52.538$ kW/°C

Outlet temperature of water at the HE for cooling H1 stream = $\frac{1495}{15.032} + 40 = 139.46°\text{C}$

Similarly, outlet temperature of water at the HE for cooling H2 stream = $\frac{5225}{52.538} + 40 = 139.46°\text{C}$

In this case, utility coolers are not required to cool H1 or H2 stream.

Next, water at 139.46°C enters the heat sink section, where water flow rate splits according to the heating duty required for C1 and C2 streams.

Heat capacity flow rate of water through HE for heating C1 stream = $\frac{3600}{6150} \times 67.57 = 39.553\,\text{kW/°C}$

Heat capacity flow rate of water through HE for heating C2 stream = $67.57 - 39.553 = 28.017$ kW/°C

Outlet temperature of water at the HE for heating C1 stream = $139.46 - \frac{3600}{39.553} = 48.44°\text{C}$

Similarly, outlet temperature of water at the HE for heating C2 stream = $139.46 - \frac{2550}{28.017} = 48.44°\text{C}$

As the water temperature after supplying full heating duty for C1 and C2 streams is greater than 40°C, one utility cooler is required. Duty of this cooler required to cool the water stream from 48.44°C to 40°C = 67.57 × (48.44 – 40) = 570 kW.

Compared to the series arrangement requiring one utility cooler and two utility heaters, parallel arrangement requires only one utility cooler. Hence, it requires lower capital cost as the number of HEs required is lower (five versus seven). Like for series arrangement, bypasses with isolation valves can be installed to enable onstream maintenance/shutdown work without affecting LTHEN loop. For providing more operational heat duty assurance when some of the HEs are taken out for maintenance, utility heater (with steam) and cooler (with cooling water) can be provided in the water circulation loop. These utility HEs (not shown in both Figs. 14.15 and 14.16) compensate the heating or cooling duties when some HEs in heat source or sink section are taken out for maintenance/shutdown.

As illustrated in Example 14.3, LTHEN can be designed easily for one or two process streams in each of heat source and sink sections. As the number of process streams increases beyond two in heat source and/or sink section, it would be difficult to arrive at the optimal LTHEN by manual calculations as it requires many iterations, substantial effort and time. In such cases, HEs in heat source and sink sections can be arranged in series or parallel or series-parallel configurations. Optimization software such as GAMS (General Algebraic Modeling System) or Frontline Analytic solver are required to solve mixed-integer nonlinear programming (MINLP) model.

Two MINLP models for developing optimal LTHEN network involving many heat sources and sinks are presented by Moorthy *et al.*[5] The first model recovers all the WH from heat source streams and transfers it to the heat sink streams. The second model recovers all the WH from heat source streams and transfers it to the heat sink streams only if it is economical to do so. This model uses utility coolers if it is not economical to recover heat from hot streams, and utility heaters if it is not economical to supply recovered heat to cold sink streams. These two models also consider both capital and operating costs for all HEs and piping, for developing the optimal solution. A case study involving five process streams in each of heat source and sink sections is available in Moorthy *et al.*[5]

14.6 Use of Pressurized Water and Steam Generation for Heat Integration

Pressurized water is frequently used as HTF in process plants, for recovering reaction heat from reactors. One example of reaction heat recovery scheme from a very high-pressure tubular polymer reactor is shown in Fig. 14.17. This arrangement uses pressurized water (at 30 bar and 160°C) in the jacket of the tubular reactor. In the jacket, water temperature increases to 200°C, which is sent to the tube side of a kettle-type HE. Boiler feed water is sent to the shell side of this HE, which generates LPS due to heat gain from hot water on the tube side. LPS generated is sent to other equipment in the plant that require a heat source.

Example 14.4: Estimate the maximum amount of saturated LPS at 4.5 bar that can be generated using the details given in Fig. 14.17.

Solution: Mass flow rate of the water at 30 bar pressure = 200,000 kg/hr

From steam tables, specific heat of water at 160°C and 200°C are 4.329 kJ/kg.K and 4.486 kJ/kg.K, respectively, and average of these two values is $\frac{(4.329+4.486)}{2}$ = 4.408 kJ/kg.K

Heat transferred by hot water at the LPS generator = 200,000 × 4.408 × (200 – 160) =35,264,000 kJ/hr

From steam tables, saturation temperature of LPS at 4.5 bar pressure = 147.91°C

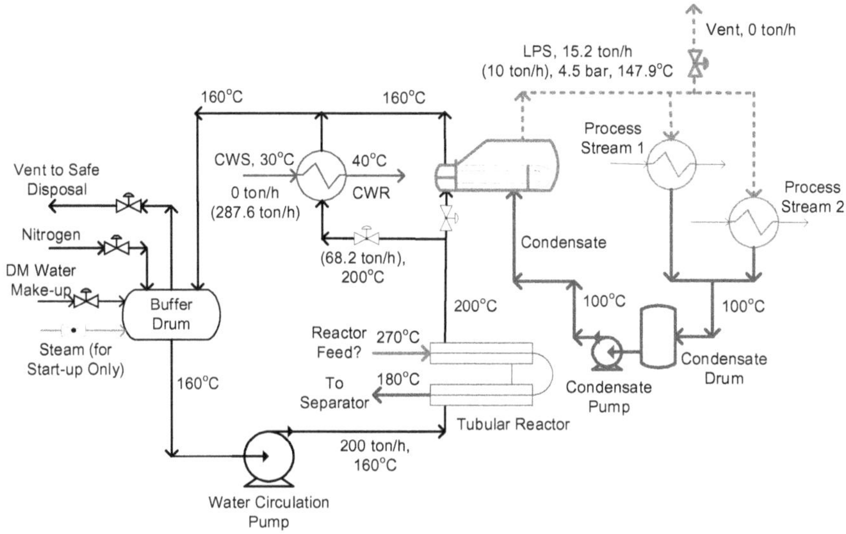

Fig. 14.17: Typical closed-loop WHR system using pressurized water as HTF and generating LPS. Bypass water cooler is for operational flexibility for use during lower LPS demand. All values without brackets are for Example 14.4, whereas those in brackets are for Example 14.5.

Steam condensate temperature to the steam generator = 100°C

Specific enthalpy of steam condensate at 100°C = 419.1 kJ/kg

Specific enthalpy of steam condensate at 147.9°C = 623.22 kJ/kg

Sensible heat change required for increasing condensate temperature from 100°C to147.9°C = 623.224 – 419.099 = 204.125 kJ/kg

Latent heat of steam at 4.5 bar = 2120.162 kJ/kg

Total heat required for LPS production from condensate at 100°C = 204.125 + 2120.162 = 2,324.287 kJ/kg

Maximum amount of LPS produced = $\frac{35{,}264{,}000}{2324.287}$ = 15,171.96 kg/hr = 15.2 ton/hr

Example 14.5: Estimate the cooling water required to cool the hot water to 160°C in the cooler shown in Fig. 14.17, if the maximum production of LPS is restricted to 10 ton/hr, due to low LPS demand in the unit/section. Cooling water supply and return temperatures are 30°C and 40°C, respectively.

Solution: From Example 14.4, the heat required to be removed from hot water = 35,264,000 kJ/hr

Amount of heat removed from hot water, for generation of 10 ton/hr of saturated LPS at 4.5 bar = 10,000 × 2324.287 = 23,242,870 kJ/hr

Here, 2324.287 is the sensible heat and latent heat change required for increasing condensate temperature from 100°C to147. 91°C and then converting condensate to LPS, as calculated in Example 14.4.

Amount of heat that needs to be transferred to cooling water = 35,264,000 – 23,242,870 = 12,021,130 kJ/hr

Specific heat of cooling water = 4.18 kJ/kg.K

Amount of cooling water required = $\frac{12{,}021{,}130}{4.18 \times (40 - 30)}$ = 287,586.84 kg/hr = 287.6 ton/hr

Amount of hot water to be diverted to the cooler = $\frac{12{,}021{,}130}{4.408 \times (200 - 160)}$ = 68177.9 kg/h = 68.2 ton/h

As the LPS demand is only 10 ton/hr, 68.2 ton/hr of the hot water from the reactor is diverted to a cooler, which cools hot water to 160°C. The cooler requires 287.6 ton/hr of cooling water. This wastes 34.1% $\left(= \frac{35{,}264{,}000 - 23{,}242{,}870}{35{,}264{,}000} \times 100\right)$ of the WH to cooling water. Instead of this wastage, one should explore potential uses of excess LPS generated (i.e., 15.2 − 10 = 5.2 ton/hr). One scenario is as follows. A neighbourhood unit/section (built at a later stage) in the same plant, imports 8 ton/hr of LPS, from an external utility company, for process heating purposes. By comprehensive review, one can find that imported LPS used can be reduced by 5.2 ton/hr if an LPS header inter-connection is made from the unit in Fig. 14.17 to this neighbourhood unit. This saves 5.2 ton/hr of LPS cost for the plant, and also 287.6 ton/hr of cooling water.

14.7 Summary

In this chapter, many closed-loop WHR opportunities, using hot oil and pressurized water as HTF, are described with examples. Main points of this chapter are as follows.

- Closed-loop WHR systems are important to increase the energy efficiency of process systems.

- Use of closed-loop hot oil HTF systems enable WHR at relatively low pressures. This option is attractive for WHR from WH sources with temperature above 150°C.
- There are many WHR opportunities for improving energy efficiency of hot oil HTF systems. These are presented in Section 14.4.
- For WH at temperatures below 150°C, use of pressurized water as HTF in closed loop, is very beneficial.
- Pinch technology is useful for establishing water circulation rate requirement in LTHEN.
- LTHEN design procedure illustrated in this chapter maximizes the inter- and intra-unit WHR; hence, it can improve energy efficiency of process plants.

Chapter 13 of the book by Hall[2] is recommended for further reading on the closed-loop heat transfer systems.

References

1. Lai SM, Wu H, Hui CW, *et al.* (2011) Flexible heat exchanger network design for low-temperature heat utilization in oil refinery. *Asia-Pac J Chem Eng* **6**: 713–733.
2. Hall SM. (2018) *Rules of Thumb for Chemical Engineers*, 6th ed. Elsevier.
3. Gamble CE. (July 2006) Cost management in heat transfer fluid systems. *Chemical Engineering Progress*, pp. 22–26.
4. Bahadori A. (2016) *Essentials of Oil and Gas Utilities: Process Design, Equipment, and Operations*. Gulf Professional Publishing.
5. Moorthy AN, Reddy CCS, Rangaiah GP. (2014) Optimization of heat exchanger networks for the utilization of low-temperature process heat. *Ind Eng Chem Res*. **53**: 17989–18004.

Acronyms and Notation

Acronyms

APH	Air Preheater
CWS	Cooling Water Supply
CWR	Cooling Water Return

FGWHR	Flue Gas Waste Heat Recovery
HE	Heat Exchanger
HTF	Heat Transfer Fluid
LPS	Low-Pressure Steam
LTWHR	Low-Temperature Waste Heat Recovery
LTHEN	Low-Temperature Heat Exchanger Network
WH	Waste Heat
WHR	Waste Heat Recovery
WHRHE	Waste Heat Recovery Heat Exchanger

Notation

M	Mass flow rate (kg/hr)
T	Temperature (°C or K)

Subscripts

C1O	Outlet of HE used for heat transfer from water to heat sink stream (C1)
C2O	Outlet of HE used for heat transfer from water to heat sink stream (C2)
H1O	Outlet of HE used for heat transfer from heat source stream (H1) to water
H2O	Outlet of HE used for heat transfer from heat source stream (H2) to water
hw	Hot water
cw	Cold water
in	Inlet
out	Outlet

Exercises

14.1 Estimate NG consumption for an HTF-fired heater operating at 85% thermal efficiency and supplying 15 MW of process heating duty. LHV of NG is 46,000 kJ/kg.

14.2 Estimate water pressure required to prevent steam generation in a closed-loop WHR system, operating with a maximum water temperature of 210°C. Assume hydraulic pressure drop of 5 bar in the system and a safety margin of 5 bar on the calculated pressure. Comment on the system operating pressure if an HTF such as SYLTHERM (density = 679.67 kg/m^3) is used (instead of pressurized water). State any other assumptions as required.

14.3 Estimate water circulation rate required to recover WH from a heat source and then transfer it to a heat sink, given in the following table. Also, estimate the heating and cooling duties that can be saved by WHR. Use an approach temperature of 10°C between water and heat source/sink streams A small heater and/or cooler may be required.

Process Streams	Heat Available, kW	T_{in}, °C	T_{out}, °C	Heat Capacity Flow Rate, kW/°C
Heat source	3250	160	35	26
Heat sink	3200	30	130	32

14.4 Repeat the calculations in the previous exercise if the approach temperature between the water and heat source/sink streams is increased to 20°C. State the effect of increasing the approach temperature from 10°C to 20°C.

14.5 Improve the solution shown in Fig. 14.15 by avoiding the steam heaters for streams C1 and C2, with use of a new steam heater on water stream, before it enters the heat sink section. Ensure that the approach temperature of 10°C is maintained for each of all heat exchangers in the network.

Chapter 15

Cooling System Options for Waste Heat Reduction

15.1 Overview

Process industries require waste heat recovery (WHR) to maximize energy efficiency and sustainability. WHR methods such as heat pumps (Chapter 5), organic Rankine cycle, Kalina cycle and thermal desalination (Chapter 13) require significant amount of cooling duties. Economic feasibility of these methods strongly depend on the cheap availability of cooling systems/equipment. Process industries use a variety of cooling systems such as once-through cooling using seawater (SW) or freshwater (FW), cooling towers (CTs), closed cooling water (CW) systems using SW and/or FW, air/fin-fan cooler (air-cooled heat exchanger, ACHE) and hybrid systems depending on the availability of water sources, geographical location, plant size and local government regulations. Installing a new cooling system for WHR often leads to significant capital cost escalation and hence makes the project economically infeasible. Hence, for WHR, cost-effective debottlenecking/retrofitting the existing cooling system is beneficial.

Optimal debottlenecking/retrofitting depends on many constraints such as space, cost, payback period, reliability, availability of FW/SW and CTs, maximum SW return temperature to water bodies. For this, a systematic review and understanding of key design features of the cooling systems are essential. Development of a robust cooling system solution is a prerequisite for sustainability of some WHR projects. Accordingly, the main purposes of this chapter are to provide a comprehensive review of various cooling systems, debottleneck/retrofit options and to present illustrative applications for selecting one of them.

This chapter begins with the detailed description and relative merits of various cooling systems employed in the process industries, in Section 15.2. The subsequent section presents SW cooling systems. Section 15.4 discusses application of FW cooling systems and an optimization strategy to reduce the CW demand. Details and application of air-cooling system and hybrid cooling systems are presented in Sections 15.5 and 15.6, respectively. Debottleneck/retrofit options for water cooling and air-cooling systems are presented in Section 15.7. This chapter ends with summary in Section 15.8.

Learning outcomes of this chapter are as follows:

1. Outline the various types of cooling systems used in the process industries.
2. Discuss the advantages and disadvantages of various cooling systems.
3. Describe the key design features and constraints of various cooling systems.
4. Identify debottlenecking/retrofitting options for various cooling systems.
5. Select the cost-effective debottlenecking/retrofitting option for a cooling system.

15.2 Process Cooling Systems

There are mainly five types of cooling systems, commonly used for meeting the required cooling in WHR equipment. Although refrigerant cooling systems are used in process industries, it is not employed in WHR projects; hence, it is not discussed in this chapter. However, one refrigerant cooling system for chilled water production is described in Chapter 7, Section 7.3.4.

1. **SW cooling system:** This cooling system is commonly used in the plants, located near the seashore. Different SW cooling systems are

presented in Section 15.3. In these systems, circulating SW removes the process heat and discards it to sea and/or air.

2. **FW cooling system:** This cooling system uses FW such as river, municipal, recycled and/or ground water. Different FW cooling systems are discussed in Section 15.4. Circulating FW removes the process heat and discards it to air and/or SW.
3. **Air cooling system:** This type of cooling system does not use any water for cooling. It uses finned tube bundles and fans to remove and discard the process heat to air. This cooling system is discussed in Section 15.5.
4. **Cooling by exchanging heat with process streams:** This type of cooling exchanges useful heat in WHR equipment with process streams. This is the best strategy, using waste heat for process purposes, before applying any additional cooling system. This is covered in Chapters 2–5, and 8–17 of this book. Hence, it is not discussed further in this chapter.
5. **Hybrid cooling system:** This type of cooling system uses a combination of some of the above four cooling techniques. Examples are the use of CW for process heat removal and subsequent use of SW to remove heat from CW, or the use of air cooling followed by CW or SW cooling.

Advantages and disadvantages of various cooling systems are summarized in Table 15.1. All the cooling systems have their own advantages

Table 15.1: Advantages and disadvantages of various cooling systems

Cooling System	Advantages	Disadvantages
SW cooling system	• Avoids use of scarce FW resources and hence operating cost for FW • Stable cooling performance • Achieves highest cooling duty, with good process design such as maintaining high velocity in plate heat exchangers (PHEs) and use of automatic backflush system (illustrated in Fig. 18.8)	• Applicable only for coastal plants • High fouling due to barnacles, if chlorine treatment is not adequate or SW velocity is <2 m/sec • Severe corrosion and hence requires expensive heat exchanger (HE) metallurgy such as titanium

(*Continued*)

Table 15.1: (*Continued*)

Cooling System	Advantages	Disadvantages
		• Pipelines require cement or polymer lining, which are difficult to maintain. Hence, they can lead to leakages and high maintenance cost. • Increase of SW temperature and water pollution due to injected chemicals, which can damage marine ecosystem
FW cooling system	• Compared to SW cooling systems, less corrosion and fouling, with proper use of adequate filtering and chemical corrosion inhibitors • Can use cheaper materials such as carbon steel (CS) • Can achieve high heat transfer by designing HEs with low approach temperatures	• Uses scarce FW and hence competes with community needs • High make-up water cost, especially in countries with high water tariffs
Air cooling system	• Does not require any water • Does not pollute water sources • Easy to revamp	• Requires significant space and hence usually installed above plant pipe racks • Can economically achieve process cooling to only 10°C–15°C above the ambient temperature • Cooling duty varies with ambient weather • Prone to air particulate fouling on external surface of finned tubes
Process cooling system	• Uses WH for process heating and hence reduces the amount of fresh fuel used • Zero air/water pollution	• Difficulty in finding suitable nearby cold process stream for heat exchange • May limit cooling duty based on process and approach temperature requirements
Hybrid CW system	• Minimizes water consumption • Provides cost-effective options for debottlenecking/revamping cooling systems	• May require additional equipment and/or space, and hence more capital cost • May require more equipment to maintain and operate

and disadvantages. One should carefully evaluate their potential for the application on hand, to establish a cost-effective option for achieving the additional cooling duty required for WHR in the plant.

15.3 Seawater Cooling Systems

There are mainly three types of SW cooling systems, commonly used in the process industries; these are shown in Figs. 15.1, 15.2 and 15.3. SW, if available, can replace FW for process cooling and hence reduce the demand for FW. The main problems associated with the use of SW in the process HEs are corrosion; scaling due to barnacles (a marine arthropod) which can foul piping, valves and HEs; carryover of sand and debris and microbiological growth.[1] Hypo or other types of chlorine injection systems are used in SW systems, for barnacles control; they are located in the SW intake station.

To prevent debris, a stationary bar screen and travelling screen(s) are used at SW intake station, as shown in Fig. 15.4. A stop gate is also provided, which facilitates the maintenance work in SW intake station. To prevent sand and other solid fines carryover, SW filter(s) with automatic backflush system is provided at the discharge of the SW pumps. For brevity, the details in Fig. 15.4 are not shown in figures (such as Figs. 15.1, 15.2 and 15.3) for SW and hybrid cooling systems of this chapter.

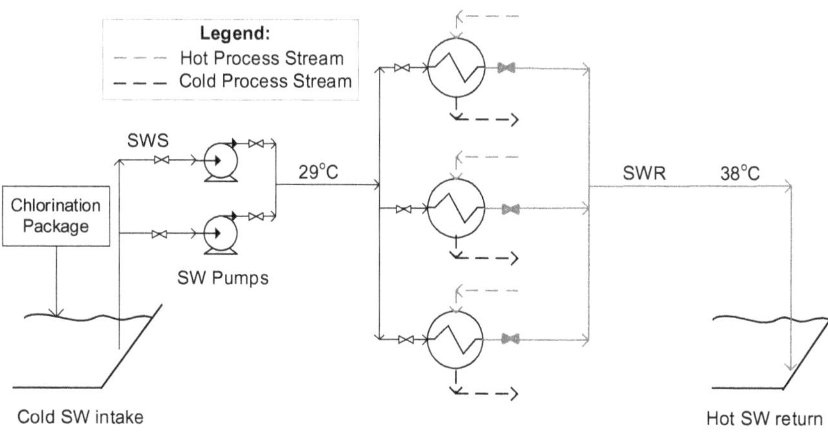

Fig. 15.1: Typical once-through SW cooling system; as shown, coolers are generally in parallel.

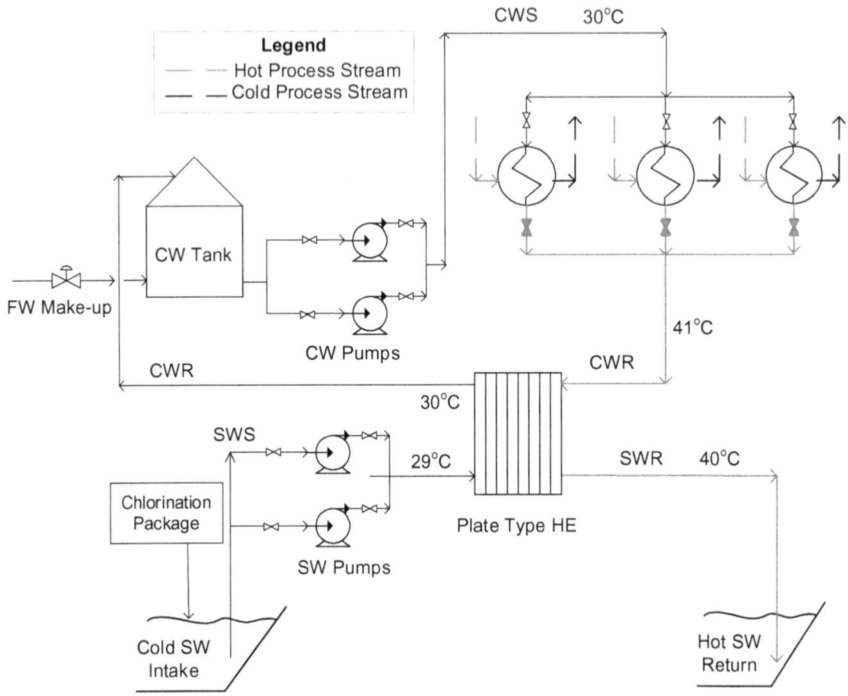

Fig. 15.2: SW (once-through)-FW (closed-loop) cooling system.

Once-through SW cooling system (Fig. 15.1) is the simplest SW cooling system. It uses maximum SW for a given cooling duty as SW is discarded after one-time use in an HE. Typical once-through SW cooling system consists of a SW pumping basin, chlorine injection package, circulating/rotating screens (for debris removal), SW filters, SW pumps, SW piping network and SW return station, as shown in Figs. 15.1 and 15.4. They are mainly used for cooling of process stream(s) to a temperature very close (i.e., 1°C–2°C approach) to the ambient temperature. Seawater supply (SWS) temperature is the SW temperature. After use in process HEs, seawater return (SWR) is sent to a separate sea basin, far away from SW intake.

SW can severely corrode CS piping. Hence cement-/epoxy-/polymer-lined CS piping is generally used. Alternatively, polymer piping can be used. For easy cleaning, SW is always used in the tube side of shell and tube HEs. Often, titanium tubes are used for minimizing corrosion to achieve long sustained operation. PHEs use titanium plates. If process

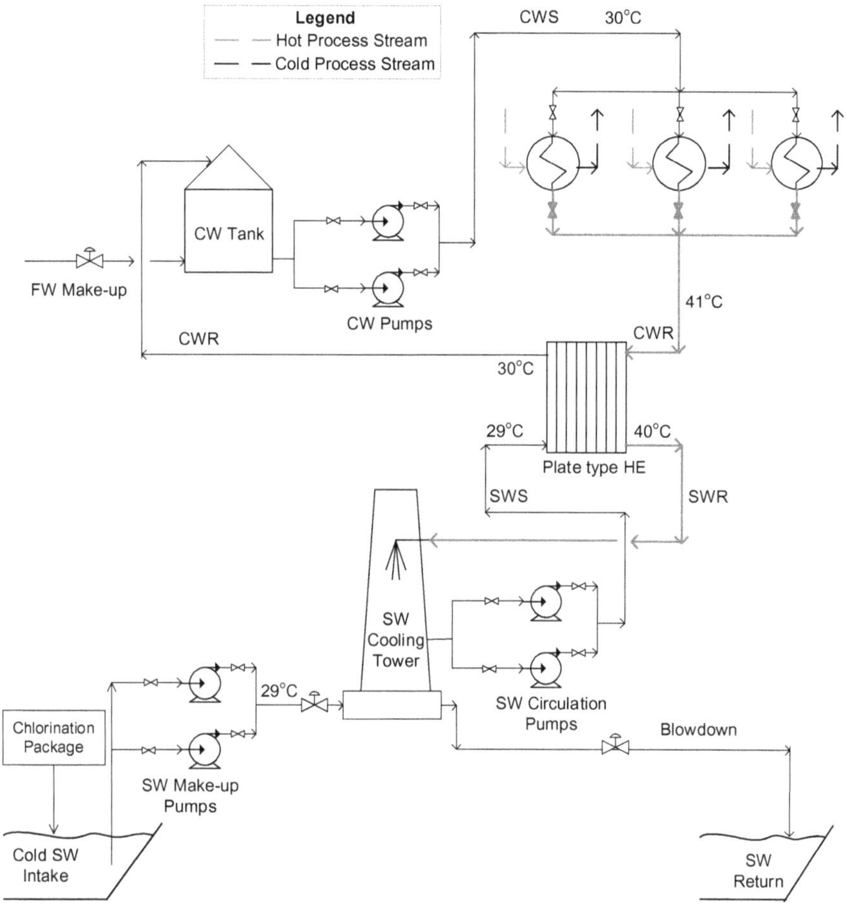

Fig. 15.3: Typical SW (closed loop) and CW (closed loop) systems.

plant has a large number of HEs, use of titanium in all HEs can be very expensive. Further, SW HEs can foul quickly if chlorine injection system is not adequate and/or upset, mainly due to barnacle growth. To overcome these problems for large cooling systems such as in a petroleum refinery, petrochemical and power plants, it may be more economical to use an **SW (once-through)-FW (closed-loop) cooling system** as shown in Fig. 15.2. This system uses CW[a] in closed loop and SW in once-through/open loop. SWR should not be at more than 43°C to avoid scale

[a] CW in this and other chapters of this book refers to circulating FW.

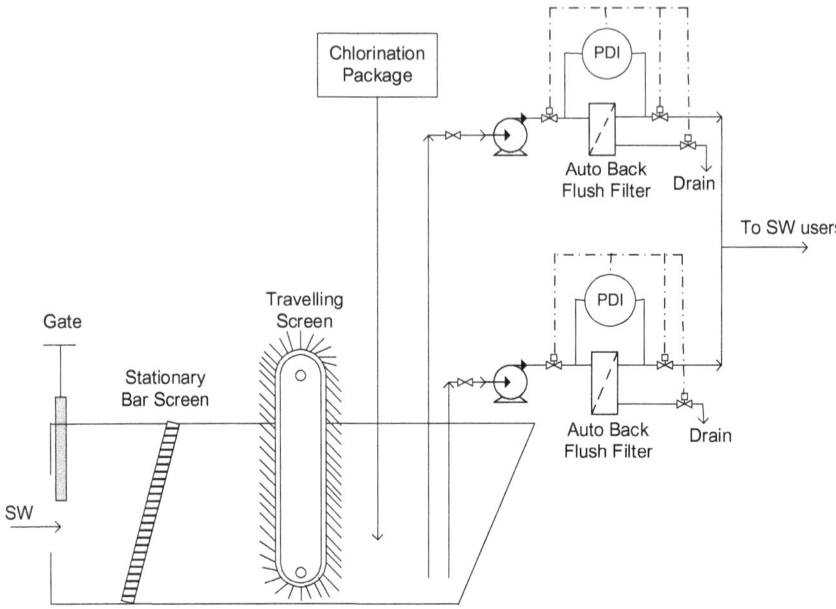

Fig. 15.4: Details of an SW intake station.

formation. Local environmental regulations also limit the maximum SWR temperature.

The SW-FW cooling system (Fig. 15.2) allows the use of cheaper CS tubes in HEs and minimize SW piping. Process heat is exchanged to circulating CW in HEs. Then, heat from hotter CW is transferred to SW using PHEs, installed with titanium plates. This cooling system requires more equipment such as circulating CW pumps, CW tank and PHEs. However, this can be economical if there are many process HEs. Also, because of the use of CW in process HEs, they can sustain process cooling duty due to lower fouling.

The SW–FW cooling system still has a high rate of usage of SW, and ultimately rejects process heat to the sea through hotter SWR. In some locations, SW temperature may be relatively hotter, especially if many companies reject WH into the same sea basin. For such locations, SW (closed-loop)–FW (closed-loop) cooling system shown in Fig. 15.3 is used. This system uses even more equipment compared to the cooling system shown in Fig. 15.1. Here, an SW CT and SW circulation pumps are additionally needed. SW intake pumps for this system are very small as they only supply SW to compensate SW loss due to evaporation, drift and blowdown (BD). In this system (Fig. 15.3), WH is ultimately rejected to air at SW CT. The only SWR stream to sea is the BD stream, which is very small in quantity. The two systems in Figs. 15.2 and 15.4 are both hybrid cooling systems.

15.4 Freshwater Cooling Systems

CTs along with FW re-circulation are widely used in the industries for process cooling, especially rotating equipment (such as pumps and compressors), where other types of cooling such as air cooling and SW cooling are not feasible/acceptable. Typical FW cooling system is shown in Fig. 15.5. A similar system is described in Chapter 7, Section 7.3.3. The principle of CT operation is the simultaneous transfer of mass and heat. Latent heat of vaporization of water provides nearly 80% of the cooling of water[2] and the rest is by sensible heat transfer, which raise both the dry- and wet-bulb temperatures of air. Heat from water is ultimately discarded to air.

Cooling water supply (CWS) is pumped to process HEs and hot CW is sent back to CT as cooling water return (CWR). CWR is introduced at

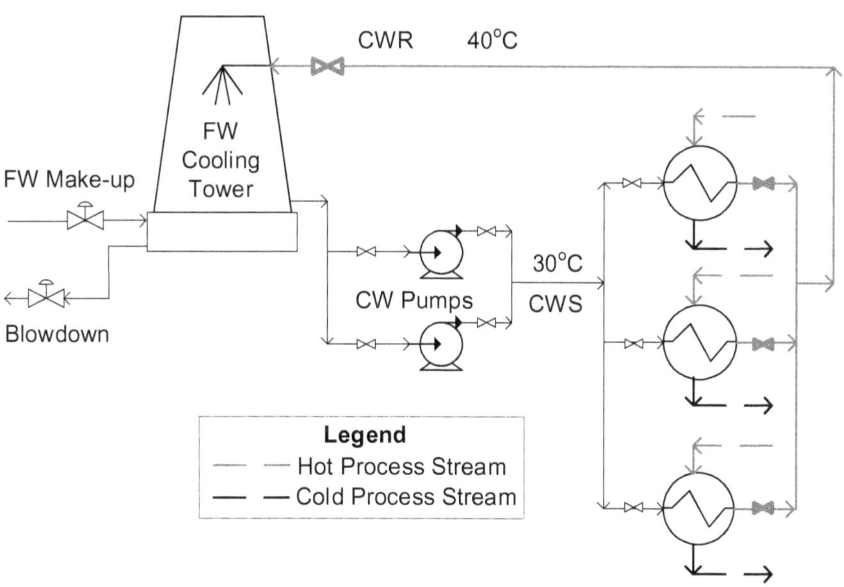

Fig. 15.5: Typical FW type CW system.

Table 15.2: Temperature approach versus relative CT size

Temperature Approach in CT, °C	Relative CT Size
3	2.4
8	1
14	0.55

the top of CT, using distribution headers and spray nozzles. Hot water flows through packing or fill of wood or plastic material, where it forms water droplets, which exchange heat to air. CT fill increases the contact between the hot water and air streams. As the water falls through CT, its temperature drops below the dry-bulb temperature of air entering the tower and approaches the wet-bulb temperature of air. For economic reasons, CWS temperature is designed 2.8°C to 5.5°C above the wet-bulb temperature. CT size increases exponentially as the approach temperature (= CWS temperature − wet-bulb temperature of air to the tower) decreases as shown in Table 15.2. The values in Table 15.2 are extracted from Table 18.5 in Raju.[2]

As a rule of thumb, for every 5.5°C of water cooling, 1% total mass of water is lost because of evaporation. Drift loss is usually 0.1%–0.3% of the circulation water rate.[3] The cooled CW is collected in the sump (or basin) of the CT, and it is typically pumped to the plant as CWS stream. BD from CT basin (Fig. 15.5) is performed to maintain CW quality, especially mineral concentration. Usually, 10% of the circulating CW is sent through 10-micron filter, installed with auto backflush provision, to remove solids from circulating CW. CWS temperature depends mainly on the wet-bulb temperature of ambient air. CWR reaching CT is usually at below 52°C, mainly to control the fouling in HEs and protect the plastic fill used in CT.

Usually, HEs are installed in parallel arrangement in water cooling systems. In such an arrangement, every HE receives CW at CWS temperature from the CT and, if following conventional design, every HE has an outlet temperature, which is 8°C–12°C higher than the inlet, which is then the CWR temperature. Such an arrangement, if matched to the total cooling demand, would result in the maximum CW flow required to satisfy that demand. For illustration, let us look at Table 15.3, which shows hot

Table 15.3: Hot process stream data

Heat Exchanger	Hot Process Stream Data			
	T_i (°C)	T_o(°C)	MC_p, kW/°C	Cooling Demand, kW
HE-1	50	40	20	200
HE-2	50	40	100	1000
HE-3	85	40	40	1800
HE-4	120	65	10	550

process stream data for a cooling system. Here, T_i and T_o are the inlet and outlet temperatures of a stream, respectively, MC_p is the product of mass flow rate and heat capacity at constant pressure, and cooling demand by the stream can be calculated as $MC_p \times (T_i - T_o)$. The sum of the cooling demands in Table 15.3 is 3550 kW. The total CW flow required for CWS and CWR temperatures of 30°C and 40°C, respectively, for all the HEs, is:

$$\frac{3550 \times 3600}{4.184 \times (40-30)} = 305{,}449 \text{ kg/hr}$$

where 4.184 kJ/kg K is the heat capacity of water at constant pressure. As mentioned earlier, this is the maximum CW flow required through this network, to satisfy the cooling demand of 3550 kW. This value can also be obtained by the reciprocal of the slope of the conventional design line in Fig. 15.6, which plots CW temperature and the amount of heat removed for different design cases.

However, by careful examination of stream fouling tendency and temperatures of process streams, one often finds that not all HEs require CW at CWS temperature and similarly not all HEs should return CW at CWR temperature of the CT. This can be seen, for example, in Table 15.4, which lists maximum allowed CWS and CWR temperatures for each exchanger listed in Table 15.3.

The CW temperatures (Table 15.4) are limited as follows: CWS temperature for HE-1, HE-2 and HE-3 is limited to 30°C as the process stream flowing through each of them requires to be cooled for storage to 40°C and the minimum approach temperature is 10°C. Process stream in HE-4 has to be cooled to 65°C (with 10°C approach temperature), based on maximum allowable feed temperature for a reactor. Maximum

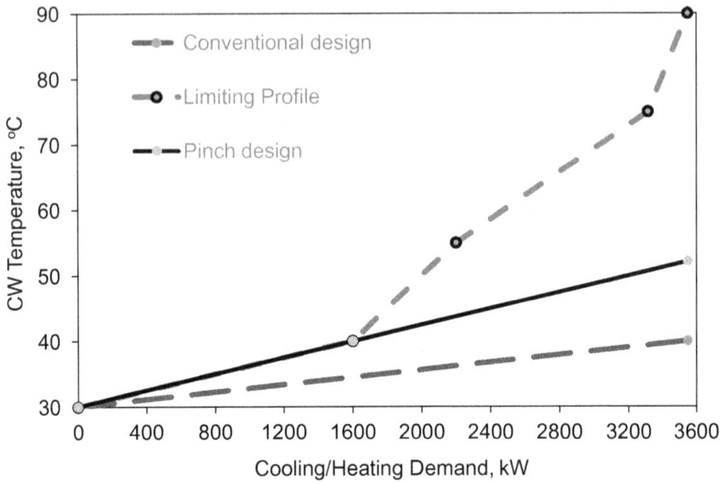

Fig. 15.6: CW temperature for conventional, limiting and water pinch designs.

Table 15.4: Limiting CW data

Heat Exchanger	Limiting CW Data			
	T_i(°C)	T_o(°C)	MC_p, kW/°C	Heating Demand, kW
HE-1	30	40	20	200
HE-2	30	40	100	1000
HE-3	30	75	40	1800
HE-4	55	90	15.7	550

allowable CWR temperatures for HE-1, HE-2 and HE-3 are established by using a 10°C approach temperature with inlet temperature of the respective process stream. Maximum CWR temperature in HE-4 is restricted to 90°C, mainly to prevent any chance of steam generation if CWR pressure drops to 1 bar during any maintenance activities.

The minimum CW flow required for the cooling system can be obtained using water pinch. Drawing a line joining CWS temperature and other lower temperature on the limiting profile curve in Fig. 15.6 provides the water pinch line. Note that this line must not cross or be above the limiting profile. The slope of water pinch line in Fig. 14.6 is $\frac{(40-30)}{(1600-0)}$ = 0.00625°C/kW.

The CWR temperature to CT for the required demand of 3550 kW, following this water pinch line, is 30 + (0.00625 × 3550) = 52.19°C, which is within the maximum CWR temperature allowed for the CT. The corresponding CW flow rate is:

$$\frac{3550\times3600}{4.184\times(52.19-30)}=137{,}652 \text{ kg/hr}$$

This is the minimum CW flow rate for this system.

Thus, in this case study, there will be a reduction of 54.93% $\left(=\frac{305{,}449-137{,}652}{305{,}449}\times100\right)$ in CW flow rate, if the water pinch target can be achieved. However, decreasing CW flow rate lowers the driving force in the coolers, and consequently increases the size and capital cost. The design of CW network meeting water pinch target requires optimization, the detailed procedure of which is beyond the scope of this book. Interested readers are referred to Reddy *et al.*,[4] who have presented MILP (Mixed Integer Linear Programming) and MINLP (Mixed Integer Nonlinear Programming) models for designing optimal CW networks and meeting water pinch target.

15.4.1 *Cooling Tower Performance Evaluation Model*

Heat removal in CTs increases with increase in CWR temperature. Similarly, with increase in SWR temperature (i.e., by re-use of SW), heat removal increases for the case of SW cooling systems. For understanding and evaluating the impact of higher CWR temperature, CT performance estimation method is presented below. It is based on Eqs. 15.1–15.8.

The Merkel equation, shown below, is used to calculate the number of transfer units (NTU) for a CT.

$$\frac{KaV}{m_{\text{w}}}=\text{NTU}=\int_{T_{\text{CWS}}}^{T_{\text{CWR}}}\left(\frac{\text{Cp}}{h_w-h_a}\right)dT \tag{15.1}$$

Here, h_w, h_a and C_{p} are enthalpy of saturated air film surrounding a water droplet (kJ/kg), enthalpy of the bulk air (kJ/kg) and heat capacity of water at constant pressure (= 4.184 kJ/kg K), respectively. Saturated air

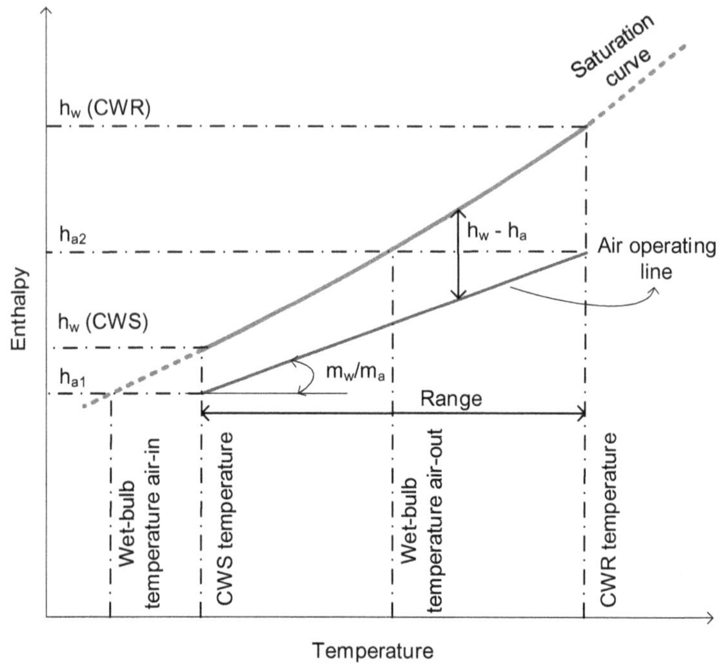

Fig. 15.7: Enthalpy versus temperature diagram for the CT; air-in and air-out temperatures are assumed to be equal to CWS and CWR temperatures, respectively.

follows a slightly non-linear curve while air operating line (through the CT tower) is a straight line, the slope of which is equal to the ratio m_w/m_a (on the enthalpy versus temperature plot in Fig. 15.7), where m_w and m_a are the mass flow rates (kg/hr) of CW and air in CT, respectively. Difference between CWR and CWS temperatures is called the range of CT. Area between the saturation curve and air operating line from CWS to CWR temperatures is proportional to the cooling duty of a CT. Fig. 15.7 indicates that lower m_w/m_a ratio (e.g., lower CW flow rate for a fixed air flow rate) and larger range lead to a higher cooling duty from the CT.

The integral on the right-hand side of Eq. 15.1 can be computed by numerical integration. For this, the range in Fig. 15.7 is divided into several equal temperature intervals. h_w is estimated using Eq. 15.6 (given later) along with CW (saturation) temperature at each temperature interval. Enthalpy of air inlet to the CT (i.e., h_{a1}) is estimated using Eq. 15.6 along with wet-bulb temperature of entering air. Enthalpies of air leaving

CT and at each temperature interval (inside the CT) are estimated using Eq. 15.7.

Humidity ratio of saturated air (W) is estimated by:

$$W = \frac{0.62198 \times pp_w}{p - pp_w} \tag{15.2}$$

Here, 0.62198 is the MW ratio of water to air, p and pp_w are the atmospheric pressure and partial pressure of water vapour in air, calculated by Eqs. 15.3 and 15.4, respectively. Atmospheric pressure (kPa) at altitude Z m is given by:

$$p = 101.325 \times (1 - 2.5577 \times 10^{-5}\ Z)^{5.2559} \tag{15.3}$$

Partial pressure (kPa) of water in air having relative humidity of φ is:

$$pp_w = p_{ws} \times \varphi \tag{15.4}$$

Here, saturation pressure of water vapour (p_{ws}, kPa) is given by:

$$\ln(p_{ws}) = [C_1T^{-1} + C_2 + C_3T + C_4T^2 + C_5T^3 + C_6\ \ln(T)] \tag{15.5}$$

Here, $C_1 = -5{,}800.2206$, $C_2 = -5.516256$, $C_3 = -4.8640239 \times 10^{-2}$, $C_4 = 4.1764768 \times 10^{-5}$, $C_5 = -1.4452093 \times 10^{-8}$, $C_6 = -6.5459673$ and T is temperature in K. Eq. 15.5 can be used to calculate p_{ws} at any temperature (T) such as dry-bulb temperature (T_{db}).

Enthalpy of saturated air (kJ/kg) is given by:

$$h_w \text{ or } h_{a1} = 1.006 \times (\mathrm{T} - 273.15) + W \times (1992.941 + 1.86 \times T) \tag{15.6}$$

Here, W is estimated using Eq. 15.2 and T is air temperature in K. For estimation of h_{a1} and h_w, wet-bulb temperature of air (at the inlet of CT) and water temperature are used, respectively (Fig. 15.7). Enthalpy of exit air from a CT is given by:

$$h_{a2} = h_{a1} + \left[\left(\frac{m_w}{m_a}\right) \times \Delta T \times 4.186\right] \tag{15.7}$$

Here, ΔT, m_w and m_a are the temperature range (K) of the CT (i.e., $T_{CWR} - T_{CWS}$), mass flow rates (kg/hr) of CW and air in CT, respectively. Eq. 15.7 is based on energy balance (i.e., energy transferred from CW to air).

CT characteristic equation from the manufacturer is:

$$\frac{KaV}{m_w} = C \times \left(\frac{m_w}{m_a}\right)^n \tag{15.8}$$

Here, $\frac{KaV}{m_w}$, C and n are the CT characteristic, coefficient and exponent, provided by the CT manufacturer.

Reddy *et al.*[4] tested the above model for the performance of three cooling towers and found it to be reasonably accurate. Right side of Eq. 15.1 is solved by numerical integration over the cooling tower range (i.e., from CWS temperature to CWR temperature). Details of NTU calculations for a CT are shown in Table 15.5; CT design conditions are m_w/

Table 15.5: Details of NTU calculations

Water Temperature °C	Water Temperature K	Partial Pressure of Water Vapour (pp_w), kPa	Humidity Ratio (W), kg water/ kg air	Enthalpy of Air Film (h_w), kJ/kg	Enthalpy of Air (h_a), kJ/ kg	Enthalpy Difference ($h_w - h_a$), kJ/kg	$\frac{1}{(h_w - h_a)}$	ΔT	$\frac{4.184 \times \Delta T}{(h_w - h_a)\text{avg}}$
32.2	305.4	4.8186	0.031	112.05	86.1	25.90	0.039		
33.7	306.9	5.2465	0.034	121.12	92.3	28.82	0.035	1.5	0.232
35.3	308.4	5.7069	0.037	130.86	98.5	32.41	0.031	1.5	0.208
36.8	309.9	6.2019	0.041	141.32	104.6	36.71	0.027	1.5	0.184
38.3	311.4	6.7337	0.044	152.55	110.8	41.79	0.024	1.5	0.162
39.8	312.9	7.3044	0.048	164.63	116.9	47.71	0.021	1.5	0.142
41.3	314.5	7.9165	0.053	177.62	123.1	54.56	0.018	1.5	0.125
42.8	316.0	8.5723	0.058	191.62	129.2	62.41	0.016	1.5	0.109
44.3	317.5	9.2743	0.063	206.72	135.4	71.35	0.014	1.5	0.095
45.9	319.0	10.0254	0.068	223.01	141.5	81.48	0.012	1.5	0.083
47.4	320.5	10.8281	0.075	240.60	147.7	92.92	0.011	1.5	0.073
48.9	322.04	11.6854	0.081	259.62	153.8	105.79	0.009	1.5	0.064
				NTU					**1.478**

m_a, wet-bulb temperature and relative humidity of inlet air, altitude, average CW density, CWS and CWR temperatures equal to 0.97, 27.22°C, 100%, 10 m, 992 kg/m^3, 32.2°C and 48.89°C, respectively). In these calculations, Eqs. 15.4 and 15.5 are used to estimate pp_w; W and h_w at every water temperature are estimated by Eqs. 15.2 and 15.6, respectively; and h_a for the CT inlet (i.e., value in the first row corresponding to CWS temperature of 32.2°C) is estimated using Eq. 15.6 and wet-bulb temperature of 27.22°C. h_a in the second row onwards is calculated based on h_a value in the respective previous row and Eq. 15.7. Calculations are continued until water temperature is equal to 48.89 °C, which is the CWR temperature for the CT. NTU of the CT is obtained by summing all values in the last column of Table 15.5.

For different values of m_w/m_a in a suitable range such as 0.5–1.5, the design NTU calculations (using Eqs. 15.1–15.7) are performed for the given operating conditions. These are: T_{CWR}, T_{CWS}, T_{wb} of inlet air; assumed altitude of 10 m; and relative humidity of inlet air (assumed 100%). This requires repetition of calculations in Table 15.5 for each value of m_w/m_a. In addition, CT characteristic, KaV/m_w is calculated using Eq. 15.8 for different values of m_w/m_a. Then, design NTU and KaV/m_w can be plotted against m_w/m_a. This is referred to as the performance plot of a CT (Fig. 15.8). For this figure, $C = 1.446$ and $n = -0.75$ are used from the CT vendor data; and T_{wb}, T_{CWS} and T_{CWR} are 27.22°C, 32.2°C and 48.89°C, respectively. Intersection of the two curves in Fig. 15.8 is the m_w/m_a ratio that can be achieved in the CT for these conditions.[b]

For a fixed m_a and given m_w/m_a ratio, water flowrate m_w through the CT can be calculated; then, heat removed by the CT can be calculated using m_w, range and specific heat capacity of water (4.184 kJ/kg.°C). For $m_a = 616{,}234$ kg/hr, $m_w/m_a = 0.97$ and range = 48.89 − 32.2 = 16.69°C, calculated $m_w = 0.97 \times 616{,}234 = 597{,}747$ kg/hr and heat removed = 597,747 × 4.184 × 16.69 = 41,741,247 kJ/hr = 11,595 kW.

[b] Instead of plotting for finding the intersection as in Figure 15.8, the implicit non-linear equation obtained by equating the right side of Eq. 15.1 with that of Eq. 15.8 can be solved for m_w/m_a using Goal Seek in Excel. For this, calculations in Table 15.5 for a particular value of m_w/m_a should be coded in Excel.

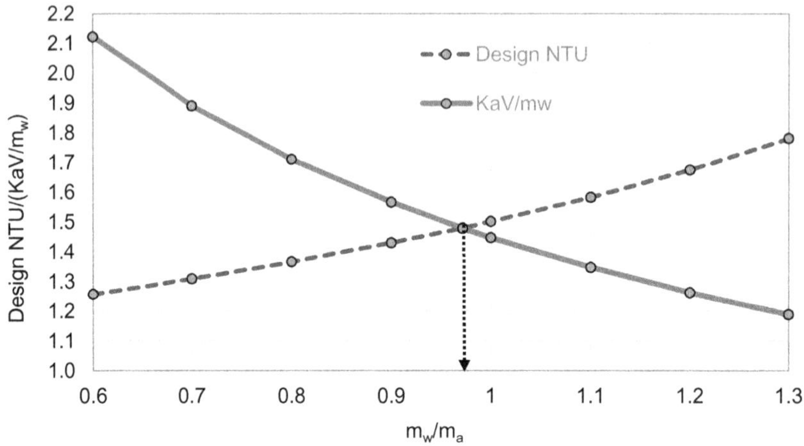

Fig. 15.8: Intersection of CT characteristic (KaV/m_w) and design NTU curves gives design m_w/m_a ratio, as shown by the vertical dotted line.

CWR temperature (assumed to be 48.89°C for Fig. 15.8) changes the design NTU curve and hence its intersection with KaV/m_w curve (i.e., m_w/m_a ratio that can be achieved in the CT). Calculations like those in Table 15.5 for Fig. 15.8 as well as heat removed can be repeated for different CWR temperatures between the design values of CWS and CWR temperature, while keeping CWS temperature constant. These results can then be plotted as in Fig. 15.9, which shows that m_w/m_a and water flow rate (since air flow rate is a given value) decrease non-linearly, whereas the heat removed in CT increases linearly with increase in CWR temperature. Specifically, heat removed increases from 6000 to 7700 kW and m_w/m_a ratio decreases from 2.2 to 1.4 steeply for CWR temperature increase from 36°C to ~40°C; then, heat removed increases from 7700 to 14,330 kW and m_w/m_a ratio decreases further, less steeply, from 1.4 to 0.8, for CWR temperature increase from ~ 40°C to 55°C. Fig. 15.9 can be used to estimate m_w/m_a and heat removal for a CWR temperature. For example, for CWR temperature of 48.89°C, m_w/m_a is 0.97 and heat removal is 11,595 kW.

CT cooling duty can be increased without any capital cost (i.e., for installing a new CT or modifying existing CT), if m_w/m_a ratio is

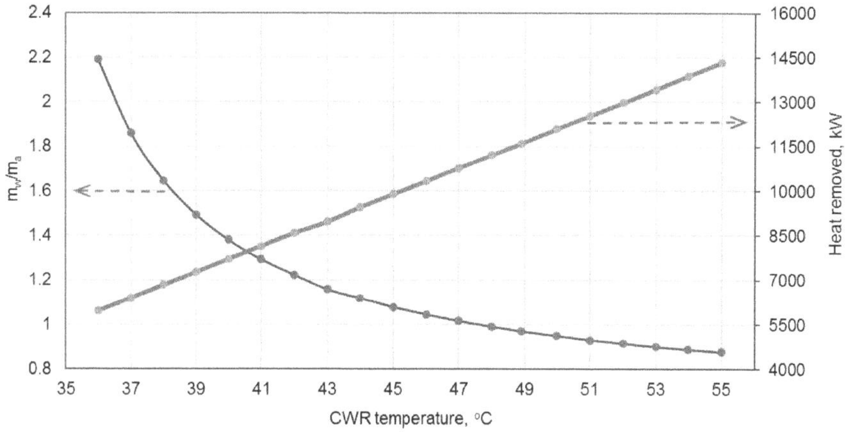

Fig. 15.9: Typical effect of CWR temperature on m_w/m_a and heat removal of CT.

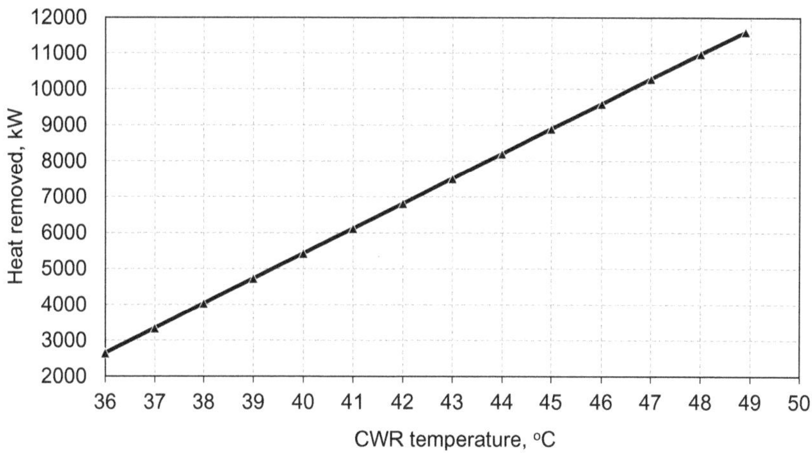

Fig. 15.10: Effect of CWR temperature on heat removal from CT, with m_w/m_a ratio of 0.97.

maintained at or below its design value and CWR temperature is lower than the design value. This is illustrated in Fig. 15.10. The values in this figure are calculated using a fixed design m_w/m_a of 0.97. Heat removal increases from 2640 kW to 11,595 kW (by 339%) as the CWR temperature increases from 36°C to 48.89°C (design CWR temperature). For this

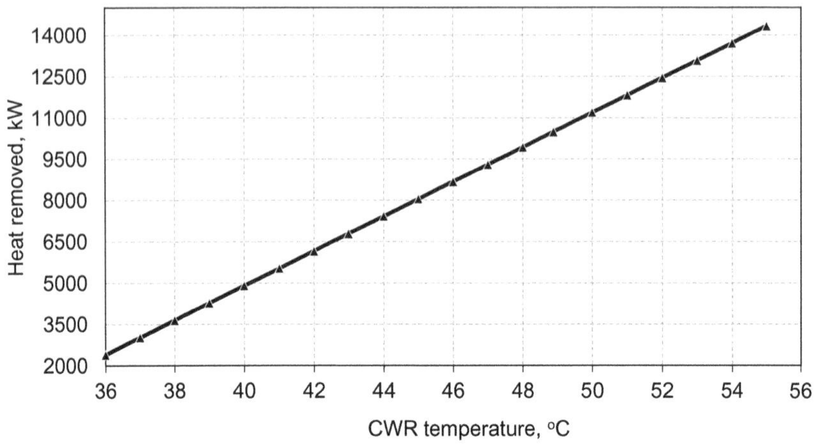

Fig. 15.11: Effect of CWR temperature on heat removal from CT, with m_w/m_a ratio of 0.8774.

case, CWR temperature cannot be increased further above the design value of 48.89°C, without increase in CWS temperature, unless m_w/m_a ratio is decreased to below the design value of 0.97.[c]

If CWS temperature is required to be kept constant, a CT can remove more heat if the operating m_w/m_a ratio is lower than the design value. In Fig. 15.11, CT is operating at $m_w/m_a = 0.8774$ (compared to 0.97 used for Fig. 15.10). In this case, NTU = KaV/m_w is 1.595 at CWR temperature of 55°C (the extreme right point in Fig. 15.9). Hence, CWR temperature can be increased up to 55°C, without increase in CWS temperature; for example, CT heat removal increases from 2388 kW to 14,328 kW (by 500%) as the CWR increases from 36°C to 55°C. As the $m_w/m_a = 0.8774$ used in this case, is lower than the design $m_w/m_a = 0.97$, heat removal at CWR of 48.89°C is lower (at 10,488 kW in Fig. 15.11) compared to the design heat removal value (11,595 kW in Fig. 15.10). However, in this case, CWR can be increased to 55°C and hence heat removal can be increased to 14,328 kW, which is 23.57% higher compared to the design heat removal.

[c]There are other ways to increase CT heat removal. One way is by increasing the air flow rate, but this requires modifications in CT. Another way is by allowing higher CWS temperature if it can be tolerated by the process/coolers; then, CT can remove more heat by increasing water flowrate. These cases are not covered in this chapter.

Maximum CWR temperature is often restricted by the CT fill/packing temperature limit, fouling and/or corrosion. Ryznar Stability Index (RSI) is recommended for quick evaluation of the fouling tendency. Details of such evaluation is available in Chapter 22 of Hall.[5]

Thermal efficiency of a CT is calculated by

$$\eta = \frac{(T_{\text{CWR}} - T_{\text{CWS}})}{(T_{\text{CWR}} - T_{\text{wb}})} \times 100 \tag{15.9}$$

Thus, efficiency of a CT increases with CWR temperature as illustrated in Fig. 15.12 for $T_{\text{wb}} = 27.22°\text{C}$.

Evaporation and BD losses of a CT are calculated using Eqs. 15.10 and 15.11, respectively.

$$E = \frac{m_{\text{w}} \times (T_{\text{CWR}} - T_{\text{CWS}}) \times 4.184}{\text{H}_{\lambda}} \tag{15.10}$$

Here, E, m_{w} and H_{λ} are evaporation rate of water (m^3/hr), circulation rate of CW across CT (m^3/hr) and heat of vaporization of water (kJ/kg), respectively; and 4.184 kJ/kg.K is the specific heat of water. Eq. 15.10 assumes that the heat removed from water is entirely by water evaporation and sensible heat gain of air is negligible.

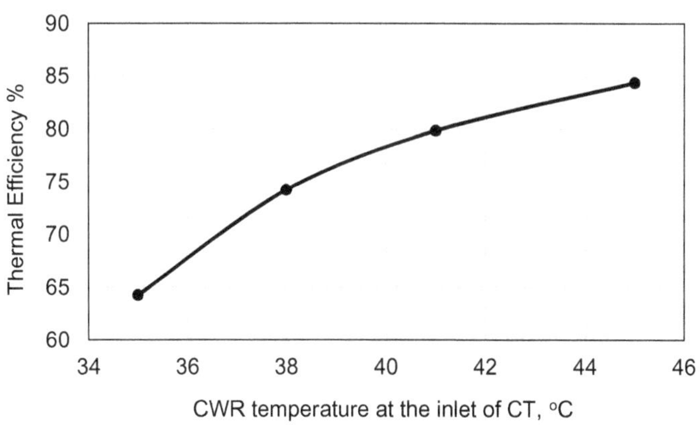

Fig. 15.12: Effect CWR temperature (at the inlet of CT) on the thermal efficiency of CT, for a wet-bulb temperature of 27.22°C and CWS temperature of 30°C.

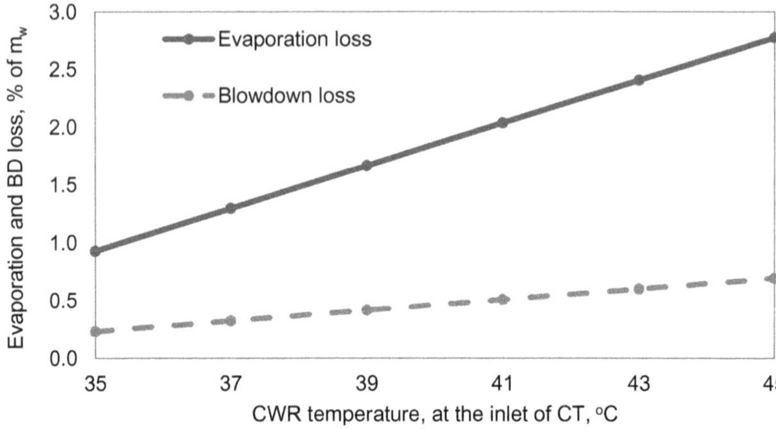

Fig. 15.13: Effect CWR temperature (at the inlet of CT) on BD and evaporation losses (both as percentage of water circulation rate through CT).

$$\mathrm{BD} = \frac{E}{\mathrm{COC} - 1} \tag{15.11}$$

Here, BD and cycles of concentration (COC)[d] are BD rate (m^3/hr) and cycle of concentration of CT. For generating Fig. 15.13 based on Eqs. 15.10 and 15.11, a value of COC of 5, CWS of 30°C, H_λ of 2260 kJ/kg and m_w of 1000 m^3/hr are used. As can be seen in this figure, both evaporation rate and BD increase with the increase in CWR temperature. However, a small increase in losses due to evaporation and BD can be tolerated, for the sake of the gain in efficiency with increasing CWR temperature and thus the ability to meet increased cooling requirements in the plant, without the additional capital cost for installing a new cooling system.

15.4.2 *Debottlenecking Options for Cooling Systems*

For meeting the increased cooling duty in the plant equipment, some or all the following options can be used to debottleneck SW/FW cooling systems.

[d] COC is the ratio of the concentration of total dissolved solids in the CT water to that in the CT make-up water.

- Install additional SW/FW pumps to provide a higher SW/FW flow rate and/or pressure head.
- Increase SW/FW piping size or install parallel SW/FW piping header, to reduce pressure drop in the piping.
- Install parallel SW/FW coolers for increasing cooling duty.
- For PHEs, increase the number of plates to the maximum allowed by the PHE frame.
- For closed-loop SW/FW cooling system, maintain COC of 5 or higher.
- Use automatic backflush filter for maintaining SW HEs with minimum fouling.
- Use the side-stream filter for closed-loop FW cooling system.

Both automatic backflush filter and side-stream filter reduce fouling in process HEs and help to sustain good cooling performance and avoid wastage of SW/CW.

15.5 Air-Cooled Heat Exchangers

ACHE is often a good alternative to a water-cooled heat exchanger (WCHE). Water shortage and/or increasing water cost, together with concerns on water pollution and CT plumes, have reduced the industry's use of WCHEs. As a result, ACHEs are increasingly used in process industries. They are advantageous for applications such as condensation of low-pressure steam and/or process vapour streams, where the condensate temperature does not need to be close to the ambient temperature.

Another advantage of ACHE over a WCHE is that CW circuit and associated equipment such as CT, circulation pumps, water treating systems, and piping are not required. ACHEs incur lower operating cost compared to WCHEs. Additionally, BD and evaporation losses at CT can be avoided. Further, ACHEs can provide partial cooling (20%–30% of the rated duty, due to natural circulation of air) during a power failure. Location of the ACHE is independent of water supply location. Maintenance cost of ACHE is less than one-third of WCHE.[2] ACHE can operate efficiently with temperature above 50°C at tube skin. There would be concerns about fouling at these temperatures, in the case of WCHEs.[6]

However, for a given duty, ACHEs are generally bigger units compared to WCHEs, mainly due to low heat transfer coefficient of air compared to water. They are usually installed on pipe racks to reduce the required ground area. Capital cost of ACHEs is usually higher than that of water-cooled shell-and-tube HEs. ACHEs are the only option in locations where CW is not available or in process plants where the CW systems have no spare capacity and therefore additional CTs would need to be installed if WCHEs were added.

Due to economic reasons (excessive capital cost for larger ACHEs), process streams can only be cooled in ACHEs to about 15°C above the ambient air temperature, compared to 3°C–5°C above the CW temperature for WCHEs.[7] Cooling duty in ACHE is subjected to seasonal variations in ambient air conditions. Hence, costly temperature control may be required if accurate temperature control of process streams is required. ACHEs may not be economical for cooling highly viscous fluids, mainly due to low tube side heat transfer coefficient. For such applications, tube inserts may be required, which are not well proven.

Following are the main parts of ACHEs.

1. One or more tube bundles. Each tube bundle includes many tubes, tube supports and header boxes. The tube length is usually defined by the pipe rack width or other platform/plot layout considerations.
2. Forced or induced draft fan(s) to circulate air on the external surface of tubes.
3. A plenum between the fan and tube bundle(s), which is a totally enclosed space that facilitates uniform airflow distribution between the fan and tube bundles. It can be designed as pyramidal or straight section.
4. A supporting structure, which consists of columns, braces, and cross beams that support the tube bundle at high enough elevation to allow the necessary volume of air to enter below the tube bundle, at an approach velocity low enough to allow unimpeded fan performance and to prevent unwanted re-circulation of hot air.
5. Platforms and ladders for maintenance and operational access.

6. Optionally, louvres above tube bundle or variable speed drive on fan motor/steam turbine driver, to regulate air flow.

To compensate for low film heat transfer coefficients on the air side, finned tubes are used in ACHEs. This increases the air-side heat transfer and hence reduces the bundle size compared to the case of plain tube bundles. There are several types of fins available for ACHEs. The most common are tension-wrapped, embedded, extruded, and footed. Tube lengths vary from 6 feet (1.83 m) to 60 feet (18.3 m). Tubes are available from 5/8 inch (15.9 mm) to 6 inches (152.4 mm) in diameter. The most used tube diameter is 1 inch (25.4 mm). Fin height can vary between ½ inch (12.7 mm) and 1 inch (25.4 mm); popular sizes are ½ and 5/8 inches. Fins are generally made of aluminium (for process stream temperatures up to 400°C) or CS (above 400°C). Finned tubes generally have 275–433 fins per metre (7 or 11 fins per inch). The ratio between the fin area and the plain-tube external area varies from 7 to 25. Fin thickness is typically 0.4–0.8 mm.

Tube bundles typically consist of 2–10 rows of finned tubes arranged on triangular pitch (2 and 2.5 tube diameters). Net-free area for air flow through bundles is ~50% of tube bundle face area. Tubes are rolled or welded into the tube sheets. The tube bundles are commonly installed in a horizontal position. However, in some cases such as vapour condensing applications, tube bundle can be arranged vertically or V-shape or A-shape; of these, A-shape arrangement is shown in Fig. 15.14. This mainly saves plot/pipe rack size required for the ACHE. The main disadvantage of A-shape and V-shape arrangements is higher fan power requirement.

Axial flow type fans are commonly used in ACHE. Fans can have 2–20 blades. Usually, the fan diameter is limited to 4 or 5 m. Blades are commonly made of aluminium or fiberglass-reinforced plastic. They can also be made of steel. Uniform distribution of air across the tube bundle is essential for achieving efficient heat transfer. This can be ensured by designing adequate fan coverage area[e] (e.g., 40% of the tube bundle's

---

[e] Fan coverage is defined as the ratio between the projected area of the fans and the projected area of the bundles they serve.

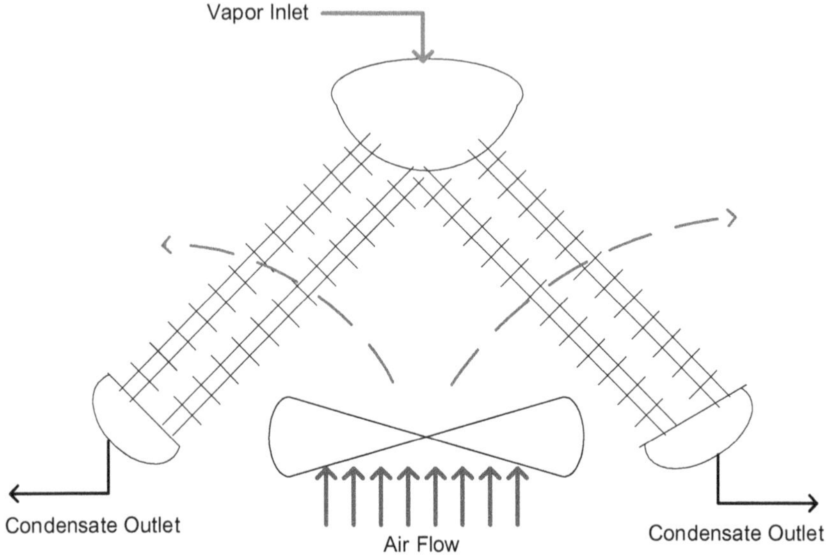

Fig. 15.14: A typical A-shape type ACHE.

projected face area) and static pressure loss across the tube bundle (at least 3.5 times the velocity pressure loss through the fan ring).

Fans are commonly driven by motors. The fan power requirement depends on the air flow rate and air pressure drop (which mainly depends on the number of tubes in the flow direction). Often, it is economical to use a larger number of fans, to avoid high motor power; for economic reasons, fan power is usually maintained below 35 kW. In some cases, fans may be driven by steam turbines, gas engines or hydraulic motors. Fans may have fixed or adjustable pitch blades. The blades' pitch can be adjusted either manually or automatically to regulate the air flow.

ACHE Configurations: ACHEs are installed in many configurations. Some of the configurations are illustrated in Fig. 15.15. The assembly of one/more tube bundles, served by one/more fans, a plenum and structure, is called an ACHE bay. When large duties/large flow rates are required, it is common for the ACHE to be divided into several bays, mainly to facilitate economic design, easy transportation and construction. One or more bundles for the same or different cooling service may be combined in one bay with one/more fans.

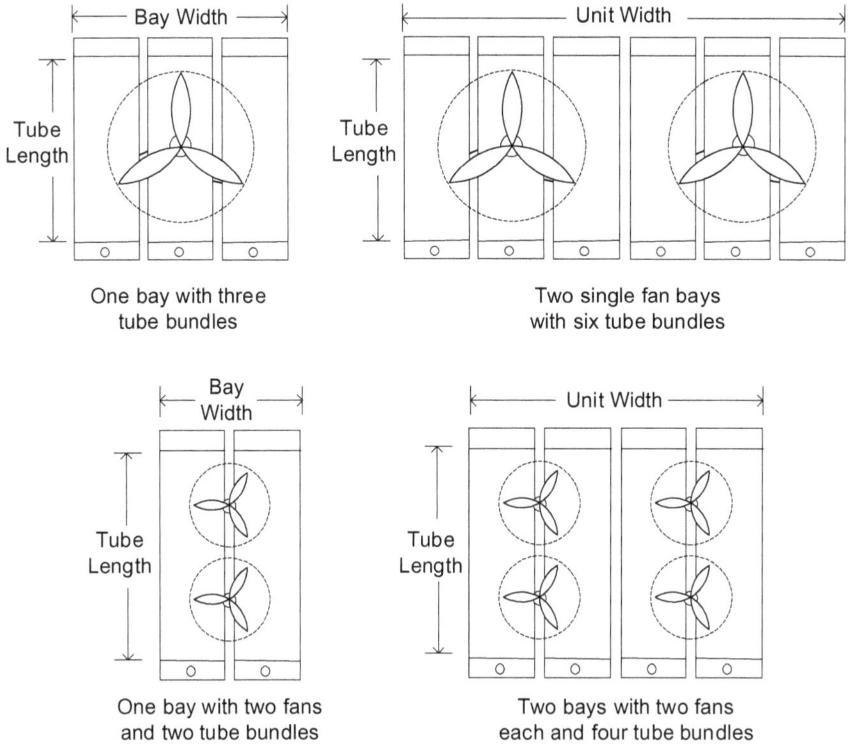

Fig. 15.15: Some arrangements of ACHE tube bundles and fans in one or more bays.

In ACHE, air and process stream flow at right angles to each other (i.e., air flows vertically from bottom to the top surface of the tube bundle, while process fluid flows through the tubes in horizontal or inclined direction) and hence result in cross flow heat transfer.[f] ACHE with forced and induced draft arrangements are shown in Figs. 15.16 and 15.17, respectively. Forced draft ACHEs are more common compared to induced draft ACHEs. However, induced draft fan is sometimes preferred for the following scenarios.

---

[f]Cross-flow configuration is employed for the following reasons: (a) ACHEs are installed on pipe racks with horizontal tubes; (b) to utilize cooling via natural circulation, especially in the case of fan trips and (c) driving forces are often large compared to shell and tube heat exchangers.

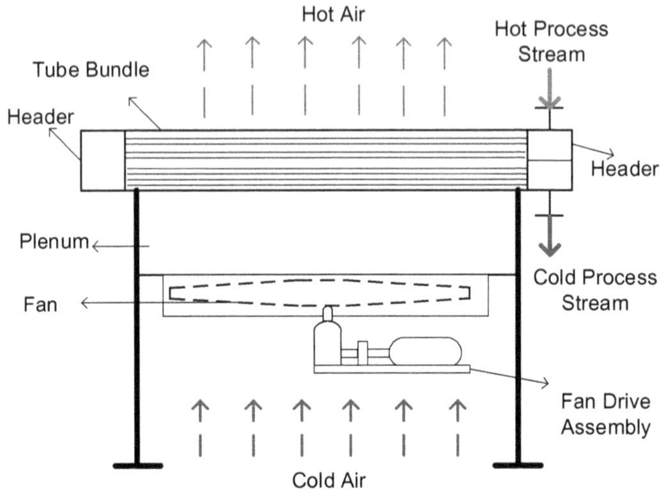

Fig. 15.16: Typical forced draft ACHE arrangement.

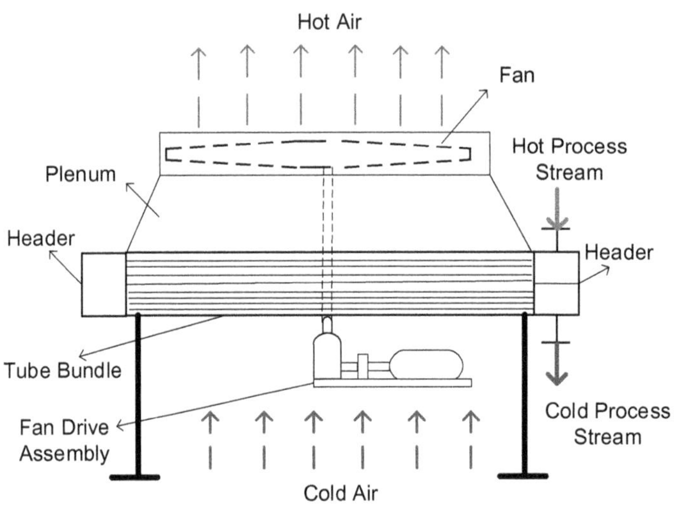

Fig. 15.17: Typical induced draft ACHE arrangement.

- To reduce the risk of hot air re-circulation, which would make the inlet hotter and therefore reduce the heat transfer driving force at the inlet.
- When approach temperature (i.e., temperature difference between process outlet and air inlet) is 11°C (20°F) or less[8] or in extremely hot climates.

- When natural circulation (due to the stack effect[g]) is important in the event of a fan failure
- Excessive cooling due to rain is a problem.
- Air side fouling is a major problem.

Key differences between forced draft and induced draft ACHEs are summarized in Table 15.6.

ACHE area (A m^2) can be calculated using $A = \frac{Q}{U \times \text{LMTD} \times F}$. Here, F is the LMTD correction factor for deviation from counter current flow, Q is the amount of cooling duty (kW), U is the overall heat transfer coefficient (kW/m^2K) and LMTD is the log mean temperature difference between air and process stream (K); note that LMTD is calculated assuming counter current flow. Value of F is equal to one if ACHE has more than three tube passes.[10] If the number of passes are three or below, F value needs to be estimated like in a shell and tube HE. Overall heat transfer coefficients for ACHEs are presented in Table 15.7. Detailed design procedure for ACHE is available in books by Eduardo[10] and Smith.[11]

ACHE cooling generally increases by increasing air flow rate. Following options can be used to vary air flow rate in ACHE.

- Use of variable frequency drives on fan motors
- Use of steam turbine drivers for fans
- Switching-off some fans in multi-fan units
- Adjustable louvres on top of bundles
- Use of two-speed fan motors
- Reducing gears or V belts
- Adjust fan pitch

The following affinity rules are useful to optimize air flow rate, static pressure and power consumption in ACHE fans.

---

[g] Stack effect is the movement of air through ACHE due to density difference between cold and hot air. Density of ambient air is higher (due to its lower temperature) than the air density at the top of the tube bundle (due to high air out temperature). This causes air buoyancy and hence moves some air through ACHE, even if the fan trips (i.e., stops working).

Table 15.6: Key differences between forced and induced draft ACHEs

Factor	Forced Draft	Induced Draft
Air distribution over entire cross section of the bundle	Poor	Better
Hot air re-circulation	Higher possibility due to low air velocity (~3 m/sec) at the exit of tube bundle and lack of any stack on top of the tube bundle[9]	Lower possibility due to high discharge velocity at the exit of tube bundle (up to 10 m/sec), which is directed upwards above the tube bundle[9]
Maintenance of fan	Easy as the fan handles cold atmospheric air	Difficult as fan is exposed to high temperature air
Maintenance access for fan and tube bundles	Tube bundle can be accessed easily as fan is located below the tube bundle	Difficult as fan is located above the tube bundle. For accessing the tube bundle, plenum needs to be removed.
Process temperature limit	No limit with proper metallurgy of tube bundle	Limited to 175°C, above which fan component failures are possible
Air temperature limit	No limit	Limited to 95°C, above which fan component failures are possible
Natural draft stack effect during fan failures	Low natural draft due to smaller stack effect	Higher natural draft due to higher stack effect
Power requirement	Lower power requirement as fan handles colder and hence high-density air; since power increases with the cube of volume flow (see example 15.1, below)	Higher power requirement as fan handles hotter and hence low-density air
Effects of climatic conditions such as heavy rain	Significantly affected	Less affected as plenum covers ~60% of the tube bundle area

Table 15.7: Overall heat transfer coefficients in ACHE (using 5/8-Inch aluminium fins on 1 inch OD tubes with 393 fins per meter)[10] for different condensation/cooling services

Condensation, W/m²K		Gas Cooling, W/m²K		Liquid Cooling, W/m²K	
Amine regenerator	570–670	Air or combustion gas at 50 psig (ΔP = 1 psi)	60	Machine CW	740–900
Ammonia	600–700	Air or combustion gas at 100 psig (ΔP = 2 psi)	112	Fuel oil	115–170
Heavy naphtha	400–500	Air or combustion gas at 100 psig (ΔP = 5 psi)	170	Reforming or platforming liquids	480
Light hydrocarbons	540–600	Hydrocarbon gas at 15–50 psig (ΔP = 1 psi)	170–220	Light gas oil	450–550
Light naphtha	450–550	Ammonia reactor outlet stream	500–600	Light hydrocarbons	510–680
Steam	800–1200	Gaseous hydrocarbons at 50–250 psig (ΔP = 3 psi)	280–340	Light naphtha	510
		Gaseous hydrocarbons at 250–1500 psig (ΔP = 5 psi)	400–500	Process water	680–820

- Air volumetric flow rate is directly proportion to fan speed (i.e., revolutions per minute).
- Static pressure developed by fan varies directly with the square of fan speed.
- Power required by the fan varies directly with the cube of fan speed.

Often in WHR projects, it is beneficial to debottleneck/revamp cooling capacity of existing ACHEs, before deciding to install a new ACHE. Table 15.8 summarizes various performance improvement adjustments/modifications for ACHE.

Table 15.8: Options for improving the performance of air coolers

Key Adjustment/Modification	Typical Improvement	Cost Impact
Replace old, low efficiency aluminium blades with high efficiency aerodynamic fan blades	30–40% more air flow	Medium
Clean fins/tubes (applicable for dirty environment/services)	5–50% more duty	Medium
Replace tubes/tube bundle with high efficiency fins	More duty up to 50%	High
Replace V belts and drives	3–4% driver efficiency	Medium
Replace inefficient drivers with high efficiency ones	Up to 10% driver efficiency	Medium to High
Use variable speed drives	Energy savings for part-load operation	Medium
Adjust blade pitch	More air flow	Low
Increase fan speed	More air flow	Low
Install seal disc/air seal or inlet bell	2–3% more air flow	Low
Limit tip clearance	2–3% more air flow	Low
Seal any leakages in fan housing	More air flow	Low

Example 15.1: Estimate the approximate percent increase in flow capacity and power consumption if increase in cooling load for a WHR project requires the ACHE fans to operate at 20% increase in speed.

Solution: Flow capacity is directly proportional to the increase in fan speed. Hence, air flow capacity will increase by 20%. Power consumption increases directly with the cube of the fan speed. Hence, increase in power consumption = $(1.2^3 - 1) \times 100 = 72.8\%$.

15.6 Hybrid Cooling Systems

Hybrid cooling systems consists of combination of FW cooling HEs, ACHEs and/or SW cooled HEs. They are particularly useful for the revamp of a cooling system to provide more cooling to the equipment in the plant. Main advantages of hybrid cooling systems are the following.

- Facilitate the development of cost-effective option for achieving additional cooling duty.
- May avoid the need for additional plot space.
- Reduce FW and/or SW consumption.
- Increase the reliability of the cooling system.

A hybrid cooling system with ACHE and WCHEs in series (for cooling one hot process stream) is shown in Fig. 15.18. Not all HEs require CW for the entire range of cooling. For example, HE-1 and HE-2 in Fig. 15.18, the base case uses CW to cool two process streams, one from 150°C to 35°C in HE-1 and another from 100°C to 35°C in HE-2. A WHR project requires the cooling of a hot stream from 70°C to 35°C, in a new HE-3 (shown in a dashed rectangle). This requires additional CW flow rate and new CT, which can make the WHR project uneconomical.

Careful examination of HE-1 indicates that it may be able to reduce CW demand in this exchanger by the addition of ACHE in series, for partially cooling the hot process stream. ACHEs can easily achieve an approach temperature of 20°C with respect to the ambient temperature. Hence, hot process stream in HE-1 can be cooled by ACHE from 150°C to 50°C and the remaining cooling from 50°C to 35°C can be carried out using CW. This saves CW at HE-1, and the saved CW can be used in HE-3, without installing an additional CT. If required, another ACHE can be added in series for HE-2 also. In the example in Fig. 15.18, for brevity, only two HEs are considered for the base case. However, in process

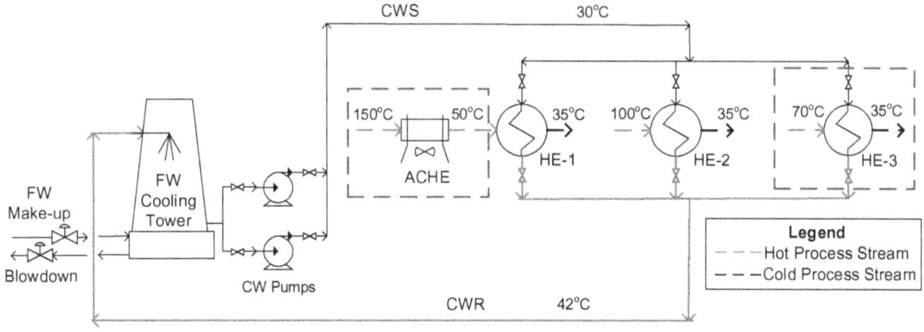

Fig. 15.18: Hybrid cooling system with ACHE and WCHEs in series (for cooling one process stream from 150°C to 35°C); new units are shown in dashed rectangles.

plants, there can be many HEs in parallel arrangement. Hence, there are many opportunities to optimize the CW flow and divert the saved CW for WHR projects.

Example 15.2: Currently, one CT serves two HEs (HE-1 and HE-2 shown in Fig. 15.18) in a chemical plant. CWR temperature = 42°C, CWS temperature = 30°C, CW flow rate = 650 ton/hr, and data on HE-1 and HE-2 are in the following table. A WHR project requires condensing 9.2 ton/hr of steam at a vacuum pressure of 0.1235 bar. Identify a hybrid cooling solution. Assume the site has sufficient pipe rack space for installing an ACHE.

	HE-1		HE-2	
Quantity	**Process Fluid**	**CW**	**Process Fluid**	**CW**
Inlet temperature, °C	150	30	125	30
Outlet temperature, °C	35	42	35	42
Average heat capacity, kJ/kg.K	2	4.2	1.012	4.2
Flow rate, kg/hr	109,600	500,000	83,000	150,000

Solution: In HE-1, heat removed from the process stream = $\frac{109{,}600\times 2\times(150-35)}{3600}$ = 7002.2 kW, and heat gained by CW = $\frac{500{,}000\times 4.2\times(42-30)}{3600}$ = 7002.2 kW. Hence, energy balance is satisfied.

Similarly, in HE-2, heat removed from the process stream = $\frac{83{,}000\times 1.012\times(125-35)}{3600}$ = 2099.9 kW, and heat gained by CW $\frac{150{,}000\times 4.2\times(42-30)}{3600}$ = 2099.9 kW. Hence, energy balance is satisfied.

Mass flow rate of steam to be condensed = 9200 kg/hr. From steam tables, latent heat and saturation temperature of steam at 0.1235 bar are 2381.974 kJ/kg and 50°C, respectively. With CWS and CWR temperatures of 30°C and 42°C, respectively, amount of CW required for condensing the steam is $\frac{9200\times 2381.974}{4.2\times(42-30)}$ = 434,805 kg/hr.

As can be seen, HE-1 is consuming 500,000 kg/hr of CW for cooling the process stream from 150°C to 35°C. Hence, ACHE can be used to cool this process stream from 150°C to 50°C (assuming atmospheric temperature of 30°C) before cooling this process stream further to 35°C using CW. in HE-1.

Heat that can be removed by ACHE $= \frac{109{,}600 \times 2 \times (150-50)}{3600} = 6088.9$ kW.

Remaining heat from 50°C to 35°C from the process stream can be removed using CW. Amount of CW required for this cooling $= \frac{109{,}600 \times 2 \times (50-35)}{4.2 \times (42-30)} = 65{,}238$ kg/hr.

Hence, at HE-1, there is a CW saving of 434,762 (= 500,000 − 65,238) kg/hr. This amount of CW is sufficient for use in the new WHR (i.e., for condensing 9200 kg/hr of steam at 0.1235 bar).

It should be noted that, due to the large reduction (87%) of CW in HE-1, velocity of CW in HE-1 tubes reduces substantially. So, HE-1 may need to be modified such as increasing the number of passes or plugging some tubes. Otherwise, excessive fouling occurs due to very low velocity of CW. Further, there will be capital cost for ACHE, which should be considered in the profitability analysis of WHR project.

In general, more complex hybrid CW systems are possible. If CT duty is maximized and ACHE addition option is fully utilized, SW pumping system and distribution network are readily available, additional cooling capability for CWR can be provided by exchanging hot CWR with fresh SW (Fig. 15.19) or used SW (Fig. 15.20), both using titanium PHEs.

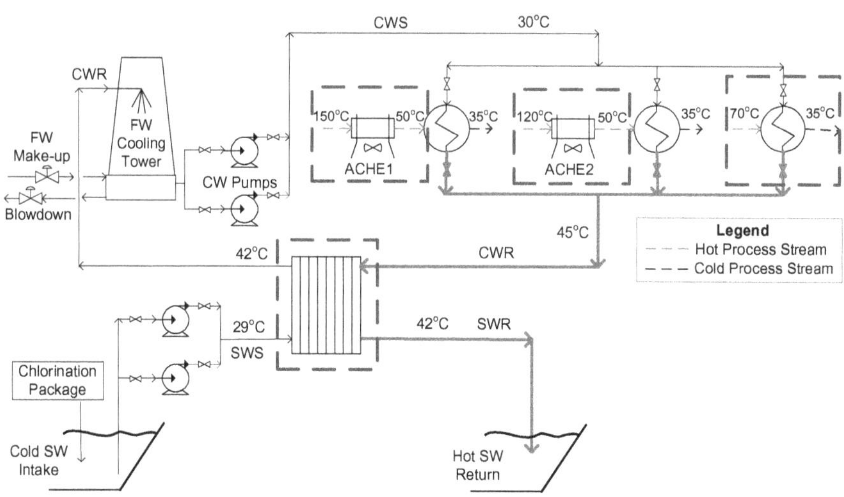

Fig. 15.19: Hybrid cooling system with ACHEs and WCHEs in series (for cooling process streams) integrated with fresh SW for CWR colling from 45°C to 30°C; new units are shown inside dashed rectangles.

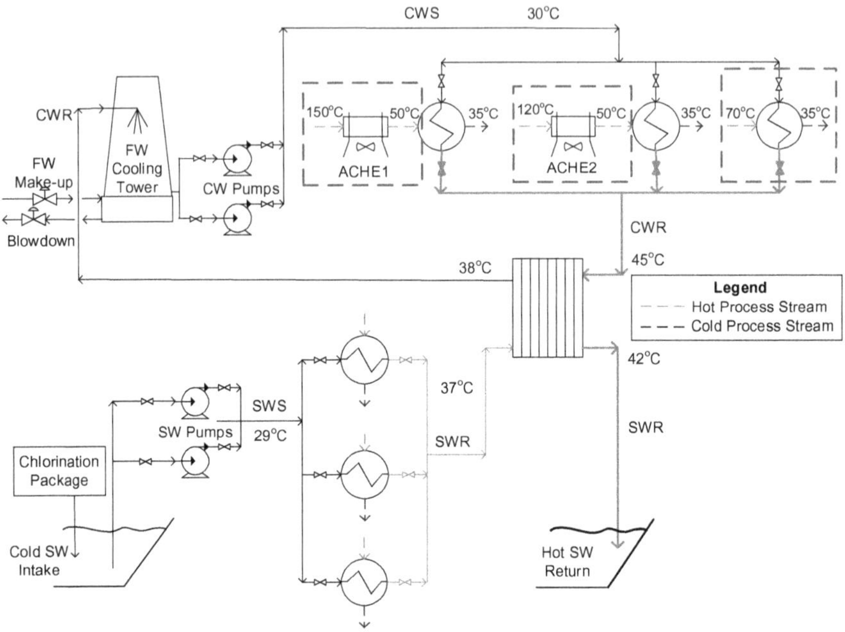

Fig. 15.20: Hybrid cooling system with ACHE and WCHEs in series (for cooling process streams) with used SW for CWR colling from 45°C to 38°C before CT. Note that SW is used for cooling some other process streams before using for CWR cooling. New units are shown in dashed rectangles.

15.7 Strategies for Optimizing Cooling Systems

Different types of CW systems are selected in projects depending on many parameters such as available type and quantities of water sources, weather conditions, accepted corrosion and fouling rates, and costs. Frequently, several options are available. Hence process engineers usually perform screening or conceptual studies to evaluate and select the best cooling option. Often, this is a tedious process. Methodology outlined in Fig. 15.21 can be very helpful to establish the best cooling solution for WHR projects. The best strategy for cooling hot process streams is exchanging heat as much as possible with colder process streams (i.e., heat integration described in Chapters 2 and 3). If this option is not feasible or already fully utilized, and if further cooling duty is required, one should review the availability of water-cooled systems.

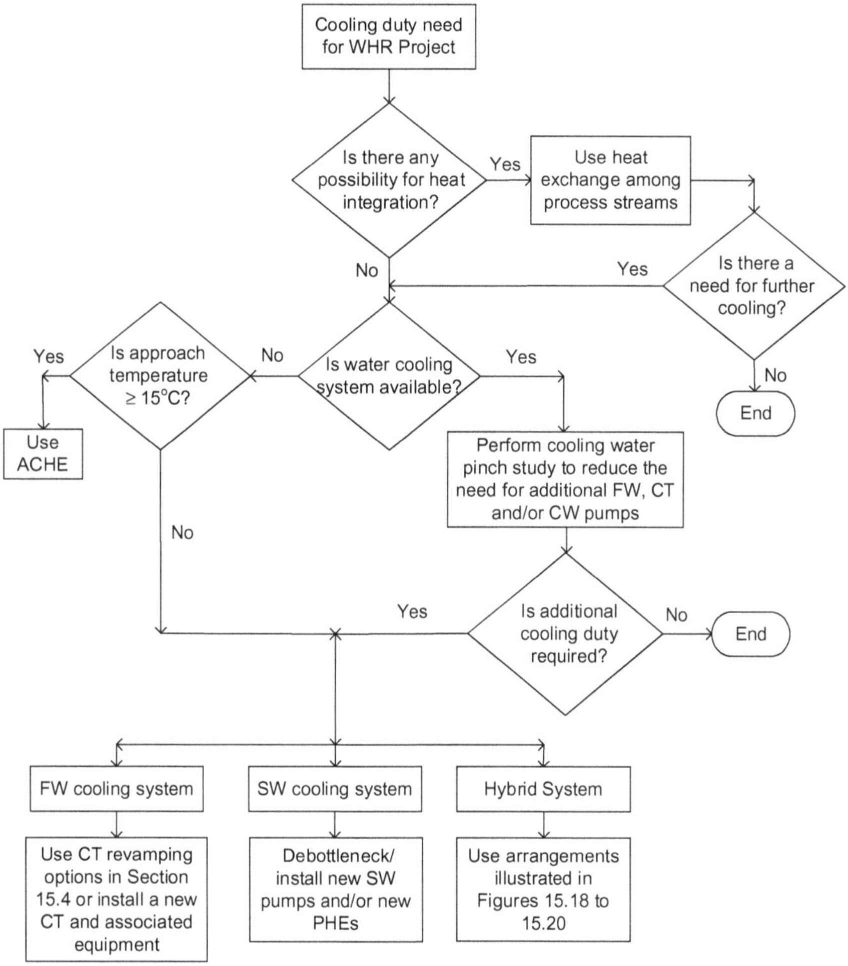

Fig. 15.21: Strategies for debottlenecking industrial cooling systems.

If the water-cooled system is possible, water pinch study needs to be performed to establish the minimum CW required (i.e., target). A suitable modification scheme can be developed to retrofit the existing cooling system to use the minimum CW and use the saved CW in a new HE for meeting the cooling duty of the new WHR project. If the water-cooled system is not feasible, one should carefully examine the approach temperature at the cold end of the (existing) coolers. If the approach temperature is >15°C, addition of ACHE before such coolers may be cost effective. If the approach temperature is ≤15°C or cooling duty is required even after CW pinch design implementation, one should explore other revamp/new equipment options for FW/SW cooling systems (outlined in Section 15.4.2) and/or use of hybrid cooling systems (outlined in Section 15.6).

15.8 Summary

This chapter presented a comprehensive review of cooling options applicable for WHR projects, which are very useful for practising engineers and researchers. Following are the key learning points of this chapter.

- WCHE and ACHEs are the commonly used cooling equipment in process industries.
- To minimize FW demand, process plants located near sea can effectively use SW cooling systems.
- CW pinch can be performed to establish the minimum CW demand.
- With appropriate modifications, one can avoid the need for installing a new CT and associated pumps, for meeting the new cooling duty required in a WHR project.
- CT can perform at higher efficiency and capacity if CWR temperature is increased and/or circulating water flow rate is decreased.
- Cooling duty of FW/SW system can be increased by many options summarized in Section 15.4.3.
- Cooling duty of ACHE can be improved by use of high-efficiency fans, and options for this are discussed in Section 15.5.
- Use of hybrid cooling systems can provide beneficial design options for WHR projects as it would minimize the need for FW, CTs and/or SW pumping systems.

Chapters 24 and 26 in Smith[11] cover many technical details of CW optimization and design of various cooling equipment. It is recommended for further reading on the topics of this chapter. Reddy *et al.*[4] presented comprehensive models with and without the use of intermediate CW header, for optimizing CW systems. These models include simultaneous hydraulic assessment to develop optimized CW system. The paper by Reddy *et al.*[4] is recommended for advanced topics related to this chapter.

References

1. Gougal M, Cham JP, Resnani HD. (May 2009) *Optimal Cooling Systems for Coastal Plants*. Chemical Engineering, pp. 45–48.
2. Raju KS. (2011) *Fluid Mechanics, Heat Transfer, and Mass Transfer: Chemical Engineering Practice*. Wiley.
3. Vengateson U. (April 2017) *Cooling Towers: Estimate Evaporation Loss and Makeup Water Requirements*. Chemical Engineering, pp. 64–67.
4. Reddy CCS, Rangaiah GP, Lim WL, Naidu SV. (2013) Holistic approach for retrofit design of cooling water networks. *Ind Eng Chem Res* **52**: 13059–13078.
5. Hall S. (2017) *Rules of Thumb for Chemical Engineers*. Butterworth-Heinemann.
6. Mandal MK. (February 2019) *Air Coolers Versus Shell and Tube Water Coolers*. Chemical Engineering, pp. 42–49.
7. Mukherjee R. (February 1997) *Effectively Design Air-Cooled Heat Exchangers*. Chemical Engineering, pp. 26–47.
8. API Std. 661. (July 2013) *Petroleum, Petrochemical, and Natural Gas Industries — Air-Cooled Heat Exchangers*, 7th ed.
9. Couper JR, Roy PW, James RF, Walas SM. (2005) *Chemical Process Equipment: Selection and Design*. Gulf Professional Publishing.
10. Eduardo C. (2010) *Heat Transfer in Process Engineering*. McGraw-Hill Education.
11. Smith R. (2016) *Chemical Process Design and Integration*, 2nd ed. Wiley.

Acronyms and Notation

Acronyms

ACHE Air-Cooled Heat Exchanger
BD Blowdown
COC Cycles of Concentration

CS	Carbon Steel
CT	Cooling Tower
CWS	Cooling Water Supply
CWR	Colling Water Return
DM	Demineralized
FW	Freshwater
HE	Heat Exchanger
LMTD	Log Mean Temperature Difference
NTU	Number of Transfer Units
PHE	Plate Heat Exchanger
SS	Stainless Steel
SW	Seawater
SWR	Seawater Return
SWS	Seawater Supply
WCHE	Water-Cooled Heat Exchanger
WHR	Waste Heat Recovery

Notation

A	Heat transfer area in the heat exchanger (m^2)
C	Coefficient in CT characteristic equation
E	Evaporation rate (kg/s)
C_p	Heat capacity at constant pressure (kJ/kg.K = kJ/kg.°C)
h	Heat transfer coefficient ($kW/m^2{\cdot}K = kW/m^2.°C$)
h_w	Enthalpy of saturated air film, surrounding a water droplet (kJ/kg)
h_a	Enthalpy of air at dry-bulb temperature (kJ/kg)
H_l	Heat of vaporization of water (kJ/kg)
$\frac{Kav}{m_w}$	CT characteristic (dimensionless)
M	Mass flow rate (kg/h)
N	Exponent in CT characteristic equation
Q	Amount of cooling duty (kW)
MW	Molecular weight (kg/kmole)
p	Atmospheric pressure (kPa)
p_{ws}	Saturation vapour pressure of water (kPa)
pp_w	Partial pressure of water vapour (kPa)

T	Temperature (°C or K)
U	Overall heat transfer coefficient ($kW/m^2.K = kW/m^2.°C$)
W	Humidity ratio of dry air
W_s	Humidity ratio of wet air
Z	CT site altitude (m)

Greek Symbols

φ	Relative humidity

Subscripts/Superscripts

a	Air
a1	Condition of inlet air to a CT
a2	Condition of outlet air from a CT
CWS	CW supply
CWR	CW return
db	Dry bulb
w	Water
wb	Wet bulb

Exercises

15.1 State the advantages and disadvantages of ACHEs compared to FW and SW cooling systems.

15.2 Assuming a wet-bulb temperature of 29°C, estimate the percent increase in blowdown loss and evaporation rate if the CWR temperature is increased from 40°C to 50°C, while maintaining the CWS temperature of 30°C.

15.3 List key considerations for re-use of CW and SW.

15.4 A process plant is using 300 ton/hr of CW with CWS temperature of 30°C from a FW CT, and CWR temperature of 36°C. Existing CT is designed to supply 300 ton/hr of CW with a CWS and CWR temperatures of 30°C and 45°C, respectively. A new project needs 200 ton/hr of CW with CWS and CWR temperatures of 30°C and

40°C, respectively. Based on detailed assessment, the new project can use CW with a maximum CWS and CWR temperatures of 40°C and 52°C, respectively. Assess whether a new CT is required for this project. Using the cooling water system pinch study (outlined in Section 15.4), develop a possible solution.

15.5 Assume that the plant in the previous exercise has a ready access to a SW system designed to supply 500 ton/hr of SW with SWS and SWR temperatures of 30°C and 40°C, respectively. This system is operating at the full design flow rate but with a SWR temperature of 36°C only. The new project now needs 250 ton/hr of CW with maximum CWS and CWR temperatures of 36°C and 52°C, respectively. Is a FW–SW hybrid solution possible (instead of a new CT) for this case? If yes, suggest one possible solution.

Chapter 16

Waste Heat Recovery in and Optimization of Steam Systems

16.1 Overview

Steam is the primary source of thermal energy (heat) below 260°C, in process plants. It has the advantage of very high heat transfer coefficient (due to phase change from vapour to liquid) over other heating media. Further, steam is relatively safer compared to other heating media such as combustion of fuels and hot oils. In addition to process heating, it is mainly used in steam turbine (ST) drivers for improving reliability of pumps/compressors, in steam jet ejectors (SJEs) as motive steam for thermal compression and for (electric) power generation. Other uses are in atomization, refrigeration, reaction feed such as in steam reforming and steam cracking, equipment cleaning, humidification and moisturization.

A steam system is present in almost all process plants. It supplies approximately 30% of the energy used in a typical petroleum refinery,[1] and contributes significantly to the operating cost of process industries. Hence, it plays a vital role for maximizing waste heat recovery (WHR) of the plant. Energy and costs saved by its optimization can substantially improve both profitability and sustainability of the plant.

The steam system of a process plant can be divided into four parts: steam generation, steam distribution, steam usage and condensate recovery system. This chapter begins with an outline of these parts of a steam system in Section 16.2. Next, Section 16.3 illustrates WHR opportunities in the steam generation system. Section 16.4 outlines WHR opportunities in steam distribution. The subsequent section presents WHR opportunities, efficient methods and strategies for steam usage. Section16.6 presents a steam network model for optimizing power generation. This chapter ends with summary in Section 16.7. Condensate recovery system, which is an important part of a steam system, is covered in Chapter 17.

Learning outcomes of this chapter are as follows.

1. Describe steam system components.
2. Evaluate blowdown (BD)/flash steam WHR opportunity.
3. Identify the WHR opportunities in steam distribution.
4. Identify the opportunities for maximizing WHR in steam usage.
5. Calculate potential savings from steam system WHR.
6. Describe the various types of ST drivers and evaluate their relative efficiency.
7. Identify the opportunities for optimizing power generation using a steam system model.
8. Discuss the true cost of steam for evaluating WHR opportunities.

16.2 Steam Systems in Process Plants

Steam is generated in plant's utility boilers, heat recovery steam generators (HRSG) in co-generation plants, waste-heat boilers (WHB) and incinerators. Some plants purchase their steam requirements from an external utility supplier. Depending on the process and power generation requirements, process plants may have steam at two to four levels of pressure. They are low-pressure steam (LPS; at 3–5 barg), medium-pressure steam (MPS; at 10–11 barg), high-pressure steam (HPS; at ~ 40 barg) and very high-pressure steam (VHPS; at ~ 100 barg). Steam systems involve large capital and operating costs. Capital cost is for boilers, STs, steam and condensate systems, demineralization plants, deaerators, condensate flash drums, and so on. Operating cost is for water make-up, fuel, electricity, chemical costs and so on. Main parts of a steam system are as follows.

Steam Generation: It includes boilers, HRSGs, WHBs, boiler auxiliaries (such as forced- or induced-draft fans, instrumentation and controls), economizers, air pre-heaters, superheaters, BD system, water treatment equipment, deaerator, feed water pumps, fuel storage and handling system, heat pumps (covered in Chapters 5 and 14) and condensate flash vessels.

Steam Distribution: It mainly consists of a number of steam headers, each at a different pressure, STs, pressure reducing stations (control valves for pressure reduction, flow control and venting), condensate drip legs, steam desuperheaters, steam accumulators (for variable steam demand) and relief valves.

Steam Usage: Examples of steam usage in process plants are for process heating in heat exchangers (including reboilers), steam stripping, vacuum generation, evaporators, dryers, thermal compression using SJEs, furnace atomization, steam jacketing/tracing, flare quench, reaction feed (e.g. methane reforming), absorption chillers, heat pumps, ST drivers and steam turbo generators.

Condensate Return System: Condensate after steam usage needs to be returned to steam generation (in order to reduce make-up water). Condensate return system consists of steam traps, control valves, condensate/flash tanks, condensate pumps (motor or ST driven), hot condensate header, cold condensate header, condensate and water recovery systems and treatment and disposal systems for contaminated condensate. It is the topic of Chapter 17.

16.3 WHR Opportunities in Steam Generation

Steam is generated at HP level in utility boilers and/or co-generation plants; later, most of it is converted to MPS and LPS by processing it through STs for generating electrical and/or shaft power to drive compressors/pumps. Some saturated MPS and LPS is generated from WHBs. Some facilities may generate VHP steam and use it for power generation in STs and eventually generate HPS, MPS and LPS.

16.3.1 *Boilers*

Boilers in the process plant generally produce HPS. Steam production rate is controlled by the pressure control of the plant's steam network.

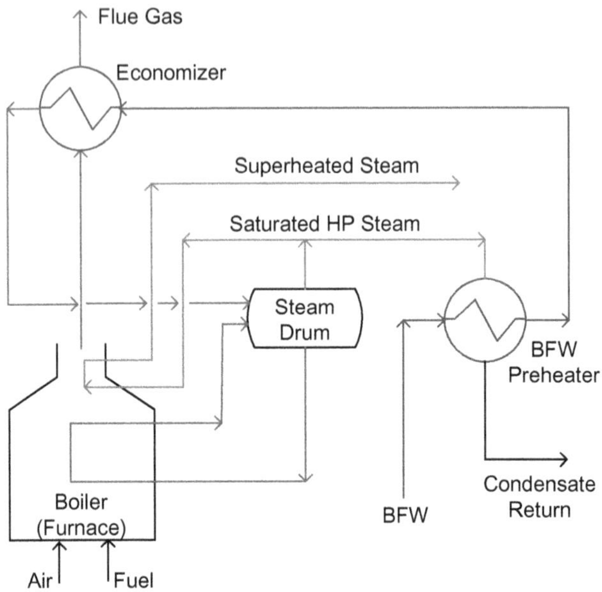

Fig. 16.1: Typical utility boiler with economizer and BFW pre-heater.

High-efficiency boilers are installed with non-condensing or condensing economizers (to recover heat from flue gases) and boiler feed water (BFW) pre-heater. As shown in Fig. 16.1, BFW pre-heater is generally used to heat BFW to a temperature such that when BFW enters economizer, it maintains a safe tube surface temperature, above acid dew point of flue gases. Thus, it prevents corrosion of economizer tubes. HPS or MPS is used in this pre-heater before the economizer.

Efficiency of a boiler (η_B) is defined as the ratio of the energy of steam produced to the energy of fuel used to generate steam. Although it varies nonlinearly with steam generation rate, constant boiler efficiency is assumed in this chapter. This assumption is valid for small changes in steam generation rates. Required energy of fuel supplied to the boiler is given by:

$$Q_f = \frac{Q_S}{\eta_B} = \frac{\Delta H_B \times M_S}{\eta_B} \quad (16.1)$$

Here, Q_f (kW) is the heat required from fuel combustion in the boiler, Q_S (kW) is the heat content of steam, ΔH_B(kJ/kg) is the specific enthalpy of steam produced and M_S (kg/sec) is the mass flow rate of steam. Amount of fuel (M_f kg/sec) required can be found by dividing Q_f by the higher heating value (HHV)[a] (kJ/kg) of fuel used in the boiler.

Opportunities for WHR in steam generation are as follows:

- Boiler BD/flash steam heat recovery
- Economizers
- Air pre-heaters
- Direct contact WHR from flue gas
- Co-generation and tri-generation: These are highly efficient to maximize WHR, and hence will reduce the cost of power and steam.

WHR using economizers, air pre-heaters and direct contact type flue gas heat recovery details are covered in Chapter 11. Co-generation and tri-generation are covered in Chapter 10. The following sub-sections are on WHR from boiler BD.

16.3.2 *Boiler Blowdown Heat Recovery*

Boiler BD is a process for the removal of some water from boiler mud drum (located at the lower portion of a boiler, into which sediments in water settle) on continuous or intermittent basis, to prevent accumulation of impurities, which can cause scaling, corrosion, carryover and other boiler integrity problems. BD ratio is defined as the ratio of BD flowrate to deaerator feed water (i.e., demineralized [DM] water make-up) flow rate, and it is determined by conductivities of deaerator feed water and BD water using:

$$\text{BD ratio (BR)} = \frac{\text{BD flow rate}}{\text{Deaerator Feed water flow rate}} \simeq \frac{\text{Deaerator Feed water conductivity}}{\text{BD water conductivity}} \quad (16.2)$$

[a] HHV refers to higher heating or gross calorific value of the fuel.

The following equations are used to estimate the mass flow rate of BD and thermal energy loss due to BD.

$$\text{Mass flow rate of BD water, } M_{\text{BD}} = \left(\frac{\text{BR}}{1-\text{BR}}\right) \times M_{\text{S}} \quad (16.3)$$

$$\text{Thermal energy loss due to BD, } Q_{\text{BD}} = M_{\text{BD}} \times (H_{\text{BD}} - H_{\text{FW}}) \quad (16.4)$$

Here, H_{BD} and H_{FW} are specific enthalpy of BD water and deaerator feed water, respectively.

$$\text{Percentage thermal energy loss due to BD} = \frac{Q_{\text{BD}}}{M_{\text{f}} \times \text{HHV}} \times 100 \quad (16.5)$$

Example 16.1: Estimate the quantity of boiler BD and consequent thermal energy loss using the following data: boiler capacity = 50 ton of steam/hr, operating pressure of boiler mud drum = 30 barg, deaerator feed/DM water temperature = 30°C, natural gas fuel (at flow rate of 3.13 ton/hr, HHV of 52,200 kJ/kg and cost of \$396/ton), conductivity of deaerator feed and BD water are 100 and 1500 μS/cm, respectively, and annual operating hours are 8760.

Solution: From steam tables, enthalpy of BD water (saturated liquid) at 30 barg is 1017.1 kJ/kg and enthalpy of deaerator feed/DM water at 30°C = 125.7 kJ/kg. From Eqs.16.2 to 16.5:

$$\text{BD ratio} = \tfrac{100}{1500} = 0.06667$$

$$\text{BD mass flow rate} = \left(\tfrac{0.06667}{1-0.06667}\right) \times 50 \times 1000 = 3571\,\tfrac{\text{kg}}{\text{h}} = 0.9921\,\tfrac{\text{kg}}{\text{s}}$$

$$\text{Thermal loss due to BD} = 0.9921 \times (1017.1 - 125.7) = 884.3\ \text{kW}$$

$$\text{Percentage thermal loss due to BD} = \frac{884.3 \times 100}{\left(\frac{3.13 \times 1000 \times 52{,}200}{3600}\right)} = 1.948\%$$

Finally, annual energy cost due to BD is $\left(3.13 \times 396 \times 8760 \times \tfrac{1.948}{100}\right) =$ \$211,558

Most of the wasted thermal energy due to BD can be recovered using a BD heat recovery system, which consists of a BD flash drum to separate flash steam and condensate, and a condensate heat recovery exchanger to pre-heat deaerator make-up water, as shown in Fig. 16.2. Flash drum usually operates at a pressure close to that of flash steam consuming

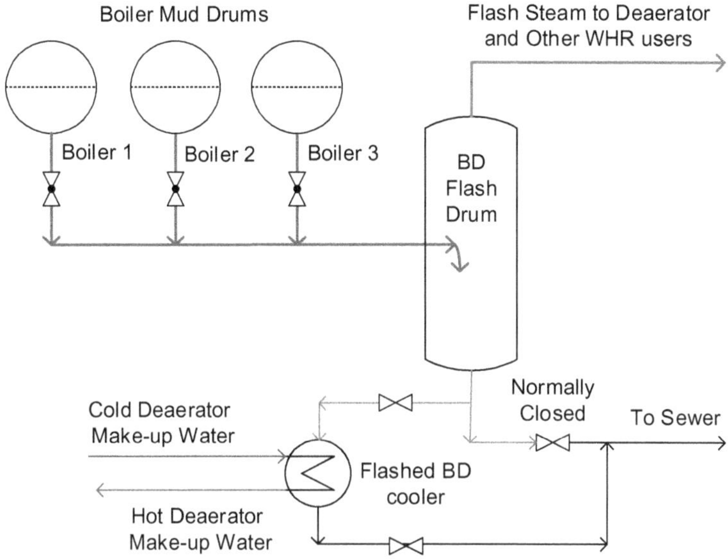

Fig. 16.2: BD WHR system showing boiler mud drums, BD flash drum and flashed BD cooler.

equipment such as deaerator. For smaller BD flow rates <1 ton/hr, condensate heat recovery exchanger is not used for economic reasons. BD WHR recovers wasted flash steam, condensate heat energy and eliminates capital and operating costs for cooling BD water using sea/raw water or cooling ponds.

Whenever BD water (i.e., high-pressure condensate) is depressurized, flash steam is generated. Flash steam generated (Flash%) from BD can be estimated using:

$$\text{Flash\%} = \frac{(Q_{\text{sh}} - Q_{\text{sl}}) \times 100}{\lambda} \tag{16.6}$$

Here, Q_{sh} and Q_{sl} (kJ/kg) are enthalpy of BD water/condensate at high pressure and flash tank (low) pressure, respectively, and λ (kJ/kg) is the latent heat of water vaporization at flash tank pressure. Table 16.1 shows the percentage of BD water/condensate converted into flash steam, at various condensate and flash drum pressures. Values in this table are found by using Eq. 16.6.

Table 16.1: Percentage of BD/condensate mass converted to flash steam in a flash tank

Condensate Pressure, barg	Flash Tank Pressure, barg						
	0	**1**	**2**	**3**	**5**	**10**	**20**
2	6.34	2.57	0.00				
4	9.82	6.14	3.63	1.66			
6	12.34	8.72	6.26	4.32	1.28		
8	14.36	10.79	8.36	6.46	3.46		
10	16.06	12.53	10.14	8.26	5.30	0.00	
12	17.55	14.06	11.69	9.83	6.91	1.68	
14	18.87	15.42	13.07	11.24	8.35	3.17	
16	20.08	16.65	14.33	12.51	9.65	4.53	
18	21.18	17.78	15.48	13.68	10.84	5.78	
20	22.21	18.84	16.55	14.76	11.95	6.94	0.00
22	23.17	19.82	17.55	15.78	12.99	8.02	1.15
24	24.07	20.74	18.49	16.73	13.97	9.04	2.23
26	24.92	21.62	19.38	17.63	14.89	10.00	3.26
28	25.73	22.45	20.23	18.49	15.77	10.92	4.23
30	26.51	23.24	21.04	19.31	16.60	11.79	5.16
32	27.25	24.00	21.81	20.09	17.40	12.62	6.05
34	27.96	24.73	22.55	20.85	18.17	13.43	6.90
36	28.64	25.43	23.27	21.57	18.92	14.20	7.73
38	29.30	26.11	23.96	22.27	19.63	14.95	8.52
40	29.94	26.77	24.62	22.95	20.32	15.67	9.29
42	30.56	27.40	25.27	23.60	20.99	16.37	10.03

Example 16.2: BD operation is performed from a boiler operating at 42 barg and a BD water flow rate of 15 ton/hr. Estimate the flash steam that can be obtained if a BD WHR system is available with flash drum pressure of 2 barg.

Solution: For BD operation from 42 barg to 2 barg, we can quickly find from Table 16.1 that 25.27% flash steam is produced. Hence, flash steam produced = 15 × 0.2527 = 3.7905 ton/hr.

16.3.3 *Sizing of Blowdown Flash Drum*

Flash drum in a BD WHR system can be designed as follows.

$$\text{Diameter of the flash vessel without demister pad, } D = \left(\frac{4 \times V_v}{\pi \times u_s}\right)^{0.5} \quad (16.7)$$

Here, V_v is the volumetric flow rate of flash steam in m^3/sec and u_S is the settling velocity of liquid drops, in m/sec. The latter is calculated by:

$$u_s = 0.07 \times 0.15 \times \left(\frac{\rho_l - \rho_v}{\rho_v}\right)^{0.5} \quad (16.8)$$

Here, ρ_l and ρ_v (kg/m^3) are the density of flashed BD water and flash steam, respectively, at the BD flash drum pressure. If demister pad is used in the flash vessel, factor of 0.15 should be removed from the above equation.[2] Assuming 10 minutes of liquid holdup time, height of the flash drum (h in m) is given by:

$$h = \left(\frac{V_l \times 10 \times 60 \times 4}{\pi \times D^2}\right) + \max(0.5\,D, 0.6) + \max(D, 1) + 0.4 \quad (16.9)$$

Here, V_l is the flashed BD liquid volumetric flow rate in m^3/sec. The last three terms on the right side of the earlier equation follow the details of the BD flash drum shown in Fig. 16.3.[2]

Example 16.3: Estimate the sizes of flash drum and condensate (i.e., Flashed BD) cooler for Example 16.2, using the following data: flashed BD water temperature at outlet of the cooler = 40°C, overall heat transfer coefficient (U) of condensate cooler = 1136 W/m^2.K (from heat exchanger design book by Serth[3]), and DM water (i.e., make-up water to deaerator) inlet and outlet temperatures are 30°C and 60°C, respectively. Cost of flash steam is US$19.18/ton. Assume no demister pad in the flash drum.

Solution: From Example 16.2, BD water flow rate = 15 ton/hr and flash steam produced = 3.7905 ton/hr. Density of flash steam (ρ_v) from steam tables is 1.658 kg/m^3. First, let's size the flash drum.

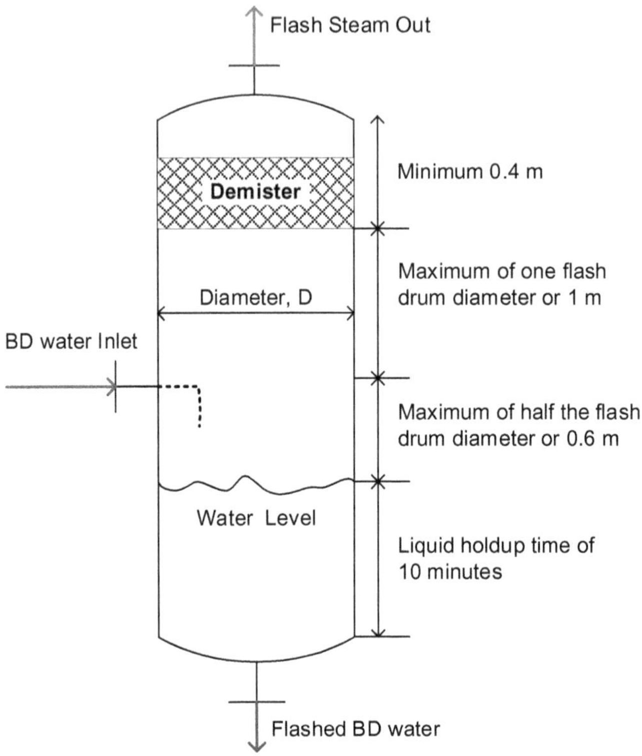

Fig. 16.3: BD flash drum design details.

Volumetric flow rate of flash steam, $V_v = \frac{3.7905 \times 1000}{1.658 \times 3600} = 0.6351 = \text{m}^3/\text{sec}$

Density of BD liquid (ρ_l) in flash drum = 931.683 kg/m^3 (from steam tables)

Volumetric flow rate of BD liquid after flashing in the vessel (V_l) = $\frac{(15-3.7905)\times 1000}{931.683 \times 3600} = 0.00334$ m^3/sec

Using Eq. 16.8, $u_s = 0.07 \times 0.15 \times \left(\frac{931.683 - 1.658}{1.658}\right)^{0.5} = 0.249$ m/s

Using Eq. 16.7, $D = \left(\frac{4 \times 0.6351}{\Pi \times 0.249}\right)^{0.5} = 1.802$ m

Using Eq. 16.9, $H = \left(\frac{0.00334 \times 10 \times 60 \times 4}{\Pi \times 1.802^2}\right) + 0.5 \times 1.802 + 1.802 + 0.4 = 3.89$ m

Thus, flash drum diameter and height are 1.802 m and 3.89 m, respectively. Let's now size the condensate (Flashed BD) cooler.

Saturation temperature of steam at 2 barg = 133.68°C (from steam tables). This is the temperature of flashed BD water at inlet to the cooler.

Hence, heat transferred by flashed BD water (heat capacity = 4.184 kJ/kg.K) to deaerator make-up water is $\frac{(15-3.7905)\times 1000\times 4.184}{3600} \times (133.68 - 40) =$ 1220.456 kW.

Log mean temperature difference (LMTD) = $\frac{((133.68-60)-(40-30))}{ln\left(\frac{(133.68-60)}{(40-30)}\right)} =$ 31.88°C

LMTD correction factor (two-shell and four-tube pass heat exchanger) = 0.863 (obtained from free online calculator available at checalc.com[b])

LMTD corrected = 31.88 × 0.863 = 27.51°C

Area of BD condensate cooler = $\frac{1220.456\times 1000}{1136\times 27.51} = 39.05\ \text{m}^2$

As can be seen in Table 16.2, payback period for the BD WHR system increases with decrease on BD water flow rate. In this table, monetary value of energy savings from flash steam is found by multiplying the amount of flash steam produced with its given cost (= $\frac{3.7905\times 19.18\times 24\times 365}{10^6}$ = 0.637 million $/year). From steam tables, enthalpy of flash steam at 2 barg = 2725 KJ/kg; based on this and given cost of flash steam, monetary value of energy of flash steam is $\frac{19.18}{1000\times 2725}$ = US$7.04 × 10^{-6}/kJ. Monetary value of energy savings due to heat recovery from flashed BD water is $\frac{(1220.456\times 3600\times 7.04\times 10^{-6}\times 24\times 365)}{10^6} = 0.27$ million $year. Capital costs of flash separator and condensate cooler are estimated using CAPCOST program (based on MS Excel), available with the design book by Turton *et al.*[4] Results for other BD water flow rates in Table 16.2 were obtained by calculations similar to the above including Exercises 16.2 and 16.3.

Table 16.2: Typical payback period for BD WHR system

BD Water Flow Rate, ton/hr	Flash Separator (without demister, 10 Min Holdup) Size: Diameter (m) × Height (m)	Condensate Cooler Area, m²	Monetary value of Energy Savings, Million $/year: Flash Steam	Monetary value of Energy Savings, Million $/year: Heat Recovered in Flashed BD Cooler	Equipment Cost, Million $: Flash Drum	Equipment Cost, Million $: Flashed BD Cooler	Payback Period, Months
15	1.81 × 3.89	39	0.637	0.27	0.086	0.09	2.3
10	1.47 × 3.39	26	0.425	0.18	0.056	0.083	2.8
5	1.04 × 2.83	13	0.212	0.09	0.033	0.077	4.4

[b] https://checalc.com/solved/LMTD_Chart.html

16.4 WHR Opportunities in Steam Distribution

A typical steam system in a large process plant is shown in Fig. 16.4. It has three boilers and three steam headers (one each for HPS, MPS and LPS), one LP condensate header (also known as hot condensate header), one cold condensate header, one DM water header, steam generators and consumers in the process, and numerous STs (although only some are shown in Fig. 16.4 for simplicity) including a few for electric power generation (of which one between HP and MP headers is shown in Fig. 16.4). To maintain good balance between steam supply and demand, some pressure control valves are installed between steam headers to depressurize steam (e.g., HP to MP, MP to LP and vent to atmosphere). Steam is distributed throughout the plant at multiple steam pressure levels (i.e., as HPS, MPS and LPS).

16.4.1 *Steam Pipe Sizing*

Right sizing of steam header is important to minimize heat losses and piping cost. Higher velocity in the header pipe leads to higher pressure drop

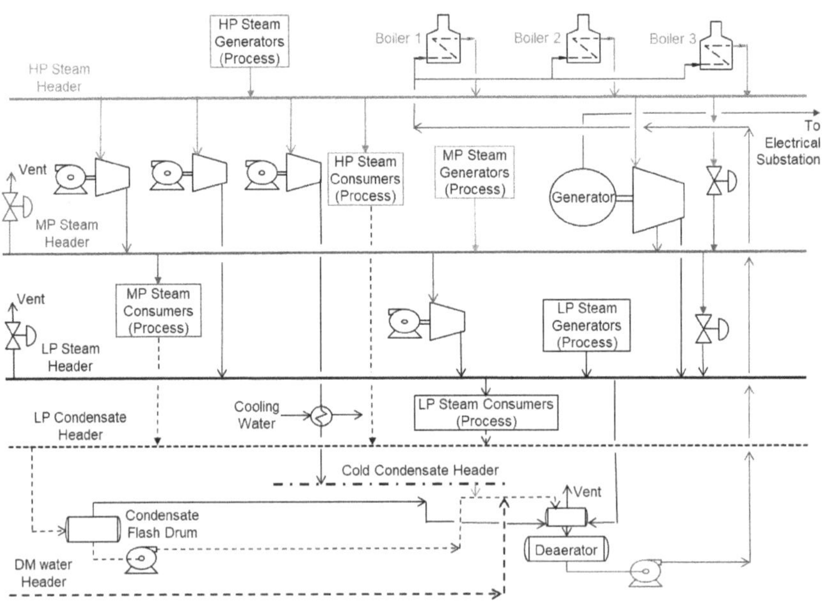

Fig. 16.4: Typical steam distribution system in a large process plant.

and erosion of pipes, whereas lower velocities increase energy losses. A conservative guideline is to limit steam velocity in piping to less than 24.4 m/sec. Maximum allowable velocity in steam pipes is 30 m/sec,[5] and steam velocity of more than 36.6 m/sec may cause piping erosion. Recommended slope for steam header in the direction of flow is 1:100 (Spirax Sarco[c]).

16.4.2 *Heat Loss Through Uninsulated Pipe and Economic Insulation Thickness*

Uninsulated (or damaged insulation at) steam/condensate pipelines leads to significant loss of thermal energy. Figs. 16.5 and 16.6 are prepared using 3Eplus®,[d] which is a free insulation evaluation software developed by the North American Insulation Manufacturers Association (NAIMA). Here, pipe temperature is assumed to be same as that of steam flowing inside the pipe. As can be seen in Figs. 16.5 and 16.6, heat loss from an uninsulated pipe increases with increase in steam temperature, pipe size and wind speed. This increase is substantial and nonlinear for larger pipe

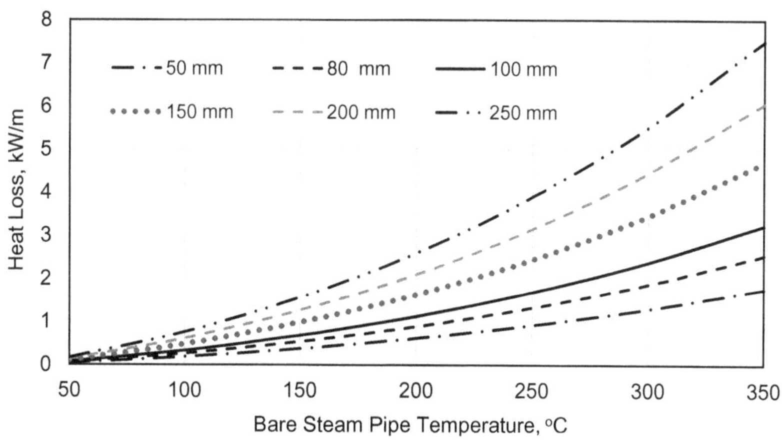

Fig. 16.5: Heat loss from bare/uninsulated pipes for various nominal pipe sizes from 50 to 250 mm, with zero wind velocity and ambient temperature of 30°C.

[c] https://www.spiraxsarco.com/learn-about-steam/steam-distribution/steam-mains-and-drainage.

[d] This software is available at https://insulationinstitute.org/tools-resources/free-3e-plus/.

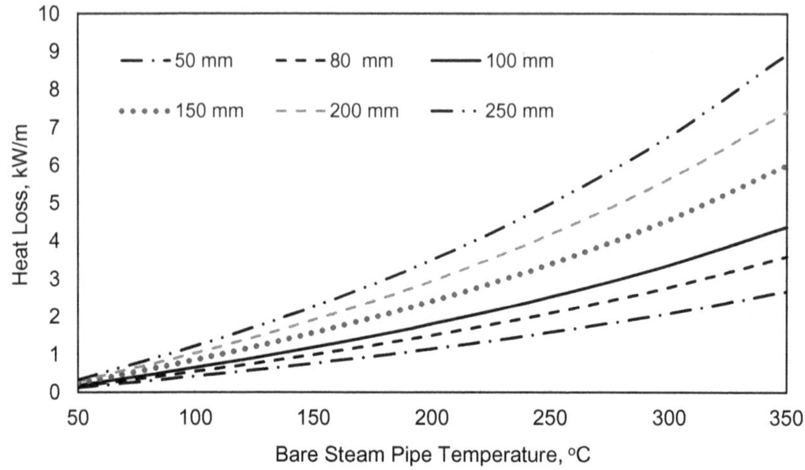

Fig. 16.6: Heat loss from bare/uninsulated pipes for various nominal pipe sizes from 50 to 250 mm, with 3 m/sec wind velocity and ambient temperature of 30°C.

sizes of 100–250 mm, which is due to its large lateral area for heat loss. Further, heat loss from a pipe of 250 mm at 200°C increases from ~ 2500 W/m in the absence of no wind to ~ 3500 W/m (by ~ 40%) in the presence of 3 m/sec (10.8 km/hr) wind. Note that wind speed depends on the location, season and time of the day, and winds in Singapore are light, with wind speed less than 2.5 m/sec.

Heat loss from a pipe (and hence wasted thermal energy) can be reduced by the application of insulation. Hence, apart from using the recommended pipe velocity, optimum insulation thickness plays a major role in reducing the heat (energy) loss from the pipe. The following two examples illustrate the calculation of heat loss from uninsulated and insulated pipes as well as reduction in heat loss and energy cost due to addition of insulation.

Example 16.4: A 10 barg steam header portion of 20 m long is uninsulated. Nominal pipe size of the header is 200 mm with a steam temperature of 200°C. Assume the header is horizontal. Ambient temperature is 30°C and wind speed is 3 m/sec. Natural gas (HHV: 52,200 kJ/kg and cost of \$396/ton) is used as fuel in the boiler operating with an efficiency of 90%. Estimate the energy loss from this uninsulated piping and its annual cost (assuming 8760 hours of operation per year).

Solution: Using 3Eplus® software or reading from Fig. 16.6, heat loss from the uninsulated pipe of 200 mm size is 2941 W/m.

Heat loss for 20 metres pipe length = 2941 × 20 = 58,820 W = 58.82 kW

Accounting for boiler efficiency, actual energy loss = $\frac{58.82}{0.9}$ = 65.36 kW

Natural gas cost = \$396/ton = $\frac{396}{1000 \times 52{,}200}$ = \$7.586 $\times 10^{-6}$/kJ

Annual cost of heat loss from the uninsulated pipe = 65.36 × 3600 × 8760 × 7.586 × 10^{-6} = \$15,636

Example 16.5: As a continuation of the previous exercise, assume the pipe has 90-mm thick calcium silicate insulation. Estimate the energy loss from this insulated steam pipe, and then annual energy/cost savings due to insulation.

Solution: Using 3Eplus® software (Fig. 16.7), heat loss from the insulated pipe = 114.8 W/m

Heat recovery efficiency = $\frac{(2941-114.8)}{2941}$ × 100 = 96.1%

Heat loss for 20 metres pipe length = 114.8 × 20 = 2296 W = 2.296 kW

Accounting for boiler efficiency, actual energy loss = $\frac{2.296}{0.9}$ = 2.551 kW

Natural gas cost = \$396/ton = $\frac{396}{1000 \times 52{,}220}$ = \$7.586 × 10^{-6}/kJ

Annual cost of heat loss from the insulated pipe = 2.551 × 3600 × 8760 × \$7.586 × 10^{-6} = \$610.3

Annual value of energy savings due to insulation = \$15636 – \$610.3 = \$15025.7

Heat loss for various insulation thicknesses for Example 16.5 (using 3EPlus® software) is shown in Fig. 16.7. These results along with cost of insulation and payback period can be used to find the optimum insulation thickness.

16.4.3 *Estimation of Steam Loss Through a Hole in a Steam Pipe*

Steam leaks in piping waste lot of energy. Napier equation[6] for estimation of steam leak in a pipe is given by:

$$M_{\text{Leak}} = 0.413 \times P_S \times D_h^2 \tag{16.10}$$

Here, M_{Leak} is the steam leak in kg/hr through the hole, P_S is the saturated steam pressure in bar and D_h is the hole diameter in millimetre.

Heat Loss Per Hour Report

Item ID:	1
Item Description:	
System Application:	Pipe - Horizontal
Dimensional Standard:	ASTM C 585 Rigid
Calculation Type:	Heat Loss Per Hour
Process Temp:	200
Ambient Temp:	30
Wind Speed:	3
NPS Pipe Size:	200

Variable Insulation Thickness	Surface Temp (°C)	Heat Loss (W/m)	Efficiency (%)
Bare	199.3	2941.00	
15.0	77.2	543.30	81.53
25.0	54.3	300.30	89.79
40.0	46.9	220.70	92.49
50.0	42.9	177.50	93.97
65.0	39.9	144.90	95.07
80.0	38.3	127.60	95.66
90.0	37.2	114.80	96.10
100.0	36.3	104.90	96.43
115.0	35.6	96.96	96.70
125.0	35.0	90.51	96.92
140.0	34.6	85.13	97.11
150.0	34.2	80.57	97.26
165.0	33.8	76.66	97.39
175.0	33.6	73.26	97.51

Fig. 16.7: Heat loss for various insulation thicknesses for Example 16.5 (using 3EPlus®).

Eq. 16.10 is used to estimate steam leak rates at various operating pressures and the results are presented in Fig. 16.8.

Example 16.6: Estimate steam leak rate through a 10-mm hole in a steam pipe operating at a pressure of 10 bar. If cost of 10 bar steam is \$22.3/ton,

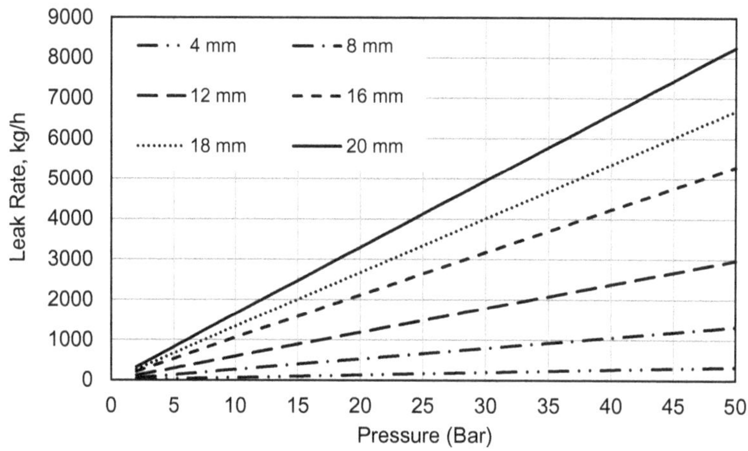

Fig. 16.8: Effect of pressure and hole diameter on leak rate in a steam pipe.

estimate the value of annual energy loss due to steam leak. Assume operating hours of 8760 per annum.

Solution

Using Eq. 16.10, steam leak rate from the pipe = $0.413 \times 10 \times 10^2 =$ 413 kg//hr

Annual steam loss due to leak = 413 × 8760 = 3617880 kg = 3617.88 ton

Annual value of energy loss due to steam leak = 3617.88 × 22.3 = \$80678.7

16.5 WHR Opportunities for Efficient Steam Usage

In many process plants, steam generation and demand vary from time to time according to changes in process operations. If they are not balanced, it may lead to problems such as steam venting, pressure relief valves lifting and drop in steam supply. Switching some ST (back pressure type) driven pumps to motor driven standby pumps can quickly decrease the demand for steam and vice versa. Along with back pressure type STs, automatic let-down valves between various steam levels are useful for balancing the steam supply and demand.

Due to WHR opportunities in process plants, there is always potential to generate LP/flash steam. However, there may not be enough demand for LP steam, in the plant. Any excess LP steam generation greater than the demand, wastes recoverable energy due to the need for venting to the atmosphere. Process plants typically address this problem by switching some ST drives generating LP steam at their exit, to electric motors, and/or reducing LP steam generation from WHR. However, these will reduce co-generation benefits as less steam flows through ST drives, to generate the power. Excess flash steam is often condensed using cooling water or air cooling, to recover condensate. However, this will lead to wastage of latent heat of flash steam. Some strategies and useful solutions for utilizing LP/flash steam are described below.

16.5.1 *Maximizing the Use of Low-Pressure/Flash Steam*

16.5.1.1 *Optimization of the deaerator pressure*

Deaerator is used for the removal of dissolved gases such as oxygen and carbon dioxide from the treated water prior to its introduction into the boiler. Schematic of a typical deaerator is shown in Fig. 16.9. Generally,

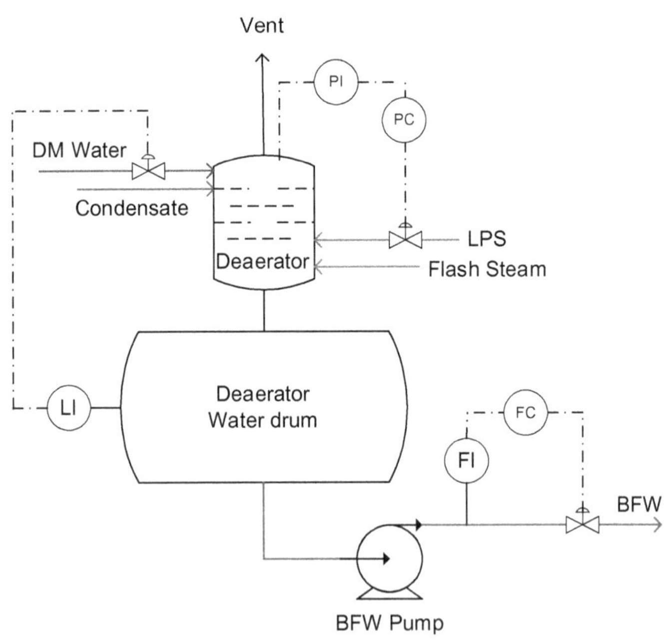

Fig. 16.9: Typical schematic of a deaerator with instrumentation and control.

mechanical design of deaerators is for a design pressure of 3.5–5 barg, whereas they typically operate at 1–1.5 barg. LP/flash steam required for heating fresh DM water increases with increase in deaerator operating pressure. One can increase deaerator operating pressure with careful assessment of safe operating pressure margin for make-up water, flash steam and condensate headers. This increases the demand for LP/flash steam.

Key benefits of higher operating pressure at deaerator are as follows:

1. Increased consumption of LP/flash steam with no new capital cost.
2. Increase WHR at the process plants.
3. Reduce/eliminate BFW heater at boilers and save fuel at boilers.
4. Reduce the demand of HPS/MPS, used for heating BFW at boilers. This steam can be used for increasing co-generation in STs.

Note that a tiny fraction (~ 0.1%) of flash steam is lost in the vent of a deaerator (Fig. 16.9).

The usefulness of deaerator pressure increase can be understood from Example 16.7, which requires the following mass and energy balances around the deaerator.

$$M_V = \propto (M_{FS} + M_{LPS}) \quad (16.11)$$

$$M_{DM} + M_{CR} + M_{LPS} + M_{FS} = M_V + M_{BFW} \quad (16.12)$$

$$M_{DM}H_{DM} + M_{CR}H_{CR} + M_{LPS}H_{LPS} + M_{FS}H_{FS} = M_VH_V + M_{BFW}H_{BFW} \quad (16.13)$$

Here, $M_{V,}$ M_{FS}, M_{LPS}, M_{DM}, M_{CR} and M_{BFW} are mass flow rates of deaerator vent flow, flash steam, LPS, make-up/DM water, condensate return and BFW, respectively; α is the fraction of LP and flash steam, vented at the deaerator; and H_{DM}, H_{CR}, H_{LPS}, H_{FS}, H_V and H_{BFW} are the specific enthalpies of make-up/DM water, condensate return, LPS, flash steam, deaerator vent and BFW, respectively.

Example 16.7: A deaerator (design pressure = 6 bar) is producing 300 ton/hr of BFW, using only make-up/DM water at 70°C and 5 bar (i.e., no condensate return). LPS is available at 5 bar, and flash steam is not available. Neglect vent flow rate. Calculate BFW temperature, LPS and make-up water flow rates by increasing deaerator operating pressure from 2 to 4 bars in steps of 0.5 bar.

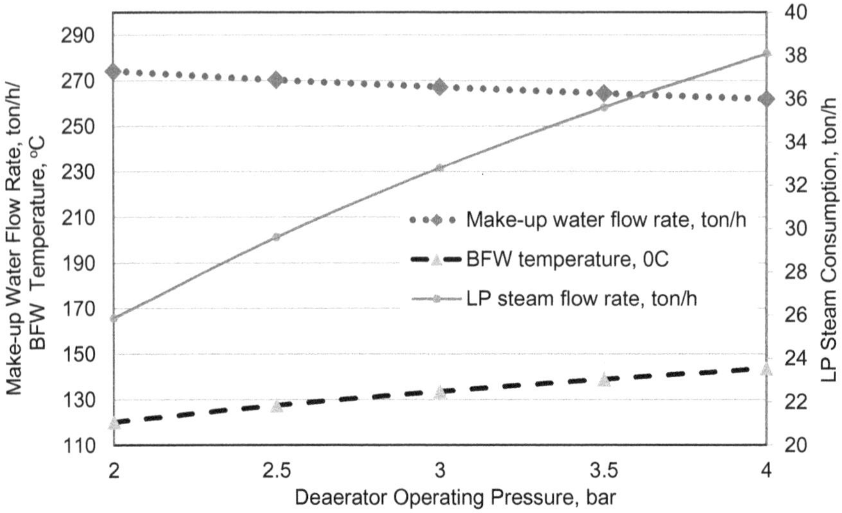

Fig. 16.10: Effect of increasing operating pressure of steam deaerator on make-up water flowrate, BFW temperature and LP steam consumption.

Solution: Chose a deaerator pressure from 2 to 4 bars in steps of 0.5 bar. Assuming BFW leaves the deaerator as saturated water, its temperature will be the saturation temperature at the chosen deaerator pressure. Find this saturation temperature and also values of H_{DM}, H_{LPS} and H_{BFW} from steam tables. Solve Eqs. 16.12 and 16.13 with $M_{CR} = M_{FS} = M_V = \alpha = 0$ and $M_{BFW} = 300$ (ton/hr) to find M_{LPS} and M_{DM}; these two equations are linear and can be solved easily. BFW temperature, M_{DM} and M_{LPS} found at different deaerator pressures are plotted in Fig. 16.10. It is clear from this figure that, as the operating pressure of deaerator increases, LPS consumption and BFW temperature increase while the make-up water flow rate decreases; owing to mass balance, increase in LPS consumption is equal to decrease in make-up water flow rate.

16.5.1.2 *Flash steam usage for combustion air heating in boilers and fired heaters*

Maximizing combustion air pre-heating at boilers/fired heaters by LP/ flash steam can reduce the fuel consumption. Often, it can also help to prevent cold-end corrosion in air pre-heaters.

16.5.1.3 *Upgrading flash steam with mechanical/thermal vapour recompression*

Steam recompression requires only 5%–10% of the energy needed to produce an equivalent amount of steam in a boiler. Flash steam (at atmospheric or slightly higher pressure) can be upgraded to low-low pressure (LLP; of 2–3 bar) and/or LP steam using mechanical vapour recompression (MVR) or thermal vapour recompression (TVR). Vapour recompression requires a mechanical compressor (in MVR) or SJE (in TVR) to increase the pressure and temperature of steam to make it usable for process heating.

MVR can be employed in applications, where the compressor inlet pressure is above atmospheric pressure. Coefficient of performance (COP) of MVR can be 10–30, depending on the compression ratio. Due to cost considerations, maximum compression ratio is restricted to 2.1 per stage.[7] Single-stage centrifugal compressor is generally used. Fig. 16.11 shows the energy requirements of MVR, to produce LLP steam at 2.5 bar, assuming compressor's polytropic efficiency of 75% and saturated steam condition at the compressor's inlet. It indicates that compressor power increases by ≈ 5 times as the flash steam pressure decreases from 2.2 to 1.2 bar.

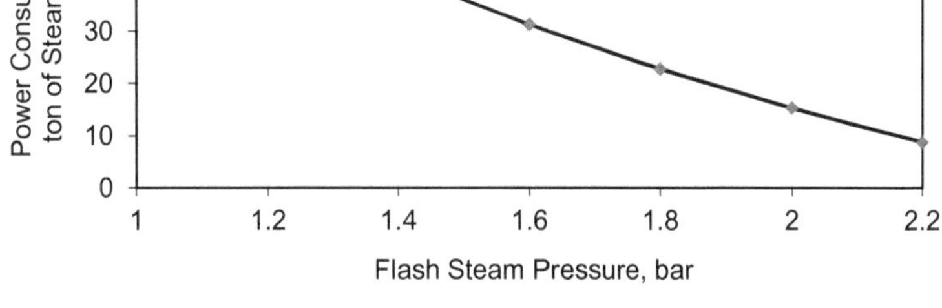

Fig. 16.11: Effect of flash steam pressure on energy requirement of MVR, for a discharge pressure of 2.5 bar.

In TVR, HPS/MPS is used as motive steam, to compress flash steam using a SJE and deliver steam at LP or an intermediate pressure. Eq. 16.14 can be used to quickly estimate the approximate quantity of motive (i.e., HP/MP) steam required for upgrading a given quantity of flash steam.[8]

$$\mathrm{RM} = 0.4 \times e^{\frac{4.6 \times \ln\left(\frac{P_D}{P_L}\right)}{\ln\left(\frac{P_M}{P_L}\right)}} \tag{16.14}$$

Here, RM is the ratio of mass flow rate of motive steam to mass flow rate of load (flash) steam, P_M is the absolute pressure of motive steam (bar), P_L is the absolute pressure of load steam (bar) and P_D is the target pressure of discharge steam (bar). This empirical equation is applicable for motive saturated steam at pressures below 20 bar and for RM between 0.5 and 6. Typical motive MPS (11.5 bar)/LPS (4.5 bar) requirements of TVR, for a discharge steam pressure of 2.5 bar, at various load steam pressures as shown in Fig. 16.12. It can be seen that RM for MPS (LPS) motive steam increases by a factor of ≈ 2 (≈ 5) when flash steam pressure decreases from 2.2 to 1.2 bar.

For upgrading smaller amounts of flash steam (e.g., in food and pharma industries), TVR is preferred as it does not require any electrical

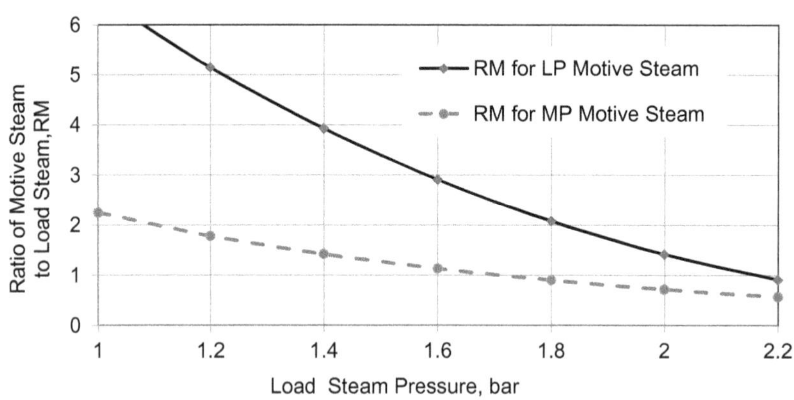

Fig. 16.12: LPS (4.5 bar) or MPS (11.5 bar) requirements for TVR for compressing load steam at different pressures to a discharge pressure of 2.5 bar.

infrastructure. MVR is generally preferred for upgrading larger quantity of flash steam in applications such as thermal desalination and heat pumps in distillation systems.

16.5.2 *Power Generation with Operating Flexibility*

Generally, steam header pressure in process plants is much higher than the required pressure on steam side in heat exchangers, due to following reasons.[9]

- Steam is generally available at two or three pressure levels only.
- Requirement of temperature driving force margin for fouled operation of heat exchangers
- Consideration for steam pressure fluctuations
- Consideration for condenser operation/duty during summer. Distillation columns will have to be operated at higher pressure as temperature of cooling water or air (in water or cooled condenser, respectively) will be higher in summer in many countries. This requires reboiler to operate at higher pressure/temperature, thus requiring higher steam pressure at the reboiler.
- Consideration for occasional peak duty requirements for process heating.

If MPS is used for services requiring the use of LPS for heating (for most of the operating time), excess pressure over LPS is removed in the steam pressure control valve. This wastes recoverable energy. In such cases, an ST as shown in Fig. 16.13 can be used for power generation. In this figure, an ST is connected to a process pump. Hence, power generated by the ST is used for driving the pump. If excess power is generated by the ST's generator (not shown in Figure 16.13) over that required for the pump, it can automatically flow into the electricity grid of the plant.[9] Thus, power is generated from steam pressure reduction. The arrangement shown in Fig. 16.13 supports export of LPS to the plant's LPS header, if reboiler duty requirement is lower, by opening control valve B. It also allows stoppage and maintenance of ST for repairs without any impact on

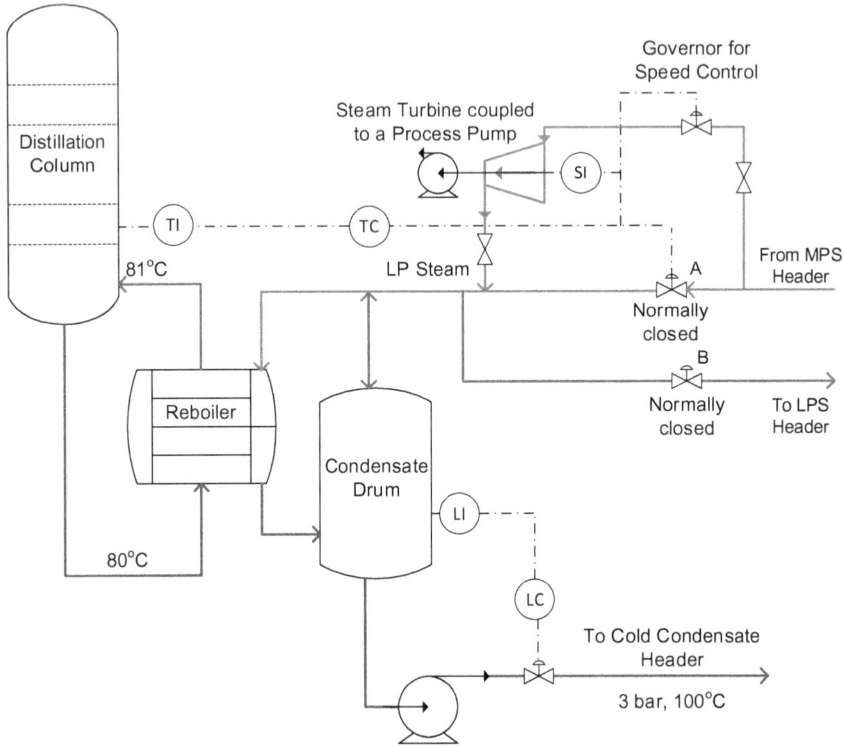

Fig. 16.13: Power generation using an ST at steam pressure reduction station.

reboiler duty, by supplying LPS through ST's bypass control valve (A). Further, if required, it can support higher reboiler duty by supplying more LPS to reboiler by partially opening bypass control valve (A).

16.5.3 *Direct Usage of Steam*

Direct usage of steam (contact steam) is more beneficial and requires lower capital cost compared to indirect steam usage. This is illustrated in the following example.

Example 16.8: A heat exchanger heats 20 kg/sec of water from 30°C to 80°C. Saturated steam at 2 bar is available for this heating. Condensate leaves the heat exchanger as saturated liquid. (a) Calculate the heat exchanger duty and the amount of steam required for this indirect heating process. (b) Calculate the required steam if it can be directly injected into water to produce 20 kg/sec of hot water at 80°C.

Solution

(a) Taking heat capacity of water as 4.183 kJ/kg.K, heat transferred to water is calculated as follows:

$$Q_W = M_W \times C_p \times (T_o - T_i) = 20 \times 4.183 \times (80 - 30) = 4183 \text{ kW}$$

Latent heat of saturated steam at 2 bar = 2201.6 kJ/kg (from steam tables)

Amount of steam required = $\frac{4183}{2201.6} = 1.90 \frac{\text{kg}}{\text{s}} = 6.84$ ton/hr

(b) For direct steam injection into water, following mass and energy balances are used.

Mass balance: $M_{W-in} + M_S = M_{W-out}$

Energy Balance: $M_{W-in} \times H_{W-in} + M_S \times H_S = M_{W-out} \times H_{W-out}$

From steam tables, enthalpy of water at 30°C ($H_{W\text{-}in}$) is 125.75 kJ/kg, enthalpy of water at 80°C ($H_{W\text{-}out}$) is 334.95 kJ/kg and enthalpy of saturated steam (H_S) at 2 bar is 2706.24 kJ/kg. With these quantities and for 20 kg/sec of hot water, mass and energy balances give:

$$M_{W-in} + M_S = 20 \text{ kg/sec} \quad \text{and}$$
$$M_{W-in} \times 125.75 + M_S \times 2706.24 = 20 \times 334.95.$$

Solution of these equations gives: $M_{W\text{-}in}$ = 18.379 kg/sec and M_S = 1.621 kg/sec = 5.836 ton/hr.

As can be seen from the earlier example, steam required for direct steam injection is lower (by 15%) than that for indirect heat exchange. This is mainly due to the complete transfer of condensate thermal energy to feed water. With direct steam injection, there is no need to recover condensate and transfer it to the condensate header. Also, as condensate mixes with water, water inlet flow rate is lower (by 8%). However, where the process stream is not water, direct steam injection leads to dilution, which may need to be taken out in order to maintain the concentration (e.g., by evaporation) at a cost.

16.5.4 *Steam Turbines*

This sub-section describes ST drivers, their advantages and disadvantages, selection between ST driver versus electric driver (motor) and ST performance.

16.5.4.1 *Steam turbine drivers*

STs are used for generating power and for driving pumps/compressors by converting a part of steam energy into power, and they can be configured in different ways. Depending on the power requirements and steam/electricity balance of a process plant, STs can be arranged between different steam pressure levels (e.g., HP to LP, HP to MP, MP to LP, HP to MP and LP, HP to MP and vacuum, HP to vacuum, MP to vacuum and LP to vacuum).

STs can be classified into two basic configurations: back pressure and condensing turbines as illustrated in Figs. 16.14 and 16.15, respectively. Outlet steam of STs is at above and below atmospheric pressure in case of back pressure and condensing type, respectively. Extraction STs produce more than one steam pressure level at their discharge; for example, back pressure type extraction ST (given in Fig. 16.16) uses HPS and produces both MPS and LPS. Power generated by the ST is proportional to the difference between the ST inlet and outlet steam enthalpy. A backpressure ST is beneficial if LPS or MPS generated by it can be used in the process. Condensing turbines (Fig. 16.15) generate more power as there is larger enthalpy difference across them. However, they are highly inefficient because of the loss of latent heat of flash steam under vacuum conditions.

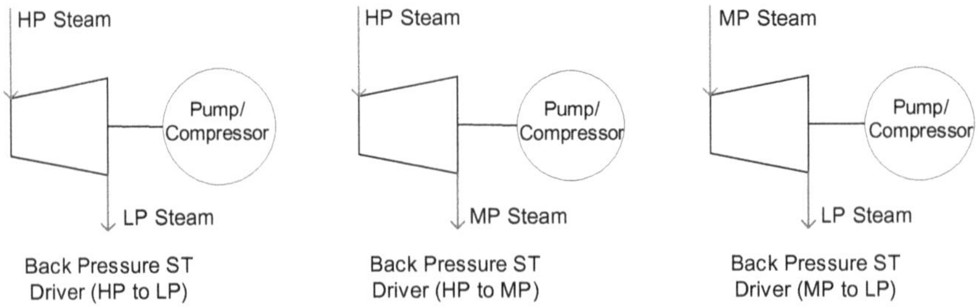

Fig. 16.14: Three examples of back pressure ST drivers.

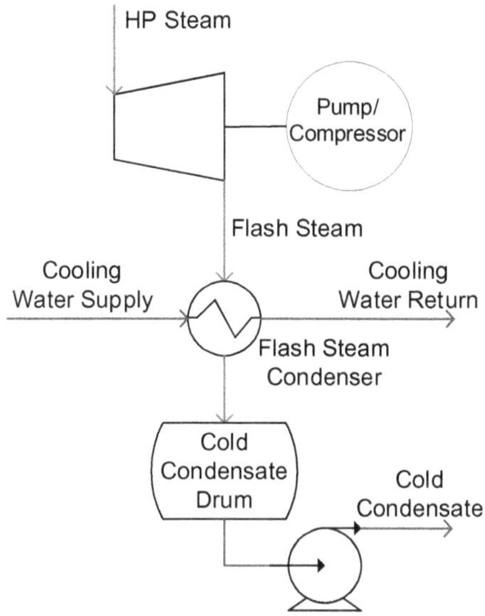

Fig. 16.15: A condensing ST driver using HPS.

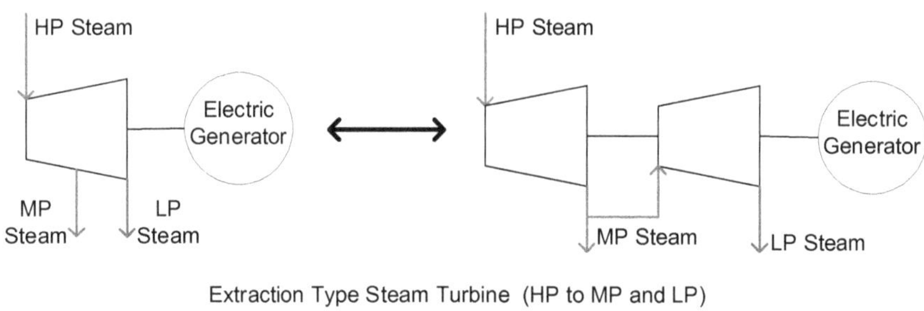

Fig. 16.16: Extraction type ST generator (left plot) and its equivalent model (right plot).

16.5.4.2 *Pros and cons of steam turbine drivers*

Advantages of ST drivers are as follows:

- Reliable operation during power failures.
- Ability to operate across a wide range of speed. So, they can act like a variable speed drive to save energy during turndown by closely matching pump/compressor curve with system head curve, without needing a throttling device such as a control valve.

- Flexibility to switch on/off to balance the steam network and prevent steam venting.
- Provide high starting torque required for constant torque loads such as positive displacement pumps.
- Installation of ST drivers does not require changes in electrical network as they are directly coupled with pumps/compressors, with or without gears.

Disadvantages of ST drivers are as follows:

- Generally, they incur more operating and maintenance costs (due to steam leaks, heat losses, wastage of some steam energy for slow rolling operation when they are on standby mode, and lower efficiency) compared to motor drivers.
- Condensing turbines require large amount of cooling water for surface condensers (Fig. 16.15).
- Lower efficiency for smaller capacities.

16.5.4.3 *Choice between steam turbine driver and electric driver*

Process pumps are generally installed in pairs with parallel arrangement (i.e., one pump in operation and the other pump on standby), for the sake of reliability. For critical services, one of the pumps is installed with ST while the other pump is installed with a motor. Switching between motor and ST drivers can lead to better balance between steam supply and demand in the plant's steam network. This is useful for maximizing cogeneration and minimizing steam venting.

The choice between ST and electric driver also depends on requirements of quick start up, pressure relief system capacity and reliability, mainly during power failures. For example, if there is excess HPS/MPS or pump operation is very critical during electricity interruptions, then an ST driver is the obvious choice. If electricity is cheaper, steam network is short of HPS/MPS, less demand for MPS/LPS and/or pump operation is not critical, then a motor driven pump is better.

Apart from critical process pumps and compressors, many equipment in utility systems such as boiler fans, boiler feed water pumps, air compressors and cooling tower pumps are usually driven by STs. Some of the non-critical pump/compressor drivers can be switched to electric motors, based on the plant's steam balance. They are called switchable STs, and provide flexibility for adjusting steam balance and co-generation potential. If ST drivers need to be always in operation for reliability and process safety considerations, they are called non-switchable STs.

16.5.4.4 *Steam turbine performance*

Enthalpy of steam is a function of temperature and pressure, and it is lower at lower temperatures and pressures. So, work produced by ST using steam at lower pressure will be low and steam consumption for the required output will be higher. In other words, at higher inlet temperature/pressure of steam, energy extraction by ST will be higher and hence steam consumption will be reduced for the required output. Typical energy utilization breakdown for various STs is shown in Table 16.3, based on steam conditions in Table 16.4.

ST efficiency in Column 2 of Table 16.3 is the overall efficiency (i.e., product of isentropic efficiency calculated using Eq. 16.15 and

Table 16.3: Energy utilization in various STs.

Turbine Type	ST Overall Efficiency	Useful Mechanical Energy, kJ/kg	Useful Heat Energy (Latent Heat Transferred for Process Heating), kJ/kg	Wasted Energy to Cooling Water, kJ/kg	Mechanical Energy as a % of Inlet Steam Enthalpy	Useful Heat Energy as a % of Inlet Steam Enthalpy	Wasted Energy to Cooling Water as a % of Inlet Steam Enthalpy
HP to LP (back pressure)	0.578	262.20	2216	0	8.4%	71.1%	0
HP to MP (back pressure)	0.442	126.45	2193	0	4.1%	70.3%	0
HP to Vacuum (condensing)	0.758	773.42	0	2153	24.8%	0	69.0%

Table 16.4: Conditions of steam used in STs presented in Table 16.3

	Steam Conditions			
Steam Level	**Pressure, Bar**	**Temperature, °C**	**Enthalpy, kJ/kg**	**Entropy kJ/Kg**
HP steam	40	360	3118.1	6.62
MP steam	12	275	2991.6	6.94
LP steam	5	200	2855.9	7.06
Flash steam (10% wet)	0.1	46	2344.7	

mechanical efficiency assumed to be unity here). For example, overall efficiency of an HP to LP ST is:

$$\frac{\left(\text{Enthapy of HPS}, 3118.1\,\frac{\text{kJ}}{\text{kg}} - \text{Enthalpy of LPS}, 2855.9\,\frac{\text{kJ}}{\text{kg}}\right)\times 1}{(\text{Enthapy of HPS}, 3118.1\frac{\text{kJ}}{\text{kg}} - \text{Enthapy of steam at LPS pressure and HPS entropy}, 2664.44\ \text{kJ/kg})} = 0.578$$

Useful mechanical energy is found by multiplying mechanical efficiency (assumed to be unity here) and enthalpy difference between ST inlet and outlet conditions; for example, for HP to LP ST, useful mechanical energy = 1 × (3118.1–2855.9) = 262.2 kJ/kg. From Table 16.3, it can be seen that back pressure type STs do not waste steam energy and are hence better for power generation while simultaneously providing LPS for process heating; on the other hand, condensing type STs waste major portion of steam energy (68.3% due to loss of latent heat to cooling water, for operation under vacuum level at their outlet) but they produce more power.

Smaller ST drivers are single-staged. They are installed with single governor control valve and one or more manually operated hand valves or automatic/horsepower valves for controlling and optimizing steam flow, when the system is operated for long periods at above/below normal capacity or operation with reduced steam pressure.[10] ST operation is a combination of inefficient isoenthalpic expansion (across governor speed control valve) and efficient isentropic expansion (across steam nozzles) in the steam chest.[11] Pressure drop across the governor valve varies greatly among smaller ST drivers, and hence it is difficult to predict the performance of these turbines, with a single model. Larger ST drivers are multi-staged and are installed with multiple automatic valves, known as steam rack, to minimize pressure drop across steam control valves.

Work produced in an ST can be calculated using the following equations.

$$W = \eta_{is} \times \eta_{mech} \times \Delta H_{is} \times M_s \tag{16.15}$$

$$\eta_{is} = \frac{H_i - H_o}{H_i - H_{is}} = \frac{\Delta H_{real}}{\Delta H_{is}} \tag{16.16}$$

Isentropic efficiency (η_{is}) changes significantly with load, whereas mechanical efficiency (η_{mech}) is typically high (at 0.95 to 0.99) and changes over a narrow range.[12] H_i and H_o are specific enthalpies of ST inlet and outlet steam, respectively. H_{is} is the specific enthalpy at ST inlet steam entropy and outlet steam pressure. Knowing inlet and outlet steam temperature and pressure, these enthalpies can be obtained from steam tables. Then, η_{is} can be computed using Eq. 16.16. With efficiency data from the vendor or assumed mechanical efficiency and a given M_s, W can be found from $\eta_{mech} \times (H_i - H_o) \times M_s$.

Smith[12] presented a predictive model for evaluating ST performance. Power generated (P) by an ST is proportional to steam flow rate (M_s) through it. This linear relationship (Eq. 16.17) over the entire range of ST operation is known as Willan's line.[12]

$$P = n \times M_s - P_{INT} \tag{16.17}$$

Here, n and P_{INT} are slope and y-axis intercept of Willan's line, respectively. For better accuracy, it is good to obtain the performance curve from the ST vendor. Fig. 16.17 shows typical ST performance from a vendor for ST inlet pressure and temperature of 11.5 bar and 254°C, and outlet pressure of 4.5 bar. Note that outlet temperature varies with steam flow rate, and rated capacity in Fig. 16.17 corresponds to normal operation.

A series combination of two back pressure type STs, one from HP to MP and another MP to LP, can be used to model HP to MP and LP extraction ST. This is illustrated in Fig. 16.16. Hence, for analysing HP to MP and LP extraction ST, Eqs. 16.15 and 16.16 can be employed for HP to MP backpressure ST, and then for MP to LP backpressure ST (using part of MP steam from the exit of HP to MP backpressure ST).

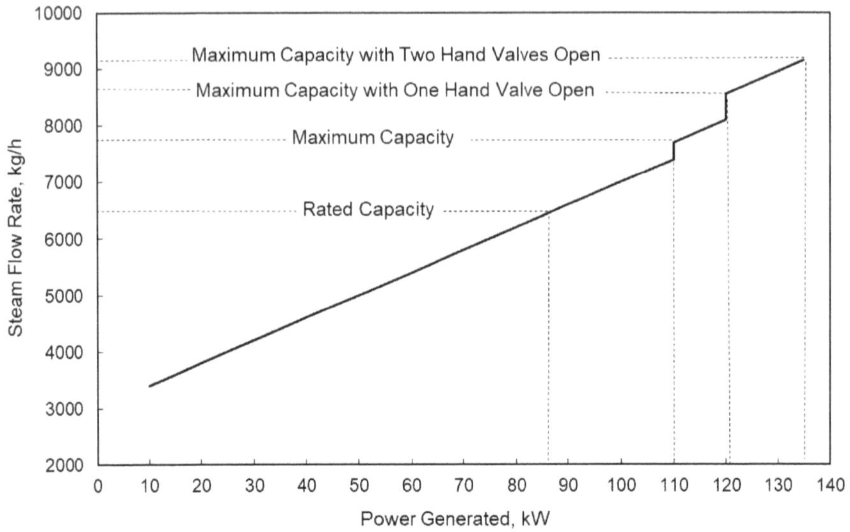

Fig. 16.17: Typical relationship between steam consumption and power generated in an ST for steam inlet at 11.5 bar and 254°C, and outlet steam at 4.5 bar.

16.5.5 *Other Strategies for Maximizing WHR in a Steam System*

Other strategies, outlined as follows, can be used to maximize WHR from a steam system.

- Minimize the venting of LPS by utilizing excess LP steam in the process, after upgrading it to MPS by absorption heat transformer or MVR or TVR; or use it for thermal desalination/chilled water production or in organic Rankine cycle.
- Maximize the use of back pressure STs for steam pressure reduction instead of pressure let-down control valves in a steam system (Fig. 16.4).
- Minimize the use of smaller STs. For example, replace several low-efficiency back pressure STs with one single high-efficiency back pressure ST; this is illustrated later in Sub-section 16.6.4.
- Use optimum setting of hand valves and governors for single-stage ST drivers. By closing hand valves, governor speed control valve will

open wider and hence consume less steam for a given power output. This is mainly due to reduction in isenthalpic effect.[11]

- Control the ST governor using pump discharge flow instead of using a control valve in the pump discharge. This will reduce steam required, due to lower pressure requirement at the pump discharge. Again, this is mainly due to avoidance of isenthalpic expansion that would otherwise happen across the flow control valve on pump discharge.[11] This will also save capital cost for pump and piping system.
- Avoid slow rolling of STs by using reliable and efficient steam condensate removal system and regular testing of auto start function of ST drive. Note that slow rolling is performed to keep STs in hot standby condition. This may consume more than 20% of rated capacity of steam flow, which does not produce any power output. Energy is lost by isenthalpic expansion from HPS/MPS to LPS.
- Recover vented steam at the STs. SJEs are used in STs to keep seals free of condensate. Motive steam used at these ejectors is generally vented to atmosphere as flash steam. If the plant has many STs, it may be economical to collect and recover this flash steam using MVR or TVR (Sub-section 16.5.1.3).
- Preferentially, use steam to supply heat duty of reboilers, rather than using fired reboilers.

16.6 Steam System Optimization

Process industries use a lot of steam in STs, mainly to improve the steam balance, reliability and availability of operations. Steam extracted from back pressure STs is used for process heating. Power generation using multiple STs is very reliable. Petroleum refineries either generate all electricity requirements internally using co-generation plants, or generate partial requirements internally using multiple STs and purchase the balance from external power supply grids. Operation of steam systems offers multiple degrees of freedom such as number of steam generators and consumers, operating pressures of steam headers, selection of drivers (multiple STs, electric motors, gas turbines, diesel engines etc.), steam pressure let-down stations and vents, which can be used for optimization.

Energy costs and environmental impact concerns have been increasing. Hence, it is important to optimize steam systems and reduce steam costs by maximizing power generation. This section presents a model for steam system optimization and then its use for several cases of optimization.

16.6.1 *Optimization Model*

Process simulators are useful for analysing steam networks, but they may be expensive and may not provide flexibility for evaluating all possible scenarios. Advanced models using specialized optimization software may not be understood by users. A third option is to use spreadsheet models such as ProSteam™ (available from KBC Energy Services), which incurs a licence fee. As an effective alternative to these, this chapter presents a spreadsheet-based mixed integer linear programming (MILP) optimization model, with built-in physical properties database; this model can be modified by users for studying energy saving schemes and arriving at accurate steam costs. Solution of this model determines the optimum operational configuration of ST drivers for maximum power generation, while satisfying constraints on steam system mass balances, ST selection and capacity limitations. MILP model is most suitable for large petrochemical plants and petroleum refineries. Hence, all the case studies in this chapter are presented and solved as MILP models. The spreadsheet-based model can easily be linked with the distributed control system of the plant, for achieving real-time optimization.

The goal of the proposed optimization model for the steam network is to maximize total power generation (which includes both the total shaft power from ST drivers used for driving pumps/compressors and the electrical power from steam turbo generators), with the consideration of the following constraints:

- Mass and energy balances
- Variable steam flow through steam turbo generator (assumed to be only one in the case study)
- Type of ST drivers

- Choice of switchable (ST or electric motor) drivers
- Capacity limitations of the selected STs
- Constraints to ensure operation of steam turbo generator within its capacity limits

Steam rate (SR, kg/kWh) of an ST is defined as the ratio of steam flow rate (kg/hr) through it, to the power produced (kW) by it. This is used for the selection of efficient STs (i.e., those with low SR). Steam flow rate through the turbo generator can be varied during optimization, using linear Willan's line (Eq. 16.17) for HP to MP and MP to LP extraction.

Reddy *et al.* [13] presented the following MILP formulation for a steam network (Fig. 16.18) with 90 ST drivers and a single steam turbo generator. Of the ST drivers, 65 are switchable and 25 are non-switchable. Significance of symbols in the following equations is given in the Nomenclature section at the end of this chapter.

Objective Function: Maximize total power generation, P = total shaft power from ST drivers + electrical power from the steam turbo generator. In equation form, the objective function is given by

$$P = \sum_{i=1}^{p}\left(\frac{M_i^{\mathrm{HL}}}{SR_i^{\mathrm{HL}}}\right)Y_i^{\mathrm{HL}} + \sum_{j=1}^{q}\left(\frac{M_j^{\mathrm{HM}}}{SR_j^{\mathrm{HM}}}\right)Y_j^{\mathrm{HM}} + \sum_{k=1}^{r}\left(\frac{M_k^{\mathrm{HC}}}{SR_k^{\mathrm{HC}}}\right)$$

$$+\sum_{l=1}^{s}\left(\frac{M_l^{\mathrm{ML}}}{SR_l^{\mathrm{ML}}}\right)Y_l^{\mathrm{ML}} + \left(n^{\mathrm{HM}}M_G^{\mathrm{HML}} - P_{\mathrm{INT}}^{HM}\right) + \left(n^{ML}M_G^{ML} - P_{\mathrm{INT}}^{ML}\right) \quad (16.18)$$

Here, the first, second and fourth terms on the right side calculate the power generated by both switchable and non-switchable HP-LP, HP-MP and MP-LP STs, respectively. The third term is the power generated by HP to vacuum (i.e., condensing) STs, which are not switchable. The last two terms on the right side calculate the power generated by the steam turbo generator (for HP-MP and MP-LP stages) by Willan's line. Binary variable Y of each switchable ST has two values: 0 if it is not operating and 1 if it is operating, whereas Y of each non-switchable ST has only one value of 1 (i.e., always operating and not changed by optimization). Note that $p + q + r + s = 90$ to include all ST drivers.

Constraints

HP steam mass balance:

$$M_{\mathrm{B}} + M_{\mathrm{HPG}} = \sum_{i=1}^{p} M_i^{\mathrm{HL}} Y_i^{\mathrm{HL}} + \sum_{j=1}^{q} M_j^{\mathrm{HM}} Y_i^{\mathrm{HM}} + \sum_{k=0}^{r} M_k^{\mathrm{HC}} + M_G^{\mathrm{HML}} + M_{\mathrm{HPC}} + M_{\mathrm{ld}}^{\mathrm{HM}} \tag{16.19}$$

MP steam balance

$$M_{\mathrm{MPG}} + M_{\mathrm{ld}}^{\mathrm{HM}} + M_G^{\mathrm{HM}} + \sum_{j=1}^{q} M_j^{\mathrm{HM}} Y_i^{\mathrm{HM}} = \sum_{l=1}^{s} M_l^{\mathrm{ML}} Y_i^{\mathrm{ML}} + M_G^{\mathrm{ML}} + M_{\mathrm{MPC}} + M_{\mathrm{ld}}^{\mathrm{ML}} + M_v^{\mathrm{MPS}} \tag{16.20}$$

LP steam balance:

$$M_{\mathrm{LPG}} + M_{\mathrm{ld}}^{\mathrm{ML}} + M_G^{\mathrm{ML}} + \sum_{l=1}^{s} M_l^{\mathrm{ML}} Y_i^{\mathrm{ML}} + \sum_{i=1}^{p} M_i^{\mathrm{HL}} Y_i^{\mathrm{HL}} = M_{\mathrm{LPC}} + M_v^{\mathrm{LPS}} \tag{16.21}$$

The term M_{LPC} in the above equation includes LPS requirement for deaerator, which is obtained, before optimization, by solving mass and energy balance equations around deaerator. Mass balance and capacity constraints for steam turbo generator (extraction type ST):

$$M_G^{\mathrm{HML}} = M_G^{\mathrm{HM}} + M_G^{\mathrm{ML}} \tag{16.23}$$

$$M_G^{\mathrm{HML}} \le M_G^{\mathrm{HMLmax}} \tag{16.24}$$

$$M_G^{\mathrm{HML}} \ge M_G^{\mathrm{HMLmin}} \tag{16.25}$$

$$M_G^{\mathrm{HM}} \le M_G^{\mathrm{HMmax}} \tag{16.26}$$

$$M_G^{\mathrm{HM}} \ge 0 \tag{16.27}$$

$$M_G^{\mathrm{ML}} \le M_G^{\mathrm{MLmax}} \tag{16.28}$$

$$M_G^{\mathrm{ML}} \ge 0 \tag{16.29}$$

Decision variables: Binary variables are Y_i^{HL} $for\, i = 1, \ldots, p;$ Y_j^{HM} $for\, j = 1, \ldots, q$ and Y_l^{ML} $for\, l = 1, \ldots s$. Continuous variables are M_G^{HML}, M_G^{HM} and M_G^{ML}. In this model, mass flow rate through let-down (M_{ld}) and vent valves (M_{v}) are set to zero.

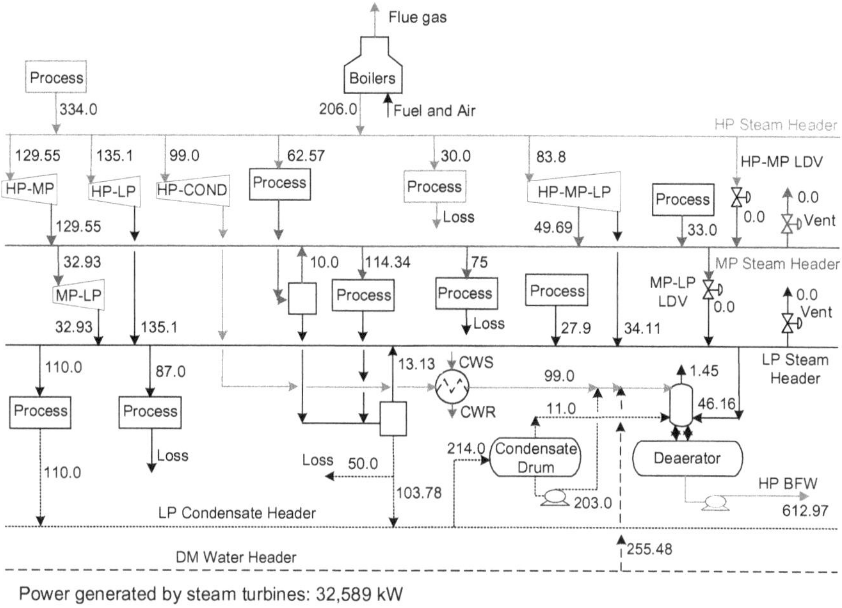

Fig. 16.18: Schematic of the steam network for the base case; HP-MP-LP is the steam turbo generator and values (without units) are steam flow rates in ton/hr.

Note that there are only three continuous variables but number of binary variables in the following case studies is 65 (for 65 switchable STs). Steam flow rate through each of switchable STs is fixed at the respective operating value if it is operating; else, it is zero. The above MILP model does not include condensate recovery system (i.e., LP condensate header, DM water header and deaerator in Fig. 16.18).

In the following sub-sections, base case, its optimization and five other cases for a typical steam system are described. The above optimization problem in various cases presented later was solved using the Solver tool in MS Excel.

16.6.2 *Base Case*

The steam network in the base case is shown in Fig. 16.18. It involves 93 ST drivers in total (of which 35 are HP to LP, 15 are HP to MP, 3 are HP to Vacuum, and 40 are MP to LP) and one steam turbo generator. Further, 65 of the ST drivers are switchable while the remaining 28 are non-switchable. Overall summary of steam generation and consumption at the process, various ST drivers and steam turbo generator is given in Fig. 16.19.

Following assumptions are made in obtaining the optimum solution for all the cases presented in this chapter.

1. Negligible pressure and temperature drop in the steam headers.
2. HPS pressure is 43 bar, and its temperature is 360°C (for ST drivers) and 380°C (for turbo generator).
3. MPS pressure is 11.5 bar and LPS pressure is 4.5 bar.
4. Steady state of HPS, MPS and LPS generation and consumption at process units, ST drivers and steam turbo generator
5. Generation from process units is saturated steam at MP or LP header pressure.
6. BD for each boiler is 2% of its steam generation rate.
7. Annual operating hours of 8760.

16.6.3 *Case A: Optimization of the Base Case*

In this case, steam flow to the steam turbo generator can be varied within the equipment limits. Steam generation and consumption in the process remain the same. Switchable turbines can be made on/off based on mass balances of steam headers. The optimization problem was solved and the optimized network for the base case is shown in Fig. 16.20. Optimal steam flows, SR and status of various ST drivers, steam flow through steam turbo generator and total power generated from the steam network are given later in Tables 16.7 and 16.8.

Optimization reduced the total steam flow for power generation at HP to LP STs from 135.1 ton/hr in the base case to 114.2 ton/hr in Case A (i.e., after optimization), increased the steam flow for power generation through the steam turbo generator (from 83.8 ton/hr in the base case to 104.66 ton/hr in Case A) and through MP to LP turbines (from 32.93 ton/hr in the base case to 47.9 ton/hr in Case A). Overall, total power generation increased from 32,589 kW to 34,554 kW. This is mainly due to higher efficiency of the steam turbo generator.

Switchable STs

ST	Y	Steam flow rate, ton/h	SR, kg/kWh
HL1	1	6	32.0
HL2	0	2.4	28.0
HL3	0	4	31.0
HL4	0	6.6	35.0
HL5	0	3.4	32.0
HL6	0	14	19.0
ML1	1	2.5	58.0
ML2	1	3.2	109.0
HL7	1	5	27.0
HL8	0	4	35.0
HL9	1	2.6	35.0
HL10	0	0.9	77.5
HL11	1	2.2	99.6
HL12	1	1.05	76.5
HL13	1	1.6	85.6
ML3	0	1.45	84.8
ML4	0	0.63	95.7
HL14	0	5	34.7
HL15	0	12	22.7
HL16	0	8.32	33.4
HL17	1	9.6	19.5
ML5	0	6.95	48.0
ML6	0	1.38	55.0
HL18	1	4.5	27.5
ML7	0	2.86	62.0
ML8	0	2.32	75.3
ML9	0	5	55.5
ML10	0	3.8	59.6
ML11	0	0.9	82.9
ML12	1	0.69	115.8
HL19	0	14.89	35.7
HM1	0	1.32	55.3

ST	Y	Steam flow rate, ton/h	SR, kg/kWh
ML13	1	7.02	59.0
ML14	0	6.3	65.6
ML15	0	3.86	56.7
ML16	0	5.3	59.0
ML17	0	4.3	72.5
ML18	1	0.69	105.6
ML19	1	0.245	196.7
ML20	0	4.89	86.7
HL20	0	4.59	22.7
ML21	0	0.96	77.8
ML22	0	0.89	96.7
ML23	0	5.4	77.4
HL21	0	5.06	15.9
HL22	0	3.9	17.5
ML24	0	2.95	62.0
ML25	0	3.86	63.0
HL23	0	4.9	28.6
HL24	1	12.22	44.5
HM2	0	5.67	73.0
HM-3	1	9.5	25.9
HM-4	1	8.88	57.9
HM5	0	7.69	56.2
HM6	1	18	27.6
ML26	0	3.26	43.6
ML27	0	4.43	75.0
ML28	0	0.88	140.0
ML29	0	3.21	57.5
ML30	0	0.65	64.5
ML31	0	8.47	59.7
ML32	1	2.07	49.6
ML33	0	0.78	144.4
ML34	1	0.49	66.8
ML35	0	2.89	66.5

Non-switchable STs

ST	Steam flow rate, ton/h	SR, kg/kWh
ML36	8	44.2
HM7	7.2	118.0
HM8	6	65.0
HL25	1.2	53.7
HL26	3.28	69.6
HL27	1.2	54.0
HL28	1.8	72.8
HL29	1.8	59.4
HL30	3	89.4
HL31	3	79.0
HL32	38	13.7
HM9	14	39.5
HM10	6.25	44.8
HM11	4.89	46.3
HL33	13.92	41
HL34	8.9	19
HL35	14.23	23.5
HM12	11.66	38.5
HM13	16.5	35.6
HM14	9.45	34
HM15	17.22	33.5
ML37	3.89	39.2
ML38	1.32	49.5
ML39	1.69	77.8
ML40	1.12	66.5

Condensing turbines

HC1	15	6.521
HC2	19	6.6
HC3	65	5.88

Fig. 16.19: (*Continued*)

Flow rates to steam turbo generator

HPS Inlet flow, ton/h	83.8
MP extraction flow, ton/h	49.69
LP exhaust flow, ton/h	34.11
Maximum steam flow to inlet, ton/h	105
Minimum steam flow to inlet, ton/h	40
Maximum MP extraction steam flow, ton/h	65
Maximum LP exhaust flow, ton/h	40

	Generation, ton/h			**Consumption, ton/h**	
	Process	Boilers	STs	Process	STs
HPS	334	206	0	92.5	447.5
MPS	43.03	0	179.24	189.34	32.93
LPS	41	0	202.1	243.1	0

Total power generation, kW	32589

Notes:

Extraction turbine power (HP to MP), kW = 284 x (HP steam flow rate at turbine inlet, ton/h) x (1000/3600) - 1382.4

Extraction turbine power (MP to LP), kW = 179 x (MP to LP steam flow rate, ton/h) x (1000/3600) - 495.5

Selection parameter (Y), for example, for HL2 and ML11 are denoted by $Y^{HL}{}_2$ and $Y^{ML}{}_{11}$, respectively.

Fig. 16.19: Input data for the base case.

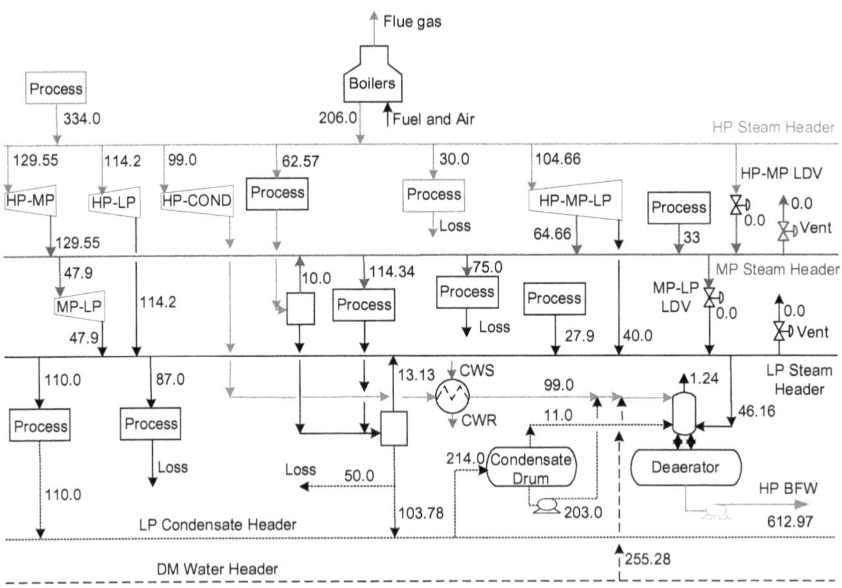

Power generated by steam turbines: 34,554 kW

Fig. 16.20: Schematic of the steam network for the optimized base case, Case A.

16.6.4 *Case B: Replacement of Smaller Steam Turbines*

ST drivers have lot of variation in efficiency, as shown by SR values given in Fig. 16.19; recall that lower SR means higher efficiency and vice versa. This is mainly due to pressure drop across the governing valves and ST windage losses. Industrial data show that small (~5 kW) to medium (~1000 kW) ST driver efficiency varies between 8% and 60% (i.e., SR between 100 and 10). Operating cost of an electric motor driver can be 30%–40% lower than that for the ST driver.[14] So, it is advantageous if low-efficiency ST drivers can be replaced by electrical drives.

However, non-switchable turbines pose problems for maximizing power generation from the steam network. With proper process safety and reliability reviews, some of these ST drivers, originally designated as non-switchable drivers, can be converted to electrical motors. This will increase the electrical power demand of the site. If additional power required is unavailable at the site, one big ST-electrical power generator can be installed to supply power for these new motors.

Case B illustrates the potential of replacing some non-switchable STs by electric motors. For this case, six non-switchable STs selected for motorization are shown in Table 16.5; their efficiency is between 8.33% and 13.79%. Further, a new high-efficiency (70%) ST power generator is considered between HP and LP headers, to generate electrical power required for the new motors. Optimization was repeated with these

Table 16.5: Details of non-switchable STs considered for motorization in Case B

Turbine	Actual Flow Rate, ton/hr	SR, kg/kWh	Shaft Power Generated (kW)	Efficiency
HL26	3.28	69.6	47.13	10.70
HL27	1.2	54.0	22.22	13.79
HL28	1.8	72.8	24.73	10.23
HL29	1.8	59.4	30.30	12.54
HL30	3.0	89.4	33.56	8.33
HL31	3.0	79.0	37.97	9.43
Total	14.08		195.91	

changes. For Case B after optimization, total power generation increased by ~ 3% to 35,681 kW from the total power generation of 34,554 kW in Case A. Detailed benefit analysis is presented in Section 16.6.7.

Condensing STs are highly inefficient due to condensation of flash steam under vacuum conditions. Motorizing large condensing turbines require modifications and investment in electrical infrastructure. However, motorizing smaller condensing turbines is relatively easier. As a sub-case of Case B (after optimization) and for illustration, motorization of condensing ST (HC1) is considered and optimized. Benefits from motorization of HC1 turbine are presented in Table 16.6. Payback period of 2.47

Table 16.6: Benefits from motorization of condensing ST (HC1)

HP Steam Operating Cost	
HC1 HP steam flow rate	15.0 ton/hr
Enthalpy of HP steam	3160.834 kJ/kg
Boiler efficiency (assumed)	0.92
Fuel energy required at boiler	51,535,339 kJ/hr
Calorific value of fuel oil	39,539 kJ/kg
Amount of fuel saved	1.303 ton/hr
Fuel cost	376.87 US$/ton
Fuel savings	491.21 US$/hr
Cooling Water Operating Cost	
HC1 outlet steam pressure	174 mm Hg abs.
Heat removed by cooling water	35,244,049 kJ/hr
Cooling water required	1052.2 ton/hr
Cost of cooling water	0.05 US$/ton
Savings due to elimination of cooling water	52.61 US$/hr
Required Power Purchase Cost	
Power generated by HC1	2300.26 kW/hr
Cost of power taken from power grid	0.15 US$/kW
Cost for imported power if HC1 is motorized	345.04 US$/hr
Net operating savings due to motorization of HC1	491.21 + 52.61 − 345.04 = 198.78 US$/hr
Annual savings due to motorization of HC1	1,741,311 US$/year
Capital cost for electric motor	358,000 US$
Payback period	2.47 months

months for this modification is very attractive. Note that payback period depends on costs of fuel and power from the grid.

16.6.5 *Case C: Using Low-Pressure Steam in Reboilers*

The best steam for heating at reboilers is dry saturated steam, which has higher film heat transfer coefficient than superheated steam. Lower the steam pressure, the higher is the latent heat of steam. Steam saturation temperature may be selected as 20°C–30°C higher than the process target temperature (at reboiler outlet), to ensure nucleate boiling. In most cases, process plants have three steam headers (HP, MP and LP). Depending on the process heating temperature in the reboiler, it may be possible to generate (via thermo compressors) and then use intermediate pressure (IP) steam from LPS/MPS, instead of HPS. Thermo compressors are essentially SJEs, which utilize energy of HPS (motive steam) to compress LPS/MPS (load), to generate IP steam. For generation of IP steam, a chart is available (Figure 4.8 in Power[15]) to estimate the motive steam required for a given load steam and pressures of motive, load and IP steam.

For Case C, two process reboilers in the base case are selected for modification; these are consuming 14.94 ton/hr and 24 ton/hr of saturated HP steam at 43 bar. Process target temperatures are 175.56°C and 185°C with reboiler inlet temperatures of 168.89°C and 176°C, respectively. For this case, IP steam at 18.8 bar (saturation temperature of 209.3°C) is sufficient to ensure temperature driving force of 33.7°C and 24.3°C at these two reboilers. The two reboilers together require 34.64 ton/hr of IP steam, whose production requires HPS and MPS for the thermo compressor, as shown in Fig. 16.21.

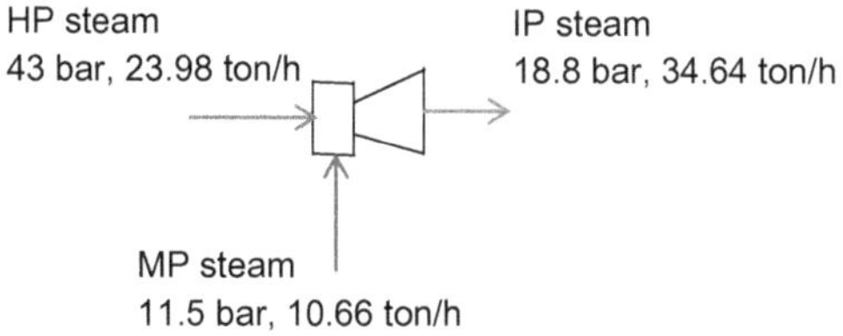

Fig. 16.21: Pressure and flow rate of inlet and outlet streams of the thermo compressor (SJE) for producing IP steam required for the two reboilers selected for modification in Case C.

The modified steam system with the thermo compressor requires a total of 14.96 (= 14.94 + 24 − 23.98) ton/hr less HPS compared to the optimized base case (Case A). However, as shown in Fig. 16.21, the thermo compressor requires 10.66 ton/hr of MPS, which can be generated by processing 10.66 ton/hr of HPS through STs. The remaining 4.3 (= 14.96 − 10.66) ton/hr of HPS can also be processed to MPS through STs. MILP model is modified with these stream flow rate changes and then solved for maximum power generation.

The optimal power generation found for Case C (i.e., base case except two reboilers using IP steam produced by the thermocompressor) is 35,069 kW, with almost all the additional steam available at HP level passing through the steam turbo generator due to its better efficiency. However, owing to the reduction in LMTD (from 77.34°C to 36.95°C and 69°C to 28.54°C) for the two reboilers, additional heat transfer area (185 m^2 and 467 m^2) and/or heat transfer enhancements are required in the existing reboilers (with area of 169 m^2 and 329 m^2, respectively). Feasibility of this substantial increase in heat transfer area in the two reboilers should be assessed carefully.

16.6.6 *Case D: Lowering Exhaust Pressure of Steam Turbo Generator*

Deaerators generally operate at 2.05–2.1 bar, whereas LPS from its header is at 4–5 bar, mainly due to process heating requirements. So, LPS is depressurized through pressure control valves, for the use at deaerators; this wastes recoverable power at STs. If LLP steam header (maintained at 3 bar) is built (for the exhaust of the steam turbo generator) to supply required steam for deaerators, additional power can be generated due to larger differential pressure across the steam turbo generator. Further, it is possible to increase steam flow rate through the turbo generator, due to higher differential pressure across it, for increasing power generation. However, for Case D, steam flow rate through the turbo generator is considered constant. Further, base case (Fig. 16.18) is assumed for the modifications in Case D.

Maximum LLP exhaust steam (at 3 bar and 141°C) from the steam turbo generator is 40 ton/hr, whereas the deaerators require more than this, that is, 7.84 ton/hr (estimated before optimization, from mass and energy balances around the deaerator), which can be supplied from LPS header. The LPS header is supplemented, in this case, by 2.47 ton/hr of LPS generation by WHR from a hot process stream in the plant, to compensate for lower heat available from LLP steam compared to LPS. Thus total LPS generation from process is 27.9 ton/hr from case A + 2.47 ton/hr from WHR = 30.37 ton/hr. Extraction turbine power (kW) for MP to LLP stage is given by kW = 252 × (MP to LLP steam flow rate, ton/hr) × (1000/3600) − 698.47, which is based on actual vendor performance of turbo generator (MP to LLP) with exhaust conditions of 3 bar and 141°C. Optimization of this case (with the use of both 40 ton/hr of LLP steam in the deaerator, and 2.47 ton/hr of additional LPS generation from a process stream) was performed after making minor changes in the MILP model presented here. After optimization, site power generation has increased to 35,164 kW, as shown in Fig. 16.22. Summary of optimization results and benefits analysis for this case are given later in Tables 16.7 and 16.8.

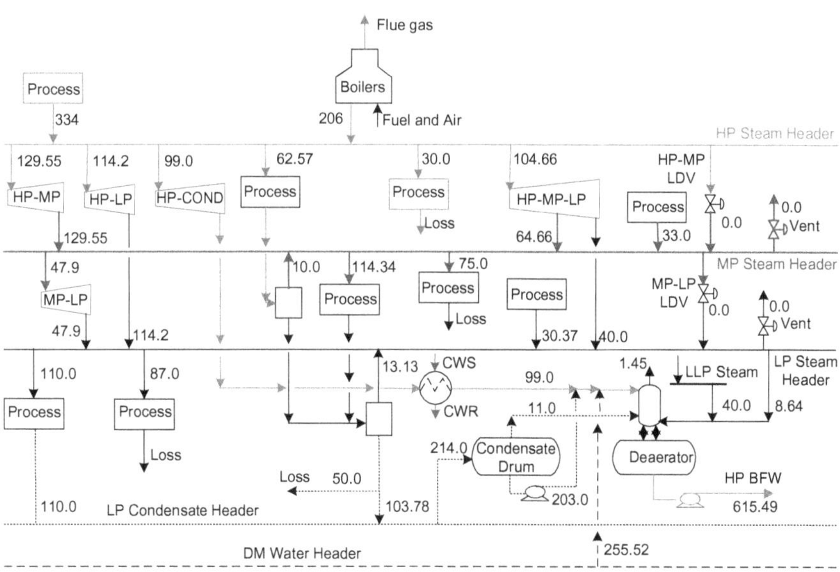

Fig. 16.22: Optimal solution for case D using both LLP steam and LPS at deaerator as well increased LPS generation by WHR in the process.

Table 16.7: Optimal status of switchable ST drivers in base case and Cases A to D

Steam Turbine	ST Selection Parameter (Y) for Case					Steam Turbine	ST Selection Parameter (Y) for Case				
	Base	**A**	**B**	**C**	**D**		**Base**	**A**	**B**	**C**	**D**
HL1	1	0	0	0	0	ML14	0	0	0	0	0
HL2	0	0	0	0	0	ML15	0	0	0	0	0
HL3	0	0	0	0	0	ML16	0	0	0	0	0
HL4	0	0	0	0	0	ML17	0	0	0	0	0
HL5	0	0	0	0	0	ML18	1	0	0	0	0
HL6	0	1	1	1	1	ML19	1	0	0	0	0
ML1	1	1	1	0	1	ML20	0	0	0	0	0
ML2	1	0	0	0	0	HL20	0	0	0	0	0
HL7	1	0	0	0	0	ML21	0	0	0	0	0
HL8	0	0	0	0	0	ML22	0	0	0	0	0
HL9	1	0	0	0	0	ML23	0	0	0	0	0
HL10	0	0	0	0	0	HL21	0	1	1	1	1
HL11	1	0	0	0	0	HL22	0	0	0	1	0
HL12	1	0	0	0	0	ML24	0	0	0	0	0
HL13	1	0	0	0	0	ML25	0	0	0	0	0
ML3	0	0	0	0	0	HL23	0	0	0	0	0
ML4	0	0	0	0	0	HL24	1	0	0	0	0
HL14	0	0	0	0	0	HM2	0	0	0	0	0
HL15	0	0	0	0	0	HM-3	1	1	1	1	1
HL16	0	0	0	0	0	HM-4	1	0	0	0	0
HL17	1	0	0	1	0	HM5	0	0	0	0	0
ML5	0	1	1	1	1	HM6	1	1	1	1	1
ML6	0	1	1	0	1	ML26	0	1	1	1	1
HL18	1	0	0	0	0	ML27	0	0	0	0	0
ML7	0	0	0	0	0	ML28	0	0	0	0	0
ML8	0	0	0	0	0	ML29	0	1	1	0	1
ML9	0	1	1	1	1	ML30	0	0	0	1	0
ML10	0	0	0	0	0	ML31	0	0	0	0	0
ML11	0	0	0	0	0	ML32	1	0	0	0	0
ML12	1	0	0	0	0	ML33	0	0	0	0	0
HL19	0	0	0	0	0	ML34	1	0	0	0	0
HM1	0	0	0	1	0	ML35	0	0	0	0	0
ML13	1	1	1	0	1						

Note: For example, ST selection parameter for HL2 and ML11 are denoted by $Y^{HL}{}_2$ and $Y^{ML}{}_{11}$, respectively.

Table 16.8: Steam flow rates for turbo generator and benefit analysis for all cases

Detail	Base Case	Case A	Case B	Case C	Case D
Turbo generator inlet flow, ton/hr	83.8	104.66	104.66	104.80	104.66
Turbo generator extraction flow, ton/hr	49.69	64.66	64.66	64.84	64.66
Turbo generator exhaust flow, ton/hr	34.11	39.99	39.99	39.95	39.99
Power from turbo generator, kW	6435	8375	8375	8384	8985
Total power generation, kW	32,589	34,554	35,681	35,069	35,164
Total savings (compared to base case), US$/year	0	2,581,883	4,063,090	3,258,606	3,383,712
Total savings (compared to optimized case), US$/year	NA	0	1,481,207	676,723	801,829
Capital cost details, US$	NA	NA	200,000 (piping) + 519,000 (ST) + 418,200 (electric motors)	154,185 (piping) + 26,071 (SJE) + 46,000 (instrumentation) + 540,000 (reboilers)	50,000 (Instrumentation) + 313,000 (WHB-kettle type)
Capital cost for modifications, US$	NA	NA	1,137,200	766,256	363,000
Payback, years	NA	NA	0.77	1.13	0.45

16.6.7 *Summary and Analysis of Results*

Optimal status of switchable ST drivers for the base case and Cases A to D are presented in Tables 16.7. Steam flow rates through the steam turbo generator and benefit analysis for all cases are summarized in Table 16.8. Total savings for each of Cases A to D in this table are calculated by the product of difference in power generation between the base case and each of Cases A to D, electricity cost of US$0.15/kWh and 8760 hrs/year. Capital costs of heat exchangers, pumps, pressure vessels are estimated using CAPCOST program (based on MS Excel), available with the book by Turton *et al.*[4] Chemical Engineering Plant Cost Index (CEPCI) value of 598.3 in July 2020 is used. Purchase costs of SJEs are derived from the chart (Figure 10.2 in Power[15]). Piping, insulation and instrumentation costs for ejectors, condensate recovery system and WHR are estimated based on industrial data.

It can be seen from Table 16.6 that the operating conditions in the base case can be optimized (in Case A), with no additional capital or operating cost, to increase total power generation by 1965 kW (6%). Except for Case A, all other cases require modifications in the steam system and hence involve capital costs. Switching from STs to motorization and use of LP steam for process heating (Cases B and C) can achieve very good savings. But their implementation requires major plant modifications such as substation works and replacement of reboilers. Case D involving lowering the exhaust pressure of steam turbo generator requires the lowest capital and the shortest payback period. For implementation of changes in Case D, turbo generator vendor should be consulted for detailed mechanical and electrical generator checks.

Payback period for all cases studied except Case C is less than a year (Table 16.6). Even for Case C, it is only 1.13 years. So, the scenarios presented are economically very attractive; they also improve energy efficiency of the steam system, which will reduce greenhouse gas emissions.

16.6.8 *True Cost of Steam*

Inaccurate steam prices affect WHR project economics and can lead to wrong business decisions. Hence, accurate steam cost is critical for reliable assessment of WHR projects. There are four approaches: enthalpy based, work based, fuel equivalent based and cost based, to estimate the cost of steam.[16] Of these, cost based method is more appropriate for economic decisions. It estimates the cost of LP steam by subtracting the credit for power generated at STs from the cost of HP or MP steam (that is estimated based on its cost of production). According to Smith,[12] cost based method reflects the true value of steam at any level and should be used to evaluate true benefits of energy conservation projects.

Here, true cost of steam for the base case is estimated using electricity cost of US$0.15/kWh, fuel heating value of 39,539.24 kJ/kg, fuel cost US$376.87/ton, boiler efficiency of 0.92, and steam flowrates and power generation values from the base case in Fig. 16.18. Let's first estimate enthalpy-based steam costs for the base case. For this, HPS (pressure 43 bar and temperature 380°C), MPS (pressure 11.5 bar and temperature 257.9°C), LPS (pressure 4.5 bar and temperature 214.9°C), and BFW (pressure 56 bar and temperature 121°C), are used.

Enthalpy-based Costs of HPS, MPS and LPS: Enthalpy-based cost of HPS is estimated by calculating equivalent fuel energy, boiler efficiency and fuel cost. Since all boilers in the present steam system have same efficiency and use the same fuel, HPS cost is estimated as $\frac{\text{Enthalpy of HPS}(3160.8\,\frac{\text{kJ}}{\text{kg}}) - \text{Enthapy of BFW}(508.04\,\frac{\text{kJ}}{\text{kg}})}{\text{Boiler efficiency }(0.92)\times\text{fuel heating value }(39539.24\,\frac{\text{kJ}}{\text{kg}})} \times$ Cost of fuel C $\left(376.87\frac{\text{US\$}}{\text{ton}}\right)$ = US$27.48/ton. Enthalpy-based cost of MPS and LPS are estimated as follows.

$$\text{MPS cost} = \frac{\text{Enthalpy of MPS}(2955.35\,\frac{\text{kJ}}{\text{kg}}) - \text{Enthapy of BFW}(508.04\,\frac{\text{kJ}}{\text{kg}})}{\text{Enthalpy of HPS}(3160.8\,\frac{\text{kJ}}{\text{kg}}) - \text{Enthapy of BFW}(508.04\,\frac{\text{kJ}}{\text{kg}})}$$

$$\times \text{Cost of HPS}\left(27.48\frac{\text{US\$}}{\text{ton}}\right) = 25.35\ \text{US\$/ton}$$

$$\text{LPS cost} = \frac{\text{Enthalpy of LPS}(2889.9\,\frac{\text{kJ}}{\text{kg}}) - \text{Enthapy of BFW}(508.04\,\frac{\text{kJ}}{\text{kg}})}{\text{Enthalpy of HPS}(3160.8\,\frac{\text{kJ}}{\text{kg}}) - \text{Enthapy of BFW}(508.04\,\frac{\text{kJ}}{\text{kg}})}$$

$$\times \text{Cost of HPS}\left(27.48\frac{\text{US\$}}{\text{ton}}\right) = 24.67\ \text{US\$/ton}$$

In summary, enthalpy-based cost of HPS, MPS and LPS are 27.48, 25.35 and 24.67 US$/ton, respectively.

True Cost for HPS: True cost of HPS is the same as enthalpy-based cost of HPS since HPS is produced in boilers and no power generation is involved. If VHPS is produced at the plant, then HPS cost will be lower than the enthalpy-based cost due to power generation credit from the ST, used for production of HPS by processing VHPS through it.

True Cost for MPS: This is estimated by weighted average cost of MPS generated by HP to MP STs, turbo generator and saturated MPS generated at the process. In the base case, a total of 213.35 ton/hr of steam is processed through HP to MP STs (129.55 ton/hr) and through steam turbo generator (83.8 ton/hr), thus producing 213.35 ton/hr of MPS and generating total power of 8720.74 kW (= 3483.53kW by HP to MP STs + 5237.21 kW by steam turbo generator). As stated, electricity cost is assumed to be US$0.15/kWh.

MPS cost after power generation credits from HP to MP STs = Cost of HPS $\left(27.48\frac{\text{US\$}}{\text{ton}}\right)-\frac{3483.53\ kW}{129.55\ \frac{ton}{h}}\times 0.15$ US$/kWh = 23.45 US$/ton

MPS cost after power generation credits from turbo generator (HP to MP) = Cost of HPS $\left(27.48\frac{\text{US\$}}{\text{ton}}\right)-\frac{5237.2\text{ kW}}{83.8\frac{\text{ton}}{\text{h}}}\times 0.15$ US$/kWh = 18.11 US$/ton

The process steam generators produce 43.03 ton/hr of saturated MPS. Cost of this MPS = $\frac{\left(\text{Enthalpy of MPS}(2782.27\frac{\text{kJ}}{\text{kg}})-\text{Enthapy of BFW}(508.04\frac{\text{kJ}}{\text{kg}})\right)\times\text{Fuel cost}(376.9\frac{\text{US\$}}{\text{ton}})}{\text{Boiler efficincy}(0.92)\times\text{Fuel heating value}(39539.24\frac{\text{kJ}}{\text{kg}})}$ = 23.56 US$/ton

Note that 34.11 ton/hr out of 83.8 ton/hr of MPS generated from steam turbo generator re-enters the steam turbo generator to produce more power by expanding to LPS. So, effectively, only 49.69 ton/hr (= 83.8 − 34.11) of MPS enters its header. This steam flow rate value is used in weighted average cost calculation for MPS, as follows.

Weighted average cost of MPS generated by HP to MP STs, turbo generator and in the process steam generators = $\frac{23.45\times 129.55+18.11\times 49.69+23.56\times 43.03}{129.55+49.69+43.03}$ = 22.28 US$/ton.

True cost for LPS: Similarly, true cost of LPS is estimated by weighted average cost of LPS generated by HP to LP STs, MP to LP STs, turbo generator and saturated LPS generated at the process. In the base case, a total of 202.14 ton/hr of steam is processed through all STs (32.93 ton/hr through MP to LP STs + 34.11 ton/hr through steam turbo generator + 135.1 ton/hr through HP to LP STs). It generates total power of 7635.07 kW (= 599.71

kW by MP to LP STs + 1197.89 kW by steam turbo generator + 5837.47 kW by HP to LP STs). As earlier, electricity cost is US$0.15/kWh.

LPS cost after power generation credits from MP to LP STs = Cost of MPS $\left(22.28\frac{\text{US\$}}{\text{ton}}\right)-\frac{599.91\ \text{kW}}{32.93\frac{\text{ton}}{\text{h}}}\times 0.15\ \text{US\$/kWh} = 19.55\ \text{US\$/ton}$

LPS cost after power generation credits from turbo generator (MP to LP) = Cost of MPS $\left(18.11\frac{\text{US\$}}{\text{ton}}\right)-\frac{1197.89\ \text{kW}}{34.11\frac{\text{ton}}{\text{h}}}\times 0.15\ \text{US\$/kWh} = 12.84\ \text{US\$/ton}$

LPS cost after power generation credits from HP to LP STs: = Cost of HPS $\left(27.48\frac{\text{US\$}}{\text{ton}}\right)-\frac{5837.47\ \text{kW}}{135.1\frac{\text{ton}}{\text{h}}}\times 0.15\ \text{US\$/kWh} = 21.0\ \text{US\$/ton}$

LPS cost after power generation credits from STs and turbo generator $\frac{(32.93\times 19.55)+(34.11\times 12.84)+(135.1\times 21)}{32.93+34.11+135.1} = 19.39\ \text{US\$/ton}$

The process steam generators produce 40.9 ton/hr of saturated LPS. Cost of this LPS = $\frac{\left(\text{Enthalpy of LPS}(2743.39\frac{\text{kJ}}{\text{kg}}) - \text{Enthapy of BFW}(508.04\frac{\text{kJ}}{\text{kg}})\right)\times \text{Fuel cost}(376.9\text{U}\frac{\text{S\$}}{\text{ton}})}{\text{Boiler efficincy}(0.92)\times \text{Fuel heating value}(39539.24\frac{\text{kJ}}{\text{kg}})}$ = 23.16 US$/ton

By mixing LPS generated by STs, turbo generator and process, overall LPS cost = $\frac{(32.93\times 19.55)+(34.11\times 12.84)+(135.1\times 21)+(40.9\times 23.16)}{243.04} = 20.02\ \text{US\$/ton}$

In summary, for the base case, true cost of MPS is 22.28 US$/ton whereas that of LPS is 20.02 US$/ton. Similarly, true cost of MPS/LPS for Cases A to D can be found, and they will be marginally lower because of optimization and/or modifications in these cases. It is clear that the true cost of MPS/LPS is lower than enthalpy-based cost of MPS/LPS at 25.35/24.67 US$/ton.

16.7 Summary

In this chapter, many WHR opportunities for steam system are presented with examples. Key points for maximizing WHR in steam generation, distribution and usage are as follows.

- BD WHR is very important and often very attractive.
- Application of economic insulation thickness is beneficial to reduce heat losses.
- Monitoring for and quick repair of steam leaks are essential to reduce steam/water losses and operating costs.

- Maximizing LP/flash steam utilization will help to maximize WHR in process plants
- Where possible, direct steam usage is more efficient compared to indirect steam usage in a heat exchanger, bearing in mind the cost of dilution, if applicable.
- Spreadsheet-based model of steam system can be employed for studying different scenarios and for optimization.
- Significant energy savings can be generated by optimizing the ST driver selection and operation.
- For maximizing co-generation effect, steam should be utilized in reboilers/heat exchangers at the lowest possible temperature/pressure level.
- True steam cost is essential for the proper evaluation of WHR and energy conservation projects.

Books by Smith[12] and Zhu[16] are recommended for further reading on the topics covered in this chapter. More details and examples on ST modelling can be found in Chapter 23 of Smith.[12] Detailed assessment on modelling steam and power system and true cost of steam can be found in Chapters 15 and 17 of Zhu,[16] respectively.

References

1. Hou A, Mita T. (May 2018) Advanced steam system optimization program. *Hydrocarb Process* 45–49.
2. Towler G, Sinnott R. (2009) *Chemical Engineering Design Principles, Practice and Economics of Plant and Process Design*, 5th ed. Elsevier.
3. Serth RW. (2007) *Process Heat Transfer: Principles and Applications*, 1st ed. Elsevier.
4. Turton R, Bailie RC, Whiting WB, Shaeiwitz JA. (2009) *Analysis, Synthesis, and Design of Chemical Processes*, 3rd ed. Prentice Hall.
5. Vengateson U. (May–June 2015) Retool heat exchanger design for different operating scenarios. *Gas Process* 23–28.
6. Branan C. (2012) *Rules of Thumb for Chemical Engineers*, 5th ed. Gulf Professional Publishing.

7. Green DW, Perry RH. (2008) *Perry's Chemical Engineers' Handbook*, 8th ed. McGraw-Hill.
8. Minton PE. (1986) *Handbook of Evaporation Technology*. Noyes Publications.
9. Lieberman NP. (2010) *Process Engineering for a Small Planet: How to Re-use, Re-purpose and Retrofit Existing Process Equipment*. Wiley.
10. Bloch HP, Singh MP. (2009) *Steam Turbines Design, Applications and Re-Rating*, 2nd ed. McGraw-Hill.
11. Lieberman NP, Lieberman ET. (2008) *A Working Guide to Process Equipment*, 3rd ed. McGraw-Hill.
12. Smith R. (2005) *Chemical Process Design and Integration*. Wiley.
13. Reddy CCS, Rangaiah GP, Naidu SV. (January 2013) Waste heat recovery methods and technologies. *Chem Eng* 28–38.
14. Tanthapanichakoon W. (January 2012) Saving energy in multilevel steam systems. *Chem Eng Prog* 27–32.
15. Power RB. (2005). *Steam Jet Ejectors for the Process Industries*, 2nd ed. McGraw-Hill.
16. Zhu F. (2014) *Energy and Process Optimization for the Process Industries*. Wiley.

Acronyms and Notation

Acronyms

BD	Blowdown
BFW	Boiler Feed Water
BR	Blowdown Ratio
CC	Cold Condensate
CEPCI	Chemical Engineering Plant Cost Index
COP	Coefficient of Performance
CWS	Cooling Water Supply
CWR	Cooling Water Return
FW	Feed Water
FS	Flash Steam
HE	Heat Exchanger
HP	High Pressure
HPS	High-Pressure Steam
IP	Intermediate Pressure
LP	Low Pressure

LLP	Low-Low Pressure
LPS	Low-Pressure Steam
MILP	Mixed Integer Linear Programming
MP	Medium Pressure
MPS	Medium-Pressure Steam
MVR	Mechanical Vapour Recompression
RM	Ratio of mass flow rate of motive steam to mass flow rate of load steam
SJE	Steam Jet Ejector
ST	Steam Turbine
SR	Steam rate for the steam turbine, kg/kWh
TVR	Thermal Vapour Recompression
VHPS	Very High-Pressure Steam
WHB	Waste Heat Boiler
WHR	Waste Heat Recovery

Notation

A	Heat transfer area of a heat exchanger (m^2)
D	Diameter
H, ΔH	Enthalpy of a stream, enthalpy change (kW)
h	Height of flash vessel (m)
i	Index for HP to LP back pressure steam turbine
j	Index for HP to MP back pressure steam turbine
k	Index for HP to Vacuum condensing steam turbine
l	Index for MP to LP back pressure steam turbine
M	Mass flow rate of a stream (kg/sec)
n	Slope of Willan's line
P	Pressure (bar) or power (kW)
p	Number of HP to LP back pressure steam turbines
P_{INT}	*Y* intercept of Willan's line
Q	Heat duty of (i.e., heat transferred from one stream to another in) a heat exchanger (kW)
q	Number of HP to MP back pressure steam turbines
s	Number of MP to LP back pressure steam turbines
T, ΔT	Temperature, temperature difference (°C or K)

U	Overall heat transfer coefficient ($W/m^2.K$)
u	Velocity (m/sec)
V	Volumetric flow rate (m^3/sec)
Y	Binary parameter
α	Fraction of flash steam and LP steam
η	Efficiency
ρ	Density (kg/m^3)

Subscripts

B	Boiler
BD	Blowdown
BFW	Boiler feed water
CR	Recovered condensate
cs	Saturated condition at condensate header pressure
D	Discharge
DM	Demineralized water
F	Fuel
G	Turbo generator or steam generation
h	Hole
FW	Deaerator feed water
HC	Hot condensate
HHV	Higher heating value
HPC	HPS consumption
HPG	HP steam generation at process units
i	Refers to inlet or inner condition(s)
is	Isentropic
L	Load steam
LH	Latent heat
M	Motive steam
MPC	MPS consumption
MPG	MP steam generation at process units
l	Liquid
ld	Pressure let-down
LMTD	Log mean temperature difference
LPC	LPS consumption

LPG	LP steam generation at process units
M	Mean value
o	Refers to outlet or outer condition(s)
S	Steam
sh	Enthalpy of high-pressure condensate/BD water
sl	Enthalpy of low-pressure condensate/BD water
v	Vent or vapour
W-in	Water in
W-out	Water out

Superscripts

HC	HP to vacuum steam (condensation)
HL	HP to LP back pressure steam turbine
HM	HP to MP back pressure steam turbine
HML	HP to MP and LP extraction type steam turbine
ML	MP to LP steam
MPS	MP steam
LPS	LP steam

Exercises

16.1. A boiler is producing 100 ton/hr of steam at 40 bar. Temperature of DM water make-up to the deaerator is 25°C. Natural gas (HHV = 52,200 kJ/kg, cost = \$300/ton) is used as fuel at the boiler, which operates at an efficiency of 85%. Conductivity of DM water and BD water are 100 and 1000 μS/cm, respectively. Estimate the quantity of BD and percentage thermal energy loss due to BD. Assume 8000 operating hours per year.

16.2. A 40 bar saturated steam header portion of 10 m long is uninsulated. Nominal pipe size of the header is 150 mm. Ambient temperature is 20°C and wind speed is 2 m/sec. Natural gas (HHV = 52,200 kJ/kg, cost = \$300/ton) is used as fuel in the boiler operating with an efficiency of 90%. Estimate the energy loss from this uninsulated piping and its annual value (assuming 8000 hours of operation per year). Use 3E-Plus program available at https://insulationinstitute.org/tools-resources/free-3e-plus/

16.3. Estimate steam leak rate through a 15 mm hole in a steam pipe operating at a pressure of 30 bar. If cost of 30 bar steam is $25/ton, estimate annual energy loss due to steam leak and its cost. Assume 8000 hours of operation per year.

16.4. Estimate the amount of MP steam (11 bar) required to compress 2 ton/hr of flash steam from atmospheric pressure to 2 bar.

16.5. An ST uses HP steam at 40 bar and 320°C, with a flow rate of 10 ton/hr. Exhaust condition of steam at turbine outlet is 4 bar and 160°C. Mechanical efficiency (η_{mech}) is 0.98. Calculate the isentropic efficiency (η_{is}) of the ST and power generated.

16.6. Estimate the true cost of MPS and LPS for Case A. For this case, HP to MP STs and steam turbo generator (HP to MP stage) produce 3483.53 kW and 6884.97 kW power, respectively. MP to LP STs and steam turbo generator (MP to LP stage) produce 947.19 kW and 1489.98 kW power, respectively. HP to LP STs produce 5515.03 kW power. Assume the electricity cost of US$0.15/kWh, fuel heating value of 39539.24 kJ/kg, fuel cost US$376.87/ton and boiler efficiency of 0.92. HPS pressure and temperature are 43 bar and 380°C, respectively. MPS (11.5 bar) and LPS (4.5 bar) temperatures are 255.92°C and 209.27°C, respectively. BFW pressure and temperature are 56 bar and 121°C, respectively. Steam flow rates through STs and steam turbo generator can be obtained from Fig. 16.20.

Chapter 17

Waste Heat Recovery in Condensate Return Systems

17.1 Overview

Steam condensate recovery (CR) is a very important part of a steam system. It is essential for maximizing energy efficiency of steam systems. Installing/modifying a CR system plays a vital role in increasing waste heat recovery (WHR) from the steam system. Steam condensate (SC) is a valuable resource due to its energy content; moreover, it has the potential to reduce freshwater demand, demand on DM (demineralized) water plants, chemical costs, blowdown (BD) requirements/losses at boilers, demand on wastewater treatment plants and hot condensate (HC) disposal costs. As SC is relatively pure compared to most of the water sources and free of oxygen, it is frequently used as wash water in process applications in petroleum refineries and petrochemical plants.

Although very important, CR in many process plants is poor. The main reasons for this are water hammering, stall problems, inadequate/improper CR system design and condensate contamination. Steam traps play an important role in CR; their proper selection, sizing and maintenance are essential for maximizing CR. This chapter begins with emphasizing the importance of CR and the main barriers for maximizing it, in Section 17.2. Next, Sections 17.3 and 17.4 present the main reasons for water hammer (WH) and solutions to prevent it, respectively. Section 17.5 describes the problem of condensate stall, and efficient strategies to prevent it. Section 17.6 discusses the importance of steam traps, their selection and management, for maximizing CR. Key strategies for reducing pressure drop in condensate header are summarized in Section 17.7. Next, Section 17.8 discusses flash steam (FS) recovery and its efficient utilization. Section 17.9 illustrates the importance of CR and WHR, for maximizing power generation, using steam network optimization model, presented in Chapter 16. Finally, this chapter ends with summary in Section 17.10.

Learning outcomes of this chapter are as follows.

1. Describe the CR return system components.
2. Identify the benefits of CR and estimate CR.
3. Identify the barriers for CR and good solutions to overcome them.
4. Evaluate the methods to avoid WH and stall problems for maximizing CR.
5. Describe the importance, selection and management of steam traps in maximizing CR.
6. Explain the design considerations for sustainable CR.
7. Evaluate the impact of CR on optimization of power generation from the steam system.

17.2 Condensate Return System

Collection and reusing HC in the deaerator saves significant energy and operating costs for freshwater make-up, BD losses and chemical treatment. Hence, building a dedicated CR system is generally attractive.

17.2.1 *Importance of Condensate Recovery*

As shown in Fig. 17.1, SC contains significant energy. At 10 bar pressure, SC contains 27.5% of the total energy in steam. This value increases to 38.8% at 40 bar pressure. To recover full SC heat, with minimum capital cost, SC should be recovered as hot as possible.

Example 17.1: In a process plant, 4 ton/hr of SC at 100°C is drained to sewer. Temperature of feed water make-up to deaerator is 30°C. Natural gas (higher heating value [HHV] of 52,220 kJ/kg and cost of $396/ton) is

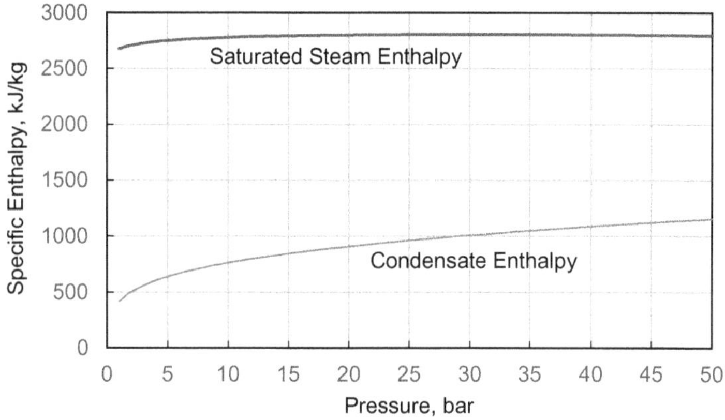

Fig. 17.1: Total energy in SC and saturated steam.

used as fuel in the boiler operating with an efficiency of 90%. Calculate the energy loss and its annual cost due to SC draining. Assume 8760 hours of operation per year.

Solution

Enthalpy of condensate at 100°C = 419.1 kJ/kg (from steam tables)

Enthalpy of feed water at 30°C = 125.7 kJ/kg (from steam tables)

Energy loss from condensate draining $= \frac{4000}{3600} \times (419.1 - 125.7) =$ 325.95 kW

Energy loss considering boiler efficiency $= \frac{325.95}{0.9} = 362.17$ kW

NG cost = \$396/ton = $\frac{396}{1000 \times 52{,}200} = \$7.58 \times 10^{-6}$ /kJ

Cost of energy loss per annum due to SC draining = 362.17 × 3600 × 8760 × 7.58 × 10^{-6} = \$86,644

17.2.2 *Estimation of Condensate Recovery*

Steam production and SC system are shown in Fig. 17.2, where the block of process units includes all units in the plant that are consuming steam. Mass and energy balances for the deaerator are given earlier in Eqs. 16.11–16.13. Mass balance for the boiler is given by:

$$M_{BFW} = M_S + M_{BD} \tag{17.1}$$

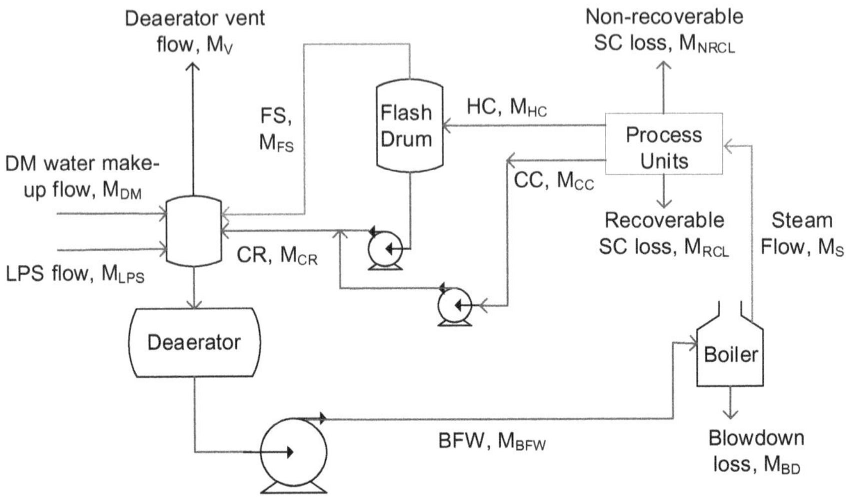

Fig. 17.2: Block flow diagram for steam production and SC system.

Here, M_{BFW}, M_S and M_{BD} are the mass flow rates of boiler feed water (BFW), steam and BD, respectively. Steam and condensate balance for the process units is:

$$M_S = M_{HC} + M_{CC} + M_{RCL} + M_{NRCL} \tag{17.2}$$

Here, M_{HC}, M_{CC}, M_{RCL} and M_{NRCL} are the mass flow rates of HC, cold condensate (CC), recoverable condensate loss (RCL) and non-recoverable condensate loss (NRCL), respectively.

RCL refers to the SC/steam flows that are not recovered due to issues like WH, stall problem, steam/FS venting and/or lack of sufficiently sized condensate headers. With proper mitigation measures described in this chapter, this SC portion can be recovered to maximize WHR and minimize freshwater consumption. NRCL refers to the SC/steam flows that are not recovered to deaerator due to the contamination of SC (e.g., generated due to the use of stripping steam in distillation columns, SC injection in process furnace tubes and the use of SC as wash water) and steam loss to atmosphere after use in the plant (e.g. atomization steam used in oil burners).

The following equations are for mass and energy balances around the SC flash drum (Fig. 17.2).

$$M_{\mathrm{HC}} + M_{\mathrm{CC}} = M_{\mathrm{CR}} + M_{\mathrm{FS}} \tag{17.3}$$

$$M_{\mathrm{HC}} \times H_{\mathrm{HC}} + M_{\mathrm{CC}} \times H_{\mathrm{CC}} = M_{\mathrm{CR}} \times H_{\mathrm{CR}} + M_{\mathrm{FS}} \times H_{\mathrm{FS}} \tag{17.4}$$

Here, M_{CR} and M_{FS} are the mass flow rates of CR and FS, respectively, and H_{HC}, H_{CC}, H_{CR} and H_{FS} are specific enthalpies of HC, CC, CR and FS, respectively. Referring to Fig. 17.2, percent CR is given by:

$$\mathrm{PCR} = \frac{(M_{\mathrm{CR}} + M_{\mathrm{FS}})}{M_S} \times 100 = \frac{(M_{\mathrm{BFW}} + M_V - M_{\mathrm{DM}} - M_{\mathrm{LPS}})}{(M_{\mathrm{BFW}} - M_{\mathrm{BD}})} \times 100 \tag{17.5}$$

Here, M_{V} is the mass flow rate of deaerator vent, which is small at ~0.1% of the BFW and can be neglected.

Example 17.2: Estimate the CR for a process plant with the following data: BFW as saturated liquid at 2.5 bar with flow rate = 500 ton/hr, DM water makeu-p to deaerator = 200 ton/hr at 30°C, deaerator operating pressure = 2.5 bar, pressure of LPS to deaerator = 2.5 bar, CR temperature to deaerator = 100°C, BD rate = 6 ton/hr at 45 bar and neglect deaerator vent flow rate. Assume FS flow rate of zero for this example.

Solution

BFW temperature = Saturated water temperature at 2.5 bar = 127°C

With negligible deaerator vent flow rate, mass balance across deaerator (Eq. 16.12) is:

$$200 + M_{\mathrm{LPS}} + M_{\mathrm{CR}} = 500$$

From steam tables, specific enthalpy of DM water make-up at 30°C, low-pressure steam (LPS) at 2.5 bar, CR at 100°C and 2.5 bar, and BFW are 125.7, 2716.5, 419.1 and 535.4 kJ/kg, respectively. Using the energy balance Eq. 16.13,

$$(200 \times 1257.7) + (M_{\mathrm{LPS}} \times 2716.5) + (M_{\mathrm{CR}} \times 419.1) = (500 \times 535.4)$$

Solving the above two equations simultaneously, $M_{\mathrm{LPS}} = 50.84$ ton/hr and $M_{\mathrm{CR}} = 249.16$ ton/hr.

Using Eq. 17.5, CR percentage can be calculated as follows.

$$\text{PCR} = \frac{(500 - 200 - 50.84)}{(500 - 6)} \times 100 = 50.43\%$$

As stated earlier, CR in many process plants is poor with recovery and re-use of recoverable condensate at 30%–70% or even lower. CR rate of more than 70% is considered good.[1] Percent CR of 50.43% in the earlier example shows potential to improve it.

17.2.3 *Barriers for Good Condensate Recovery*

Barriers for good CR are the following:

- WH in condensate header (discussed in Section 17.3)
- Improperly sized condensate headers for two/single phase flow
- Lack of condensate headers in remote areas
- Overdesign of heat exchangers (HEs), which can result in stall problem (discussed in Section 17.5)
- Use of high differential pressure (DP) type steam and/or SC control valves at HEs
- Contamination of SC. Contributing factors for SC contamination and solutions for them are summarized in Table 17A-3.
- Improperly designed controls for SC drainage (discussed in Section 17.5.2)
- Damaged steam trap (i.e., blockage or steam blowing problem) and/or control valve in SC removal system. A blowing steam trap can reduce heat transfer by 50% in reboilers and HEs[1] (discussed in Section 17.6).
- Steam locking of steam traps, which occurs if steam collects at the inlet of steam trap and hence prevents it from opening. It is caused mainly due to long piping, many bends and/or vertical rise in piping between HE and steam trap.

Operators generally deal with the above problems by draining the SC and by opening manual bypass valves across steam traps and control valves to the sewer. Such a draining wastes energy and water, increases the load of wastewater plant and can also cause operational instabilities. The following sections discuss the above problems in detail and present practical solutions to prevent them.

17.3 Water Hammer[a]

WH/steam hammer is mainly caused by sudden conversion of liquid velocity into pressure. This causes a pressure surge inside the piping and can damage piping, its accessories and/or result in personal injuries/fatalities. Specifically, dangers of WH are the following:

- Damage to internals of steam traps
- Cracking of steam traps and valve bodies
- Damage to control valves
- Damage to piping and HE tubes
- Failure of pipe supports

Four conditions that can cause WH are outlined below.

Hydraulic Shock: This is mainly caused by the sudden start/stop of SC flow to the SC header. This type of WH is commonly seen in SC headers involving positive displacement type of SC pumps (especially if proper flow dampening devices are not installed) and on–off operation of SC pumps.

Thermal Shock: This type of WH frequently happens in CR headers when CC below 100°C is mixed into the two-phase SC header, without a proper mixing device. In SC headers, FS can occupy significant volume even if it is present in small quantities due to its very large specific volume compared to SC. As can be seen in Fig. 17.3, the ratio of FS vapour to SC liquid volume decreases as SC header pressure increases. When FS contacts CC entering SC header, it will rapidly condense by transferring latent heat to SC and hence create a localized low pressure/vacuum region. In turn, this will quickly accelerate surrounding liquid from all directions[2] towards the vacuum space, thus creating WH.

---

[a]Note that WH in this chapter is the acronym for water hammer (and not waste heat in other chapters).

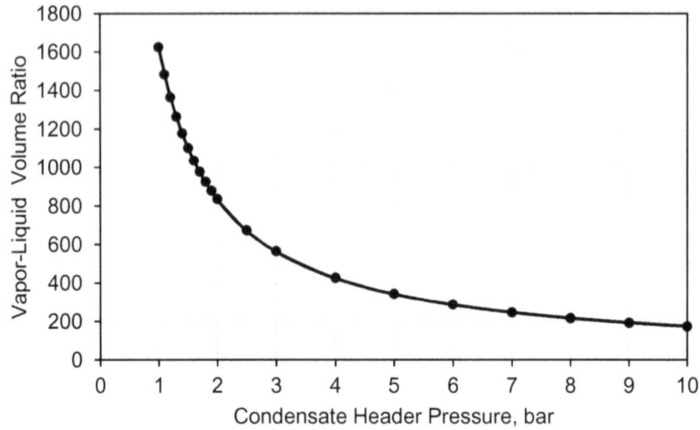

Fig. 17.3: FS–SC (vapour to liquid) volume ratio as a function of SC header pressure.

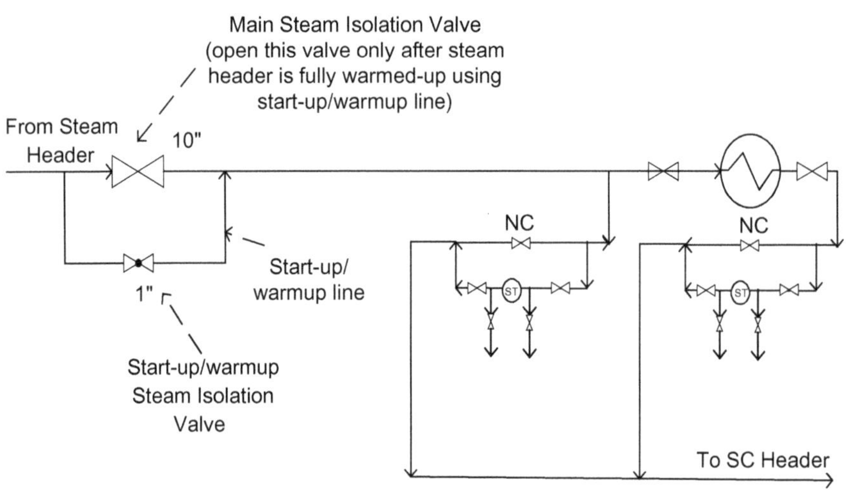

Fig. 17.4: Steam header start-up pipe arrangement to avoid the flow induced WH; ST inside the circle is a steam trap and NC refers to a normally closed valve.

Flow Induced Shock: Flow shock is commonly caused by lack of SC removal before steam isolation or control valve. During start-up operation, if the steam pipe is not warmed gradually using small start-up pipe (shown in Fig. 17.4), cold steam pipe rapidly condenses steam and leads to a big wave of SC, accelerated by steam. When this big SC slug approaches a closed isolation or control valve, it is required to stop and reverse the direction. This causes WH and can break the isolation/control valve. If the isolation/control valve is open, SC slug impinges with a very high velocity on the equipment (e.g., HE tubes), which can cause equipment damage and also WH.

Differential shock: This type of WH can happen in two-phase SC headers. FS is always present in SC system due to SC flashing. Sometimes, float of the steam trap may be stuck at open position and hence releases steam directly to SC header. Occasionally, steam trap bypass valve allows some steam to flow through or the bypass valve of the steam trap is opened by operators during stall conditions. Steam traps whose internals have been damaged by previous WH incidents, may release live steam to SC header. In short, all these conditions can lead to lot of steam in SC header, which can cause high SC header back pressure and/or WH.

SC can be accelerated by steam moving at velocities of 30–100 m/sec in the steam header. Steam and SC phases travel in the same header at different velocities; often steam velocity is more than 10 times SC velocity. SC flow at the bottom of the pipe forms growing waves as they progress along the pipe. Eventually, the liquid wave fills the pipe and blocks the steam flow temporarily. This creates a pressure difference across the liquid seal/slug and accelerates it with high velocity. Momentum builds as the liquid seal/slug volume increases. When this liquid seal/slug approaches a pipe bend, a tee connection or isolation valve, it stops abruptly and/or changes direction leading to WH.

Potential for WH on mixing of SC and steam/FS can be evaluated by examining a ratio, $R_{c/s}$,[3] which is defined as follows:

$$R_{c/s} = \frac{\text{Condensing capacity of the SC flow}}{\text{Heating capcity of steam flow}} = \frac{M_C \times C_p \times \Delta T}{M_S \times \lambda} \qquad (17.6)$$

Here, M_C is the mass flow rate of subcooled SC (kg/hr), C_p is the heat capacity of subcooled SC (J/K.kg), ΔT is degrees of sub-cooling below the saturation temperature (°C or K), M_S is the mass flow rate of FS (kg/hr) and λ is the latent heat of vaporization of FS, J/kg.

If $R_{c/s}$ is much less than 1.0, the condensing capacity of SC is not sufficient to condense all the FS/steam flow entering or existing in the SC header. Hence, there will be two-phase flow in the SC header. However, if the SC header is properly designed (i.e., without loops in vertical plane but with a continuous slope of 1:100 towards SC collection drum) for two-phase flow, WH will not result. If $R_{c/s}$ is much more than 1.0 and there is perfect mixing of steam/FS and SC, the condensing capacity of SC is more than sufficient to condense all the FS/steam flow, entering or existing in the SC header. Hence, there will be only liquid phase flow in the SC header and WH will not result if liquid velocity is kept within the general design limit. If $R_{c/s}$ is slightly above or below 1.0 with imperfect mixing or stratification of flows, some steam bubbles may persist temporarily and WH is possible.

17.4 Measures to Prevent Water Hammer

Various measures to prevent WH are detailed in this section.

1. Avoid the mixing of CC and HC by maintaining separate SC headers for CC and HC as illustrated in Fig. 17.5.

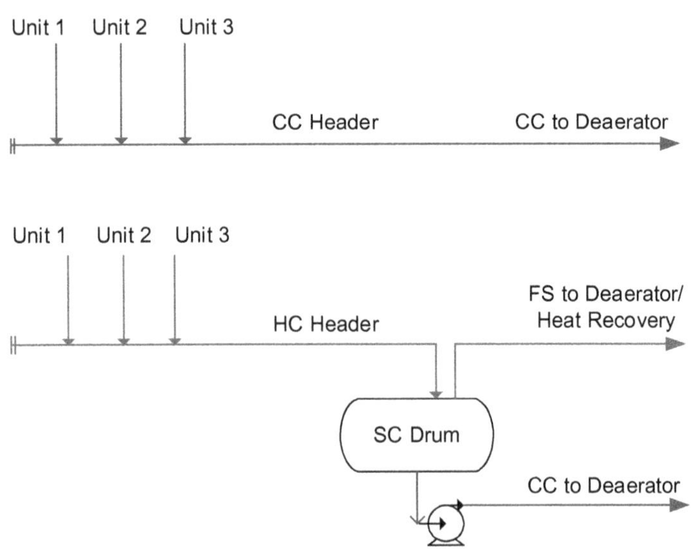

Fig. 17.5: Avoid WH in SC recovery header by using separate CC and HC headers.

2. In large petrochemical/refinery complexes, individual process units may be very far from the central utility area. In such cases, flash the HC in individual plants to atmospheric pressure, separate FS and pump CC to the central SC collection drum, as shown in Fig. 17.6. FS can be upgraded to LPS using thermo vapour recompression (TVR).

   For avoiding WH problem and minimizing SC header sizing, arrangement shown in Fig. 17.7 is recommended. In process plants, there will be short pipes for high pressure (HP) SC and medium pressure (MP) SC, to remove SC from HEs using high-pressure steam (HPS) and medium-pressure steam (MPS). As shown in Fig. 17.7, HP SC should be flashed to MPS pressure level, MP SC should be flashed to LPS pressure level and finally low pressure (LP) SC or HC should be flashed to low-low pressure (LLP)/deaerator pressure level, using flash drums. MP/LP FS obtained by flashing HP/MP SC can be mixed with MPS/LPS headers. LLP level FS generated in LP SC flash drum can be used directly in deaerator or upgraded to LPS by TVR or mechanical vapour recompression (MVR), for process use. LLP SC/CC can be pumped to the deaerator.
3. Use continuous type and correctly sized steam trap and maintain them with regular inspection and/or wireless online monitoring. Repair a steam trap if found passing or damaged to ensure the timely removal of SC and prevent steam leak to SC header.
4. Avoid bypassing of steam traps.
5. Avoid wide opening of SC drain control valves. Repair leaking control valves.
6. If pressurized CC needs to be mixed with HC, use direct contact steam heaters to heat CC before mixing with HC.
7. If DP of steam trap permits, increase pressure in SC header. This will reduce FS generation in the header and hence reduce WH. However, this needs to be done carefully while watching for any potential stall operation of the equipment.
8. Avoid SC piping loops in vertical plane. High points collect steam, whereas low points accumulate SC slugs. Any thermal expansion loops can be provided in horizontal plane. SC drain should be installed with a positive slope (1:100) towards SC collection drum.
9. Always connect SC into SC header from top side.

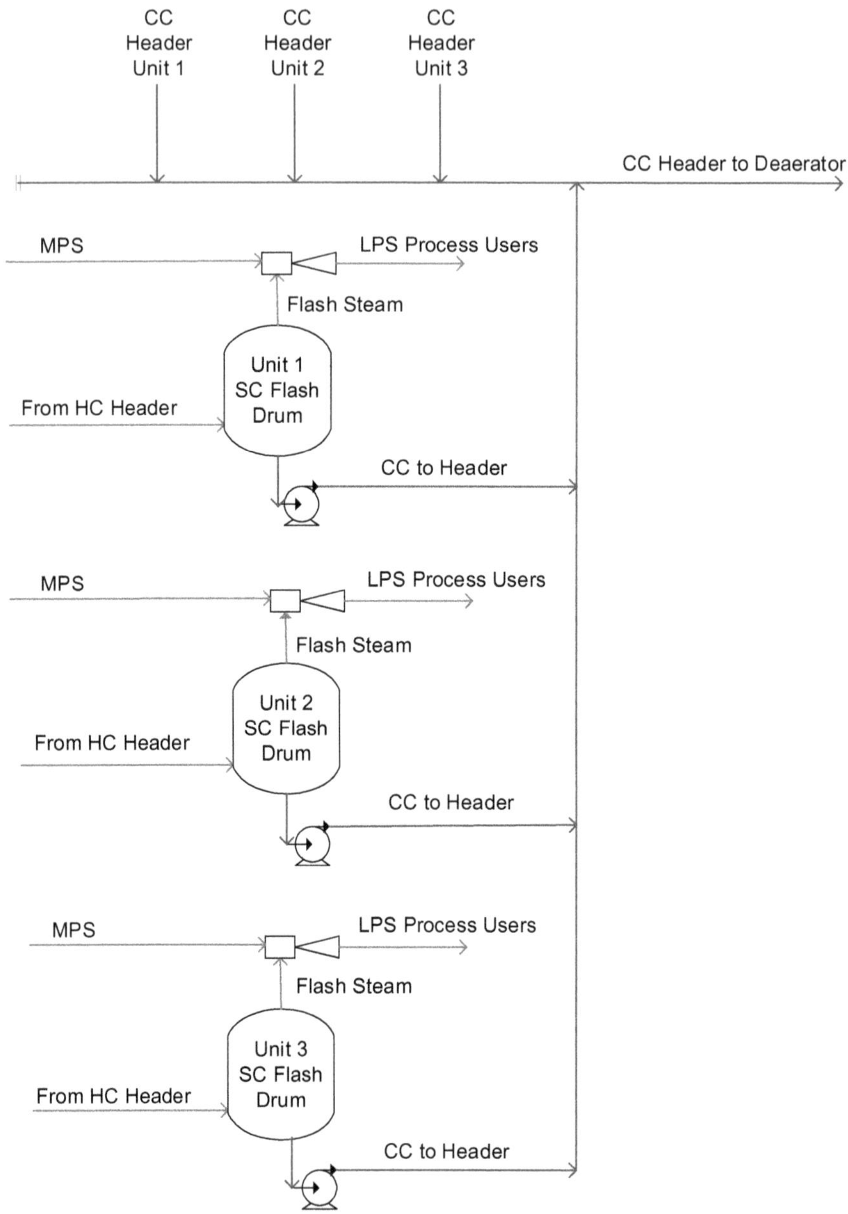

Fig. 17.6: Avoiding WH in CR header by flashing HC and sending CC to the CC header.

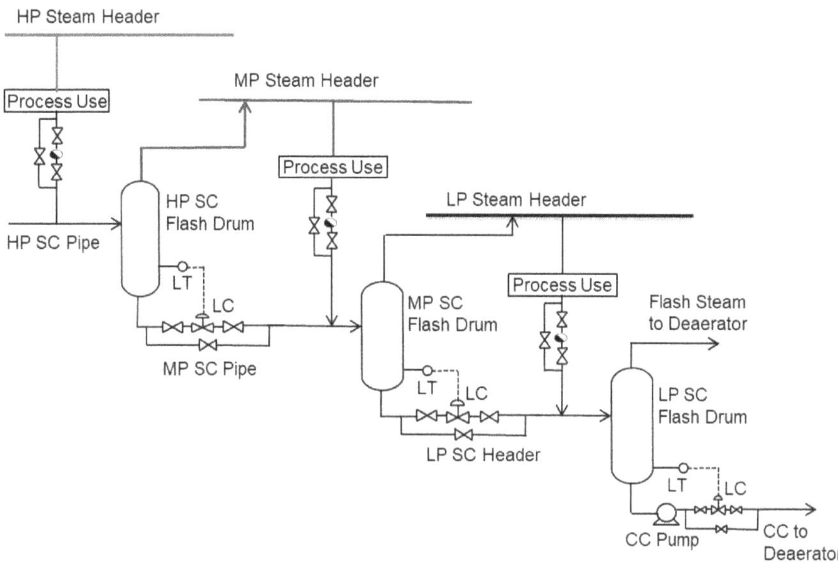

Fig. 17.7: Efficient CR system.

10. Accurately estimate the amount of flashing in the SC header and size this header adequately for two-phase flow.
11. Maintain SC flow as steady as possible by installing a variable flow SC pump (as opposed to on/off control).
12. Encourage the use of modulating steam traps such as float traps and thermostatic traps (as opposed to intermittent blast discharge traps such as inverted bucket traps).
13. Non-slam-type check valves can be used to prevent WH in SC sub-headers.
14. Have proper start-up/shutdown procedures for the steam system.
15. Implement proper training for operators.
16. Use start-up lines (Fig. 17.4) for steam pipes larger than 2″ (50 mm) in diameter. Do not use main isolation valve to warm up the steam pipe.

Contributing factors for WH and solutions for them are summarized in Table 17A-1.

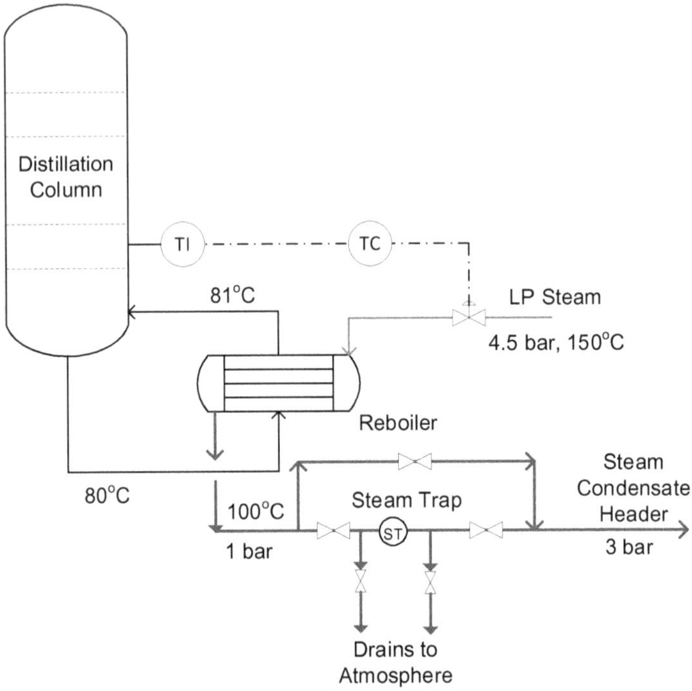

Fig. 17.8: Reboiler/HE stall problem; ST inside the circle is the steam trap and the valve on the line bypassing steam trap is normally closed; in this case, due to stalling operation, SC needs to be drained from upstream drain valve of ST.

17.5 Stall Condition

If back pressure in SC return header is more than the steam pressure in the HE or process equipment, a phenomenon, known as a 'stall' condition occurs. It causes SC to build up and sub-cool in the HE, reducing heat transfer rate. Due to stalling, SC will not be recovered into SC header. Stalling is mainly due to negative pressure drop across the steam trap, as shown in Fig. 17.8. In this case, steam chest pressure of the reboiler (HE) is 1 bar, pressure drop across steam trap is 2 bar and SC cannot return to its header. During stall situations, it is common for the plant operator to open steam trap bypass valve and/or drain SC to atmosphere; this usually results in the loss of live steam and/or SC.

Stall situation typically arises when LPS pressure is modulated by a control valve based on process temperature/duty control. HEs are typically oversized due to the consideration of fouling and design allowances for future duty additions. Such HEs may cause stalling problem even at 100% design duty, which is covered in the following sub-section. The only reliable solution to recover SC in such stall cases is to use a condensate pump.

17.5.1 *Evaluating Stall Conditions*

Stall conditions can be quickly verified using a chart[4] shown in Fig. 17.9. It is developed based on the following equations.

$$Q = M \times C_p \times (T_i - T_o) = M_S \times \lambda_S = U \times A \times \text{LMTD} \simeq U \times A \times (T_s - T_m) \quad (17.7)$$

$$T_m = \frac{(T_i + T_o)}{2} \quad (17.8)$$

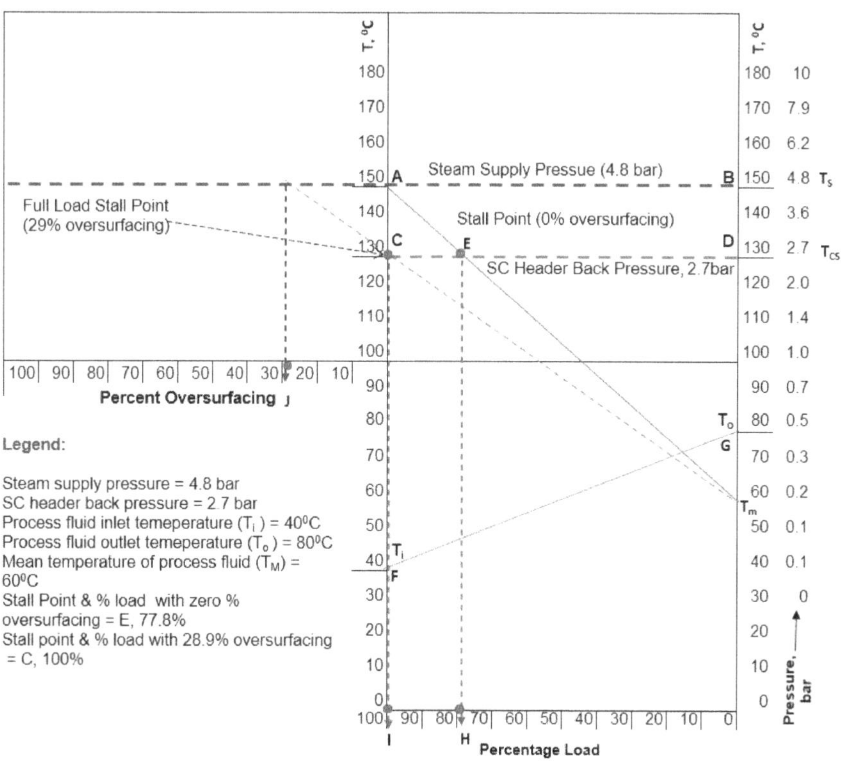

Fig. 17.9: HE stall conditions.

Here, U, T_s and T_m are the overall heat transfer coefficient, steam supply temperature and mean process fluid temperature, respectively, and M, C_P, T_i and T_o are respectively mass flow rate, specific heat capacity, inlet and outlet temperatures of process fluid.

In Fig. 17.9, steam pressure/temperature and process stream temperature are plotted on y axis. Percentage HE load/duty is shown on lower x axis, whereas percentage over surface (compared to the surface required for design capacity) is shown on upper x axis. Steam supply pressure/temperature to the HE is indicated by the line AB. SC header saturation temperature (T_{CS}) corresponds to the SC header pressure and is indicated by the line CD. Line FG joins inlet and outlet temperatures of the process stream. As the steam supply to the HE is controlled by T_o, steam pressure in the HE reduces linearly with the required duty as shown by the steam supply line AT_m.

With zero excess/overdesign area, intersection of lines CD and AT_m indicates the stall point (i.e., steam pressure in the equipment is equal to constant back pressure in the SC header). For this scenario, stall point is indicated by point E (77.8% load). This means, if the HE's load is less than or equal to 77.8%, SC cannot flow into the SC header. If sufficient excess design area is available, HE can stall even at 100% design duty. For this example, if 28.9% overdesign area is available, stall point occurs at point C (i.e., 100% duty, intersection of lines CD and CT_m). For this case, if the HE's load is less than or equal to100%, SC cannot flow, without pumping, into the SC header. So, it requires a SC pump for CR. If overdesign area is more than 28.9%, stall will occur at duty > 100%. Hence, CR is not possible with steam trap and requires a SC pump at all loads below or equal to stall point. In summary, use of high overdesign factor in HE design can create CR problems with steam traps under turndown operations.

The stall chart in Fig. 17.9 is highly useful to decide whether SC pump is required in place of steam trap. Note that stall operation can occur if saturated steam temperature (T_{cs}) at SC header back pressure is between steam supply temperature (T_s) and mean process fluid temperature (T_m), which is the usual situation. Without using the chart, stall duty (as percentage) at zero excess area in HE (PD_{S0}) can be easily estimated using:

$$\mathrm{PD}_{\mathrm{S0}} = \frac{\left(T_{\mathrm{cs}} - T_{\mathrm{m}}\right) \times 100}{\left(T_{\mathrm{S}} - T_{\mathrm{m}}\right)} \tag{17.9}$$

For the illustration in Fig. 17.9, $T_{\mathrm{m}} = \frac{(40+80)}{2} = 60°\mathrm{C}$, $T_{\mathrm{cs}} = 130°\mathrm{C}$ and $T_{\mathrm{S}} = 150°\mathrm{C}$. Hence, $\mathrm{PD}_{S0} = \frac{(130-60)\times 100}{(150-60)} = 77.8\%$, which is same as that obtained from the chart.

Over design area (A_{S100}), at which HE stalls at 100% duty or full load, can also be estimated using:

$$A_{\mathrm{S100}} = \frac{\left(T_{\mathrm{S}} - T_{\mathrm{m}}\right) \times 100}{\left(T_{\mathrm{cs}} - T_{\mathrm{m}}\right)} - 100 \tag{17.10}$$

For the illustration in Fig. 17.9, $T_{\mathrm{m}} = 60°\mathrm{C}$, $T_{\mathrm{cs}} = 130°\mathrm{C}$ and $T_{\mathrm{S}} = 150°\mathrm{C}$. Hence, $A_{\mathrm{S100}} = \frac{(150-60)\times 100}{(130-60)} - 100 = 28.9\%$, which is same as that obtained from the chart. Given an over design area, stall duty (PD_{S}) can be estimated by using:

$$\mathrm{PD}_{\mathrm{S}} = \frac{\left(T_{\mathrm{cs}} - T_{\mathrm{m}}\right) \times \left(100 + A_{\mathrm{S100}}\right)}{\left(T_{\mathrm{S}} - T_{\mathrm{m}}\right)} \tag{17.11}$$

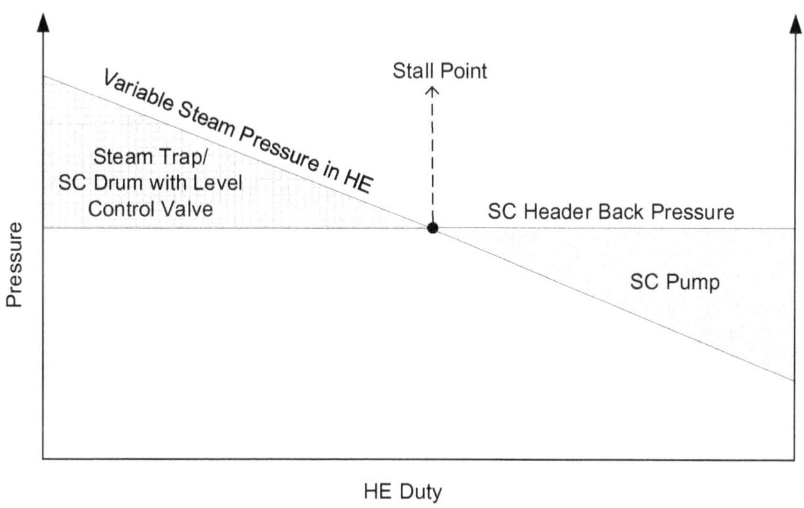

Fig. 17.10: Operating envelop for steam traps and SC pumps.

For this example, stall duty of HE with 20% over design is $PD_s = \frac{(130-60)\times(100+20)}{(150-60)} = 93.3\%$. Equations 17.9–17.11 are based on the interpolation of relevant linear equations for the straight lines in Fig. 17.9.

Steam traps can remove SC, above stall point only. Below stall point, SC pump is required, as shown in Fig. 17.10. If stall operation is likely to occur, then the only options are to use an SC pump or operate HE under partial SC flooded mode (which decreases HE duty). For partial SC flooded mode, heat transfer area available for latent heat transfer reduces and hence reduces steam flow rate and HE duty. As the SC temperature increases, SC saturation pressure at the HE outlet is more than the SC header pressure and hence CR can be achieved by using a steam trap or SC drum with level control valve, as highlighted in Fig. 17.10.

17.5.2 *Stall Condition Prevention*

Control valve manipulating heat supply to HE can be arranged on steam line (at HE inlet) and/or SC line (at HE outlet). Each of these has its own advantages and disadvantages.[5] Comparison of the two arrangements is presented in Table 17.1

If steam system chemistry (i.e., purity of DM make-up water and steam) is well maintained as well as carbonic acid corrosion and thermal stress problems are minimum, then stall operation may be avoided economically by partial flooded operation of HE/reboiler. For HEs operating on variable heat duties and using LPS (3–4 barg) as heating source, SC control valve is recommended instead of steam inlet control valve,[1] as illustrated in Fig. 17.11. This scheme maintains high pressure in HE by operating SC control valve on temperature control, with level control override, which minimizes chances of liquid seal loss. Equalization pipe between HE and SC drum is normally opened (NO) and equalization line to LPS header is NC. SC pump (either motor driven or pumping trap[b]) may or may not be required, depending on the worst case SC pressure in

[b] https://www.armstronginternational.com/products-systems/steam-condensate/condensate-recovery-equipment/pressure-operated-mechanical-pumps/pumping-traps. A pumping trap is a mechanically driven condensate pump, which uses steam as motive force instead of electricity, to pump SC to the SC header. It can pump SC even under vacuum condition.

Table 17.1: Comparison of control valve on steam inlet pipe and SC flow, for HE operation

Item	Control Valve on Steam Inlet Pipe (at HE Inlet)	Control Valve on SC Outlet Pipe (at HE Outlet)
Variable affecting heat duty	Temperature driving force, ΔT (if SC drum/pot is not used for HE level control)	HE area (A)
SC condition in reboiler/HE	Saturated liquid or slightly subcooled	Subcooled. SC goes to CC header if its temperature is lower than 100°C.
Dynamic response	Fast	Slow
Impact of SC seal loss	Rarely an issue	Leads to loss of HE duty and causes WH in SC header
Energy efficiency	Only good if DP across the steam control valve is small	Superior
Cost of control valve	Generally higher	Lower
Thermally degradable/ Fouling service	Superior	Inferior
Corrosion and/or thermal stress	Generally, no problem	Problematic unless temperature changes are small, and CO_2 is vented regularly from HE
Start-up/turndown operation	Problematic but it can be solved using SC pump or SC drum/pot for level control in HE	Generally good
Stalling problem	Generally, problematic for heating with LPS unless an SC drum/pot and pump are used. Not a problem with higher pressure steam if SC saturation pressure in HE is greater than SC header pressure.	Very good for stall prevention
Potential to cause WH	Generally, no WH unless steam trap or SC control valve is damaged, or SC is sent to CC header	Generally, not an issue unless liquid seal is lost
Operation with oversurfaced HE	Problematic, unless SC drum/pot is used for level control in HE	Generally, no problem

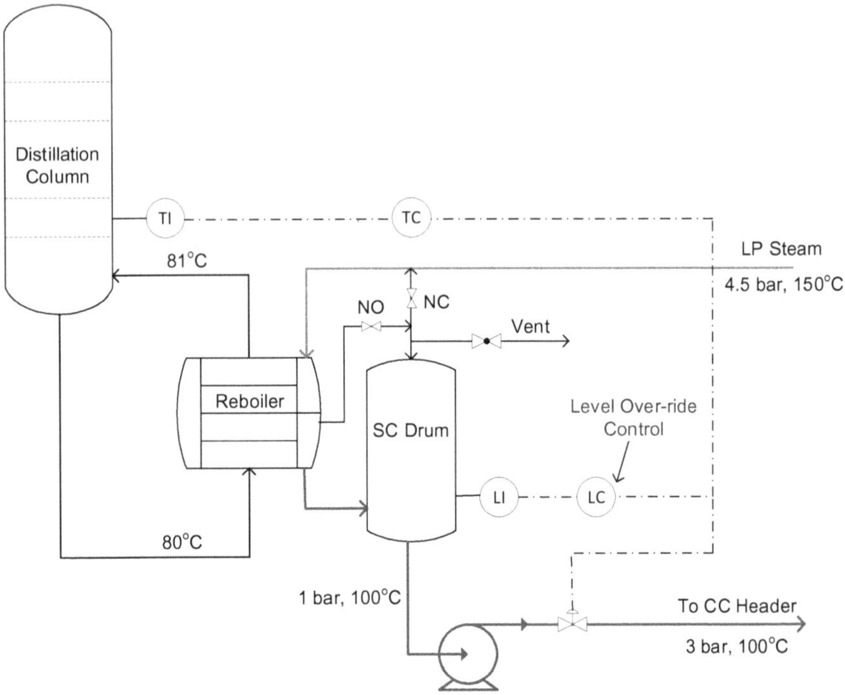

Fig. 17.11: Avoiding stall condition by partial flooding of HE/Reboiler.

HE. Pump is required if the SC header pressure is greater than SC pressure in the HE.

Alternatively, if LPS control valve at HE inlet is required for fast control of duty and other considerations given in Table 17.1, a CR system shown in Fig. 17.12 is required; it involves a control valve in LPS supply pipe to the reboiler and a pump for sending SC to CC header. The systems in Figs. 17.11 and 17.12 can recover SC to CC header without any WH issues. If this SC is sent to HC header, it may cause WH and the arrangement in Fig. 17.13 can be used to avoid WH. In this arrangement, pressurized CC is heated by a direct injection steam heater, controlled by a temperature controller, to control SC temperature close to the HC header temperature.

Heating arrangements for a vertical reboiler with SC control and with steam and SC control are discussed in detail by Kister.[5] Typical arrangements for preventing stall and SC header WH in case of SC control and

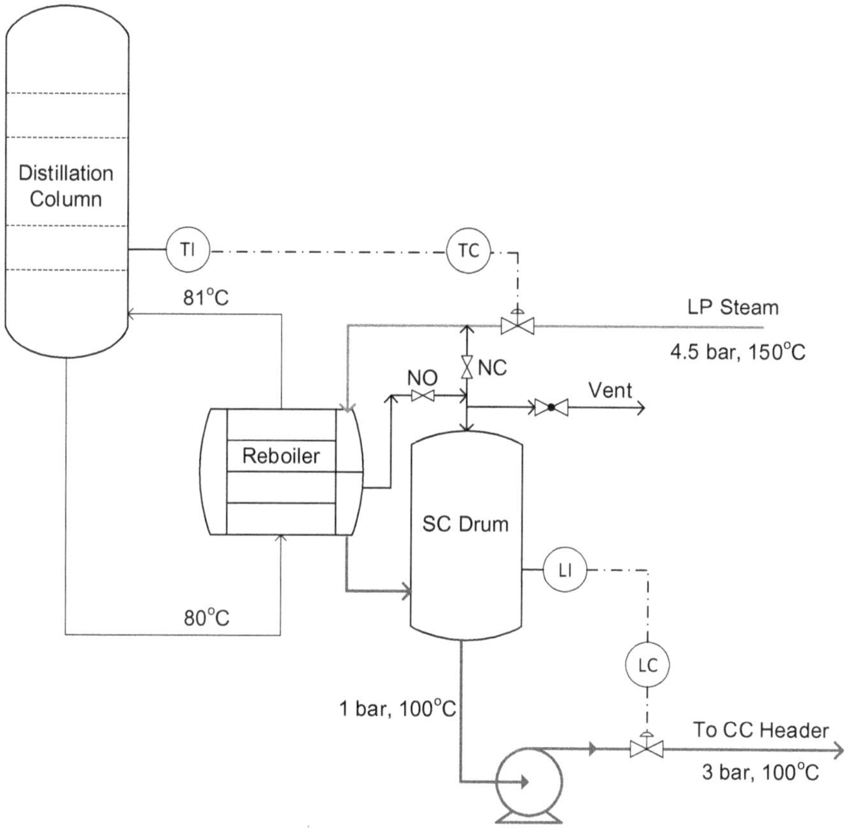

Fig. 17.12: Mitigating stall problem in case of control valve on steam inlet pipe, by pumping CC to CC Header.

steam & SC control are illustrated in Figs. 17.14 and 17.15, respectively. For a vertical reboiler, HE corrosion due to CO_2 or carbonic acid accumulation can only be managed by occasional SC blow to sewer, using a drain valve (NC) at inlet of SC drum/pot.[6]

SC backup and/or blowing steam trap/control valve can cause HE to lose its capacity. SC backup/stall operation can be verified by opening the manual bypass valve across the steam trap and monitoring the HE duty. If duty increases/decreases, the problem could be due to SC backup or blowing steam trap. Blowing steam trap can be further verified by partial closing of the manual valve at the downstream of the steam trap and see

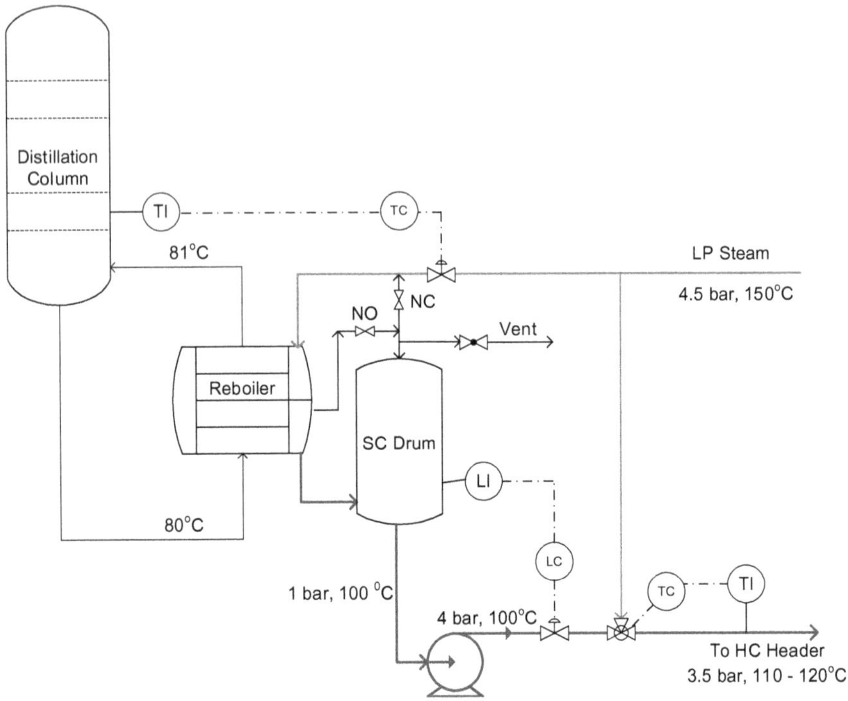

Fig. 17.13: Mitigating stall problem in case of control valve on steam inlet pipe; CC is pumped to HC Header after LPS injection for direct heating.

whether HE duty increases. If the duty increases, steam trap blowing is confirmed.[1] SC backup can be monitored by the level indicator installed on the SC drum. Contributing factors for stalling problems and solutions for them are summarized in Table 17A-2.

17.5.3 *Condensate Drum/Pot Sizing Guidelines*

Design and operation guidelines for condensate drum/pot sizing to avoid or minimize stall problem and/or WH are as follows.

- SC drum/pot can be horizontal or vertical depending on the HE orientation. Generally, vertical SC drums/pots are used with vertical HE/reboilers, and horizontal SC drums are used with horizontal HE/

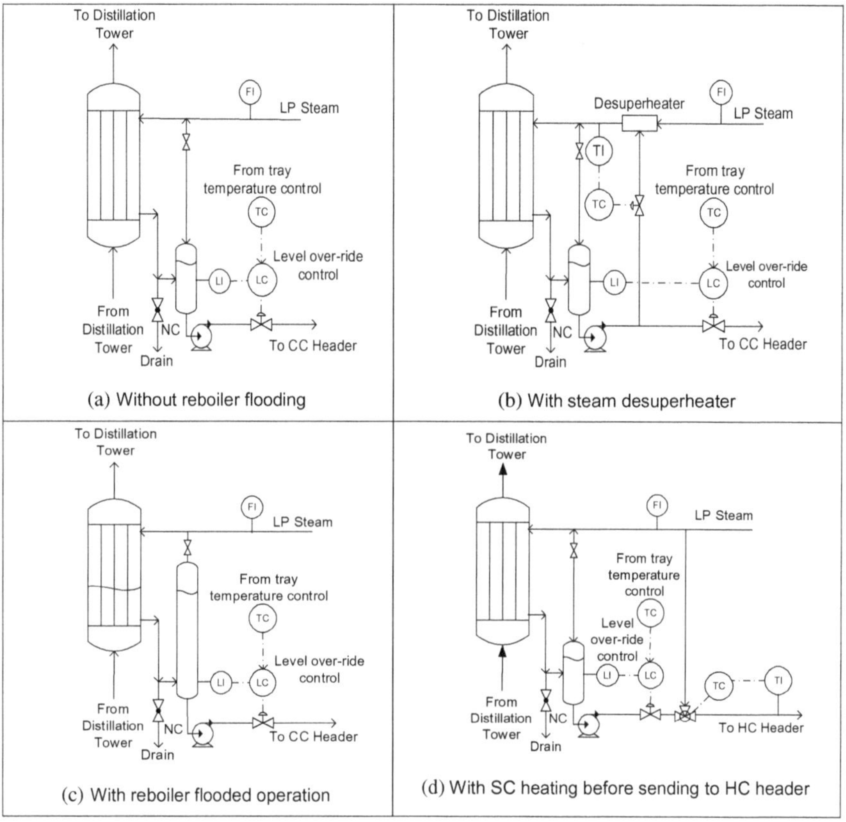

Fig. 17.14: Different possibilities to avoid stall and WH problems for a vertical reboiler with only SC level control.

reboilers. A vertical SC drum/pot can be used with a horizontal HE in certain cases, especially if the HE diameter is very large.

- Pressure equalizing line between the steam pipe/below-pass partition plate and condensate drum/pot top section should be provided, with a line size of 1″ and an isolation valve for throttling purposes.
- Always allow SC to enter the SC drum/pot at its bottom, below normal liquid level.
- SC pipe between HE and SC drum/pot should be designed for gravity flow with nearly zero pressure drop.

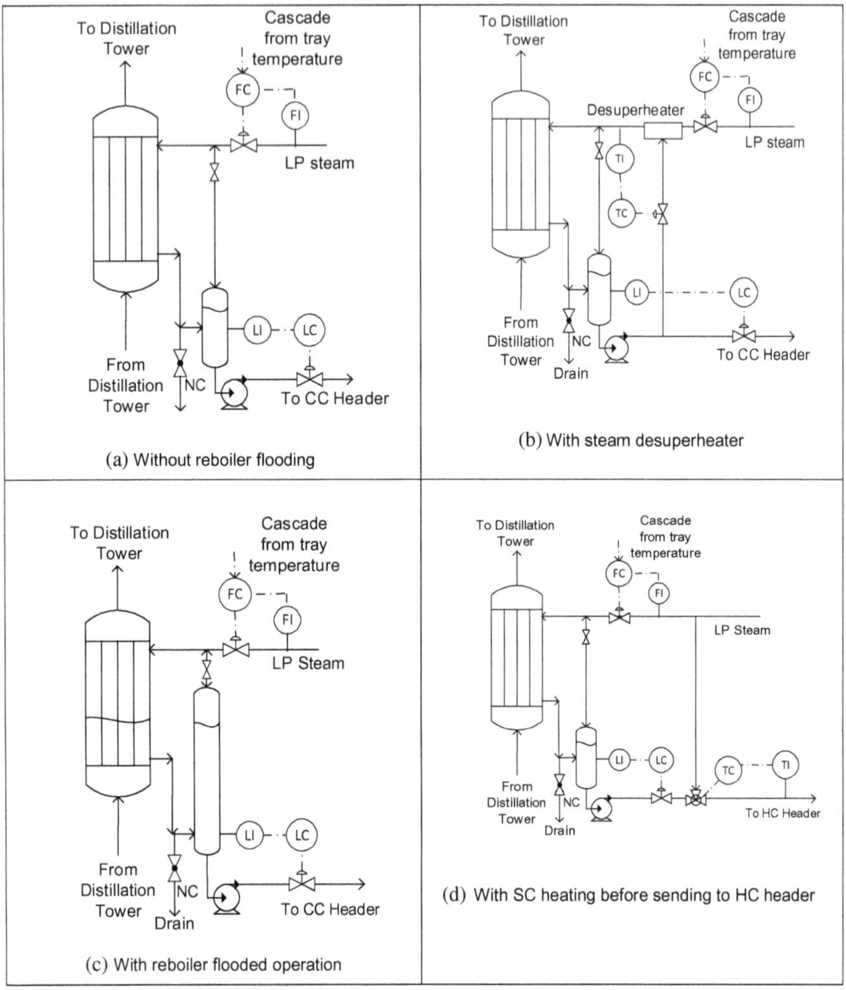

Fig. 17.15: Different possibilities to avoid stall and WH problems for a vertical reboiler with both steam and SC level control.

- For non-flooding and pressure balanced SC drum/pot design, bottom SC nozzle of HE should be sufficiently elevated above the maximum level in the SC drum/pot. This elevation height should be at least sufficient to overcome pressure drop in the SC line between HE and SC drum/pot. Generally, an elevation of 2 feet (600 mm) is used.
- For flooded SC drum/pot design, maximum level in the SC drum should be set based on the required maximum level in the HE.

- Vapour space in the SC drum/pot above maximum liquid level should be at least 1 foot (300 mm). Distance between lowest level and bottom tangent line of SC drum/pot should be at least 6 inches (150 mm).
- Surge volume between maximum and minimum levels is maintained at 1–3 minutes of residence time based on maximum SC flowrate or maximum flow rate through SC control valve. Lower (higher) value is used for vertical (horizontal) SC drum/pot.
- Height (distance between tangents) of the SC drum/pot is usually selected based on the following criteria.
    - For horizontal reboiler/HE: 2 feet (600 mm) + reboiler/HE diameter
    - For vertical reboiler/HE: 2 feet (600 mm) + reboiler/HE height

17.6 Steam Traps

Steam traps are an important element of CR/WHR systems. However, they are least reliable.[5] Properly working steam traps separate SC, after heat exchange in the process equipment/heat tracing. Passing/leaking steam traps result in steam loss and also create WH and/or stall problems. Purchase cost of steam trap is only a small fraction of its operating cost, which includes energy loss, water loss, maintenance and emissions costs. In a plant without proper steam trap monitoring and maintenance, 25%–50% of steam traps will be either leaking steam or not working properly. According to Hou and Mita,[7] an average of 4.6% steam reduction can be achieved by good steam trap management, in large refinery and petrochemical plants.

Branan[8] presented a method for rough estimation of steam losses through a fully blowing steam trap, based on the C_v (flow coefficient). This method requires rated capacity of the steam trap, SC pressure and temperature at the steam trap inlet. If the rated capacity of steam trap is not available, Eq. 16.10 can be used for rough estimation of steam loss. As every steam trap has a restriction orifice (RO), value of D_h in Eq. 16.10 can be taken as the steam trap's RO size, which can be found easily from the vendor. C_v and steam trap orifice hole size (D_h in inches) are approximately related by

$$C_v = 22.1\, D_h^2 \qquad (17.12)$$

Ganapathy[9] presented the following equation for the estimation of steam loss using orifice size.

$$M_{\text{Leak}} = \frac{50 \times A_{\text{h}} \times P_{\text{s}}}{1 + 0.00065 \times (T_{\text{s}} - T_{\text{sat}})} \tag{17.13}$$

Here, A_{h}, P_{s}, T_{s}, and T_{sat} are hole area (in^2 based on D_{h}), steam pressure (psi), steam temperature (°F) and saturated steam temperature (°F), respectively. Hicks[10] presented another equation for estimation of steam loss using orifice size as shown below.

$$M_{\text{Leak}} = k \times A_{\text{h}} \times (P_{\text{s}} \times \rho_{\text{s}})^{0.5} \tag{17.14}$$

Here, k, A_{h}, P_{s}, and ρ_{s} are constant (1085 for saturated steam and 1138 for superheated steam), hole area (in^2), steam pressure (psi) and steam density (lb/ft^3), respectively. Comparison of methods of Branan, Ganapathy and Hicks, and also use of Eq. 16.10 are illustrated in the following example.

Example 17.3: A steam trap on 10.34 bar (150 psi) SC line is blowing live steam. Rated capacity of the steam trap is 2268 kg/hr (5000 lb/hr) of SC at saturation temperature of 10.34 bar steam. C_v of the steam trap is 2.38. Estimate the live steam blowing using Eq. 16.10. Compare it with the value of 238.14 kg/hr (525 lb/hr) given in Branan's book,[8] and values estimated by Ganapathy's and Hicks' methods.

Solution

Using Eq. 17.12, $D_{\text{h}} = \left(\frac{C_v}{22.1}\right)^{0.5} = \left(\frac{2.38}{22.1}\right)^{0.5} = 0.328 \text{ in} = 8.33 \text{ mm}$

Using Eq. 16.10, $M_{\text{Leak}} = 0.413 \times P_{\text{s}} \times D_{\text{h}}^2 = 0.413 \times 10.34 \times 8.33^2 =$ 296.4 kg/hr

Using Ganapathy's method, saturated steam condition, $T_{\text{s}} = T_{\text{sat}}$ and hence Eq. 17.13 simplifies to

$M_{\text{Leak}} = 50 \times \frac{\Pi}{4} \times D_{\text{h}}^2 \times P_{\text{s}} = 50 \times 0.7857 \times 0.328^2 \times 150 = 633.97 \frac{\text{lb}}{h} =$ 287.6 kg/hr

Using Hicks' method with $k = 1085$ (for saturated steam) and density of 150 psi saturated steam (from steam tables) = 0.3316 lb/ft^3,

$M_{\text{Leak}} = 1085 \times \frac{\Pi}{4} \times D_{\text{h}}^2 \times (P_{\text{s}} \times \rho_{\text{s}})^{0.5} = 1085 \times 0.7857 \times 0.328^2 \times$ (150 × 0.3316)$^{0.5}$

$$= 646.82 \frac{\text{lb}}{h} = 293.4 \text{ kg/hr}$$

Steam leak values estimated by Eq. 16.10, Ganapathy's and Hicks' methods are close. However, steam blowing value given by these three methods is higher than that in Branan,[8] which may be due to the approximate relationship in Eq. 17.12.

Note that not all steam trap failures result in full-bore live steam blowing. Based on industrial practise, a leaking steam trap releases only 25% of the value estimated for the blowing steam trap. Companies such as Armstrong International, TLV, Yarway and Spirax Sarco provide efficient online steam trap monitoring systems, using remote sensors. This can be very beneficial for minimizing steam leaks and CR problems.

Inadequate SC removal from stripping steam lines is the leading cause of SC-induced pressure surges that lead to damage of distillation tower internals. Poor quality steam (i.e., with minute water droplets) can damage critical equipment like steam turbines (STs), HEs and steam jet ejectors (SJEs). Inadequate heating (due to faulty/passing steam traps) in reciprocating compressor suction piping handling saturated vapours (with saturation temperature above atmospheric temperature), can result in liquid carryover and hence liquid compression. This can damage the compressor and/or result in fire/explosions.

17.6.1 *Steam Condensate Removal and Steam Trap Problems*

Well-maintained team traps help to provide reliable process heating by efficient condensate removal. However, if they are damaged, they can lead to more equipment reliability problems and maintenance costs as summarized below.

- More maintenance cost: defective steam traps may result in WH, which can destroy steam/SC headers, manual/control valves and so on
- Increased safety risk: WH can injure personnel and/or cause fatalities
- HE damage due to temperature cycling

- Liquid compression damage in reciprocating compressors handling saturated vapour temperature more than atmospheric temperature
- Higher erosion/corrosion at HEs and associated piping
- Damage to SJE nozzles due to impingement of high-velocity condensate droplets
- Poor atomization in case of oil burners
- Damage to distillation trays and column internals
- ST blades damage and/or loss of power due to low turbine efficiency
- Loss of heating duty as SC can transfer only sensible heat which could lead to loss of product yield
- Loss of product heating, causing plugging in highly viscous services such as sulphur and asphalt
- Damage to flare tip in flare stacks
- Poor process control: SC flooding of HEs can lead to control swings
- Increase in volume of effluent to its treatment plant
- Increased consumption of freshwater as make-up at deaerators due to condensate loss
- Increase in cost of water treatment for steam generation as the chemical cost for demineralization of freshwater and heating cost at deaerator are significant

Failed steam trap can result in huge financial consequence. Risco[11] reported typical consequences of defective steam traps as follows: US$0.75 million for flare tip replacement, US$1 million for plant shutdown due to analyzer failure, US$3.6 million for gas compressor failure and US$ 20 million for ST failure. Main reasons for steam trap failures are listed in Tables 17.2. As per Risko,[12] steam trap failures can be grouped into cold and hot failures. A cold failure refers to blockage of steam trap with debris or operating at abnormally low internal temperature due to stalling problem, whereas a hot failure refers to steam leaking along with condensate draining.

Proper selection and sizing of steam traps are very important for maximizing and sustaining WHR from CR. SC load of a stream trap (= mass flow rate of SC to be handled = mass flow rate of steam, M_S) is established by the following energy balance for HE.

Table 17.2: Reasons for steam trap failures

Cold Failures	Hot Failures
• Negative pressure differential across the steam trap • Undersized steam trap • Operating under a stall condition • Steam locking due to lift or distance • Steam trap is installed in a group • Double trapping • Steam trap blockage • Incorrect low temperature setpoint • High condensate header backpressure • Backwards installation by mistake • Inability to discharge air • Isolated steam trap	• Backwards installation • Incorrect orientation • Steam trap is oversized • SC header backpressure is high • Inability to handle superheat • Incorrect high temperature setpoint

$$Q = M \times C_p \times (T_i - T_o) = M_S \times \lambda_S \tag{17.15}$$

Here, M, T_i and T_o are mass flow rate, inlet temperature and outlet temperature of the process fluid through the HE, respectively, and λ_S is the latent heat of steam. For steam trap sizing, M_S value given by Eq. 17.15 is multiplied with a safety factor, as many steam traps do not discharge SC continuously. Paffel[13] suggested a safety factor of 2 for float and thermostatic steam traps, and 3 for inverted bucket, thermodynamic and thermostatic steam traps. Steam trap characteristics for selection are presented in Table 17.3, which are mostly derived from Turner and Doty.[14] There are more types of steam traps than those presented in this table. One can find their details at websites of steam trap vendors.

17.6.2 *Benefits of Good Steam Trap Management*

Benefits of good steam trap management are the following:

1. Energy savings from reduced steam losses
2. Sustained maximum steam CR

Table 17.3: Characteristics of different types of steam traps

Characteristic	Inverted Bucket	Float and Thermostatic	Thermodynamic (e.g., Disk)	Thermostatic (e.g., Bellows, Bimetallic)
Operation	Intermittent	Continuous	Intermittent	Continuous
Time in service	Excellent	Good	Poor	Fair
Resistance to wear	Excellent	Good	Poor	Fair
Resistance to hydraulic shock or WH	Excellent	Poor	Excellent	Poor
Vent air and CO_2 at steam temperature	Yes	No	No	No
Operation against back pressure	Excellent	Excellent	Poor	Excellent
Ability to handle start-up loads	Fair	Excellent	Poor	Excellent
Performance at low loads	Excellent	Excellent	Poor	Excellent
Ability to handle dirt	Excellent	Poor	Poor	Fair
Physical size	Large	Large	Small	Small
Failure position	Open	Closed	Open	Closed

3. Reduced risk of production loss
4. Reduced maintenance costs
5. Reduced operating cost due to lower freshwater intake: chemical cost, energy cost, emissions cost, and wastewater treatment cost.
6. Improved plant reliability

Strategies for maintaining healthy steam traps are the following:

1. Check sizing of all steam traps to ensure that they are adequately sized to provide proper condensate drainage.
2. Review the types of traps in various services to ensure that the most efficient steam trap is being used for each application.
3. Implement a regular steam trap survey (preferably online using wireless sensors and diagnosis) and proactive maintenance program.

17.6.3 *Influence of Human Factors on Steam Trap Reliability*

API RP 581[21] defines the probability of failure (PoF) of an equipment as:

$$\mathrm{PoF} = gff \times D_{\mathrm{f}}(t) \times F_{\mathrm{MS}} \tag{17.16}$$

Here, *gff* is failure frequency of the equipment, $D_{\mathrm{f}}(t)$ is damage factor, which accounts for the relevant damage mechanism and inspection effectiveness, and F_{MS} is management systems factor. F_{MS} accounts for the effectiveness of the organization's management system and its influence on the overall equipment integrity. PoF value is increased by a weak management approach. Human aspects of plant operation such as steam trap inspection, fixing the cold and/or damaged steam traps, keeping records and continuous staff training significantly improve the reliability of steam trap system and entire steam system.[7] If an operating process plant does not have a proactive steam trap management program, up to 50% of traps could be defective (cold/hot failure).[15]

17.6.4 *Condensate Drip Legs*

Condensate can be efficiently separated from steam headers by installing properly designed condensate drip legs (CDLs), as shown in Fig. 17.16.

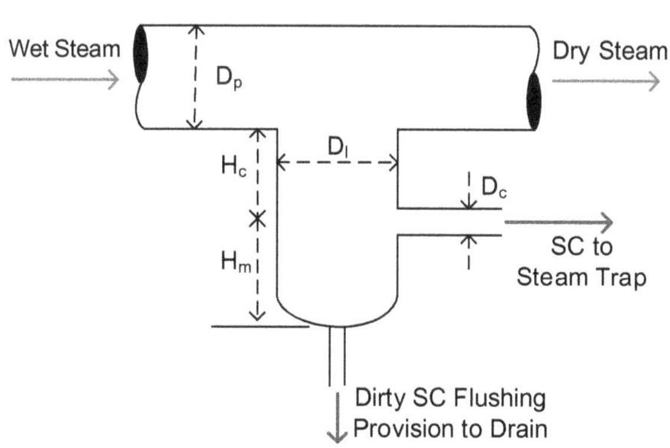

Fig. 17.16: Schematic showing CDL details; see Table 17.4 for values of dimensions.

Table 17.4: Minimum dimension requirements for CDLs; see Fig. 17.16 for significance of symbols

		H_c, mm			
D_p, mm	D_l, mm	Manual, supervised	Automatic	H_m, mm	D_c, mm
20	20	300	700	100	20
25	25	300	700	100	20
32	32	300	700	100	20
40	40	300	700	100	20
50	50	300	700	100	20
65	65	300	700	100	20
80	80	300	700	100	20
100	100	300	700	100	25
150	100	300	700	100	25
200	100	300	700	100	25
250	150	380	700	165	25
300	150	450	700	165	25
350	200	530	700	165	25
400	200	600	700	165	25
500	250	760	700	165	25

CDLs are typically spaced at 30–45 m apart in the steam header. They are required to be installed before raisers, at drops and before expansion loops. Dimension requirements for CDLs are given in Table 17.4. Values in Table 17.4 are extracted from Table 1 in Risco.[16]

17.6.5 *Steam Condensate Header Sizing*

Adequate SC header sizing is important for good CR. SC piping design methods are shown in Table 17.5.[17] Recommended slope for SC pipe is 1:100 towards SC tank/drum. To minimize flashing in SC header, velocity of SC in the pipe should not be more than 0.6 m/sec. Preferred velocity is ~0.3 m/sec.[18] SC should always be added at the top of the SC collection header.

Table 17.5: SC piping sizing methods

Type of Condensate Pipe	Sizing Method
SC pipe from equipment to steam trap	Liquid phase at saturated temperature corresponding to the pressure at the equipment outlet
SC pipe from steam trap outlet to SC header pipe	Two (i.e., vapour and liquid) phases
Main SC header pipe	Two (i.e., vapour and liquid) phases
Pumped SC pipe	Liquid phase at sub-saturated temperature and pressure at the SC pump outlet

17.7 Strategies for Reducing Pressure Drop in Steam Condensate Headers

Following strategies can be used to reduce the pressure drop in SC headers; they are useful for maximizing the CR.

- Flash the SC, and pump the resulting SC liquid to the condensate header
- Increase SC header pipe size by replacing the existing header
- Add another SC header parallel to the existing SC header
- Reduce SC demand on the SC header by shifting SC demand to another SC header
- Use dedicated HC and CC headers (i.e., do not mix HC and CC)
- Upsize control valves
- Eliminate any flow restrictions in the SC headers

The optimum SC system design is a trade-off between the cost of dedicated HC return system and cost of having a CC receiver/flash tank/ drum (having vent connected to a SJE/mechanical compressor or vent to atmosphere) and CC pumping system. If SC quantity is large, lost FS energy (in case of FS vented to atmosphere) can be significant compared to the capital cost for FS recovery using a SJE/mechanical compressor; hence, this favours capturing SC and returning it back to the deaerator, with the highest possible thermal energy.

Contributing factors for SC piping erosion/corrosion and solutions for them are summarized in Table 17A-5.

17.8 Flash Steam Recovery

FS is produced whenever BD water or SC at HP, MP or LP is depressurized to a lower pressure. It is mainly due to differences in enthalpy and vaporization temperature of SC at higher and lower pressures. The enthalpy of FS is almost comparable to the enthalpy of HP steam. Hence, there is a considerable potential for WHR by recovering and reusing FS.

FS generated by SC flashing can be calculated using Eq. 16.6. It can be used in the following ways: directly in deaerator (covered in Chapter 16); for upgrading it to LPS using absorption/mechanical heat pump (covered in Chapter 5, shown here in Fig. 17.17); for generating power using Organic Rankine Cycle (ORC, covered in Chapter 9, shown here in Fig. 17.18); for upgrading to LPS using thermo-compressor or mechanical compressor (Fig. 17.19); to generate chilled water at 5°C–7°C using single-stage lithium bromide absorption chiller (Fig. 17.20); and/or for the generation of freshwater using low temperature thermal desalination (covered in Chapter 13) (Fig. 17.21). The last option is particularly attractive in locations where water is scarce.

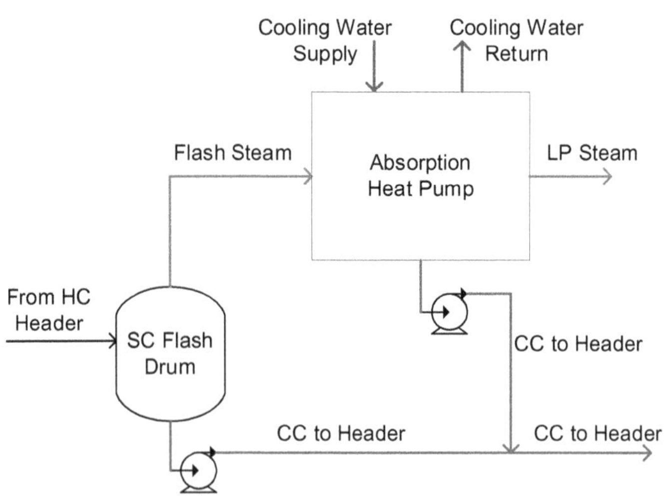

Fig. 17.17: Upgrading FS to LPS using a heat pump.

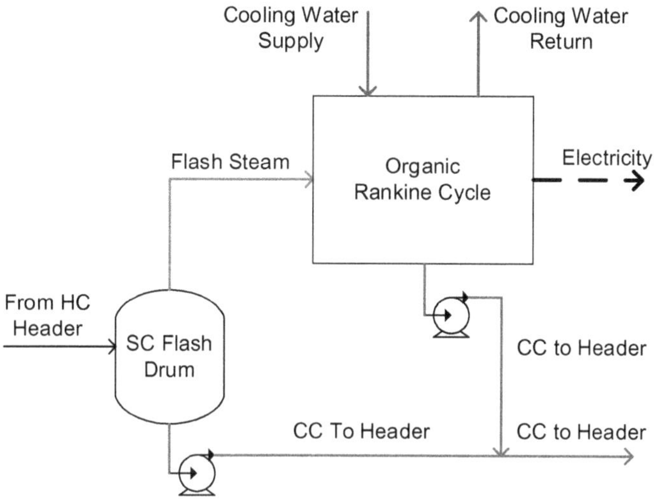

Fig. 17.18: Generation of power from FS using ORC.

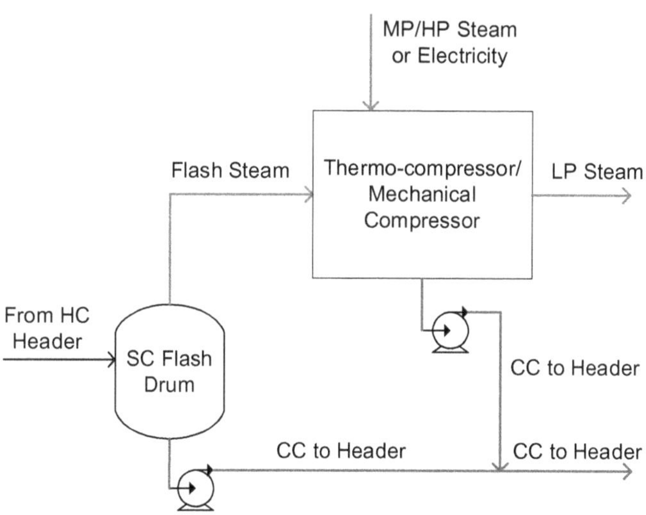

Fig. 17.19: Upgrading FS to LPS using thermo-compressor or mechanical compressor.

For power generation, although ORC efficiency is lower than (steam) Rankine cycle, it is still useful as waste FS is used as heat source. Efficiency of ORC increases with increase in FS temperature/ pressure.

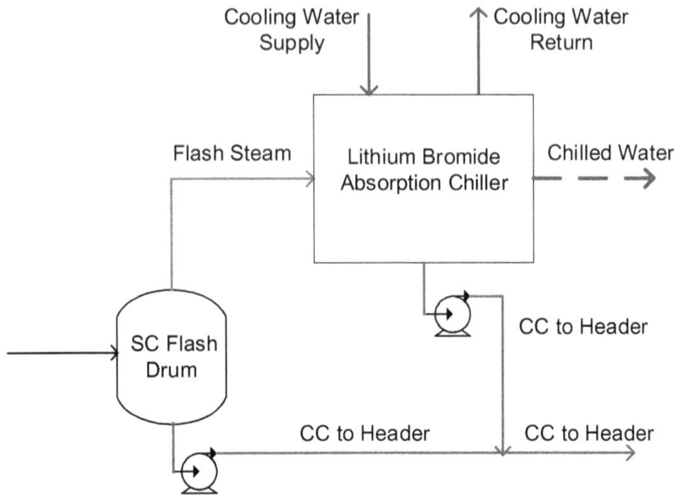

Fig. 17.20: FS usage in lithium bromide absorption chiller to produce chilled water.

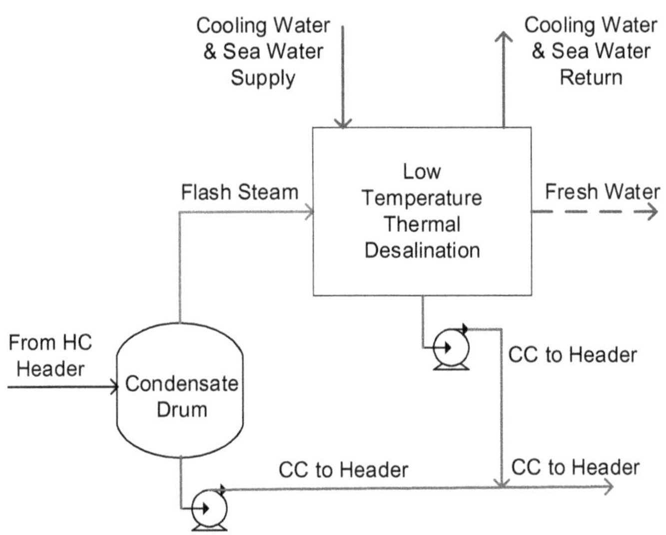

Fig. 17.21: Use of FS for low temperature thermal desalination.

FS energy can be fully recovered by heating cold DM water using the methods shown in Figs. 17.22 and 17.23. These methods fully recover waste heat from FS by indirect contact (shown in Fig. 17.22) or direct contact (shown in Fig. 17.23) of FS and cold DM water. HC flashes into FS and

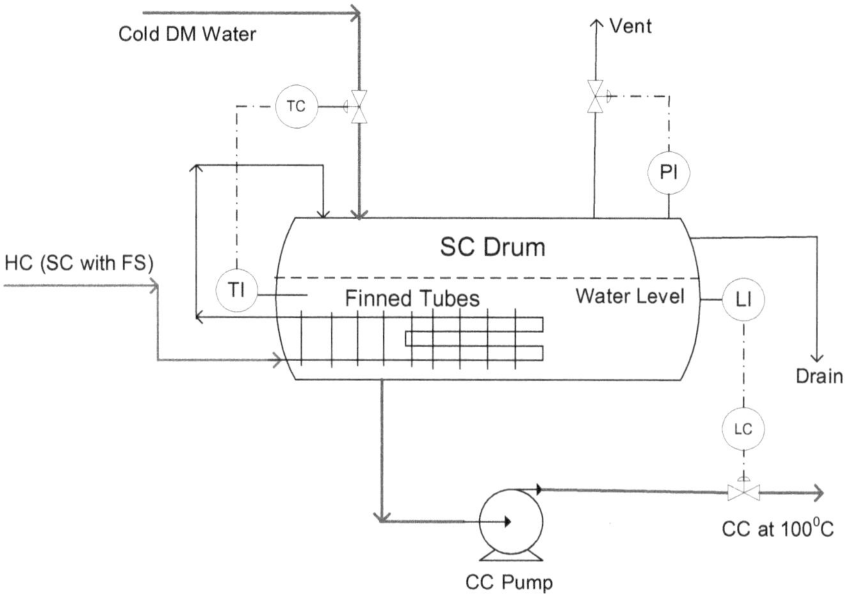

Fig. 17.22: FS recovery by condensation using finned tube HE and DM water cooling.

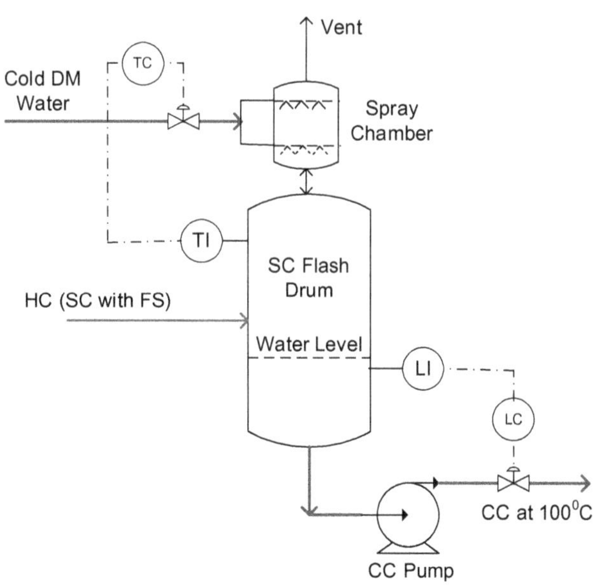

Fig. 17.23: FS recovery by condensation using DM water quench system.

SC in SC drum. Figure 17.22 illustrates FS condensation and SC cooling by heat exchange with DM water. A finned tube HE is used to condense FS and cool the SC by indirect contact of DM water with SC mixture, before sending it to the SC drum. Figure 17.23 illustrates FS condensation by direct contact with cold DM water spray system, arranged at the top of the SC flash drum. DM water flow is regulated to minimize uncondensed FS escaping through the vent into atmosphere.

Although both direct and indirect contact methods are used in industry to achieve full condensation of FS, the direct method requires lower capital cost as it uses cheaper DM water spray nozzles for FS condensation, compared to use of finned tube HE in the indirect method. Also, as SC flashes inside SC flash drum in the direct method, WH problems can be fully eliminated.

Contributing factors for poor FS recovery and solutions for them are summarized in Table 17A-4.

Example 17.4: A boiler operates at a steam pressure of 40 barg for 8000 hours annually. BD from the boiler is 4 ton/hr. Calculate FS that can be obtained from flashing this BD in a flash drum operating at 1 barg. Recovered FS can be used in the deaerator by replacing LPS as the heating medium. Assume cost of LPS is \$19.18/ton and cost of installing BD drum and FS piping is \$100000. Calculate payback period of BD FS recovery system.

Solution

Sensible heat of condensate at 40 barg = 1094.7 kJ/kg (from steam tables)

Sensible heat of condensate at 1 barg = 505.6 kJ/kg (from steam tables)

Latent heat of condensate at 1 barg = 2201 kJ/kg (from steam tables)

Using Eq. 16.6, Flash% $= \frac{(1094.7-505.6)\times 100}{2201} = 26.8\%$

Amount of FS produced from 4 ton/hr of BD $= 4 \times 0.268 = 1.1$ ton/hr

Annual energy savings from FS recovery and its use in the deaerator (instead of LPS) $= 1.1 \times 19.18 \times 8000 = \$164{,}277$

Payback period for BD FS recovery $= \frac{100{,}000}{164{,}277} = 0.61$ years or 7.3 months.

17.9 Steam System Optimization with Condensate Recovery and WHR

Steam system optimization model presented in Chapter 16 (Eqs. 16.18–16.29) is used along with STs and steam load data in Fig. 16.19 (Base Case), for exploring the benefits of CR and WHR in Sections 17.9.1 to 17.9.3. Base Case and Case A in the following sections refer to the respective cases in Chapter 16.

17.9.1 *Case CR1: Increasing Condensate Recovery*

To study the effect of CR increase on steam system economics, additional CR of 99.5 ton/hr is considered in Case CR1 (besides CR of 214 ton/hr in the Base Case). This additional CR is achieved by installing two headers, one for HC (≈ 1 km long and 8″ diameter) and another for DM water (≈ 1 km long and 6″ diameter), one SC flash drum (with DM water quench to condense FS, as shown in Fig. 17.24) and SC pump (to pump the resulting CC to the deaerator). In effect, CR increases from 51.8% (in the Base Case) to 82.8% (in Case CR1). Since process demand for LPS and MPS are constant, reduction of LPS consumption at the deaerator (due to additional CR) requires reduction of HPS generation at the boilers by the same quantity.

For Case CR1 with additional CR, LPS requirement for deaerator is calculated by performing mass and energy balances for the Base Case and CR1 Case using the following equations:

Mass balance: $M_{HC} + M_{CC} + M_{DM} + M_{LPS} = M_{BFW} + M_v$ (17.17)

Energy Balance: $M_{HC} \times H_{HC} + M_{CC} \times H_{CC} + M_{DM} \times H_{DM} + M_{LPS} \times H_{LPS} = M_{BFW} \times H_{BFW} + M_v \times H_v$ (17.18)

Data for the abovementioned calculations and results for the CR1 case are given in Table 17.6. In this table, energy input to and output from the deaerator refer the left and right side of the above energy balance, respectively. As can be seen in Table 17.6, LPS demand reduction (46.16 ton/hr

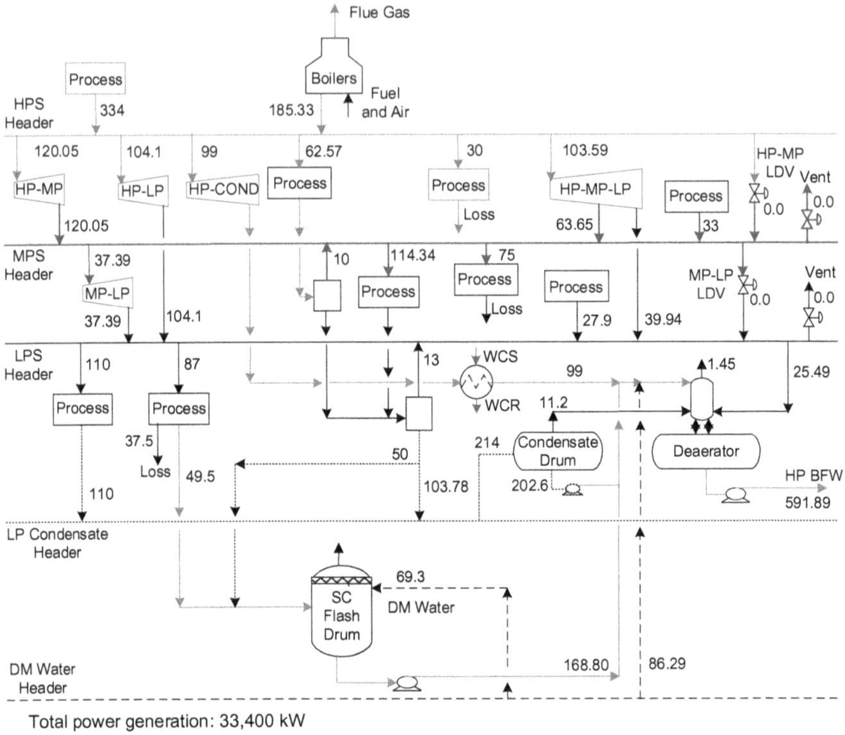

Fig. 17.24: Optimal solution for Case CR1 with increased CR.

in the Base Case to 25.49 ton/hr in CR1 Case) results in lower HPS generation, which saves fuel at boilers. Due to 20.67 ton/hr of lower LPS demand, HPS generation at boilers and the HPS flow rate through STs are reduced by the same amount. These results and Table 17.6 are before optimization (i.e., LPS and HPS demand reduction are established by mass and energy balances around the deaerator).

Next, optimization was performed with flow rate changes in HPS generation and LPS consumption, to maximize power generation from the steam system. Optimal solution found thus for Case CR1 is shown in Fig. 17.24. Optimal power production for Case CR1 is 33,400 kW, which is lower than 34,554 kW in the optimized Case A in Chapter 16 but more than 32,589 kW in the Base Case (shown in Table 17.7). However, due to HPS reduction, total savings for Case CR1 is US$3.46 million/year

Table 17.6: Mass and energy balances for deaerator in base case and CR1 case

Base Case		**CR1 Case**	
Deaerator Inlet Streams		**Deaerator inlet streams**	
HC flow rate (LP condensate), M_{HC} ton/hr	213.78	HC flow rate (LP condensate), M_{HC} ton/hr	313.28
Enthalpy of HC, M_{HC} kJ/kg	623.2	Enthalpy of HC, M_{HC} kJ/kg	623.2
CC flow rate, M_{CC} ton/hr	99.0	CC flow rate, M_{CC} ton/hr	99.0
Enthalpy of CC at 40°C, H_{CC} kJ/kg	167.5	Enthalpy of CC at 40°C, H_{CC} kJ/kg	167.5
DM water flow rate, M_{DM} ton/hr	255.48	DM water flow rate, M_{DM} ton/hr	155.58
Enthalpy of DM Water at 30°C, H_{DM} kJ/kg	125.75	Enthalpy of DM Water at 30°C, H_{DM} kJ/kg	125.75
LPS flow rate, M_{LPS} ton/hr	46.16	LPS flow rate, M_{LPS} ton/hr	25.49
Enthalpy of LPS at 214.91°C and 4.5 bar, H_{LPS} kJ/kg	2889.9	Enthalpy of LPS at 207.3°C and 4.5 bar, H_{LPS} kJ/kg	2873.9
Deaerator Outlet Streams		**Deaerator Outlet Streams**	
BFW flow rate, M_{BFW} ton/hr	612.97	BFW flow rate, M_{BFW} ton/hr	591.81
BFW enthalpy at 121°C, H_{BFW} kJ/kg	508	BFW enthalpy at 121°C, H_{BFW} kJ/kg	508
Vent flow rate, M_V ton/hr	1.45	Vent flow rate, M_V ton/hr	1.45
Enthalpy of vent flow at 121°C, H_V kJ/kg	2707.4	Enthalpy of vent flow at 121°C, H_V kJ/kg	2707.4
Mass Balance		**Mass Balance**	
Flow rate of deaerator inlet streams, ton/hr	614.42	Flow rate of streams to deaerator, ton/hr	593.35
Flow rate of deaerator outlet streams, ton/hr	614.42	Flow rate of streams out of deaerator, ton/hr	593.35
Energy Balance		**Energy Balance**	
Energy input to deaerator, MW	87.59	Energy input to deaerator, MW	84.62
Energy output from deaerator, MW	87.59	Energy output from deaerator, MW	84.62

compared to Case A (based on HPS cost of US\$27.48/ton and electricity cost of US\$0.15/kW). For Case CR1, CR requires additional SC drums (with DM water quenching of FS) at the process units, SC pumps, two ≈ one km long headers, one for DM water (6″ diameter) and another for SC

Table 17.7: Steam flow rates for turbo generator and benefit analysis for base, A, CR1 and CR2 cases

Quantity	Base Case	Case A	Case CR1	Case CR2
Turbo generator inlet flow, ton/hr	83.8	104.66	103.59	103.99
Turbo generator extraction flow, ton/hr	49.69	64.66	63.65	64.20
Turbo generator exhaust flow, ton/hr	34.11	39.99	39.93	39.78
Total power generation, kW	32,589	34,554	33,400	32,865
Power from turbo generator, kW	6,435	8,375	8,287	8,312
HP steam savings, ton/hr	0.00	0.00	20.67	30.27
Total savings (compared to optimized case), US$/year	NA	0	3,459,664	5,066,585
Capital cost for modifications, US$	0	0	1,565,586	3,145,586
Payback period, years	0	0	0.45	0.62

(8″ diameter). Costs for these modifications are estimated to be US$1.57 million. Summary of optimization results and benefits analysis are given later in Tables 17.7 and 17.8.

17.9.2 *Case CR2: Increasing Condensate Recovery and WHR*

A significant part of wasted energy in process units is low temperature heat rejected to atmosphere, mainly through cooling water, fin-fan coolers and flue gases. Usually, distillation column overhead streams at 100°C–200°C reject heat by fin-fan coolers, and streams at < 100°C reject heat to cooling water. In Case CR2, in addition to the CR increase considered in Case CR1, WHR from a process stream, shown near second process block of the left side of Fig. 17.25) is used for generation of 9.6 ton/hr of LPS. As there is no change in process demand for LPS, LPS generated by WHR and sensible heat of CR will reduce the demand for LPS (from its header) at the deaerator. As explained in Case CR1, lower LPS demand reduces HPS consumption at STs (by 20.67 + 9.6 = 30.27 ton/hr).

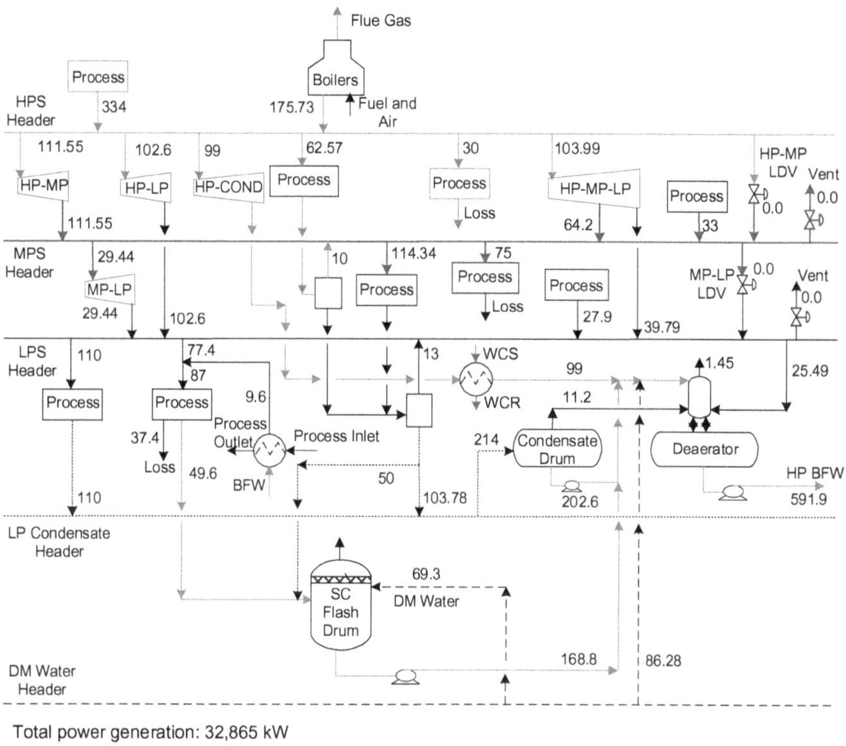

Fig. 17.25: Optimal solution for Case CR2: Effect of CR increase and WHR (generating 9.6 ton/hr of LPS) from a process stream.

The results in the previous paragraph are before optimization (i.e., LPS and HPS demand reduction are established by mass and energy balances around the deaerator). Optimization was performed with changes in HPS generation and LPS consumption flow rates, to maximize the power generation from the steam network. Optimal solution for Case CR2 is shown in Fig. 17.25. Optimal power production for this case is 32,865 kW, which is lower than 34,554 kW in the optimized Case A in Chapter 16, and also slightly lower than 33,400 kW for Case CR1 given earlier. However, total savings for Case CR2 is US$5.07 million/year compared to Case A, due to HPS reduction. Summary of optimization results and benefits analysis are given later in Tables 17.7 and 17.8.

Table 17.8: Optimal status of switchable ST drivers in base, A, CR1 and CR2 cases

	ST Selection Parameter for Case Studies					ST Selection Parameter for Case Studies			
ST	**Base**	**A**	**CR1**	**CR2**	**ST**	**Base**	**A**	**CR1**	**CR2**
HL1	1	0	0	0	ML14	0	0	0	0
HL2	0	0	0	1	ML15	0	0	0	0
HL3	0	0	0	0	ML16	0	0	0	0
HL4	0	0	0	0	ML17	0	0	0	0
HL5	0	0	0	0	ML18	1	0	0	0
HL6	0	1	0	0	ML19	1	0	0	0
ML1	1	1	0	0	ML20	0	0	0	0
ML2	1	0	0	0	HL20	0	0	0	0
HL7	1	0	0	0	ML21	0	0	0	0
HL8	0	0	0	0	ML22	0	0	0	0
HL9	1	0	0	0	ML23	0	0	0	0
HL10	0	0	0	0	HL21	0	1	1	1
HL11	1	0	0	0	HL22	0	0	1	0
HL12	1	0	0	0	ML24	0	0	1	0
HL13	1	0	0	0	ML25	0	0	0	0
ML3	0	0	0	0	HL23	0	0	0	0
ML4	0	0	0	0	HL24	1	0	0	0
HL14	0	0	0	0	HM2	0	0	0	0
HL15	0	0	0	0	HM3	1	1	0	1
HL16	0	0	0	0	HM4	1	0	0	0
HL17	1	0	0	0	HM5	0	0	0	0
ML5	0	1	1	1	HM6	1	1	1	0
ML6	0	1	0	0	ML26	0	1	1	1
HL18	1	0	0	0	ML27	0	0	0	0
ML7	0	0	0	0	ML28	0	0	0	0
ML8	0	0	0	0	ML29	0	1	0	0
ML9	0	1	1	0	ML30	0	0	1	1
ML10	0	0	0	0	ML31	0	0	0	0
ML11	0	0	0	0	ML32	1	0	0	0
ML12	1	0	0	0	ML33	0	0	0	0
HL19	0	0	0	0	ML34	1	0	0	0
HM1	0	0	0	0	ML35	0	0	0	0
ML13	1	1	0	0					

Note: For example, selection parameter for HL2 and ML11 are denoted by YHL2 and YML11, respectively.

17.9.3 *Summary and Analysis of Results*

Optimal status of switchable STs for the Base Case, Case A, Cases CR1 and CR2 are summarized in Table 17.8. Steam flow rates through the steam turbo generator and benefit analysis for these cases are presented in Table 17.7. Capital costs of HEs, pumps and pressure vessels are estimated using CAPCOST program (based on Excel spreadsheet), available with the book by Turton *et al.*[19] Piping, insulation, instrumentation, and electrical cost are estimated based on industrial data.

Cases CR1 and CR2 require modifications in the steam and CR system. As given in Table 17.7, total capital cost for modifications in Case CR1 is US$1,565,586 (= 407,586 for piping and insulation + 603,000 for SC flash drum + 55,000 for two SC pumps + 200,000 for instrumentation and electric works + 300,000 for civil and structural works), whereas it is US$3,145,586 (= 407,586 for piping and insulation + 603,000 for SC flash drum + 55,000 for two SC pumps + 300,000 for instrumentation and electric works + 1,780,000 for kettle type steam generator + 400,000 for civil and structural works) in Case CR2. Thus, total capital cost for modifications is more for Case CR2 compared to that for Case CR1.

Annual cost savings for CR1 compared to Case A are due to changes in power generation and HPS savings; they are calculated based on HPS cost of US$27.48/ton, electricity cost of US$0.15/kW and 8760 hours of operation per year. Accordingly, annual cost savings for Case CR1 compared to Case A is $[20.67 \times 27.48 + (33{,}400.362 - 34{,}554.18) \times 0.15] \times 8760 =$ US$3,459,664. Similarly, cost savings for Case CR2 compared to Case A is US$5.07 million. Thus, implementation of CR and WHR (i.e., Case CR2) produces more cost savings by nearly 50% but it also requires nearly double the capital cost, compared to Case CR1. In any case, payback period for Cases CR1 and CR2 is 0.45 and 0.62 year, respectively. So, CR and WHR scenarios studied in the two case studies are economically very attractive and improve energy efficiency of the plant, which reduces the greenhouse gas emissions.

17.10 Summary

In this chapter, CR problems and potential solutions are presented with examples. Key points for maximizing CR and WHR are summarized as follows:

- SC has significant portion of steam energy and its recovery is essential for maximizing WHR.
- WH, stall problem, SC contamination and inadequate SC header sizing are the main barriers for maximizing CR.
- WH problem can be resolved by many methods such as (a) maintaining separate CR collection systems for HC and CC, (b) flashing HC to atmospheric conditions and using CC pump, (c) heating CC before mixing it with HC, and (d) repairing leaking steam traps and SC control valves.
- Stall problem can be resolved by using SC pumping system or flooded operation of reboiler/HE.
- Improper condensate header sizing can be avoided by following the guidelines provided in this chapter.
- Online steam trap monitoring and repair plan are essential for minimizing steam losses and maximizing SC WHR.
- FS generated by flashing HC can be used for (a) direct water heating in the deaerator, (b) production of chilled water, (c) fresh water using thermal desalination, and/or (d) upgrading to LP/LLP steam using AHP, TVR or MVR.
- Increased CR may result in power generation reduction from steam network but HPS required to be generated may be lowered, which is a benefit. Hence, steam network optimization is very important for the proper evaluation of benefits of CR projects.

Books by Snow[20] and Turner and Doty[14] are recommended for further reading on the topics covered in this Chapter.

References

1. Lieberman NP, Lieberman ET. (2008) *A Working Guide to Process Equipment*, 3rd ed. McGraw-Hill.
2. Fleming I. (2010) Optimizing Steam Systems: Part I. PTQ, pp. 47–53, Q2.
3. Krisner W. (August 2012) Water Hammer in Condensate Return Lines. *Chem Eng* 33–37.
4. Risko JR. (November 2004) Steam heat exchangers are under worked and over-surfaced. *Chem Eng* 58–62.
5. Kister HZ. (July 2020) Controlling reboilers heated by condensing steam or vapour. *Chem Eng* 22–32.
6. Lieberman N. (2016) Improving Heat Transfer in Reboilers and Condensers. PTQ, pp. 39–43, Q1.
7. Hou A, Mita T. (May 2018) Advanced Steam System Optimization Program. *Hydrocarb Process* 45–49.
8. Branan C. (2012) *Rules of Thumb for Chemical Engineers*, 5th ed. Gulf Professional Publishing.
9. Ganapathy V. (2002) *Industrial Boilers and Heat Recovery Steam Generators: Design, Applications, and Calculations*. CRC Press.
10. Hicks TG. (2006) *Handbook of Mechanical Engineering Calculations*, 2nd ed. McGraw-Hill.
11. Risko JR. (February 2013) Beware of the Dangers of Cold Traps. CEP, pp. 50–53.
12. Risko JR. (April 2015) My Steam Trap is Good. Why Doesn't It Work. CEP, pp. 27–34.
13. Paffel K. (September 2013) How to properly size a steam trap. *Chem Eng* 58–61.
14. Turner WC, Doty S. (2013) *Energy Management Handbook*, 8th ed. Fairmont Press.
15. Risko JR. (November 2006) Handle steam more intelligently. *Chem Eng* 44–49.
16. Risko JR. (January 2019) Allocate new plant focus to steam system design-part 1. *Hydrocarb Process* 39–43.
17. Fleming I. (2010) Optimizing Steam Systems: Part II. PTQ, pp. 54–62 Q3.
18. Vengateson U. (May/June 2015) Retool heat exchanger design for different operating scenarios. *Gas Process* 23–28.
19. Turton R, Bailie RC, Whiting WB, Shaeiwitz JA. (2009) *Analysis, Synthesis, and Design of Chemical Processes*, 3rd ed. Prentice Hall.

20. Snow D A. (2001) *Plant Engineer's Reference Book*, 2nd ed. Elsevier.
21. API 581 (April 2019), Risk-Based Inspection Methodology, 3rd ed. American Petroleum Institute.

Acronyms and Notation

Acronyms

BD	Blowdown
BR	Blowdown Ratio
CC	Cold Condensate
CDL	Condensate Drip Leg
CR	Condensate Recovery
DM	Demineralized
DP	Differential Pressure
FS	Flash Steam
HC	Hot Condensate
HE	Heat Exchanger
HHV	Higher Heating Value
HP	High Pressure
HPS	High-Pressure Steam
LLP	Low-Low Pressure
LP	Low Pressure
LPS	Low-Pressure Steam
MP	Medium Pressure
MPS	Medium-Pressure Steam
MVR	Mechanical Vapour Recompression
NC	Normally Closed
NO	Normally Open
NRCL	Non-Recoverable Condensate Loss
PCR	Percentage Condensate Recovery
PD	Percentage Stall Duty
PoF	Probability of Failure of an Equipment
RO	Restriction Orifice
SC	Steam Condensate
SJE	Steam Jet Ejector
ST	Steam Turbine
TVR	Thermo Vapour Recompression

WH	Water Hammer
WHR	Waste Heat Recovery

Notation

A	Heat transfer area of an HE (m^2)
C_p	Heat capacity (specific heat) of a stream (kJ/kg.K or kJ/kg.°C)
C_v	Flow Coefficient
$D_f(t)$	Damage factor
F_{MS}	Management safety factor
gff	Failure frequency of the equipment
LMTD	Log mean temperature difference (°C or K)
M	Mass flow rate of process fluid (kg/sec)
Q	Heat duty of (i.e., heat transferred from one stream to another in) an HE (kW)
T	Temperature (°C or K)
ΔT	Temperature difference (°C or K)
U	Overall heat transfer coefficient of an HE ($kW/m^2.K$ or $kW/m^2.°C$)

Subscripts

BD	Blowdown
BFW	Boiler feed water
CC	Cold condensate
CR	Condensate recovered
cs	Saturated condition at condensate header pressure
DM	Demineralized
HC	Hot condensate
i	Refers to inlet or inner condition(s)
M	Mean value
NRCL	Non-Recoverable Condensate Loss
o	Refers to outlet or outer condition(s)
RCL	Recoverable condensate loss
S	Steam or stall
sh	Sensible heat of high-pressure condensate or BD water
S0	Stall at 0% excess area
S100	Staff at full load
v	Vent

Exercises

17.1 Estimate the annual credit/revenue from recovering steam generated by flashing 25 ton/hr of saturated condensate from 30 to 2 bar. Assume FS cost of $10/ton and 8500 hours of operation per year.

17.2 A process plant needs 300 ton/hr of BFW by operating a deaerator. DM water is available at 25°C for make-up rate of 180 ton/hr. LP steam is available for deaerator operation at 2 bar. Condensate return from the process plant is at 90°C. Boiler BD rate is 5 ton/hr at an operating pressure of 45 bar. Assume deaerator vent flow rate of 0.15 ton/hr. Perform mass and energy balances for the steam and condensate system, and calculate the CR for the plant.

17.3 An HE operates with a steam supply pressure of 4 bar, to heat a process stream from 50°C to 90°C. Condensate header back pressure at the outlet of the exchanger is 2.5 bar. HE has 30% oversurface area. Evaluate the possibility of stall problem; if yes, estimate the HE duty at stall point. State methods for overcoming stall problem to achieve sustainable CR.

17.4 A process generates 10 ton/hr of cold condensate at 90°C. A nearby condensate header is handling 150 ton/hr of two-phase condensate with 5 wt% FS at 2.5 bar. Review if it is safe to send this cold condensate to the two-phase condensate header directly. Suggest safe methods for sending cold CR to the two-phase header.

17.5 Estimate the annual steam loss through a fully blowing steam trap, operating at 30 bar. Consider steam trap orifice size of 5 mm and 8760 hours of operation per annum.

Appendix

Table 17A-1: Summary of contributing factors for WH and solutions for them (in italics in brackets)

- Mixing of CC and HC (*use dedicated CC and HC recovery headers*)
- Passing/damaged steam traps (*conduct regular steam trap inspection and repair them timely*)
- Steam trap bypass valves kept open and/or damaged control valves (*close any steam trap bypass valves to the CR header and/or repair damaged control valves*)
- High two-phase velocity (> 15 m/sec) in SC header due to the addition of excess SC above the design capacity of the header or under-sized SC header (*install a new parallel SC header or replace the existing header with a new bigger SC header, to reduce the two-phase velocity to < 15 m/sec; install flash drums and SC pumps to minimize flashing inside SC header*)
- Loops in vertical plane in the condensate header (*if possible, reroute the SC header to remove vertical loops*)
- Use of intermittent discharge type steam traps (*replace them with continuous discharge type*)
- SC added at the bottom of the SC header (*modify the SC pipe to join the main SC header from the top*)
- Use of slam check valves and pumps with on-off operation (*use non-slam check valves, use slower closing valves; operate condensate pumps on continuous mode by using dedicated minimum flow protection system*)

Table 17A-2: Summary of contributing factors for stalling and solutions for them (in italics in brackets)

- Modulating steam control valve, especially in LP steam heating (*install control valve on SC removal pipe from SC drums and remove steam control valve; install CR pump*)
- Excessive overdesign of HE or minimum turndown operation (*operate HE under partial flooded mode by using SC pot level control*)
- Steam trap strainer plugged (*clean steam trap strainers*)
- Under-sized steam trap (*replace it with correctly sized steam trap*)
- High back pressure of condensate header due to overload, many passing steam traps, live steam entry into the SC header (*repair steam traps, close steam trap bypass valves, stop any live steam entry into SC header; install a new parallel SC header or replace the existing header with a new bigger SC header to reduce two-phase velocity to < 15 m/sec*)

Table 17A-3: Summary of contributing factors for contaminated SC and solutions for them (in italics in brackets)

- HE tubes leak due to WH (*see the solutions for WH in Table 17A-1; if possible and if process stream contamination is tolerable, maintain process stream pressure lower than SC pressure*)
- Thermal shock during start-up/shutdown operation (*incorporate proper startup/shutdown procedures*)
- Stalled HE operation (*see the solutions for stalled operation in Table 17A-2*)
- Use of steam as stripping steam (*recover the heat content from the contaminated SC and send it to wastewater treatment plant or treat it in foul water stripper for re-use as process wash water*)

Table 17A-4: Summary of contributing factors for poor FS recovery and solutions for them (in italics in brackets)

- SC depressurization to atmosphere (*install flash tanks to separate FS and mix it with steam header of appropriate pressure level*)
- Pressure of FS is very low for process use (*increase FS pressure or utilize its heat value with the following techniques:*
    - *TVR (SJE) or MVR*
    - *Heat pumps for upgrading temperature or heat*
    - *Single stage absorption chiller to produce chilled water*
    - *Generate electricity using ORC*
    - *Low temperature thermal desalination for producing freshwater*
    - *Direct use in deaerator*)
- Passing/damaged steam traps (*conduct regular steam trap inspection and repair them timely*)
- Steam trap bypass valves kept open or steam passing due to damaged control valves (*close any steam trap bypass valves to the CR header or repair damaged control valves*)

Table 17A-5: Summary of contributing factors for SC piping erosion/corrosion and solutions for them (in italics in brackets)

- High two-phase velocity (> 15 m/sec) in SC header due to addition of excess SC above the design capacity of the header or under-sized SC header (*install a new parallel SC header or replace the existing header with a new bigger SC header, to reduce two-phase velocity to < 15 m/sec; install flash drums and SC pumps to minimize flashing inside SC header*)
- Sharp elbows in piping (*use long radius elbows*)
- Piping loops in vertical plane (*avoid piping loops in vertical plane; thermal expansion loops can be provided in horizontal plane*)
- SC addition at bottom of the SC header (*always add SC at the top of the SC header*)
- Passing steam traps or SC control valve or open bypass valves across steam traps (*repair passing steam traps or control valves, or close any steam trap bypass valves*)

Chapter 18

Sustainability of Waste Heat Recovery Projects

18.1 Overview

Energy consumption and environmental emissions generally increase with the production capacity of the process plant. Process industries continuously strive to achieve competitive advantage by reducing energy usage, environmental emissions, operating and maintenance costs. Waste heat recovery (WHR) addresses these goals in a cost-effective manner. WHR is a key to achieve triple bottom-line benefits of economic, environmental and social improvements for sustainable development.

It is a truism that waste heat (WH) results in lower efficiency. It also, of course, causes environmental damage. A gaseous WH stream exiting a process carries material waste in the form of particulate pollutants (e.g., from unburnt fuel) and various gaseous substances, which pollute the atmosphere. A liquid WH stream discharged into water bodies increases the temperature in water bodies and may seriously affect the aquatic ecosystem. WHR not only directly captures the thermal energy and pollutants released into the atmosphere, it also reduces the fuel consumption and consequently CO_2, NO_x, SO_x and/or water pollution. Thus, WHR leads to better air and water quality, with benefits and social impact in the community. Economic growth of a company as a result of efficiency gains through WHR, could create more jobs. It is in this way that WHR projects improve profitability, environmental compliance and social development.

This chapter begins with the discussion on sustainability aspects for WHR projects in Section 18.2. Next, Section 18.3 describes common damage mechanisms of WHR equipment and potential solutions to overcome them. These are important since damaged equipment leads to more maintenance/replacement costs and loss of energy efficiency, thus affecting WHR sustainability. Sections 18.4 and 18.5 present strategies to avoid acid dew point corrosion in air pre-heater (APH) and economizers, and design solutions to avoid WHR equipment and piping problems, respectively. Section 18.6 discusses material of construction requirements for heat exchangers (HEs) to ensure trouble-free operation. Section 18.7 presents HE fouling and its mitigation strategies. Section 18.8 summarizes piping velocity considerations. Section 18.9 outlines the importance of damage mechanism reviews (DMR), risk-based inspection (RBI) and integrity operating windows (IOW). This chapter ends with summary in Section 18.10.

Learning outcomes of this chapter are as follows.

1. Describe the sustainability requirements for WHR projects.
2. Discuss common damage mechanisms and potential solutions.
3. Calculate acid dew point temperature for assessing acid corrosion and identify the solutions to prevent it.
4. Identify safer solutions for WHR equipment hazards.
5. Select the material of construction for HEs.
6. Identify HE fouling, corrosion problems and effective solutions to minimize them.
7. Recommend fluid velocities to prevent erosion and settling problems.
8. Discuss the importance of DMRs, RBI and IOW limits, for safe and sustainable operation of WHR equipment.

18.2 Sustainability Model for WHR

Sustainability of a WHR project refers to its ability to deliver the performance and benefits over its entire life cycle. For achieving this, WHR projects should address the triple bottom-line factors, namely, environmental, economic and social factors, as shown in Fig. 18.1. WHR reduces the consumption of natural resources such as fossil oils, natural gas, coal and/or water. It also reduces air and water pollution by reducing fuel

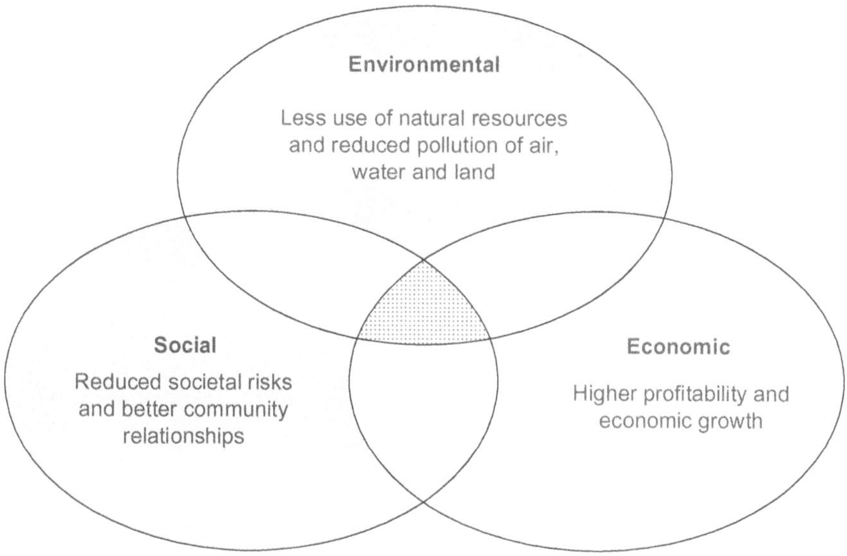

Fig. 18.1: Sustainability requirements for WHR projects.

consumption and/or minimizing wastewater generation. Thus, WHR leads to better environmental management.

WHR projects can potentially reduce both capital (e.g., avoiding capital cost required for carbon capture, utilization, pollution control and waste disposal system such as flue gas stacks) and operating costs, and hence improve the profitability and economic growth for the company. Their design should also be robust enough to minimize the maintenance costs, and the cost for environmental management (e.g., CO_2 capture or sequestration) over the entire life cycle of the project.

Lastly, WHR projects need to address social factors. They must not introduce new risks to the community, and should try to reduce existing risks. For example, a WHR project to recover energy from flue gas, thus reducing its exit temperature, can lead to reduced heat radiation into the surroundings, lower pollution and shorter flue gas stack height leading to lower 'visual' pollution. This will improve the image of the company among the community. Another example would be the economic growth of a company through WHR improving job prospects and quality of life, which would improve community support for the industry.

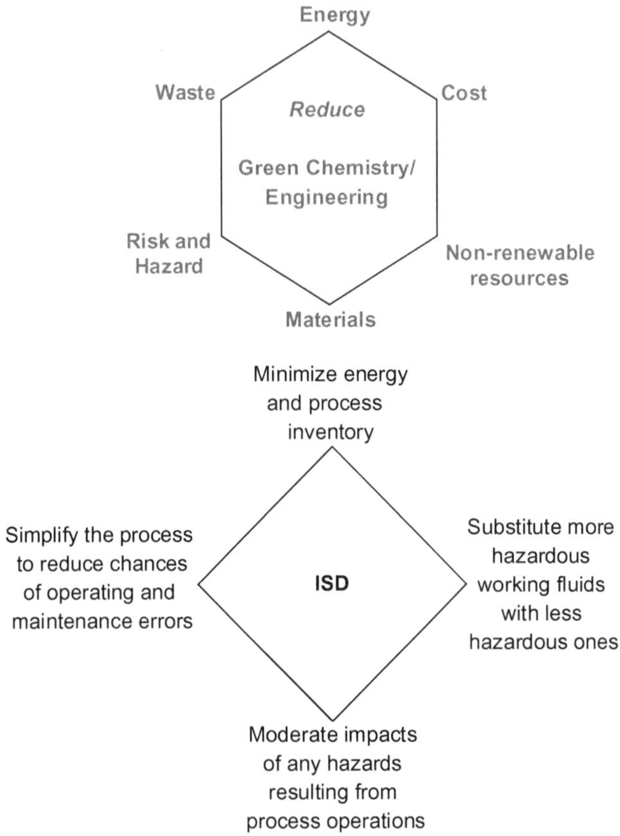

Fig. 18.2: Key features of green chemistry/engineering (top image) and ISD (bottom image).

In general, new WHR projects should adopt inherently safer design (ISD) and process safety improvement features in their project design. ISD is based on minimization/intensification, substitution, moderation and simplification principles. ISD concepts and process safety are described in detail in Chapter 20. Environmental, economic and social goals of sustainability are fully aligned with the key features of ISD and green chemistry/engineering as illustrated in Fig. 18.2. Green chemistry/engineering principles aim to reduce energy, cost, consumption of non-renewable resources, materials, risks, hazards and wastage generation. Contreras and Bravo[1] have highlighted many useful green engineering ideas, applicable for design and operating phases of projects. ISD also

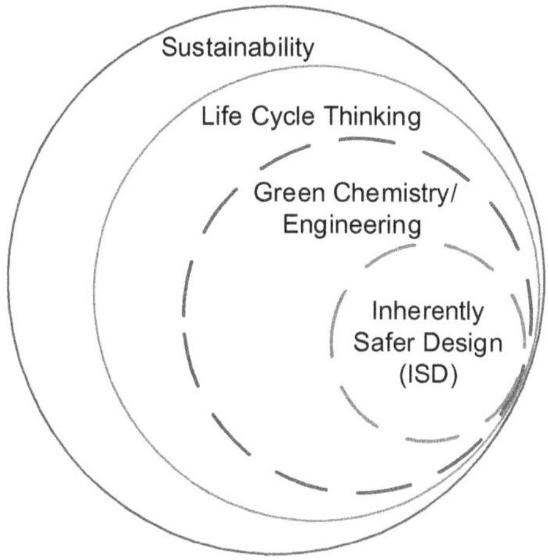

Fig. 18.3: Relationship of ISD, green chemistry/engineering, life cycle thinking and sustainability.

aims to minimize energy, inventory, risks, hazards and costs. Hence, it contributes strongly to the sustainability of WHR projects.

As illustrated in Fig. 18.3, ISD and green chemistry/engineering are in fact sub-sets of sustainability. Moreover, the concept of sustainability needs to cover the life cycle. Hence in addition to adopting ISD and green chemistry/engineering principles at the design stage, a company should develop a culture that encourages their application through the entire life cycle of the WHR project.

Environmental and economic aspects can be further classified into higher energy efficiency, efficient resource usage, lower costs, lower emissions, better energy security, better design and analysis, as shown in Fig. 18.4, which is developed based on Figure 3 in Dincer and Acar[2]. These are outlined below.

Higher Energy Efficiency: Energy efficiency can be increased by (maximizing) WHR. This can be achieved by recovering/re-using WH directly in the process or by generating utilities. Some examples are the following:

- Use of heat exchanger network (HEN) to maximize heat integration in the process.
- Use of APH to recycle WH back to boiler or fired heaters.

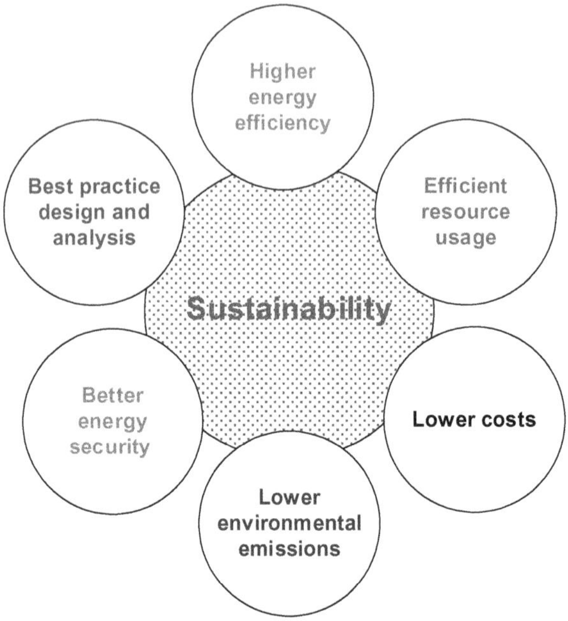

Fig. 18.4: Sustainability of WHR projects.

- Chilled water generation using WH-powered LiBr (lithium bromide) heat pump.

Many strategies and examples for WHR and re-use are presented in Chapters 3 and 8 to 17.

Efficient Resource Utilization: WHR and re-use reduces the use of non-renewable energy resources. Additionally, recovered WH can be combined with renewable and clean energy sources for reducing the demand for fresh energy further. This reduces dependence on resources that are not locally available and/or affordable. WHR from condensate recovery, described in Chapter 17, reduces fresh fuel demand of boilers. Re-use of used cooling water (CW) as discussed in Chapter 15 reduces the demand for new cooling towers. WHR can reduce capital costs by eliminating equipment required for downstream pollution abatement such as flue gas scrubbers, flue gas desulfurization units and flue gas stacks. These examples are presented in Chapter 11. Pinch analysis can establish WHR opportunities to simultaneously reduce hot and cold utilities, as explained in Chapter 3.

Lower costs: Lower costs can be achieved by the efficient use of resources and maximizing energy efficiency through WHR, process design practices such as ISD and industry codes, and re-using existing equipment. WHR can reduce CO_2 capture and storage costs. Lower capital and operating costs by WHR are illustrated in Chapters 3, 8 to 17.

Lower environmental emissions: Maximizing WHR reduces global warming and air and/or water pollution. For example, WHR using APH reduces boiler/fired heater fuel consumption and hence reduces CO_2, SO_X, particulate and NOx emissions. In once-through CW systems, WHR can reduce water discharge temperature to river/sea and hence can avoid damage to the aquatic system.

Better Energy Security: Exploiting the significant WHR potential in process industries can reduce excessive dependence on fossil fuels and also assist in the transition to alternative, reasonably priced energy sources. In this way, implementation of WHR can provide better energy security. Lower energy consumption and hence better energy security is demonstrated in many examples in Chapters 3 and 8 to17.

Best Practice Design and Analysis: Good WHR design minimizes heat losses and increases efficiency, and hence reduces wastage of resources over a long period. This can be aided by heat integration strategies such as pinch analysis, ISD principles, diverse layers of protection in process design, industry codes and best practice to deliver sustained benefits for the whole life cycle of plant operations. Comprehensive WHR design should include life cycle analysis and exergy analysis (which considers both quantity and quality of WH), as well as the usual energy analysis (considering only quantity of WH). In addition to better process design, the design team should do a thorough analysis of potential WHR equipment design problems and damage mechanisms in order to sustain the improvement from WHR. Proper documentation of these reviews is very useful for quick rectification of any equipment problems, during plant/equipment life cycle.

Many of the abovementioned factors are covered in detail in the previous chapters. Hence, the remaining sections of this chapter focus on damage mechanisms of WHR equipment, sustainable design solutions, equipment integrity monitoring, inspection, and maintenance strategies.

18.3 Damage Mechanisms and Potential Solutions

This section summarizes common damage mechanisms in WHR equipment, and potential solutions. Equipment damage increases maintenance cost and takes up scarce resources (e.g., maintenance and contractor staff, materials). It negates WHR, increasing the demand on fresh energy, and eventually increasing pollution. In some cases, (e.g., severe corrosion of plate-type APH), the only option is to completely replace the equipment. Thus, WHR equipment damage can increase both operating and maintenance/capital costs, impacting on the sustainability of the WHR project.

Comprehensive coverage of 60 damage mechanisms is available in API RP 571[3]; although these are in petroleum industry, many of them are applicable for other industries. Common damage mechanisms in WHR equipment and potential solutions are compiled in Table 18.1.

Table 18.1: Common damage mechanisms in WHR equipment and potential solutions

Damage Mechanism	Explanation	Potential Solutions
Erosion/ Corrosion	• Erosion is a physical phenomenon. It causes accelerated removal of metal surface material due to rubbing of surfaces and/or high-velocity impingement of solids, liquids and/or vapour on metal surfaces. • Corrosion is a chemical phenomenon. It contributes to erosion by removing protective films or scales on metal surfaces. In turn, erosion exposes metal surfaces to further corrosion.	• Avoid reaching erosion.velocities in process equipment. • If possible, adjust temperature, pH, deaeration, concentration of corrosive substances and injection of corrosion inhibitors to avoid corrosion. • Avoid cavitation in pumps and control valves. • Design improvements involving changes in shape, geometry and/or materials selection (e.g., increasing pipe diameter for reducing fluid velocity; use of mixing devices, impingement plates or rods; increasing wall thickness of tubes; use harder alloys for surface hardening, corrosion-resistant materials or coatings).

Table 18.1: (*Continued*)

Damage Mechanism	Explanation	Potential Solutions
		• Erosion-resistant refractories can be used in solids handling applications.
Vibration-Induced Fatigue	• It results in metal cracks due to vibration, water hammer or unstable fluid flow such as pressure surges. • Damage depends on the amplitude and frequency of vibration as well as on the fatigue resistance of the metal components. • Lack of necessary supports for piping, tubes or equipment, and flow induced vibrations are the main reasons for this type of damage mechanism.	• This type of damage cannot be solved by material upgrades. • Vibration-induced fatigue can be eliminated or reduced through proper design, use of supports and/or vibration dampening equipment. • Cavitation in pumps can be avoided by providing adequate net positive suction head (NPSH) margin over and above the required NPSH. • Cavitation in control valves (CVs) can be avoided or minimized by the use of anti-cavitation trim or use of restriction orifice (ROs) and CV in series arrangement. • Vortex shedding[a] can be minimized at the outlet of CVs and PRVs (pressure relief valves) through proper sizing of side branches and flow stabilization techniques.
Cooling Water Corrosion	It is caused in carbon steel (CS) and other metals by dissolved salts, oxygen, corrosive substances (solid, liquid and gaseous substances) and microbiological activity. • Fluid temperature, CW type (fresh, brackish, salt water), cooling system type, O_2 content and fluid velocities are important factors.	• Low velocity of CW causes solids settling problem and very high velocity causes erosion-corrosion. CW velocity of 2 to 3 m/s is beneficial. • Maintain good water chemistry using anti-corrosion and anti-microbial additives. • Metallurgy upgrades and anti-corrosion coatings for piping and HE components can be useful,

(*Continued*)

[a] Vortex shedding is an oscillating flow that takes place when a fluid flows past a bluff body (like control valve plug or PRV seat) at a velocity in the critical range.

Table 18.1: (*Continued*)

Damage Mechanism	Explanation	Potential Solutions
	• Increasing CW outlet temperature and/or process stream inlet temperature tends to increase corrosion rate and fouling tendency. • Increasing O_2 content tends to increase CS corrosion rate. • If process stream is above 60°C, scaling potential exists.	especially for CW with high chloride content, low velocity, and high process temperatures. • Use side-stream filters for cooling towers. • For effective HE corrosion control, place CW on the tube side.
Caustic Corrosion	• This type of corrosion occurs under evaporative or high heat transfer applications, which can cause localized high concentration of caustic or alkaline salts. Boilers and waste heat boilers (WHBs) are prone to this type of corrosion. • General corrosion can also occur in process systems depending on alkali or caustic solution strength used.	• Use proper process design to minimize caustic concentration in HEs. • Reduce the amount of free caustic by water dilution. • Avoid concentrated caustic impingement on hot metal surfaces. • Use proper burner management system (BMS)[b] to minimize hot spots on heater tubes. • Minimize ingress of alkaline producing salts into condensers. • Use alloys suitable for handling high-strength caustic solutions.
Steam Blanketing	• In steam generators such as fired boilers and WHBs, high heat flux can lead to a condition known as departure from nucleate boiling. This causes individual vapour bubbles on the hot tube surfaces to join and form	• Prevent flame impingement on tubes by use of proper BMS. • Maintain high water circulation rate. • Maintain heat flux well below critical heat flux for boiling situations.

[b]A BMS is a safety instrumented system (SIS), used for safe start-up, operation, and shut down of combustion systems, involving multiple burners such as boilers and/or fired heaters.

Table 18.1: (*Continued*)

Damage Mechanism	Explanation	Potential Solutions
	a steam blanket. This results in rapid tube rupture, usually within a few minutes, due to overheating. • This type of failure can also occur in superheaters and reheaters during start-up due to condensate blockage of steam flow.	
Flue gas corrosion	See Section 18.4.	See Section 18.4.

18.4 Acid Dew Point Corrosion

On cooling, acidic gases and water vapour condense and cause severe damage in APHs, economizers, WHBs and HEs. Based on the acid gas concentration and water partial pressure, dew point temperature (T_{dp} K) can be estimated using the following equations.[4]

Sulphuric acid:

$$\frac{1000}{T_{dp}} = 2.276 - 0.0294 \times \ln P_{H_2O} - 0.0858 \times \ln P_{H_2SO_4} + 0.0062 \times (\ln P_{H_2O} \times \ln P_{H_2SO_4}) \quad (18.1)$$

Sulfurous acid:

$$\frac{1000}{T_{dp}} = 3.9526 - 0.1863 \times \ln P_{H_2O} - 0.000867 \times \ln P_{SO_2} - 0.000913 \times (\ln P_{H_2O} \times \ln P_{SO_2}) \quad (18.2)$$

Nitric acid:

$$\frac{1000}{T_{dp}} = 3.6614 - 0.1446 \times \ln P_{H_2O} - 0.0827 \times \ln P_{HNO_3} + 0.00756 \times (\ln P_{H_2O} \times \ln P_{HNO_3}) \quad (18.3)$$

Hydrochloric acid:

$$\frac{1000}{T_{dp}} = 3.7368 - 0.1591 \times \ln P_{H_2O} - 0.0326 \times \ln P_{HCl} + 0.00269 \times (\ln P_{H_2O} \times \ln P_{HCl}) \quad (18.4)$$

Hydrobromic acid:

$$\frac{1000}{T_{dp}} = 3.5639 - 0.1350 \times \ln P_{H_2O} - 0.03981 \times \ln P_{HBr} + 0.00235 \times (\ln P_{H_2O} \times \ln P_{HBr}) \quad (18.5)$$

Here, P is the partial pressure (in mm Hg) of the component in the subscript.

Example 18.1: A flue gas at 6 inH_2Og pressure has the following volumetric composition: 15% water vapour, 0.01% HCl, 0.025% SO_2 and rest nitrogen, CO_2 and oxygen. Assume percent conversion of SO_2 to SO_3 is 1%. Calculate HCl and sulphuric acid dew point temperatures.

Solution

Flue gas pressure = 6 inH_2Og = 6 × 0.00245586 = 0.0147352 atm g = 1.0147352 atm abs = 1.0147352 × 760 mm Hg = 771.1987 mm Hg

Partial pressure of H_2O = Flue gas pressure × Volume fraction of H_2O = 771.1987 mm Hg × $\frac{15}{100}$ = 115.6798 mm Hg; and $\ln P_{H_2O}$ = ln (115.6798) = 4.7508

Partial pressure of SO_3 = Flue gas pressure × Volume fraction of SO_2 × fraction of SO_2 converted to SO_3 = 771.1987 mm Hg × $\frac{0.025}{100}$ × $\frac{1}{100}$ = 0.001928 mm Hg; and $\ln P_{SO_3}$ = ln (0.001928) = −6.2513

Partial pressure of SO_3 and sulphuric acid are same. Hence, $\ln P_{H_2SO_4}$ value is same as $\ln P_{SO_3}$.

Partial pressure of HCl = Flue gas pressure × Volume fraction of HCl = 771.1987 mm Hg × $\frac{0.01}{100}$ = 0.07712 mm Hg; and $\ln P_{HCl}$ = ln (0.07712) = −2.5624

From Eq. 18.1 for sulphuric acid dew point,

$$\frac{1000}{T_{dp}} = 2.276 - 0.0294 \times 4.7508 - 0.0858 \times (-6.2513)$$
$$+ 0.0062 \times 4.7508 \times (-6.2513) = 2.4886$$

Hence, T_{dp} of sulphuric acid = $\frac{1000}{2.4886}$ = 401.84 K = 128.69°C
From Eq. 18.4 for HCl dew point,

$$\frac{1000}{T_{dp}} = 3.7368 - 0.1591 \times 4.7508 - 0.0326 \times (-2.5624)$$
$$+ 0.00269 \times 4.7508 \times (-2.5624) = 3.0317$$

Hence, T_{dp} of HCl = $\frac{1000}{3.0317}$ = 329.845 K = 56.7°C

Following strategies can be used for avoiding/minimizing acid dew corrosion.

- Use fuels that are cleaner and free from corrosive substances.
- Maintain metal skin temperature at least 10°C higher than acid dew point temperature.
- For boiler economizers used for heating boiler feed water (BFW) with flue gas WH, it is beneficial to pre-heat BFW with low-pressure steam (LPS)/WH to ~ 10°C higher than acid gas dew point, before BFW enters the economizer.
- For APHs used for heating cold combustion air with flue gas WH, it is beneficial to heat the cold combustion air with LPS/WH to ~ 10°C higher than the acid gas dew point.
- Use of glass-coated CS tubes (Fig. 18.5), plates (Fig. 18.6) or glass tubes in the cold section of the APH where metal skin temperature is expected to be less than or equal to acid gas dew point temperature.
- Use of SS plates/tubes to reduce corrosion rate compared to CS plates/tubes; note that use of SS plates/tubes does not eliminate corrosion.
- Use isolation and water wash system for APHs and economizers.
- Do not partially bypass the flue gas side of APH, using any bypass dampers; this will reduce shear stress and cause fouling on the flue gas side.
- Bypass some of the cold air for APH.

Fig. 18.5: Glass-coated tubes for preventing APH cold end corrosion (courtesy of SHINHAN APEX Corp.)

Fig. 18.6: Glass-coated plates for preventing APH cold end corrosion (courtesy of SHINHAN APEX Corp.)

- Use of polymer APH first in series with metal APH, for heating cold combustion air.

18.5 WHR Equipment Hazards and Safer Solutions

Implementing safer design solutions using diverse protection layers (including ISD, active, passive and procedural) is essential for sustainability of WHR equipment/project. Some key design solutions for common WHR project equipment (namely, HEs, rotating equipment and pressure vessels) and piping are given in Tables 18.2–18.5. These tables are developed using detailed tables available in Ref. 5 in conjunction with the industrial experience of the first author.

18.6 Material of Construction (Metallurgy) Selection for HEs

Proper material selection is important for the sustained operation of WHR equipment and associated piping. Table 18.6 is a quick guide for selecting material of construction for HEs, which are common in WHR projects.

18.7 Heat Exchanger Fouling and Cleaning

HE fouling prevention is important for sustaining WHR project benefits. For this, we need to address environmental, material and stress related factors as shown in Fig. 18.7. These factors are extracted from Figure 13.4 in Shaw and Sekulic.[6] Other than process design factors, proper material selection and fabrication, testing methods are important to ensure good reliability and sustainability of HE operations.

Environmental or process related factors include fluid velocity, pH, corrosive substances, oxygen and dissolved corrosive gases. Good control of them is necessary. Nesta and Bennett[7] have presented a 'No-foul' HE design method. Key design features of this design practice are the following:

1. Velocity criteria: tube-side minimum velocity of 2 m/sec for 19 mm and 25.4 mm tubes, and 2.2 m/sec for 31.75 mm and 38.1 mm tubes and shell-side minimum crossflow stream velocity of 0.6 m/sec.

Table 18.2: Hazard and failure scenarios of heat exchangers and their solutions

S. No.	Hazard and Failure Scenarios	Solutions		
		ISD	**Active**	**Procedural**
1	Overpressure due to blocked outlet	• Increase design pressure to withstand the blocked outlet conditions • Trim the pump impeller to stay within design pressure of HE, even for its blocked outlet conditions	• Use PRV on low pressure (LP) side • Use safety instrumented system (SIS) to trip close high pressure (HP) side upon overpressure detection on LP side • Use of automatic pressure control	• Lock open HE outlet valves • Operating procedures to prevent blocked outlet
2	Overpressure due to corrosion and/or erosion of HE tubes and/or tube sheet, leading to leakage of HP side fluid to LP side	• Design fluid velocity below erosion velocity • Design LP side for the same design pressure as HP side • Use double tube sheets • Use better corrosion and erosion-resistant materials • Avoid excess over design of heat transfer area • Use higher corrosion allowance for tubes and tube sheet • Seal welding of tubes to the tube sheet • Use indirect heat transfer method using secondary heat transfer fluid	• Use of PRV on LP side • Use of SIS to close HP side upon overpressure detection on LP side	• Corrosion detection system • Periodic inspection of HE and timely repair • Online analysis of LP side for leakage • Operating procedures to control corrosion and/or erosion

3	Overpressure due to thermal expansion and contraction of tubes resulting in tube leakage/failures	• Use U-tube bundle or floating head tube bundle • Use expansion bellows on shell side • Use plate or spiral HEs instead of shell and tube HEs • Design LP side for the same design pressure as HP side	• Use of PRV on LP side • Use of process control to keep the rate of temperature change within allowed limits • Use SIS to close HP side upon overpressure detection on LP side	• Operator training on procedures to control heating or cooling of tubes in HE • Periodic inspection of HE and repair • Analysis of LP side for leakage during start-up and shutdown • Lock open cold side isolation valves
4	High tube vibrations, causing tube leakage and overpressure on LP side	• Increase the number of baffles to minimize vibration • Use impingement devices such as plates or rods • Use twisted tubes • Use rod baffles, EMBaffle® or helical baffles • Use other types of HEs such as plate and spiral HES • Design LP side for the same design pressure as HP side	• Use PRV on LP side • Use SIS to close HP side upon overpressure detection on LP side	• Periodic inspection of HE and timely repair
5	Overpressure due to excessive heat input causing leakage in tubes	• Limit temperature of heat source • Design LP side for the same design pressure as HP side	• Use PRV on LP side • Use SIS to close HP side upon overpressure detection on LP side • Automatic heat input control	• Operating procedure to control heat input

(*Continued*)

Table 18.2: (*Continued*)

S. No.	Hazard and Failure Scenarios	Solutions: ISD	Solutions: Active	Solutions: Procedural
6	Overpressure due to the loss of cooling duty (e.g., because of CW failure or fouling in vapour condensers)	• Design LP side for the same design pressure as HP side • Design with a high cooling medium velocity to minimize fouling • Use a design pressure that can withstand pressure generated by loss of cooling	• Use PRV on LP side • Use SIS to close heating source upon overpressure detection on LP side • Supply CW/cooling medium with automatic backup from multiple CW pumps/systems • Set up operating and standby pumps with different drivers such as electric motors and steam turbine drivers • Automatic venting system on LP side	• Monitor and clean HEs as needed, based on cooling duty requirements • Venting of LP side manually • Use chemical injection to reduce fouling • Manual isolation of heating medium (e.g., to reboiler of a distillation column) or supply of alternative cooling medium
7	High temperature due to external fire or excessive heat input	• Use high design temperature and pressure conditions • Locate HE outside or above fire zone • Use volatile substances inside tubes • Limit the heat input	• Install PRV • Use fire-resistant insulation to HE • Automatic activation of fire sprinklers and/or foam	• Manual firefighting activities • Emergency response procedures

		• Provide a quick drainage system for draining any spills/leaks away from HE • Apply fire proofing to HE and support structures	systems based on fire detection • Automatic heating control system for HE (for non-fire cases)	
8	High process temperature due to fouling	• Use high velocity to minimize fouling • Use filters before HEs to remove/minimize solids • Use plate or spiral HEs • Use twisted tube HE • Use tube inserts • Use rod baffles, EMBaffle or helical baffles	• Use automatic backflush for HE and strainers in water streams	• Mechanical or chemical cleaning of HE • Operating procedures
9	Loss of containment due to gasket failure because of external fire (applicable to plate HE with gasket)	• Locate HE outside fire zone • Use fully welded plate HE • Use fire-resistant insulation to HE • Provide a quick drainage system for draining any spills/leaks away from HE	• Automatic activation of fire sprinklers and/or foam systems based on fire detection • Use of PRV on LP side	• Manual firefighting activities • Emergency response procedures

Table 18.3: Hazard and failure scenarios of rotating equipment and their solutions

S. No.	Hazard and Failure Scenarios	Potential Solutions: ISD	Potential Solutions: Active	Potential Solutions: Procedural
1	Overpressure due to blockage of pump or compressor outlet by closure of block, CV or emergency shutdown valve (ESV)	• Design equipment's discharge piping and equipment to withstand deadhead pressure • Use dedicated minimum flow recirculation lines to pump suction vessel with automatic recirculation valve (ARV)/RO	• High discharge pressure and/or temperature shutdown interlocks • Low suction/discharge flow shutdown interlock • Use PRV on downstream piping/equipment • Use CV for automatic minimum flow recirculation line	• Operating procedures to prevent blocked operation • Remote stopping of pump/compressor, based on high discharge pressure or temperature alarm
2	Overpressure due to process conditions (temperature, pressure and composition) change in pump or compressor	• Design for maximum expected pressure • Use dedicated minimum flow recirculation lines to pump suction vessel with ARV/RO	• Use PRV on discharge piping • Automatic pump or compressor shutdown on high discharge pressure detection • Use CV for automatic minimum flow recirculation line	• Operating procedures to prevent deviations from design process conditions
3	Lubrication failure resulting in bearing failure in a compressor	• Use lube oil accumulator	• Use spare lube oil pump with auto cut-in based on low lube oil pressure • Shutdown interlocks for high bearing temperature and low lubrication pressure/level	• Operator corrective actions based on high bearing temperature alarm

4	Compressor damage due to high discharge temperature caused by loss of upstream/ interstage cooling	• Select proper material of construction to withstand the maximum temperature conditions	• High suction and/or discharge temperature shutdown interlock • Low coolant flow shutdown interlocks	• Operator action to adjust coolant flow based on high compressor discharge temperature alarm • Remote stopping of compressor based on high discharge pressure/ temperature alarm
5	Centrifugal pump seal leak due to cavitation	• Avoid flow restrictions in pump suction	• Shutdown interlocks for low discharge flow and high pump vibration • Install differential pressure transmitters with DCS alarm to indicate plugging of suction strainers	• Operator action based on low suction/discharge flow and/or vibration alarms • Remote stopping of pump
6	Compressor damage due to surge, caused by reduced flow through it	• Use compressors other than centrifugal type	• Automatic anti-surge control system • Shutdown interlocks for low discharge flow and high compressor vibration	• Remote stopping of compressor, based on low discharge flow alarm

(*Continued*)

Table 18.3: (*Continued*)

S. No.	Hazard and Failure Scenarios	Potential Solutions: ISD	Potential Solutions: Active	Potential Solutions: Procedural
7	Pump seal damage due to reverse flow from discharge to suction	• Use dedicated minimum flow protection system • Use seal-less pumps	• Use double check valves at pump discharge • Automatic isolation valve on discharge, activated on pump motor trip or high-high suction pressure	• Operating procedures for isolation of standby pumps
8	Loss of containment due to backflow from discharge to suction side of a centrifugal compressor and causing overpressure rupture	• Design compressor suction side piping and equipment to withstand the settle-out pressure[c]	• Use double check valves on compressor discharge • Automatic isolation valve on discharge, activated on pump motor trip or high-high suction pressure • PRV on suction piping and vessel for overpressure protection	• Remote stopping of compressor, based on high suction pressure alarm

[c] After a compressor shutdown, high-pressure gas at the compressor outlet can flow back to suction side and cause the suction-side piping and vessels pressure to increase. This equalized pressure throughout compressor loop (i.e. suction and discharge sides) is called settle-out pressure.

9	Pump seal damage due to particulate matter in inlet stream, and hence loss of containment	• Double or tandem seals • Use pumps that can accommodate solids (e.g., diaphragm type)	• Automatic pump trip on high vibration alarm • Automatic pump trip on detection of loss of seal fluid • Automatic back-flushing or duplex strainer in the pump suction	• Manual cleaning of pump suction filters • Provide seal leak detection system with alarm • Provide remotely operated isolation valves at inlet and outlet with manual activation from safe distance • Periodic inspection of shaft seals
10	Compressor damage due to liquid compression (due to liquid in the inlet stream)	If possible, use liquid ring compressor	• Provide a suction knock out (KO) drum with demister pad and automatic liquid removal and high-high level interlock to trip the compressor • Avoid condensation in suction piping by steam tracing and insulation • Online vibration monitoring with automatic shutdown	• Operation procedure to check steam trace heating of suction piping • Periodic manual checking of KO drum levels • Inspection procedures to check integrity of KO drum internals during maintenance period

Table18.4: Hazard and failure scenarios of pressure vessels and their solutions

S. No.	Hazard and Failure Scenarios	Potential Solutions		
		ISD	Active	Procedural
1	Overpressure due to liquid overfill in vessels or tanks	• Use a vessel design pressure that can withstand the maximum supply pressure • Restrict vessel inlet flows with ROs and/or minimum possible pipe sizes	• High-high level to automatically trip close the vessel inlets • Design sufficiently sized vent pipes with proper disposal system	• Operating procedures to prevent vessel overflow • Emergency handling
2	Overpressure due to external fire	• Use buried/underground tanks (e.g., for liquified natural gas [LNG] and liquified petroleum gas [LPG] storage) • Use fireproof insulation • Install vessels above 7.62 m elevation or outside fire affected zone • Provide good drainage with proper slope to quickly remove any spilled material away from the vessel • Follow spacing guidelines for vessels and tanks	• Use PRV • Fixed fire protection water spray (deluge) and/or foam systems activated by flammable gas, flame, and/or smoke detection devices	• Manual firefighting • Emergency response procedures

3	Vacuum creation due to failure of vacuum system control or excessive liquid withdrawal rate	• Design vessel for full vacuum rating • Restrict liquid withdrawal rate with CV and/or RO	• Vacuum relief device • Use blanket gas such as nitrogen • Trip close liquid withdrawal upon vacuum pressure alarm	• Manual vacuum breaking on indication of high vacuum alarm • Operating procedures to manually limit liquid withdrawal rate
4	Low temperature due to depressurization of vessel containing liquified gases	• Use impact tested low temperature CS for vessel and piping • Use closed blowdown liquid drain system, equipped with steam heated blowdown drum (to trap and vaporize liquefied gases) and connection to flare system	• Interlock to trip close depressurization valve at 0°C temperature • Provide steam jacket, heated by LPS, for depressurization pipes	• Operating procedures to drain out liquid before depressurization • Control depressurization rate by observing any ice formation on the vessel and/or piping
5	Loss of containment due to corrosion from process fluid	• Use corrosion-resistant materials of construction • Use protective coatings and paints • Use double-walled tank/pipe design	• Automatic addition of corrosion inhibitor	• Regular corrosion monitoring and repairs • On-line corrosion monitoring with alarm

Table 18.5: Hazard and failure scenarios of piping and their solutions

S. No.	Hazard and Failure Scenarios	Potential Solutions: ISD	Potential Solutions: Active	Potential Solutions: Procedural
1	Overpressure caused by blockage of piping, valves and/or flame arresters due to solid deposition	• Avoid solids deposition by use of right pipe sizes to ensure fluid velocity is above solids' settling velocity • Piping designed for maximum expected pressure • Eliminate flame arresters as much as possible	• Use PRV • Removal of solids using filters • Steam tracing/heating of vessels and piping to minimize solid deposition	• Periodic manual removal of solids at filters • Blowdown process vessels and KO drums • Periodic pipe internal cleaning using pigging • Use parallel switchable flame arresters with high DP (pressure drop) alarm to DCS
2	Overpressure and potential pipe leak/rupture caused by pressure surge	• Perform pressure surge analysis and select closing speed of actuator of CV or ESV, to prevent pressure surge • Limit closing rate for pneumatic actuator using adequately designed actuator torque and/or RO in air supply pipe • Use slow closing manual valves such as gate valves instead of ball valves	• Use variable speed motors and/or programmable electrical actuators with timers to adjust valve closing time • Use pressure surge arresters	• Operating procedures to close valves slowly
3	Overpressure and potential pipe leak/rupture caused by thermal expansion of blocked liquid	• Eliminate potential for blocking by removing unnecessary valves and blinds	• Use of thermal relief valves	• Procedures for draining and emptying all blocked-in lines during shutdown

4	Overpressure and potential pipe leak/rupture caused by inadvertent full opening of CV, leading to high pressure on the downstream piping and equipment	• Design all downstream piping and equipment to withstand the maximum upstream pressure • Install RO in series with CV to restrict the flowrate or pressure	• Use PRV on the downstream piping and equipment • Use SIS to isolate high pressure side by sensing high pressure on LP side	• Operating procedures to prevent excess opening of CV
5	Overpressure and potential vessel leak/rupture due to closure of PRV/rupture disk (RD) inlet and outlet isolation valves	• Eliminate all block valves in the relief path • Provide three-way block valve at inlet and outlet piping of dual PRV/RD installation		• Car-seal[d] open or lock open all block valves at inlet and outlet of PRV/RD
6	Loss of containment due to solids and/or two-phase fluid erosion	• Use adequate pipe size to avoid erosion velocities • Use proper material to resist erosion • Use higher pipe wall thickness at tees, elbows and other high abrasion points • Use tees and long (instead of short) radius elbows in abrasive solid service	• Control fluid velocity below erosion velocity	• Operating procedures to control flow rates below maximum allowed • Periodic inspection of piping and fittings

(Continued)

[d]Car-seal is the industry common term for sealing devices, which keep the valve in open or closed position.

Table 18.5: (*Continued*)

S. No.	Hazard and Failure Scenarios	Potential Solutions		
		ISD	Active	Procedural
7	Loss of containment due to erosion at high differential pressure CVs	• Reduce DP across CV by using ROs in series • Locate CV as close as possible to the vessel inlet • Use anti-cavitation trim at CV • Use proper supports for CV and piping	• Control operations to operate within design DP of CVs	• Periodic inspection of piping and fittings
8	Loss of containment due to breakage of sight glasses and glass rotameters because of overpressure, thermal stress or physical impact	• Eliminate sight glasses as far as possible • Use magnetic level indicators • Use orifice flow meters instead of rotameters. • Provide 3 mm size ROs in sight glass connections • Use high design pressure (> maximum working pressure) for sight glass	• Use of excess flow check valves in sight glass connections. This will avoid excess leakage to atmosphere if sight glass breaks	• Procedures to line-up sight glasses only during level checking and isolate them immediately afterwards
9	Loss of containment from piping due to piping connection leak or valve leak or piping rupture due to collision impact or inadequate piping supports	• Avoid threaded joints and hose connections by couplings • Avoid expansion bellows in hydrocarbon piping • Avoid unnecessary flanges and fittings	• Use SIS for quick isolation of piping on detection of high low or low pressure or gas detector signal • Use fusible link valves for automatic closure under fire conditions	• Remote pipe isolation from safe areas • Periodic inspection of piping systems • Procedural restrictions for vehicles movement

		• Minimize small diameter pipes (< 1″ size) • Use minimum pipe diameter for mechanical strength • Use proper pipe supports and expansion loops • Avoid dead spaces and liquid slug formation • Avoid the use of underground piping • Use double-walled pipe for toxic services • Use physical collision barriers for piping near vehicle access areas		
10	Loss of containment due to excessive thermal stress	• Perform stress analysis and provide proper piping supports and expansion loops	• Control the temperature to prevent excessive thermal stress	• Periodic inspection of piping systems

Table 18.6: Material of construction selection for HEs[6]

Material	Service and Temperature Range
Refractory lined steel	Non-corrosive; 500°C–2000°C
Alloy steels	Up to 1000°C
Stainless steel (SS)	Up to 540°C
Carbon steel (CS)	Non-corrosive, mildly corrosive fluids, tempered CW; 0°C–400°C (impact tested CS for up to −45°C)
Ferritic carbon-molybdenum and chromium-molybdenum alloys	Moderately corrosive
Ferritic chromium steels	Oils with high sulphur or hydrogen at high temperatures; > 300°C
Austenitic chromium-nickel steel	Highly corrosive
Aluminium	Moderately corrosive
Copper	Up to ~ 200°C
Copper alloys: admiralty, aluminium brass and cupronickel	CW and sea water; Up to ~ 300°C
High nickel-chromium-molybdenum alloys	Mineral acids
Titanium	Seawater
Borosilicate glass- or glass-coated CS tube/plates	APH
Coatings: aluminium and epoxy resins	Sea and brackish water
Linings: lead and rubber	Seawater
Linings: austenitic chromium-nickel steel	Corrosive services

2. Maximum temperature at the tube wall < 300°C.
3. Use single segmental baffles with 20% cut, oriented horizontally for TEMA type E and J shells. Impingement rods are recommended for impingement protection.
4. Use up to 20% excess surface area when both streams are within the scope of this design practice, without using any fouling factors.

As fouling is not always possible to avoid, next alternative approach is to control the rate of fouling. Various types of fouling and potential solutions for preventing/minimizing them are summarized in Table 18.7.

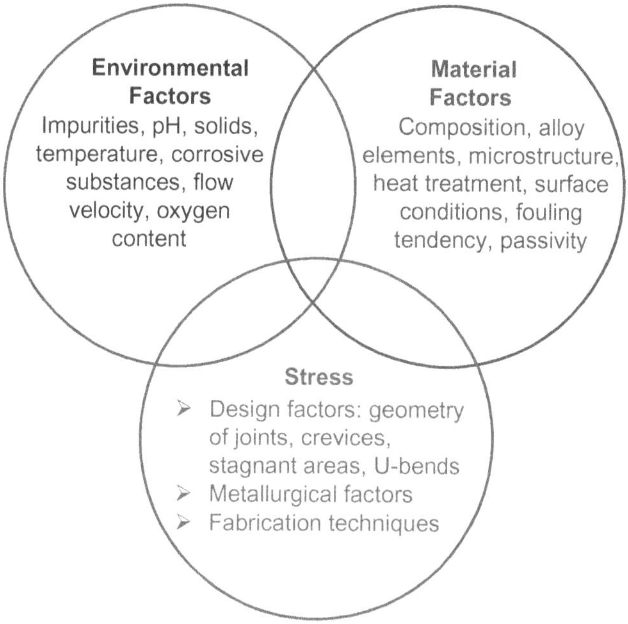

Fig. 18.7: Fouling and corrosion prevention.

Other strategies for reducing HE fouling are as follows:

- Increase the number of passes
- Increase the number of baffles and/or baffle cut
- Use shell and tube HEs with twisted tubes or tube inserts, or plate or spiral HEs
- Install filters on process streams
- Use automatic/manual backflush system for water streams as illustrated in Fig. 18.8. Sea water filters through <1000 micron filtration size. Usually, auto backflush filters in parallel arrangement are preferred. During normal operation, valves A and B are open and valves C and D are closed. During backflush, valves C and D are opened first and valves A and B are closed. This reverses the direction of water in plates and hence cleans the HE. Backflush is performed a few times in a day and hence valves are usually automated using a programmable logic controller.

Table 18.7: Common fouling problems and potential solutions

Fouling Problem	Fouling Mechanism	Susceptible Service	Potential Solutions
Particulate or solids fouling	Accumulation of finely divided solids suspended in process fluids, on heat transfer surface	Any stream– liquid, gas or two-phase	Maintain sufficient tube-side velocity to keep particulates from settling (2 to 3 m/sec to maintain high shear stress)
Salt crystallization, scaling or precipitation fouling	Crystallization of a pure liquid or one component from the liquid phase on a subcooled heat transfer surface	Stream containing water phase	Manage concentration, skin and bulk temperatures, and pH.
Chemical reaction fouling such as polymerization and coking	Deposit formation on heat transfer surface by chemical reaction	Heavy hydrocarbon liquids and streams contaminated with catalyst	Keep temperature below reaction activation, or control reaction rate
Biological fouling	Attachment of microorganisms to heat transfer surface	Water or water containing streams	Kill or control rate of biological growth by proper selection of additives, materials, operating temperature and flow velocity
Corrosion	Accumulation of corrosion products on to heat transfer surface	Practically any process stream	Adjust process conditions to prevent corrosion and select appropriate materials to resist corrosion

- Use anti-fouling chemicals and solvents
- If possible, use onstream water wash
- Use tubes with smooth surfaces
- Use coatings such as ceramic, glass, epoxy and other coatings
- Limit heat fluxes in reboilers and WHBs

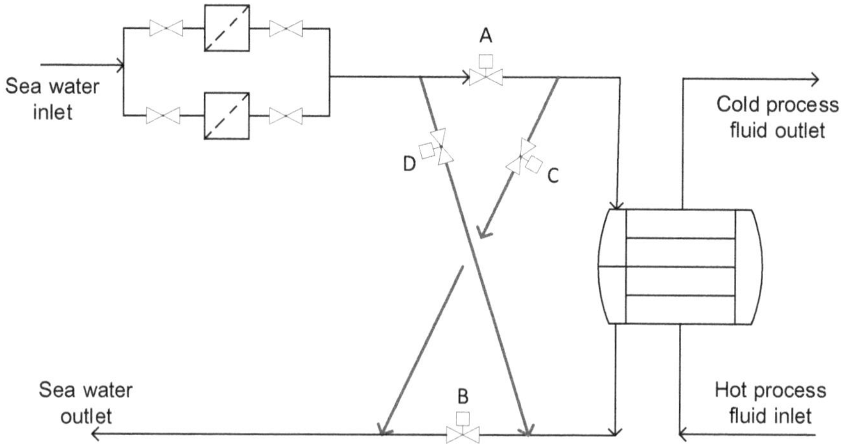

Fig. 18.8: Automatic backflush system for HE onstream cleaning.

Fouling of APH can be minimized by

- Maintaining good mechanical integrity of furnace/boiler and flue gas refractory
- Using flue gas filter screens at the inlet of APH
- Installing provisions for chemical and hot water jet cleaning, upon detection of high DP across flue gas
- Choosing bigger gap between plates (e.g., 6–8 mm).

Techniques for HE cleaning: HE cleaning methods are summarized in Table 18.8.

18.8 Recommended Fluid Velocity in Pipes

Designing and maintaining process piping with proper fluid velocity to prevent erosion or particulate settling is very important for sustainability of WHR projects. Good fluid velocities for this are summarized in Table 18.9, based on the values in Coker[8]

Erosion velocity: For preventing equipment/piping damage, fluid velocity should be maintained below erosion velocity. Erosion velocity formula published in API 14E is given by:

Table 18.8: Techniques for cleaning of HE

Online Methods	Offline Methods
Use of additives such as inhibitors, anti-scalants, dispersants and acids	Manual cleaning using liquid/steam/air jets
Cleaning using sponge balls, brushes, sonic horns, chains and scrapers and air bumping	Mechanical cleaning using drills and scrapers
	Chemical cleaning

Table 18.9: Recommended fluid velocities

Fluid	Velocity, m/sec
Air	20.3
Ammonia (liquid)	1.8
Ammonia (gas)	30.5
Boiler feed water	1.2–3.7
Natural gas	30.5
Oils	1.8
Oxygen	9.1
Seawater	1.5 to 2.4
Sodium hydroxide (0%–30%)	1.8
Sodium hydroxide (30%–50%)	1.5
Sodium hydroxide (50%–73%)	1.2
Sodium chloride solution (no solids)	1.5
Steam (0–2 bar), saturated	20.3–30.5
Steam (2–10 bar), saturated or superheated	30.5–50.8
Steam (> 10 bar), superheated	33
Sulphuric acid	1.2
Water	0.9–2.4
Water in pump suction lines	0.3–1.5

$$V = \frac{C}{\rho^{0.5}} \tag{18.6}$$

Here, V is velocity (m/sec), ρ is gas/liquid mixture density (kg/m^3) and C is a coefficient that depends on piping material and quality of the fluid. For steel pipe, clean fluid (no sand), C is 121 and 151.25, respectively, for continuous and intermittent service; and for steel, clean fluid, non-corrosive and continuous service, C is 181.5–242. According to API 14E, minimum velocity for two-phase flow should be about 3 m/sec.

18.9 Damage Mechanism Review, Risk-Based Inspection and Integrity Operating Windows

A holistic asset integrity management (AIM) program including DMR is essential for achieving sustainable WHR project operation and benefits. DMR is a systematic analysis (usually performed on each equipment and piping using P&IDs) to determine credible damage mechanisms for fixed equipment such as pressure vessels, HEs, piping, WHB, fired boilers, and process heaters. A team of multi-disciplinary subject matter experts (usually material specialists, inspection, operations and process engineers) discusses, using the knowledge of mechanical, corrosion, reliability and materials engineering, process data, inspection and maintenance reports, to identify potential damage mechanisms associated with WHR equipment and establish preventive solutions, inspection methods and testing frequencies. DMR results are also very useful for process safety management such as HAZOP, PSI, root cause analysis, incident investigations, pre-start-up safety review (PSSR) and management of change (MOC)[9] and for establishing RBI plan as shown in Fig. 18.9.

As per API RP 580,[10] RBI is an assessment methodology that prioritizes and manages equipment inspection program based on risk assessment. DMR review results are very useful for risk assessment process used in RBI, a flowchart for which is shown in Fig. 18.10. RBI starts with data collection for the equipment (HEs, WHBs, pressure vessels, piping and its components, pressure-containing parts of rotating equipment, storage vessels, boilers and fired heaters). Next, consequence and frequency of equipment failures are used to establish the risk. Based on the risk, the inspection frequency, procedures and inspection re-assessment criteria are set. Re-assessment is typically based on conditions that may change because of deterioration or damage of equipment. The

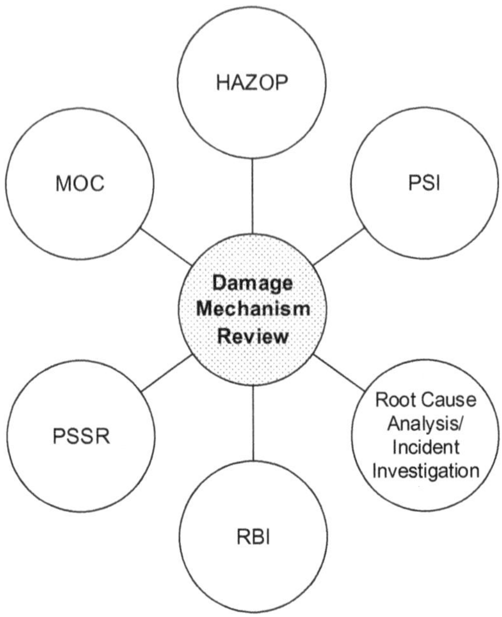

Fig. 18.9: Importance of DMR.

inspection program will continue based on the established inspection plan. If re-assessment is required, it will go through the entire RBI with the new conditions.

Maintaining asset integrity is important for the whole life cycle of plant operations. However, over time, apart from mechanical condition (e.g., degradation) of the equipment, operational parameters may change. This causes the original AIM program to become ineffective. To ensure the integrity of equipment, a comprehensive process safety management (PSM) system, including IOWs and MOC programs, is necessary.[11] Equipment subject to significant process variable fluctuations can benefit from real time monitoring and timely corrective actions, using an IOW program. The MOC process is covered in detail in Chapter 20.

According to API RP 584,[12] an IOW refers to established limits for process variables in order to maintain equipment integrity/reliability, for process operation within the window. The main goal of IOWs is to lower

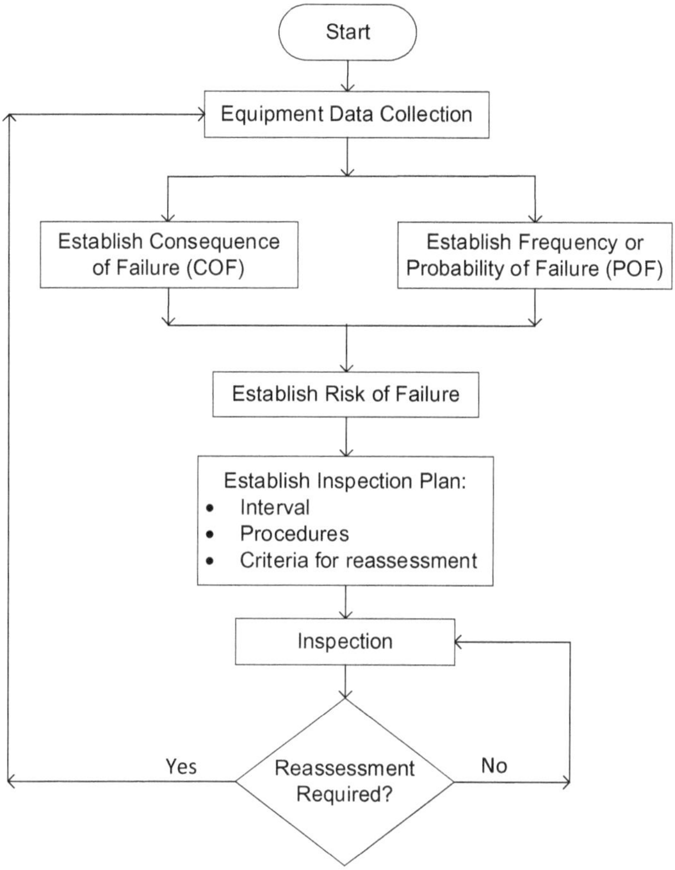

Fig. 18.10: Flowchart for RBI.

the safety risk of operating process equipment. IOWs help to continuously monitor and control process variables in such a way as to prevent equipment failure. Based on process criticality/risk, three levels of IOWs are defined,[12] as given in Table 18.10.

Based on IOW limits categorization, typical process operating limits are shown in Fig. 18.11. Within IOW information/target high and low limits, operation is very reliable. Operation within IOW standard high and low limits is safe. Operation between standard and critical IOW limits

Table 18.10: IOW classification of process risk and its significance

Process Risk	IOW Level	IOW Significance	Required Action If IOW Limit Is Exceeded	Example(s)
High	Critical limit	Exceeding this IOW level even for a **fairly short time** can lead to rapid deterioration of process equipment causing extensive damage, loss of containment and eventual catastrophic process safety incident.	Operator **must urgently** return the process to a safe condition by returning process variables to within IOW limits.	In WHB, loss of BFW level could quickly cause boiler tube rupture. Over-pressuring HE/pressure vessel can result in quick loss of containment.
Medium	Standard limit	Exceeding this IOW level **over a specified amount of time** may lead to increased corrosion rates, cracking or other types of equipment damages, causing damage to equipment, loss of containment and/or eventual catastrophic process safety incident.	Operator must return the process, **over a specified time**, to a safe condition by returning process variables to within IOW limits.	Metallurgical creep due to elevated temperature on a WHB tube skin could lead to tube failure over time.
Low	Information limit	These are mainly the limits used by subject matter experts (e.g. process engineer, inspector and corrosion specialist) to monitor, predict and/or control long-term integrity and reliability issues of equipment.	These IOWs do not generally require operator responses but are utilized for planning future testing and/or repairs during turnarounds or shutdowns	Revamped throughput exceeding original design value

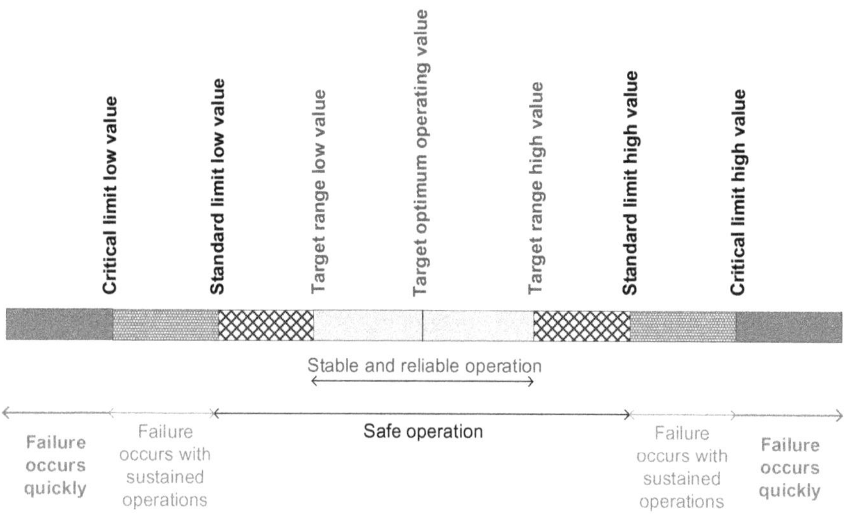

Fig. 18.11: Typical operating zones including target ranges, standard and critical limits.

would require corrective action in the specified time. Operation beyond critical IOW limits is very dangerous and require urgent corrective action.

Lastly, condition based equipment maintenance can help to maintain sustained benefits by minimizing the equipment downtime and maintenance costs. Many digitalization technologies[13] such as equipment monitoring (e.g., steam traps and PRV damage/leakage monitoring using wireless sensors), artificial intelligence, data analytics (e.g., HE fouling monitoring and prediction) and machine learning tools can be effectively used to improve sustainability and/or maximize WHR equipment performance.

18.10 Summary

This chapter presented sustainability considerations useful for WHR projects. It provides a concise summary of sustainability aspects for practising engineers and also students. Key considerations are as follows.

- Application of ISD and green engineering principles in WHR project design is a pre-condition for achieving WHR project benefits for many years. Application of these techniques during the whole life cycle of the plant/equipment is essential.

- Corrosion in flue gas WHR equipment can be minimized/avoided without compromise on WHR, using techniques such as glass-coated CS tubes/plates, pre-heating combustion air with LPS/FS steam/WH before APH and polymer APH at cold end of APH.
- Evaluating potential WHR equipment problems such as overpressure, corrosion–erosion and loss of containment, during HAZOP sessions is also essential to develop sustainable WHR project.
- Use of proper fluid velocities in pipes and material of construction is very important.
- It is very important to understand WHR equipment damage mechanisms and potential solutions at the design stage. This will help to choose the material of construction and operating conditions to enable sustained WHR benefits during the entire life cycle of the plant.
- RBI is very important to enable timely inspection and repair.
- Setting proper IOW limits helps to effectively monitor and maintain safe and reliable WHR project operations.

Hall[14] is a holistic book that provides many useful and concise design guidelines and rules of thumb for designing process equipment and systems. Chapters 4, 5–9, 12, 13, 19–22 and 25 in this book are recommended for further reading. Also, the book by CCPS[5] is recommended for deeper understanding of equipment design problems and potential solutions.

References

1. Contreras CD, Bravo F. (August 2011) Practice green chemical engineering. *Chemical Engineering*, pp. 41–44.
2. Dincer I, Acar C. (2017) Smart energy systems for a sustainable future. *Applied Energy*, pp. 225–235.
3. API RP 571. (April 2011) *Damage Mechanisms Affecting Fixed Equipment in the Refining Industry*, 2nd ed. American Petroleum Institute.
4. Ganapathy V. (2015) *Steam Generators and Waste Heat Boilers for Process and Plant Engineers*. CRC Press.

5. CCPS. (1998) *Design Solutions for Process Equipment Failures*. American Institute of Chemical Engineers, New York.
6. Shah RK, Sekulic DP. (2003) *Fundamentals of Heat Exchanger Design*. Wiley, Hoboken, New Jersey.
7. Nesta J, Bennett CA. (July 2004) Reduce fouling in shell and tube heat exchangers. *Hydrocarbon Processing*, pp. 77–82.
8. Coker KA, Ludwig EE. (2007) *Applied Process Design for Chemical and Petrochemical Plants: Volume 1*, 4th ed. Gulf Professional Publishing.
9. Rehmat TS, Montgomery R, Whittle D. (December 2017) Address damage mechanism reviews in a process hazard analysis. *Hydrocarbon Processing*, pp. 81–83.
10. API RP 580. (May 2016) *Risk Based Inspection*, 3rd ed. American Petroleum Institute.
11. Russ D. (May 2017) Implementing an "integrity operating window" program. *Chemical Engineering*, pp. 42–48.
12. API RP 584. (May 2014) *Integrity Operating Windows*, 1st ed. American Petroleum Institute.
13. Agnihotri RB. (July 2018) Digitalization for the refinery and plant of the future. *Hydrocarbon Processing*, pp. 31–36.
14. Hall SM. (2018) *Rules of Thumb for Chemical Engineers*, 6th ed. Elsevier Inc.

Acronyms and Notation

Acronyms

AIM	Asset Integrity Management
APH	Air Pre-Heater
ARV	Automatic Recirculation Valve
BFW	Boiler Feed Water
BMS	Burner Management System
COF	Consequence of Failure
CV	Control Valve
CS	Carbon Steel
CW	Cooling Water
DCS	Distributed Control System
DM	Demineralized
DMR	Damage Mechanism Review

DP	Differential Pressure/Pressure Drop
FS	Flash Steam
HAZOP	Hazard and Operability
HE	Heat Exchanger
HEN	Heat Exchanger Network
HP	High Pressure
IOW	Integrity Operating Window
KO	Knock Out
LiBr	Lithium Bromide
LNG	Liquified Natural Gas
LP	Low Pressure
LPG	Liquified Petroleum Gas
LPS	Low-Pressure Steam
MOC	Management of Change
P&ID	Piping and Instrumentation Diagram
POF	Probability of Failure
PRV	Pressure Relief Valve
PSI	Process Safety Information
PSM	Process Safety Management
PSSR	Pre-Start-Up Safety Review
RBI	Risk-Based Inspection
RD	Rupture Disk
RO	Restriction Orifice
SIS	Safety Instrumented System
SS	Stainless Steel
TEMA	Tubular Exchanger Manufacturers Association
WH	Waste Heat
WHB	Waste Heat Boiler
WHR	Waste Heat Recovery

Notation

T	Temperature (°C or K)
P	Pressure (mm Hg)

Subscript

dp Dew point

Exercises

18.1 Fuel oil consumed in a fired heater generates a flue gas with the following conditions: flue gas pressure of 1.018 atm, and composition (volume %) of water vapour = 12, SO_2 = 0.03, and rest are N_2, CO_2, CO and O_2. Assuming a 0.5%–2% conversion of SO_2 to SO_3, in steps of 0.5%, estimate sulphuric acid dew point for different conversions. What can be concluded about the effect of concentration of SO_3 in flue gas on sulphuric acid dew point?

18.2 In the previous exercise, change water concentration from 5% to 20% in steps of 5% while keeping SO_2 to SO_3 conversion fixed at 1%. Estimate corresponding sulphuric acid dew point temperatures. What can be concluded about the effect of water vapour on sulphuric acid dew point temperature?

18.3 A WHR project involves recovering WH from a hazardous process stream by LPS generation. Identify green engineering and ISD ideas applicable for designing the new HE for this heat recovery.

18.4 Briefly discuss the importance of erosion velocity. Identify possible solutions to avoid erosion in HEs and piping involving two-phase flow.

18.5 Estimate the erosion velocity of a mixture of hot water and flash steam, having a density of 800 kg/m^3.

Chapter 19

Project Management for Waste Heat Recovery Projects

19.1 Overview

Waste heat recovery (WHR) projects are essential for addressing global and/or regional business challenges such as increased competition, higher energy prices and significant environmental concerns. Engineers need to explore WHR opportunities, develop and/or manage projects to implement the selected WHR strategies/technologies. The scope of a WHR project varies from simple to complex, and hence may take more evaluation and development time as many options are available to achieve the project objective(s).

WHR projects may involve greenfield[a] or brownfield[b] projects. For the case of brownfield projects, successful design and implementation of WHR projects involves comprehensive and systematic evaluation of interactions between proposed additions and the existing/operating plant. Technological advances are making WHR projects feasible and more attractive. The main objective of WHR projects is to recover waste heat (WH) for re-use and hence reduce the energy consumption and emissions. Often, they can provide other benefits such as increased plant capacity,

---

[a] Greenfield projects are totally new without any existing/operating plant in the project area.

[b] Brownfield projects are those executed in the existing/operating plants (e.g., as additions and/or modifications to the existing plant).

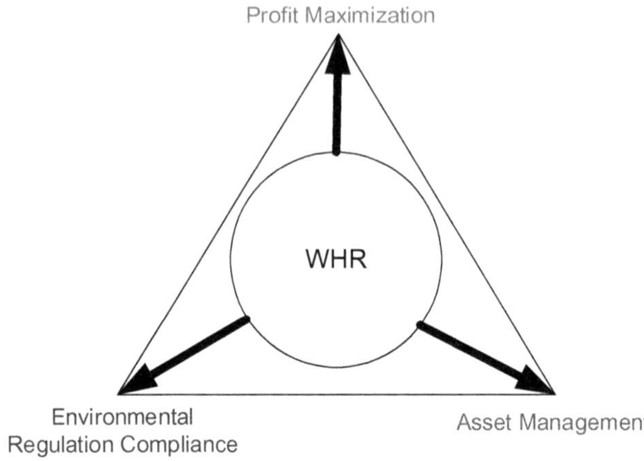

Fig. 19.1: Strategic role of WHR projects from three perspectives.

improved plant reliability, safety and reduction in WH disposal costs (e.g., smaller height of flue gas stack and lower cooling cost). These benefits are classified into three main areas as shown in Fig. 19.1. WHR can increase the profitability by reducing fuel consumption and hence operating cost. Also, it reduces emissions and hence emissions treatment costs and/or emission tax (e.g., carbon tax) where applicable. In general, WHR sometimes can help to manage the plant in a better way.

This chapter begins with highlighting the key differences between brownfield and greenfield WHR projects in Section 19.2. Next, Section 19.3 discusses the main stages in WHR project and methodology to perform a pre-feasibility study. Section 19.4 outlines pre-front end engineering design (pre-FEED) and FEED study development strategy. Sections 19.5 and 19.6 present the details of pre-FEED and FEED study stages, respectively. Section 19.7 discusses Engineering, Procurement and Construction (EPC) stages in detail. Section 19.8 highlights key project deliverables. Main elements of project management are described in Section 19.9. This chapter ends with summary in Section 19.10.

Learning outcomes of this chapter are as follows.

1. Delineate the key differences between two types of WHR projects (i.e., greenfield and brownfield projects).
2. Develop comprehensive WHR project development methodology.

3. Describe the importance of WHR project activities in pre-FEED, FEED and EPC stages, including value improvement practices (VIPs).
4. Outline WHR business drivers and barriers.
5. Discuss the development of project charter and project scope documents.
6. Identify the project deliverables for WHR project stages.
7. Discuss project risk, schedule, resources and key stakeholders' management methods.
8. Identify project scope and cost control methods.

19.2 Differences between Brownfield and Greenfield Type WHR Projects

For improving energy efficiency, business competitiveness and environmental compliance, process/chemical engineers often perform WHR studies and develop WHR projects for successful implementation. These projects can be brownfield or greenfield type. Brownfield projects generally require more effort, resources and time, compared to greenfield projects. Key requirements and considerations for both type of projects are compiled in Table 19.1.

Table 19.1: Comparison of brownfield and greenfield type WHR projects

Item	Brownfield WHR Projects	Greenfield WHR Projects
Project scope	Project scope should be developed by considering various alternatives and their integration with the existing plant equipment.	Project scope is mainly decided by the EPC company and/or process engineers in the owner company.
Conceptual design	Design data need to be calibrated with actual plant data. Steam/power savings should be verified by the careful evaluation of plant's steam/power network balances and economics. Generally, there are more restrictions.	No data calibration is required. Design can be performed offline from contractor's office. There are minimal restrictions from existing steam and power balances.

(*Continued*)

Table 19.1: (*Continued*)

Item	Brownfield WHR Projects	Greenfield WHR Projects
Expertise of process/design engineers	Process designers/engineers should have good knowledge of existing plant technology, operating experience and economic trade-offs.	These projects may be designed by process engineers with minimal operating experience and with little supervision.
Interactions with existing plant equipment and system	Complex interactions may be present with the existing equipment in the plant and also in neighbouring plants. Generally, there will be significant interactions with the fuel, steam, electric power and pressure relief systems.	Generally, there are no interactions.
Design margin allowances	Design margins of existing equipment will be utilized in the revamp. Hence, revamped plants may have to operate with minimal/zero design margins.	Usual design margins are generally provided and will be available for operating the new plant.
Process safety risks	Relatively higher process safety risks due to interactions with the existing plant equipment.	Lower safety risks as design can use inherently safer design methods/ practices.
Project team size	Project team is generally very lean as most of the activities are completed by in-house project team.	Often, a large project team to handle the entire project.
WHR project implementation timeline	Implementation timeline is influenced by factors such as equipment delivery time and next turnaround/ stoppage of the plant	Implementation timeline is independent of turnaround/stoppage of the plant
Plot space limitations	Plot space may be limited, which often restricts the extent of WHR. Considerations for maintenance, operational and emergency handling accesses can be challenging.	Generally, there are no plot space limitations
Engineering cost	Due to complex interactions and constraints, generally higher at 15%–20% of project cost.[1]	Generally lower at ~10% of the project cost.[1]
Project cost	Generally, less compared to greenfield WHR project due to the availability of operating plant's infrastructure such as control rooms, pipe racks, steam and power system.	Generally, more compared to the brownfield WHR project

19.3 Developing WHR Projects and Pre-Feasibility Study

A comprehensive project development strategy is essential for developing successful and sustainable brownfield WHR projects. WHR project opportunities vary from plant to plant depending on several factors such as process technology, energy prices and governmental regulations.

A WHR project consists of mainly four stages, namely Defining, Planning, Executing, and Closing stages, as illustrated in Fig. 19.2. Here, the defining phase involves screening/pre-feasibility studies. Planning phase includes pre-FEED/Conceptual study and FEED. Execution phase is the EPC. Closing stage involves training, pre-commissioning, commissioning, project start-up, performance testing, project evaluation and documentation.

A systematic methodology for WHR pre-feasibility/screening study shown in Fig. 19.3 can be used as the first step in WHR project development. It starts with setting the WHR evaluation team to collect and validate data before detailed analysis. Once data analysis is ready and WHR strategy is finalized, WHR project objectives and basis (e.g., energy, utility prices, economic criteria and end use for the WHR) are set. Subsequently, a roadmap of WHR opportunities with low/no, medium and high capital cost solutions are developed for prioritizing their implementation. Naturally, low/no cost solutions are prioritized for quick development and implementation. Economics of medium cost solutions are evaluated (typically with ±50% accuracy) and attractive projects are identified for further development and implementation. Unattractive medium cost solutions and high cost solutions are set aside for future consideration and development. After implementation of the selected projects, performance and lessons learned are sent to the WHR team.

19.4 Pre-FEED and FEED Development Strategy

Once a WHR project pre-feasibility study (shown in Fig. 19.3) is completed, the next step is to check whether WHR objectives are satisfied and seek approval from the management. After the approval is obtained, conceptual design (also known as feasibility or pre-FEED study) begins. For

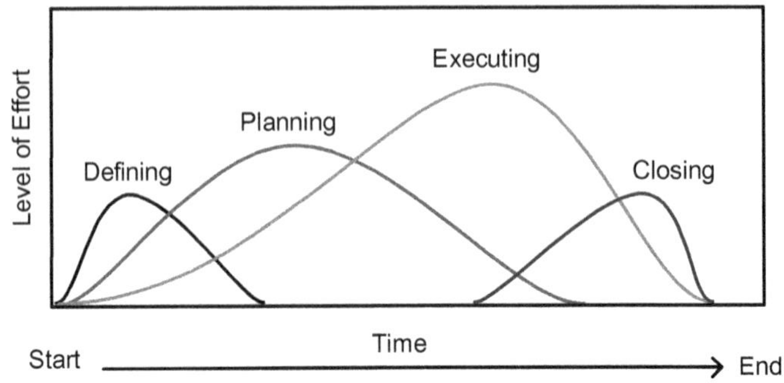

Fig. 19.2: Project life cycle.

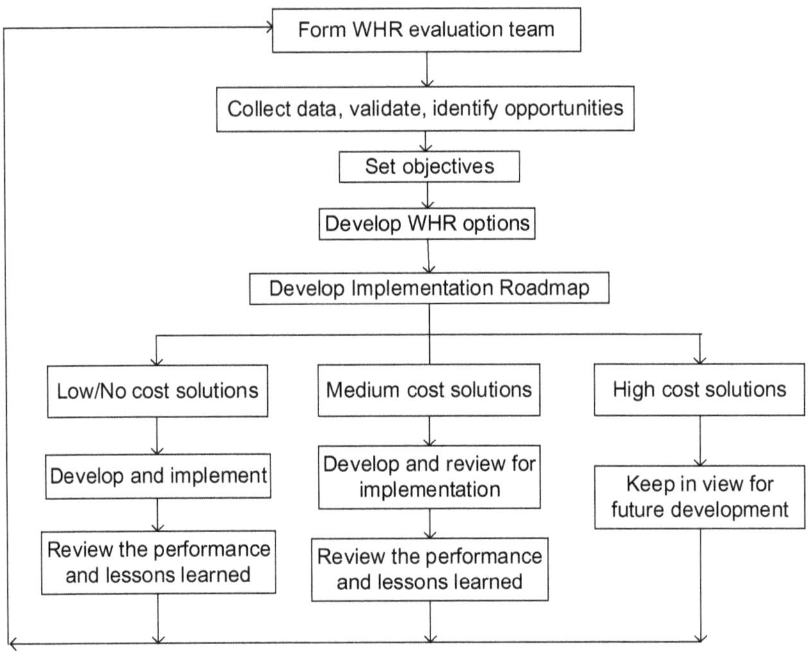

Fig. 19.3: Pre-feasibility study for WHR project development.

brownfield projects, plant design and operating data are collected and validated with plant performance tests and/or process simulations. With the validation assessment, equipment duty/sizing limitations or bottlenecks are established. (These steps are not required for greenfield projects.) The next step is to perform detailed assessment to develop many feasible WHR options for achieving project objectives.

As shown on the left side of Fig. 19.4, VIP studies are required for comprehensive development of options. After developing detailed WHR options, preliminary cost estimation (typically with ±30% accuracy) is performed for each option and a techno-economic analysis is prepared. Based on this analysis, the most attractive WHR option is selected and presented to the management for approval. After pre-FEED approval, the WHR project moves to FEED stage for full project development, to prepare detailed project deliverables and cost suitable for the EPC stage.

During the FEED stage, detailed Piping and Instrumentation Diagrams (P&IDs), cause and effect charts, equipment process/mechanical data sheets and drawings, equipment layout, piping isometric/layout drawings, instrument and electrical datasheets/drawings, civil design drawings, safety plan, risk assessment and mitigation plan, cost estimation (typically with ±10% accuracy), hazard and operability (HAZOP) and layers of protection analysis (LOPA) report, reliability, availability and maintainability (RAM) study report, 3D Model and constructability report and so on are developed. For smaller projects, some of the FEED study items and deliverables can be omitted based on the detailed assessment.

The finalized FEED is used for obtaining EPC/detailed engineering tender bid. Then, bids received are evaluated and one of them is selected. The finalized FEED study, with bid analysis and bid selection, are presented to the management again for approval. Once management approval is obtained, the WHR project proceeds to EPC stage. Pre-FEED, FEED and EPC stages are elaborated in the next three sections.

19.5 Pre-FEED, Feasibility or Conceptual Design

WHR projects mainly focus on maximizing WHR and hence maximizing profitability and better compliance of environmental regulations. For this, one needs to explore various WH utilization and upgrade options. The following three main strategies, in order of preference, are useful for maximizing the direct WH usage and WHR benefits.

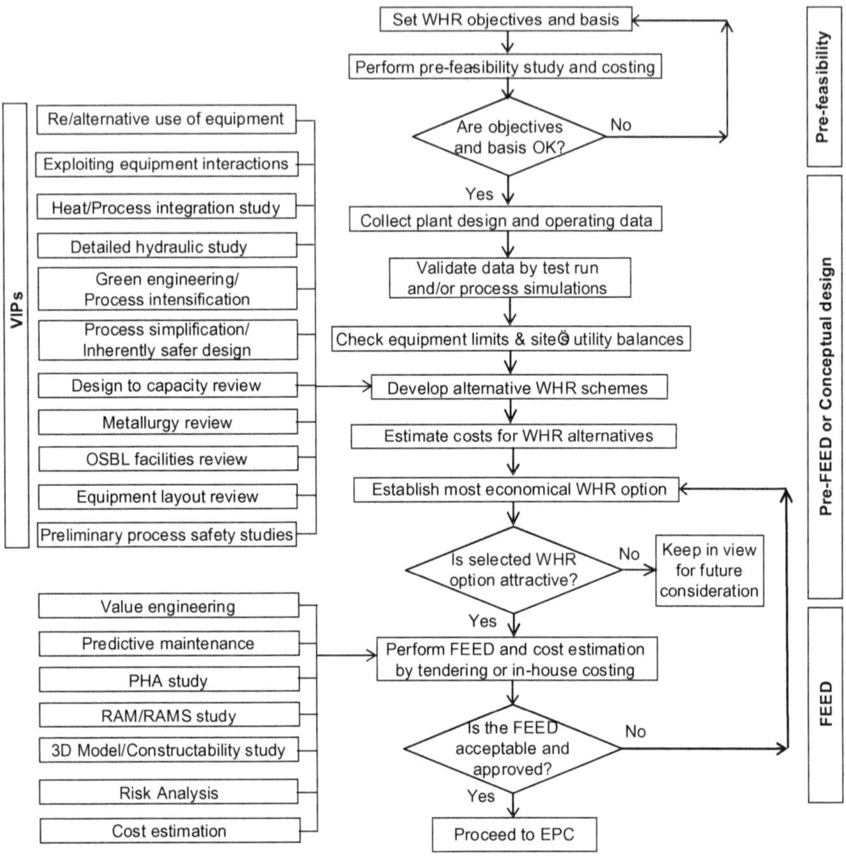

Fig. 19.4: A detailed and systematic methodology for WHR projects: Pre-feasibility study to FEED completion (Reproduced with permission from John Wiley & Sons Inc.).[3]

- First option is to re-use WH within the process or equipment itself. This is the most economical and effective method of using WH (e.g., in the boiler or fired heater). The principal advantages are that the source and sink are generally close together and there are no problems in matching heat availability with demand.
- Second option is to use WH in other equipment within the unit/section boundary itself. This may require direct use of WH in other nearby equipment or WH in the generation of plant utilities/hot oil system

(if other equipment is far away). This helps to reduce capital expenditure for piping as well as operational and maintenance issues such as shutdown of different units at different times.

- Third option is to consider WH to generate steam, hot oil or power that can be utilized in other units within the facility. This may arise because the WH is at an insufficiently low or high temperature for re-use in the originating process unit, or because of process requirements such as the need for precise control of heat input.

Fig. 19.5 illustrates the direct WH usage and various upgrade options for WH to higher grade/quality heat, generation of electrical power and chilled water. A single WH source can be directly used, WH at multiple temperatures can be combined or WH energy can be combined with other energies such as fuel and solar energy.

During pre-FEED study, the main objective and any other minor objectives need to be identified and documented. For example, reducing the energy consumption is the main objective, and other related objectives could be reduction in emissions, lower WH disposal cost, improved reliability and/or process safety. In addition to the main objective, other minor objectives are very useful for getting management approval.

A list of project objectives, key WHR project business drivers, basis of design (e.g., energy, and utility prices) and assumptions (e.g., design assumptions such as plant infrastructure, cost assumptions, project

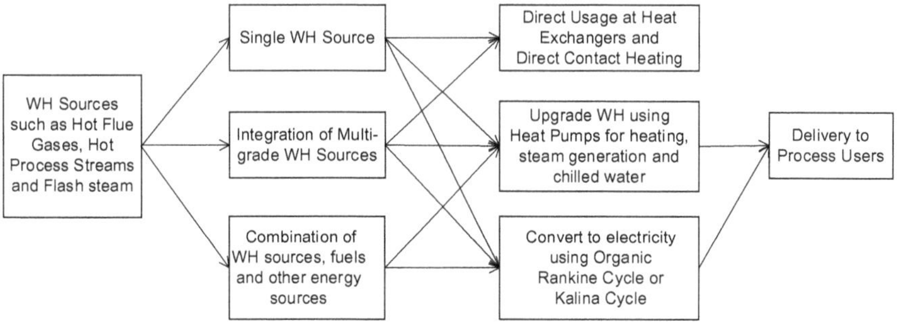

Fig. 19.5: WH utilization strategies.

interfaces and any government approvals) should be prepared. Useful project framing information is presented in the following sub-sections.

WHR Business Drivers: Key WHR project drivers are as follows:

- Sustainable development
- Conventional equipment with little technology risk
- Reduce the use of fossil fuels
- Reduce the energy cost
- May reduce the capital cost for WH disposal equipment
- May increase productivity and competitiveness
- Generation of power by Rankine cycle (RC), organic Rankine cycle (ORC) or Kalina cycle (KC)
- Reduce greenhouse gas emissions and hence air/gas emission treatment cost
- Better reputation among shareholders and society at large
- May improve the product quality in some cases

Barriers for WHR: Although WHR projects are very beneficial, they often encounter the following barriers:

- WHR projects are typically considered as discretionary by process industries.
- Cost Related: In each WHR method, only a portion of WH is recovered; hence, it may not have economies of scale and may have longer payback periods. Low energy prices and high maintenance costs make WHR projects even more challenging.
- Temperature and material restrictions: There may not be a viable end use for the recovered WH at the recovery temperature. There may be high WHR costs due to the requirement of expensive material for very high temperature usage. Low-temperature constraints, thermal cycling issues, low heat transfer rates and so on, may limit the extent of WHR.
- Chemical Composition Related: These barriers can impose restrictions on process temperatures, heat transfer rates, material of construction and flue gases/waste streams disposal to atmosphere. They

can increase operation and maintenance costs and make WHR projects unattractive. For example, it may be challenging to dispose wastewater used to wash dirty off-gases, for WHR. Wastewater in a wet scrubber contains 1000–10,000 mg/L of suspended solids. WHR system components such as recuperators often experience corrosion due to high temperatures and presence of particulates.

- Other Barriers
  - Power and steam typically outside core business
  - High capital cost for some WHR technologies such as power generation with ORC and KC.
  - Some custom equipment with little salvage value
  - Retrofit may be difficult for some WHR equipment
  - Poor payback period may not be attractive for process industries
  - Technical challenges such as plot constraints, dirty flue gas and temperature limitations
  - Lack of supporting systems such as necessary piping headers, pipe rack space and substation infrastructure
  - Batch operations and variable energy demand at the site
  - Viability heavily tied to energy prices

The WHR project developer needs to evaluate the main barriers and their mitigation measures. Following proven VIPs/methodologies[4] are useful to develop holistic WHR design options and select the best WHR design.

19.5.1 *Re-Use of Existing Equipment*

For developing profitable WHR projects, it is required to critically review, debottleneck/revamp and re-use the existing plant equipment as much as possible. Successful re-use of equipment for revamp/retrofit is highlighted by Litzen and Bravo,[5] Martin and Cheatham,[6] Higley,[7] and Bhargava and Sharma.[8] Pumps and heat exchangers (HEs) are typically installed with a design margin of 10%. Control valves and separators are usually installed with a design margin of 20%. Motors may have a design margin of ~15%. Pumps generally have allowance in the casing to increase the impeller diameter as they are installed with impeller size lower than the maximum size. So, pump capacity can often be increased by the replacement of the impeller to the bigger size, as supported by the pump casing and motor.

If sufficient DP (differential pressure/pressure drop) allowance is available, HE passes can be increased to increase the duty. Higher heating duty of a shell and tube HE can be obtained by replacing the tube bundle in the shell with a twisted tube bundle. This modification re-uses the existing HE shell and all the associated piping and valves. More importantly as the equipment size is the same, no additional civil/structural work is required. Hence, it can save lot of capital cost. Generally, operating temperature and pressure of process equipment are lower than the conditions used for mechanical design. This provides an opportunity to operate the existing equipment at higher temperature and pressure (as long as they do not cause safety valves to chatter), without requiring new equipment. If the plot space permits, new equipment such as HEs and pumps can be added in parallel and/or series to the existing equipment, for increasing duty/capacity.

19.5.2 *Equipment Interactions*

Equipment interactions are important for developing holistic WHR options. For example, consider the addition of a new air pre-heater (APH) to heat cold combustion air by transferring heat from hot flue gases. Increased temperature of cold combustion air requires re-checking the burners' performance. In extreme cases such as high combustion air temperature, greater than that recommended by the burner vendor, replacement of burners may be required. If an existing APH is required to be replaced with a bigger APH for higher WHR, additional DP is required both on flue gas and combustion air streams. Hence, this project requires checking sizes of induced and forced draft fans and, if insufficient, modifying or replacing them. Replacement of an HE with a bigger unit for higher heat duty may require changing the feed pump due to the additional DP requirement for the new HE. Consider another WHR project with the main objective of generating low-pressure steam (LPS); this requires checking the facility's steam balances and also need for a separate steam header. In summary, it is very important to check the system interactions due to proposed modifications before finalizing scope for any equipment.

19.5.3 *Heat/Process Integration Study*

Pinch analysis is well proven for quickly estimating WHR potential by heat integration. It is useful to obtain targets for utilities, heat transfer area and emissions,[9,10] and can be used to explore WHR potential within the unit and/or across the whole site. Fundamentals of heat integration and pinch analysis are covered in Chapters 2 and 3, respectively. Several industrial case studies illustrating pinch analysis application are available in the open literature.[10,11] Optimization techniques can also be used for WHR by heat integration. For example, Moorthy *et al.*[12] employed optimization for WHR using a low-temperature heat exchanger network.

19.5.4 *Detailed Hydraulic Study*

Detailed hydraulic study is essential for successful development of WHR projects. Maximizing available DP in HE maximizes heat recovery, minimizes HE capital cost and also helps to sustain the heat recovery over longer periods. Neglected hydraulic assessment may lead to use of very conservative (low) DPs across piping and HEs. This leads to oversized piping and HEs, and eventually to excessive fouling/plugging; these in turn lead to more capital, maintenance, and operating costs. On the other extreme, very high DP usage can lead to more expensive pumps/compressors, high design pressures, erosion damage (in piping and HEs) and/or excessive process relief flows. Hence, a good hydraulic study is required to maintain sufficient velocity (usually 3 m/sec for liquid services) for sustained and economical WHR by avoiding additional/costly equipment and damage mechanisms.

19.5.5 *Green Engineering/Process Intensification*

Green engineering for WHR projects aims to reduce fresh energy/fuel consumption and WH generation by operating processes at as low temperature as possible. Use of inherently safer design practices, process intensification techniques[13,14] and pinch analysis, which come under green engineering practices, are useful for improving the sustainability of WHR projects. There are many green engineering methods in practice.[15,16] Some of the useful green engineering practices for WHR projects are as follows.

- Choose operating conditions to maximize WHR and reduce energy costs.
- Use pinch and/or optimization methods for heat integration.
- Utilize compact equipment such as plate HEs, heat pipes and dividing-wall columns.
- WHR from wastewater.
- Maximize the re-use of existing equipment and wastewater (e.g., in petroleum refineries, treated fouled water is used for washing HE tubes and desalting).
- Utilize high-efficiency equipment (e.g., the use of dry vacuum pumps instead of steam jet ejectors, back pressure turbine, heat of compression air dryers in place of desiccant dryers and heat exchangers instead of fired heaters)
- Use advanced control systems and real-time optimizers for maximizing WHR.

19.5.6 *Inherently Safer Design*

Inherently safer design[4,17] refers to a systematic process methodology to improve safety and reduce capital/operating costs by eliminating non-essential equipment and process steps. An inherently safer plant relies on robust design and/or operating methods for improving process safety rather than relying on instrumented/procedural systems such as safety instrumented system, process control system, alarms, interlocks and emergency procedures.[18] Process simplification and inherently safer design practices such as process intensification, substitution, attenuation and limitation of effects, are very useful for designing safer and sustainable WHR projects. These techniques are covered in Chapter 20.

19.5.7 *Design-to-Capacity Review*

This review is useful to decide the right (and not over/under) sized WHR equipment. Note that WHR equipment and piping are frequently oversized due to the following reasons.

- Not performing detailed hydraulic studies
- Use of high DP across control valves instead of across HEs.
- Allowances for future duty/capacity creep
- Allowances for fouling, especially to operate continuously between plant turnarounds
- Double inclusion of safety margins by the designers and equipment manufacturers

Unreasonable/high design margins can decrease WHR equipment performance. For example, if very high design margin and fouling factors are used for an HE design, it will foul and plug up very fast in operation (due to low velocity), and hence its performance will deteriorate quickly. Oversized pumps/compressors require higher capital and operating costs. Also, they require extra pressure to be dropped (when HE is in clean condition) across high DP control valves and hence may create erosion/corrosion problems. For sustaining WHR benefits only reasonable design margins should be used.

19.5.8 *Metallurgy Review*

Proper selection of metallurgy (i.e., material of construction) is crucial for sustainable WHR. In addition to operating and design conditions, it should include special conditions such as start-up, shutdown and any other abnormal conditions. Proper metallurgy selection can minimize damage mechanisms such as erosion, corrosion, high temperature sulfidation, cold/hydrogen embrittlement and chloride/caustic stress cracking.

If a metallurgy upgrade is not economical, other methods such as the use of weld overlays, rubber lining, special coatings, refractory and insulation are widely used. Sometimes, process operating conditions can be controlled by the addition of additives. Metallurgy selection and metal protection methods are discussed in detail in Chapter 18.

19.5.9 *Outside Battery Limit Facilities Review*

Detailed review of existing outside battery limit (OSBL) facilities listed in Table 19.2 is very beneficial for optimizing WHR project cost and making it economically more attractive. Instrument air is required for control valves. Plant air is required for instrument panel purges. Most WHR techniques such as heat pumps (for heating and cooling) and ORC require significant amount of cooling water. Any additional steam generation requires more BFW and hence more deaeration, demineralization plant capacity and condensate recovery equipment. Significant additional equipment and piping costs are required if existing OSBL piping and equipment are not able to serve these additional demands.

For minimizing capital cost, it is essential to review and utilize/modify the existing OSBL infrastructure as much as possible. Detailed check of steam and power balances, utility and energy costs are required to verify and confirm the true savings of any WHR project. Zhu[19] highlighted this with a site-wide energy optimization study. Reddy *et al.*[20] presented a mixed integer linear programming (MILP) model for optimizing power generation from a steam system, using a spreadsheet. This can be used for quick verification of benefits from WHR projects involving

Table 19.2: List of OSBL facilities

Steam and Condensate Network	Cooling and Chilled Water Systems
Deaerators	Demin plants and water makeup
Air systems	Pressure relief and flare systems
Substation and electrical distribution system	Cogen plants and utility boilers
Wastewater treatment plants	Power import system
Distributed control system (DCS)	Fire water network
Programmable logic controller (PLC) and safety instrumented system (SIS)	Fire alarm system

additional steam generation. Some useful case studies to increase power generation from a steam system are presented in Chapters 16 and 17.

If pressure relief to flare system is required in a WHR project, it should be integrated with the existing flare system along with a detailed evaluation of its capacity. Proper checks on substations, DCS, PLC, electrical and instrumentation system, fire water, fire and gas monitoring systems are important for effective integration of a WHR project. If OSBL requirements are not correctly evaluated, it is not uncommon to incur 20%–30% cost escalation during the project implementation.

19.5.10 *Equipment Layout Review*

Equipment layout review is important to meet fire and safety distance guidelines as required by government regulations and company standards. It is also useful for minimizing piping, civil, structural and pressure relief costs. Additionally, it is essential to review and meet maintenance, operational and emergency access requirements. Instruments and electrical motors need to follow the area classification required for the new equipment area.

19.5.11 *Preliminary Process Safety Studies*

Process safety studies are very important to ensure that WHR project design meets requirements for safe design and reliable operation, for the whole life cycle of the equipment usage. Based on the nature of the WHR project, process safety studies can use simple tools such as What–if checklists, to complex tools such as HAZOP and/or LOPA studies. Detailed explanation of these studies is given in Chapter 20.

In a pre-FEED study, detailed process calculations/simulations, block/process flow diagrams, pipe/equipment sizes, cost estimation (±30% accuracy) are generated for each WHR option. Cost estimation can be performed by any of the methods such as expert judgement, analogous estimating, parametric modelling, cost curves, six-tenths rule, ratio, bottom-up estimating, three-point estimating, and computerized tools, as per the project need. Some of these cost estimation methods are discussed in Chapter 6. A comparative discussion on cost estimation methods is available in Feng and Rangaiah[21] and Hastak.[22] With detailed analysis of the project economics and the pros and cons of each option, the final recommended WHR option is selected and presented to the management for approval. After management approval, project proceeds to FEED stage for complete development. Note that pre-FEED can be completed by in-house teams for smaller projects, whereas engineering contractors are generally used for bigger projects.

A typical project charter or summary, illustrated in Fig. 19.6, is very useful for WHR project presentation to the management.

<table>
<tr><td>Project</td><td>New Deaerator Project</td><td>Project #</td><td>2018/06</td></tr>
<tr><td>Project manager</td><td>John</td><td>Sponsor</td><td>Edwards</td></tr>
<tr><td></td><td></td><td>Updated</td><td>8/1/18</td></tr>
<tr><td>Background</td><td colspan="3">Boiler feed water (BFW) system's capacity is not sufficient to meet the additional BFW requirement of 200 ton/hr for new projects. Flash steam (FS) of 29.63 ton/hr can be generated from a WHR project, which can be used in a new deaerator to produce the required additional demand for BFW. Since poorly deaerated BFW causes corrosion in boilers and heat exchangers, installation of a new deaerator of 200 ton/hr capacity is required to meet the additional BFW capacity.</td></tr>
<tr><td>Business drivers</td><td colspan="3">Support additional plant capacity expansions and achieve energy cost savings of US$3.9 million per annum, from WHR.</td></tr>
<tr><td>Objectives</td><td colspan="3">To achieve WHR by generating 29.63 ton/hr of FS and utilizing it in a new deaerator (200 ton/hr BFW capacity), to supply BFW for the new expansion projects and improve the reliability of BFW system</td></tr>
<tr><td rowspan="3">Scope</td><td>In Scope</td><td colspan="2">Out of Scope</td></tr>
<tr><td>Design and installation of a new deaerator</td><td colspan="2">Repair/replacement of existing deaerators</td></tr>
<tr><td>Diverting FS from WHR to the new deaerator</td><td colspan="2">Mechanical integrity check of the existing BFW system</td></tr>
<tr><td rowspan="3">Key considerations</td><td>Assumptions and Constraints</td><td>Category</td><td>Risk Description</td></tr>
<tr><td>BFW requirement for new projects is 200 ton/hr and required FS can be generated by WHR from a process stream</td><td>Construction</td><td>Very limited ground level plot space is available. However, elevated deaerator can be considered as risk mitigation</td></tr>
<tr><td>BFW system is part of key utilities; hence, its reliability is vital for undisrupted facility operation</td><td>Schedule</td><td>Shutdown window of only 14 days is available for construction. Hence, it is planned to pre-fabricate the equipment skids with all structures for quick installation.</td></tr>
<tr><td rowspan="5">Success criteria</td><td colspan="3">1. Successfully install a WHR heat exchanger to produce 29.63 ton/hr of FS</td></tr>
<tr><td colspan="3">2. Successfully support BFW for the plant growth</td></tr>
<tr><td colspan="3">3. Execute the project within the estimated budget and schedule</td></tr>
<tr><td colspan="3">4. Dissolved oxygen in BFW should be less than 10 ppb wt</td></tr>
<tr><td colspan="3">5. Execute project without any safely incidents</td></tr>
<tr><td>Budget</td><td colspan="3">Project cost: US$7,500,000 (± 30% accuracy)</td></tr>
<tr><td>Key milestone Schedule</td><td colspan="3">Pre-FEED: Dec. 2018; FEED: April 2019; Execution: May 2019 to June 2020; Commissioning and testing: July 2020; Post-project review and operations: Sept. 2020</td></tr>
<tr><td>Stakeholders</td><td colspan="3">Daniel (ABC), Kiran (PQR), Smith (XYZ)</td></tr>
</table>

Fig. 19.6: A WHR project charter for presentation to the management.

19.6 Front End Engineering Design

A comprehensive FEED study is required to fully develop the selected WHR option, reduce the project scope creep[c,23,24] and hence reduce the project risk significantly. Although FEED study costs ~ 5% of the project cost,[25] it defines nearly 80% of the capital cost and 98% of the operating cost, and hence it plays a major part in the project success.[15] Typical data required for a WHR project FEED study are listed in Table 19.3.

As a first step in FEED study, all the plant operating/test run data, preferably DCS data (i.e., operating data for as long as available) and process simulations need to be collected. Operating data should be validated so that they satisfy mass and energy balances. This can be facilitated with process simulation, and measured flow rates and/or stream temperatures adjusted via data reconciliation, in order to satisfy mass and energy balances. Equipment inspection data need to be reviewed for understanding damage mechanisms. Detailed hydraulic study is essential to validate

Table 19.3: Typical data required for a WHR project FEED study

Process Flow Diagrams	Process and Instrumentation Diagrams	Cause and Effect Charts
Data sheets and drawings of existing equipment	Existing piping/ducting isometrics	Company/local/ international standards applicable to the project
Plant plan	Area classification drawings	Existing utility balances
Pressure and temperature profiles of existing plant	Data of process and flue gas streams	Facility's safety management system
Existing utility pipe sizes	Electrical power generation/ import and distribution details	Existing relief system
Existing instrument datasheets	Capabilities of substation, DCS and safety instrumented system	Firefighting system
Plant turnaround or stoppage plans	Existing plant control system narrative	Equipment inspection and maintenance reports
Process simulations	Operating procedures	Operating envelope

[c] Scope creep can be defined as the tendency for a project to go beyond its initial boundaries.

the pressure profile and establish the allowable DP for the new/modified WHR equipment. Similarly, detailed review of the site's utility balances and OSBL facilities (Table 19.2) is essential for optimizing the project scope by exploiting the existing infrastructure as much as possible.

As part of FEED study, review of laser scan integrated with existing 3D plant models[26] will be useful especially for offline design from office. However, a plant walk down is very beneficial and recommended for understanding plot space limitations, physical condition of plant equipment, instrumentation, fire/emergency escape routings and so on. Detailed review of piping layout is useful for minimizing WHR equipment size, remove any liquid slugs (for gas/vapour services), installation costs and also for correct sizing and installation of control valves.

Initially, process calculations/simulations and pipeline sizes are checked and confirmed to meet the project requirements. Process controls, cause and effect charts, SIS, pipe sizes (if changes are required), process flow diagrams (PFDs) and area classification diagrams are updated. Once PFDs and design calculations are approved by the client, P&IDs are updated with the new pipe sizes, manual valves, automatic control valves, emergency shutdown valves, process control loops, process alarms and interlocks, piping tie-in points,[d] new/modified WHR equipment, piping and equipment demolitions. New/modified WHR equipment dimensions are finalized and plot plan is updated for the changes. Pressure relief calculations are performed and modifications to the plant's relief system are prepared. All the equipment, instrument and electrical (I/E) equipment datasheets are prepared and sent to vendors for quotations. Operating manual and process control narratives are prepared. Line, equipment, instrumentation, electrical equipment, cable sizing, input/output (I/O), SIS, firefighting equipment, gas and fire detection equipment, utilities and list of wastewater streams are prepared. FEED study can be completed in-house or by EPC contractor, respectively, for smaller or bigger WHR projects; the latter involves complete project management team (PMT), including engineers from all disciplines.

---

[d]A tie-in point is the place in the existing piping or a flange on equipment, where new piping needs to be connected.

The following studies are essential as part of FEED, to make the WHR project safer and sustainable.

19.6.1 *Reliability, Availability and Maintainability Study*

This is a systematic statistical study to ensure that a WHR project is designed to be reliable, available and maintainable, for a specified period, by aiming for the following goals:

- WHR equipment will perform reliably according to project objectives, under specified conditions, for a specified period such as operation between plant turnarounds.
- WHR equipment is available to perform the required function for a specified period.
- When a specified procedure is followed, failed WHR equipment can be repaired to its operation effectiveness, within a specified period (i.e., it is maintainable).

As reliability and safety are interrelated, RAM and safety studies are sometimes performed together with RAMS software.[27] Many commercial programs are available for performing RAM/RAMS studies. If reliability data is available, simple spreadsheets can be used for smaller projects. RAM/RAMS study uses equipment's failure data (in-house data, if available, is preferred) such as mean time between failures (MTBF), mean time to repair (MTTR), mean time to failure (MTTF) and mean down time (MDT).

19.6.2 *Constructability Study*

The main objective of a constructability study is to review the WHR project model (preferably 3D model) and ensure that the following safety, operational and maintenance requirements are met in the project design:

- No problems with construction issues such as lifting, excavation and installation.
- Maintenance, operational and emergency accesses are available and not partially blocked/restricted.

- Fire safety distance gaps between equipment and equipment, equipment and pipe racks, equipment and plant boundary and so on, are fully complied.
- No obstruction for firefighting access.
- Instruments and valves are accessible, without any ergonomic issues.
- Equipment can be easily removed from the plant, for maintenance.
- Safe routing of piping.
- Proper orientation of instruments, steam traps, control valves, on-off valves, check valves and so on.

19.6.3 *Predictive Maintenance Study*

This study uses digital tools such as sensors (e.g., corrosion probes, pH probes, DP measurement, pressure/temperature/flow sensors, vibration sensors and acoustic sensors) controls and/or algorithms for monitoring operation of WHR equipment and predicting maintenance requirements. For example, fouling of APH and other HEs can be monitored with DP and temperature measurements. Maintenance cleaning can be scheduled to perform acid or hot water cleaning. Many data analytic software tools are now available for achieving digitalization benefits.

19.6.4 *Value Engineering*

Value engineering is an important FEED activity, mainly performed to minimize/simplify the project cost/scope, without compromising project objectives. It is often carried out by multi-disciplinary experts or peer review, by brainstorming the project scope.

- Maximize the re-use of existing equipment and plant infrastructure
- Remove any overspecifications for new equipment and instrumentation
- Reduce excessive overdesign margins
- Review and avoid the use of expensive metals by design changes
- Elimination of unnecessary valves and instruments
- Use of alternative equipment and/or vendor
- Remove need for any customized design of equipment

- Reduce equipment sizes by following inherently safer design practices and green engineering principles
- Use of SIS system to reduce pressure relief requirements

19.6.5 *Process Hazard Analysis*

PHA is performed during FEED stage, mainly to review changes made after pre-FEED safety studies. Generally, HAZOP and LOPA are used in FEED stage for holistic review of safety and operability issues. Detailed explanation of PHA is presented in Chapter 20.

19.6.6 *Risk Assessment*

As every project has inevitable risks, PMT needs to evaluate and list all possible project risks such as long-lead equipment delivery issues, energy prices uncertainty, congested plot space, contractual issues, shortage of contractor resources and anticipated delays in government approvals. All possible project risks should be identified and recorded in the project scope document along with proper mitigation measures. Since all risks are not critical, risk mitigation measures can be prioritized using the strategy in Fig. 19.7, which is developed based on Figure 2.5 in Harold.[28] Using this strategy, risks with high strategic importance and added/rework costs should be mitigated quickly. Risks with low strategic importance and added/rework costs do not require any attention. If sound mitigation measures for project's critical risks are not identified and their costs are not included by the completion of FEED stage, the project cost may increase exponentially. Cost of project changes with project timeline is illustrated in Fig. 19.8.

WHR risk management methodology is illustrated in Fig. 19.9. Once risks are identified, proper measures to mitigate them are generated for quick implementation. Risk management is a continuous process. Hence, as the project progresses from pre-FEED to FEED and subsequently to EPC, risks and mitigation measures need to be revisited, and efforts should be made to implement the mitigation measures promptly. Feedback is obtained from key stakeholders on the effectiveness of the risk mitigation, and their comments are used to assess whether any changes in the

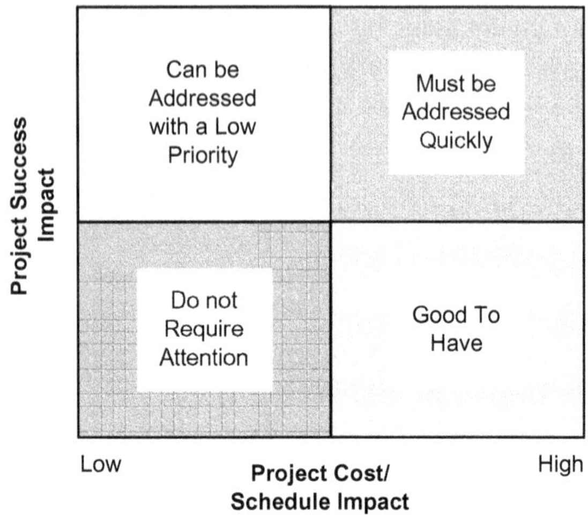

Fig. 19.7: Strategy to reduce the WHR project risk.

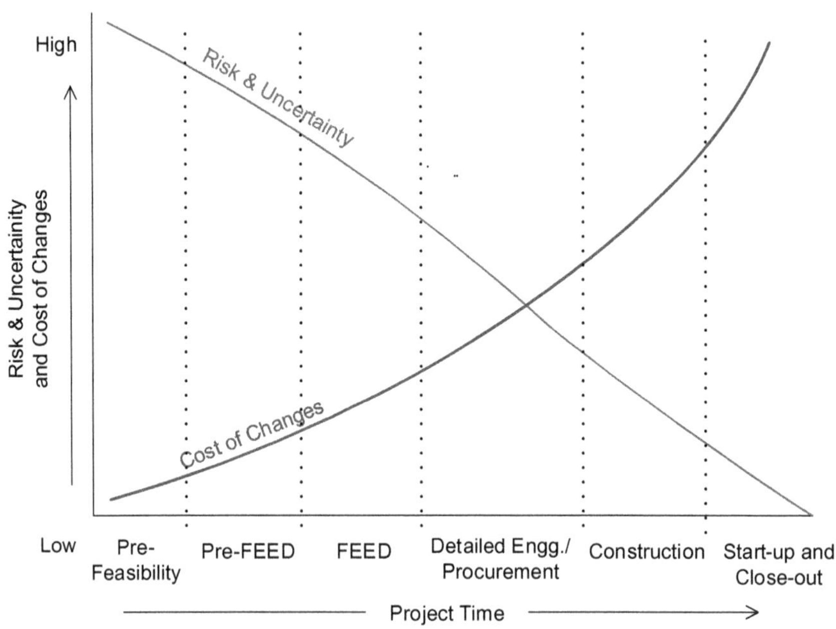

Fig. 19.8: Variation of risk/uncertainty and cost of changes with project time.

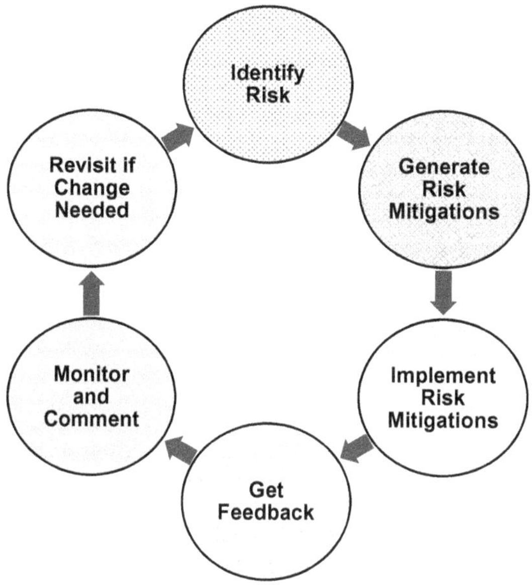

Fig. 19.9: Risk management.

risk mitigation measures are required. If required, suitable changes in risk mitigation measures are implemented and their effectiveness is monitored regularly during the entire project duration. Risk assessment continues throughout WHR project development and implementation until residual risks are acceptable.

19.6.7 *Cost Estimation*

Action items from all studies in Sections 19.6.1–19.6.6 need to be closed with proper explanation and/or re-design, in the FEED stage itself. After finalizing the FEED, vendor quotations are obtained from potential contractors and engineering companies. EPC scope is discussed in Section 19.8. After reviewing FEED equipment costing and vendor quotations, the project team prepares cost estimation (with ±10% accuracy) and recommends the EPC contractor. Based on this cost, economic analysis of the project is prepared, as illustrated in the following example. Economic evaluation methods are described in Chapter 6.

Example 19.1: A process plant needs to install a new deaerator (capacity: 200 ton/hr of BFW production), for plant expansion. A WHR project at the plant can generate 29.63 ton/hr of FS at 2.1 bar, which can be used in the new deaerator. PMT estimates a capital cost of \$6.45 million (±10% accuracy) for this WHR and deaerator project. Assuming an FS price of \$15/ton, estimate the project payback period and internal rate of return (IRR). Perform a sensitivity study for FS price variation from \$5/ton to \$20/ton, in steps of \$5/ton. Assume a plant life of 15 years, annual operating hours of 8760 and discount rate of 10%.

Solution: An Excel spreadsheet can be used for calculating simple payback period and IRR as shown in Table 19.4, where each scenario is for a different FS price. Excel's net present value (NPV) and IRR functions are used with discount rate of 10%. Payback period is calculated by dividing the capital cost of the project with annual energy savings. Annual FS energy savings are estimated by multiplication of FS flow rate (ton/hr) with FS price (US\$/ton) and annual operating hours. For example, at US\$15/ton FS cost, FS energy savings = 29.63 × 15 × 8760 = US\$ 3,893,382.

From the assessment in Table 19.4, the project base scenario results in an IRR of 60% and payback period of 1.7 years. Hence, the proposed WHR project is very attractive. For presentation to management, often an economic sensitivity study such as that in Table 19.4 is required. It can be seen from this table that IRR reduces, whereas payback period increases with decrease in FS/energy prices. For the case of low FS cost of US\$5/ton, IRR drops to 19% and payback period increases to 5 years. On the other hand, if FS price increases to US\$20/ton, IRR increases to 80% and payback period is shorter at just 1.2 years. Final project scope with cost, economic analysis and EPC recommendation are presented to the plant management. Complete project scope document (illustrated in Fig. 19.10), based on the completed FEED study is very useful for the presentation to the management for approval. This document presents firm project scope, cost (±10% accuracy) and detailed project milestone schedule compared to preliminary details given in the project charter (Fig. 19.6). After management approval, the project proceeds to EPC stage.

Table 19.4: Economic and sensitivity analysis for example 19.1

Quantity	Scenario 1	Project Base Scenario	Scenario 2	Scenario 3
FS Flow rate, kg/hr	29.63	29.63	29.63	29.63
Annual operating hours	8760	8760	8760	8760
FS price, US$/ton	20	15	10	5
FS annual savings, US$	5,191,176	3,893,382	2,595,588	1,297,794
Capital cost, US$	6,450,000	6,450,000	6,450,000	6,450,000
Year	**Capital cost/ annual energy savings US$**	**Capital cost/ annual energy savings, US$**	**Capital cost/ annual energy savings, US$**	**Capital cost/ annual energy savings US$**
0	−6,450,000	−6,450,000	−6,450,000	−6,450,000
1	5,191,176	3,893,382	2,595,588	1,297,794
2	5,191,176	3,893,382	2,595,588	1,297,794
3	5,191,176	3,893,382	2,595,588	1,297,794
4	5,191,176	3,893,382	2,595,588	1,297,794
5	5,191,176	3,893,382	2,595,588	1,297,794
6	5,191,176	3,893,382	2,595,588	1,297,794
7	5,191,176	3,893,382	2,595,588	1,297,794
8	5,191,176	3,893,382	2,595,588	1,297,794
9	5,191,176	3,893,382	2,595,588	1,297,794
10	5,191,176	3,893,382	2,595,588	1,297,794
11	5,191,176	3,893,382	2,595,588	1,297,794
12	5,191,176	3,893,382	2,595,588	1,297,794
13	5,191,176	3,893,382	2,595,588	1,297,794
14	5,191,176	3,893,382	2,595,588	1,297,794
15	5,191,176	3,893,382	2,595,588	1,297,794
Total savings	77,867,640	58,400,730	38,933,820	19,466,910
NPV@10%	30,031,361	21,057,612	12,083,862	3,110,113
IRR	**80%**	**60%**	**40%**	**19%**
Payback period Years	**1.2**	**1.7**	**2.5**	**5.0**

19.7 Engineering, Procurement and Construction

Detailed EPC activities are shown in Fig. 19.11. Any remaining engineering work from FEED study is completed in EPC stage. This activity completes all the detailed engineering and procurement work (i.e., purchase requisition of equipment and 3D model with all the piping, equipment, I/E, fire

<table>
<tr><td>Project</td><td>New Deaerator Project</td><td>Project #</td><td>2018/06</td><td>Updated</td><td>7/1/19</td></tr>
<tr><td>Project manager</td><td>John</td><td>Sponsor</td><td colspan="3">Robert</td></tr>
<tr><td>Project scope</td><td colspan="5">1. Installation of a new WHR exchanger to produce 29.63 ton/hr of FS
2. Installation of a new deaerator to produce 200 ton/hr of BFW, with new BFW pumps
3. New piping (for FS, BFW, plant air, instrument air, chemical injection, demineralized (DM) water and fire water), electrical and instrumentation works
4. New DCS graphics
5. Civil foundation and structural works</td></tr>
<tr><td>Business drivers</td><td colspan="5">Support additional plant capacity expansions and achieve energy cost savings of US$3.9 million/annum from WHR</td></tr>
<tr><td>Project objectives</td><td colspan="5">To achieve WHR by generating 29.63 ton/hr of FS, and utilize it at a new deaerator (200 ton/hr BFW capacity)</td></tr>
<tr><td rowspan="2">Project exclusions</td><td colspan="5">Repair/replacement of existing deaerators</td></tr>
<tr><td colspan="5">Mechanical integrity check for the existing BFW system</td></tr>
<tr><td rowspan="16">Key milestone schedule</td><td colspan="3">Milestone</td><td colspan="2">Due Date</td></tr>
<tr><td colspan="3">Generation of alternatives</td><td colspan="2">Sept 18</td></tr>
<tr><td colspan="3">Costing of alternatives</td><td colspan="2">Oct 18</td></tr>
<tr><td colspan="3">Selection of alternatives</td><td colspan="2">Nov 18</td></tr>
<tr><td colspan="3">Pre-FEED approval</td><td colspan="2">Dec 18</td></tr>
<tr><td colspan="3">Developing selected alternatives and HAZOP study</td><td colspan="2">Apr 19</td></tr>
<tr><td colspan="3">VIP studies</td><td colspan="2">Apr 19</td></tr>
<tr><td colspan="3">FEED costing</td><td colspan="2">Apr 19</td></tr>
<tr><td colspan="3">FEED and EPC contractor selection and approval</td><td colspan="2">July 19</td></tr>
<tr><td colspan="3">Detailed engineering</td><td colspan="2">Oct-19 to May 20</td></tr>
<tr><td colspan="3">Constructability review</td><td colspan="2">Apr-20</td></tr>
<tr><td colspan="3">Procurement of equipment</td><td colspan="2">Sept 19 to Feb 20</td></tr>
<tr><td colspan="3">Equipment Installation</td><td colspan="2">May 20</td></tr>
<tr><td colspan="3">Commissioning</td><td colspan="2">June 20</td></tr>
<tr><td colspan="3">Performance Testing</td><td colspan="2">July 20</td></tr>
<tr><td colspan="3">Operate and evaluate</td><td colspan="2">Aug 20</td></tr>
<tr><td rowspan="3">Key considerations</td><td>Assumptions and Constraints</td><td>Category</td><td colspan="3">Risk Description</td></tr>
<tr><td>BFW requirement for new projects is 200 ton/hr and required steam can be generated by WHR</td><td>Construction</td><td colspan="3">Very limited ground level plot space available. Hence, elevated deaerator can be considered for risk mitigation.</td></tr>
<tr><td>BFW system is part of key utilities and hence its reliability is essential to support undisrupted facility operation</td><td>Schedule</td><td colspan="3">Shutdown window of only 14 days is available for construction. To minimize this risk, it is planned to pre-fabricate equipment skids with all necessary structures</td></tr>
<tr><td rowspan="3">Success criteria</td><td colspan="5">1. Successfully implement WHR equipment to recover 29.63 ton/hr of FS</td></tr>
<tr><td colspan="5">2. Successfully support BFW for the refinery growth projects</td></tr>
<tr><td colspan="5">3. Execute project within the estimated budget and schedule</td></tr>
<tr><td rowspan="2"></td><td colspan="5">4. Dissolved oxygen in BFW should be maximum 10 ppb wt.</td></tr>
<tr><td colspan="5">5. Execute project without any safety incidents</td></tr>
<tr><td>Budget</td><td colspan="5">Project cost = S$6,450,000 (± 10% accuracy)</td></tr>
<tr><td>Stakeholders</td><td colspan="5">Daniel (ABC), Kiran (PQR), Smith (XYZ)</td></tr>
</table>

Fig. 19.10: Project scope document.

and safety works) for procurement of the equipment, generation and finalization of piping isometric drawings, finalization of plot plan, construction activities (i.e., site preparation, foundation and structural works, equipment installation, field I/E works such as installing instruments, motors and lighting, and cabling works including DCS and substation works), obtaining necessary government approvals, and pre-commissioning, commissioning and start-up activities of the project. All the design calculations are verified by the PMT from the owner company.

Final constructability study is conducted with the multi-disciplinary teams from both owner company and EPC contractor, to finalize the plot plan, piping, equipment (including I/E items) locations, firefighting facilities, location of fire and gas detectors, safe access space for valves and equipment, maintenance and emergency access space, pipe routings, equipment lifting plans and so on. All the recommendations are incorporated into the final 3D model, which is used for the project construction. The Engineering phase (in EPC) generates 'issued for construction' (IFC) documents. Procurement activities proceed during the engineering phase, once sufficient details are available. Plant construction will start progressively during EPC stage from site preparation to foundation, erection of structures and installation of equipment, as and when necessary details are available.

After completion of construction activities, several plant walks are conducted by multi-disciplinary teams to check mechanical completion of the project. Training sessions are conducted for the operations and maintenance teams. Pre-start-up safety review (PSSR) is conducted to check operation readiness. Any punch list[e] items generated are closed by corrective actions/modifications by the EPC contractor. After checking the closure of the punch list items, pre-commissioning activities begin. At this stage, all the As-built documents are made available.

Pre-commissioning includes the following activities:

- Cleaning and flushing of all the piping, its accessories and equipment
- Blowing any refractory dust
- Checking all the instrumentation, control checks and motor rotation checks

[e]Punch list refers to the list of work items not completed or implemented, as per the approved design. They may include some minor repair of damaged items during construction also.

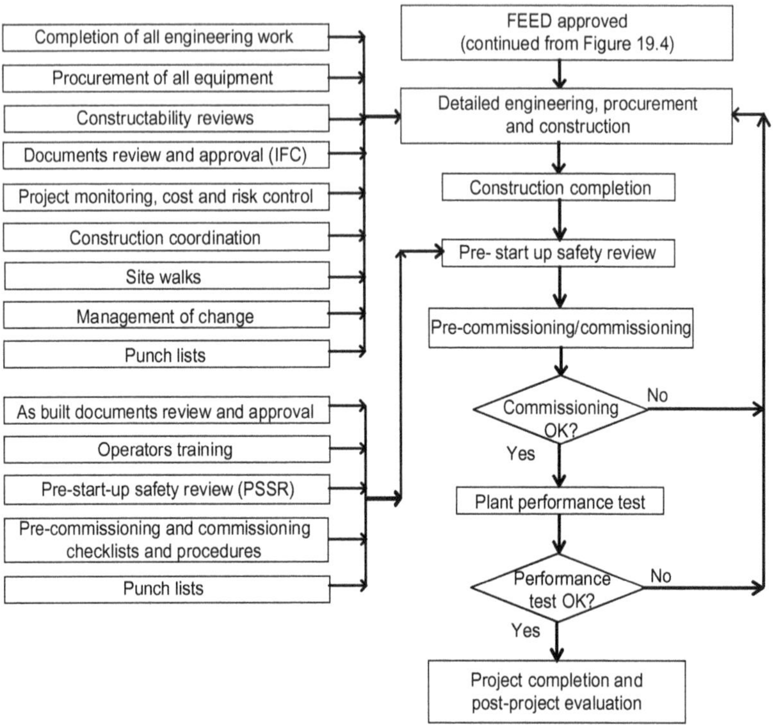

Fig. 19.11: Overview of project EPC activities (Reproduced with permission from John Wiley & Sons Inc.).[3]

- No-load rotation check of pumps, fans and compressors
- System leak tests
- Checking fire/gas detection and firefighting equipment
- Filling necessary catalysts and adsorbents

After successful completion of pre-commissioning activities, commissioning begins. This includes the following activities:

- Check each system/subsystem with motors energized and/or dynamic tests to ensure that it fully meets the design intent
- Complete chemical and/or steam cleaning of pipelines and equipment
- Equipment and pipelines drying, air freeing and leak testing
- Checking interlocks, PLC and DCS systems
- Refractory dry-out

- Commissioning the utilities
- Filling and circulating pipelines and equipment with process fluid/temporary fluid

After successful commissioning, plant operation and performance test follow.

Performance Test: After reaching stable plant operation, a performance test is conducted, generally for 24 hr. If the performance test is not successful, detailed evaluation is carried out to establish the root cause(s) for this. If the root cause is due to operating conditions, they are adjusted to perform a second performance test. Otherwise, the EPC contractor needs to implement corrective modifications to fully address the root cause(s). This process continues until the performance test meets the performance guarantees.

The performance test report is circulated to all key stakeholders. Based on this, the project manager/engineer performs post-project evaluation, with assistance from process and other engineers. The performance test report not only reflects the plant/equipment performance but also achievement of project objectives, lessons learned, safety performance, project changes and their reasons, construction and equipment photos, and opportunities for future revamp/debottlenecking opportunities. An example of a simplified project performance summary, highlights and lessons learned is presented in Tables 19.5 and 19.6.

Finally, the EPC contractor hands over the project documents including lessons learned to the plant owner. Typical summary of purpose, resources, cost estimation requirements and project deliverables for various project stages are shown in Table 19.7. Actual number of project stages may be lower or some of the stages can have sub-stages, depending on the project size/complexity and/or company's project management policies.

19.8 Final Project Deliverables

At the end of the construction phase, the EPC vendor provides the entire project documentation, which is generally arranged separately into engineering and vendor-related document dossiers. A typical list of

Table 19.5: Project performance summary

Item Detail	Yes	No	Actual Performance and/or Remarks
Project Performance			
Commissioning of WHR and new deaerator (capacity: 200 ton/hr) by July 2020		√	September 2020
Zero safety incidents	√		
Project cost not to exceed US$ 6.45		√	US$6.82 Million
Pre-feasibility study by July 2018	√		July 2018
Pre-FEED by December 2018	√		December 2018
FEED by April 2019		√	May 2019
Award EPC by May 2019		√	August 2019
Project completion by July 2020		√	September 2020
Project Objectives			
To provide additional 200 ton/hr of BFW for the plant expansion	√		New deaerator is operating reliably. Also, it is integrated with the facility's existing BFW system.
WHR for generation of 29.63 ton/hr of FS for the new deaerator	√		All the FS generated from WHR is consumed by the new deaerator.
To complete the project within budget and schedule		√	Late design change of increasing residence time in deaerator BFW storage vessel increased the project scope and cost.
To complete the project without any safety incident	√		

deliverables is given in Table 19.8. Any comments from PMT needs to be quickly addressed by the EPC contractor. Accurate project documentation is essential for equipment operation, maintenance, process safety assessments and future revamps.

19.9 Key Elements of Project Management

WHR project development and implementation mainly depend on three ingredients, namely, time, scope and resources. If any of these are out of balance, the project fails to meet some of the objectives or fail completely.

Table 19.6: Project highlights and key lessons learned

No	Item	Highlights	Key Lessons Learned	Follow-Up Actions: Description	Follow-Up Actions: Responsible Person	Follow-Up Actions: Completion Date
1	Project schedule	Project completion is delayed by 2 months	• Good teamwork between PMT and contractors • Late project changes increased the project cost significantly	Share the project success and lessons learned to all in the whole organization	Nick	April 2020
2	Project cost	Actual project cost is US$0.37 million more than the original approved budget due to: • Increased deaerator BFW storage vessel cost due to increase of residence time at EPC stage • Increase in civil structural works • Equipment expediting costs	Late project changes increased the project cost significantly	Share the lessons learned to all the stakeholders	Nick	June 2020

Hence, it is very important to control scope, time and resources throughout the WHR project. Some key reasons for a failed project are given below.

- Poor cost and schedule estimates
- Poorly defined goals and objectives
- Lack of project sponsorship
- Poor scope, cost, risk and schedule management
- Insufficient resources
- Poor communication
- Lack of stakeholder engagement
- Change in business environment

Actual reasons for a failed WHR project can differ significantly from one project/place to another project/place. As per a global survey of project

Table 19.7: Summary of WHR project stages, resources required, cost estimation and key deliverables (Reproduced with permission from John Wiley & Sons Inc.)[3]

	Pre-Feasibility	Pre-FEED	FEED	EPC	Project Start-up
Purpose	To set the project objectives, basis, assumptions and preliminary economics	To generate alternative options for WHR and establish the best WHR option	To fully develop the WHR option	To complete engineering, procurement and construction	To measure success of the WHR project and document key highlights and lessons learned.
Resources required	Process and operations engineers	Process and operation engineers; engineering contractor (if required); **Ad hoc basis:** Project engineer, mechanical engineer, electrical engineer, civil engineer, environmental engineer, laboratory analyst and document controller	Process and operations engineers, project engineer, mechanical engineer, civil engineer, electrical engineer, safety engineer, reliability engineer, environmental engineer, engineering contractor (if required) and document controller	Project manager/engineer, EPC contractor with his own project team, process and operations engineers, mechanical engineer, inspection engineer, reliability engineer, safety engineer, civil engineer, environmental engineer, electrical engineer, construction and operations supervisors	Process and operations engineers, project engineer, operations supervisors/operators, engineering contractor, equipment vendors, and laboratory staff
Cost estimation method; typical accuracy[f]	Cost estimate using in-house (past projects) and published cost data; ±50%	Cost estimate based on quotations of major equipment in pre-FEED study; ±30%	Cost estimation based on finalized quotations; ±10%	NA	NA
Deliverables	• Project note highlighting objectives • Project framing (basis, assumptions and boundary conditions)	• Review report of plant design, operation, inspection and maintenance data • Plant test report	• PFDs and P&IDs • Cause and effect charts • Equipment process and mechanical data sheets • Equipment and piping lists • Material selection diagrams	• EPC weekly and monthly progress reports • Complete project as-built documents (see Table 19.8)	• Summary of project highlights and lessons learned • Summary of WHR opportunities in the future

[f] A comprehensive discussion on cost estimation methods is available in Green and Perry.[4]

<table>
<tr>
<td>• Preliminary project economics
• Block flow diagram
• Resources and activities planning for next phase
• Preliminary project schedule</td>
<td>• Summary of equipment bottlenecks, retrofit and debottleneck opportunities
• Equipment rating study reports
• WHR options and calibrated plant simulations for all options.
• Summary of OSBL modifications
• PFDs for all options
• Summary of VIPs used for establishing WHR options
• Summary of WHR options, benefits and drawbacks
• Cost estimation for all WHR options
• Major equipment quotations
• Recommended WHR option
• Summary of project risks and mitigation plan
• Resources and activities planning for next phase
• Updated project schedule</td>
<td>• Instruments list
• DCS I/O point list
• List of tie-in points
• Request for quotations for equipment
• Vendor quotations for equipment, civil, Instrumentation, electrical works, EPC contract etc.
• Pipe layout drawings
• Plot layout and equipment layout drawings
• Interlock lists
• PHA and RAM reports
• Control philosophy
• Constructability report
• Critical long-lead equipment list
• Detailed instrument and cable sizes and routing drawings
• Civil, structural, control room and shelter drawings
• Soil survey report
• Plant lighting drawings
• Quotation for EPC work
• Firefighting system design
• Revised project risks and mitigation plan
• Revised cost estimation
• Resources and activities planning for next phase
• Updated project schedule</td>
<td>• Operating procedures
• Operations and maintenance training manuals
• All the construction, PSSR, pre-commissioning and commissioning punch lists with action items closed
• Summary of project scope variations and cost impact
• Management of change documents
• Plant performance test report
• Mitigation plans if a performance gap is identified</td>
</tr>
</table>

Table 19.8: Final project documents from EPC vendor

General

- Project specifications
- Major equipment specifications and data sheets
- Material take-offs
- Training material

Mechanical Engineering

- Equipment list
- Material selection diagrams
- Mechanical data sheets for the equipment
- Equipment mechanical design calculations
- Equipment inspection reports
- Insulation, fireproofing and painting specifications
- Spare parts and consumables list
- Equipment fabrication and testing reports
- Equipment drawings
- Equipment installation drawings
- RAM study report
- Vendor manuals
- Equipment factory acceptance test (FAT) reports
- Hydrostatic test and other tests report

Piping Engineering

- Line list
- Piping and supports design calculations
- Stress analysis report for piping
- Piping layout and isometric drawings
- Pipe rack drawings
- Miscellaneous piping accessories
- Hydrostatic test and other tests report
- Tie-in list

Electrical Engineering

- Single line diagrams
- Electrical load summary
- Electrical equipment (such as motors, heaters, VFDs and transformers) specifications and data sheets

Process Engineering

- Design basis for the WHR design
- Process block diagrams
- PFDs
- P&IDs
- Cause and effect charts
- Tie-in and line lists
- Utility summary for the project
- Relief devices summary and calculations
- Relief system design calculations
- Utilities flow diagram
- Process design calculations including process simulations
- Hydraulic calculations
- Equipment sizing calculations
- Equipment, I/E data sheets
- List of process alarms and interlock set points
- Operating manual
- PHA report and compilation of PSI data
- Operations assurance report

Fire Protection

- Firewater system design basis
- Firewater piping drawing
- Gas detectors layout diagram
- Fire sprinklers, fire monitors and deluge system design details
- Details of other firefighting systems such as fire extinguishers

Instrumentation

- Instrument index
- Specification and datasheets for both new and modified control valves and other instruments

Table 19.8: (*Continued*)

General	Process Engineering
• Electrical cable schedule • Voltage drop calculations • Electrical area classification • Electrical instrument interface drawings • Underground electrical cable and cable trench arrangement • Electrical equipment layout drawings • Motor control centre and power layout • Load and short circuit analysis • VFDs configuration • Plant lighting plan • Grounding and lightening arrestor plan	• Instrumentation, controls and metering philosophy • Instrument cable layout plan and tray drawings • Instrument air piping and tubing tie-in • PLC architecture and control philosophy narrative • Logic diagram for key shutdowns and interlocks • Junction box termination drawings • Cabinet layout and termination drawings • Loop drawings • Conduit schedule • DCS input/output list
Civil and Structural Engineering	**Project performance guarantees**
• Soil investigation report • Equipment foundation design • Survey and preliminary grading plan • Piling and structural design report • Key plot plan • Plan of underground facilities including drainage • Sewerage connection drawing • Road, pavement plan and access review • Design of buildings, equipment rain shelters	

management leaders, conducted by PWC[g] in 2012, 30% respondents voted for poor cost estimates/missed deadlines as the main reason for a failed project.

19.9.1 *Project Execution and Progress Monitoring*

A project schedule chart is very useful to schedule and monitor the progress of project activities and deliverables. The number of activities can vary from few to several hundred, depending on project size. For smaller

[g] https://www.pwc.com/gx/en/capital-projects-infrastructure/pdf/pwc-correcting-the-course-of-capital-projects-v3-pdf.pdf.

projects, a simple Gantt chart using Excel, PowerPoint or Visio can be used. A typical high-level schedule using Excel is shown in Fig. 19.12. For large projects, project management software such as Microsoft Project and Primavera can be used to track critical paths, risk assessment, and to understand detailed interactions of project activities.

WHR project execution should be robust and closely follow the project schedule. Project progress is monitored regularly (generally on weekly basis). Leading and lagging activities (detailed engineering deliverables, procurement activities, long-lead equipment deliveries, construction activities, pre-commissioning, and commissioning requirements) compared to the project schedule, are discussed during the weekly meetings. If there are any lagging issues, appropriate corrective actions are identified, for quick implementation.

Critical path and long-lead items should be ordered as early as possible. Close monitoring and, if required, expedition of their delivery is essential for successful project completion. The schedule prepared at the beginning of the project is updated from time to time, with careful review by all stakeholders, as more details become available during the project development stages. Any schedule risks are reviewed from time to time and regular updates are provided to PMT members and key stakeholders, for expediting any lagging activities. The project schedule in FEED stage is fully developed by the EPC contractor to include all the activities of detailed engineering, procurement, construction (including construction subcontractors' schedule), pre-commissioning, commissioning and project start-up.

Advanced planning is beneficial as the WHR project development and execution often require more than a year. Although a WHR project is fully completed during the plant stoppage/turnaround duration, some of the works (such as piping, ducting, civil foundations, structural works, some equipment, instrument, electrical cable laying and scaffolding) can be completed before plant stoppage/turnaround. This helps to better manage the project completion as there may be insufficient resources and/or time during the plant stoppage/turnaround.

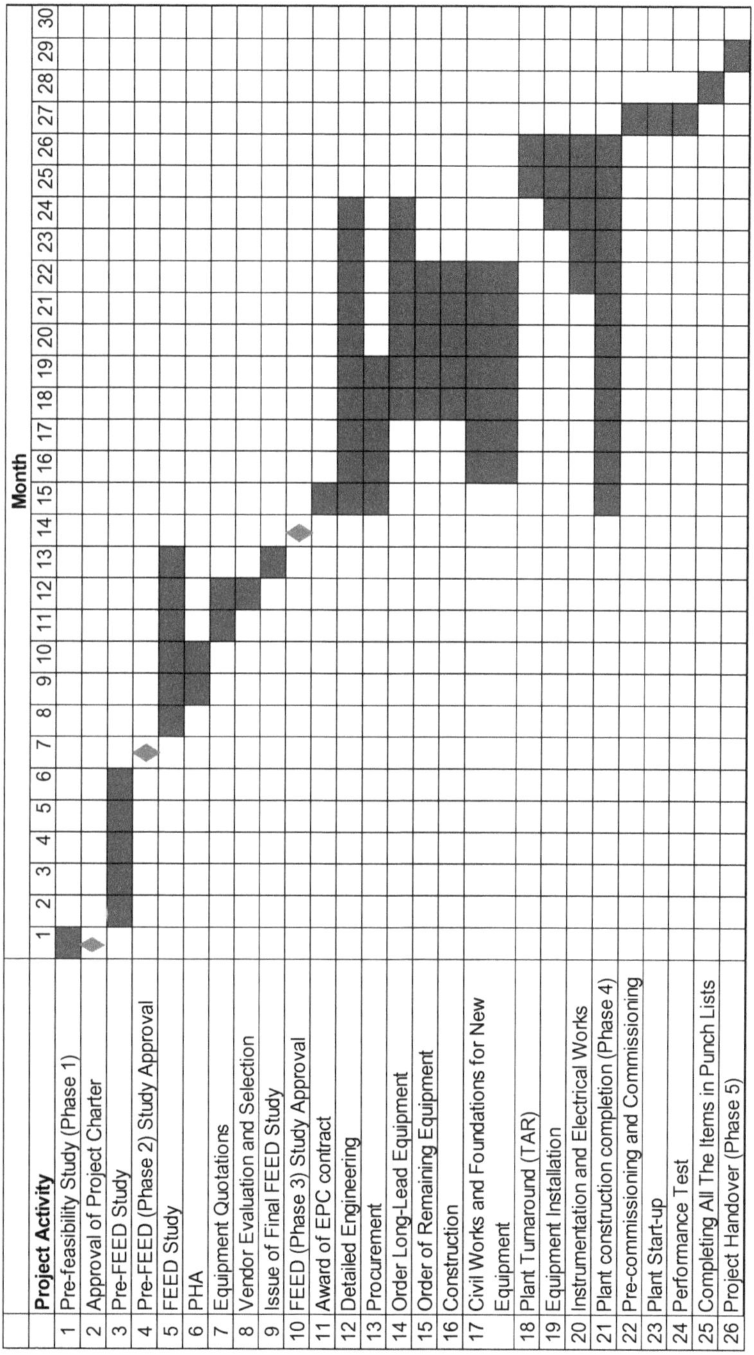

Fig. 19.12: Typical schedule of a WHR project.

19.9.2 *Resource Management*

Appropriate resources are vital for a successful WHR project development and implementation. Typical resource and cost requirements during project stages are illustrated in Fig. 19.13, which is prepared based on Figure 2.8 in PMBOK Guide.[2] Project cost and staffing levels are low in defining and planning stages. They gradually increase to a maximum during EPC stage and then reduce to zero towards the end of project implementation.

WHR project success depends on expertise of a cross-functional team of experienced engineers and managers from plant and engineering contractor. Typical organization chart of a WHR project is shown in Fig. 19.14. Project team members are selected based on the project size and nature, at the beginning of the project. Actual project team composition may change as the project progresses from one stage to another, depending on project nature and complexity. Teamwork and good coordination of engineers from multiple disciplines are required to holistically develop and analyse WHR options during the pre-FEED and FEED stages. Similarly, good coordination between EPC contractor and PMT is essential to fully develop and construct the new equipment and/or modify the existing equipment, and finally deliver a successful and sustainable WHR project.

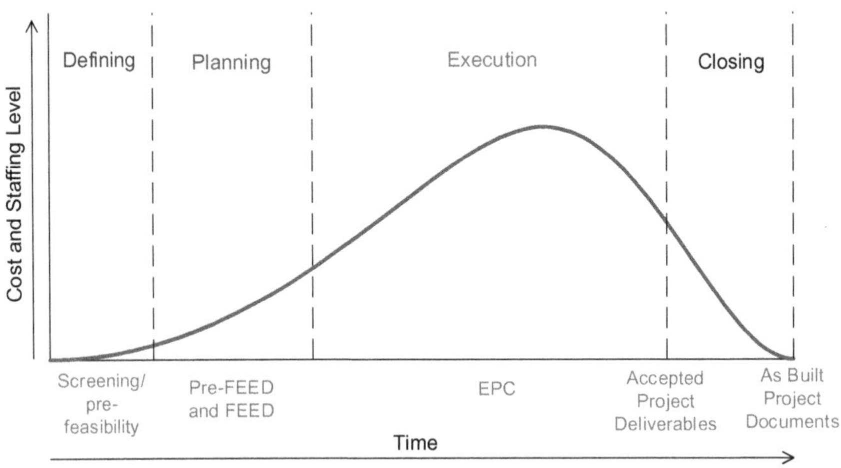

Fig. 19.13: Typical cost and staffing levels across a generic project life cycle.

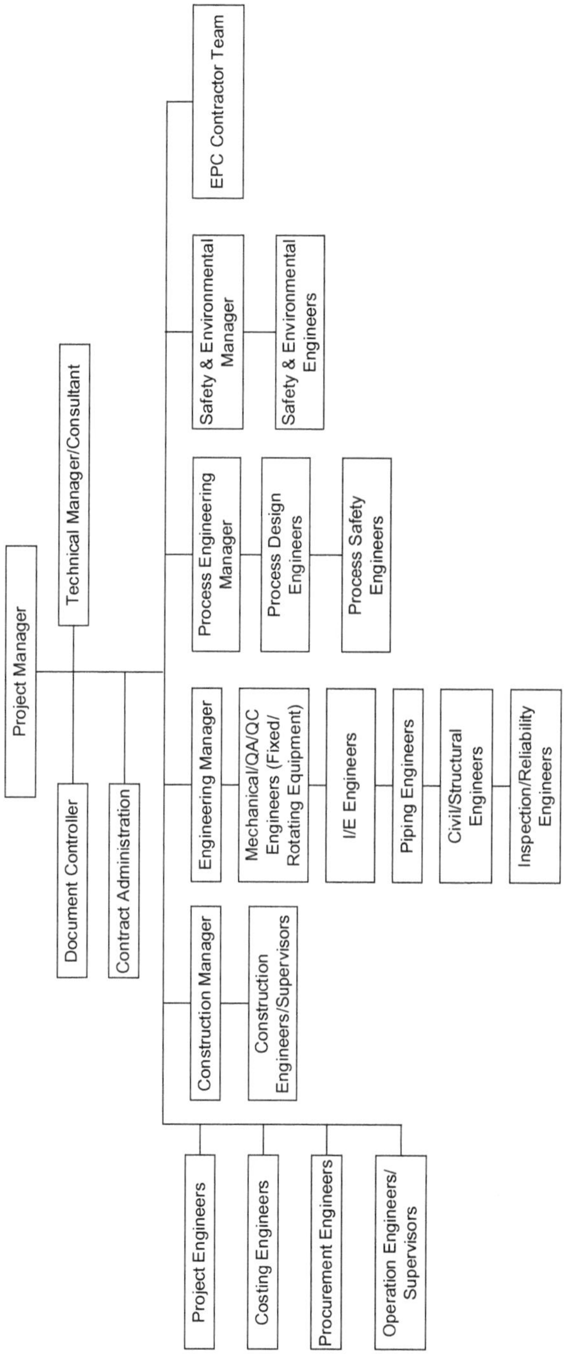

Fig. 19.14: Typical organization chart of a WHR project.

Once the project manpower resource plan for various project stages is finalized, it is documented in LACTI (lead, approve, consult, team member, inform) chart, such as that in Table 19.15. LACTI chart is used to assign specific roles to PMT members and also for project activities such as approval, consultation, review, follow-up and communication (Fig. 19.15).

Work breakdown structure (WBS) document is also generally used to assign specific project activity for each of the PMT members. A simplified WBS is shown in Fig. 19.16. For a bigger project, WBS can have few thousand items, for close monitoring of the project deliverables.

19.9.3 *Project Scope Management Plan*

Project scope management plan (PSMP) is very important for the successful execution of any project as it is the most influential factor affecting the project cost and schedule. Key components of PSMP[2] are Plan Scope Management, Collect Requirements, Define Scope, Create WBS, Validate Scope and Control Scope.

A major contributor to unsuccessful projects is poor understanding, definition, and management of the project scope.[29] PMBOK[2] defines scope as the sum of the products, services and results to be provided by the project. Failure to identify the correct number of new equipment in the project scoping will increase the project cost in detailed engineering stage and leads to budget overrun. If scoping is not performed properly, project scope can expand by 30%–70% as the project proceeds to the next stages.[23] Money spent in project scoping results in reduced project cost, schedule and risk.[30] Lagace[24] has presented a checklist for good scope definition. Pre-FEED typically costs less than 5% of the project cost even if it is performed by an EPC contractor.[25]

Proper scope development requires design and analysis of all alternatives for choosing the most attractive option. This approach can reduce capital cost by 5% to 10% and keep scope changes to no more than 1% of capital cost.[31] For refinery projects, scoping stages dictate nearly 80% of the capital cost and 98% of the operating cost,[15] and eventually the project success.

DELIVERABLE	Project Manager	Rotating Equipment Specialist	Technical Manager	Engg. Manager	Construction Manager	I/E Manager	HSE Manager/ Engineer	Operations Engineer	Process Engineer	Costing Engineer	Mechanical Engineer	Civil Engineer	I/E Engineer	Project Engineer	Contracts and Procurement Engineer
	John	David	Jimmy	Henry	Mike	Rich	Alan	Robert	Isaac	Selva	William	Tom	Ivan	Nick	Bob
Documentation	A	I	I	I	I	I	I	I	I	I	I	I	I	L	I
Project coordination procedure	A	I	T	I	I	I	I	C	C	I	C	C	C	L	C
Project monthly report	A	I	T	I	I	I	I	C	C	C	C	C	C	L	I
Project scope management	A	C	C	C	C	I	C	C	C	C	C	C	C	L	C
Project risk management	A	T	T	C	C	I	C	I	I	I	I	I	I	L	I
Cost estimation	A	I	C	C	C	I	C	C	C	L	C	C	C	C	C
Project schedule management	A	I	T	I	C	I	C	I	C	C	I	I	I	C	C
Process design	A	I	A	C	C	I	I	C	L	C	I	I	I	I	I
Mechanical design	A	C	C	A	C	I	I	I	I	C	L	C	C	C	I
Instrumentation design	A	C	C	C	C	A	I	C	C	C	C	I	L	C	I
Electrical system design	A	C	C	C	C	A	I	I	I	C	I	I	L	C	I
Civil design	A	C	C	A	C	I	I	I	I	C	I	L	I	I	I
Fire safety design	A	C	C	C	C	I	A	I	C	C	C	C	C	C	I
Award of contracts and procurement services	A	C	C	C	C	I	C	C	C	C	C	C	C	C	L
Permits	A	I	I	I	C	I	C	C	I	I	I	I	I	I	I
Develop & maintain the project quality assurance program	A	C	C	A	C	I	C	I	I		I	I	I	I	I
Perform acceptance testing	A	I	A	I	I	I	I	L	C	I	I	I	C	C	I
Prepare lessons learned	A	I	I	I	I	I	I	I	I	I	I	I	I	L	I

L - Lead - Person responsible for the deliverable and for consolidation of comments from all parties before issuing for approval.
A - Approve - person with appropriate authority to approve the document.
C - Consult - person that is critical to review the document and provide technical comments.
T - Team member - person assigned to assist with document development
I - Inform - Person that should be informed or document should be shared with.

Fig. 19.15: LACTI chart.

Project Title:		*New Dearerator Project*			
Prepared by:	*John*		*Date / Control Number:*	May-19	
Element Number	*WBS Elements: Activity, Task or Sub-Task Name*	*Definition of Activity or Task (Description)*	*Responsible Person or Group*	*Estimated (E) or Actual (A) Cost, S$*	*Project Phase*
1	**New deaerator Project**	**Entire Project**	**John**	**6,450,000**	
1.1	**Pre-FEED**	**Entire Pre-Feed**	**Jimmy**	**500,000**	**Pre-FEED**
1.1.1	Design basis	Deaerator & WHB specifications	Isaac	100000	Pre-FEED
1.1.2	Alternatives development	Generate various project options	Isaac	300000	Pre-FEED
1.1.3	Alternatives costing	Cost estimation	Selva	50000	Pre-FEED
1.1.4	Finalizing the best alternative	Final project option selection	Isaac	50000	Pre-FEED
1.1.5	Pre-FEED presentation for management approval	Pre-FEED approval	Nick		Pre-FEED
1.2	**FEED**	**Entire FEED**	**John**	**800,000**	**FEED**
1.2.1	Process	Process design	Isaac	150000	FEED
1.2.2	Mechanical	Mechanical design	Willam	200000	FEED
1.2.3	I/E	I/E design	Ivan	150000	FEED
1.2.4	Civil	Civil design	Tom	150000	FEED
1.2.5	Fire/safety	Fire/safety design	Alan	50000	FEED
1.2.6	HAZOP	HAZOP	Jimmy	50000	FEED
1.2.7	VIP studies	Design to capacity, design for reliability and energy optimization study	Jimmy	50000	FEED
1.2.8	EPC contractor quotations	EPC cost	Bob		FEED
1.2.9	FEED presentation for management approval	FEED approval	Nick		FEED
1.3	**EPC phase**	**EPC**	**John**	**5,150,000**	**EPC**
1.3.1	EPC contractor selection	Contractor selection	John		EPC
1.3.2	Detailed engineering	Design finalization	EPC contractor	700,000	EPC
1.3.3	Constructability review	Finalizing construction method	Mike	50,000	EPC
1.3.4	Procurement	Purchase of equipment	EPC contractor	4,150,000	EPC
1.3.5	Design verification	Process design	Isaac		EPC
1.3.6	Design verification	Mechanical design	Willam		EPC
1.3.7	Design verification	I/E design	Ivan		EPC
1.3.8	Design verification	Civil design	Tom		EPC
1.3.9	Design verification	Fire/safety design	Alan		EPC
1.3.10	Govt. Permits	Govt. Permits	EPC contractor	100,000	
1.4	**Pre start-up safety review**	**Preparation for start-up**	**Robert**		**Evaluate & Operate**
1.4.1	Pre-commissioning	Preparation for start-up	Robert	50,000	Evaluate & Operate
1.4.2	Commissioning	Preparation for start-up	Robert	100,000	Evaluate & Operate
1.4.3	Performance test	Start-up	Robert		Evaluate & Operate
1.4.4	Post evaluation report	Project performance and lessons learned report	Nick		Evaluate & Operate
1.5	**Project closure**	**Project closure**	**John**		**Project closure**

Fig 19.16: Work breakdown structure.

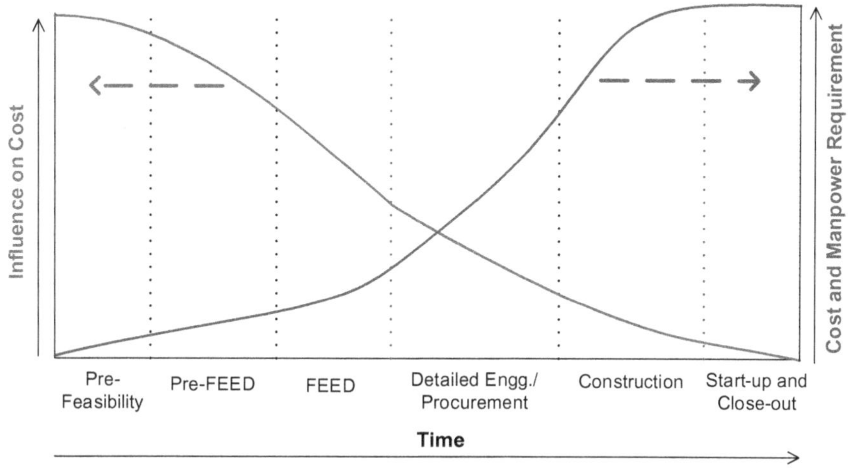

Fig. 19.17: Influence of various stages on project cost.

Figure 19.17 emphasizes the importance of project scoping decisions in early stages of the project as they decide most of the project cost, dictate the economic profitability of the project and also have long-term effects on operations and maintenance costs.[25,32] Further, Fig. 19.17 indicates that the proportion of cost/manpower required for proper scope development is very little compared to that required for other stages of the project.

Improper project scope leads to scope creep and may result in project failure. It can have a cascading effect on many project deliverables and can lead to failure of project objectives, cost overrun, poor project performance, stakeholder/customer dissatisfaction and/or loss of reputation. Main reasons for project scope creep are given as follows:

- Insufficient allocation of resources and funding in the scoping phases.
- Lack of good communication between stakeholders and PMT.
- Poorly documented technical specifications.
- No formal risk management process.
- Poorly defined initial requirements and allowing requirement changes later.
- Poor understanding of stakeholders' requirements prior to scope definition and/or contract signing.

- Project deliverables are not fully defined upfront.
- No formal review and approval process for project scope changes. No proper change management.
- Improper or no work breakdown structure.
- Lack of proper project roles and responsibilities.
- Misinterpretation of what is contained in the scope, contract or Statement of Work.
- Business environment changes such as management change.
- Willingness to accept whatever the client/stakeholder says.
- Thinking that one little change won't matter.
- Allowing staff, who don't execute the work, to accept the changes.

Use of VIPs for developing the project scope, reduces later changes in the project scope, controls the project cost and also significantly improves the profitability of the project.[4] Poor control of project scope changes is one important factor for escalating costs and hence its failure.[33] Once the project scope is fixed, the project team should minimize further changes. Any project scope changes should be carefully reviewed and approved/rejected by the responsible project engineers/managers. Typical control of project scope change is illustrated in Fig. 19.18.

19.9.4 *Stakeholders Management*

For development of project scope and executing a WHR project, it is very important to identify all stakeholders for the project at the beginning stage itself. Stakeholders have a key role in the project management as they positively or negatively impact the success of the project. They can be internal and/or external. Internal stakeholders include PMT members and other key company employees, who can affect the key decisions in the project. External stakeholders are those who are not directly involved in the project but can have significant influence on the project activities through many ways. Examples of external stakeholders are government agencies, surrounding communities, neighbouring companies and political groups.

Project scope should be finalized by validating the scope developed by the project team, with all stakeholders. For this, the project manager

No.	Decision Date	Owner	Decision Makers	Change Request	Category	Action Plan	Comments	Decision Basis	Cost Impact	Final Decision	Project Sponsor Endorsement	Key Stakeholder Endorsement
1	15-Jan-20	John	Edwards	Increase residence time in deaerator BFW storage vessel from 10 to 20 minutes	Operations Management	Project team to change the deaerator specifications in the terms of deaerator BFW storage vessel residence time	Operations management requested for higher residence time in deaerator BFW storage vessel	Higher residence time ensures reliable plant operation	+US$500,000	Recommended	Endorsed	Daniel (ABC), Kiran (PQR), Smith (XYZ)
2	8-Feb-20	John	Edwards	Remove one BFW pumps and associated piping and I/E works by re-using existing pumps	Scope Optimization	Project will need to install one pipeline to connect to the existing pumps' suction pipe	Technical specialist has verified this value engineering proposal and recommended it.	To avoid addition of one new BFW pump and maximize usage of existing pumps	-US$140,000	Recommended	Endorsed	Daniel (ABC), Kiran (PQR), Smith (XYZ)
3	9-Jul-20	John	John	Provide additional staircase for the new deaerator	Safety	Project team to install one additional access staircase to deaerator	Operator safety	Pre-start-up safety review recommendation	+US$$10,000	Recommended	Endorsed	Daniel (ABC), Kiran (PQR), Smith (XYZ)

Fig. 19.18: Project scope control document.

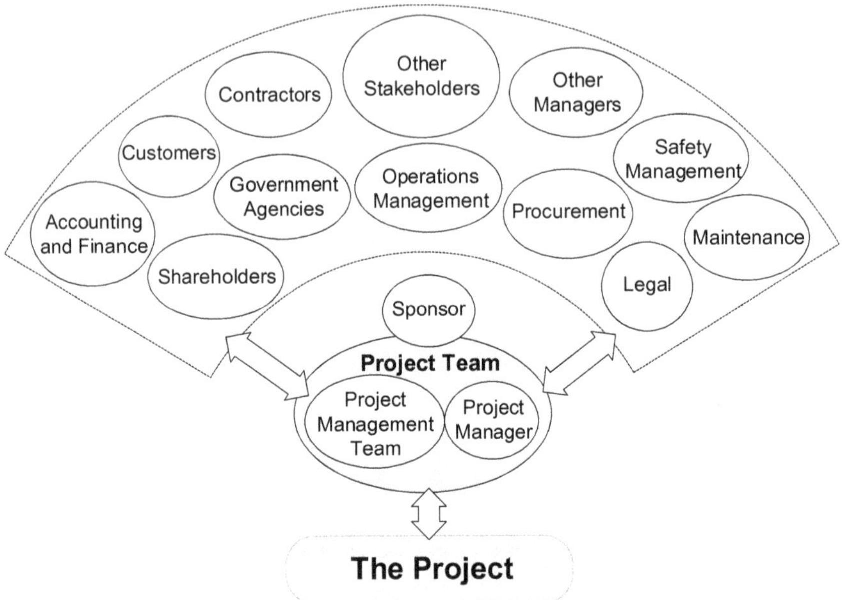

Fig. 19.19: Typical project stakeholders.

(PM) should identify all stakeholders such as those shown in Fig. 19.19, and then analyse to identify their requirements. The next step is to engage stakeholders and get their inputs on the project. Once agreements are reached, PM should follow them. Project progress along with any deviations from the agreements should be communicated to the stakeholders at least once a month. During project execution, stakeholders may change. In such a case, PM should identify the new stakeholder and repeat the whole process. Typical stakeholder relationship management is depicted in Fig. 19.20.

19.9.5 *Project Cost Management*

A cost management plan can be used to establish how staff and resources will be measured in monetary terms, level of cost estimation accuracy (required for various project phases), links with WBS, cost variance thresholds and also rules for cost performance measurement.[2] Typical cost estimation accuracy at various project phases, used in process plants is

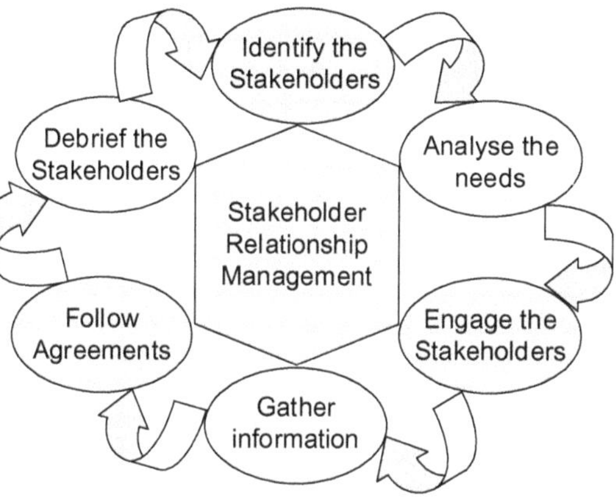

Fig. 19.20: Stakeholder relationship management.

shown in Table 19.7 given earlier. It also includes other cost considerations such as contingencies (e.g., design developments and minor variations within the scope, minor price fluctuations, changes in market and environmental conditions, and errors in project planning and cost estimations)[22]; and project allowances (e.g., allowance for changes in material selection, allowance for undefined items and anticipated but unknown full work scope).

Project cost can be effectively monitored using spreadsheets, with graphs such as project S curves and planned versus actual costs. Value analysis and cost–benefit analysis are useful to efficiently manage the project cost. A good pre-FEED includes detailed technical and economic analysis of various alternative WHR options and the selected best WHR option. This reduces capital cost by 5% to 10% and keeps scope changes to less than 1% of capital cost.[31] Further, FEED study to fully develop the selected WHR option using VIPs, defines the project scope firmly and hence helps to control the project cost. Use of experienced engineers, managers and contract management staff, and good teamwork culture are vital for good cost management. Project scope change control is a key factor for preventing cost overruns and project failures.[33] Reasons for cost overrun are shown in Fig. 19.21. Some of these are related to project scope creep given earlier in Section 19.9.3. Cognition of many causes in Fig. 19.21 and Section 19.9.3, and a good cost management plan, are essential to avoid cost overrun.

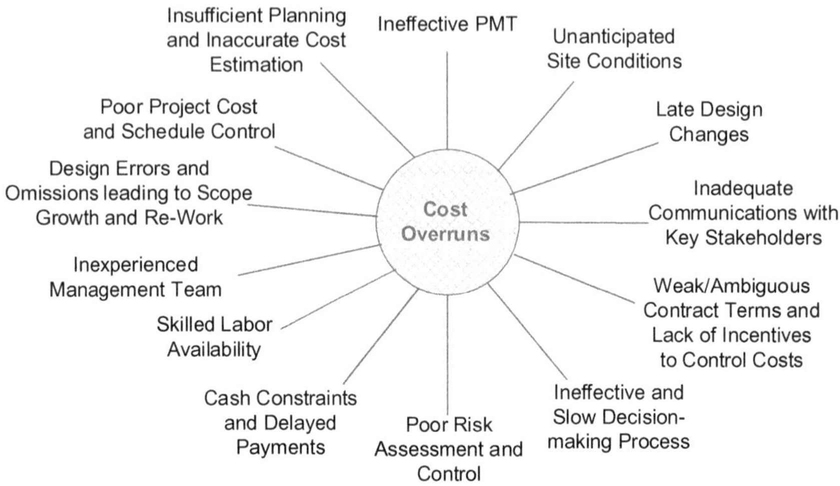

Fig. 19.21: Causes of cost overrun.

19.10 Summary

This chapter presented a holistic methodology for developing and executing WHR projects, in detail. Essential ingredients for successful WHR projects are follows.

- Identify WH source, WHR method and intended use of WH.
- For brownfield/revamp projects, collect and validate the data with plant performance tests and/or process simulations. For greenfield projects, this step is not required.
- Develop realistic WHR project scope by performing the pre-feasibility study.
- Study various WHR and utilization options including VIPs, in the pre-FEED stage, for final selection of best WHR and utilization option.
- Develop complete project scope by performing comprehensive FEED study for establishing project costing (±10% accuracy) of the selected option for WHR.
- Follow scope, resource, project, risk and cost control/management plans for effective project execution and control.
- Perform thorough pre-commissioning and commissioning activities for smooth project start-up. Evaluate the project success by conducting a plant performance test.

- Prepare and share the post-project evaluation including project performance, lessons learned and future revamp options.
- Verify the accuracy and completeness of the entire project documents and save them for future reference.

PMBOK guide[2] is a holistic book on project management. It is recommended for further reading on the topics of project management. Also, the book by Tayntor[34] is recommended for deeper understanding of project management concepts and applications.

References

1. Elshout R, Gracia D. (July 2009) Revamps: Strategies for a smooth turnaround. *Chem Eng* 34–39.
2. PMBOK Guide. (2013) *A Guide to the Project Management Body of Knowledge (PMBOK guide)*, 5th ed. Project Management Institute.
3. Reddy CCS. (2016) Project engineering and management for process retrofitting and revamping. In: GP Rangaiah (ed), *Chemical Process Retrofitting and Revamping*. Wiley.
4. Green DW, Perry RH. (2008) *Perry's Chemical Engineers' Handbook*, 8th ed. McGraw-Hill.
5. Litzen DB, Bravo JL. (March 1999) Uncover low-cost debottlenecking opportunities. *Chem Eng Prog* 25–32.
6. Martin GR, Cheatham BE. (Summer 1999) Keeping Down the Cost of Revamp Investment. PTQ, pp. 99–107.
7. Higley T. (2015) Enhanced Heat Transfer for Improved Throughput and Energy Consumption. PTQ, pp. 45–49, Q4.
8. Bhargava M, Sharma AP. (2020) Dividing Wall Technology in Distillation Columns. PTQ, pp. 41–45, Q1.
9. Smith R. (2005) *Chemical Process Design and Integration*. Wiley.
10. Kemp IC. (2007) *Pinch Analysis and Process Integration*, 2nd ed. Elsevier Butterworth-Heinemann.
11. Milosevic M, Rudman A, Brown R. (June 2013) Are you using pinch technology effectively in your daily operations: Parts 1 and 2. *Hydrocarb Process* 99–102 and pp. 77–81.
12. Moorthy AN, Reddy CCS, Rangaiah GP. (2014) Optimization of heat exchanger networks for the utilization of low-temperature process heat. *Ind Eng Chem Res* 53: 17989–18004.

13. Stankiewicz AI, Moulijn JA. (January 2000) Process Intensification: Transforming Chemical Engineering. *Chem Eng Prog* 96: 22–34.
14. Stankiewicz AI, Moulijn JA. (2004) *Re-Engineering the Chemical Processing Plant: Process Intensification.* Marcel Dekker.
15. Mendez M. (December 2007) The rate of chemical engineers in green engineering, what we can do to support its goals. *Chem Eng* 41–44.
16. Contreras CD, Bravo F. (August 2011) Practice green chemical engineering. *Chem Eng* 41–44.
17. Koolen JLA. (2002) *Design of Simple and Robust Process Plants.* Wiley-VCH Verlag GMBH.
18. CCPS. (1993) *Guidelines for Engineering Design for Process Safety.* AIChE, New York.
19. Zhu F. (2014) *Energy and Process Optimization for the Process Industries.* Wiley.
20. Reddy CCS, Naidu SV, Rangaiah GP. (December 2013) Optimization of a steam network. *Chem Eng* 54–59.
21. Feng Y, Rangaiah GP. (August 2011) Evaluating capital cost estimation programs. *Chem Eng* 22–29.
22. Hastak M. (2015) *Skills & Knowledge of Cost Engineering*, 6th ed. AACE International.
23. Barletta T, Martin GR, Visser A. (June 2002) Consider comprehensive CPD efforts to cut costs. *Hydrocarb Process* 53–62.
24. Lagace Jr. JC. (February 2006) Project success builds on a well-defined scope. *Chem Eng* 36–38.
25. Golden SW, Moore T, Nigg JM. (September 2003) Optimize revamp projects with a logic based approach. *Hydrocarb Process* 75–83.
26. Farrow G. (2013) Laser Surveying a Revamp. PTQ Revamps, pp. 29–31.
27. Sikos L, Klemes J. (2010) Reliability, availability and maintenance optimization of heat exchanger networks. *Appl Therm Eng* 30: 63–69.
28. Harold K. (2017) *Project Management Metrics, KPIs, and Dashboards: A Guide to Measuring and Monitoring Project Performance.* Wiley.
29. Mirza MN, Pourzolfaghar Z, Shahnazari M. (2013) Significance of Scope in Project Success. Procedia Technol 9: 722–729.
30. Gebert T, Villarreal R, Whitsura FR. (2004) Revamping Automation Systems. PTQ Revamps & Operations, pp. 9–12.
31. Brown T, Singh S. (July 2014) Project optimization through engineering. *Chem Eng* 51–58.
32. Dipaolo CR. (April 2007) Target costing. *Chem Eng* 66–69.
33. Harding J. (December 2012) Avoiding project failures. *Chem Eng* 51–54.
34. Tayntor CB. (2010) *Project Management Tools and Techniques for Success.* CRC Press.

Acronyms

APH	Air Pre-Heater
DCS	Distributed Control System
DP	Differential Pressure/Pressure Drop
EPC	Engineering, Procurement and Construction
FAT	Factory Acceptance Test
FEED	Front End Engineering Design
FS	Flash Steam
HAZOP	Hazard and Operability
HE	Heat Exchanger
I/E	Instrumentation and Electrical
IFC	Issued for Construction
I/O	Input/Output
IRR	Internal Rate of Return
KC	Kalina Cycle
LACTI	Lead, Approve, Consult, Team Member, Inform
LOPA	Layers of Protection Analysis
LPS	Low-Pressure Steam
MDT	Mean Down Time
MTBF	Mean Time between Failures
MTTF	Mean Time to Failure
MTTR	Mean Time to Repair
NPV	Net Present Value
ORC	Organic Rankine Cycle
OSBL	Outside Battery Limit
PFD	Process Flow Diagram
PHA	Process Hazard Analysis
P&ID	Piping and Instrumentation Diagram
PLC	Programmable Logic Controller
PMT	Project Management Team
PSI	Process Safety Information
PSMP	Project Scope Management Plan
PSSR	Pre-Start-Up Safety Review
RAM	Reliability, Availability, and Maintainability
RAMS	Reliability, Availability, Maintainability, and Safety
RFQ	Request for Quotation

SIS	Safety Instrumented System
TAR	Turnaround
VFD	Variable Frequency Drive
VIP	Value Improvement Practice
WBS	Breakdown Structure
WH	Waste Heat
WHB	Waste Heat Boiler
WHR	Waste Heat Recovery

Exercises

19.1 WHR survey at a plant identifies an opportunity for recovering 5 ton/hr of saturated steam at 2.2 bar. This plant purchases LPS (at 4 bar) and medium-pressure steam (at 11 bar) as well as electricity from an external utility company, with purchase costs of US$ 15/ton, US$ 20/ton and US$ 0.10/kWh, respectively. Presently the plant utilizes 20 ton/hr of LPS, out of which a reboiler uses 3 ton/hr of LPS (by reducing its pressure to 2 bar). Assume there are no space constraints for WHR project. Using flash steam direct usage/upgrade options presented in Chapter 16 and cost estimation techniques in Chapter 6, generate WHR options and suggest the best WHR option.

19.2 If there are space constraints, how will the selected WHR option in Exercise 19.1 change? Give valid reasons for the new WHR option.

19.3 Identify VIPs for pre-FEED study of the selected WHR option in Exercise 19.1.

19.4 Summarize the good practices for successful development and implementation of a WHR project.

19.5 Briefly discuss the importance of hydraulic study for a heat integration project, involving installation of four new shell and tube heat exchangers.

Chapter 20

Process Safety in Waste Heat Recovery Projects

20.1 Overview

Process safety is crucial for the entire life cycle of WHR project development, execution, and operation phases. Also, sustainability of a WHR project is strongly dependant on process safety. If proper safety considerations are not considered in the design stage, WHR projects can pose process safety threats and result in project failures. Lack of proper process safety hazard assessment and mitigation right from the design stage and all the way up to project operation, are the main reasons for many severe safety incidents.

Although process safety incidents occur less frequently, they can cause huge consequences. They can lead to great negative impact on stakeholders, employees, and society with the loss of property, revenue, lives and environmental damage. A company can lose its reputation; additionally, it may be subjected to huge penalties from government agencies. Safe and sustained project operation will boost the company image, morale of the employees and encourage shareholders for further investments in WHR projects. This chapter discusses the main process safety techniques and methodologies applicable for development and sustained operation of WHR projects.

First, the key lessons from process safety incidents related to some past heat exchanger (HE) failures, which are important for WHR projects, are highlighted in Section 20.2. Next, Section 20.3 discusses the process safety reviews applicable to WHR projects. Section 20.4 presents the risk assessment methodology, risk mitigation hazard categories and risk matrix. Preliminary hazard reviews during conceptual design, namely, inherently safer design (ISD), what-if (WI) study, plot plan review, hazardous area classification, relief systems review, and fire safety consideration reviews are presented in Section 20.5.The subsequent section presents other process hazard analysis (PHA) procedures such as hazard and operability (HAZOP) study, failure modes and effects analysis (FMEA), instrumented protection system (IPS), fault tree analysis (FTA), event tree analysis (ETA) and layers of protection analysis (LOPA). Section 20.7 discusses the life cycle management of a safety instrumented system (SIS). Importance of process safety information (PSI) is highlighted in Section 20.8. Management of change (MOC) and pre-start-up safety review (PSSR) procedures are described in Sections 20.9 and 20.10, respectively. Finally, this chapter ends with summary in Section 20.11.

Learning outcomes of this chapter on process safety in WHR projects are the following.

- Explain the importance of process safety aspects for reducing and/or managing the risk.
- Outline the key process safety aspects for developing sustainable WHR projects.
- Discuss the importance of ISD in developing sustainable WHR projects.
- Identify the key methods for risk assessment and risk mitigation using as low as reasonably practicable (ALARP) concepts.
- Describe PHA methods (e.g., WI, FMEA, HAZOP, FTA, ETA, and LOPA).
- Identify the key requirements for SIS design and life cycle.
- Discuss the importance of PSI, MOC and PSSR in process safety management.

20.2 Lessons from Past Process Safety Incidents

Lessons learnt from the past process safety incidents are very valuable to recognize the importance of process safety aspects during the entire life cycle of the WHR project. They will help design engineers and operators to adapt and implement key process safety features. According to Crowl and Louvar,[1] piping system failures account for the largest number of industrial accidents, followed by reactors and storage tanks. Based on the analysis of 100 industrial accidents resulting in largest losses during 1972–2001, they have reported the main contributing causes as mechanical failures (53%), operator errors (18%) and design issues (10%).

An HE explosion at Good Year Tire and Rubber Company, Houston, Texas, on 11 June, 2008, revealed several gaps in process safety. As per the report of U.S. Chemical Safety Board (CSB),[a] isolation valve (V-1) before the rupture disk (RD, located on the shell side, at top location of the HE in Fig. 20.1) was closed for RD replacement. However, it was not opened back after RD replacement (due to human error). Next day, valve (V-2) before pressure control valve (PCV) was closed (contrary to operating procedure) while steam was introduced to clean the reactor charge pipe, through HE tubes in Fig. 20.1. Due to thermal energy from steam, isolated liquid ammonia on the shell side of the HE was heated and resulted in overpressure. As both the valves at the inlet of RD/pressure

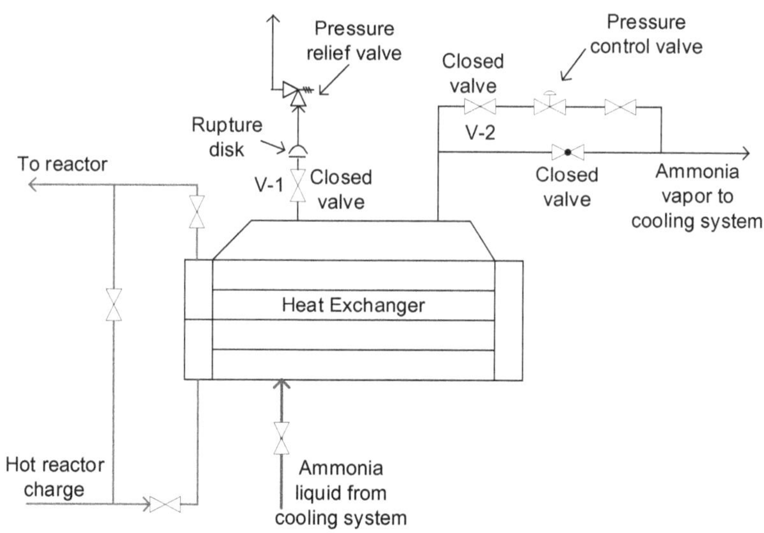

Fig. 20.1: Process flow diagram of ammonia heat exchanger based on details and diagram in CSB report.

[a] https://www.csb.gov/goodyear-heat-exchanger-rupture/.

Fig. 20.2: Ammonia heat exchanger after rupture (https://www.csb.gov/goodyear-heat-exchanger-rupture/).

relief valve (PRV) and pressure control valves were closed,[b] overpressure on the shell side caused the HE to rupture (Fig. 20.2), killing an employee of the company.

In this case, there was no problem with the design of the overpressure protection system of HE. As per American Society of Mechanical Engineers (ASME) Boiler and Pressure Vessel Code, overpressure protection shall be continuously provided for the pressure vessel (HE in this case) if there is any chance of overpressure by external heating (steam heating in this case) and/or external mechanical pressure. However, the correct procedure was not followed during the *maintenance work* (i.e., to check and ensure that V-1 is open before steam heating, and also during steam heating to clean reactor charge pipe), which resulted in fatality and HE rupture damage.

CSB[c] reported a brazed aluminium HE failure in 2016, due to thermal fatigue,[d] which resulted in huge financial loss of US$10.4 million for fire

---

[b] The valves at inlet of RD/PRV and PCV will be in open position during normal operation. However, during the time of accident occurrence, they were kept in closed position leading to overpressure and HE rupture.

[c] https://www.csb.gov/assets/1/6/final_case_study_-_enterprise.pdf

[d] Thermal fatigue is mainly caused by increased stress created by different expansion and contraction of HE parts

response activities and another US$ 7.1million non-cash loss. It also reported an HE failure due to high temperature hydrogen attack[e] at Tesoro's refinery[f] on 2 April, 2010, which resulted in huge damage and death of seven employees.

CCPS[2] has presented a good summary and analysis of many industrial accidents. From these incidents, we can learn an important lesson to understand the importance of safety layers and maintain them as per design intent, during the whole life cycle of the plant/equipment.

20.3 Process Safety Reviews for WHR projects

Process safety studies/reviews in Table 20.1 are recommended during the WHR project execution stages. In particular, for sustainable benefits and safety performance, it is highly recommended to adopt ISD principles in the WHR project design right from the conceptual design phase to project

Table 20.1: Recommended Process Safety Studies during WHR Project Stages

WHR project stage	Process safety study or review
Pre-feasibility study	ISD review and WI
Pre-FEED or conceptual study	i. ISD review ii. WI iii. Plot plan and layout review iv. Area classification review v. Pressure relief system considerations vi. Fire safety design considerations
Front-end engineering design (FEED)	i. HAZOP ii. FMEA iii. FTA iv. ETA v. LOPA vi. SIS
Engineering, procurement and construction (EPC)	i. SIS detailed design, validation ii. MOC iii. PSSR

[e] This is a damage mechanism in carbon steel in hydrogen service at high temperatures and pressures, which causes fissures and cracking.

[f] http://www.csb.gov/assets/1/7/tesoro_anacortes_2014-may-01.pdf

operation. A complete set of process safety information (PSI) should be available before any process safety study/review. Some of the studies in Table 20.1 are elaborated later in this chapter.

20.4 Risk Assessment and Risk Mitigation

All WHR project activities (like any plant activities) involve risks. It is neither practical nor economical to remove all risks. However, they should be reduced to an acceptable level with necessary design and/or process technology changes. Typical risk assessment methodology is shown in Fig. 20.3. It begins with the project's process description and identifying the hazards by hazard assessment tools such as WI, HAZOP, LOPA and FMEA. For this, hazard event likelihood/frequency and consequence severity are assessed. Risk is estimated as the product of severity and likelihood/frequency. If the risk is acceptable or in as low as reasonably practicable (ALARP) region, the project can be implemented without any changes. In the ALARP region, which lies between intolerable and tolerable risk levels (RL; Fig. 20.4), resources required become grossly disproportionate to the additional risk reduction obtained (as illustrated in Fig. 20.5). Once ALARP is demonstrated, the risk may be acceptable.

If the risk is totally intolerable, the project may need to be stopped or redesigned using more safety protection layers or using a less hazardous technology. With the addition of sufficient safety design layers or using a less hazardous technology, risk can be reduced to ALARP region. Effective safety protection layers with low cost should always be incorporated for risk reduction, irrespective of their estimated frequencies. Below the ALARP region, residual risk is tolerable and hence further risk reduction may not be required.

The hazard review leader decides the hazard assessment method, documentation requirements and team selection. After brainstorming, the hazard review team evaluates the process risks. If the risk is not acceptable, appropriate risk mitigation recommendations are made. If the risks are significant, quantitative risk assessment (QRA) tools such as LOPA, ETA and FTA are used to evaluate whether the recommended risk mitigation measures are adequate to bring the risk to an acceptable level set by the corporate/local authority, or relevant codes, whichever is more

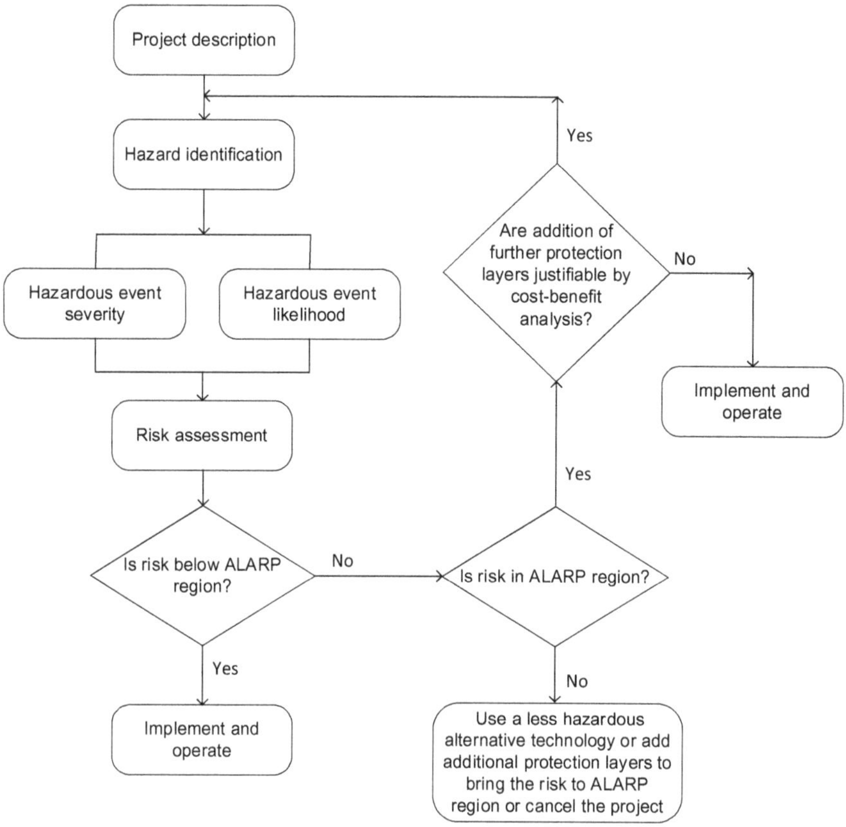

Fig. 20.3: Flowchart of a typical risk assessment methodology.

stringent. SIS is frequently used if the risk is not tolerable even after incorporating economical design of mitigation measures such as ISD practices, basic process control systems (BPCS), pressure relief systems and administrative layers. SIS design satisfies the Safety Integrity Level (SIL) or the maximum allowable average probability of failure on demand (PFD_{avg})[g] to bring the overall risk to well within the target. Detailed explanation of SIS and SIL concepts are given in Section 20.6.3. SIS design is performed and verified during EPC phase.

---

[g] PFD_{avg} is the average probability of failure of a SIS to perform its safety function when required.

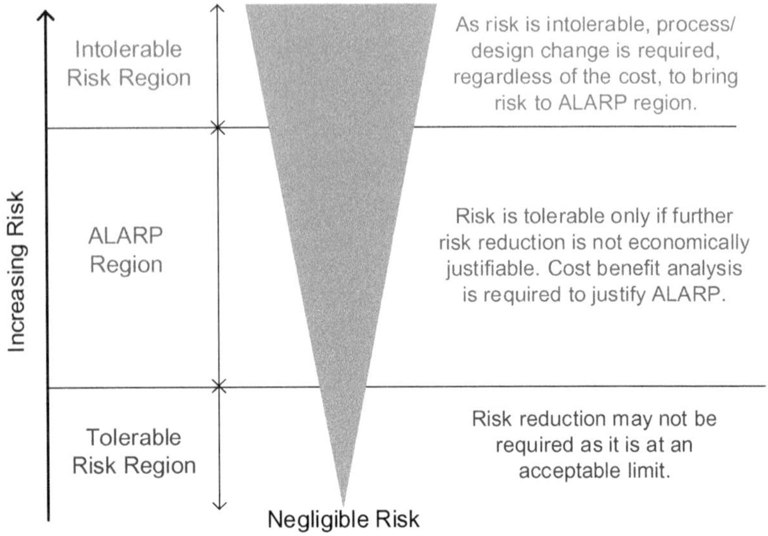

Fig. 20.4: ALARP principle.

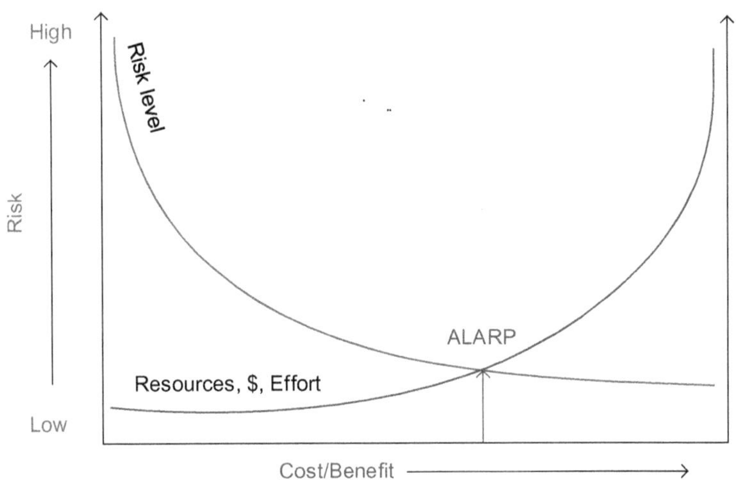

Fig. 20.5: ALARP illustration.[h]

[h] With quantitative cost–benefit analysis, ALARP can be represented by a point where cost is equal to benefit.

Safety walks and PSSR are performed prior to the commissioning and start-up of the WHR project to ensure all design- and safety-related actions are completed, documented and communicated. Any critical punch list items identified during safety walks and/or PSSR requires immediate rectification. One important but often ignored activity in WHR projects is PSI updates, particularly, revision of standard operating procedures and training the operators on the revised procedures. MOC is an essential safety work process to be followed diligently in all project execution phases, as every overlooked or new change in design has some safety implications. Detailed explanation for MOC is provided in Section 20.9.

Major hazard categories (i.e., fire, explosion and toxic release) and their sub-categories are listed in Table 20.2. This information is very useful for developing hazard scenarios and during hazard analysis. The hazard analysis study report is very useful not only for developing and implementing safer projects but also for getting project permits from local authorities.

Risk Matrix: This is a qualitative tool, mostly used in process industries for quick risk assessment. It lists the probability/likelihood/frequency and consequence/severity of hazards in a matrix form. Risk matrix size and categories vary (from 3 to 6) depending on industry nature and risk management practices. Based on the likelihood and consequence, which are often judged by the hazard review team using process technology, operations and/or industry experience, overall RL can be established from the matrix. In the typical 4 × 4 risk matrix in Table 20.3, four levels of probabilities (likelihood) and four severity levels are categorized into five RL). H3 risk should be addressed quickly as it is the most serious RL. Risk mitigation follows the order H2, H1M and L, with decreasing order of priority. Events with severity category 1 and likelihood category N (i.e., last row in Table 20.3) should be analysed with QRA methods, which are beyond the scope of this chapter. Interested readers can refer to the book by CCPS.[3] Apart from RL, risk matrix can also be utilized to determine SIL requirement for a safety instrumented function (SIF).[i] Likelihood and consequence of typical process plant accidents are listed in Table 20.4.

[i] SIL and SIF are explained in detail in Section 20.6.3.

Table 20.2: Details of Hazard Categories

Hazard		
Category	**Sub-Category**	**Cause of Hazard**
Fire	Pool fire	Ignition of spilled flammable liquid on the ground or water
	Flash fire	Ignition of flammable vapour/gas cloud in air, without any overpressure. If vapour/gas is not ignited immediately after release, it can disperse and ignite after finding an ignition source and result in flash fire. Flame front can travel back through the vapour/gas cloud to the release location.
	Jet fire	Ignition of pressurized flammable fluid close to the release point. It results in very high thermal radiation that could be hazardous to life and property, away from the flame's visible boundary.
Explosion	Boiling liquid expanding vapour explosion (BLEVE)	An explosion that occurs during failure of a pressure vessel, containing a liquid at a temperature significantly above its boiling point and at atmospheric pressure. It releases large quantities of vapour and expanding liquid, which ignite immediately. One example is propane vessel bursting under fire attack.
	Vapour cloud explosion	An explosion resulting from ignition of a flammable gas, vapour or mist in air, when flame front travels through the flammable mixture at sufficiently high speed due to confinement and/or a high degree of turbulence, to cause overpressure. It can cause severe damage.
	Pressure vessel burst	Uncontrollable pressure build-up above test pressure and/ or mechanical failure of a pressure vessel.
Toxic	Toxic gas cloud	Leak/failure of equipment containing toxic fluid. It can cause injuries or fatalities if released toxic substance concentration is above safe levels.

20.5 Preliminary Hazard Review during Pre-FEED or Conceptual Design

This hazard review focuses on identifying, assessing and controlling process hazards. This helps WHR project team to understand major on/off-site risks, potential solutions, and cost implications. Hazard reviews in

Table 20.3: Typical 4 × 4 Risk Matrix for Qualitative Risk Judgement

Likelihood \ Severity	Multiple fatalities, facility damage >1 million US$, significant long-term environmental impacts and community evacuation, shelter in place (Level 1)	Multiple serious injuries to employees, facility damage of 0.5 to 1 million US$, significant short-term environmental impacts, some minor injuries to community (Level 2)	Serious Injury to employees, facility damage of 0.05 to 0.5 million US$, serious environmental impact, community complaints such as smell, smoke and noise (Level 3)	Minor Injury to employees/contractors, business reputation issues, minor environmental Impact but no community issues (Level 4)
HIGH (H) "Possible" (more frequent than 1 incident every 100 years)	H3[a]	H2[b]	H1[c]	L[e]
MEDIUM (M) "Occasional" (1 incident every 100 to 10,000 years)	H2[b]	H1[c]	M[d]	L[e]
LOW (L) "Seldom" (1 incident every 10,000 to 1,000,000 years)	H1[c]	M[d]	L[e]	L[e]
NIL (N) "Unlikely" (less frequent than 1 incident every 1,000,000 years)	Perform QRA	L[e]	L[e]	L[e]

[a] One SIL3 or one SIL2 + one SIL1 or two SIL2 SIFs would be sufficient for risk reduction.
[b] One SIL2 SIF would be sufficient for risk reduction.
[c] One SIL1 SIF would be sufficient for risk reduction.
[d] This indicates ALARP region. SIL SIFs are not generally required unless there is a justification to do so.
[e] SIL SIFs are not required.

Table 20.4: Likelihood and Consequence of Typical Process Plant Accidents

Type of Accident	Likelihood	Potential for Fatalities	Potential for Asset Damage
Fire	High	Low	Intermediate
Explosion	Low to intermediate	High	High
Toxic release	Low	High	Low

feasibility and Pre-FEED stages focus to evaluate the major hazards with process technology and/or preliminary process design.

Two types of barriers or protection layers, namely, protective and reactive/mitigative type, are used in process design to address the hazards, as illustrated in Fig. 20.6. Protective layers prevent the hazard, whereas reactive layers mitigate the hazard effects after it happened. Hence, in the conceptual design, more emphasis shall be given for protective layers. Each of these protective and reactive layers/barriers may have some weaknesses or 'holes'. If these holes in all layers are aligned, an unsafe

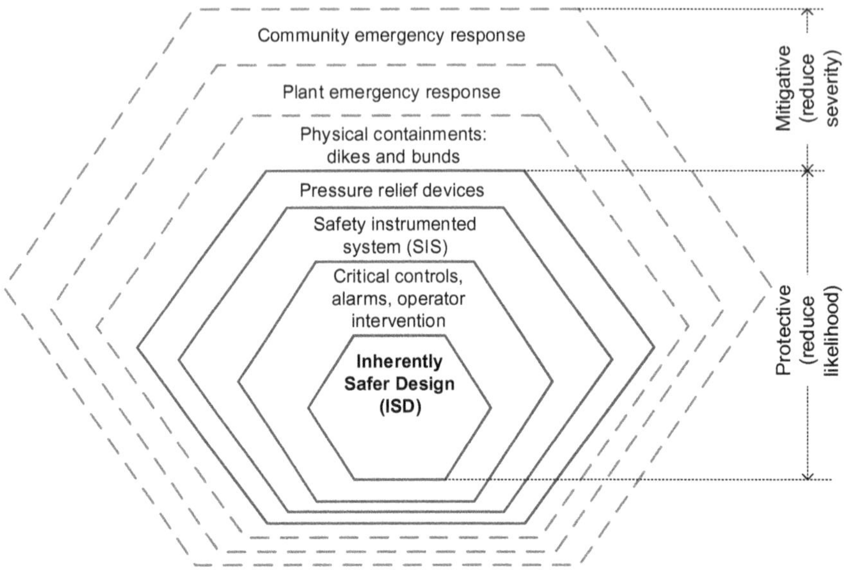

Fig. 20.6: Layers of protection for a process plant.

event can propagate through them and an accident can happen as shown in Fig. 20.7. Various hazard reviews/studies typically conducted in pre-FEED design stage are elaborated in the following sub-sections.

20.5.1 *Inherently Safer Design Review*

At the conceptual design stage of a new WHR project, there is the greatest opportunity to achieve the lowest residual risk by proper evaluation of possible design options. By choosing the less hazardous technology/ design, the process can be made inherently safer. This avoids the need for dependence on safety protection layers such as safety valves, SIS and controls. ISD is a fundamentally different way of thinking about the design of chemical processes and plants.[4] It is a philosophy for addressing safety issues in the design and operation of facilities that use or process hazardous chemicals.[5] Many ISD design ideas are presented by Kletz and Amyotte.[6]

Conventional process design uses a reactive approach in dealing with hazards (i.e., it accepts the existence and magnitude of hazards in a

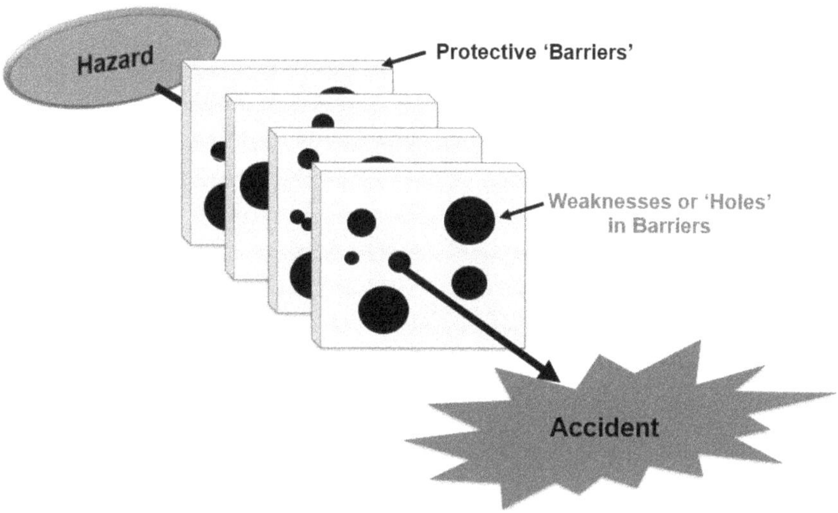

Fig. 20.7: Swiss cheese model on how the hazard penetrates through holes of different barriers.

process, and incorporates engineering and administrative controls to reduce process risks by using safety protection layers). On the other hand, ISD provides a proactive, more robust and reliable risk reduction approach that addresses safety, health, environmental and loss prevention issues in a strategic and integrated manner by dealing with the hazards at source. It has the potential to make the process design simpler and more economical by eliminating the need for expensive safety systems and/or procedures.

Projects that are more tolerant of failures and/or human errors have lower consequences, if any failure occurs. Process safety strategies for risk reduction are based on reducing frequency and/or consequence severity of potential unsafe incidents or accidents. They are classified into four categories: inherent, passive, active and procedural,[5,7] which are outlined below. Their assurance and reliability are shown in Fig. 20.8. Design solutions for some of the WHR equipment using the following strategies are summarized in Section 18.5 of Chapter 18.

Inherent Strategies: These eliminate or greatly reduce a hazard by using equipment, materials and/or process conditions, which are much less

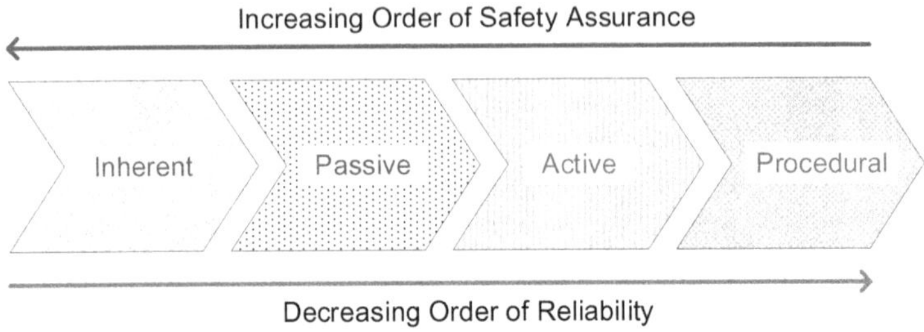

Fig. 20.8: Process safety strategies and their assurance and reliability.

hazardous, or are non-hazardous (e.g., use of water instead of a flammable and/or toxic compound in WHR application).

Passive Strategies: These minimize a hazard by equipment design features such as engineering aspects, without the need to depend on active functioning of any safety device or protection layers. Passive methods can reduce frequency and/or consequence of a hazard; for example, use of higher design pressure for equipment can eliminate the need for non-fire case pressure relief requirements specified in API Standard 521.[8]

Active Strategies: These mainly use SIS or engineering controls such as distributed control system (DCS), safety interlocks and emergency shutdown systems, to detect any unsafe conditions and automatically apply corrective actions (e.g., a pump that is shut off by a low-low pressure switch in the suction piping).

Procedural Strategies: These utilize mainly administrative control procedures such as operating and maintenance procedures, shutdown procedures, firefighting procedures, safety rules, administrative checks, emergency response and other management approaches, to minimize/prevent accidents and/or to minimize the consequence of the accident.

CCPS[7] has categorized ISD strategies into four groups: Substitution, Minimization, Moderation and Simplification. Typical techniques/examples for these groups are presented in Table 20.5.

Example 20.1: A steam-heated reboiler tube rupture leads to excessive heat input to distillation column and hence causes overpressure hazard to

Table 20.5: ISD Groups with Typical Techniques/Examples

ISD Group Description	Typical Techniques/Examples
Substitution: Use less hazardous working fluids, materials, chemistry and/or processes.	• Use water as the working fluid • Use stainless steel or other corrosive-resistant alloys for corrosive services • Use double mechanical seal for hazardous services • Use welded pipe, avoid threaded joints and minimize pipe flanges • Use chemicals with higher flash point, auto ignition temperature and boiling point • Use non-toxic fluids
Minimization or Intensification: Use minimum quantities and equipment inventory of hazardous materials, and avoid operation under hazardous conditions	• Reduce storage inventory of toxic and/or hazardous substances such as ammonia • Any examples of intensified equipment such as plate HE, dividing-wall column and reactive distillation.
Moderation or attenuation and limitation of effects: Reduce hazards by dilution, refrigeration or process alternatives that operate at less hazardous conditions.	• Use lowest allowable operating pressure to reduce boiling point • Inject water to wash/dilute corrosive salts in HEs • Reduce process pressure and temperatures as far as possible • Reduce pump/compressor pressure by reducing system pressure drops • Avoid non-fire case pressure relief by designing the equipment for blocked outlet scenarios • Maintain good safety margin (~ 20%) for the design pressure over maximum operating pressure • Design equipment for full vacuum • Use refrigerated storage for hazardous material • Use mounded bullets[j] for storing liquified gases • Place equipment away from the process control room by following adequate siting and spacing guidelines

(*Continued*)

[j] Mounded bullets are horizontal bullets located above ground and covered with a mound of earth or suitable inert material. This will prevent heating of bullet during any nearby fires and hence dramatically reduce the pressure relief requirements.

Table 20.5: (*Continued*)

ISD Group Description	Typical Techniques/Examples
Simplify: Eliminate unnecessary complexity, and design user-friendly or error-tolerant plants.	• Design for safe operation, maintenance, and emergency access • Maintain piping system and important valves with clean labelling • Select the equipment that requires less maintenance • Use equipment with high reliability • Design and implement safety protection system that is easy to understand and use • Design equipment that is inherently safer and hence requires no or minimal human intervention during plant upsets • Use fireproofing on equipment structures, and fire- and explosion-resistant barricades in the plant

the reboiler, column and condenser. Identify safer design solutions to minimize the overpressure relief requirements, using the ISD concept.

Solution: Following possible solutions based on ISD concept can be used.

1. Increase the condenser heat duty to handle maximum heat input case involving tube rupture. In this case, overpressure relief is not required but it is an expensive solution and often not economical.
2. Use restriction orifice (RO) and/or smaller steam piping to limit the maximum steam flow to the reboiler. Although this is a cheaper solution, RO shall be labelled critical and staff needs to be trained on its importance. Procedures shall be in place to ensure that RO is not removed during plant operation.
3. Reduce steam pressure to the lowest allowable value by careful consideration of reboiler size and thermal duty requirements. This can be achieved by the use of pressure control valve and downstream relief valve on the steam supply pipe to the reboiler. Lowering steam pressure can reduce steam leak in case tube rupture occurs.

Solution 3 is often the preferred solution in this example.

ISD concepts are useful for the entire life cycle of WHR project, from pre-feasibility stage to EPC stage and project operation stages. For

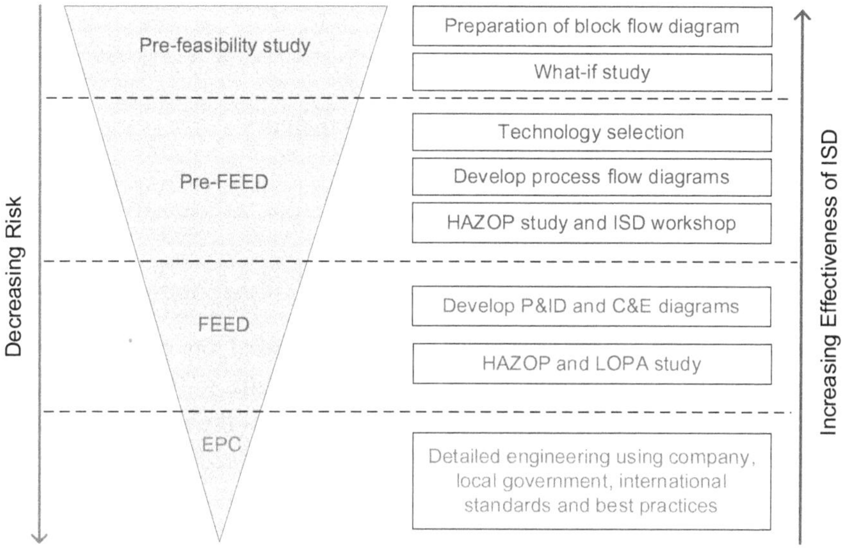

Fig. 20.9: Use of ISD concept in WHR project stages.

achieving maximum safety and cost benefits, ISD workshops and reviews are recommended to be conducted in the early phases of WHR project development, as illustrated in Fig. 20.9. Use of ISD concepts in pre-FEED and FEED stages can significantly reduce the risks and dependence on unnecessary/costly protection layers for the whole life cycle. In addition to improved process safety, ISD can simplify the process and hence reduce the project cost.

ISD checklist is summarized in the Appendix. Although preferred, ISD may not always be fully achieved due to process technology and/or economic reasons. A combination of ISD, engineering and administrative controls are always required to adequately manage all process risks.[5]

20.5.2 *What-If Check or Process Safety Checklists*

WI or WI-check method uses a list of questions about initiating events that can lead to hazardous scenarios. WI team brainstorms the project and generates these questions based on their experience and/or accident history available in the company and/or public domain. During a WI study, the following main activities are performed.

1. Description of the process
2. Develop WI questions
3. Identify hazards
4. Identify likelihood and severity of consequences
5. Identify safeguards
6. Perform risk assessment
7. Develop recommendations as required
8. Record the recommendations

For example, a WI question can be generated on exceeding the design pressure of proposed condensate pot (or drum, V-3) in the separation process (Fig. 20.10), initiated by tube rupture of the reboiler (HE-2). For this question, adequacy of safeguards are analysed and risk assessment is performed using the risk matrix (in Table 20.3) and necessary

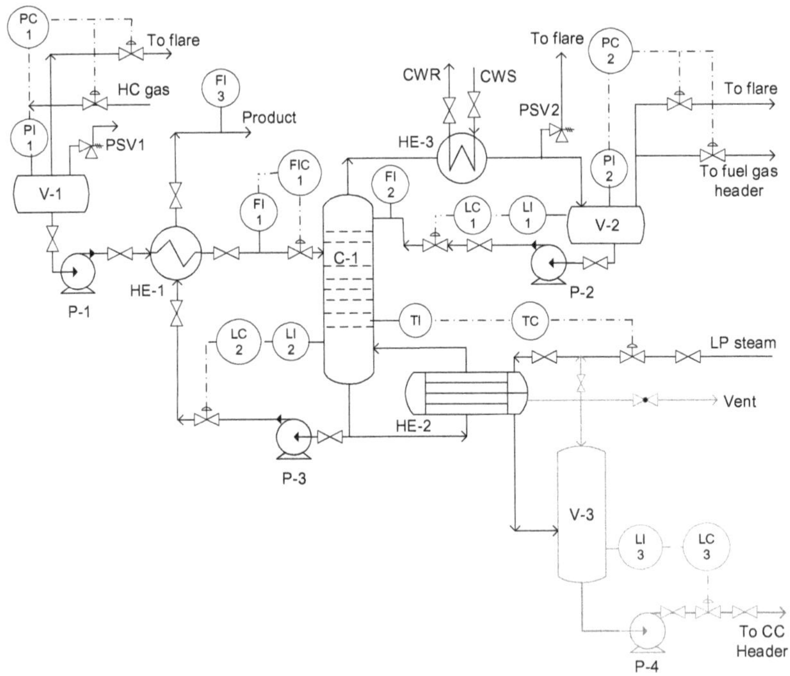

Fig. 20.10: P&ID of a WHR project to recover condensate (i.e., installation of V-3 and P-4 at the bottom right).

recommendations are made. A sample WI check list is presented in Table 20.6 for the condensate recovery section at the bottom right in Fig. 20.10.

20.5.3 *Plot Plan and Layout Review*

WHR equipment siting is very important as it can greatly impact process safety and project cost. Most common type of design error leading to industrial accidents is poor layout.[9] Proper project site layout reduces the likelihood of an accident. During plot plan layout, not only WHR equipment location but also the surrounding plant layout assessment is required. This can be conducted by a multi-disciplinary team of process engineers, project management, maintenance, operations, instrumentation, electrical, occupational and process safety representatives.

Depending on WHR project nature, some or all of the following should be considered during plot plan layout review.

- Review of WHR equipment location considering the existing equipment in the plant
- Review of existing pipe racks, utility, and process piping tie-ins
- Review of piping, instrument, and electrical cable routings
- Operations, maintenance, and emergency access
- Review of fire, thermal radiation, toxic and explosion contours from existing as well as new WHR project
- Review of electrical area classification of the new and existing equipment
- Proximity of firefighting and sewerage systems
- Close access to electrical sub-stations
- Close access to process, utilities, flare network and product storage area tie-ins
- Requirement for piling, new support structures and platforms. Project economics favour sites with minimum piling and additional structural requirements.
- Ability to support safe and efficient project construction

Table 20.6: Example of WI Worksheet for Preliminary Hazard Analysis

WI Safety Review Log Sheet									**Page of**
Study Name:	Steam condensate recovery from the distillation column (Fig. 20.10)	Team Leader: Charles							
Node:	X	Team Members:	John, Robert, Ravi, Edwards, Nick, Rosy						
Date:	15/10/2020	Ref. #:							
WI Question	**Cause**	**Consequence**	**Severity**	**Likelihood**	**RL***	**Safeguards**	**Rec#**	**Recommendations**	
Pressure of steam condensate pot (V-3) exceeds its design limit	Rupture of tube in the reboiler (HE-2)	Rupture of steam condensate pot, LPS and condensate piping (due to high-pressure fluid from the shell side of reboiler entering tube side) and resulting fire or explosion due to the loss of containment of hydrocarbon material. This may cause fatalities and asset damage.	2	H	H2	None	1.1	1. Consider installation of a pressure relief valve on the condensate recovery pot (V-3) with safe disposal of the relief. 2. Provide adequate safe design margin between maximum operating pressure and design pressure of HE-2 (on tube side), V-3 and associated piping system. 3. Establish proper program for regular inspection of HE-2 tubes	

*RL is based on the Risk Matrix classification as per Table 20.3.

- Noise generated at the surrounding facilities
- Any future plant expansion provisions

Potential flammable, toxic chemical and/or particulate matter release points shall be sited downwind of potential ignition sources. Careful consideration of topography in WHR project site selection and layout is very important as it can have great impact on potential consequences of an event.[10] Proximity of new WHR project site to existing plant, utilities and other supporting systems reduces the project cost and the inventory of hazardous substances by minimizing the piping runs. New WHR project shall be free of any flooding issues. While selecting the project site, careful review shall be conducted on the flammable, toxic and/or particulate releases from upwind equipment. For example, cooling towers and fin-fan coolers installed next to particulate release areas, may get fouled very fast and impair their cooling performance; toxic chemical releases will harm operators, and flammable releases may cause fire/explosion.

Another important consideration is electrical area classification. Locating WHR equipment in the existing classified area may require new equipment to meet the existing area classification or more stringent area classification, depending on the nature of the WHR equipment.

Proper safe access is essential for normal operation, maintenance and emergency operations. New WHR project site shall not have any obstruction for these activities. WHR equipment when installed on ground level shall have adequate surface slope to facilitate quick drainage of any leaked flammable material away from the equipment. This can prevent accumulation of any spilled material and prevent chances of equipment fire.

Easy access to existing fire hydrants, firefighting equipment and fire monitors can save project cost. High noise generating equipment shall be installed with noise suppression devices and also be located at a safe distance away from human access areas as required by company and local regulations.

New WHR facilities shall be installed at places sufficiently far away (allowable values can be obtained from API Standard 521[8] local government, and company standards) from thermal radiation sources such as flare and flue gas stacks. Equipment spacing needs to comply to local government and company standards.

20.5.4 *Area Classification Reviews*

Area classification review is a systematic methodology using standards such as API RP 500,[29] API RP 505,[30] National Fire Protection Association (NFPA) publication 70,[31] NFPA publication 497, National Electric Code (NEC), International Electrotechnical Commission (IEC) and local government standards, to identify the locations where explosive/hazardous environment may be present. It is based on three main criteria: type of hazard (groups), auto ignition temperature (AIT) of the hazardous material (temperature or T rating) and likelihood of the hazard being present in flammable concentrations (zones). Currently, there are two main systems used to classify hazardous areas. They are the class/division system (NEC) and the zone system (IEC). The class/division system is mainly used in United States and Canada, whereas rest of the world generally uses the zone system. A comprehensive comparison of both the systems is presented by Hall.[11]

Codes and standards help to select electrical equipment and instrumentation that will safely operate in a hazardous environment. An explosive environment may be created by a gas, vapour, liquid, dust or even fibres. Late project changes due to improper area classification in early project stages can lead to significant cost and schedule overrun. In some cases, installation of new equipment can result in change requirement of area classification for the existing equipment, which then requires additional large investment for replacing the existing equipment. Using appropriate standards, the project team can identify classified areas surrounding the plant equipment, based on characteristics and probability of leaked/released substances. Accordingly, they can choose electrical and instrument equipment that can safely operate in the hazardous environment by complying with the required area classification. Area classification review results are useful in conducting PHA, MOC and buildings/equipment siting studies, and hence it is an essential part of process safety.

For WHR project area classification review, area classification drawings for the existing plant need to be reviewed first. Subsequently, detailed analysis shall be conducted using the standards to identify the area classification for new and existing plant areas. Accordingly, the project team can develop/modify area classification diagrams for the new/existing project sites. Sometimes, the project layout plan may need to be changed or modified to control the project cost.

20.5.5 *Pressure Relief System Considerations*

Pressure relief system is a very important layer of protection in process design. It is a passive safeguard to protect WHR equipment from overpressure and hence prevent fire, explosion and toxic release hazards. API Standard 520[12] Parts 1 & 2 and API Standard 521[8] provide comprehensive guidelines for the design of PRVs and relief systems. Several books such as Parry[13] and Mannan[14] provide good discussion on this topic. Although a PRV does not prevent overpressure in WHR equipment, it releases the overpressure for safe disposal. ISD of process, controls, critical process alarms, operator supervision and automatic action using SIS and ESD (emergency shutdown) systems can avoid or minimize the pressure and/or vacuum relief. By implementing some or all of these, avoid or reduce the pressure relief requirements as much as possible.

The list of standards pertaining to PRVs are the following:

- ASME Section I Power Boilers
- ASME Section IV Heating Boilers
- ASME Section VIII, Division 1 Pressure Vessels
- ASME Section VIII, Division 2 Pressure Vessels — Alternative Rules
- B31.3, Chapter II, Part 3 Power Piping — Safety and Relief Valves
- B31.3, Chapter II, Part 6 Power Piping — Pressure Relief Piping

Overpressure occurs in process equipment, mainly due to excessive heat input and/or excessive fluid pressure entering the equipment. Its relief assessment is required for all WHR equipment in pre-FEED and FEED stages. Detailed explanation for 17 possible overpressure causes and comprehensive coverage of PRV sizing calculation methods, relief headers and flare stack sizing methods are available in API Standard

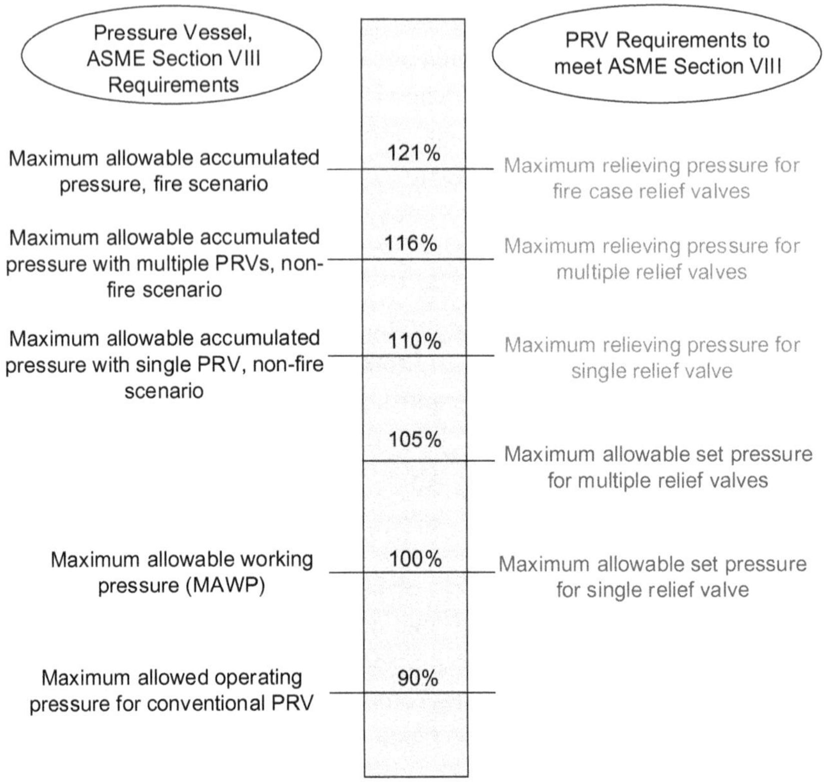

Fig. 20.11: ASME Section VIII — pressure vessel and PRV requirements.

521.[8] One case study using the overpressure scenarios is presented by Balajee and Reddy.[15] Based on the project nature, relevant overpressure scenarios from API Standard 521[8] can be selected for detailed assessment. Relief valve set pressure guidelines as per ASME section VIII are available in API Standard 520[12] Part 1. PRV set and relieving pressure requirements to meet ASME Section VIII clauses are compiled in Fig. 20.11.

Relief system design methodology, shown in Fig.20.12, starts with defining the relief system, which may contain one or more equipment connected by piping. All the pressure vessels, HEs, steam turbines, compressors and positive displacement (and not centrifugal) pumps require

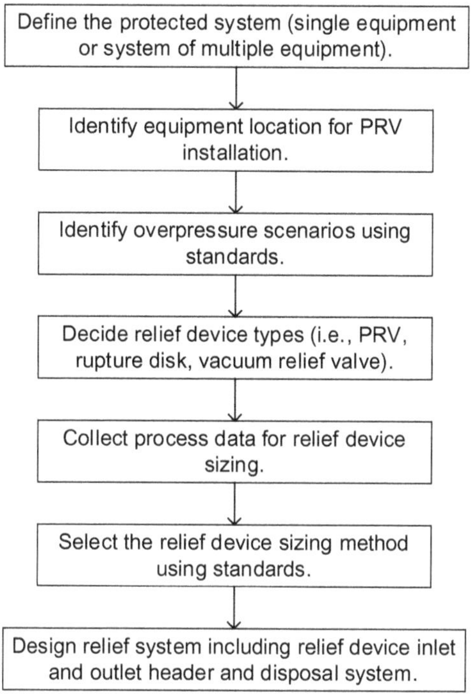

Fig. 20.12: Relief system design methodology.

relief. Piping containing light hydrocarbons or accidently receiving pressure more than maximum allowable working pressure (MAWP) also require relief.

Next step is to choose the location where the relief device can be installed (Fig. 20.12). Depending on the WHR equipment/system, identify relief cases using standards such as API Standard 521.[8] Identify the type of relief device (e.g., PRV, RD, PRV-RD combination and/or vacuum relief valve) that is appropriate for the equipment. See Fig.2.13 in Parry[13] for a useful flowchart for selecting relief devices. Next, select the relief set pressure and allowable accumulated pressure or relief pressure from Fig. 20.11. Collect properties of fluids involved and select the appropriate relief device sizing equation from API Standard 521.[8] Finally, design relief device inlet and outlet piping, and relief disposal system (i.e., to

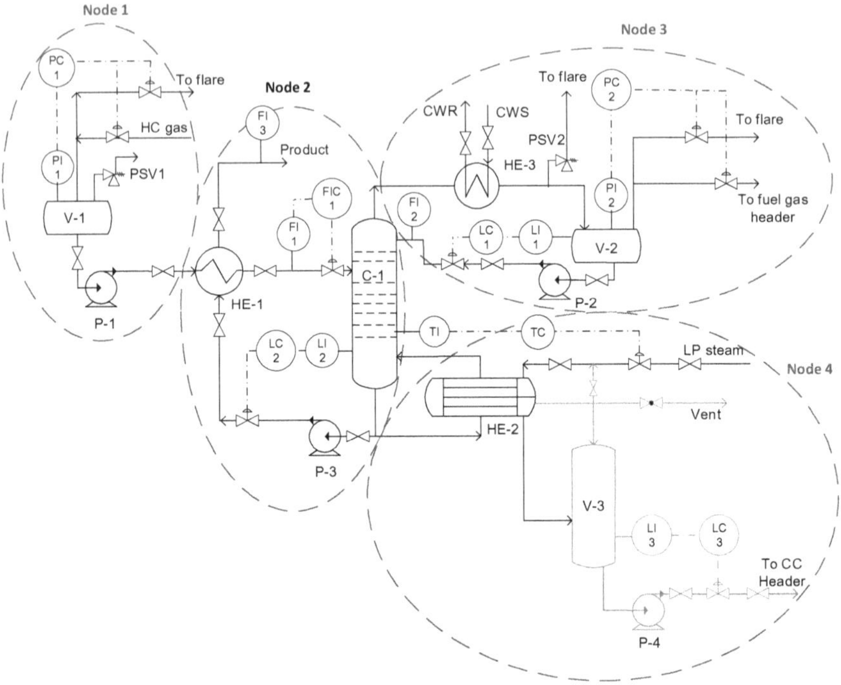

Fig. 20.13: P&ID with nodal classification for a WHR project to recover condensate (i.e., addition of V-3 and P-4).

flare system or atmospheric safe location) as per API Standard 521[8] guidelines. Some useful guidelines for WHR project relief system design are described in the following paragraphs.

Sometimes, existing PRVs can be re-used if the WHR project involves debottlenecking or retrofitting an existing equipment. In such cases, capacity of the existing relief valve and margin between MAWP and maximum operating pressure shall be reviewed. If the required relieving capacity of the new PRV is greater than the existing PRV capacity and/or the margin between MAWP and PRV set pressure is lower than that allowed for the existing PRV type, replace the existing PRV with a new one. It is also important to check and confirm the PRV inlet and outlet pipe pressure drops. PRV inlet pipe pressure drop shall be lower than 3% of

PRV set pressure. Pressure drop in the PRV outlet pipe shall be less than 10% and 30%–50%[k] of PRV set pressure for the conventional and balanced bellow type PRVs, respectively. If the backpressure[l] is very high (> 50% of PRV set pressure) and/or inlet pipe pressure drop is high (> 3% of PRV set pressure), pilot-operated PRV can be used. Pilot-operated PRVs are also used when operating pressure is more than 90% of the PRV set pressure.

Care must be taken while evaluating a relief system involving several process vessels and/or HEs. If some of the equipment are modified or replaced, new pressure relief conditions may result in scenarios exceeding MAWP of some of the vessels.

Vacuum relief requirement can be avoided by designing the equipment for full vacuum. New WHR equipment installed above 7.6 m from the grade/ground level do not require pressure relief devices (PRDs) for fire case scenario. Hence, it may be advantageous to place vessels on platforms at > 7.6 m if fire case relief load is highest among the relief loads (readers may refer to API Standard 521[8] for details of various relief load requirements) and/or the existing flare system has relief capacity constraint. Fire case relief can also be reduced, irrespective of the vessel elevation, by applying fire-resistant insulation or refractory as per the guidelines in API Standard 521.[8] Fire circle or zone is defined as maximum ground level area surrounding any equipment, that can be affected by fire; API Standard 521[8] defines this area as 230–460 m^2. All the equipment in the fire zone will simultaneously relieve under fire condition. Addition of new process equipment inside existing fire circle may increase the fire circle size. Hence, care shall be taken to review the fire circle size with each equipment addition as it may significantly increase the peak relief load from that zone during the fire scenario.

---

[k] Actual value depends on the vendor; user shall check the limit with the vendor, for each application.

[l] PRV back pressure includes total system pressure drop from PRV outlet piping to atmospheric release point. For example, if PRV is connected to the flare system, back pressure will include pressure drops in PRV outlet pipe, flare main header, flare KO drum liquid seal, flare stack and flare tip.

Changes in PRV set pressure need to be reviewed carefully. Increasing/decreasing PRV set pressure can increase/decrease PRV capacity. Increase of PRV set pressure, if allowable by consideration of MAWP, can increase PRV inlet and outlet piping pressure drops, which may violate the allowable pressure drop limits.

A WHR project often requires an increase in the capacity and/or head of an existing centrifugal pump (by increasing impellor size if pump casing has allowance for a bigger impeller), replacement of an existing pump or installation of more pumps in parallel/series. During such modifications, it is important to check MAWP and relieving capacities of the downstream piping and equipment. Changes in pump drivers can have significant impact on the relief system. For example, changing steam turbine driver to electrical motor can increase relief load during power failure scenario. Also, it is always essential to check the impact of driver changes on the pump downstream equipment.

HE tube rupture relief can be neglected when the hydraulic test pressure on the low-pressure side is greater than or equal to MAWP of the high-pressure side. Removal or increasing size of critical RO in piping may significantly increase the flare load. Hence, it is very important to maintain critical ROs as per the design intent. Addition of isolation valves on equipment (not installed with PRD) may require new PRDs due to blocked outlet and/or fire case scenarios. Increasing WHR may require more relief in the downstream equipment.

Change of existing control valves (to reduce pressure drop and/or increase capacity) and/or opening their bypass valves may overpressure the downstream equipment and hence increase relief requirements. Hence, care must be taken with such modifications unless changes are verified to be safe by detailed hydraulic calculations. If operator response time of 10–30 minutes is available, as per API Standard 521,[8] relief credit (i.e., no requirement of pressure relief) can be taken in relief system design, depending on the complexity of the plant.

Addition of new PRDs in the WHR project may increase the back-pressure of some existing PRDs and/or may exceed velocity and temperature limits in some relief headers in the flare network. Complex flare network design and analysis can be performed with commercial software such as Aspen flare analyzer, Flaretot, and Visual Flare. A simplified rough method for quick estimation of PRV manifold pressure drops is given by Branan.[16] Relief headers sizing should be checked to ensure that backpressure does not exceed the allowable backpressure limit for the PRDs. It is general practice to limit velocity at the flare stack tip to 0.5 Mach for peak, short-term and infrequent flow, and 0.2 Mach for normal conditions. It is very important to check the size of flare knockout drum to ensure that proper vapour–liquid separation takes place to prevent carryover of big liquid droplets (> 300 μm in size) to the flare stack.

Overpressure relief capacity requirements and relief piping, flare tip pressure drops, and flare knockout drum capacity need to be evaluated for each utility failure, fire cases and any other combination scenarios. Pressure relieving system must meet the design guidelines for all such scenarios. If the flare quantity increases substantially in the project case beyond acceptable values, QRA of flare system and/or dynamic simulation can be performed to establish credible flare load, which can be lower than the flare load established by the steady state analysis. If flare tip Mach number is still higher, flare tip and/or flare stack modification or replacement maybe needed, to minimize thermal radiation (for peak flaring), and flammable and/or toxic substances dispersion (for the case of flare flame-off scenarios) effects. A case study on revamping a flare network is presented by Madlani.[17] PRD and entire relief system calculations shall be documented, for future reference, as they are important documents for PHA studies.

20.5.6 *Fire Safety for WHR Projects*

Fire hazards can be minimized by adapting ISD concepts in pre-FEED and/or FEED stage. Before finalizing the WHR project scope, designers/ project team should evaluate the impact of the project on the existing firefighting system of the plant. Some of the important fire safety aspects applicable for WHR projects are summarized below.

Design according to standards: Use local and international standards such as API, NFPA, ASME, ASTM, National Association of Corrosion Engineers (NACE), IEC, and NEC in designing new WHR projects.

Siting, spacing and access: New WHR equipment should be installed as per local and international siting and spacing regulations and guidelines. Spacing between equipment should allow safe operational, maintenance, emergency, and firefighting access.

Fire-resistant ESD system: New equipment handling flammable substances should be installed with fire-resistant ESD system, for remote quick isolation, in case of any fire.

Inventory disposal: install automatic de-inventorying of hydrocarbon material from storage vessels to remote disposal systems, during fire emergencies.

Drainage: Install adequate slope for the drainage system to quickly drain off any spilled hydrocarbon away from process equipment. Drainage capacity shall be designed to handle combined flow of worst-case spill/leak rate and fire water rate (used for the firefighting).

Minimize leak points: Connect level gauges, instrumentation and sampling point drains to flare system by using closed piping. Avoid manual operations such as hose transfers and atmospheric venting. Design should eliminate any escape of flammable fluid to atmosphere. Where vapour release is necessary, sufficient mitigations such as dilution, release at very high locations using flame arrestor and/or guard against ignition sources are required.

Control of ignition sources: Ignition sources (e.g., vehicles and smoking) should be spaced as far as possible from process equipment. Sources of any potential ignition sources in hazardous locations should be identified, and reasonable safeguards should be provided to minimize fire/explosion risks.

Use of fireproofing and insulation: Process structures and equipment supports should be fire proofed. Equipment shall be insulated to minimize the effects of potential fires.

Avoid congestion: New project should avoid congestion where any leaked hazardous substance can accumulate and cause potential fire or explosion. Sufficient air circulation and ventilation shall be ensured by engineering design.

Accumulation of electrostatic charges: Avoid the accumulation of static charges by adequate design practices. Applicable design methods are well presented by Pratt.[18]

Flammable vapour, fire and smoke detectors: Flammable vapour detectors should be considered in locations where there is potential for leakage of flammable vapour such as pump and compressor leaks or areas involving manual activities. Fire detection systems and alarms should be considered in hazardous locations. These include gas, thermal, smoke detectors, thermal detectors and IR/UV flame scanners.

Firefighting system: Fire water system with foam injection facilities, fire monitors, hose reels, fire water sprinkler systems, fire extinguishers (dry chemical, CO_2, halogen etc.) should be installed as per local regulations, company and/or international standards such as NFPA. It is essential to ensure uninterrupted supply of sufficient fire water all the time by proper selection and installation of fire water pumps. Deluge systems, which operate by automatic fire water opening on detection of smoke or heat, are useful for automatic firefighting.

Fire safety design consideration for some WHR equipment are listed in Table 20.7. A comprehensive coverage of many fire safety design and implementation guidelines is available in Nolan.[19]

Table 20.7: Fire Safety Consideration for Some WHR Equipment

Equipment	Fire Safety Consideration
Pumps handling flammable material	• Do not locate them below pipe racks. • Use double mechanical seals, with leak detectors. • Use dedicated minimum flow protection and low suction pressure trips. • Use steel or alloy material as per requirement. • Install deluge system. • Install remotely operated emergency shut off switch and isolation valves. • Pump base shall be designed to prevent oil accumulation. • Install gas detectors at pump seals. • Install vibration monitoring system to trip the pump on high vibrations. • Install deluge fire water system.
Compressors	• Avoid liquid compression in a reciprocating compressor. • Distance piece vents[m] of reciprocating compressors shall be routed to a safe location. • Minimize small diameter piping in compressors. • Prevent blocked outlet for reciprocating compressors. • Provide anti-surge and stonewall protection systems for centrifugal compressors. • Install adequately designed PRDs. • Install remote isolation valves and tripping switches. • Install gas detectors at compressor seals. • Install vibration monitoring system to trip the compressor on high vibrations.
Process furnaces	• Avoid overheating of furnace tubes. • Install spare feed pumps with auto cut-in provision in case duty pump stops. • Install remote isolation valves for stopping feed and fuel during fire scenarios. • Design automatic shutdown of furnace on loss of feed and low fuel pressure as per API RP 556 guidelines. • Install deflagration or detonation arrestors in low-pressure waste gas systems.
Pressure vessels and piping	• Protect with PRDs as per requirement. • Always use proper MOC to prevent corrosion, erosion, cold and hydrogen embrittlement, chloride stress cracking, high temperature sulfidation and other damage mechanisms. • Avoid flexible bellows in piping and HEs.

[m] A distance piece refers to an open or enclosed casing through which the piston rod travels, separating the compressor cylinder from the crankcase.

Table 20.7: (*Continued*)

Equipment	Fire Safety Consideration
	• Design to operate below corrosion and erosion velocities. • Use proper supports. • Perform stress analysis to avoid excessive stresses. • Minimize number of pipe flanges and joints. • Avoid threaded joints in piping handling hazardous substances. • Avoid pressure surge in pipes by reducing fluid velocity and avoiding quick close on-off valves. • Install deluge fire water system for flammable substance storage vessels.
Relief system	• Route PRD discharges to the flare system. • PRD discharge to atmosphere, if any, should be routed to a safe location with a safe disposal system such as sufficiently sized knockout drums (to prevent liquid release), steam or nitrogen dilution.
Cooling towers	• Locate them far away from process units. • If possible, maintain cooling water pressure more than process stream pressure, at the process coolers. • Install gas detectors at the top of cooling towers, and automatic shutdown system for cooling tower fans, in case of gas detection.
Boilers	• Locate them far away from process units. • Install boiler management system. • Install deflagration or detonation arrestors in low-pressure waste gas systems.
Storage tanks	• Consider nitrogen blanketing as needed. • Install safety systems to prevent overfill and formation of explosion mixtures during tank emptying or draining. • Install pressure/vacuum protection devices and flame arrestors in vents. • Install proper grounding system to prevent lightning strikes. • Install bunds to contain any leaked material. • Locate pumps and ignition sources out of tank bund walls. • Use double seals for the roof of floating-roof tanks. • Install frangible[n] tank shell-to-roof welding for API 650 tanks. • Install fire water and foam injection system.
Cabling	• Install critical power, control, instrument and alarm wiring, and pneumatic tubing outside the fire-exposed envelope. • Cable trays and conduit banks should be routed underground or on the upper levels of elevated pipe racks at least 7.62 m above the ground and outside the drainage path of flammable material spills. • Install fire proofing of critical systems and equipment if they are inside the fire-exposed envelope.

[n] A frangible roof is a roof to shell joint or junction that is weaker than the rest of the tank and will preferentially fail if the tank is over pressurised.

20.6 Process Hazard Analysis

PHA is an essential part of WHR project management for enhancing process safety. It is an experience-based evaluation process conducted by a multi-disciplinary team to identify process hazards, estimate associated risks, and make risk mitigation recommendations. PHA may contain some or all of the following studies/activities: WI analysis, HAZOP, FMEA, LOPA, FTA, ETA, QRA, SIS analysis, toxic/flammable substance dispersion and thermal radiation modelling, review of any past PHAs, PSI verification, human factors study, facility siting reviews, and past incidents review. Some of these studies/activities are outlined in the following subsections. Based on the WHR project scope, the hazard review leader will decide PHA study scope and duration. Team composition may vary according to the studies included in PHA.

20.6.1 *Hazard and Operability Study*

HAZOP is a qualitative and most used technique in process industries, during pre-FEED and FEED stages. It uses a systematic review of process, nodes (small sections of the plant) and guide words (deviations from design intent), to identify potential hazards associated with operation and maintenance of the process. It frequently uses a risk matrix to assess the risk (includes safety, environmental, asset and often societal risks).

High quality PSI is essential for conducting a productive HAZOP study. Updated PSI information, especially piping and instrumentation diagram (P&ID), PFD and cause and effect (C&E) charts, alarm and trip summary with set points, relief device specification and sizing documents are required for conducting HAZOP study. If WHR project is revamp type, then past MOC records, past HAZOPs of the plant, inspection and accident history can be used beneficially for the new HAZOP study. HAZOP leader will choose team composition and duration depending on the complexity of the WHR project. HAZOP study is conducted by a multi-disciplinary team led by a person knowledgeable in HAZOP technique; typical team consists of HAZOP team leader, scribe, process, operations, process safety, maintenance, instrumentation, electrical and environmental engineers. Subject matter experts or technology licensor may be invited for clarification on some technical matters on ad-hoc basis. Team leader facilitates all HAZOP discussions and team members actively participate in brainstorming hazardous scenarios, their likelihood and severity.

The HAZOP leader breaks the P&ID into nodes such that HAZOP technique can be applied efficiently. Sequence of nodes will be decided by the team leader, often starting from process nodes followed by utility nodes. Node boundaries usually occur where process parameters change significantly (e.g., change in design temperature, pressure or phase). For better productivity, all the design information, session date, skid name, node number, node description, associated drawing numbers and node intent can be updated, and nodes are clearly marked on P&IDs, ahead of the HAZOP session. If risks are not acceptable, appropriate recommendations are made to bring them to ALARP or even lower level. Recommendations are assigned to the responsible staff by the HAZOP leader to follow-up and close/complete them. A comprehensive coverage of HAZOP technique and its application are presented by Dave.[20]

Figure 20.13 illustrates nodal classification for a distillation system modified with a condensate recovery system shown in Fig. 20.10. The feed tank (V-1) up to the feed pre-heat HE-1 is classified as node 1, HE-1 system and distillation column (C-1) as node 2, distillation column (C1) overhead section as node 3, and LPS, new condensate recovery system and reboiler (HE-2) as node 4.

HAZOP Methodology: The technical expert (usually the technology licensor or process engineer) will explain the system design and operational philosophy at the beginning of each nodal review. Deviations for specific parameters are listed in Table 20.8.

A sample HAZOP worksheet for node 4 in Fig. 20.13 is presented in Table 20.9. The scribe will follow the HAZOP team leader to fill up the worksheet as HAZOP study progresses. Significance of HAZOP worksheet items/columns is as follows.

Table 20.8: List of Typical Deviations from Intended Operations for Specific Process Parameters

Parameter	Deviation from Intended Operation	Parameter	Deviation from Intended Operation
Composition	Low, high, contaminated or wrong composition	Containment	Partial or total loss of containment
Corrosion/erosion	High corrosion and/or erosion	Flow	No, low, high, reverse or misdirected flow
Level	No, low or high level	Maintenance	Special maintenance procedures
Mixing	No, less or more mixing	Phase	Loss or additional phases
Pressure	High or low pressure	Procedure	Omitted or additional step
Reaction	No, low, high, reverse or additional reaction	Shutdown	Improper shutdown
Start-up	Special start-up procedures	Temperature	Low or high temperature
Utilities	Loss of utilities	Vibration	High vibration

1. Deviation: Process parameter, which deviates from the design intent.
2. Cause: Reason for the deviation.
3. Consequence: Potential undesired effects of the deviation, which can include fatalities, injuries, production loss, asset, environmental and/or societal damage due to fires, explosions, pressure vessel rupture and/or toxic releases.
4. Category: Typical categories of consequences such as safety, reliability, operability and equipment protection.
5. Safeguards: Existing protection layers designed to prevent causes and/or mitigate consequences.
6. Severity/Likelihood: Severity/Likelihood of an unwanted event based on the risk matrix in Table 20.3.
7. RL: The combination of likelihood and severity. It can be obtained from Table 20.3.
8. SIL of a SIF: Integrity level of SIF. It can be obtained from Table 20.3. Section 20.6.3 covers the methodology of SIS design.

Table 20.9: Sample HAZOP Worksheet for Node 4 in Fig. 20.13

HAZOP Worksheet

Date & Session:	20/10/2020
Skid:	Steam condensate pot
Node Number:	4
Node Description:	LPS pipe to reboiler (HE-2), condensate pot (V-3) and steam condensate pipe to the header, process inlet and outlet pipes for HE-2
Drawings:	ABC.000.xx Sht 1 Rev. 3
Intention:	LPS at 145°C heats process stream in HE-2 from 90 to 95°C. Condensate formed in HE-2 is drained into condensate pot (V-3) and subsequently removed to condensate header by condensate pump (P-4)

Deviation	Cause	Consequence	Category	Safeguards	Severity	Likelihood	RL	SIF	Recommendations	By	Remarks
High Pres sure	Reboiler (HE-2) tube rupture	Rupture of the steam condensate pot (V-3), LP steam and condensate piping (due to high pressure fluid from the shell side of reboiler entering tube side) and resulting fire or explosion due to loss of containment of hydrocarbon material. This may cause fatalities and asset damage.	S	None	2	H	H2	SIL2	1. To consider installation of a pressure relief valve on the condensate recovery pot (V-3) with safe relief disposal.	Process Engineer (John)	
								SIL0	2. To provide adequate safe design margin between maximum operating pressure and design pressure of HE-2 (tube side), V-3 and associated piping system.	Process Engineer (John)	
								SIL0	3. Establish proper inspection program for inspecting the HE tubes	Inspection Engineer (Robert)	

*RL and SIF requirements are classified as per the risk matrix in Table 20.3.

9. Recommendation: Corrective action(s) to reduce the unwanted event's likelihood and/or severity to reduce the risk to acceptable level.
10. 'By' column: Responsible person to follow up and close the recommendation.

Severity and likelihood help to make a (semi-)quantitative judgement during the hazard review, to trigger a requirement of QRA or SIL assessment and verification as per IEC61511. SIL assessment and verification are detailed in the Sections 20.6.6 and 20.7. QRA is not covered in this chapter; for a good coverage of this topic, interested readers can refer to CCPS (2000).[3]

20.6.2 *Failure Modes and Effects Analysis*

FMEA is a systematic and proactive method to identify, prevent and/or mitigate process safety, environmental and asset integrity related problems. It is well suited in process development stages such as pre-FEED and FEED. It can also be employed for improving safety of existing processes. FMEA uses different failure modes and their effects to analyse the process for potential hazards, as well as likelihood and severity to evaluate the risk using a risk matrix table. If the risk is not acceptable, appropriate recommendations are made to bring the risk to an acceptable level. FMEA is very useful for proper assessment of highly instrumented or electrical systems. Like HAZOP, it is also conducted by the team leader along with the scribe and members from different disciplines. Involving technical experts at least on part-time basis can be very useful.

FMEA procedure is as follows. As for HAZOP, it requires updated PSI.

- Review the process using PFDs and P&IDs.
- Brainstorm and identify all possible failure modes of process and instrumentation and electrical (I&E) equipment. See the example in Table 20.10.
- List potential effects or consequences of all failure modes.
- List the existing safeguards.

Table 20.10: Example of FMEA Worksheet

FMEA Safety Review Log Sheet

Date:	15/11/2020		
Study Name:	Condensate recovery project	Team Leader:	Charles
Node:	4	Team Members:	John, Robert, Ravi, Edwards, Nick, Rosy
Node Description:	LPS pipe to reboiler (HE-2), condensate pot (V-3) and steam condensate pipe to the header, process inlet and outlet pipes for HE-2		

Tag No.	Failure Mode	Consequence	Safeguards	Severity	Likelihood	RL	Rec#	SIF	Recommendations	By
Reboiler (HE-2)	Rupture of tube(s) in reboiler (HE-2)	Rupture of the steam condensate pot (V-3), LP steam and condensate piping (due to high pressure fluid from the shell side of reboiler entering tube side) and resulting fire or explosion due to loss of containment of hydrocarbon material. This may cause fatalities and asset damage.	None	2	H	H2	1.1	SIL2	1. To consider installation of a pressure relief valve on the condensate recovery pot (V-3) with safe relief disposal.	Process Engineer (John)
								SIL0	2. To provide adequate safe design margin between maximum operating pressure and design pressure of HE-2 (tube side), V-3 and associated piping system.	Process Engineer (John)
								SIL0	3. Establish proper inspection program for inspecting the HE tubes	Inspection Engineer (Robert)

- Establish the likelihood of cause detection using Table 20.10. This is an assessment of likelihood that the current safeguards will detect the cause of failure mode or failure mode itself.
- Establish the severity of the effect(s) using a risk matrix.
- Assess the RL for each effect or consequence based on the classification in the risk matrix. If the risk is very significant, QRA and/or SIS studies may be required.
- Develop recommendations to address the failure modes with high RL. Assign responsibility and target completion date for each of these actions.

FMEA in Table 20.10 illustrates an overpressure scenario of LPS and condensate collection system due to tube rupture in the reboiler (HE-2 in Fig. 20.13). This can generate unacceptable risk for the plant and people. At least SIF of SIL2 configuration is necessary to mitigate this risk.

20.6.3 *Instrumented Protective Systems*

IPS is an important layer of protection in process plants to prevent abnormal operations/situations that can potentially result in serious safety consequences, by using the action of instrumentation. Depending on the risk reduction achieved, they can be classified as independent protection layer (IPL) or non-IPL type (e.g., DCS alarm or interlock). SIF is an IPL-type IPS designed and maintained as per technical specifications and guidance available in relevant standards such as IEC 61508 and IEC61511. It uses life cycle management concept to achieve the target risk reduction. IPS are considered as safeguards during PHA studies.

SIF is defined as an instrumented function that contains a collection of sensors, logic solver and final elements (Figure 20.14), which together detect a specific hazard and bring the process to a safe state. It provides a defined level of risk reduction (or SIL) for a specific hazard by using automatic instrumentation action. Each SIF must have a specified SIL, necessary to achieve the desired risk reduction factor (RRF).The SIL of the SIF can be derived by considering the risk reduction provided by that function. An SIS system contains one or more SIFs. The target PFD_{avg},

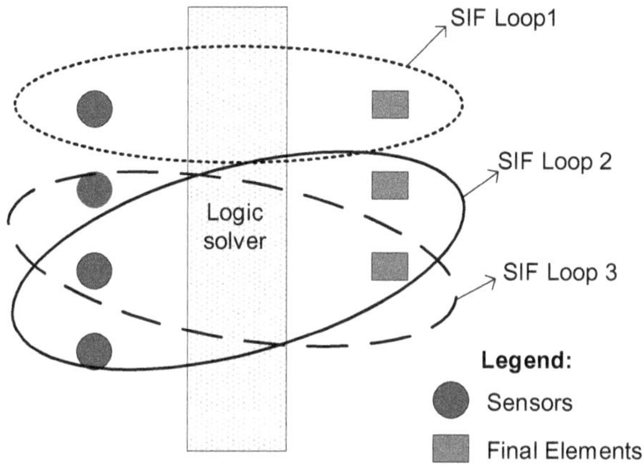

Fig. 20.14: Configuration of SIF.

Table 20.11: Details of Different SILs

SIL	Target PFD$_{avg}$	Target RRF	Target Frequency of Dangerous Failures to Perform the SIF (per Hour)
4	$\geq 10^{-5}$ to $< 10^{-4}$	> 10,000 to ≤ 100,000	$\geq 10^{-9}$ to $< 10^{-8}$
3	$\geq 10^{-4}$ to $< 10^{-3}$	> 1000 to ≤ 10,000	$\geq 10^{-8}$ to $< 10^{-7}$
2	$\geq 10^{-3}$ to $< 10^{-2}$	> 100 to ≤ 1000	$\geq 10^{-7}$ to $< 10^{-6}$
1	$\geq 10^{-2}$ to $< 10^{-1}$	> 10 to ≤ 100	$\geq 10^{-6}$ to $< 10^{-5}$

RRF and the target frequencies of dangerous failures for each SIL type are listed in Table 20.11. Several risk analysis methods, from qualitative to fully quantitative, are available (Table 20.12) for evaluating SIL requirements. Selection of a particular method will depend on the severity and complexity of the project scope. SIL4 requirement is not expected in process industries.

20.6.4 *Fault Tree Analysis*

FTA uses a backward and deductive logic. It is mainly used to develop causes and/or estimate frequency or probability for the top event. It is carried out via top-down approach, starting with a top or hazardous event or

Table 20.12: Risk Analysis Methodologies

Risk Analysis Method	**Type**
Risk matrix	Qualitative
LOPA	Semi-quantitative
FTA	Quantitative
ETA	Quantitative
Consequence analysis	Quantitative

hazard, and moving downwards through intermediate events to various initiating or basic events at the bottom of the tree. Depending on the complexity of the system, the number of intermediate and basic events in the tree can vary. However, there is only one hazardous event at the top. FTA uses a clear graphical representation of the system, which makes it very easy to understand the logic leading to the top event, while considering interactions to individual events. Events in an FTA can include process and I/E equipment, software, human, and environmental factors.

Procedure for constructing an FTA is as follows.

1. Place top event or an accident at the top of the page.
2. Identify the intermediate events that can contribute to the top event.
3. If all these intermediate events must occur together for the top event to happen, use 'AND' gate to connect them to the top event.
4. If any of these intermediate events occurring can enable the top event to happen, use 'OR' gate to connect them to the top event.
5. Consider any one of the new intermediate events and identify what events must occur using 'AND' or 'OR' gates and connect them to the intermediate event.
6. Repeat Step 5 for all the intermediate events.
7. Continue building FTA downwards until branches terminate at initiating or basic events.

Detailed assessment of FTA is available in Crowl and Louvar.[1] FTA for an overpressure event of a pressure vessel (V-1) shown in Fig. 20.15, is depicted in Fig. 20.16.

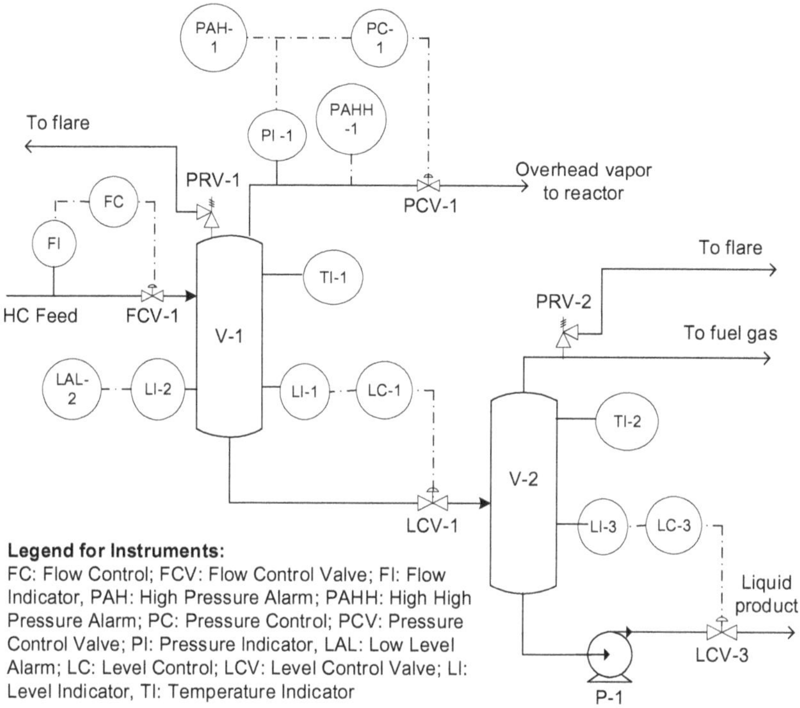

Fig. 20.15: Details of an HP and LP flash drums (V-1 and V-2, respectively).

20.6.5 *Event Tree Analysis*

ETA is a form of fault propagation modelling, which is divergent in nature. It starts with an initiating event, which is usually a process upset or failure of process or I&E equipment and propagates the branches by including intermediate events, and finally generates one or more outcomes. Some of the outcomes show hazard frequencies. Branches of the event tree follow different event sequence paths to result in various outcomes. Number of outcomes depends on number of events in each branch. ETA establishes logic for occurrence of a failure/accident and its probability.

Generally, FTA is used to find the frequency/probability of a hazardous event, whereas ETA to study how a hazardous event propagates to various consequences. ETA is also used to analyse cost-effective barriers/ intermediate events/solutions for reducing the frequency of unsafe

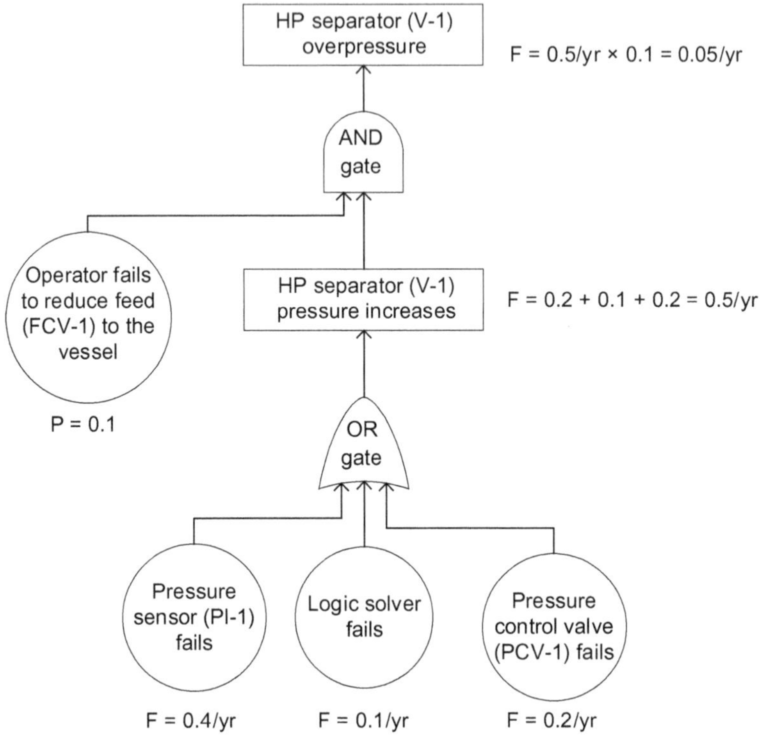

Fig. 20.16: FTA for over-pressurization of HP separator (V-1) in Fig. 20.15.

consequences. Both methods require probability or frequency values of intermediate events, can be used to calculate the frequency/probability of various consequences, and require considerable process knowledge. Their understanding requires basic knowledge of statistics. Hence, the usage of these methods is generally restricted for analysing high consequence events.

Event tree branches are usually complementary events. For example, a branch event could be the failure of a control valve. The event set contains two complementary events (e.g., control valve fails and control valve operates). Industrial accidents occur only when various safety protection layers/systems fail to prevent the hazard from propagating. Hence, whether an accident will occur or not depends on the success or failure of each of these safety systems. ETA depicts this information clearly in a diagram.

Typical steps in an ETA are as follows:

1. Start with an initiating event of interest.
2. Identify the safety protection layers/systems designed to prevent propagation of the initiating event.
3. Consider the success and failure of layers.
4. Construct the ETA to show various outcomes including accident event scenarios.
5. Assign necessary numerical values (frequencies/probability of the initiating event and intermediate events/safety protection layers) to calculate the frequency of unwanted accidents.

Thus, the sequence of events that lead to potential accidents/outcomes is developed considering both the success and failure of intermediate events or safety functions or protective layers, as the event progresses from the initiating event to the outcome. Every branch of the tree indicates a possible sequence of events. Probability of each outcome is calculated using probability multiplication of events in the branch as ETA uses 'AND' logic only. A sample ETA with these calculations is shown in Fig. 20.17; initiating event for this ETA is overpressure in HP separator (V-1) in Fig. 20.15.

20.6.6 *Layers of Protection Analysis*

LOPA, a variation of ETA, is an effective and commonly used risk assessment method; it analyses the effect of adding additional protective layers for reducing frequency of the unwanted consequence and hence eventual risk reduction. Also, process industries use LOPA to identify critical high consequence scenarios and initiate a quantitative study such as QRA for such scenarios. LOPA uses IPLs as safeguards for risk reduction. It is conducted for one scenario at a time, which includes a single initiating event and a resulting consequence only.

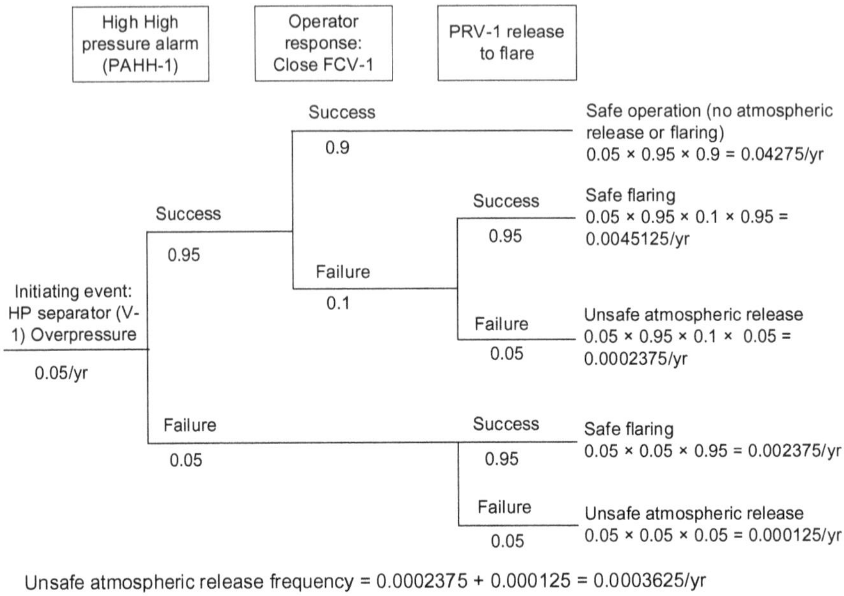

Fig. 20.17: ETA for HP separator (V-1) rupture in Fig. 20.15.

LOPA investigates the reduction in frequency of an initiating event by considering both success and failure of each IPL. Top and bottom images in Fig. 20.18 illustrate the reduction in frequency (as indicated by arrow thickness) of the initiating event. Lower thickness means lower frequency) with the success of each IPL. Note that, although consequence frequency is reduced, the hazard can still occur with the same consequential severity. However, risk of the process system reduces to the acceptable level. If all IPLs succeed (fail), it will lead to safe outcome (unwanted consequence) as shown in the bottom image of Fig. 20.18. Process system can be made safer by implementing good design beginning with ISD (thus, reducing the frequency and/or consequence of the initiating event) and continuing with as many IPLs as possible.

IPL requirements are as follows.

1. Every IPL must be specific and must be designed to prevent or mitigate the consequences of one potentially hazardous event (e.g., a fire, an explosion or a toxic material release).

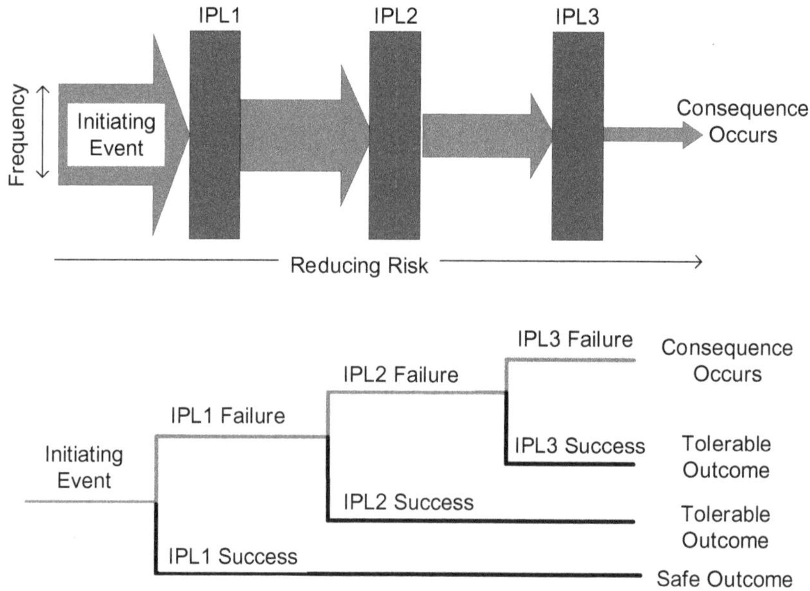

Fig. 20.18: Graphical representation of LOPA (top image) and generic LOPA logic (bottom image).

2. Each IPL must be completely independent of the initiating event and other IPLs. Equipment in one IPL cannot be shared with other IPLs. A common mistake is to share the same field equipment for both IPL and DCS.
3. IPL must be dependable to achieve its intended design all the time. Both random and systematic failure modes must be addressed in the IPL design.
4. Lastly, IPL needs to be auditable. It must be tested and maintained so that target risk reduction is always achieved. It must be designed to enable regular validation of its protective functions (e.g., proof of testing and maintenance of IPLs are necessary to maintain their integrity).

Depending on the abovementioned requirements, it is important to recognize that not all safeguards can be qualified as IPLs in LOPA. In other words, all IPLs are safeguards but not all safeguards are IPLs. Selection of valid IPLs, LOPA and its applications are comprehensively covered in CCPS.[21]

Steps to determine the maximum allowable PFD_{avg} for a SIF using LOPA methodology are presented below. Frequencies of initiating event and probability of failure on demand (PFD_{avg}) can be obtained from company and/or industrial databases.

1. Document the initiating events from HAZOP/WI/FMEA/FTA study.
2. Document frequencies of all initiating events. Begin with one initiating event.
3. List the PFD_{avg} of each IPL that can be used to reduce the risk of the selected initial event. Typical PFD_{avg} of some common protection layers are[3]: BPCS control loop = 0.10; Operator's response to alarm = 0.10; Relief safety valve = 0.01–0.001; and vessel failure probability at maximum design pressure = 10^{-4}.
4. Estimate the risk and compare the calculated risk with the tolerable risk target.
5. If the estimated risk is higher than the tolerable risk target, calculate the required RRF by dividing the former with the latter. With RRF value, identify the PFD_{avg} necessary for a new SIF to achieve the tolerable risk target. This will establish required SIL level.
6. Repeat steps 3 to 5 using the PFD_{avg} value of the proposed SIL modification, to re-calculate the overall incident frequency and confirm it meets the risk target.
7. Repeat Steps 3 to 6 for all initiating events.
8. Summarize all the proposed SIL recommendations in the LOPA report.

Example 2: Perform LOPA for the scenario of a level sensor (LI-1, installed on V-1) blockage in Fig. 20.15, resulting in malfunction of level control valve (LCV-1), causing potential overpressure in the LP flash drum (V-2) and loss of containment of hydrocarbon vapour, which can cause fire/explosion damages. Existing PRV-2 is undersized to handle the required relief flow for this scenario. Assume the target mitigated likelihood as 10^{-5} events/year.

Solution: LOPA for this scenario is illustrated in Table 20.13. V-1 is installed with level sensor (LI-1), which serves to operate LCV-1. Level sensor (LI-2) generates low level alarm (LAL-2). Loss of level sensor (LI-1) will fail to activate LCV-1. LAL-2 alarm provides protection for

Table 20.13: LOPA Solution for the Initial Design

	1	2	3	4	5	6	7	8	9	10
				Protective Layers						
#	Initial Event Description	Initiating Cause	Cause Likelihood	Process Design	BPCS	Alarm (LAL-2)	SIS	Old PRV-2	Mitigated Likelihood	Notes
1	High pressure in LP separator (V-2)	Connection (tap) for level sensor LI-1 becomes plugged	0.1	0.1	1.0	0.1	1.0	1.0	0.001	PRV-2 is not considered a safe guard as it is undersized to handle the required relief flow rate for this scenario
	Mitigated likelihood frequency is higher than the target! Need to improve the design.									

Table 20.14: LOPA Exercise: Solution for the Enhanced Design

	1	2	3	4	5	6	7	8	9	10
				Protective Layers						
#	Initial Event Description	Initiating Cause	Cause Likelihood	Process Design	BPCS	Alarm	SIS	New PRV-2	Mitigated Likelihood	Notes
1	High pressure in LP separator (V-2)	Connection (tap) for pressure sensor PI-1 becomes plugged	0.1	0.1	1.0	0.1	1.0	0.01	0.00001	New PRV-2 with sufficient capacity is considered in design. Hence, it is can be taken as a safeguard.
	The enhanced design achieves the target likelihood by including an independent pressure sensor with alarm and also a new PRV.									

this initiating cause. No PRV credit can be taken as PRV-2 is undersized to handle the required relief flow, and no SIS protection is provided in the design. So, no credit can be applied for them and hence probability '1' is assigned for BPCS, SIS and PRV in columns 5–8 in Table 20.13. Apart from the failure frequency of level sensor, credit by virtue of process design and low level alarm in V-1, we cannot take any other instrumented or non-instrumented IPL credits in the initial design.

For the initial design, mitigated likelihood is 0.001 (by multiplying values in columns 3–8 in Table 20.13), which is not acceptable compared to the target value of 10^{-5} events/year. Hence, additional risk mitigation magnitude of 0.01 (= $10^{-5}/10^{-3}$) is required. This can be achieved by replacing the existing PRV-2 with a new PRV-2, with the required relief handling capacity, installed on V-2 overhead pipe (PFD_{avg} = 0.01). LOPA for the new design enhancement is shown in Table 20.14; the PRV-2 replacement leads to the change in column 8, and consequently in column 9. As can be seen, new mitigated likelihood = 0.00001 (by multiplying columns 3–8 in Table 20.14) meets the target mitigated likelihood (10^{-5}).

20.7 Safety Instrumented System Life Cycle

SIS life cycle involves three main phases,[22] namely, analysis, realization and operations, as shown in Fig. 20.19. Analysis phase involves hazard identification (for which HAZOP and LOPA are frequently used), risk assessment and risk mitigation using non-SIL and SIL safeguards and establishing SIL targets (if new SIL recommendations are required).If new SIS is not required, design can explore non-instrumented safeguards. Finally, safety requirements specification (SRS) document is prepared to develop appropriate cost-effective designs for SIS to achieve optimal test and maintenance strategies for the lifespan of the plant. This completes analysis phase.

Realization phase uses SRS document to perform conceptual SIS design. For this, SIS technology, architecture and test intervals are

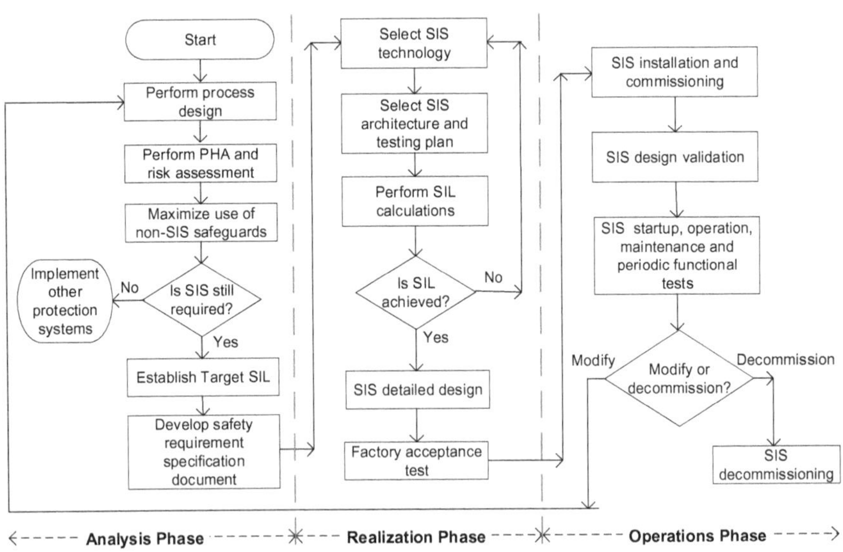

Fig. 20.19: Three phases of SIS life cycle.

selected. SIL calculations are performed to verify whether the proposed SIS can achieve target SIL or requires further risk reduction. If target SIL is not achieved, conceptual SIS design is modified until target SIL level is achieved. SIS detailed design is performed subsequently, and factory acceptance test is completed to ensure SIS is designed as per design intent.

Finally, operations phase involves SIS installation, commissioning, and SIS validation to ensure that the installed SIS and its associated SIFs achieve requirements as specified in the SRS document. Qualified and experienced SIL experts are essential to validate the results of SIS. Once validation is completed, SIS will be in operation. It needs to be tested and maintained to ensure it always meets the target SIL. If SIS needs modification anytime during the plant life, analysis, realization and operations phase assessments shall be repeated with the proposed modifications.

20.8 Revision of Process Safety Information and Operator Training

PSI and operator training are important ingredients for incident-free plant operations. Typical PSI data is shown in Table 20.15.They need to be updated before WHR project commissioning. Operators need to be trained on normal operation control, start-up, shutdown, material safety data sheet and emergency handling procedures. Lack of proper PSI and operator training is an important contributing factor for many industrial accidents.[23]

Table 20.15: PSI Information

Process Description	PFDs
P&IDs	C&E charts
Electrical area classification drawing	Process equipment and piping design conditions
Plot plan layout	Operating procedures, including start-up/ shutdown and emergency handling
Operating envelope	Design basis for pressure relief and emergency systems and devices
Equipment inventories	Maintenance records*
Any past PHA records*	

* Applicable for brownfield projects.

Operating procedures play a major role in enhancing the understanding of operator-dependent safeguards in the event of process upsets. They should be updated during EPC phase of WHR project (Section F of OSHA PSM 1910.119[o]) by addressing the following aspects as the minimum:

- Critical controls
- Critical alarms and interlocks
- Changes in normal operating, start-up, shutdown, and emergency handling
- Changes in operating envelope
- Safety and health considerations for hazardous chemical handling
- SIS and their functions

20.9 Management of Change

MOC procedure is a systematic assessment to examine whether any of the proposed process modifications in the design pose new hazards or increase magnitude of existing hazards.[24] Like-to-like replacement of equipment is known as replacement in kind (RIK), and it is not considered a change. The well-known Flixborough disaster in the United Kingdom is a result of a change in plant configuration, without using proper MOC procedure (CCPS, 2008).[25] Sanders[26] presented many industrial accident cases that clearly emphasize the importance of MOC for avoiding them altogether. MOC is applicable for whole life of the plant. In WHR projects, it is mainly used to manage any design changes (in EPC, pre-commissioning, commissioning, start-up and operating phases), after HAZOP and LOPA are completed in FEED stage. A comprehensive coverage and checklists are available in CCPS.[27]

For WHR project modifications, following are the important causes requiring MOC procedure, post-HAZOP and/or LOPA study:

- Modifications in specifications of process equipment, piping, instrumentation and/or electrical components
- Plot plan changes

[o] https://www.ecfr.gov/cgi-bin/text-idx?SID=b1288696e6d827e1feebe00cb0ccc1eb&mc=true&node=se29.5.1910_1119&rgn=div8 (accessed on 16th Jan 2022).

- Modifications in specification of PRVs, RDs and/or relief system piping
- Changes in critical process control systems
- Changes in SIS design and implementation, critical safety alarm settings, interlocks and process trips
- Any changes in material of construction
- Deviations from codes and standards
- Changes in operating envelope

MOC uses detailed technical review and authorization by a multidisciplinary team (usually technical, operations, maintenance, inspection, reliability, occupational safety, process safety and plant managers) before any change is implemented in the process. A typical flowchart of MOC is shown in Fig. 20.20. MOC form can be raised by any staff (e.g., process engineers/operators, maintenance, project management, inspection, reliability, and safety staff) in the company. MOC coordinator will review the MOC requirements and assess whether the change involves any RIK activities. If change proposed is RIK type, MOC is not required and it will be communicated to the MOC initiator.

If the change is not RIK type, MOC coordinator appoints a review team to conduct PHA. Type of PHA (e.g., WI, FMEA, HAZOP and LOPA) will be decided by the MOC leader. PHA team will perform the PHA and, if required, generate recommendations to bring the risk to ALARP or lower. PHA team leader reviews the PHA report, approves the change and assigns responsible person(s) to close the recommendations in a specific time frame. Once all the recommendations are closed, PSSR will be conducted. All applicable PSI documents need to be updated and operator training needs to be conducted on the changes. Any punch list generated is required to be closed quickly. MOC will be closed once the plant is started and operation is assessed to be safe. Completed MOC will be filed for future safety studies and reference.

20.10 Pre-Start-Up Safety Review

PSSR refers to pre-checks of the process for any safety hazards before start-up operation. It uses a systematic process of field verification to ensure that all the project works are built as per the design intent,

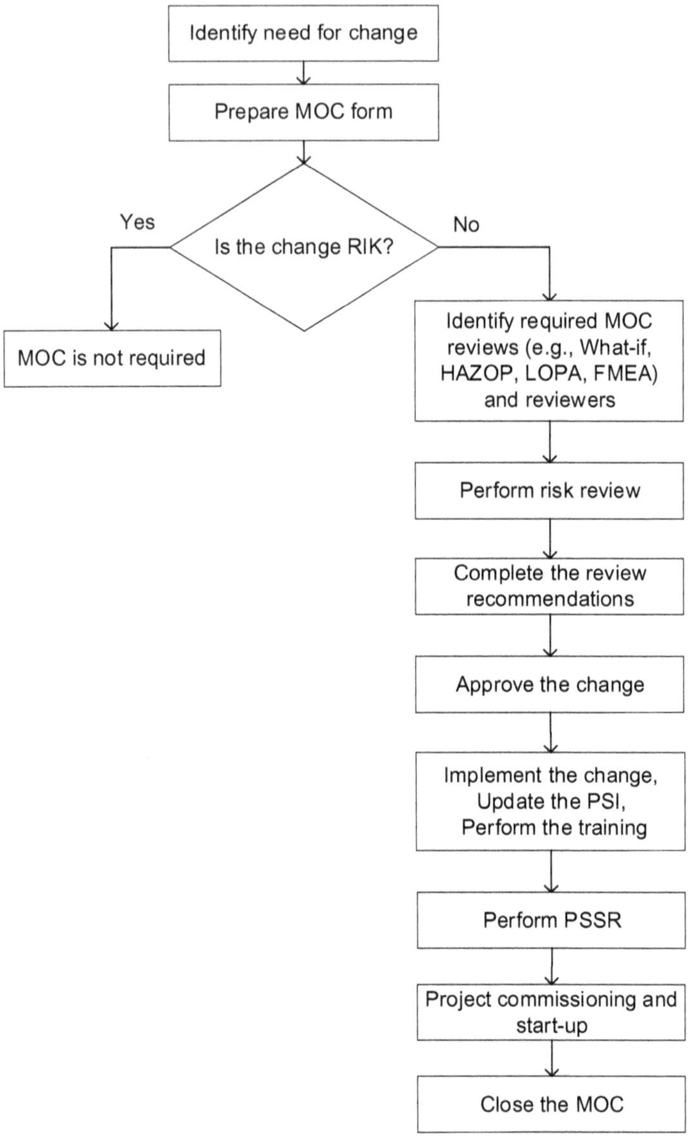

Fig. 20.20: MOC flowchart.

construction, and testing standards. PSSR is essential to ensure that the new/modified facility will perform as intended.

WHR project engineer/manager organizes the PSSR timing and composition of the review team after verification of the field works is completed. PSSR team members are selected based on the project nature. They use pre-defined checklists to ensure the following PSSR goals are met.

- Project/modification is built as per the design intent by following all the applicable design, construction and testing codes and standards.
- All the procedures including safety, operating, maintenance and emergency aspects are in place and adequate.
- All the PHA action items are fully resolved and implemented.
- All the PSI is fully updated.
- All the operator trainings are completed.

PSSR best practices and several detailed PSSR checklists are available in CCPS.[24]

20.11 Summary

This chapter presented an overview of important process safety considerations applicable for WHR projects. It provides a brief and handy reference for practicing engineers and students. Main studies, reviews and considerations for designing a safer and sustainable WHR project are as follows.

- Maximize utilization of ISD principles in pre-FEED and FEED stages for safer and sustainable WHR projects. Often, this will reduce the project cost.
- Comprehensive risk assessment and mitigation are essential for successful WHR projects.
- Proper selection and use of PHA methods (such as WI, HAZOP, FMEA, LOPA, ETA and FTA) in pre-FEED and FEED stages will

make the process safer by addressing the process risks and will result in cost-effective designs.

- Plot plan, layout, hazardous area classification, pressure relief and fire safety design reviews are essential for addressing plant-level safety risks.
- LOPA is a powerful technique to examine the adequacy of existing protection layers and determine the necessity for additional layer(s).
- Non-SIL protection layers shall be fully utilized before making SIL recommendations. This will reduce unnecessary dependence on SIS.
- SIS achieves the required risk reduction by proper design, FAT, validation, periodic inspection and maintenance. Hence, its integrity should be maintained without any compromise during the whole lifespan of plant operation.
- PSI update and accuracy are very important for process safety studies such as PHAs.
- Implementation of MOC procedure is essential to manage additional risks that can arise due to any changes in the design after PHA study.
- PSSR is critical to make sure WHR project is ready for safe start-up by addressing all PHA, MOC action items and construction punch list items.

Mannan[14] and Crowl and Louvar[1] are holistic books on process safety and are recommended for further reading on the topics of this chapter. The book by Kletz and Amyotte[6] is recommended for deeper understanding of ISD concepts and applications.

References

1. Crowl DA, Louvar JF. (2011) *Chemical Process Safety Fundamentals with Applications*, 3rd ed. Prentice Hall International Series.
2. CCPS (1998) *Guidelines for Design Solutions for Process Equipment Failures*, AIChE, New York.
3. CCPS (2000) *Guidelines for Chemical Process Quantitative Risk Analysis*, 2nd ed. AIChE, New York.

4. Bollinger RE, Crowl DA. (1997) *Inherently Safer Chemical Processes: A Life Cycle Approach*. Wiley-AIChE.
5. Hendershot DC. (February 2011) *Inherently Safer Design an Overview of Key Elements*, pp. 48–54.
6. Kletz TA, Amyotte P. (2010) *Process Plants: A Handbook for Inherently Safer Design*. CRC Press.
7. CCPS (2009) *Inherently Safer Chemical Processes: A Life Cycle Approach*, 2nd ed. Wiley.
8. API Standard 521. (June 2020) *Pressure-Relieving and Depressuring Systems*, 7th ed. American Petroleum Institute.
9. Moran S. (September 2016) Process plant layout — becoming a lost art? *Chem Eng* 71–76.
10. CCPS. (2003) *Guidelines for Facilities Siting and Layout*, AIChE, New York.
11. Hall S. (2018) *Rules of Thumb for Chemical Engineers*, 6th ed. Elsevier.
12. API Standard 520. (October 2020) *Sizing, Selection and Installation of Pressure-Relieving Devices in Refineries, Parts 1&2*, 10th ed. American Petroleum Institute.
13. Parry C. (1994) *Relief Systems Handbook, Institution of Chemical Engineers*. VCH Publishers.
14. Mannan S. (2005) *Lee's Loss Prevention in the Process Industries*, 3rd ed. Elsevier Butterworth-Heinemann.
15. Balajee R, Reddy CCS. (2016) Project safety in revamp projects. In: Rangaiah GP (ed), *Chemical Process Retrofitting and Revamping*, Wiley.
16. Branan C. (2012) *Rules of Thumb for Chemical Engineers*, 5th ed. Gulf Professional Publishing.
17. Madlani D. (May 2012) Flare for Design, Hydrocarbon Engineering (http://www.ivisionindia.com/Flaretot%20Hydrocarbon%20Engineering%20Article%20May%202012.pdf).
18. Pratt TH. (2000) *Electrostatic Ignitions of Fires and Explosions*. AIChE, New York.
19. Nolan DP. (2019) *Handbook of Fire and Explosion Protection Engineering Principles for Oil, Gas, Chemical, and Related Facilities*, 4th ed. Elsevier.
20. Dave M. (2004) *Practical HAZOPs, Trips and Alarms*, 1st ed. Elsevier, Burlington.
21. CCPS. (2001) *Layer of Protection Analysis: Simplified Risk Assessment*, AIChE, New York.

22. Marszal EM, Scharpf EW. (2002) *Safety Integrity Level Selection: Systematic Methods Including Layer of Protection Analysis*. Instrumentation, Systems, and Automation Society, Research Triangle Park, NC.
23. Abu Bakar HT, Siong PH, Yan CK, et al. (2017) Analysis of main accident contributor according to process safety management elements failure. *Chem Eng Trans* **56**: 991–996.
24. CCPS. (2007) *Guidelines for Performing Effective Pre-startup Safety Review*, AIChE, New York.
25. CCPS. (2008) *Guidelines for the Management of Change for Process Safety*, Wiley.
26. Sanders RE. (2005) *Chemical Process Safety-Learning from Case Studies*, 3rd ed. Elsevier Butterworth-Heinemann.
27. CCPS. (1995) *Plant Guidelines for Technical Management of Chemical Process Safety*, AIChE, New York.
28. IEC61511 Parts 1 to 3 (2016) Functional Safety: Safety Instrumented Systems for the Process Industry Sector (https://webstore.iec.ch/searchform&q=IEC%2061511).
29. API RP 500. (2012) *Recommended Practices for Classification of Locations for Electrical Installations at Petroleum Facilities Classified as Class I, Division I and Division 2*, 3rd ed.
30. API RP 505. (2018) *Recommended Practice for Classification of Locations for Electrical Installations at Petroleum Facilities*, 2nd ed.
31. NFPA 70. (2020) *National Electric Code*, National Fire protection Association.
32. NFPA 497. (2021) *Recommended Practice for the Classification of Flammable Liquids, Gases, or Vapors and of Hazardous (Classified) Locations for Electrical Installations in Chemical Process Areas*. National Fire Protection Association.

Acronyms

AIT	Auto Ignition Temperature
ALARP	As Low As Reasonably Practicable
ANSI	American National Standards Institute
API	American Petroleum Institute
ASME	American Society of Mechanical Engineers
BLEVE	Boiling Liquid Expanding Vapour Explosion
BPCS	Basic Process Control System

CCPS	Centre for Chemical Process Safety
CIMAH	Control of Industrial Major Accident Hazards
COMAH	Control of Major Accident Hazards
CSB	Chemical Safety Board
C&E	Cause and Effect
DCS	Distributed Control System
ETA	Event Tree Analysis
EPC	Engineering, Procurement and Construction
ESD	Emergency Shutdown
FTA	Fault Tree Analysis
FEED	Front-End Engineering Design
FMEA	Failure Modes and Effects Analysis
HAZOP	Hazard and Operability
HE	Heat Exchanger
HSE	Health and Safety Executive
I&E	Instrumentation and Electrical
IEC	International Electrotechnical Commission
IPL	Independent Protection Layer
IPS	Instrumented Protection System
ISD	Inherently Safer Design
LOPA	Layers of Protection Analysis
LPG	Liquefied Petroleum Gas
MAWP	Maximum Allowable Working Pressure
MOC	Management of Change
NACE	National Association of Corrosion Engineers
NEC	National Electric Code
NEPA	National Fire Protection Association
P&ID	Piping and Instrumentation Diagram
PFD	Process Flow Diagram
PFD_{avg}	Average Probability of Failure on Demand
PHA	Process Hazard Analysis
PRD	Pressure Relief Device
PRV	Pressure Relief Valve
PSI	Process Safety Information
PSM	Process Safety Management
PSSR	Pre-Start-up Safety Review

QRA	Quantitative Risk Assessment
RD	Rupture Disc
RO	Restriction Orifice
RIK	Replacement in Kind
RRF	Risk Reduction Factor
RL	Risk Level
SIF	Safety Instrumented Function
SIL	Safety Integrity Level
SIS	Safety Instrumented System
SRS	Safety Requirements Specifications
WI	What-If

Exercises

20.1. Perform LOPA for the scenario of 'loss of liquid level in V-2, resulting in pump P-1 seal failure and eventual fire and explosion incident', for the system shown in Fig. 20.15. Assume the probability of occurrence is of high category (likelihood of ~ 0.01/year) and severity is category 2 as per risk matrix in Table 20.3.Propose the minimum IPLs required in the design to meet the target mitigated likelihood below 10^{-5} event/year.

20.2. Perform ETA for the scenario in Exercise 20.1.

20.3. Compare the results obtained from Exercises 20.1 and 20.2. What conclusions can be made about these methods?

20.4. Identify ISD options for HEs. Comment whether implementing ISD for HE designs will be beneficial. Explain your reasons.

20.5. What is the importance of ALARP in risk mitigation?

Appendix: Inherently Safer Design Checklist

Minimize

1. Can alternative equipment with reduced hazardous material inventory requirement be used? Examples are:

   a. Compact HEs (higher heat transfer area per unit volume such as spiral, plate and frame, and plate and fin) in place of shell-and-tube HE
   b. Continuous in-line mixers (e.g., static mixer) in place of mixing vessels
   c. Assigning more hazardous material to the tube side in shell-and tube HEs
   d. Use water or other non-flammable heat transfer medium such as a vapour-phase medium or a medium below its boiling point
2. Has the length of hazardous material piping been minimized?
3. Has hazardous material piping been designed for minimum pipe diameter?
4. Can hazardous inventory be reduced by minimizing new storage drums/vessels?
5. Can pipeline inventory be reduced by using the hazardous material as a gas rather than a liquid?
6. Can process conditions/technology be changed to reduce hazardous conditions?
7. Has energy assessment been performed to ensure minimum energy is used?
8. Are deadhead locations in piping eliminated?

Substitute

1. Is it possible to substitute less hazardous raw materials? Some examples are the following:
   - Use of non-combustible, less volatile, less reactive/more stable, and less toxic substance
   - Use of steam rather than combustible or toxic heat transfer fluid

Moderate

1. Is it possible to design operating conditions to avoid any decomposition or freezing of working fluids?
2. Can process conditions be changed to avoid handling flammable liquids above their flash points or auto ignition temperatures?

3. For processes handling flammable materials, is it possible to design the layout to minimize the number and size of confined areas, and to limit the potential for serious overpressure in the event of a loss of containment and subsequent ignition?
4. Can two-phase flow be avoided? Are velocities in pipes below erosion velocities?
5. Can process units be located to eliminate or minimize the following?
   - Adverse effects from adjacent hazardous installations
   - Off-site impacts
   - On-site impacts on employees and plant facilities
6. Can process units be designed to limit the extent of the following process deviations?
   - Maximum feed rate may be limited by pipe size to be within safe limits for gravity fed systems
   - Maximum feed rate may be limited by pipe size and/or safety-critical RO for pressurized systems
7. Are maximum flow rates restricted by pump/compressor design and/or use of safety-critical ROs in the piping?
8. Can liquid spill of hazardous material be prevented from entering drainage system/sewer?
9. For flammable materials, can spills be directed away from the storage vessel to reduce the risk of a BLEVE in the event of a fire?
10. Passive protection layers are preferred over active protection layers; examples of passive protection layers are: use of fire proofing/insulation, refractory, secondary containment, use of flare knockout/cold blowdown drums, permanent bonding and grounding systems, use of inert gas blanketing for storing flammable and explosive substances.
11. Are knock-on or domino effects avoided by ample layout spacing, equipment orientation, fail-safe shutdown, and open construction?
12. Is minimum flow protection for pumps/compressors implemented?
13. Are compressors installed with anti-surge and stonewall protection system?

Simplify

1. Can equipment be designed such that it is difficult or impossible to create a potential hazardous situation due to an operation or maintenance error? Examples are the following:
   - Easy operation of valves designed to prevent inadvertent errors
   - Simplified control displays
   - Use corrosion resistant materials for the process
   - Design temperature-limited heat transfer equipment
   - Operate at as low pressure as possible to limit any hazardous release quantity
   - Avoid cryogenic conditions by design methods such as controlling depressurization rate of volatile hydrocarbons by use of safety-critical RO and use of steam jacketing in the depressurizing flare piping.
   - Operate at higher temperature to eliminate cryogenic effects such as embrittlement failures
   - Operate equipment well below design temperature and pressure
   - Use passive rather than active controls
   - Use buried or shielded tanks for storing materials such as liquefied petroleum gas (LPG)
   - Eliminate unnecessary cross-connection piping.
   - Use fail-safe controls if utilities are lost
   - Limit the complexity in instrumented system design
   - Implement redundant modules for critical control system
   - Use refrigerated storage versus pressurized storage
   - Provide electrical supply from independent and/or emergency sources
   - Reduce wall area to minimize corrosion/fire exposure
   - Install hydrocarbon storage vessels above 7.6 m height
   - Minimize connections, paths, and number of flanges in hazardous processes
   - Use fewer bends in piping
   - Use expansion loops in piping rather than bellows
   - Develop the equipment isolation mechanisms for maintenance works

- Limit manual operations such as filter cleaning, manual sampling, and hose handling operations
- Design vessels for full vacuum, which eliminates the risk of vessel collapse
- Design both shell and tube sides of HEs to contain the maximum attainable pressure, thus eliminating the need for pressure relief due to tube rupture
- Design the equipment to make incorrect assembly impossible
- Use equipment that clearly identifies status:
  - o Check valves with easy to identify direction of flow
  - o Gate valves with rising spindles to clearly indicate open or close position
  - o Spectacle blinds instead of slip plates

2. Design the equipment with sufficient strength to contain the maximum pressure generated, even if 'worst credible event' occurs?
3. Was assessment performed to use safer materials?
4. Can passive leak-limiting technology be used to limit potential loss of containment? Examples include the following:
   - Blowout resistant gaskets
   - Increasing wall strength
   - Using fewer seams and joints
   - Providing extra corrosion/erosion allowance
   - Reducing vibration
   - Minimizing the use of open-ended, quick-opening valves
   - Eliminating open-ended, quick-opening valves in hazardous service
   - Improving valve seating reliability
   - Eliminating unnecessary expansion joints, hoses, and rupture disks
   - Eliminating unnecessary sight glasses/glass rotameters
5. Has process simplification such as avoiding congested piping, complex process logics/interlocks been performed?
6. Is equipment designed to withstand maximum operating conditions?
7. Are expansion bellows avoided in HEs and pumps by adjusting operating conditions and/or critically evaluating the requirement for steam cleaning?
8. Are human errors minimized by clear tagging of critical valves and also by making operation controlled by automatic action rather than human action?

Index

Q

T

Lightning Source UK Ltd.
Milton Keynes UK
UKHW020754120522
402872UK00003B/152

9 789811 248399